Functional Anatomy of the Vertebrates

An Evolutionary Perspective

Second Edition

Warren F. Walker, Jr.
Emeritus Professor, Oberlin College

Karel F. Liem
Henry Bryant Bigelow Professor of Biology, Museum of Comparative Zoology, Harvard University

SAUNDERS COLLEGE PUBLISHING
Harcourt Brace College Publishers

Fort Worth   Philadelphia   San Diego
New York   Orlando   San Antonio
Toronto   Montreal   London
Sydney   Tokyo

Text Typeface: New Caledonia
Compositor: University Graphics, Inc.
Acquisitions Editor: Julie Levin Alexander
Developmental Editor: Christine Rickoff
Managing Editor: Carol Field
Project Editors: Nancy Lubars, Linda Boyle
Copy Editor: Patricia Daly
Manager of Art and Design: Carol Bleistine
Art Director: Anne Muldrow, Susan Blaker
Text Designer: Julie Anderson
Cover Designer: Robin Milicevic
Text Artwork: Hudson River Studio
Director of EDP: Tim Frelick
Production Manager: Joanne Cassetti
Cover: A radiograph of the skate, *Raja erinacea,* courtesy of William M. Winn and Joseph Eastman

Printed in the United States of America

FUNCTIONAL ANATOMY OF THE VERTEBRATES:
An Evolutionary Perspective, second edition

ISBN 0-03-096846-1

Library of Congress Catalog Card Number: 93-085819

9 0 1 2 3 4 5 6   069   13 12 11 10 9 8 7 6 5

To our former teachers and mentors,

the late Alfred Sherwood Romer of Harvard University

and Hobart Muir Smith of the University of Illinois.

Preface

PURPOSE

Professor Karel F. Liem of Harvard University has joined the senior author in preparing the second edition of this book. Dr. Liem, who has coauthored other books at both the introductory and advanced college level, brings to this book the perspectives of an active researcher and teacher in functional anatomy. Our purpose and basic approach in preparing the second edition of *Functional Anatomy of the Vertebrates* remains the same as for the first edition: to help students understand vertebrate form and function by tracing the major morphological changes that have occurred during the 500 million years of vertebrate evolution and to relate these changes both to the functions the organs perform and to the adaptations of the animals to their environment. Classic comparative anatomy emphasizes the morphological changes, but structures **do** things; they perform functions that must be integrated with the functions of other body parts, with the animal's life style, and with the environment in which the animal lives. In a sense, form and function are opposite sides of the same coin; one does not exist without the other. Functional anatomy is not only the description of form and function, it also asks why and how the form-function changes we see came about. By studying form and function together, students will see how the threads of form, function, and ecology weave together to form a coherent tapestry. Functional anatomy explores the relationship between form, function, and adaptation. It is one of today's most challenging and dynamic fields of research.

We assume that students taking a course in functional anatomy will have had a college course in biology or zoology, and therefore will be familiar with the basic principles of cell and developmental biology, physiology, evolutionary theory, and ecology. We have not developed these subjects, but rather summarized them and then gone on to build on them. The physics necessary to understand functional anatomy is presented as we develop the subject.

DISTINCTIVE FEATURES AND APPROACH

The most distinctive feature of this book is the discussion of form and function together to the extent that this is possible. The first four chapters deal with preliminary topics that are essential to understanding vertebrate evolution: the integration of organ size with body size (scaling), the basis of constructing and interpreting phylogenies (homology and homoplasty), the origin of vertebrates and their relationship to other chordates, vertebrate diversity, and the early embryonic development of vertebrates. Thereafter, we consider, in turn, the functional anatomy of each organ system. Usually our discussion of form and function can be included in the same chapter, but some aspects of functional anatomy cannot be appreciated until the anatomy of more than one organ system is understood.

In discussing the organ systems, we utilize the systems approach. This approach facilitates understanding the evolutionary changes that have occurred in the organ systems, but it also makes it easy to lose sight of the integration of the organ systems within an entire organism. We have tried to ameliorate this by making it clear how changes in organ systems relate to the animal as a whole, to the environment in which it lives, and to its mode of life. The sequence in which we discuss organ systems is conventional, except that we have placed the discussion of the nervous system and the introduction to endocrine glands after our consideration of the muscles and locomotion. It is logical to introduce these control systems as close as possible to the muscles, the primary effectors being controlled. We can also utilize information on control systems as we consider the organ systems that deal with metabolic processes.

Several pedagogic devices will help students study this material. Structures have names, and effective communication necessitates that students learn them. Since most of the terms are a description, in Latin or Greek, of some aspect of the structure, some knowledge of the terms' derivation helps students learn and understand them. When introducing an important term that is likely to be unfamiliar, we have given its classical derivation so students can see that the term is descriptive of the structure. Since most of the classical roots are used repeatedly, students will soon recognize them, and learning new terms will become easier. Giving the derivations when terms are introduced is a natural way for students to learn anatomical terminology.

Each chapter is preceded by a precis and a chapter outline that will give students an overview of what is to come and its importance. Chapter summaries highlight the main points of each chapter.

NEW TO THE SECOND EDITION

Advances in functional anatomy have led to new interpretations of structure. Discussing these and introducing more functional anatomy has necessitated reorganizing and rewriting much of the book. A few examples will illustrate the scope of the changes we have made. We have expanded the discussion of scaling and allometry, and applied these concepts to embryonic development as well as to adult structure. A discussion of the role of homeobox genes in controlling the vertebrate body plan is included in the chapter on embryology. Our consideration of the properties and mechanics of structural materials has been removed from other chapters and collected in a single chapter. The structure and evolution of the vertebral column and appendicular skeleton are dealt with in a single chapter on the postcranial skeleton. The treatment in this chapter is largely descriptive, and functional aspects of support and locomotion are considered in a new chapter on the functional anatomy of locomotion, which follows the chapter on the muscular system. New discoveries on the development of cranial muscles are incorporated and have led to a regrouping of cranial muscles. We have rewritten our discussion of cranial nerves to incorporate recent studies on head segmentation and the multiple origin of the cranial nerves. The enteric system is recognized as a distinctive part of the autonomic nervous system. We recognize in our discussion of the mammalian brain that it evolved independently from that of contemporary reptiles and birds. We have split the chapter on the digestive system into one on the oral cavity and feeding mechanisms and one on other parts of the digestive system. Gas exchange in fish gills and in early lungs is described in more functional terms, as is the double circulation through the lungfish heart. The chapter on the excretory system and osmoregulation has been extensively reworked. As we revised chapters, we simplified terminology and tried to make the material more accessible and user friendly.

Many chapters include Focus Essays in which we expand the discussion of selected topics mentioned in the text. Cladograms (phylogenetic diagrams) at the end of many chapters summarize important events in the evolution of the organ systems. Many figures have been added and many others modified or replaced to make them clearer. Color continues to be used in the chapters on embryology and circulation, and color has been added to the chapter on the skull to help students identify and follow changes in the major skull components. References are placed at the end of each chapter rather than being collected at the end of the book. General references that apply to more than one organ system are placed at the end of the first chapter. A glossary has been added; it includes the phonetic pronunciation of major terms, and their classic derivation and meaning.

Acknowledgments

We are indebted to many for help in preparing this edition. Richard Estes, San Diego State University; Stanley Hillman, Portland State University; Allan H. Savitsky, Old Dominion University; Donald O. Straney, Michigan State University; Bedford M. Vestal, University of Oklahoma; John R. Winkelmann, Gettysburg College; and one anonymous person reviewed the first edition. The following instructors have reviewed different portions of the manuscript for the second edition: William E. Bemis, University of Massachusetts at Amherst; François Chapleau, University of Ottawa; John W. Hermanson, Cornell University; Stan Hillman, University of Wisconsin at Madison; Larry Miller, Gannon University; Gail R. Patt, Boston University; Alan H. Savitzky, Old Dominion University; Kurt Schwenk, University of Connecticut; G. G. E. Scudder, University of British Columbia; Margaret Simpson, Sweet Briar College; James F. Thompson, Huntingdon College; John P. Tramontano, Orange County Community College N.Y.; and Alexander J. Werth, Hampden-Sydney College. All have made suggestions that have helped guide us in preparing this edition. We thank them all for taking time from busy schedules to help us in this way. We, of course, take full responsibilities for any errors or shortcomings that remain. We hope that users of this edition will bring errors to our attention and make suggestions that will improve a future edition. We also appreciate the help of the following colleagues with whom we have frequently consulted: E. Blaban, E. Brainerd, A. W. Crompton, K. Hartell, F. A. Jenkins, C. Souza.

The staff of Saunders College Publishing has, as always, been most helpful. Julie Levin Alexander, our Senior Editor, encouraged and supported us as we planned this edition. Christine Connelly and Christine Rickoff, our Developmental Editors, guided us as we prepared the manuscript. Patricia Daly, our Copy Editor, helped us attain a consistent and uniform style throughout the book. Nancy Lubars, our Project Editor, and her assistant, Linda Boyle, helped us with the numerous phases of production. New and revised art for this edition was prepared under the supervision of Anne Muldrow and Susan Blaker, our Art Directors. John Norton drew the chapter opening art. Tensy Walker, wife of the senior author, has continued to lend encouragement and help in the proofreading.

Warren F. Walker, Jr.
Karel F. Liem

Contents

1 | Introduction

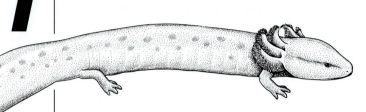

PRECIS

Important introductory concepts that you must know to understand the evolution of vertebrate structure and function are introduced in this chapter.

Outline

Anatomical and Functional Interpretation

Since the publication of Charles Darwin's *On the Origin of Species* (1859) and the general acceptance of the theory of organic evolution, biologists have studied the course of evolution primarily by comparing the anatomy of different species. **Comparative anatomy** has served us well; it has enabled us to trace the evolution of organisms and organs and to see ourselves in a broader context.

Using recently developed technologies, biologists can now investigate structure, and structural transformations that occurred during evolution, by analyzing structure in a functional context. **Functional anatomy** examines how structures within organisms, such as cells, tissues, organs, organ systems, and other complex functional units, perform specific functions. The form of an organ is linked to its function just as the design of a musical instrument is tied to its purpose. In a sense, form and function are two sides of the same coin; one cannot exist without the other. It is therefore important to study both the form and function of the various components that make up the vertebrate body. Functional anatomy can reveal how animals adjust to different modes of life. Structures can only be described anatomically in a dead animal, preserved material, or museum specimens; a living animal must be analyzed to find out how the structures work in an intact organism.

Functional anatomy is not only the description of form and function; it also asks how, why, and "How did it come about?" To answer these questions, it is necessary to be familiar not only with the anatomy and function of various organ systems, but with the comparative method. By comparing the functional anatomy of different vertebrates, we can gain insights into how each of the vertebrate groups has solved its adaptive problems in its natural environment over time. Comparisons of living and fossil vertebrates often result in discoveries about evolutionary changes in anatomical design and function. We will learn that new anatomical designs evolve only by changing pre-existing structures and their functions in the ancestral species. In the course of evolution a structure can take on a new form and maintain its function, or it may lose its original function and acquire a new one.

To understand why certain designs and functions have changed in a particular way during the course of evolution, it is necessary to become familiar with the changes in form and function during development. The nature and extent of the modifications in form and function of an adult structure are often dependent on the pattern and mechanism of its development in the embryo.

Evolution of form and function of various components of the vertebrate body depends on three factors: (1) the nature of the pre-existing structures in the ancestors; (2) the pattern and mechanism of development during the transformation from embryo to adult; and (3) the sum of the prevailing environmental factors in the animal's habitat, which act as agents of selection. Unlike an engine, which is turned off when mechanical adjustments are made, evolutionary changes in the machinery of the vertebrate body must take place while the engine is running (Frazzetta, 1975). In this book we will discuss the functional anatomy of musculoskeletal systems, locomotion, reception of stimuli, neuronal and hormonal

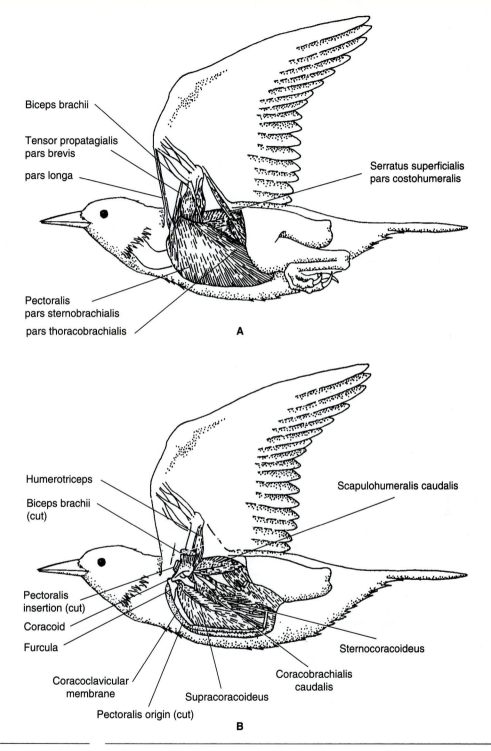

Biceps brachii

Tensor propatagialis
pars brevis

pars longa

Serratus superficialis
pars costohumeralis

Pectoralis
pars sternobrachialis

pars thoracobrachialis

A

Humerotriceps

Biceps brachii
(cut)

Scapulohumeralis caudalis

Pectoralis
insertion (cut)

Coracoid

Furcula

Sternocoracoideus

Coracoclavicular
membrane

Coracobrachialis
caudalis

Supracoracoideus

Pectoralis origin (cut)

B

FIGURE 1–1   Lateral views of the muscles of the shoulder of the European starling, *Sturnus vulgaris.* **A**, Superficial muscles. **B**, deeper muscles after resection of most of the pectoralis major. (After Dial, Golsow, and Jenkins, 1991.)

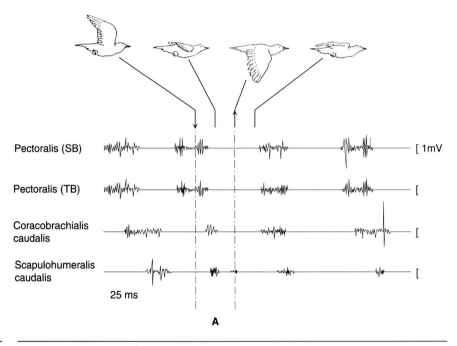

Pectoralis (SB)

Pectoralis (TB)

Coracobrachialis
caudalis

Scapulohumeralis
caudalis

25 ms

A

FIGURE 1–2   **A,** Representative traces of electromyograph signals recorded from shoulder muscles of the European starling during flight. The strength of the signals in millivolts is shown on the vertical axis and their duration in milliseconds on the horizontal axis. SB = sternobrachial part of the pectoralis, TB = thoracobrachial part. Sketches of flying birds above key the signals to phases of flight. The first figure is the start of the downstroke; the third figure, the start of the upstroke.

integration, feeding, digestion, respiration, circulation, excretion, and reproduction. It is with these systems that vertebrates interact with each other and with their environment. Functional adjustments in one or more of these systems are often necessary to cope with changes in environmental factors.

As an example of how functional anatomy is practiced, we present a brief summary of research on bird flight by Dial, Goslow, and Jenkins (1991). The major flight muscles and associated skeletal elements were dissected in dead starlings, and their anatomy was described and illustrated (Fig. 1–1). A thorough knowledge of the anatomy is an absolute prerequisite for the next step, functional analysis of the living bird. Starlings were trained either to fly down a 50-m-long hallway to a landing platform, or to fly in a wind tunnel. The major flight muscles were then surgically implanted with electrodes. The activity of the muscles (electromyography) was recorded synchronously with a motion picture of wing actions of the flying bird taken at 400 frames per second (Fig. 1–2**A**). The activity of the flight muscles was then studied in relation to wing movements (Fig. 1–2**B**). The results indicate that some muscles are active during the downstroke of the wing, while others are responsible for the upstroke. Some overlapping muscle action occurs during the transitions between the downstroke and upstroke. (We will return to this study when we discuss flight on p. 389.) With these and other powerful ana-

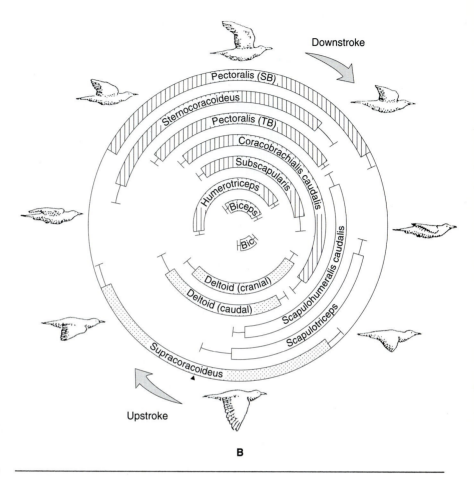

Downstroke

Upstroke

B

FIGURE 1–2   *continued*

B, A summary of the mean onset and offset of the activity of the shoulder muscles of the European starling during a wing cycle of 72 ms. Hatched muscles are used in the downstroke; stippled ones, during the upstroke; and those not shaded, during the transition between downstroke and upstroke. Thin lines extending beyond the range of mean muscle activity indicate the standard error. (After Dial, Goslow, and Jenkins, 1991.)

lytical tools, functional anatomists are making many new discoveries. Recent findings in functional anatomy have revolutionized the study of form, function, and adaptation.

Scaling and Allometric Growth

The anatomy and function of organs are dependent partly on body size. Since vertebrates vary greatly in size (ranging from fishes that weigh a fraction of a gram to the 90-metric-ton blue whale), biologists use scaling analysis to study the structural and functional consequences of changes in size among otherwise similar organisms. Scaling analysis can be illustrated by examining the effects of doubling the size of a cube (Fig. 1–3). If the linear dimension of the cube (e.g., the length of a side) is doubled, the small and large cubes are said to be geometrically similar or **isometric** (Gr. *isos* = equal + *metron* = measure). The surface area and volume of a cube, however, do not change in proportion to its linear dimensions. If a linear dimension of a cube is doubled (Fig. 1–3), its surface area, which is a two-dimensional quantity, increases by the square of the linear increase, or four times (2^2 = 4) and its volume, which is a three-dimensional quantity, increases by the cube of the linear increase, eight times (2^3 = 8). If the linear dimensions of the cube were quadrupled, the surface area would increase 16 times (4^2) and volume 64 times (4^3). Stated in other terms:

$$\text{surface area} = \text{height}^2$$
$$\text{volume} = \text{height}^3$$
$$\text{surface area} = \text{volume}^{2/3}$$

The last equation, which is derived from the first two, states that as geometrically similar objects increase in size, their surface area increases more slowly than their volume (at only ⅔ the power of the volume). When surface area is plotted against volume on an arithmetic plot, the slope of the curve decreases as volume increases (Fig. 1–4**A**). When the same data are plotted on logarithmic coordinates, the curve is a straight line with a slope of 0.67 (Fig. 1–4**B**).

The same rules apply to animals that increase (or decrease) in size during a phylogenetic or ontogenetic series. In the cube we discussed, each dimension was doubled (that is, changed in the same proportion or isometrically). However, the linear dimensions of comparable parts of animals usually do not change in the same proportion but in different proportions, or **allometrically** (Gr. *allos* = different). Allometric scaling is the analysis of the change in one body part (e.g., limb diameter), or rate of activity (e.g., metabolism), relative to another (usually body mass as measured by weight). The general formula of the allometric equation is $y = a \times x^b$ where

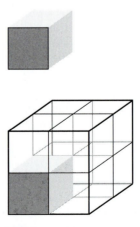

FIGURE 1–3   Changes in the surface area and volume of a cube when its linear dimensions are doubled.

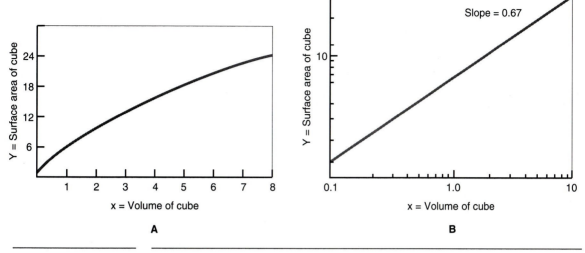

FIGURE 1–4   Graphs of the relationship between surface area and volume in cubes of increasing size. **A,** Arithmetic coordinates. **B,** logarithmic coordinates. (After Schmidt-Nielsen.)

y = the dependent variable (size or activity of some part)
a = the proportionality coefficient, which is the intercept of the regression line on the y-axis when $x = 1$
x = the independent variable (usually body mass)
b = the body mass exponent, or slope of the regression line.

The exponent b (the slope) is not always ⅔ (which describes the relation of surface area to volume), but has different values for different structures and functions. A slope of 1 indicates that the size of the part or function in question increases at the same rate as the body mass, so there is no proportional change (Fig. 1–5**A,** dashed line). An example is the size of the heart in mammals. Larger mammals have larger hearts than small ones, but in all cases the heart is about 0.6 percent of body mass. A slope greater than 1 indicates that the part in question increases at a faster rate than body mass, so its relative size becomes greater as the animal increases in size (Fig. 1–5**A,** solid line). An example is skeletal mass, which is relatively greater in large animals than small ones. A slope less than 1 means that the part in question increases at a slower rate than body mass as the animal increases in size, so it becomes relatively smaller (Fig. 1–5**A,** dotted line). This is the relationship of surface area relative to body mass, and the relationship of metabolic rate relative to body mass. A negative slope means that the part or function in question decreases with increasing body mass (Fig. 1–5**B**). This is the case for heart rate in mammals. An elephant's heart rate is only about ⅛ that of a mouse's.

Allometric ratios describe structural and functional correlations, but the challenge is to find the underlying reasons for these correlations. Often the reason for a relative increase in the size of a structure is found in the changing surface to

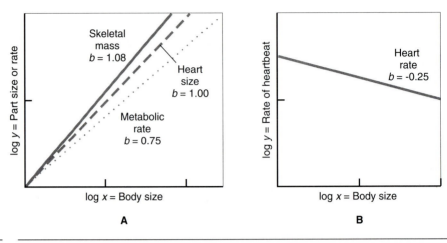

FIGURE 1–5   Graphs on logarithmic coordinates of the relationship between the size of an organ, or rate of activity, and body size. **A**, Skeletal mass (solid line), heart size (dashed line), and metabolic rate (dotted line). **B**, Rate of heart beat. (After Schmidt-Nielsen.)

volume ratio. As bridges, buildings, or animals become larger, more weight needs to be supported or moved, but weight increases as the cube of the increase in linear dimension. The strength of supporting materials (e.g., bones) and the strength of muscles, however, increase in proportion to their cross-sectional area, a square function. If all body parts grew isometrically, the bones and muscles of large animals would not be strong enough to support and move the animal. As animals become larger, the surface area across which food is absorbed, gases exchanged, or waste products eliminated decreases relative to the mass requiring these processes. To overcome problems of this sort, three parameters can change:

1. The relative dimensions of certain structures can change. For instance, large stone buildings have relatively thicker walls than small ones, and the supporting bones of large animals have a relatively greater diameter than those of small ones.
2. The nature of the structural materials may change. Steel may replace bricks, bone may replace cartilage, or bones may become denser.
3. A fundamental design change may be needed to accomplish the task. For instance, to cross a wide river, a suspension bridge with a more complex structure is used instead of a simple stone arch. Protochordates, which are relatively small animals, exchange gases across their body surface by diffusion. Diffusion across the body surface is not adequate in vertebrates, where body surface is smaller relative to body mass in these larger and more active animals. Aquatic vertebrates have evolved gills (a design change that greatly increases surface area) and a method of pumping water across the external surface of the gills and blood through them.

Although changes in surface-volume relationships help explain many phenomena, they do not explain others. It is well known that small mammals have higher metabolic rates than large ones. The allometric formula for metabolic rate is

$$\text{metabolic rate} = 70\ M_{\text{body}}^{0.75}$$

In this equation, M = mass. Since the body mass exponent (0.75) approaches that of surface area, which scales to body mass to the power of 0.67, biologists first proposed that the higher metabolic rate of small mammals compensates for the heat lost through their relatively larger surface area. However, the ratio of metabolic rate to body mass in mammals is significantly higher than would be expected if heat loss through the surface were the explanation. Many alternative explanations have been explored. A model developed by McMahon (1973) is currently favored by many investigators. McMahon showed that the maximum power output (work per unit of time) of a muscle is proportional to its cross-sectional area and scales to body mass as follows:

$$P_{\text{max}} = M^{0.75}$$

If this formula applies to all the metabolic variables involved in supplying muscles with energy and oxygen, as is likely, metabolism would also scale to body mass to the power of 0.75. Power output of muscles relative to body mass appears to explain metabolic rates better than surface-volume relationships.

Since the metabolic rate of mammals decreases as body mass increases, the relative need for oxygen decreases. Relative heart size remains the same (about 0.6 percent of body mass), so the relatively decreased oxygen need is met by a decrease in heart rate. Heart rate scales to body size to the power of -0.25.

Scaling analysis helps us understand many evolutionary changes in form and function, and we will see other examples as we consider the functional and structural evolution of vertebrates. However, not all factors are explained as the result of simple mathematical or geometric relationships. The pull of gravity on an animal, the laws of thermodynamics, and similar physicochemical factors do not change with body size. Similarly, the properties of many biological materials are independent of body size. Because of the nature of muscle architecture and contraction (p. 309), the force of muscle contraction per unit of cross-sectional area is the same for a mouse and an elephant.

Functional Anatomy and Development

Growth and Changing Proportions of Body Parts

Studies of allometric growth in animals help us understand changing proportions of body parts among species on an evolutionary scale. Allometric growth also occurs during the development and growth of individual animals. Body size increases rapidly during embryonic development. If all parts of the embryo increased in size in the same proportion, surface-volume relationships would soon result in the exchange surface becoming too small to supply gas, nutrients, and other products to the increasing mass. According to Gilbert (1991), "The most successful growth strategy to circumvent the surface-volume problem is invagination." The digestive tract and lungs are invaginated regions that increase surface area relative to mass. As size increases further, the exchange surfaces become more distant from the cells to be supplied, but the development of a circulatory system

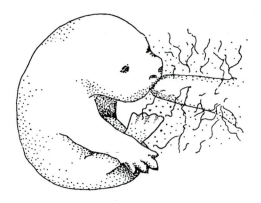

FIGURE 1–6   A recently born marsupial attached to a nipple in its mother's pouch. Note the relatively large head and pectoral appendage.

shortens diffusion distance by bringing materials to be exchanged closer to the cells. New organisms must function as they develop, and this phenomenon requires continuing allometric changes and remodeling of exhange surfaces, the circulatory system, the skeleton, and other organ systems.

Body proportions as a whole are also affected during embryonic development. Intrauterine development in marsupial mammals is very short (only about two weeks for the American opossum). A newborn marsupial is essentially an embryo, but its pectoral appendages, face, and jaws have grown at a much faster rate than the rest of the body (Fig. 1–6). The embryo can pull itself into the mother's pouch, or marsupium, where it attaches itself to a nipple. Pouch life is several times longer than intrauterine life, and "embryonic" development is completed here.

Paedomorphosis

Allometric growth helps us understand many changes that occur during the development of individuals. Other changes occur as a result of a change in the timing of the development of a part of an embryo relative to the timing of its development in the ancestral species. This general phenomenon is known as heterochrony, and we discuss it further later (p. 135). One aspect of heterochrony of interest now involves differential growth rates in the embryo between reproductive organs and other parts of the body. Such differences in growth rates have led to the evolution of new forms that retain certain larval or juvenile features in otherwise sexually mature individuals. Larvae and juveniles possess body shapes and other features that differ considerably from their adult counterparts. The retention of larval or juvenile features is known as **paedomorphosis** (Gr. *paido*, from *pais* = child + *morphosis* = form). Paedomorphosis can be a result of two different developmental mechanisms. During the development of an animal, the differentiation and maturation of the reproductive organs can proceed at an accelerated pace relative to the rest of the body. The result is a sexually mature individual with the body appearance of a larva or juvenile. This phenomenon is known as **progenesis.** In sharp contrast, the second developmental mechanism leading to paedomorphosis involves a delay or decelerated development of nonreproductive body parts relative to the maturation of the reproductive organs. The result is also a sexually mature individual with many other body parts in the larval or juvenile condition. This phenomenon is called **neoteny.** It is sometimes difficult to distinguish between progenesis and neoteny unless comparisons can be made with related species that have normal development.

It is hypothesized that progenesis is found in species living in unpredictable environments (Gould, 1977). Food and other resources may be abundant only for very short periods of time. An acceleration of the development and maturation of the reproductive organs would give the species a high reproductive potential and enable it to capitalize on the short periods of abundant resources. Some miniature fish species with a juvenile appearance living in unpredictable environments may have evolved by progenesis.

Neoteny is found in species that inhabit predictable and stable environments. Species exploit these stable environments by retaining larval or juvenile features, while sexual maturation proceeds at the rate found in the nonneotenic ancestor. Many salamanders are neotenic and remain aquatic throughout their life cycle rather than metamorphosing to terrestrial adults, as do other salamanders. The juvenile (neotenic) features, such as external gills and a finned tail, are adaptations to the stable aquatic habitats and resources. In this way neotenic salamanders have escaped the vicissitudes of the harsh terrestrial habitats surrounding the water in which they live. A well-known example of neoteny is the mudpuppy, *Necturus,* which is often studied in comparative anatomy laboratories.

Paedomorphosis is important in evolution because it produces new morphological combinations of juvenile and adult characters that can be acted on by natural selection. The potential exists for a paedomorphic population to take an evolutionary direction that would not have been possible for an adult population that was highly adapted to a restricted mode of life. Paedomorphosis may explain the large morphological gaps we sometimes see between major groups of animals. The prevalent theory for the origin of vertebrates invokes paedomorphosis (p. 47).

Comparative Interpretation

Phylogeny and Classification

To see order in the great diversity of organisms and to present a general overview, biologists today use a system of classifying organisms that was developed by Karl von Linnaeus in 1758. The individual kinds, or species, of organisms are arranged into a nested hierarchy of progressively inclusive groups or taxa. For instance, lions, tigers, and closely related cats are placed in one genus *(Panthera);* the domestic cat and wild cat of the Old World are in the genus *Felis;* and the lynx and other cats are in yet other genera. All cats are grouped into a cat family (Felidae), which together with the dog family (Canidae), bear family (Ursidae), and related families constitutes the order of carnivores (Carnivora). Linnaeus based his groups on degree of similarity; contemporary biologists attempt to arrange their groups on the basis of their evolutionary relationship.

Evolutionary relationships of organisms can be graphically represented in a **phylogeny** (Gr. *phylon* = tribe + *genesis* = origin) (Fig. 1–7). Diverging taxa are

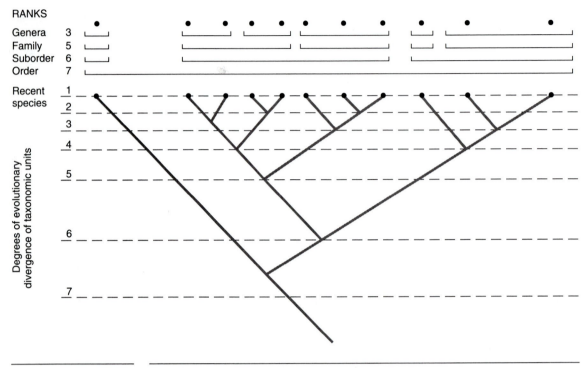

FIGURE 1–7 A hypothetical phylogeny based on monophyly. Note the dichotomous branching pattern of the taxa. The grouping of species into a nested hierarchy of progressively inclusive groups is diagrammed at the top of the figure.

represented by a dichotomous branching pattern. Each pair of lineages is represented by lines that diverge from their most recent common ancestor. The oldest divergence forms the base, or root, of the phylogenetic tree, while the tips of the branches represent the most recently evolved taxa. In a carefully constructed phylogeny, each taxon or phylogenetic unit (genus, family, order, and so on) contains all the species derived from a common ancestor. A taxon that contains one or more lineages and the most recent common ancestor from which the lineages originated is said to be **monophyletic.** Thus we believe that all cats form a monophyletic group, and that cats, dogs, and bears, who share a more remote common ancestor, form a larger monophyletic group. A phylogeny is a hypothesis of evolutionary relationships between groups of organisms based on monophyly. New discoveries test hypotheses of evolutionary relationship and may require modifications in the phylogenetic tree.

Homology, Homoplasty, and Analogy

As a first step in deducing the evolution of organisms and organ systems, biologists compare living and extinct organisms and search for similarities that suggest a relationship. Structural, functional, molecular, genetic, and other features may be compared. All contribute to our understanding of relationships among animals,

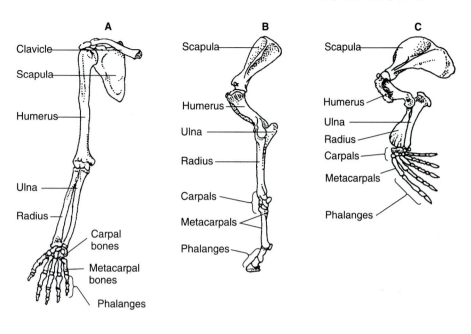

FIGURE 1–8 The bones of the left pectoral girdle and limb of a human (**A**), horse (**B**), and seal (**C**) illustrate the similar topographical relationships of the major elements despite differences in limb function: grasping, running, and a stabilizing keel, respectively. The human limb is viewed from the front with the hand in the prone position; the others are from a lateral view.

but an anatomist emphasizes the structural ones. In making comparisons, an investigator must clearly understand the types of features being compared and their biological significance. The concepts of homology, homoplasty, and analogy help investigators do this.

There are several kinds of homology. We will be concerned primarily with **evolutionary homology.** Structures in two or more species are homologous if they can, in principle, be traced back to a common, ancestral precursor. When structures in two species have a fundamental resemblance to each other, we have reason to believe that these species are descended from a common, ancestral precursor, or that one has transformed into the other. The fundamental resemblance between homologues is not based on shape or other superficial features alone but is usually manifested by similar topographical relationships and connections to surrounding parts. An example is the bones in the forelimbs of a human, a horse, and a seal (Fig. 1–8). Since the limbs are used in different types of locomotion, we should not expect them to be identical, yet the bones have very similar topographical relationships. Sometimes the resemblance is less obvious, as in the case of the auditory ossicles of mammals and certain visceral arches of fishes. The resemblance becomes evident when one examines stages in the transformation (p. 242). The ossicles and visceral arches arise from nearly identical embryonic precursors. Homologous organs, then, may or may not have similar functions. Function is not a criterion for homology.

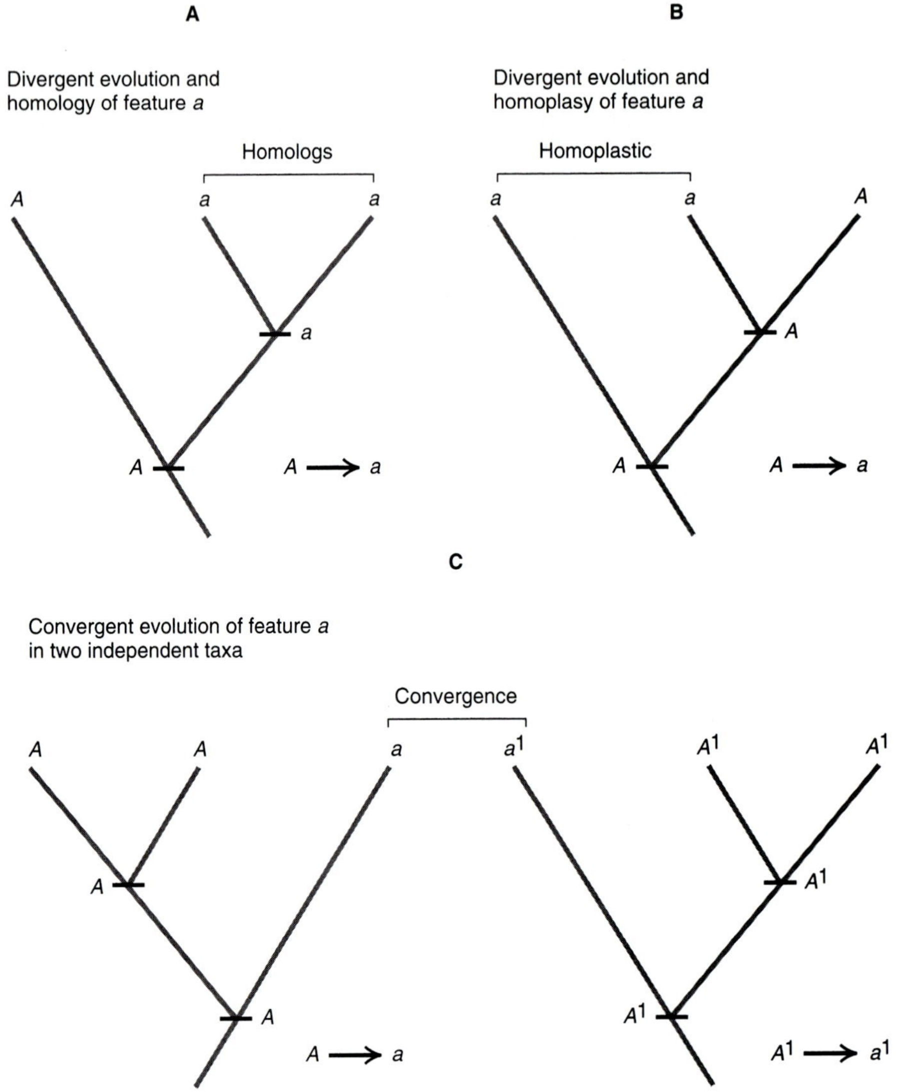

A

Divergent evolution and
homology of feature *a*

B

Divergent evolution and
homoplasy of feature *a*

C

Convergent evolution of feature *a*
in two independent taxa

FIGURE 1–9 Diagrams of phylogenies showing different types of relationships of structures symbolized by *a*. **A**, Divergent evolution and homology of features *a*. **B**, divergent evolution and homoplasty of features *a*. **C**, convergent evolution.

Serial homology refers to entities in different parts of the same individual that can be traced back to elements of a precursor that are repeated in linear sequence. The individual vertebrae are serially homologous because each develops from the same part of a linear series of embryonic body segments. **Sexual homology** refers to entities in different sexes of the same species that can be traced back to the same embryonic source (e.g., the glans penis of the male and the glans clitoridis of the female, which both derive from the phallus of an embryo).

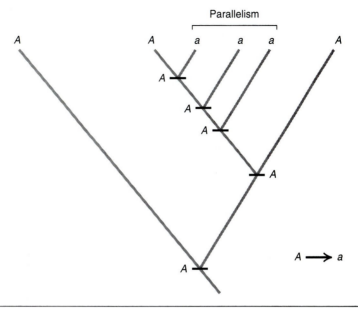

D

Parallel evolution and parallelism
of feature *a* in related taxa

FIGURE 1–9   *continued*
D, parallel evolution. Features *a* are homoplastic in both convergent and parallel evolution.

Evolutionary homologues are organs that have a common evolutionary origin (Fig. 1–9**A**). However, organs that are structurally similar do not necessarily originate from a common ancestral condition. If structural similarity evolved in organisms with different ancestors, the condition is called **homoplasty,** and the structurally similar organs are said to be **homoplastic** (Fig. 1–9**B**). An example of homoplasty is the wings of birds and bats (Fig. 1–10). Since both are airfoils and perform the same function—flight—they necessarily resemble each other superficially. As wings they are homoplastic and not homologous because each wing evolved independently.

A few biologists still use the term **analogy** for nonhomologous organs with similar functions. Homoplasty is a subset of analogy in which there is some degree of resemblance between the organs, but many analogous organs do not resemble each other. All that is required is a similar function. One example is the gills of a fish and the lungs of a terrestrial vertebrate. They share the same function—gas exchange—but do not resemble each other because gas exchange occurs in different media (water or air).

Divergent, Convergent, and Parallel Evolution

Biologists reconstruct the evolutionary history of organisms and organ systems by comparing homologous organs (Focus 1–1). Identifying homologous organs is not

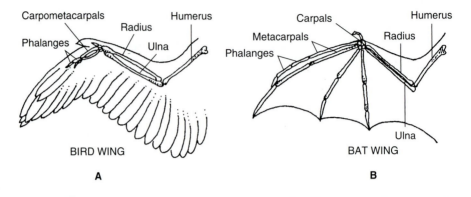

FIGURE 1–10   Various types of wings: **A**, bird; **B**, bat. As wings, both are homoplastic because each type of wing evolved independently. However, at the level of comparison of appendages, the bird and bat wing are homologous because each represents a modification of a terrestrial vertebrate forelimb.

always easy because of the potential confusion with homoplastic ones. When organs in different species are identified as homologous, the implication is that the species arose from a common ancestor by **divergent evolution** (Fig. 1–9**A**). Organs are identified as homoplastic when species arose not by divergence from a common ancestor but by convergent or parallel evolution (Fig. 1–9**C** and **D**). **Convergent evolution** is the independent evolution of homoplastic structures among species that do not share a recent common ancestry. Birds and bats converged as they both adapted to a flying mode of life and evolved wings (Fig. 1–10). The wings *as wings* are homoplastic. They are both airfoils and look alike superficially, but one type of wing did not evolve from the other. That they are homoplastic and not homologous is indicated by the nature of the airfoil surface. A bat wing is essentially an enlarged, webbed hand; a bird wing is composed of feathers that grow out from the caudal border of the limb. However, the *bones* in their limbs are homologous, and this implies that birds and bats descended from some remote common ancestor (an early terrestrial vertebrate) in which the pattern of limb bones had become established. Birds and bats evolved wings independently of each other, but both types of wings are specialized forelimbs. **Parallel evolution** is the independent evolution of homoplastic structures among animals that have had a more recent common ancestry (Fig. 1–9**D**). The independent evolution of quills by American and African porcupines is an example of parallel evolution. Because the difference between convergent and parallel evolution is the degree of relationship between the animals concerned, a sharp distinction sometimes is difficult to make (compare Fig. 1–9**C** and **D**).

Divergent, convergent, and parallel evolution reflect the adaptive responses of organisms to environmental factors. Natural selection often determines the direction of change in form and function of a structure. Biotic and abiotic factors in the environment are potentially capable of affecting the course of evolutionary change. The environment acts like a filter; it preserves individuals that are better adapted to environmental conditions and eliminates less well-designed individuals.

FOCUS 1–1   Reconstructing Phylogenies

Biologists reconstruct the evolutionary history, or **phylogeny,** of groups by trying to determine the points, or nodes, at which lineages branch and diverge. To do this, investigators undertake a careful analysis of the distribution of homologous organs among the species being studied, and evaluate which homologous characters evolved recently and which a long time ago. If two species or groups share a recently evolved, homologous character that is not present in any other group, it is a reasonable hypothesis that they have diverged from a recent common ancestor. Such a character is called a **derived** or **advanced character** or, technically, an **apomorphic character** (Gr. *apo* = separation + *morphe* = form). A character that evolved at a more distant time would be more widely distributed and found in more groups. Such a character is described as a **primitive** or **ancestral character** or, technically, a **plesiomorphic character** (Gr. *plesios* = near, i.e., close to the ancestor).

The way primitive and derived characters are used in constructing a phylogeny can be seen by looking at the evolution of mammals (see the figure). Mammals, birds, and reptiles share a cleidoic egg, or some modification of such an egg, in which the embryo develops an amnion and other extraembryonic membranes needed to survive in a terrestrial environment until it hatches as a miniature adult. The cleidoic egg is a primitive character, but it is so widespread that it cannot be used to separate mammals from other groups. Hair, warm bloodedness (endothermy), and other features of ancestral mammals are derived characters that define mammals. However, since all mammals have them, they are of no use in sorting out the lines of mammalian evolution. To do this, an investigator must search for more recently evolved derived characters. The retention of embryos in a uterus for at least part of embryonic development is a derived character that separates marsupials and eutherian mammals from the egg-laying monotremes. A type of placenta that allows a long intrauterine life is a derived character of eutherian mammals that distinguishes them from marsupials. These examples illustrate the essential aspects of character analysis, but character analysis is a large and growing field of biology. We refer those who wish to pursue the topic further to books by Eldredge and Cracraft (1980) and Wiley (1981).

Two groups of biologists treat the data on branching points somewhat differently (Charig, 1982). **Traditional** or **evolutionary systematists** sometimes group more than one lineage into a higher taxon based on shared structural and functional characters, and they may exclude a closely related lineage with a different suite of characters. For example (see accompanying figure), a traditional systematist would include turtles, lizards and snakes, crocodiles, and the extinct mammal-like reptiles that gave rise to mammals as one class, the Reptilia. All share the cleidoic egg, and all are scaled. (Note that the cleidoic egg is a derived character for a reptile, since it helps define the group, but it is a primitive character from the perspective of a mammal.) Birds, which are more closely related to crocodiles than crocodiles are to turtles, are excluded from the reptiles and placed in their own class, the Aves, based on the presence of feathers, endothermy, and other shared adaptations for flight.

Phylogenetic systematists, or **cladists** (Gr. *clados* = branch), insist that all groups be monophyletic and include the ancestral group and all of its descendants. In their view, reptiles is an unnatural, **paraphyletic** group because it excludes certain reptile descendants (i.e., the birds and mammals). Cladists try to avoid such groupings. To them, turtles would be one group, lizards and snakes another, crocodiles and birds another, and mammals another. In preparing this book, we have tended to favor cladistics, but we retain some traditional groupings (such as reptiles and birds) because they have been used for so long and are so familiar; however, we recognize their artificial nature.

continued

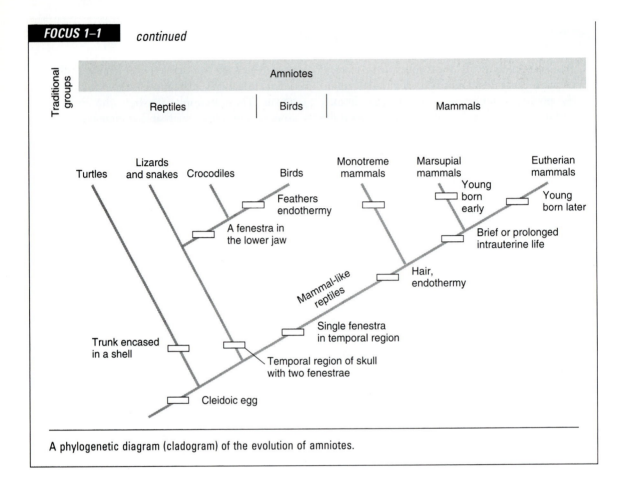

A phylogenetic diagram (cladogram) of the evolution of amniotes.

Adaptations are hereditary adjustments in the design of organisms. Divergent evolution shows how divergent environmental factors have resulted in a variety of adaptations (Fig. 1–9**A** and **B**). Convergent and parallel evolution reflect how unrelated and related animals evolve similar solutions (adaptations) to common environmental challenges (Fig. 1–9**C** and **D**).

Phylogeny and Comparative Biology

Ideally, a phylogenetic tree (Fig. 1–7) is the final summation of all data gathered in a comparative study (Kluge, 1977). It provides a framework from which new hypotheses and generalizations can be formulated. For example, one can ask whether a functional anatomical adaptation Z (Fig. 1–11**A**) has played a role in producing divergent evolution leading to the evolution of taxa A, B, C, D, and E while the primitive taxon (X) not possessing the adaptation has not diversified. It is a hypothesis that relates the possession of the adaptation Z to the generation of diversity. To test this hypothesis, one can compare an unrelated lineage that possesses a structurally identical, but convergent, homoplastic adaptation Z^1 and

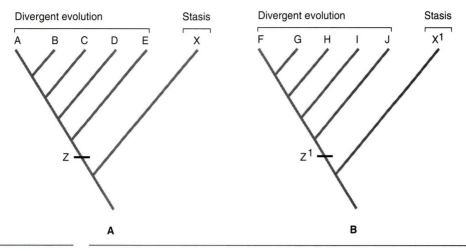

FIGURE 1–11   A phylogenetic analysis of the effect of the evolution of feature Z on divergence. **A**, Group being analyzed. **B**, the unrelated comparison group.

examine the effects of Z^1 (Fig. 1–11**B**). If the lineage with Z^1 has undergone comparable divergent evolution (F, G, H, I, and J) while the lineage without Z^1 has not diversified (X^1), we can draw the conclusion that the presence of the convergent adaptations Z and Z^1 are both causally related to divergent evolution. Research in functional anatomy within the framework of phylogeny provides for feedback, counterchecks, and testing of functional anatomical and theoretical approaches to the evolution of form (Lauder, 1981).

Anatomical Terminology

Before continuing with a structural and functional analysis of the evolution of the organ systems of vertebrates, you should understand the basis for the terminology that anatomists use.

Over the years many authors have used different terms for the same organs. For example, the vagus nerve that carries parasympathetic fibers to the thoracic and abdominal organs was at one time called the pneumogastric nerve. Medical anatomists recognized such confusion many years ago and agreed on a code of anatomical nomenclature known as the *Nomina Anatomica*. This has been updated periodically and has been extended to include embryological and histological terms. In the meantime veterinary anatomists have agreed on a *Nomina Anatomica Veterinaria* that applies human terms to quadruped mammals as far as possible. Major differences relate to differences in posture (e.g., "superior" in human beings becomes "cranial" in a quadruped). Bird anatomists have agreed on a *Nomina Anatomica Avium*. These codes are not binding, but they are helpful, and we have used them as a guide for the terminology in this book. Unfortunately, no guide, other than conventional usage, is available for structures in fishes and amphibians.

Most anatomical terms are derived from Latin or Greek, and they often are written in Latin in scientific papers, as are the names of animals. Our biceps muscle is technically known as *Musculus biceps brachii*, for example, but anatomical terms

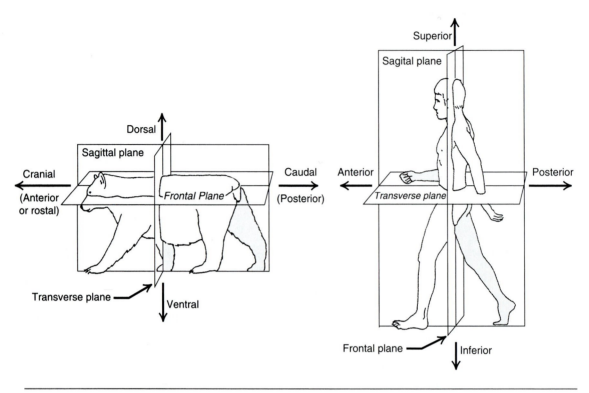

FIGURE 1–12   Diagrams to show the terms for direction and for the planes of the body in a quadruped and a biped.

are placed in the vernacular in most writing (biceps brachii muscle). The derivation of many widely used terms from their Greek or Latin roots will be indicated in this book. The roots are repeated in many combinations, and learning the vast anatomical terminology will become easier and easier if derivations are studied carefully.

Since terms that define positions within the body are needed to describe the location of organs, they must be understood at the outset. Most should be clear from Figure 1–12. Notice the differences in terminology between a quadruped and a biped. The terms *dorsal* and *ventral* are not used in an adult biped, and *anterior* and *posterior* are used differently in adult bipeds and quadrupeds. Use of the terms *anterior* and *posterior* is therefore often avoided in adult quadrupeds, and *cranial* and *caudal* are used instead. *Rostral* may be used for *cranial*, especially for structures within the head.

SUMMARY

1. Functional anatomy examines how the various components of organisms perform specific functions, and the linkage between form and function. It examines why and how changes in form and function came about. Recent findings in functional anatomy have revolutionized the study of form, function, and adaptation.

2. Biologists use scaling analysis to study the structural and functional consequences of changes in size among otherwise similar organisms. As the linear dimensions of an object or animal increase, the surface area (a square function) increases at a slower rate than the mass or volume (a cubic function). This often poses problems because the strength and power of supporting bones and muscles, which change in proportion to their cross-sectional area, do not keep pace with the increasing mass, and the surface area through which physiological exchanges occur also does not keep pace with the mass to be supplied. Such problems are met in organisms by differential (allometric) growth of various parts of the body, by changes in structural materials (cartilage to bone), or by design changes in the structures (evolution of gills).

3. Allometric growth occurs as animals change in size during an evolutionary lineage, and during the embryonic development and growth of an animal to maturity. The invagination of surfaces is one device that helps the surface area keep pace with the enlarging mass of an embryo, but most organ systems and the proportions of different parts of the body are affected.

4. A differential growth rate during development between reproductive organs and other parts of the body has led to the evolution of paedomorphosis, the retention of some larval or juvenile features by otherwise sexually mature individuals. Paedomorphosis may occur by progenesis or neoteny. Paedomorphosis may have been a factor in adapting populations to different environmental conditions, and in major evolutionary transitions.

5. Our system of classifying organisms groups them into a nested hierarchy of progressively inclusive groups, or taxa.

6. The evolutionary relationship among the taxa can be depicted by a phylogeny, in which each pair of lineages is displayed by lines that diverge from their most recent common ancestor. In a well-constructed phylogeny, each taxon is monophyletic and contains one or more lineages and their most recent common ancestor.

7. The underlying basis for constructing phylogenies is to search for similarities. In searching for similarities, investigators must distinguish between similarities due to inheritance from a common ancestor (homology) and ones that have evolved independently among unrelated species (homoplasty).

8. When organs in different species are identified as homologous, the implication is that the species arose from a common ancestor by divergent evolution. Homoplastic organs indicate that the species arose by convergent or parallel evolution. Divergent, convergent, and parallel evolution reflect the adaptive responses of the organisms to environmental factors.

9. Ideally, phylogenies summarize all of the data that have been gathered during a comparative study. They are hypotheses about relationships, and they can be used as a framework from which new hypotheses can be formulated and tested.

GENERAL REFERENCES

Our references include books and articles cited in the text and others that will lead students to important additional sources of information. We have been selective and emphasized material published in the last decade or two. Older works are included only if they are cited in the text, are classics, or are of particular interest. References listed in this section are works of general interest and works that apply to more than one organ system. Those related to the subject of particular chapters are placed at the ends of the chapters.

Alexander, R. McN., 1974: *Functional Design in Fishes*, 3rd edition. London, Hutchinson and Company Limited.

Alexander, R. McN., 1981: *The Chordates*, 2nd edition. Cambridge, Cambridge University Press.

Alexander, R. McN., 1983: *Animal Mechanics*, 2nd edition. Oxford, Blackwell Scientific Publications.

Andrews, S. M., Miles, R. S., and Walker, A. D., editors, 1977: *Problems in Vertebrate Evolution*. Linnean Society Symposium, No. 4. London, Academic Press.

Bellairs, A. d'A., 1969: *The Life of Reptiles*. London, Weidenfeld and Nicolson.

Bellairs, A. d'A., and Cox, C. B., 1976: *Morphology and Biology of the Reptiles*. London, published for the Linnean Society by Academic Press.

Bemis, W. E., Burggren, W. W., and Kemp, N. E., 1986: *The Biology and Evolution of the Lungfishes*. New York, Alan R. Liss, Inc.

Bolk, L., and others, editors, 1931–1938: *Handbuch der vergleichenden Anatomie der Wirbeltiere*. 6 vols. Reprinted: Amsterdam, A. Asher and Company, 1967.

Bond, C. E., 1979: *Biology of Fishes*. Philadelphia, W. B. Saunders Company.

Carroll, R. L., 1988: *Vertebrate Paleontology and Evolution*. New York, W. H. Freeman and Company.

Corliss, C. E., 1976: *Patten's Human Embryology*. McGraw-Hill Book Company.

Dawson, T. J., 1983: *Monotremes and Marsupials: The Other Mammals*. Southampton, Edward Arnold.

Dorst, J., 1974: *The Life of Birds*. Translated by J. C. J. Galbraith. New York, Columbia University Press.

Duellman, W. E., and Trueb, L., 1985: *Biology of the Amphibia*. New York, McGraw-Hill Book Company.

Dullemeijer, P., 1974: *Concepts and Approaches in Animal Morphology*. Assen, The Netherlands, Van Gorcum and Company.

Fawcett, D. M., 1986: *Bloom and Fawcett, A Textbook of Histology*. Philadelphia, W. B. Saunders Company.

Feder, M. E., and Burggren, W. W., editors, 1992: *Environmental Physiology of the Amphibians*. Chicago, University of Chicago Press.

Feduccia, A., and McCrady, E., 1991: *Torrey's Morphogenesis of the Vertebrates*, 5th edition. New York, John Wiley and Sons, Inc.

Ferguson, M. W. J., editor, 1985: *Structure, Development and Evolution of Reptiles*. Symposia of the Zoological Society of London, no. 52. London, Academic Press.

Gans, C., and others, editors, 1969–1992: *Biology of the Reptilia*. New York, Academic Press.

Gans, C., 1974: *Biomechanics, An Approach to Vertebrate Biology*. Philadelphia, J. P. Lippincott Company. Reprinted 1980, Ann Arbor, University of Michigan Press.

Goin, C. J., Goin, O. B., and Zug, G. R., 1978: *Introduction to Herpetology*, 3rd edition. San Francisco, W. H. Freeman and Company.

Goodrich, E. S., 1930: *Studies on the Development and Structure of Vertebrates*. London, MacMillan and Company.

Grassé, P. P., editor, 1948–1973: *Traité de Zoologie*. Volumes 11–17 deal with protochordates and vertebrates. Paris, Masson et Cie.

Gregory, W. K., 1951: *Evolution Emerging*. New York, The Macmillan Company.

Hardisty, M. W., 1979: *Biology of Cyclostomes*. London, Chapman and Hall.

Harrison, R. J., editor, 1972: *Functional Anatomy of Marine Mammals*. New York, Academic Press.

Hildebrand, M., 1988: *Analysis of Vertebrate Structure*, 3rd edition. New York, John Wiley and Sons.

Hildebrand, M., Bramble, D. M., Liem, K. F., and Wake, D. B., editors, 1985: *Functional Vertebrate Morphology*. Cambridge, Harvard University Press.

Hoar, W. S., 1975: *General and Comparative Physiology*, 2nd edition. Englewood Cliffs, N.J., Prentice-Hall.

Hoar, W. S., and Randall, D. J., editors, volumes 1–12, 1969–1992: *Fish Physiology*. New York, Academic Press.

Jarvik, E., 1980: *Basic Structure and Function of Vertebrates*. London, Academic Press.

Jollie, M., 1973: *Chordate Morphology*. Huntington, N.Y., Robert E. Krieger Publishing Company.

Kemp, T. S., 1982: *Mammal–like Reptiles and the Origin of Mammals*. London, Academic Press.

Kent, G. C., 1991: *Comparative Anatomy of the Vertebrates*, 7th edition. St. Louis, C. V. Mosby Company.

King, A. S., and McLelland, J., 1979: *Form and Function in Birds*. London, Academic Press.

Kluge, A. G., Frye, B. E., Johansen, K., Liem, K. F., Noback, C. R., Olsen, I. D., Waterman, A. J., 1977: *Chordate Structure and Function*, 2nd edition. New York, Macmillan Publishing Company.

Moy-Thomas, J. A., and Miles, R. S., 1971: *Paleozoic Fishes*. Philadelphia, W. B. Saunders Company.

Moyle, P. B., and Cech, J. J., Jr., 1982: *Fishes: An Introduction to Ichthyology*. Englewood Cliffs, N.J., Prentice-Hall.

Neal, H. V., and Rand, H. W., 1936: *Comparative Anatomy*. Philadelphia, P. Blakiston's Son & Company.

Piveteau, J., editor, 1952–1968: *Traité de Paleontologie*. Paris, Masson et Cie.

Porter, K. R., 1972: *Herpetology*. Philadelphia, W. B. Saunders Company.

Portmann, A., 1948: *Einfuhrung in die Vergleichende Morphologie der Wirbeltiere*. Basel, Benno Schwabe & Co. Verlag.

Pough, F. H., Heiser, J. B., and McFarland, W. N., 1989: *Vertebrate Life*, 3rd edition. New York, Macmillan Publishing Company.

Prosser, C. L., editor, 1991: *Environmental and Metabolic Animal Physiology, Comparative Animal Physiology*, 4th edition. New York, Wiley-Liss.

Prosser, C. L., editor, 1991: *Neural and Integrative Animal Physiology, Comparative Animal Physiology*, 4th edition. New York, Wiley-Liss.

Romer, A. S., and Parsons, T. S., 1986: *The Vertebrate Body*, 6th edition. Philadelphia, Saunders College Publishing.

Schmalhausen, I. I., 1968: *The Origin of Terrestrial Vertebrates*. New York, Academic Press.

Schmidt-Nielsen, K., Bolis, L., Taylor, C. R., Bentley, P. J., and Stevens, C. E., editors, 1980: *Comparative Physiology: Primitive Mammals*. Cambridge, Cambridge University Press.

Schmidt-Nielsen, K., 1983: *Animal Physiology: Adaptation and Environment*, 3rd edition. Cambridge, Cambridge University Press.

Shuttleworth, T. V., editor, 1988: *Physiology of Elasmobranch Fishes*. Berlin, Springer-Verlag.

Smith, H. M., 1960: *Evolution of Chordate Structure*. New York, Holt.

Starck, D., 1978–1982: *Vergleichende Anatomie der Wirbeltiere*. Berlin, Springer-Verlag.

Sturkie, P. D., editor, 1986: *Avian Physiology*, 4th edition. New York, Springer-Verlag.

Taylor, C. R., Johansen, K., and Bolis, L., editors, 1982: *A Companion to Animal Physiology*. Cambridge, Cambridge University Press.

Tyler, M. J., 1976: *Frogs*. Sydney, Collins.

Tyndale-Biscoe, H., 1973: *Life of Marsupials*. London, Edward Arnold, Australia.

Vaughan, T. A., 1988: *Mammalogy*, 2nd edition, Philadelphia, W. B. Saunders Company.

Wake, D. B., and Roth, G., 1989: *Complex Organismal Function: Integration and Evolution in Vertebrates*. New York, John Wiley and Sons.

Wake, M. H., editor, 1979: *Hyman's Comparative Anatomy*, 3rd edition. Chicago, University of Chicago Press.

Walker, W. F., Jr., and Homberger, D. G., 1992: *Vertebrate Dissection*, 8th edition. Philadelphia, Saunders College Publishing.

Welty, J. C., and Baptista, L., 1988: *The Life of Birds*, 4th edition. Philadelphia, Saunders College Publishing.

Williams, P. L., Warwick, R., Dyson, M., and Bannister, L. H., editors, 1989: *Gray's Anatomy*, 37th British edition. Edinburgh, Churchill Livingstone.

Wolfe, R. G., 1991: *Functional Chordate Anatomy*. Lexington, Mass., D. C. Heath and Company.

Young, J. Z., 1981: *The Life of Vertebrates*, 3rd edition. Oxford, Clarendon Press.

Young, J. Z., and Hobbs, M. J., 1975: *The Life of Mammals*, 2nd edition. Oxford, Clarendon Press.

Zug, G. R., 1993: *Herpetology: An Introductory Biology of Amphibians and Reptiles*. San Diego, Academic Press.

REFERENCES FOR CHAPTER 1

Alexander, R. McN., 1985: Body support, scaling, and allometry. *In* Hildebrand, M., Bramble, D. M., Liem, K. F., and Wake, D. B., editors: *Functional Vertebrate Morphology*. Cambridge, Harvard University Press.

Baumel, J. J., King, A. S., Lucas, A. M., Breazile, J. E., and Evans, H. E., 1979: *Nomina Anatomica Avium*. London, Academic Press.

Calder, W. A. III, 1984: *Size, Function, and Life History*. Cambridge, Harvard University Press.

Charig, A., 1982: Systematics in biology: A fundamental comparison of some major schools of thought. *In* Joysey, K. A., and Friday, A. E., editors: *Problems of Phylogenetic Reconstruction*. New York, Academic Press.

DeBeer, G. R., 1958: *Embryos and Ancestors*, 3rd edition. London, Oxford University Press.

Dial, K. P., Goslow, G. E., Jr., and Jenkins, F. A., Jr., 1991: The functional anatomy of the shoulder of the European starling *(Sturnus vulgaris)*. *Journal of Morphology*, 207: 327–344.

Eldredge, N., and Cracraft, J., 1980: *Phylogenetic Patterns and the Evolutionary Process*. New York, Columbia University Press.

Frazzetta, T. H., 1975: *Complex Adaptations in Evolving Populations*. Sunderland, Mass., Sinauer Associates Inc.

Ghiselin, M. T., 1976: The nomenclature of correspondence: A new look at "homology" and "analogy." *In* Masterton, R. B., Hodos, W., and Jerison, H., editors: *Evolution, Brain and Behavior*, vol. 2. Hillsdale, N.J., Lawrence Erlbaum and Associates.

Gilbert, S. F., 1991: *Developmental Biology*, 3rd edition. Sunderland, Mass., Sinauer Associates Inc.

Gould, S. J., 1966: Allometry and size in ontogeny and phylogeny. *Biological Reviews, Cambridge Philosophical Society*, 41: 587–640.

Gould S. J., 1977: *Ontogeny and Phylogeny*. Cambridge, Harvard University Press.

Huxley, J., 1972: *Problems of Relative Growth.* Reprinted in 1932: New York, Dover Publications.

International Committee on Veterinary Anatomical Nomenclature, 1983: *Nomina Anatomica Veterinaria,* 3rd edition. Ithaca, N.Y., Department of Veterinary Anatomy, Cornell University.

Kluge, A. G., Frye, B. E., Johansen, K., Liem, K. F., Noback, C. R., Olsen, I. D., and Waterman, A. J., 1977: *Chordate Structure and Function,* 2nd edition. New York, Macmillan Publishing Co., Inc.

Lauder, G. V., 1981: Form and Function: Structural analysis in evolutionary morphology. *Paleobiology,* 7: 430–442.

McMahon, T., 1973: Size and shape in biology. *Science,* 179: 1201–1204.

Mayr, E., 1969: *Principles of Systematic Zoology.* New York, McGraw-Hill Book Company.

Schmidt-Nielsen, K., 1984: *Scaling, Why Is Animal Shape So Important?* Cambridge, Cambridge University Press.

Thompson, D'Arcy W., 1961: *On Growth and Form,* abridged edition. Edited by J. T. Bonner, Cambridge, Cambridge University Press.

Wiley, E. O., 1981: *Phylogenetics: The Theory and Practice of Phylogenetic Systematics.* New York, John Wiley and Sons.

2 | The Protochordates and the Origin of Vertebrates

PRECIS

In this chapter we explore the nature of animals that show an affinity to the vertebrates. This enables us to understand the position of vertebrates in the Animal Kingdom, and to understand the evolution of their distinctive, derived characters. This leads us to consider hypotheses about the origin of vertebrates.

We are interested in the evolution of vertebrates, but vertebrates are only one subphylum of a larger group, the phylum **Chordata.** Chordates also include two subphyla of small, soft-bodied, marine animals, the sea squirts and amphioxus. Sea squirts and amphioxus are collectively called the **protochordates.** Protochordates are slow-moving or sessile creatures that spend their adult lives sifting minute food particles from sea water. Sac-shaped sea squirts live attached to the ocean floor; amphioxus is a superficially fish-shaped animal that burrows shallowly in the sand. The different subphyla of chordates contain quite different kinds of animals, but at some stage of their development all chordates share four distinctive, derived characters: (1) a longitudinal stiffening rod of turgid cells along the dorsal part of their body that is called a notochord, (2) a single nerve cord that contains a central canal and is located in the back dorsal to the notochord, (3) pharyngeal pouches that grow laterally from the pharynx and often open to the surface as pores or slits, and (4) a groove in the pharynx floor known as the endostyle, or a thyroid gland derived from part of the endostyle. To understand the signficance of these derived characters, and to understand hypotheses concerning the evolutionary origin of vertebrates, we must first look at the protochordates. We must also consider the acorn worms and their relatives in the phylum **Hemichordata** because hemichordates, although not chordates, have some characters that resemble those of chordates.

The Phylum Hemichordata

The most familiar hemichordates are the approximately 80 species of acorn worms that belong to the class **Enteropneusta** (Fig. 2–1A). These worm-shaped animals range in length from a few centimeters to more than 2 meters. The body is divided into a **proboscis, collar,** and **trunk,** each of which contains a separate coelomic cavity. The proboscis and collar coelom connect by pores to the outside, so they can be inflated with sea water. Enteropneusts move through the sand by the ciliary action of epidermal cells assisted by the actions of the proboscis and collar. Once the inflated collar anchors the wormlike body, the proboscis is pushed forward. Next, the proboscis can be inflated and anchored and the rest of the body pulled up behind it. The proboscis is also a food-gathering organ. Food particles falling on it are trapped in mucus and carried by cilia into the mouth, which is located ventrally just in front of the collar. Excess water escapes from the pharynx through many paired pharyngeal pores or slits located dorsolaterally along the anterior part of the trunk. These pores increase in number during development, as in amphioxus, by the formation of secondary tongue bars between the primary ones. In some species folds of the body wall extend dorsally and form an atrium-like space that partly covers the pharyngeal slits. An intestine extends caudally from the pharynx to a terminal anus. The nervous system resembles that of starfish and other echinoderms. A subepidermal network of neurons becomes condensed dorsally and ventrally into solid nerve strands. The dorsal strand contains a few cavities in the collar region, but these do not qualify it as a tubular nerve cord characteristic of chordates. The base of the proboscis contains a pharyngeal diverticulum known as the **stomochord.** Its wall is composed of vacuolated cells that resemble, in

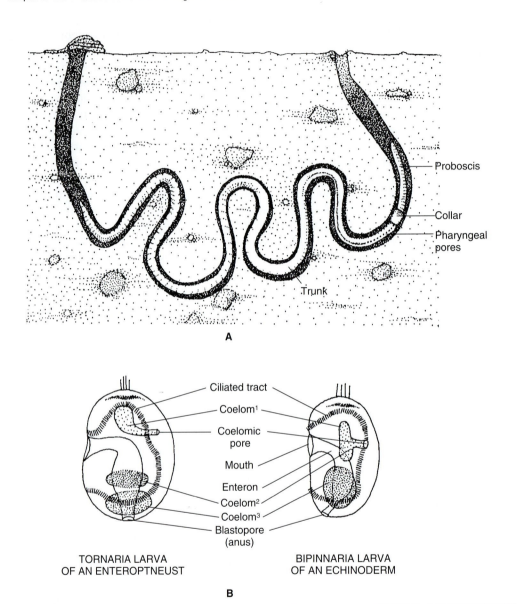

A

B

- Proboscis
- Collar
- Pharyngeal pores
- Trunk

- Ciliated tract
- Coelom¹
- Coelomic pore
- Mouth
- Enteron
- Coelom²
- Coelom³
- Blastopore (anus)

TORNARIA LARVA
OF AN ENTEROPTNEUST

BIPINNARIA LARVA
OF AN ECHINODERM

FIGURE 2–1   Hemichordates: **A**, An adult acorn worm in its burrow. **B**, The tornaria larva of a hemichordate (left) compared with the bipinnaria larva of a starfish (right).

some respects, those that make up the distinctive notochord of chordates. Some investigators consider the stomochord and notochord to be homologous, but this is doubtful because their positional relationships and embryonic development are quite different.

Enteropneusts are of interest because the presence of pharyngeal slits suggests an affinity to the chordates. The nervous system suggests an affinity to the

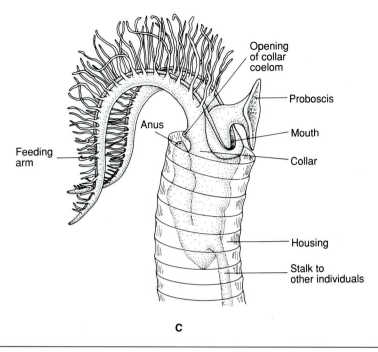

FIGURE 2–1   *continued*
C, One individual of a pterobranch colony, *Rhabdopleura*. (**B,** from Dorit, Walker, and Barnes. **C,** redrawn from Feduccia and McCrady.)

echinoderms. The **bipinnaria larva** of many echinoderms and the enteropneust **tornaria larva** are also similar (Fig. 2–1**B**). In addition to a similar shape, gut cavity, and surface ciliated bands used in locomotion, both larval types have a tripartite coelom.

The remaining few species of hemichordates are placed in the class **Ptero-branchia** (Fig. 2–1**C**). They are small, colonial animals that superficially look like plants. Each individual, is lodged in a little cup of a branching, trunklike house that is secreted by members of the colony as they multiply by budding. Individuals may or may not remain connected to each other. Each individual has a distinct proboscis, collar, and trunk, each with its own coelom. The collar is extended into one or more pairs of branching, ciliated feeding arms that collect food. Species in one genus, *Cephalodiscus,* have a single pair of pharyngeal slits; other ptero-branchs have none. The intestine forms a U-shaped loop, so the anus opens near the feeding arms.

The Tunicata

Turning now to the phylum **Chordata,** we will first consider the subphylum **Tun-icata,** which contains approximately 2,000 species. All but approximately 100 are sessile sea squirts belonging to the class **Ascidiacea.** The remaining species, placed in the class **Thaliacea,** are pelagic. Most sea squirts live attached to the ocean bottom in shallow water. Many sea squirts are colonial; others are solitary (Fig. 2–2**A** and **B**).

Adult Ascidian Structure

An adult sea squirt is a small, sac-shaped animal encased in a protective and supportive **tunic**. The tunic is composed of a mucopolysaccharide containing fibers of **tunicin,** a carbohydrate similar to cellulose. It is secreted by the epidermis of the thin body wall, or **mantle,** and by ameboid cells that move through the tunic.

Water and food particles enter a large pharynx through an apical, **incurrent siphon**. Food is trapped in mucus secreted by a ciliated groove in the pharynx floor known as the **endostyle**. Excess water escapes through hundreds of minute pharyngeal slits into a chamber, the **atrium,** that surrounds the pharynx laterally and ventrally. The atrium discharges through an **excurrent siphon** located on one side of the sac beside the incurrent one. Food passes through an intestine that also empties into the atrium. Digestive enzymes are secreted primarily by the first part of the intestine, sometimes called the stomach. Digestion is entirely extracellular, and absorption occurs throughout the intestine.

The circulatory system consists of a small **heart** and blood vessels that extend from it to the various organs and into the tunic. The vessels extending from the heart form a continuous system of channels, but they are not completely lined with endothelial cells. Blood is pumped by peristaltic contractions of the heart in one direction for several minutes, and then the flow reverses. Most excretion occurs by diffusion.

A **ganglion** lies in the mantle between the siphons. Nerves extend from it to the siphons, mantle muscles, and some viscera. A **subneural gland** is located ventral to the ganglion.

The Ascidian Larva

Tunicates can reproduce asexually by budding and also sexually. Although individuals are hermaphroditic, they are rarely self-fertilizing. The eggs develop quickly into small, motile larvae. After a few days, during which the population is dispersed, the larvae attach to the bottom and metamorphose into sessile adults.

The distinctive chordate pharyngeal slits and endostyle are present in the adult, but the other chordate characters, which are related to locomotion, can be seen only in the larvae. The ascidian larva is tadpole shaped, with an expanded front end and a slender tail (Fig. 2–2**C**). The tail is a locomotor organ and contains a notochord of approximately 40 vacuolated and turgid cells. Longitudinal muscle fibers flank the notochord but are not segmentally arranged. A characteristic chordate tubular nerve cord lies dorsal to the notochord and integrates locomotor movements. The anterior end of the nerve cord expands into a **cerebral vesicle,** but this is not comparable to a brain because it contains only a simple eyespot and a balancing statolith, which the larva needs to find a suitable habitat on which to settle. The pharynx and atrium begin to develop ventral to the cerebral vesicle, but the larva does not feed. At metamorphosis the larva attaches to the substratum

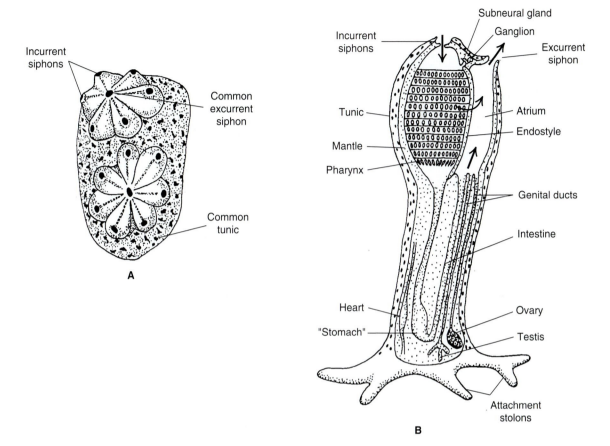

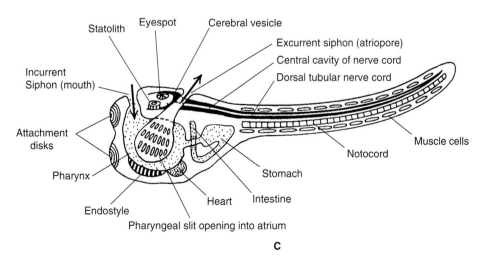

FIGURE 2–2   Representative ascidians: **A**, *Botryllus,* a colonial species. **B**, An individual of *Clavelina.* **C**, Diagram of an ascidian larva.

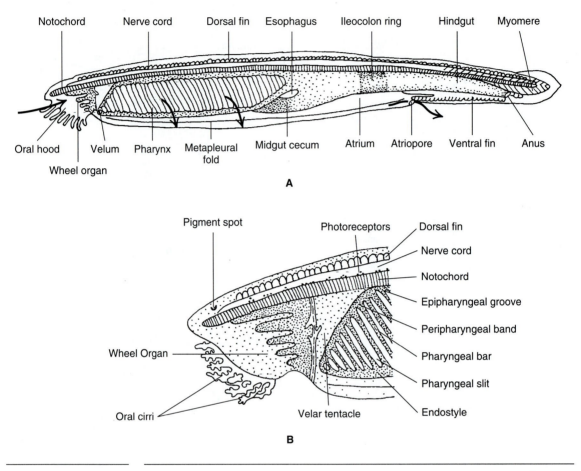

FIGURE 2–3   The structure of amphioxus: **A**, Lateral view drawn as though it were transparent. **B**, Details of anterior end. (From Walker and Homberger)

by adhesive suckers located at its anterior end. The tail is absorbed, and part of the cerebral vesicle becomes the ganglion.

Chordate Characters and Their Evolution

Sea squirts display some of the derived characters that define the chordates and distinguish them from all other phyla in the adult stage and others only in the larva. All show particularly well in the adult of amphioxus (Figs. 2–3 and 2–4). All of these characters evolved in connection with a unique method of feeding and pattern of locomotion. The **pharyngeal pouches** evaginate embryonically from the lateral walls of the pharynx. They open to the body surface as pharyngeal slits in protochordates and fishes. Gills are not present in the pharyngeal pouches of protochordates as they are in fishes; rather the slits of protochordates are part of

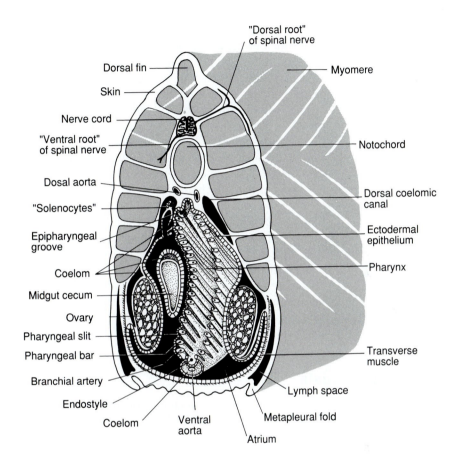

"Dorsal root" of spinal nerve

Dorsal fin

Skin

Nerve cord

"Ventral root" of spinal nerve

Dosal aorta

"Solenocytes"

Epipharyngeal groove

Coelom

Midgut cecum

Ovary

Pharyngeal slit

Pharyngeal bar

Branchial artery

Endostyle

Coelom

Ventral aorta

Myomere

Notochord

Dorsal coelomic canal

Ectodermal epithelium

Pharynx

Transverse muscle

Lymph space

Metapleural fold

Atrium

FIGURE 2–4   Drawing of a cross section of amphioxus through the posterior part of the pharynx. (From Walker and Homberger.)

their feeding mechanism. This was probably their primitive function. A current of water is drawn by cilia through the mouth into the pharynx. Cells in a ciliated groove in the floor of the pharynx, known as the **endostyle,** secrete mucus that spreads over the inside of the pharynx wall. The minute food particles suspended in the water are trapped in this mucus, and excess water escapes through the pharyngeal slits. These animals are filter feeders. Filter feeding is widespread among invertebrates, but the combined use of pharyngeal slits and an endostyle in the filtering mechanism is a unique chordate character. Certain cells in the endostyle also bind proteins with iodine, and many believe that these cells are homologous to the secretory cells of the thyroid gland of vertebrates.

Other derived chordate characters relate to locomotion. All chordates have at some stage of their life history a nerve cord of a unique type that integrates the contraction of the locomotor muscles in the body wall. The nerve cord develops embryonically as a pair of ectodermal folds that meet and close on the dorsal surface of the body (p. 125), so the **nerve cord** is **dorsal, single,** and **tubular.** When present in other phyla, nerve cords are solid structures that are located more ventrally and are usually double strands.

A **notochord,** which all chordates have at some stage of their development, is also a part of the locomotor system. It lies just ventral to the nerve cord. The notochord is a hydroskeleton that consists of a longitudinal column of cells, each of which contains a large, central vacuole full of liquid. These cells are surrounded by a sheath of dense connective tissue that maintains their turgidity. The primary function of the notochord is to resist compression and so prevent the body from shortening, or telescoping, in the manner of an earthworm when the longitudinal muscle fibers in the body wall contract. Muscle contraction causes instead a lateral bending from side to side, and this propels the animal.

Pharyngeal slits, an endostyle or its derivative (the thyroid gland), a unique type of nerve cord (single, dorsal, and tubular), and a notochord are the derived chordate characters that evolved in ancestral species. They are retained in the larvae or embryos of all chordates, but they are not always present in the adult stage.

The Cephalochordata

The subphylum Cephalochordata comprises approximately 45 species, most of which belong to the genus *Branchiostoma,* commonly called amphioxus.° Amphioxus (Gr. *amphi* = both + *oxys* = sharp) is widely distributed along the ocean coasts of the world. It is a small animal, about 3 cm to 5 cm long, and pointed at each end. It has a distinct anterior end but not a well-developed head. Amphioxus is more active than other protochordates and resembles a vertebrate in some ways.

The notochord of amphioxus extends from the anterior end of the body to nearly the tip of the tail (Fig. 2–3A). Its extension to the front of the body is the basis for the name of the subphylum, Cephalochordata (Gr. *kephale* = head + *chorde* = cord). A very long notochord stiffens each end of the animal and facilitates pushing into the sand in either direction. Each cell of the notochord forms a highly vacuolated, disk-shaped plate whose flat surface lies in the transverse plane of the body. The whole notochord resembles a stack of checkers. Electron microscope studies have shown that the notochordal cells contain striated muscle fibrils that resemble those in the paramyosin muscles that hold the two shells of a clam

---

°The name *Branchiostoma,* Costa, 1834, has priority over *Amphioxus,* Yarrell, 1836, but *Amphioxus* had been so widely used before the synonymy was recognized that it is retained as the common name for the animal.

together. Slender cytoplasmic processes from the notochordal cells extend into the nerve cord and synapse with motor nerve cells. Stimulation of the notochordal cells causes them to contract and increases their internal fluid pressure. Thus the notochord of amphioxus is not only a flexible hydroskeleton but its turgidity can be regulated.

Although amphioxus burrows in the sand most of the time, it also swims. Small **caudal, dorsal,** and **ventral fins** and paired lateral fins provide some stability for swimming (Fig. 2–3**A**). The lateral fins, called **metapleural folds,** are continuous structures that extend along the anterior two thirds of the body (Fig. 2– 4). They unite caudally on the ventral side. The longitudinal muscle fibers in the body wall are not continuous but are grouped into a long series of <-shaped muscle segments known as **myomeres**. Separate waves of contraction can begin at one end of the body (either anterior or posterior) and sweep to the other end, causing a series of lateral undulations. Effective swimming requires that the functional posterior end of the body oscillate more than the other end. The great length of the notochord and the ability to control its rigidity differentially along its length are adaptations that allow the animal to swim and burrow in either direction. The segmentation of the muscle fibers imposes a segmentation, or **metamerism,** on much of the rest of the body because blood vessels that supply the myomeres, and nerves that pass between them, are also segmented.

The unpigmented skin is thin and consists of an **epidermis** of only a single layer of columnar epithelial cells resting on a basement membrane of collagen fibrils. Amphioxus has no organized respiratory structures, but its small size and compressed shape allow gas exchange through the skin.

When feeding, amphioxus lies partly buried in the sand with only its anterior end protruding (Fig. 2–5). Filter feeders spend a great deal of time feeding, and the digestive organs make up much of the body. The mouth lies deep within the **oral hood** and is surrounded by **velar tentacles** (Fig. 2–3**B**). The oral hood is fringed with **oral cirri** that strain out coarse particles, and it is lined by ciliated bands known as the **wheel organ**. Water is drawn into the pharynx by the action of the cilia on the wheel organ and in the large pharynx. The pharynx wall is perforated by many vertically elongated, pharyngeal slits separated by fibrocartilaginous pharyngeal bars. The number of slits and bars increases as the animal grows because secondary, tongue-shaped bars extend downward from the dorsal wall of the pharynx between the first-formed or primary ones. Cells in the endostyle in the pharyngeal floor secrete mucus that entraps minute food particles (Fig. 2– 4). The mucus and food are carried by ciliary action dorsally along the pharyngeal bars to an **epipharyngeal groove.** Some endostylar cells also secrete iodinated proteins.

Excess water taken in with the food escapes through the pharyngeal slits into a chamber, the **atrium,** around the pharynx (Fig. 2– 4). An **atriopore** carries water to the outside (Fig. 2–3**A**). Contractions of muscles in the floor of the atrium expel the water. The atrium protects the numerous, delicate pharyngeal bars from abrasion in a burrowing animal. Amphioxus also gains oxygen and discharges carbon dioxide from the water flowing through the pharynx even though gills are not present.

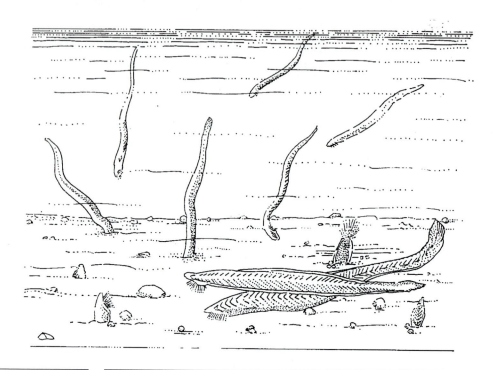

FIGURE 2–5   Amphioxus in its habitat.

The pharynx narrows caudally into an **esophagus**. A **midgut** and **hindgut** follow, and the intestine opens by an anus that is located on the left side of the caudal fin. There is a short, postanal tail, as in most vertebrates. A ciliated band, the **ileocolon ring,** moves the mucus-food cord caudally, within the gut tube. There are no muscle fibers in the gut wall. A large, hollow **midgut cecum** extends from the floor of the midgut forward into the atrium along the right side of the pharynx. Minute food particles that break off the mucus-food cord are carried by ciliary action into the cecum, where both extracellular and intracellular digestion and some absorption occur. Absorbed food is stored as glycogen and lipid in some cecal cells. The cecum thus has some functions resembling those of a vertebrate liver (storage) and others not present in a liver (enzyme secretion).

A **ventral aorta** carries blood forward beneath the pharynx, and paired **branchial arteries** carry it dorsally through each pair of pharyngeal bars (Fig. 2–4). Blood is collected from the branchial arteries by a pair of dorsal aortae. These unite caudal to the pharynx into a single **dorsal aorta** that distributes blood to the tissues. This pattern of blood flow, forward ventrally and caudad dorsally, resembles that in a vertebrate but is opposite that in other invertebrates with well-organized circulatory systems. Blood flows through tissue spaces, for there are no true capillaries. It is collected by a system of body wall veins and a **subintestinal vein** that converge on a chamber known as the **sinus venosus,** located at the caudal end of the ventral aorta. There is no heart. Blood is propelled by the slow contraction of many vessels, including muscular **branchial bulbs** at the bases of the branchial arteries. The blood contains amebocytes but no other cells. Oxygen-binding blood pigments are lacking, so oxygen is carried in physical solution.

A pair of **dorsal coelomic canals** lies above the atrium on either side of the pharynx (Fig. 2–4). They contain the excretory organs, often called **solenocytes,** which remove waste products from the blood and discharge them into the atrium (Focus 2–1).

Amphioxus lacks highly organized sense organs, such as nose, eyes, and ears, but receptor cells occur in those parts of the body on which stimuli impinge. Simple **tactile receptors** occur in the skin, atrium, and part of the gut wall. **Chemoreceptors** on the body surface and on the velar tentacles detect changes in the chemical environment and provide information on the type of particles entering the pharynx. Noxious particles are rejected. Many **photoreceptors** lie in the floor of the central canal of the nerve cord (Fig. 2–3**A**). Each consists of a single cell shielded ventrally by pigment, so it responds only to light coming from above through the thin integument. Several photoreceptors are aggregated to form a pigment spot at the front of the nerve cord. Sensory neurons from the receptor cells pass between the myomeres to the nerve cord. As in vertebrates, these fibers contribute to the **dorsal roots** of the spinal nerves (Fig. 2–4), but the cell bodies of these nerve cells lie within the spinal cord and not in peripheral spinal ganglia, as in vertebrates. Motor neurons that are visceral in function leave the nerve cord and travel in the dorsal roots to part of the gut and atrial walls. None of the neurons have a myelin sheath. The separate **"ventral roots"** of the nerves are composed of the processes of myomere muscle cells, not neurons. Although the central canal expands slightly at the anterior end of the nerve cord, the cord does not, and there is no brain. Amphioxus has a very limited behavioral repertoire that consists of little more than feeding and escape movements, and reproductive behavior.

The sexes are separate in amphioxus. The **gonads,** testes or ovaries, develop in a series of segmented coelomic sacs that bulge from the ventral part of the body wall into the atrium (Fig. 2–4). Mature gametes are discharged into the atrium, and the gonads develop anew in the next reproductive season. Fertilization and development are external.

The Vertebrata

The subphylum **Vertebrata** is clearly a member of the phylum chordata because all vertebrates share the derived chordate characters. As embryos, all have pharyngeal pouches, which have perforated the body wall as gill slits in fishes and larval amphibians. As adults, all have a thyroid gland, which develops in lampreys from an endostyle-like subpharyngeal gland. A dorsal, single, and tubular nerve cord develops in the embryos and persists in adult vertebrates. All possess an embryonic notochord, and some fishes and a few amphibians and reptiles retain it as adults.

The Excretory Organs of Amphioxus

Excretion in amphioxus occurs via clusters of highly specialized cells often called **solenocytes,** which lie in the dorsal coelomic canals (figure **A**). Although called solenocytes, these cells are certainly not homologous to the solenocytes of primitive worms. The solenocytes of amphioxus are truly remarkable since they have both filtration and pumping effects. Each cell has a group of footlike processes that wrap around an arterial branch from the dorsal aorta (figure **B**). Each solenocyte also has a pump consisting of a slotted, tubelike structure that extends through the

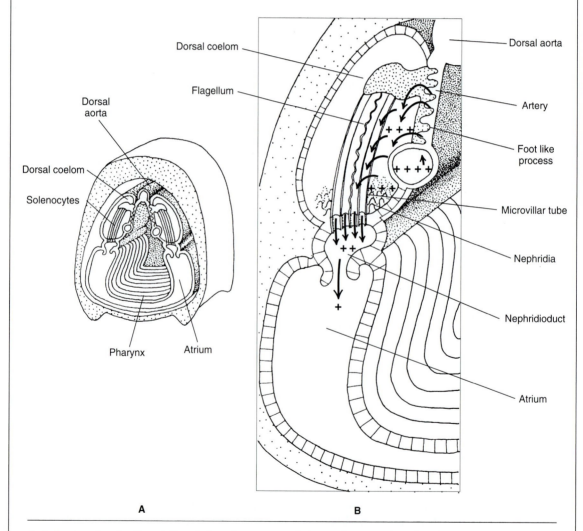

A   B

A, A diagram showing the relation of the solenocytes to the arterial system, dorsal coelomic canals, and atrium. The pharynx has been removed. **B**, Details of the relationships of one solenocyte to surrounding structures. (**B**, modified after Brandenburg.)

FOCUS 2–1 *continued*

dorsal coelomic canal to enter a **nephridioduct.** The nephridioduct opens into the atrium. The slotted tube wall consists of ten long microvilli that extend from the cell body of the solenocyte (figure **B**). A long flagellum lies within the tube, and its movements pump and propel liquid. Hydrostatic pressure in the artery causes filtration by forcing water and various wastes through the arterial wall, between the footlike processes of the solenocyte, and into the dorsal coelomic canal. Once in the dorsal coelomic canal, water and waste products are drawn through the slots and into the cavity of the microvillar tube by the movements of the flagellum, and then pumped into the nephridioduct. Water and wastes are then driven into the atrium. The loss of water that this mechanism entails is not a problem because amphioxus has the same salt concentration in its body fluid as in sea water. Lost water readily diffuses back into the body.

The evolution of the distinctive, derived characters of vertebrates is correlated with their activity and size. An active animal needs a concentration of sense organs and nervous tissue anteriorly because the front of the body is the part that first encounters new environments. Vertebrates have become **cephalized.** They have a distinctive head that, in addition to housing the mouth and pharyngeal pouches, contains well-developed **eyes, ears,** and a **nose** (Fig. 2–6). The lens of the eye and the nose and ears develop embryonically from **sensory placodes.** Sensory placodes, which are unique to vertebrates, are thickened areas of surface ectoderm in an early embryo that invaginate later in development. Other placodes, adjacent to the auditory one, form the **lateral line system** that ramifies over the head and extends caudally along the side of the trunk and tail. This system provides fishes and larval amphibians with important information about water movements and pressure changes, and also gives rise to specialized electroreceptors in many aquatic vertebrates. The front end of the neural tube is enlarged and forms a **brain** that receives and integrates the information from the newly evolved sense organs. The vertebrate brain is tripartite, consisting of a **forebrain** that in early vertebrates integrates olfactory information from the nose, a **midbrain** that processes information from the eyes, and a **hindbrain** whose development is related in large part to the ear and lateral line. The sense organs and brain are encased and protected by a skeletal **cranium** of cartilage or bone. **Bone** is a calcium phosphate compound with a complex histological organization that adds greatly to its strength (p. 184). It occurs only in vertebrates and is another of their derived characters, although it has been lost secondarily in some lineages. Ossicles in the body wall of many invertebrates differ in being calcium carbonate compounds.

An active animal also needs an efficient locomotor system. Ancestral aquatic vertebrates swim by lateral undulations of their trunk and tail. These movements are generated, as in amphioxus, by the sequential contraction of a series of seg-

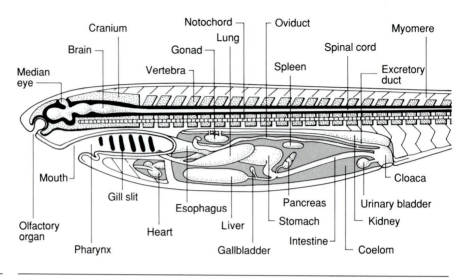

FIGURE 2–6   Diagram of a generalized vertebrate showing the major derived features. (From Dorit, Walker, and Barnes.)

mental myomeres. The action of the myomeres and the larger size of most fishes require a stronger strut than a notochord to resist compression as the animal pushes through the water. A **vertebral column** does this. Vertebrae develop around the nerve cord and notochord, and the vertebral bodies largely replace the notochord in most adult vertebrates.

Increased activity and size are not possible without increased metabolic activity. More oxygen and carbon dioxide must be exchanged with the environment, a larger volume of food must be digested and absorbed, and materials must be distributed efficiently throughout the body. **Gills** develop in the pharyngeal slits, and they are supported partly by a series of **visceral arches** of cartilage or bone that develop in the wall of the pharynx between the pharyngeal slits. The gut wall also becomes muscularized, which it is not in the protochordates. Specialized muscles become associated with the visceral arches, and their action helps expand and compress the pharynx. These movements draw a respiratory current of water into the mouth and expel it across the gills. The earliest vertebrates captured prey by means of the pumping action of the pharynx. Later in vertebrate evolution, an anterior visceral arch contributed to the formation of jaws. This enabled vertebrates to feed on more types of food. Muscles in the wall of the rest of the digestive tract move food effectively through a series of chambers specialized for the successive storage, digestion, and absorption of food. A **liver** evolved that, among its many functions, stored considerable energy as glycogen or lipid. The efficiency of the circulatory system increased with the evolution of a muscularized, multichambered **heart,** closed capillaries connecting arteries and veins, and the respiratory pigment hemoglobin that binds reversibly with oxygen. The evolution of distinctive

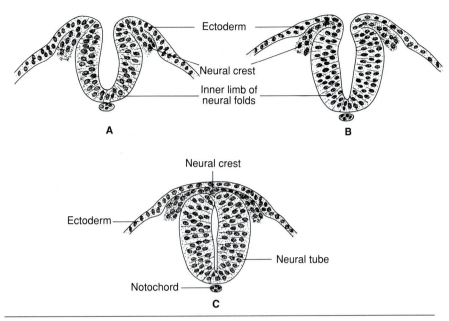

FIGURE 2–7   Diagrams through the back of a vertebrate embryo showing the formation of the neural tube and neural crest. (Redrawn from Corliss.)

kidney tubules helped eliminate nitrogenous wastes and control water balance in the freshwater environment, which vertebrates entered early in their evolutionary history. The evolution of all these features will be explored more thoroughly in later chapters.

The functional movements of the gut, heart, and other visceral organs are facilitated by a surrounding body cavity, or **coelom.** Primitively, the coelom is divided into a **pericardial cavity** around the heart and a **peritoneal cavity** around most of the digestive organs. The coelom divides the vertebrate body into two distinctive parts, separating a muscularized gut tube and heart from the muscularized body wall. Skeleton, muscles, and nerves in the body wall constitute the **somatic** part of the body. The somatic body is associated in ancestral vertebrates with locomotion. Skeleton, muscles, and nerves in the gut tube and heart form the **visceral** part of the body. This part is associated primarily with metabolic functions.

The most distinctive features of vertebrates correlate with the appearance in an early embryo of a group of cells that separate from the developing nerve cord to form a **neural crest**. The nerve cord develops by the elevation and joining of a pair of longitudinal, ectodermal folds (Fig. 2–7**A**). As these meet, neural crest cells are "pinched off" (Fig. 2–7**B** and **C**) and form ridges along the dorsolateral wall of the nerve cord. The nerve cord forms in a similar way in protochordates, but a neural crest does not develop. Cells from the neural crest spread to many parts of the body, give rise to many important structures, and appear to exert a regulatory role on many aspects of development. Some cells give rise to the distinctive visceral skeleton of vertebrates. Some form certain of the motor nerve cells that innervate the visceral tube. Others give rise to the sensory neurons of vertebrates. These neurons arise in protochordates directly from the nerve cord. Some neural crest cells give rise to cells that form the dentine of teeth, and the dentine, in turn, induces the formation of the enamel. Dentine and enamel not only are found in teeth, but also characterize the bony scales and plates of early fishes. We will see many other derivatives of the neural crest as we consider the organ systems, for it is an important derived vertebrate feature that underlies the improved functions and performance so characteristic of vertebrate life.

The Origin of Chordates and Vertebrates
Chordate Affinities

The protochordates are soft-bodied animals and have not left a fossil record, but chordates must be an ancient group. Fragments of early vertebrate scales have been found in late Cambrian deposits, and this indicates that chordate origins go back more than 500 million years. In the absence of a clear fossil record, investigators have tried to identify by other means the phyla to which protochordates and vertebrates show the closest affinity. Because vertebrates are active animals, early investigators sought an affinity between them and such active invertebrates as annelids and arthropods. All are metameric animals, but the nature of the metamerism and the basic body organization of a vertebrate are different from those of annelids and arthropods. The metamerism of the annelid-arthropod evolutionary line affects not only the body wall muscles but also the coelom, which is divided primitively into a series of compartments. (These compartments are secondarily reduced in arthropods.) Contractions of the body wall can vary the fluid pressure in the coelomic compartments, and the coelom acts as a flexible hydroskeleton. The nerve cord of annelids and arthropods is a solid, ganglionated, double strand on the ventral surface of the body. The direction of blood flow in annelids and arthropods is opposite that of vertebrates, moving posteriorly on the ventral surface of the body and anteriorly on the dorsal side. It would be necessary to turn an annelid or arthropod upside down before it could transform into a vertebrate. Early investigators had good imaginations, and some ingenious hypotheses were proposed. All beg the question, however, of the origin of protochordates, other than regarding them as degenerate vertebrates.

Later investigators sought the origin of vertebrates not among the highly specialized annelids and arthropods but among filter-feeding protochordates and early, filter-feeding echinoderms. Echinoderms, hemichordates, and chordates do show a close affinity in many ways. We have seen that the hemichordates resemble echinoderms in having a subepidermal nerve net, tripartite coelomic division, and a larva with ciliated bands. Enteropneust hemichordates resemble protochordates in having a filter-feeding mechanism that includes pharyngeal slits. The slits of hemichordates multiply during development when distinctive tongue bars form, as they do in amphioxus.

The distribution of certain phosphate compounds used in muscle contraction reinforces the notion that chordates have a closer affinity to echinoderms than to annelids and arthropods. Muscle contraction utilizes a great deal of adenosine triphosphate (ATP) as an energy source. ATP is eventually resynthesized from the metabolism of food products, but some ATP is reformed very rapidly by the breakdown of other phosphate compounds. This alternative energy source is **phosphoarginine** in nearly all invertebrate groups, including protozoa, cnidarians, primitive worms, arthropods, and molluscs. The compound is **phosphocreatine** in vertebrates and cephalochordates. Some species of tunicates and hemichordates have one of these compounds, and other species have the other compound. This distribution suggests an affinity between chordates and echinoderms, but the evidence is not conclusive. Hemichordates retain the more primitive arginine system, and a few species of annelids have evolved phosphocreatine.

The most convincing evidence for an affinity among chordates, hemichordates, and echinoderms comes from basic similarities in their early embryonic development. The patterns are different from those of other advanced invertebrates, including the annelids, arthropods, and molluscs. Details of embryonic development will be considered in a later chapter, but in brief, in annelids, arthropods, and molluscs the division, or cleavage, of the fertilized egg into a sphere of cells has a precise, spiral pattern, and the parts of the body to which cleavage cells give rise is determined very early. In echinoderms, hemichordates, and chordates cleavage is neither spiral nor so determinate. As development continues, the sphere of cells is converted into a double-layered gastrula that contains a simple gut cavity, or **archenteron,** that opens to the surface through a **blastopore** (Fig. 2–8**A**). In annelids, arthropods, and molluscs the embryonic blastopore becomes or at least contributes to the adult mouth (Fig. 2–8**B**), for which reason these animals are called **protostomes** (Gr. *protos* = first + *stoma* = mouth). Echinoderms, hemichordates, and chordates are called **deuterostomes** (Gr. *deuteros* = second) because the mouth arises not from the blastopore at the caudal end of the gastrula but from a second invagination at the anterior end of the larva that pushes into the archenteron (Fig. 2–8**C**). The blastopore becomes the anus, or is located near the future site of the anus, in many species. In primitive members of all deuterostome phyla the middle layer of cells, known as the mesoderm, and the coelom develop from paired enterocoelic pouches that pinch off from the endodermal archenteron (Fig. 2–8**F** and **G**). At some point in development the coelom has a connection with the gut cavity. This pattern of development is modified in advanced vertebrates, but the coelom in all deuterostomes is considered an **enterocoele**. The mesoderm arises by an earlier separation of cells in protostomes, and the coelom develops by a cavitation or splitting of originally solid bands of mesoderm (Fig. 2–8**D** and **E**). There is never a connection to the gut cavity. The coelom of protostomes is thus referred to as a **schizocoele** (Gr. *schizein* = to cleave).

The pattern of early development is a good indication of relationship, and it is unlikely that animals with quite different patterns can be closely related. There appears to have been a major dichotomy in the evolution of coelomate animals. Protostomes, in general, have a spiral and determinate cleavage, the blastopore forms or contributes to the mouth, and the coelom is a schizocoele. The

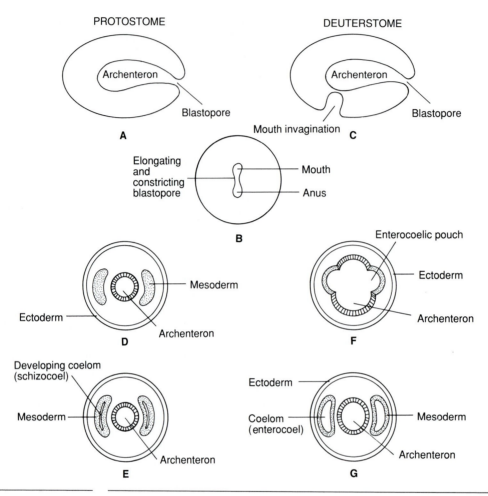

PROTOSTOME

DEUTERSTOME

Archenteron

Blastopore

A

Archenteron

Blastopore

Mouth invagination   **C**

Elongating and constricting blastopore

Mouth

Anus

B

Ectoderm

Mesoderm

Archenteron

D

Enterocoelic pouch

Ectoderm

Archenteron

F

Developing coelom (schizocoel)

Mesoderm

Archenteron

E

Ectoderm

Coelom (enterocoel)

Mesoderm

Archenteron

G

FIGURE 2–8   Developmental features of protostomes and deuterostomes: **A**, A lateral view of a protostome gastrula. **B**, End view of a later stage of a protostome in which the blastopore is
elongating and constricting to form the mouth and anus. **C**, Lateral view of a later deuterostome gastrula in which a secondary invagination is forming the mouth. **D** and **E**, Cross
sections of two stages in the formation of the mesoderm and coelom in protostomes. **F** and
G, Cross sections of two stages in the formation of the mesoderm and coelom in
deuterostomes.

deuterostomes have a nonspiral and indeterminate cleavage, the blastopore does
not form the mouth, and the coelom is an enterocoele.

Although considerable evidence points to an affinity among echinoderms,
hemichordates, and chordates, we know little about their course of evolution. It
is clear that hemichordates and chordates did not diverge from echinoderms
resembling contemporary starfish, sea urchins, or sea cucumbers. They must have
diverged from early echinoderms or pre-echinoderm ancestors. The early echinoderms of the Cambrian period were sessile and often stalked animals that
resembled modern sea lilies. They were filter feeders, but they must have sepa-

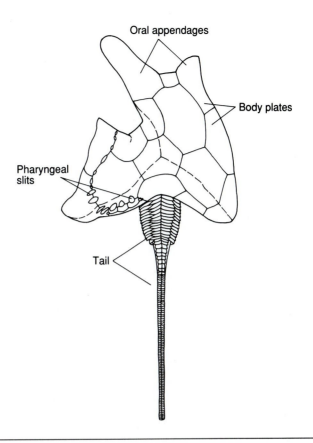

FIGURE 2–9   A reconstruction of the Cambrian fossil *Ceratocystis*. (Redrawn from Jefferies.)

rated food particles from debris and water before the food entered the mouth because they had no pharyngeal slits.

Chordate and Vertebrate Origins

The ancestors of modern echinoderms, hemichordates, and chordates may have resembled those earliest echinoderms. Beyond that we enter a realm of speculation. Jefferies (1975) has proposed that chordates arose from Cambrian animals known as cornuates (Fig. 2–9). These creatures are usually regarded as stalked echinoderms related to the sea lilies, but Jefferies believes they were ancestral chordates that lay on the bottom and crawled with a motile stalk or tail. They appear to have had pharyngeal slits, but the evidence that the stalk was motile and contained a nerve cord, notochord, and muscles is less certain.

Most investigators favor a hypothesis originally suggested by Bateson (1886) and later developed and modified by Garstang (1928), Berrill (1955), Northcutt and Gans (1983), and others. Chordate ancestors probably resembled primitive, stalked echinoderms that used ciliated arms in food gathering (Fig. 2–10). Some pterobranch hemichordates presumably added a pair of pharyngeal slits to the

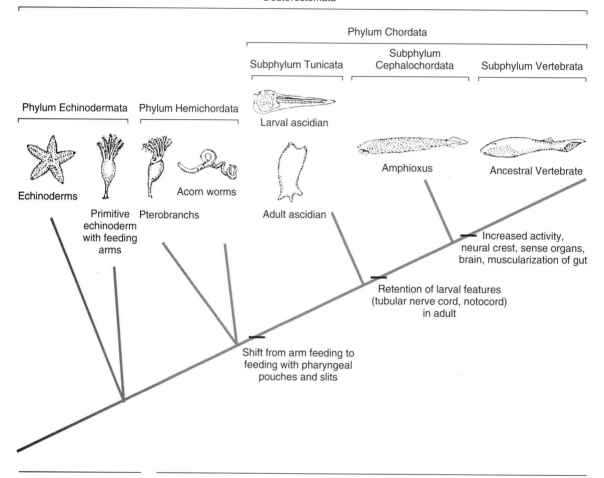

FIGURE 2–10 A cladogram showing a hypothesis on the probable major stages of evolution from echinoderms to vertebrates.

feeding arm filter-feeding mechanism. We see this in one contemporary genus of pterobranchs, *Cephalodiscus*. Pharyngeal slits apparently were an effective way to eliminate excess water and concentrate food. Selection appears to have favored them over the feeding arm system. Slits increased in number in all other hemichordates and chordates, and feeding arms were reduced and eventually lost. Enteropneust hemichordates adapted to a burrowing mode of life. The body elongated, the U-shaped gut of pterobranchs straightened out, and an elongated proboscis became part of both the feeding and the burrowing mechanism. Other pterobranch-like creatures with multiple pharyngeal slits and reduced feeding arms remained sessile. They lost the proboscis and collar, kept the U-shaped gut, improved the pharyngeal slit filtering mechanism, and added a protective atrium around the slits. These animals became the ascidian tunicates.

Sessile echinoderms and hemichordates rely on small, ciliated larvae to disperse the population. Tunicates evolved larger larvae. Because mass is a cubic function and surface area a square function (p. 6), surface ciliated bands would have been inadequate to propel a large larva. As larval size increased, then, muscular action must have become more important in locomotion. The larval tail with its muscles, a nerve cord, and a notochord may have evolved in this context, along with a simple eye and a balancing organ, or statolith, at the anterior end. Such a larva certainly is better adapted to disperse the population and find a suitable habitat in which to settle. The cephalochordates and vertebrates may have evolved by paeodomorphosis (p. 10), that is, by retaining the larval tail in the adult. Cephalochordates became burrowing filter feeders capable of only limited movement. Vertebrates became more mobile and larger animals that could exploit a wider range of habitats. Increased activity entailed the evolution of better navigational equipment (nose, eyes, and ears); a brain that processed information from the sense organs; and the metabolic machinery to support increased activity, including a muscularized gut, gills, a heart, and a unique type of kidney tubule. Many of these features develop embryonically, as pointed out earlier (p. 41), from the neural crest and placodes. Northcutt and Gans postulate that the neural crest and placodes evolved from the primitive subepidermal nerve net. We are uncertain about what selective forces led to increased activity. The ancestral vertebrates were probably marine or estuarine. Berrill has proposed that increased activity evolved as ancestral vertebrates entered rivers and ascended to fresh water. Northcutt and Gans believe that it evolved as ancestral adult vertebrates, although jawless, shifted from filter feeding to predation upon small, soft-bodied animals that could be scooped up and crushed in the muscularized pharynx. The concept outlined above has little direct supporting evidence because soft-bodied animals rarely leave a fossil record. However, this hypothesis does not violate any morphological principles or known evolutionary mechanisms.

SUMMARY

1. Chordates are characterized by the presence at some stage of their development of a notochord; a single, dorsal, and tubular nerve cord; pharyngeal pouches that often open to the surface as slits; and an endostyle or thyroid gland (an endostyle derivative). The significance of these structures can be seen in the protochordates and hemichordates.

2. Although not chordates, acorn worms and pterobranchs of the phylum Hemichordata have an echinoderm-like nervous system and chordate-like pharyngeal slits. They bridge the gap to some extent between echinoderms and chordates.

3. Adult ascidians of the chordate subphylum Tunicata have pharyngeal gill slits and an endostyle, but the notochord and nerve cord occur only in the motile larva.

4. The derived characters of chordates evolved in connection with unique methods of feeding and locomotion.

5. The subphylum Cephalochordata is represented by amphioxus. Its structure and mode of life clearly illustrate the use and significance of the derived chordate characters.

6. Members of the subphylum Vertebrata display the distinctive derived characters of chordates at some stage of their life. In addition, vertebrates have a number of derived characters of their own that correlate with the evolution of increased body size and level of metabolism. Among these are a brain and special sense organs, usually a vertebral column that replaces the notochord in the adult, a muscularization of the gut tube, gills, more regional specialization of the digestive tract, a liver, and a heart.

7. Many of the derived characters of vertebrates result from the evolution of a unique group of cells, the neural crest, that separate from the neural tube early in embryonic development, and the evolution of ectodermal placodes that give rise to certain sensory structures and nerves.

8. Echinoderms, hemichordates, and chordates appear to be related phyla for they share many early developmental features not found in other phyla: indeterminate and nonspiral cleavage, the formation of a mouth independently of the blastopore, and primitively the formation of mesoderm and the coelom from a series of enterocoelic pouches.

9. The pattern of evolution among this group of phyla is uncertain. Hemichordates and tunicates may have arisen from early echinoderm-like filter feeders by using pharyngeal slits as a method of concentrating food particles rather than ciliated feeding arms.

10. Tunicates evolved a larva with a motile tail containing a notochord and distinctive nerve cord as a means of dispersing the colony.

11. Cephalochordates and vertebrates may have evolved from tunicate larvae by retaining the motile tail into the adult. Vertebrates became particularly active animals early in their evolution. This was correlated with the evolution of sensory organs and a brain, a powerful locomotor apparatus, and the metabolic machinery needed to support increased activity.

REFERENCES

Barrington, E. J. W., 1965: *The Biology of the Hemichordata and Protochordata.* Edinburgh, Oliver and Boyd.

Barrington, E. J. W., and Jefferies, R. P. S., editors, 1975: *Protochordates.* Symposia of the Zoological Society of London, no. 36. Published for the society by Academic Press.

Bateson W., 1886: The Ancestry of the Chordata. *Quarterly Journal of the Microscopical Society,* 26: 535–571.

Berrill, N. J., 1955: *The Origin of Vertebrates.* Oxford, Clarendon Press.

Bone, Q., 1972: *The Origin of Chordates.* Oxford Biology Readers. London, Oxford University Press.

Brandenburg, J., and Kummel, G., 1961: Die Feinstruktur der Solenocyten. *Journal of Ultrastructure Research,* 5: 437–452.

Flood, P. R., 1968: Structure of the segmental trunk muscles in amphioxus with notes on the course and "endings" of the so-called ventral root fibers. *Zeitschrift für Zellforschung und mikroskopische Anatomie,* 8: 389–416.

Flood, P. R., 1975: Fine structure of the notochord of amphioxus. *In* Barrington and Jefferies, *q.v.*

Gans, C., and Northcutt, R. G., 1983: Neural crest and the origin of vertebrates: A new head. *Science,* 220: 268–274.

Garstang, W., 1928: The morphology of the Tunicata and its bearings on the phylogeny of the Chordata. *Quarterly Journal of Microscopical Science,* 72: 51–185.

Jefferies, R. P. S., 1975: Fossil evidence concerning the origin of the chordates. *In* Barrington and Jefferies, *q.v.*

Northcutt, R. G., and Gans, C., 1983: The genesis of neural crest and epidermal placodes: A reinterpretation of vertebrate origins. *Quarterly Review of Biology,* 58: 1–28.

Ruppert, E. E., and Smith, P. R., 1988: The functional organization of filtration nephridia. *Biological Reviews,* 63: 231–258.

Webb, J. E., 1973: The role of the notochord in forward and reverse swimming and borrowing in the amphioxus, *Branchiostoma lanceolatum. Journal of Zoology* (London), 170: 325–338.

Welsch, U., 1975: The fine structure of the pharynx, cryptopodocytes, and digestive caecum of amphioxus *(Branchiostoma lanceolatum). Symposium of the Zoological Society of London,* 36: 17–41.

3 | The Diversity of Vertebrates

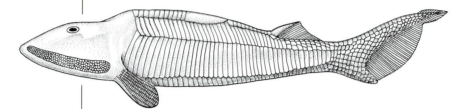

PRECIS

An overview of vertebrate evolution is followed by a brief discussion of the major vertebrate groups, their distinctive characters, and their modes of life. We will frequently refer to these groups as we consider the structural and functional evolution of the organ systems.

Outline

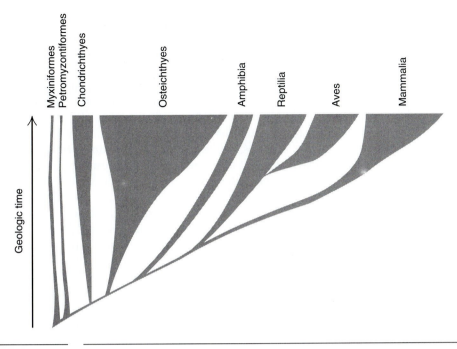

FIGURE 3–1   The eight extant classes of vertebrates, their general interrelationships, distribution in time, and relative abundance.

The subphylum Vertebrata contains nearly 43,000 extant species, and probably far more species have become extinct during the 500 million years of vertebrate evolution. We must consider the relationship among these species and sort out their main lines of evolution to have a framework for discussing their structural and functional evolution. In dealing with the evolution of vertebrates we are on firmer ground than when speculating about their origin, since their skeletons have left a fossil record. Although not complete, the fossil record provides a great deal of information against which to test hypotheses derived from a character analysis of living species. Vertebrates are usually grouped into eight classes (see Fig. 3–1 and the synopsis of chordates at the end of this chapter). Four of these classes are aquatic and are often grouped together as the **Pisces** (L. *piscis* = fish). The other four, largely terrestrial classes, constitute the superclass **Tetrapoda** (Gr. *tetra* = four; Gr. *podos* = foot), although legs are reduced or lost in some species.

Ectothermy

Central to an understanding of vertebrate evolution and of the adaptations of vertebrates to their environments is a realization that their metabolic rate, in common with all chemical processes, is greatly affected by their body temperature. An increase or decrease in body temperature of 10°C can double or halve metabolic rate and hence the activity of the animal. Vertebrate body temperature is determined by the rate at which heat is produced internally by metabolic processes and the rate at which it is lost to, or gained from, the environment. Since fishes have no insulating layer at their body surface and are surrounded by water, which is a reasonably good conductor of heat, their body temperature is largely determined by the water in which they live. Fishes lose metabolic heat in cold water and gain heat from tropical water. Tunas and some rapidly swimming sharks generate a great deal of heat from their muscular activity and conserve some of it. They are "warm blooded" to a degree, but most fishes cannot long maintain an overall body temperature that is different from the huge heat sink, or heat reservoir, in which they live. Fishes are said to be **ectothermic.** "Cold blooded" is a loose synonym, but it is not an accurate term because a tropical fish may have a rather high body temperature. Water, however, has a high thermal stability, and much energy is required to change its temperature. Diurnal temperature fluctuations are not great, and the more extensive seasonal changes occur slowly. Although water temperature changes considerably with latitude and with depth, an individual fish typically lives in a relatively stable thermal environment. Its enzyme systems maintain a rate of metabolism that is no higher than need be to sustain its activities. This minimizes the amount of food and other resources it must obtain. A fish can be active in water to which it is adapted at a moderate energy cost. It cannot tolerate rapid temperature changes, but its enzyme systems do adapt to the slower seasonal ones. An acclimatized fish may have the same rate of metabolism at 5°C in the winter as at 25°C in the summer.

Ectothermic metabolism, the ancestral vertebrate condition, restricts the range of environmental conditions under which a particular species can live. Ectotherms are usually not effective at night in cold climates when solar energy is absent. Ectothermy is continued in early terrestrial vertebrates, but the more derived tetrapods gain more control over their body temperature, as we shall see.

Class Agnatha

Ostracoderms

The most primitive vertebrates are the extant cyclostomes and extinct ostracoderms, both of which belong to the class **Agnatha.** The term derives from the absence of jaws in these primitive fishlike vertebrates (Gr. *a* = without + *gnathos* = jaw). Cyclostomes and most ostracoderms also lack paired pectoral or pelvic appendages. Like other fishes, cyclostomes are ectothermic, and the ostracoderms must have been as well. **Ostracoderm** is a collective term for several distinct groups of small aquatic vertebrates, usually under 20 cm in length. They had shell-like bony plates in their skin (Gr. *ostrakon* = shell + *derma* = skin). Fragmented plates have been found in late Cambrian deposits 500 million years old (Fig. 3–2). The Cambrian and Ordovician fragments were marine or estuarine. This habitat can be determined from the echinoderm and other marine invertebrate fossils found in the same deposits and from an analysis of the rock matrix itself. Early vertebrates, then, like their protochordate ancestors, were marine. Most of our knowledge of these ancestral vertebrates comes from more complete fossils of

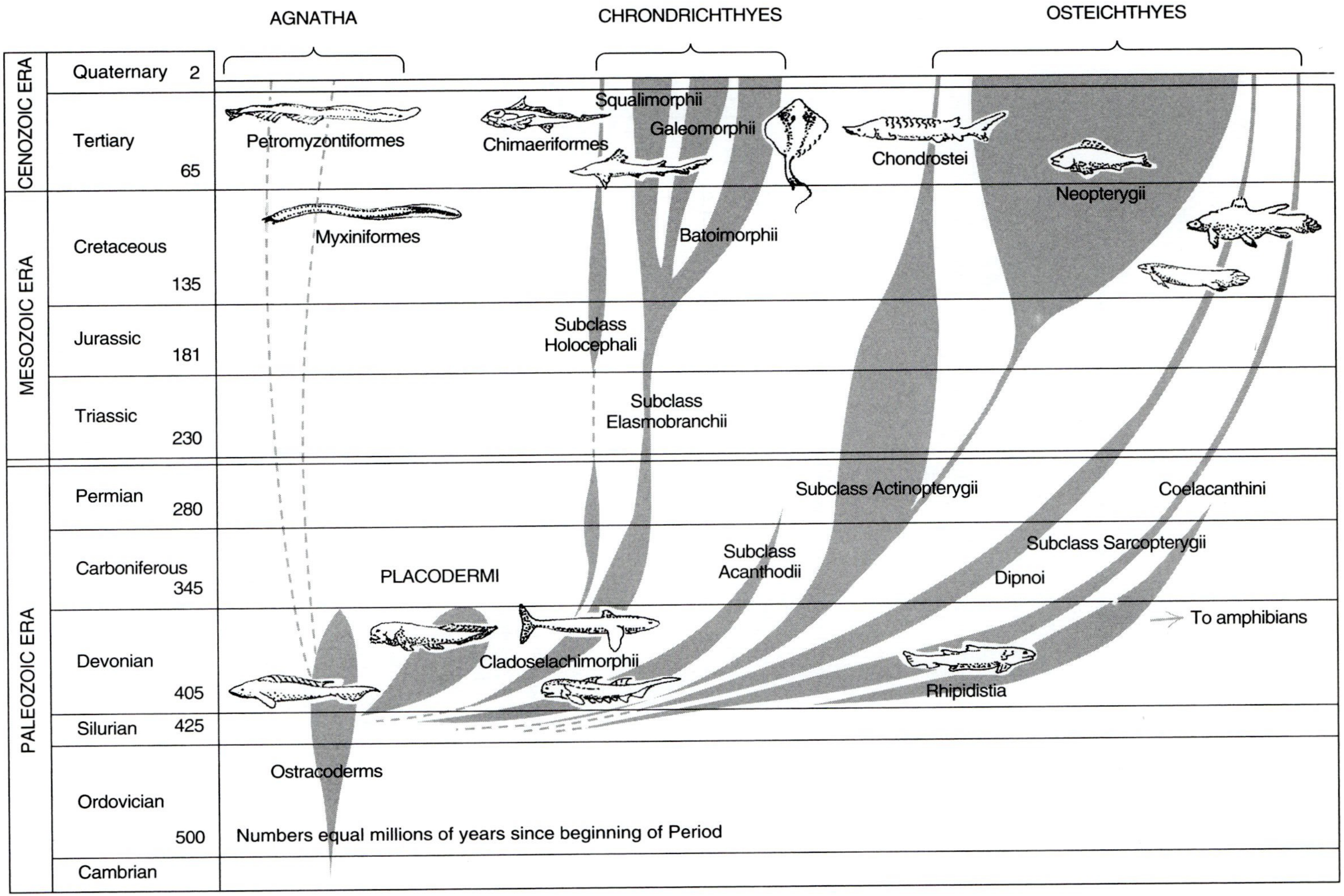

FIGURE 3-2 The evolutionary history of fishes, their distribution in time, and relative abundance.

53

Silurian and Devonian periods. By this time most vertebrates had entered fresh water, and fresh water appears to have been the center for their subsequent radiations, which took some groups back to the sea and some onto the land.

The cephalaspids, belonging to the order **Osteostraci,** are among the best known ostracoderms (Fig. 3–3**A**). A cephalaspid had a large, solid head shield and large plates along the trunk. The shield and plates were composed of bone overlain by dentine and enamel-like material. The surface bone may have protected the ostracoderms from predacious aquatic arthropods, which were abundant, or it may have been an important metabolic reservoir for calcium and phosphate, or it may have encased and insulated electroreceptors. Electroreceptors are modified parts of the lateral line system, a series of sense organs in the skin with which primitive aquatic vertebrates detect water currents. Electroreceptors enable an animal to locate prey by detecting the action potentials generated by the contraction of their muscles. Considerable ossification occurred within the connective tissue of the head, and this has enabled investigators to determine the shape of the brain and inner ear and the course of the cranial nerves. Other parts of the internal skeleton were not ossified and presumably were cartilaginous. The flattened ventral surface and dorsally directed eyes of cephalaspids indicate that they moved along the bottom of ponds and streams. The tail had a large dorsal lobe that was stiffened by an extension of the trunk. This type of tail, known as **heterocercal** (Gr. *heteros* = other + *kerkos* = tail), is often found in heavy-bodied fishes. It is related to their method of swimming and buoyancy control (p. 350). A single dorsal fin and a pair of pectoral flaps probably helped stabilize the animal as it swam. The structure of the mouth and pharynx indicates that these fishes were predators of small, soft-bodied animals or possibly filter feeders. The mouth was jawless and lay at the front of the underside of the head. The large pharynx behind it was floored with many small plates that gave the pharynx considerable mobility (Fig. 3–3**B**). The pharynx apparently contracted and expanded by muscular action, as it does in living fishes, and this suction drew in water and food. The food was trapped and probably crushed in the pharynx, and the water escaped through many pairs of saccule-shaped gill pouches that opened to the surface through pore-shaped gill slits. Some species had as many as 15 pairs of pouches. The use of a muscular pump to draw in a respiratory and feeding current of water is a derived vertebrate character and is more efficient than the ciliary mechanism used by the protochordates. This certainly was a factor in the increase in size and level of activity of early vertebrates. A single median or pineal eye lay between the lateral eyes on the top of the head; a single median nostril was rostral to it. Three groups of small plates on the top of the head were underlain by canals containing branches of cranial nerves. The function of these areas is uncertain, but they may have been sensory fields similar to the lateral line sensory system of contemporary fishes, or

---

FIGURE 3–3 Representative ostracoderms: **A,** Dorsolateral view of a cephalaspid, *Hemicyclaspis.* **B,** Ventral view of the head of *Hemicyclaspis.* **C,** An anaspid, *Pterolepis,* in presumed feeding posture. **D,** A heterostrachean, *Pteraspis.* (**A** and **B,** after Stensiö. **C,** after Ritchie. **D,** after Heintz, 1939.)

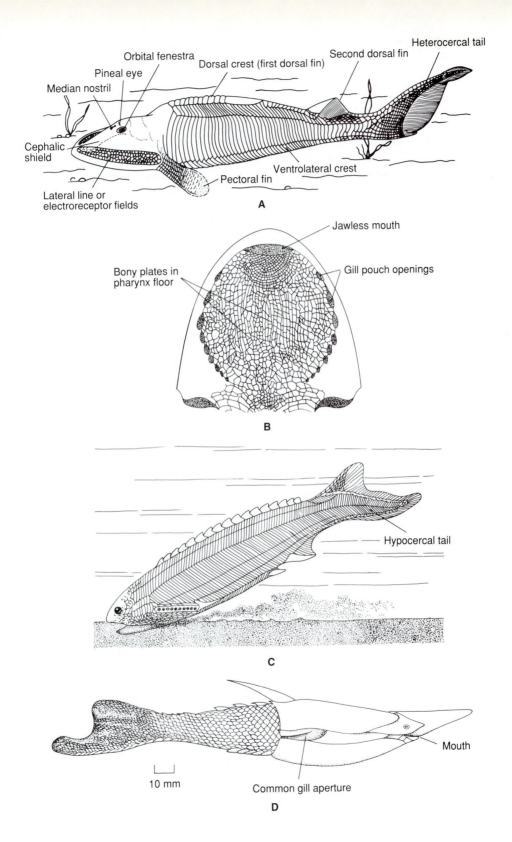

A

Median nostril
Pineal eye
Orbital fenestra
Dorsal crest (first dorsal fin)
Second dorsal fin
Heterocercal tail
Cephalic shield
Lateral line or electroreceptor fields
Pectoral fin
Ventrolateral crest

B

Jawless mouth
Bony plates in pharynx floor
Gill pouch openings

C

Hypocercal tail

D

10 mm
Common gill aperture
Mouth

electroreceptors. If cephalaspids had electroreceptors, they must have been predacious. The inner ear contained only two semicircular ducts.

Ostracoderms were successful animals until their extinction near the end of the Devonian. They underwent an extensive diversification and specialization for somewhat different modes of life as they exploited the many available habitats. We call this an **adaptive radiation.** Four additional orders are recognized. **Heterostraceans** (also called pteraspids), **coelolepids, galeapsids,** and **anaspids** had **hypocercal** tails in which the ventral lobe was larger and stiffened. Such a tail would propel the animal as it swam head downward, scooping up food from the bottom (Fig. 3–3**C**). Although their terminal mouths lacked jaws, some heterostraceans had small bony plates along the margin of the mouth. These may have functioned as teeth and helped the fish scrape food or nibble soft material. Heterostraceans were heavily armored, and some had elaborate spines (Fig. 3–3**D**). Armor was reduced in many other ostracoderms. Anaspids, galeapsids, and cephalaspids had a single nostril, so they are placed in the subclass **Monorhina** (Gr. *rhin-* = nose). Heterostraceans and coelolepids appear to have had a pair of nostrils and are placed in the subclass **Diplorhina.**

Lampreys and Hagfishes

Extant lampreys and hagfishes (collectively known as cyclostomes) share with the ancient ostracoderms an absence of jaws and paired appendages, the presence of a pineal eye, an inner ear that lacks a horizontal semicircular duct, and many pouch-shaped gill chambers that open to the surface by pores. These features suggest that they are direct descendants of one or more of the ostracoderm groups and belong in the class Agnatha. The single nostril points to an affinity to a monorhine group, possibly the anaspids. One Carboniferous fossil lamprey is known, but apart from this the fossil record is very meager. Both lampreys and hagfishes have lost the ancestral armor and become eel shaped, and the adults have specialized for unique modes of feeding. Because lampreys and hagfishes have a round, jawless mouth, they are commonly called **cyclostomes** (Gr. *kyklos* = circle + *stoma* = mouth), but they are different in so many other ways that they must have had a long independent evolution. They are now placed in separate orders.

About 40 species of lampreys, order **Petromyzontiformes,** are known (Fig. 3–4**A**). Many, including most populations of *Petromyzon marinus*, live in the ocean and enter rivers to spawn; others live entirely in fresh water. Lampreys attach to other fishes by a round, suctorial buccal funnel and feed on their blood. The buccal region and pharynx are highly specialized for this mode of life. A tongue that is covered with horny, toothlike structures can be protruded to rasp a hole in the prey. Buccal glands secrete an anticoagulant, and the sucking pharynx is divided longitudinally into a dorsal food passage (called the "esophagus") and a ventral respiratory tube that ends blindly posteriorly. Many other aspects of the internal anatomy, including a notochord that persists into the adult and rudimentary vertebrae around the spinal cord, may reflect the primitive condition for all vertebrates.

Adult lampreys die after they have spawned. The eggs hatch in about two weeks into **ammocoetes larvae** that live and grow as filter feeders in the stream

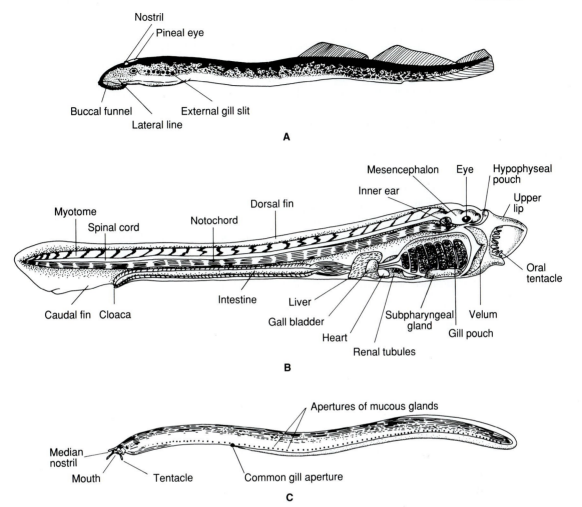

Nostril
Pineal eye
Buccal funnel
External gill slit
Lateral line

A

Mesencephalon Eye Hypophyseal pouch
Inner ear Upper lip
Myotome Dorsal fin
Spinal cord Notochord
Oral tentacle
Caudal fin Cloaca
Intestine Liver
Gall bladder Subpharyngeal gland Velum
Heart Gill pouch
Renal tubules

B

Apertures of mucous glands
Median nostril
Mouth Tentacle Common gill aperture

C

FIGURE 3–4 Contemporary agnathans: **A**, A lamprey, *Petromyzon*. **B**, The ammocoetes larva of a lamprey. **C**, A hagfish, *Myxine*.

bottom for as long as six or seven years. After metamorphosis into an adult, *Petromyzon marinus* normally returns to the ocean, where it lives for a year or two. Some stream lampreys lead very short adult lives. They do not feed but reproduce soon after metamorphosis and then die. The ammocoetes larva (Fig. 3–4**B**) lacks the specializations of the adult lamprey and may provide clues about the nature of ancestral vertebrates. Their filter-feeding mode of life is certainly similar to that of protochordates. Superficially, an ammocoetes larva resembles amphioxus. The notochord and myomeres are well developed. Its pharynx is not so large as in amphioxus and is not surrounded by an atrium. Unlike the mechanism in amphioxus, water is drawn into the pharynx by an expansion of the pharynx and by the action of a pair of muscular flaps, the velum. Food is entrapped in mucus that is secreted by an endostyle-like subpharyngeal gland. The distinctive, derived characters of vertebrates are all present, including a nose, eyes, ears, a brain, gills, a ventral liver, a chambered heart, and renal tubules.

The approximately 30 species of hagfishes (order **Myxiniformes**) are exclusively marine. Like marine invertebrates but unlike all other extant vertebrates, they have the same overall salt concentration in their body fluids as sea water. The iso-osmotic condition indicates that the hagfishes have been marine throughout their evolution. They presumably diverged from ancestral marine vertebrates before vertebrates entered fresh water and adapted to a lower salt concentration in their body fluids. Hagfishes typically burrow in the mud along the bottom of the continental shelves. Their eyes are degenerate, and they find food with three pairs of sensory tentacles around the mouth and with the single terminal nostril (Fig. 3–4**C**). They feed primarily on polychaete worms and dead fishes, tearing off bits of soft flesh with a pair of horny tooth plates in their jawless mouths. Some species of *Bdellostoma* have as many as 14 pairs of gill pouches, each of which opens by a pore on the body surface. *Myxine* has fewer, and branchial ducts lead from them to a common branchial aperture far back on the trunk. All hagfishes have many large, integumentary glands that secrete a copious protective mucus.

Gnathostomes: The Evolution of Jaws and Paired Appendages

Fishes with jaws and paired appendages appear in the fossil record during the late Silurian and early Devonian periods. Jaws evolved from an anterior visceral arch that lay close to the mouth opening and may primitively have supported gills. This arch is called the **mandibular arch** in all jawed vertebrates, or **gnathostomes** (Gr. *gnathos* = jaw + *stoma* = mouth). The arch is composed of cartilage or of bone that replaces the cartilage during embryonic development. This type of bone is called **cartilage replacement bone** (see Fig. 6–8). Plates of bone in or beneath the dermis of the skin overlie the mandibular arch in most groups and contribute to the jaws. These plates are composed of **dermal bone,** which develops directly in connective tissue without cartilaginous precursors. The second visceral arch, known as the **hyoid arch,** frequently contributes to the suspension of the mandibular arch from the cranium (see Fig. 7–4). The gill pouch that in agnathans lies between the homologues of the mandibular and hyoid arches is lost or reduced to a small passage known as the spiracle. The incipient and intermediate stages of the evolving jaws must have had significant functional advantages resulting in a sustained trend toward full differentiation of the jaws. Intermediate stages under such a selective pressure would not persist. Once the direction of evolution of jaw formation was established, natural selection acted in that direction until completion of the trend, resulting in an all-or-none pattern of jaw evolution. Paired appendages appear to have evolved from flaps of flesh associated with lateral spines.

Jaws and paired appendages increased greatly the evolutionary potential of fishes. Jaws enabled them to seize or to suck in a much wider range of food. Paired appendages provided stability and maneuverability and enabled fishes to swim more efficiently. A great adaptive radiation of fishes accompanied the evolution of these structures. At least six distinct evolutionary lines appeared by the early Devonian (Fig. 3–2). The success of these fishes probably was a factor that led to the extinction of the ostracoderms and restricted the lampreys and hagfishes to very specialized modes of life. The presence of jaws and paired appendages in all these lines implies a common origin, possibly from one of the diplorhine ostracoderms, since all gnathostomes have paired nasal passages. There are, however, many differences among these groups, so a common origin is by no means certain. Many vertebrate biologists sort the numerous lines of jawed fish evolution into three groups: the extinct placoderms, the cartilaginous fishes (sharks and their relatives), and the bony fishes (lungfishes, coelacanths, garpikes, salmon, minnows, perch, and the like).

Class Placodermi

Half a dozen orders of Devonian fishes are placed in the class **Placodermi** (Gr. *plak-* = plate + *derma* = skin). Only two, the arthrodires (order **Arthrodira**) and the antiarchs (order **Antiarchi**), are known well enough to warrant consideration here (Fig. 3–5). These fishes resembled ostracoderms in having extensive armor on the head and anterior part of the trunk, but bony scales on the rest of the body were small and sometimes reduced to minute denticles embedded in the skin. The internal skeleton was cartilaginous. Apparently, a notochord persisted in the adult because vertebral centra were absent, yet vertebral arches covered the spinal cord. The tail was heterocercal. The lower jaw could, of course, be lowered, but an unusual joint between the cephalic and the thoracic plates also allowed the skull to be raised. This mechanism was particularly well developed in the arthrodires (Gr. *arthron* = joint + *deire* = neck). Teeth were absent in most arthrodires, but the dermal bones overlying the mandibular arch had sharp cutting edges and resembled meat cleavers. Small pectoral spines in arthrodires were probably associated with a pectoral fin. Antiarchs had long, jointed pectoral spines that were covered with dermal plates and superficially resembled an arthropod limb. These spines may have elevated the front part of the body and held the fish stationary in flowing water. Pelvic fins have been found in some species, and in one species parts of the fin were modified as an intromittent organ for internal fertilization. One antiarch species shows impressions of lunglike organs. Placoderms appear to have been bottom-dwelling fish. Most were predacious, and they probably laid in wait for prey. Early species lived in fresh water, but most secondarily became marine. One arthrodire, *Dunkleosteus* from marine shale deposits in Cleveland, attained a length of 6 meters. Some specimens in the Cleveland deposits are exceptionally well preserved. Myomeres can be recognized, and microscopic examination reveals striated muscle fibers. All placoderms became extinct early in the Carboniferous.

Class Chondrichthyes

Characteristics and Relationships of Cartilaginous Fishes

There are approximately 700 living species of sharks, skates, rays, chimaeras, and related fishes that are placed in the class **Chondrichthyes** (Gr. *chondros* = cartilage + *ichthys* = fish). As in all vertebrates, their internal skeleton is cartilaginous during early embryonic development, but none of the cartilage is replaced by bone. The cartilage of Chondrichthyes is often strengthened by calcification, which

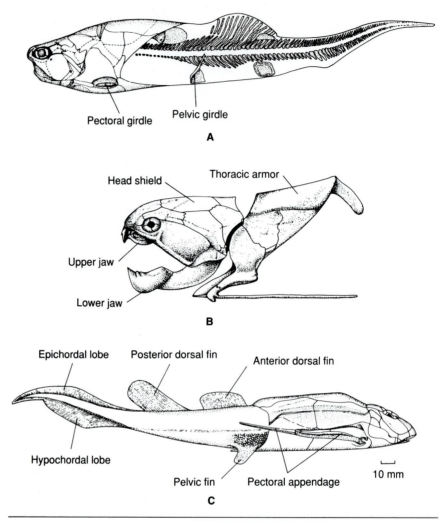

Pectoral girdle   Pelvic girdle

A

Head shield   Thoracic armor

Upper jaw

Lower jaw

B

Epichordal lobe   Posterior dorsal fin   Anterior dorsal fin

Hypochordal lobe

Pelvic fin   Pectoral appendage

10 mm

C

FIGURE 3–5   Representative placoderms: **A,** A small primitive arthrodire, *Coccosteus.* **B,** Cephalothoracic armor of the giant arthrodire *Dunkleosteus,* which attained a length of 6 meters. **C,** The small antiarch *Bothriolepis.* (**A,** after Miles and Westol. **B,** after Heintz, 1931. **C,** after Stensiö.)

is deposited as a superficial layer of prismatic plates. However, calcified cartilage is not the same as bone, which has a distinctive histological structure usually consisting of alternating layers of bone-forming cells and matrix (p. 184). The extensive dermal armor in the skin of early fish groups is reduced in chondrichthyans to minute **dermal denticles,** or **placoid scales,** and to spines that lie at the anterior border of the median fins of some species. Since there are no dermal plates, the jaws are formed entirely by the mandibular arch and are supported by the hyoid arch. The first gill pouch is reduced to a spiracle or lost. Part of each pelvic fin of the male is modified as an intromittent organ known as the clasper, and fertilization is internal. Most species are ovoviviparous or viviparous, but others, such as some skates, lay large yolky eggs enclosed in horny egg cases.

Internal fertilization is related to the production of a small number of eggs, each of which has a better chance of survival. Most cartilaginous fishes have a hetero-cercal tail. They are heavy-bodied fishes without lungs or swim bladders (air filled sacs within the body that control the buoyancy of the fish), although the large amount of lipid in the liver provides considerable buoyancy. The reduction of the ancestral dermal armor and the retention of cartilage, which is not so dense as bone, also lighten the body and facilitate swimming. Most species rest on the bottom when not swimming. Elasmobranchs have an extraordinarily enhanced sensory apparatus including an extensive network of electroreceptors. In conjunction with the increased sophistication of the sensory apparatus and locomotory and feeding modes, some modern sharks, such as the hammerheads *(Sphyrna),* have the highest known brain weight to body weight ratio among fishes and amphibians, approaching that of birds and mammals. Cartilaginous fishes are first found in Devonian marine deposits but they may extend back into the late Silarian. The group has been primarily marine throughout its history, and only a few species have adapted secondarily to fresh water.

Cartilaginous fishes appear to have an affinity to the placoderms. Both groups have internal skeletons of cartilage, both have a reduction of the dermal armor (at least on the trunk and tail), and (except for one antiarch species) neither group has lungs or swim bladders. Possibly, cartilaginous fishes diverged from a group of marine placoderms.

The Adaptive Radiation of Cartilaginous Fishes

Elasmobranchs.   Sharks, skates, and rays constitute the subclass **Elasmobranchii.** The term derives from the thin plates of tissue that bear the gills and separate the vertically elongated gill pouches (Gr. *elasmos* = thin plate + *branchia* = gills). Each pouch opens independently on the body surface. The upper jaw of an elasmobranch is not fused to the cranium. *Cladoselache* and other primitive Devonian sharks had a terminal mouth, long jaws, and paired fins, each of which attached to the trunk by a broad base (Fig. 3–6A). The notochord persisted in the adult. These early sharks are placed in the superorder **Cladoselachimorphii.**

Early sharks underwent several evolutionary radiations in the late Silurian and early Devonian. One line led to the freshwater pleuracanths, which were successful for millions of years but became extinct during the Triassic. Modern sharks and rays appeared in the Jurassic. They differ from cladoselachimorphs in having a subterminal mouth supported by shorter and more powerful jaws. The upper and lower jaws can be raised and lowered relative to the braincase and move forward and backward. Sharks that feed on large prey can gouge out large pieces, while species feeding on small prey use their mobile jaws for suction. Cartilaginous vertebral centra largely replace the notochord and provide a stronger vertebral column. Extant sharks are grouped into three major lineages, which we consider superorders. The superorders are the Squalimorphii, Galeomorphii, and the Batoidimorphii. Among the **Squalimorphi** is the order **Squaliformes** which includes the familiar spiny dogfish with fin spines and no anal fin. Squaliform sharks tend to live near the bottom of the ocean. Other Squalimorphs are the angel sharks (order **Squatiniformes**) and saw sharks (order **Pristophoriformes**). The **Galeomorphii** represent the second lineage of living sharks. Among the galeomorphs

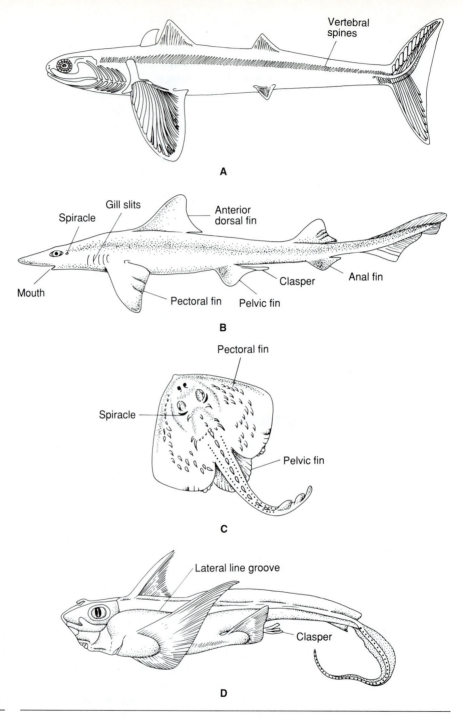

A

B

C

D

FIGURE 3–6
Representative cartilaginous fishes: **A**, An ancestral Devonian shark, *Cladoselache*. Note the absence of vertebral bodies. **B**, The smooth dogfish *Mustela*. **C**, A skate, *Raja*. **D**, A holocephalan, *Chimaera*. (**A**, after Dean. **B** and **C**, after Garman. **D**, after Dean.)

are the **Hexanchiformes,** which represent the most primitive living sharks possessing six or seven gill slits. Other galeomorphs are the requiem (order **Carchariniformes**), mako (order **Lamniformes**), carpet (order **Orectolobiformes**), and the bullhead (order **Heterodontiformes**) sharks. Most sharks have sharp, triangular cutting teeth and are predaceous.

A derivative of the sharks are the highly specialized skates and rays, which are grouped together as the superorder **Batoidimorphii (Batoidea),** a surprising diverse group with more than 400 known species. The flat body form and dorsally directed eyes of the batoids are adaptations for a bottom-dwelling life (Fig. 3–6**C**). The enormous pectoral fins have a broad attachment that extends from the head along most of the trunk. These fishes swim over the bottom by undulatory waves that travel caudally along the pectoral fins. The respiratory current enters the pharynx through a pair of enlarged spiracles located behind the eyes rather than through the mouth, which often is in the sand. Water leaves through the gill slits on the ventral surface. Most skates and rays feed on molluscs and crustaceans, which they crush with their specialized flattened teeth.

Holocephalans.   The subclass **Holocephali** includes about 30 species of peculiar marine fishes known as chimaeras (Fig. 3–6**D**). Their long evolutionary history goes back to the Devonian, but they have never been abundant. Unlike the elasmobranchs, they have an upper jaw firmly united to the cranium, and the gill pouches on each side are covered by a fleshy flap, called the operculum, that has a single opening just anterior to the pectoral fin. They, too, feed on molluscs and other shelled invertebrates with crushing and grinding tooth plates in both upper and lower jaws.

Class Osteichthyes

Characteristics of Bony Fishes

The remaining 22,000 or more species of extant fishes belong to the class **Osteichthyes** (Gr. *osteon* = bone + *ichthys* = fish). A few species retain considerable cartilage in their internal skeletons, but bone always replaces some of it; most have well-ossified skeletons. The heavy, bony surface armor of ancestral fishes is reduced, but most species retain thin, bony scales embedded in the skin. The spiracle is retained in early bony fishes but is lost in most advanced species. A **bony operculum** that covers the gills is a distinctive derived character (Fig. 3–7). Pelvic claspers are never present, but in a few species the male has an intromittent organ that develops from the anal fin. **Lungs,** or the **swim bladder** that evolved from lungs, are a derived feature of bony fishes (although one placoderm may have had lungs). One or the other is present in all species except a few deep-sea and bottom-dwelling forms, in which it has been secondarily lost. Lungs are primarily accessory respiratory organs that permit a fish to live in water with reduced amounts of oxygen while simultaneously providing buoyancy; the swim bladder is a hydrostatic organ. By controlling the amount of air in the swim bladder, many bony fishes can regulate their buoyancy and maintain their depth in the

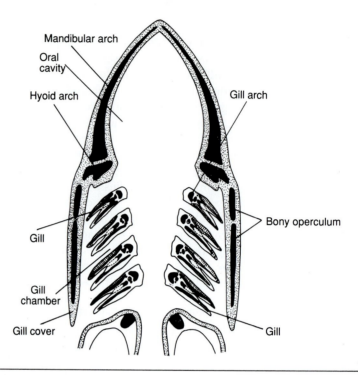

Mandibular arch

Oral cavity

Hyoid arch

Gill arch

Gill

Bony operculum

Gill chamber

Gill cover

Gill

FIGURE 3–7   A frontal section through the oral cavity and pharynx of a gnathostome. (After Jarvik.)

water with little muscular effort. Bony fishes are an ancient group that can be traced further back into the Silurian than the placoderms. Three distinct evolutionary lines, usually given the status of subclasses, can be recognized. Early members of each line were freshwater fishes.

Subclass Acanthodii

The **acanthodians** were small fishes that lived from the Silurian into the Permian (Figs. 3–2 and 3–8). Most were freshwater species, but some later ones have been found in marine deposits. They are often called spiny sharks because their heterocercal tail and shape give them a sharklike appearance, and spines precede their fins (Gr. *akanthodes* = spiny). Unlike sharks, they were covered by small, thick bony scales overlain by dentine. They had a partly ossified internal skeleton, and a bony operculum covered the gills. In specimens in which the bony head scales and operculum have been removed, it is evident that the mandibular arch was a modified visceral arch. Accessory paired spines lay between the pectoral and pelvic fins. Some species had as many as five additional paired spines. The partly ossified skeleton and bony operculum suggest that acanthodians are related to bony fishes, and many paleontologists treat them as a subclass of the Osteichthyes. Others elevate them to the rank of a distinct class.

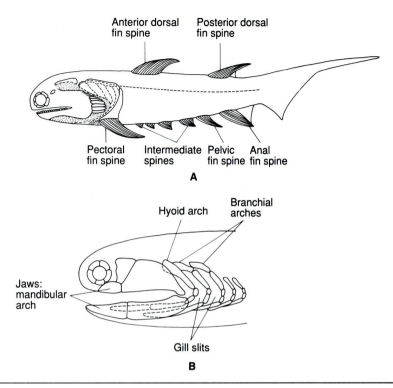

FIGURE 3-8   The structure of an acanthodian, *Climatius:* **A**, Lateral view. **B**, The visceral skeleton. Superficial scales are not shown in **A**. (After Watson.)

Subclass Actinopterygii

Except for three species of lungfish and one coelacanth, all recent bony fishes belong to the subclass **Actinopterygii** (Gr. *aktin-* = ray + *pteryg-* = wing or fin). They are characterized by paired fins that are supported primarily by superficial, segmented, and branching bony rays that evolved from rows of minute bony scales. The rays attach to deeper elements, but ray-finned fishes do not have a skeletal axis that extends far into the fin. Movements of the fins are brought about almost entirely by muscles within the body wall. Primitive members of the group have **ganoid scales,** in which many layers of enamel, or ganoine, cover the deeper dentine and bone (see Fig. 5–6). Each nasal sac has an entrance and an exit on the body surface through which an olfactory current of water flows. There are never internal nostrils opening into the buccal cavity. Primitive actinopterygians have a single dorsal fin.

Actinopterygians can be divided into two major infraclasses: **Chondrostei** and **Neopterygii,** which represent successive radiations (Fig. 3–2). The chondrosteans are a primarily Paleozoic and Triassic group, with only several surviving genera. They possess simple scissor-like jaw actions, and the jaw suspension does not allow much lateral movement. The neopterygians made their first appearance in the late Permian, and one of their lineages gave rise to a great radiation, which resulted in the vast majority of recent species of bony fishes, the **Teleostei.** In the past, more primitive neopterygians (e.g., *Amia* and *Lepisosteus*) were included in a group termed the "Holostei."

Chondrosteans.   Chondrosteans are the oldest group of actinopterygians and can be traced back to the Devonian. They radiated widely and were the dominant group of fishes during the Paleozoic era. Early species inhabited fresh water, but many later ones became marine. Ancestral chondrosteans, the **palaeoniscoids,** were long-bodied fishes with heterocercal tails, single dorsal fins, and pelvic fins set far back on the trunk (Fig. 3–9**A**). The rays of the paired fins spread out from a small fleshy lobe at the fin base. A small spiracle was retained. They had rather heavy ganoid scales. Probably, like their contemporary descendants, they had lungs that grew out of the floor of the pharynx and remained connected to the pharynx by an air duct. The maxillary bone that formed most of the margin of the upper jaw was firmly connected to adjacent bones and could not move.

The chondrosteans are represented today only by the freshwater bichirs (*Polypterus*), reedfish (*Erpetoichthys*), sturgeons (*Acipenser*), and paddlefishes (*Polyodon*). (Fig 3–9**B** and **C**.) Both the bichirs (*Polypterus*) and reedfishes (*Erpetoichthys*) are characterized by a long dorsal fin that is divided into a series of shorter elements, each bearing a sharp anterior spine (Fig. 3–9**B**). The caudal fin is nearly symmetrical. The swim bladder is in the form of paired ventral lungs. Bichirs will drown if prevented from breathing air. Bichirs and reedfishes retain a small spiracle. The fleshy appearance of the pectoral fin is a specialization of the group and does not reflect affinities with another subclass of bony fishes the sarcopterygians, that also have fleshy appendages.

The sturgeons, which produce caviar, retain the ancestral heterocercal tail, spiracle, and some other primitive features, but they have lost the ganoid scales. Only several rows of bony plates remain. Sturgeons are all predators. The paddlefish from eastern North America is an open-water filter feeder using long gill rakers to trap plankton.

Neopterygians.   Neopterygians include all actinopterygians other than the chondrosteans and constitute the vast majority of living marine and freshwater fishes. In all extant Neopterygii except gars, the scales are reduced to thin, flexible elements, the **cycloid** and **ctenoid** scales, enabling the body to be bent more rapidly and powerfully. The number of fin rays is also reduced, enabling the whole fin to be raised and lowered. Portions of it can be moved from side to side. The shape of the caudal fin in neopterygians is superficially symmetrical, or **homocercal.** The spiracle is lost and major changes can be seen in the feeding apparatus. In comparison with chondrosteans, neopterygians have shorter jaws that can move forward and backward to form a round mouth opening. The jaw suspension allows for extensive lateral movements. This lateral movement brings about a considerable increase in the volume of the oral cavity, while the round mouth opening is brought forward resulting in a powerful sucking action not unlike that of a pipette (p. 578). In this way neopterygians have developed very versatile feeding modes. The swim bladder is fully differentiated as a hydrostatic organ with which the fish can regulate its buoyancy. Living neopterygians comprise the gar, *Lepisosteus*, which is placed in the division **Ginglymodi,** and the bowfin, *Amia*, which are placed in the division **Haleocomorphi.** All other neopterygians, which is grouped as the division **Teleostei.** Because of their great numbers (Fig. 3–10)—over

(Text continues on p. 70)

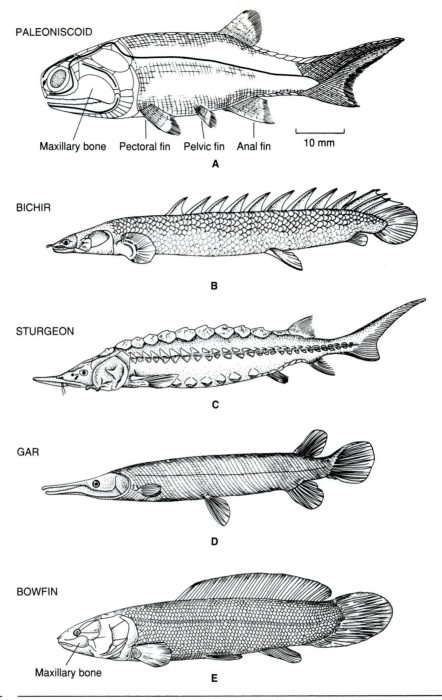

PALEONISCOID

Maxillary bone Pectoral fin Pelvic fin Anal fin

10 mm

A

BICHIR

B

STURGEON

C

GAR

D

BOWFIN

Maxillary bone

E

FIGURE 3–9 Representative chondrosteans (**A–C**) and early neopterygians (**D** and **E**): **A**, The palaeoniscoid *Moythomasia*. **B**, The bichir *Polypterus* (Chondrostean). **C**, A sturgeon, *Acipenser* (Chondrostean). **D**, A garpike, *Lepisosteus* (Neopterygian). **E**, The bowfin *Amia* (Neopterygian). (**A**, after Moy-Thomas and Miles. **B**, after Dean. **C** and **D**, after Jordan and Evermann. **E**, after Jarvik.)

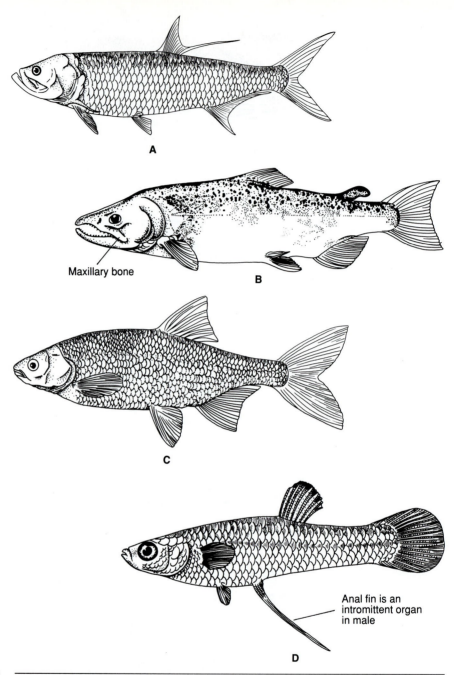

Maxillary bone

Anal fin is an intromittent organ in male

FIGURE 3–10 Representative teleosts: **A**, The tarpon *Megalops* is a primitive teleost. **B**, The salmon *Oncorhynchus*. **C**, A minnow, *Lavinia*. **D**, A male mosquito fish, *Gambusia*. **E**, The perch *Perca* is an advanced teleost. **F**, A deep-sea anglerfish, *Melanocetus*. **G**, The seahorse *Hippocampus*. **H–J**, Three stages in the development of the halibut *Hippoglossus* to show the formation of its flattened body. (**F**, after Regan and Trewavas. **H–J**, after Thompson and Van Cleve. Others after Jordan and Evermann or after Moyle and Cech.)

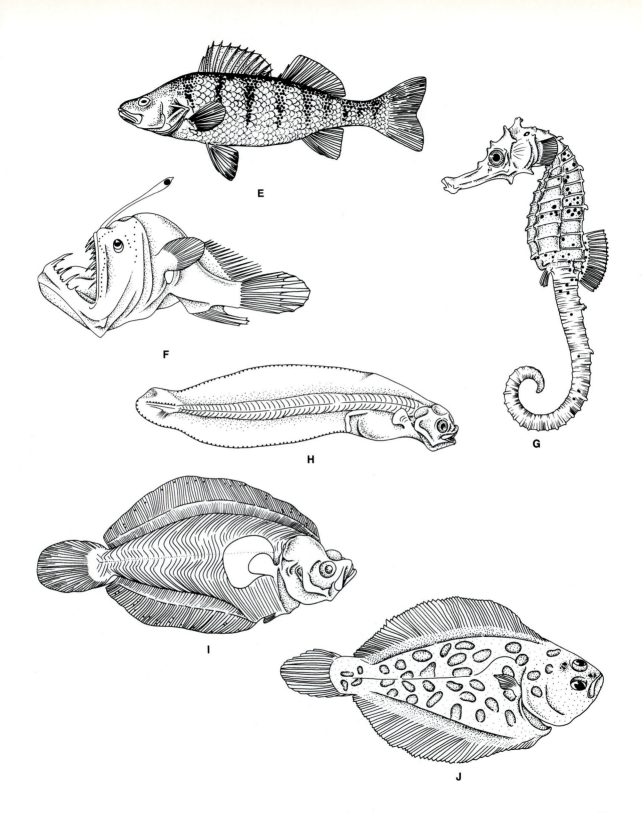

E

F

G

H

I

J

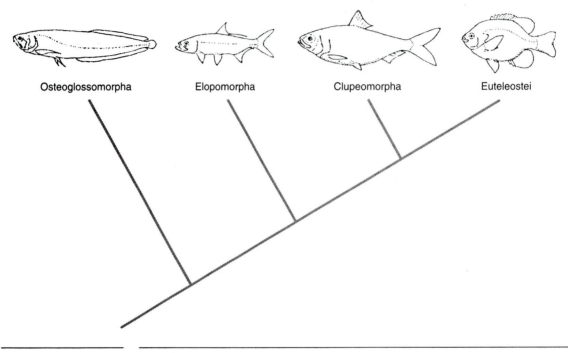

FIGURE 3–11   A cladogram showing the interrelationships among the superorders of living Teleostei.

species in 35 orders and 409 families—it is difficult to define the Teleostei, but most ichthyologists accept that all teleosts have modified neural arches in the tail vertebrae to ensure equal stiffness of the dorsal and ventral half of the caudal fin. The evolution and classification of teleosts are beyond the scope of this book, although the relationships of the contemporary superorders is shown in Figure 3–11 and included in the synopsis. The mooneye, arowana (**Osteoglossomorpha**), the tarpon and eels (**Elopomorpha**), and herring (**Clupeomorpha**) are among the more primitive teleosts. The **Euteleostei** include the primitive salmon and such derived forms as the sunfish, seahorse, flounder, and pufferfish. In this scheme such divergent forms as tarpons and eels are grouped into a mono-phyletic assemblage (Elopomorpha) because all members share a highly special-ized and very distinct leaflike larva, the leptocephalus larva. It is highly unlikely that such a larva would have arisen independently in unrelated groups. Among the Euteleostei, the perchlike fishes have undergone one of the greatest adaptive radiations in the history of life (Fig. 3–12). Members of this group are distributed throughout the world and have penetrated deep seas, open oceans, fresh waters, coral reefs, and benthic habitats, and some live in deoxygenated waters or migrate over land. As they penetrated and exploited these diverse environments, an amaz-ing array of body and fin shapes, specialized jaws, unusual sense organs, radically different respiratory systems, and behavioral mechanisms evolved. Those diver-sifications evolved so rapidly that it is difficult to determine the precise ancestral

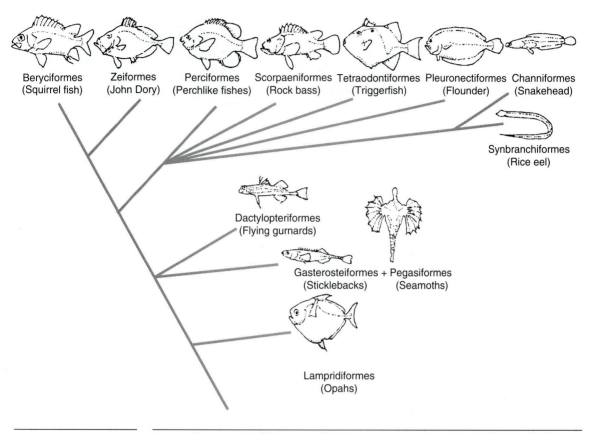

FIGURE 3–12   Adaptive radiation of the major groups (orders) of the perchlike fishes. (Modified from Lauder and Liem.)

and intermediate stages, although much headway has been made in understanding phylogenetic relationships of major groups in recent years. The vast numbers of species and morphologies have made it a challenge for zoologists to unravel the precise relationships and evolutionary history of the most derived teleosts, the spiny rayed Acanthopterygii.

Subclass Sarcopterygii

Actinopterygians are of interest because of their tremendous adaptive diversity, but they gave rise to no other classes of vertebrates. The piscine ancestors of terrestrial vertebrates are to be found among the fleshy-finned fishes of the subclass **Sarcopterygii** (Gr. *sarkodes* = fleshy + *pterygion* = wing or fin). This subclass includes three orders: the lungfishes (order **Dipnoi,** Fig. 3–13), the coelacanths **(Coelacanthini),** and the extinct **Rhipidistia.** The last two orders have been referred to as the crossopterygians in the past (Fig. 3–14).

Lungfishes, coelacanths, and **rhipidistians** have been distinct groups since they appeared in the Devonian, but they share features indicating that they prob-

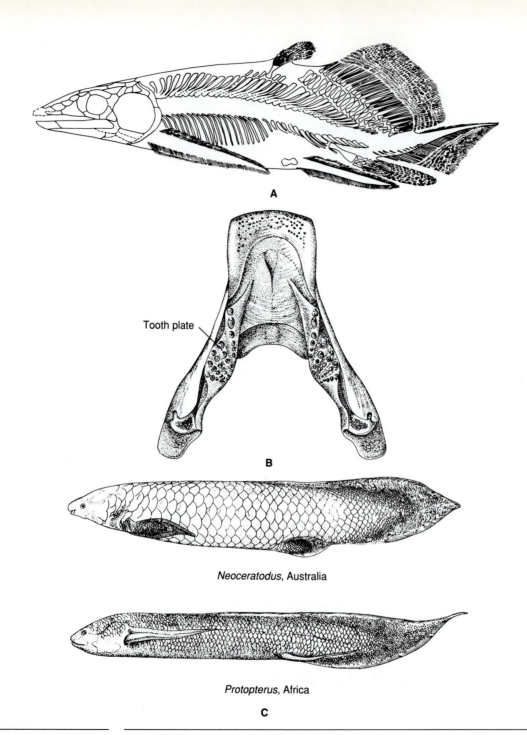

A

Tooth plate

B

Neoceratodus, Australia

Protopterus, Africa

C

FIGURE 3–13   Lungfishes: **A**, A Devonian species, *Fleuratia*. Note the incompletely ossified vertebral column and appendicular skeleton. **B**, Dorsal view of the lower jaw of the extinct *Rhinodipterus* to show the pair of crushing tooth plates. **C**, Two species of contemporary lungfishes. (**A**, after Graham-Smith and Westol. **B** and **C**, after Jarvik.)

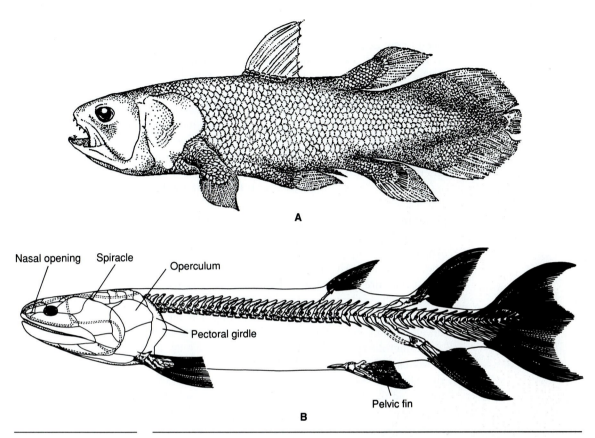

FIGURE 3–14   **A**, *Latimeria,* a surviving coelacanth. **B**, The rhipidistian *Eusthenopteron* was close to the ancestry of terrestial vertebrates. (**A**, after Millot, **B**, after Jarvik.)

ably had a common ancestor distinct from the ancestor of other bony fishes. Unlike the ray fins of actinopterygians, each of their paired fins has a distinct skeletal axis surrounded by muscles that extends far into the fin. The muscular nature of the fins may have evolved as an adaptation to life on or near the bottom, where they could be used not only as paddles but also to push against the substratum. Early lungfishes, coelacanths and rhipidistians had **cosmoid scales** rather than the ganoid scales that characterize early actinopterygians. Although both cosmine and ganoine are present in both kinds of scales, cosmoid scales include more dentine-like cosmine (see Fig. 5–6).

Common to the coelacanths, rhipidistians, dipnoans, *Polypterus,* sturgeons, and paddlefishes is the presence of a special electroreceptive sensory system associated with the superficial layers of the dermal skeleton. These receptors resemble the electroreceptors of the Chondrichthyes. They may have been used to detect prey and avoid predators and obstacles in turbid waters where vision is limited. Early sarcopterygians had two dorsal fins instead of one. The primitive heterocercal tail became symmetrical internally as well as externally (**diphycercal tail**). Both groups have been primarily freshwater fishes, although an early lungfish has been found in marine deposits and one side line of coelacanth evolution adapted to the marine habitat. Lungfishes, coelacanths, and rhipidistians were abundant during the Paleozoic, but they are nearly extinct today.

Lungfishes.   Even though Devonian lungfishes were partly ossified, extant lungfishes have relatively little ossification of their internal skeletons. Reduced ossification and the loss of premaxillae and maxillae, which ordinarily make up the upper jaw, may well be related to paedomorphosis or evolutionary juvenilization. As discussed on p. 10, paedomorphosis involves the retention of juvenile characters in the adult stages. Lungfishes have triangular tooth plates on their palate and lower jaw with which they crush the shells of invertebrates (Fig. 3–13**B**). In lungfishes the anterior nares are situated inside of the posterior edge of the upper lip, while the posterior nares are within the oral cavity. Only three forms survive in tropical rivers and ponds: *Neoceratodus* (Australia), *Lepidosiren* (South America), and *Protopterus* (Africa). The paired appendages of the South American and African species are reduced to little more than tendrils. Although the weak skeleton and specialized dentition of lungfishes preclude their being the ancestors of terrestrial vertebrates, many features of their soft anatomy represent the level of development that we would expect to see in the ancestors of terrestrial vertebrates.

Lungfishes possess well-developed lungs. Lungs appear to have evolved in the Devonian in bony fishes as accessory respiratory organs. The prevalent view is that these fishes lived in warm freshwater habitats where an accessory respiratory organ would supplement gill breathing and offer a selective advantage. Warm water holds less oxygen than cold water, and some bodies of fresh water frequently become stagnant and deoxygenated. Lungfishes live in such an environment and supplement aquatic respiration by taking air into their lungs. Many investigators agree with the late Alfred S. Romer that the bodies of fresh water in which Devonian fishes lived were subject to stagnation and periodic drying up. An African lungfish survives the drying up of its ponds not by trying to escape but by burrowing and forming a cocoon of mucus and dried mud about itself, lowering its metabolism, and breathing air through a small hole in the cocoon.

Rhipidistians.   Early rhipidistians had stronger and more thoroughly ossified internal skeletons than lungfishes. Many skeletal details found in rhipidistians also occur in the first amphibians. Rhipidistians had an unusual transverse joint in the cranium that allowed skull movements during feeding, and a suture that represents the remains of this joint is found in the early amphibians. The pattern of bones in the skull roof and in the paired appendages of rhipidistians is very similar to that in early amphibians. Both rhipidistians and early amphibians had a median eye. Both had conical teeth with a characteristic infolding of the enamel (labyrinthodont tooth). The list could go on. There can be little doubt that the rhipidistians and amphibians shared a common ancestor (Schultze, 1991). Amphibians evolved from freshwater rhipidistians that became extinct during the Carboniferous (Fig. 3–14**B**). We know nothing directly about the soft anatomy of the rhipidistians. We believe they had internal nostrils and lungs or lung remnants that are found in their close relatives, the coelacanths, lungfishes, and primitive actinopterygians.

Coelacanths.   Although once more abundant, coelacanths are today represented by a single species that lives off the Comoro Islands near Madagascar (Fig. 3–14**A**). *Latimeria* has fleshy paired fins, two dorsal fins, cosmoid scales, and lacks internal nostrils. A small vestige of the lung is present, although it is largely filled with fat.

Class Amphibia

Amphibians represent the first vertebrates to invade terrestrial habitats successfully. They exhibit numerous structural, functional, as well as behavioral adaptations that are correlated with their bimodal life. We will begin our discussion of the amphibians with the transition from water to land, followed by the behavioral and ecological factors that may have triggered aquatic vertebrates to abandon water and invade land, and the hypothesis of the origin of the first tetrapods.

The Transition from Water to Land

The different physical and chemical properties of water and air have affected the evolution of vertebrates. Air is less dense than water and does not provide the buoyancy of water, but it offers less resistance to movement. Sound waves travel more slowly in air than in water. Fewer wavelengths of light are absorbed by the atmosphere than by water. Oxygen is more abundant in air. Water is in short supply in many terrestrial environments. Air does not have the thermal stability of water, so ambient temperatures on land can fluctuate rapidly and widely. Because of the number and magnitude of the differences between water and land, it should not be surprising that the transition of vertebrates from water to land extended over millions of years. Changes occurred in all organ systems that enabled vertebrates to meet the challenges of terrestrial life and take advantage of its opportunities. Indeed, the major theme for terrestrial vertebrate evolution has been their continuing adaptations for using more and more fully the habitats and resources available on land.

Behavioral and Ecological Factors.   Because the transition from water to land occurred during the Devonian, it is impossible to study the behavioral and ecological factors directly. One way to answer the question, "Which ecological and behavioral factors triggered the invasion of land?" is by studying recent animals that engage in a similar transition from water to land. Reasoning by analogy from the recent to the past, it is hypothesized that the shift from aquatic to terrestrial life was triggered by declining resources for food, opportunities for reproduction, and increasing intraspecific and interspecific competition in the Devonian waters. At the same time, food resources on land became more and more abundant. Recent air-breathing fishes such as the chondrostean *Polypterus*, the teleost *Clarias* (the walking catfish), and *Anabas* (the climbing perch) regularly engage in terrestrial excursions in response to naturally occurring or experimentally induced dwindling food resources or increasing population densities (Liem, 1989). Rhipidistians are hypothesized to have engaged in similar terrestrial sojourns to escape fierce competition, or in search of abundant prey and new mating habitats. Superior air-breathing capacities with enlarged lungs, sturdy muscular fins, and a strong skeleton must have been greatly advantageous during these terrestrial excursions. By the late Devonian the land was becoming a hospitable environment for vertebrates, with continuously warm and humid climates. There was a rich cover of primitive plants, and insects and other terrestrial invertebrates were becoming abundant. Thus the selection pressures for the evolution of terrestriality must have become progressively stronger during the Devonian.

Origin of Tetrapods.   The transformation from fish to amphibian is considered one of the major evolutionary events in the history of vertebrates. After all, it is the amphibians that gave rise to reptiles, which in turn gave rise to birds and mammals. Yet the origin of the first amphibian tetrapods is still being debated, although a consensus is emerging. Are dipnoans or the rhipidistians more closely related to the first amphibians? Proponents for a close relationship between lungfishes and amphibians argue that lungfishes and amphibians have both developed **choanae,** openings in the palate that connect the mouth cavity with the external environment via the nasal sacs, and external nostrils or **nares** (Rosen et al., 1981; Forey, 1987). However, it has been shown that the internal nares of lungfishes differ significantly from those of amphibians (Fig. 3–15). Lungfishes have a unique nasal sac in which both external and internal nares lie inside the mouth. In tetrapods the external nostrils (nares) are outside the mouth, and the internal nostrils (choanae) are characteristically surrounded by the vomer, palatine, maxilla, and premaxilla. Members of the rhipidistians (e.g., *Panderichthys*) possess external nostrils (nares) outside the mouth and internal nostrils (choanae) surrounded by exactly the same bones as in tetrapods. Consequently, in rhipidistians and tetrapods the nasal passage can serve as a passage for air from outside to the inside of the mouth cavity while the mouth is closed, while in lungfishes such a passage is not possible because both "external" and "internal" nares lie within the mouth cavity. Because of these and other anatomical features, it is hypothesized that the earliest tetrapods are more closely aligned with the rhipidistians than with the lungfishes (Fig. 3–15). The earliest amphibians, the subclass **Labyrinthodontia,** share many features with the rhipidistians, especially with *Panderichthys.* It is now generally accepted that labyrinthodonts have evolved from rhipidistians like *Panderichthys* and that they have given rise to the modern amphibians. If this current hypothesis is correct, amphibians represent a monophyletic assemblage.

Selection pressures for the perfection of the terrestrial adaptations of the rhipidistian continued during the Devonian as new habitats in a continuously humid climate persisted and terrestrial food resources and mating sites proliferated. Such sustained selection pressure regimes often result in evolutionary trends, as exhibited in the structural and functional transformations of the locomotory apparatus in the rhipidistians.

Amphibians continued the transition. The amphibian's most obvious derived character for life on land is its four legs, although legs have been secondarily lost in some groups of tetrapods. Amphibians are reasonably well adapted to land in their other organ systems as well, but only to life in damp and humid microhabitats. Most amphibians must return to the water to reproduce because their larvae are still aquatic. The term *amphibian* (Gr. *amphi* = double + *bios* = life) refers to their life in both water and land. However, many are fully terrestrial.

Additional terrestrial adaptations evolved in reptiles. The aquatic larval stage

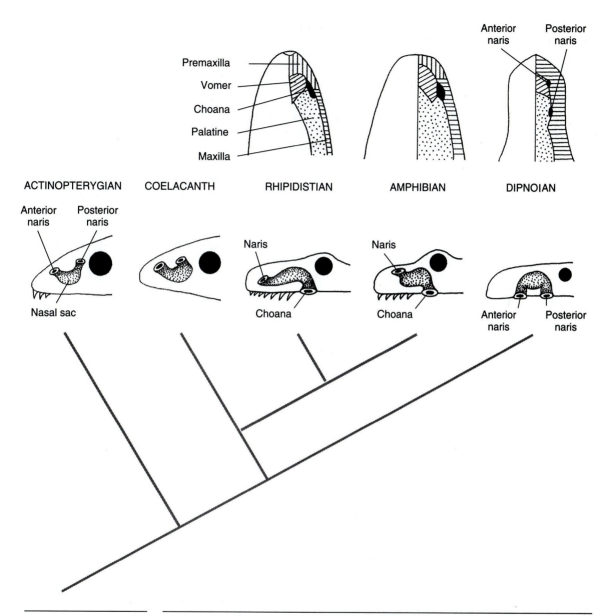

FIGURE 3–15 Rhipidistian (osteolepiform)—Tetrapod interrelationships. Top row of figures shows one half of the roof of the mouth (palate). Lower row of figures shows the location of the anterior and posterior nares and choanae, and the nasal sacs. (Modified from Schultze.)

has been suppressed, and reptiles reproduce on land. Many are sufficiently well adapted to terrestrial conditions to live in deserts. Most reptiles are limited in their ability to exploit fully the terrestrial resources by their inability to maintain the high body temperature needed to sustain their metabolism at night or in cold climates. They become torpid under these conditions. Endothermy evolved in birds and mammals and has permitted them to be active under a far wider range of environmental conditions than other tetrapods.

Characteristics of Amphibians

Most frogs and salamanders have reasonably sturdy skeletons and four well-developed legs. They tend to have flatter and broader skulls than other tetrapods, and the pelvic girdle of most species attaches to the vertebral column only by a single pair of sacral ribs and one sacral vertebra. The sense organs of adult amphibians are adapted to receive stimuli in the terrestrial environment. Most amphibians retain lungs inherited from their rhipidistian ancestors, and gills are lost in metamorphosed adults. Osteoderms, which are small nodules of bone embedded in the skin, are absent in most living species. The thin, moist, and vascular skin of amphibians serves as an accessory respiratory membrane. Amphibians are well adapted to life beside or in ponds and streams, under rocks and logs, in damp forests, and in other humid habitats. However, many tropical species are fully terrestrial in humid habitats. A few frogs live in deserts (p. 733), where they survive by burrowing into the soil and becoming torpid (**estivating**), except during brief rainy periods. Most amphibians reproduce in the water and pass through an aquatic larval stage. However, many species can lay eggs in very damp places on land.

Amphibians are ectothermic, and terrestrial ectotherms face more complex problems than fishes. Ambient temperatures in a given habitat may fluctuate rapidly and greatly. Electromagnetic heat waves, which are absorbed by water, reach a terrestrial animal directly. The evaporative loss of body water through the respiratory passages and skin, which is not a problem for a fish, involves the conversion of water to water vapor. Since 584 calories are needed for the evaporative conversion of 1 gram of water to water vapor, the amount of body heat lost in this way is considerable; over five times the number of calories required to drive off water vapor by bringing 1 gram of water from 0°C to its boiling point at 100°C. The terrestrial environment, however, is a mosaic of microclimates ranging from exposed areas where solar radiation and temperatures vary greatly to less stressful humid and shady habitats where conditions are more stable. Amphibians tend to limit their activity to the less stressful habitats. Even so, ambient temperatures in the temperate regions of the world can fall in winter far below the level that amphibians can tolerate. They burrow into soft ground or retreat to the bottoms of ponds where the temperature remains at freezing or higher and enter a dormant state known as **hibernation.** Metabolic rates drop, and the animal lives from food reserves stored in its body.

The Adaptive Radiation of Amphibians

Labyrinthodonts. The earliest amphibians were the ichthyostegalians (order **Ichthyostegalia**), found in late Devonian deposits (Fig. 3–16A). They had a peculiar mixture of piscine and terrestrial characteristics. Their four legs enabled them to move about land, but they retained a well-developed, fishlike caudal fin that must have been used in swimming. The lateral line canals on their skulls were parts of an aquatic sensory system. Probably ichthyostegalians spent most of their time in the water and only made brief excursions to the land. Ichthyostegalians are one order of a large subclass of amphibians (subclass **Labyrinthodontia**) that

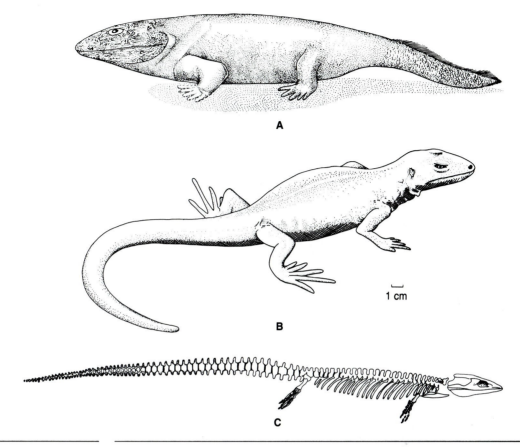

FIGURE 3–16   Some early amphibians: **A,** The labyrinthodont *Ichthyostega* is the earliest known amphibian. **B,** *Gephyrostegus,* a late anthracosaur. **C,** A lepospondyl, *Urocordylus* (**A,** after Jarvik. **B,** after Carroll. **C,** after Steen.)

were abundant during the Carboniferous and Permian periods (Fig. 3–17). All became extinct by the end of the Triassic. Some were quite large animals, more than a meter long. Each vertebral body, or centrum, of labyrinthodonts was a double unit composed of anterior and posterior arcs of bone that surrounded and largely replaced the notochord. Labyrinthodonts were a diverse lot, and two additional orders are recognized. Temnospondyls (order **Temnospondyli**) included a wide variety of forms, both aquatic and terrestrial. Anthracosaurs (order **Anthracosauria**) were becoming reptile-like in vertebral structure and in other ways (Fig. 3–16**B**).

Lepospondyls.   Another group of Paleozoic amphibians (subclass **Lepospondyli**) (Fig. 3–16**C**) is characterized by a single, spool-shaped vertebral body that developed around and partly constricted the notochord. Most lepospondyls were small, aquatic species. They were an evolutionary sideline that did not give rise to any living group.

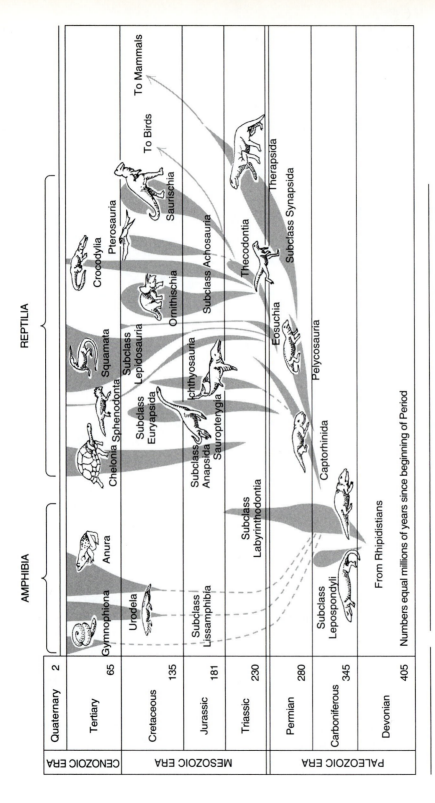

AMPHIBIA

Gymnophiona
Anura
Urodela

Subclass
Lissamphibia

Subclass
Labyrinthodontia

Subclass
Lepospondyli

From Rhipidistians

Numbers equal millions of years since beginning of Period

REPTILIA

Chelonia Sphenodonta
Squamata
Crocodylia
Pterosauria

Subclass
Anapsida

Subclass
Lepidosauria

Subclass
Euryapsida

Ornithischia
Saurischia

To Birds

To Mammals

Ichthyosauria
Sauropterygia

Subclass Achosauria

Thecodontia

Therapsida

Subclass Synapsida

Eosuchia

Pelycosauria

Captorhinida

		CENOZOIC ERA	Quaternary	2
			Tertiary	65
		MESOZOIC ERA	Cretaceous	135
			Jurassic	181
			Triassic	230
		PALEOZOIC ERA	Permian	280
			Carboniferous	345
			Devonian	405

FIGURE 3–17 The evolutionary history of amphibians and reptiles, their distribution in time, and relative abundance.

80

Lissamphibians. The nearly 3000 species of contemporary amphibians are grouped into three orders. Salamanders (order **Urodela** or **Caudata**) have well-developed tails and the general shape of labyrinthodonts, but they are small, secretive animals (Fig. 3–18**A** to **C**). Although most species are terrestrial as adults, many salamanders have become paedomorphic through neoteny. Neotenic species remain aquatic and fail to complete their metamorphosis. Frogs (Ranidae) and toads (Bufonidae) (order **Anura** or **Salientia**) are the most abundant of living amphibians. They are adapted for jumping and swimming by powerful thrusts of their long hind legs (Fig. 3–18**D**). The trunk is short, and the tail has been lost in adults of all species. The tropical caecilians (order **Gymnophiona** or **Apoda**) are adapted for burrowing (Fig. 3–18**E**). They have long, wormlike trunks, short tails, and reduced eyes, and they have lost their legs. Contemporary orders share many derived features and probably had a common origin in some labyrinthodont group, probably one of the temnospondyls. Juveniles of certain temnospondyls share with the adults of modern amphibians an unusual tooth type that is slightly bicuspid and is attached to the jaw by a narrowed pedicle. The ribs of modern amphibians are short and sometimes fused to the vertebrae. Modern amphibians never have more than four toes in their front feet, their wrist and ankle bones are incompletely ossified, and they share certain distinctive peculiarities in the ear (p. 423). For these reasons they are grouped together in the subclass **Lissamphibia** (Gr. *lissos* = smooth). Just when lissamphibians separated from the labyrinthodonts is unknown, for their fossil record goes back only to the Triassic.

The phylogenetic relationships of the Lissamphibia are depicted in Fig. 3–19. It is generally accepted that the Lissamphibia are monophyletic, sharing a common ancestor with *Trematops* and the Dissorophidae among the labyrinthodonts. During their evolutionary history, the lissamphibians inherited a tooth replacement pattern that proceeds from the medial side to the lateral side of the jaw. During subsequent evolution, two bones in the palate, the pterygoid and vomer, became separated (see Fig. 7–7**B** and **D**, p. 217). Loss of the supratemporal and jugal bones in the skull, and the interclavicle bones in the pectoral girdle led to the specialized Gymnophiona inhabiting subterranean habitats. Further losses of the postfrontal in the skull and surangular and splenial bones of the lower jaw occurred in the urodeles. In addition to all these losses of bony elements, the anurans developed a fusion of two skull bones to form the frontoparietal and a very elongate ilium, a bone in the hip, and the loss of a tail as adaptations for jumping.

Class Reptilia

Characteristics of Reptiles

Turtles, lizards, snakes, crocodiles, and other reptiles (L. *reptare* = to crawl) are well adapted to life on land. They have a stronger skeleton than amphibians, and their pelvic girdle attaches to the vertebral column by at least two sacral vertebrae. The skull is deeper and narrower than that of an amphibian. The outer layer of the skin is composed of dry, **horny scales** that offer protection.

The major adaptive feature and derived character of reptiles has been the evolution of a **cleidoic egg** that enables them to bypass the aquatic larval stage.

FIGURE 3–18 Contemporary amphibians: **A**, The tiger salamander *Ambystoma*. **B**, The mudpuppy *Necturus*. **C**, *Siren*. **D**, The tree frog *Hyla*. **E**, A caecilian, *Ichthyophis*, brooding eggs. (**A–D**, after Noble. **E**, after Gadow.)

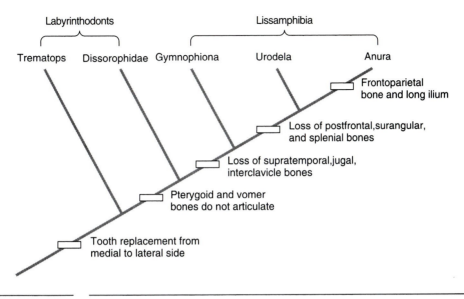

Labyrinthodonts   Lissamphibia

Trematops   Dissorophidae   Gymnophiona   Urodela   Anura

Frontoparietal
bone and long ilium

Loss of postfrontal, surangular,
and splenial bones

Loss of supratemporal, jugal,
interclavicle bones

Pterygoid and vomer
bones do not articulate

Tooth replacement from
medial to lateral side

FIGURE 3–19   The phylogenetic relationships of the Lissamphibia, with key characters. (Modified from Trueb and Cloutier.)

The egg contains a large store of yolk that is suspended from the embryo in a yolk sac, an **amnion** that forms a liquid cushion around the embryo, and other membranes needed to protect the embryo and provide for gas exchange and the elimination of nitrogenous wastes (p. 131). These extraembryonic membranes are surrounded by albumen and by a porous, parchment-like or calcareous shell, which are secreted by the oviduct. This type of egg must be deposited on land or retained within the oviduct; it cannot be laid in water, for the embryo depends on gas exchange with the air or mother. Reptiles hatch or are born as miniature adults. The cleidoic egg is a self-contained egg (Gr. *kleid-* = key). In a figurative sense it has been the key to terrestrial life. The cleidoic egg with its amnion is retained by birds and, in modified form, by mammals. Reptiles, birds, and mammals collectively are called **amniotes** for this reason. In contrast, fishes and amphibians are **anamniotes.**

Most reptiles have no body insulation and are ectotherms, but mechanisms have evolved that give them more control over their body temperature than amphibians have. When the sun rises, the body temperature of a reptile increases to some species-specific optimum level, metabolic rate rises, and the animal becomes active. Features that maximize the rate at which reptiles warm up, minimize the rate at which they cool off, and enable them to maintain a relatively high and constant body temperature during their period of activity differentiate them from amphibians. An Andean lizard can maintain a body temperature close to 35°C, about the same as in many mammals, even when the air temperature is 0°C to 10°C. In the cool of the night the lizard frequently burrows shallowly into the ground, which retains some heat. When the sun rises, the lizard first warms up by heat conduction through the soil, and then it emerges and gains heat rapidly by solar radiation. The animal increases the radiant energy it receives by dispersing a dark pigment in certain skin cells known as chromatophores. When the animal is warm enough, the pigment is retracted. A reptile detects the amount of solar radiation received, often by the median parietal eye (p. 430), and regulates its behavior accordingly. It moves in and out of the shade as needed or changes its orientation to the sun's rays. It lies perpendicular to the sun's rays to maximize heat absorption, and parallel to them to minimize heat absorption. A reptile's horny skin reduces loss of heat through the evaporation of water better than an amphibian's thin skin. A reptile also has some ability to control heat loss or gain by varying its heart rate and hence the amount of blood flowing through the skin. Early in the morning, when body core temperature is low, heart rate automatically increases and heat is transferred rapidly from the body surface being warmed by the sun to deeper body tissues. Heart rate slows down when core temperature reaches the optimum level or begins to fall. Because of the importance of solar radiation in their thermoregulation, reptiles are sometimes called **heliotherms** (Gr. *helios* = sun + *thermos* = heat).

These sorts of temperature control mechanisms do not require the expenditures of a great deal of energy. Reptiles can maintain their high and relatively constant body temperature on sunny days at a relatively low metabolic cost. They can enter more stressful environments than amphibians and be active under a wider range of conditions. But reptiles are still limited in the ecological niches that they can exploit by their inability to maintain their body temperature at optimum levels in the absence of warm ambient temperatures. Most reptiles are diurnal. Only a few tropical species can be active at night, and temperate species must hibernate during the winter. Very large species, such as crocodiles and giant tortoises, have a certain thermal stability that derives from their body size, since a large mass with its relatively small surface area changes temperature slowly.

Bakker (1971) has proposed that the extinct dinosaurs were endothermic and maintained a high body temperature by an increased metabolic rate. Among other lines of evidence, he points out that an endothermic, carnivorous dinosaur must eat more food than an ectothermic reptile to sustain its metabolism, so a given habitat cannot support as many endotherms as ectotherms. Bakker believes that the ratio of dinosaur remains to their prey organisms supports this notion. Many paleontologists do not believe that the samples from the deposits are large enough statistically to justify this argument. The large size of most species may have resulted in relative thermal stability because of the favorable small surface area.

The Adaptive Radiation of Reptiles

Stem Reptiles, Turtles.   The over 6000 living species of reptiles are an important part of the contemporary vertebrate fauna. Reptiles, however, were much more numerous during the Mesozoic era, when they were the dominant terrestrial vertebrates. The ancestral, or stem, reptiles, order **Captorhinida,** appear in the fossil record in the Carboniferous (Fig. 3–17). Captorhinids were modest-sized, superficially lizard-like reptiles similar to the anthracosaur labyrinthodonts from which they evolved. They differed from anthracosaurs in having a higher and narrower skull (Fig. 3–20**A**). All of their living descendents have an amnion and other char-

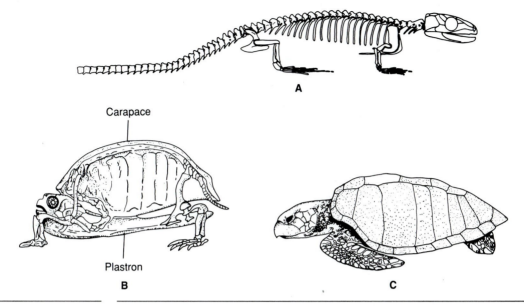

Carapace

Plastron

A

B

C

FIGURE 3–20   Anapsid reptiles: **A**, A captorhinid, *Hylonomus*. **B**, A turtle skeleton. **C**, A sea turtle. (**A**, after Colbert. **B**, after Gregory.)

acteristic extraembryonic membranes of the cleidoic egg, so they must have this type of egg too. Fossil-shelled eggs, which resemble cleidoic eggs, have been found in deposits containing early reptiles. The captorhinids are placed in the subclass **Anapsida** because a solid bony roof lay over the jaw muscles in the temporal region of the skull (Gr. *an* = without + *apsid* = loop or bar; i.e., there are no bars of bone between temporal openings). In this respect they resembled labyrinthodonts and many bony fishes.

As reptiles exploited the resources available to them, they underwent a rapid and extensive adaptive radiation from this basal stock. They diverged in feeding mechanisms, which affected skull morphology, and in patterns of locomotion, which affected limb and foot structure. Two primary lines of evolution can be traced back to the earliest stem reptiles or amniotes. One line, the **Synapsida,** became more and more mammal-like in their characteristics, and culminated in mammals. The other line, the **Sauropsida,** led to all of the other reptiles and to birds.°

°It follows from this that reptiles, as they are usually thought of, are not a monophyletic group. A monophyletic group by definition includes all of the descendants of a common ancestor. Reptiles had a common origin in the earliest amniotes, but the class does not include all of the descendants of these ancestral amniotes (birds and mammals are excluded). Technically reptiles are a paraphyletic, not a monophyletic, group.

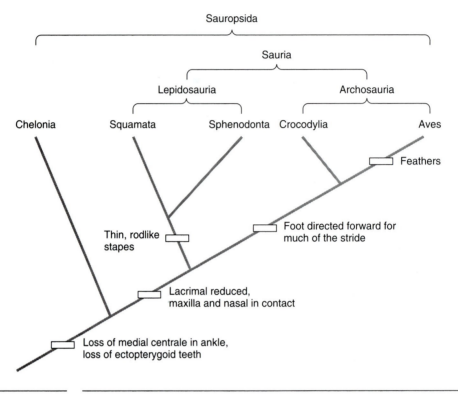

Sauropsida

Sauria

Lepidosauria   Archosauria

Chelonia   Squamata   Sphenodonta   Crocodylia   Aves

Feathers

Thin, rodlike stapes

Foot directed forward for much of the stride

Lacrimal reduced, maxilla and nasal in contact

Loss of medial centrale in ankle, loss of ectopterygoid teeth

FIGURE 3–21   A hypothesis of the phylogenetic relationships of living reptiles.

Turtles, order **Chelonia,** diverged very early from captorhinids. They retain the anapsid skull and so are often included with the captorhinids in the subclass Anapsida. In common with other sauropsids, they have lost one of the small, central bones (centralia, Fig. 8–21A) in their ankle, and teeth on the ectopterygoid bone of their palate (Fig. 3–21). Turtles, in fact, have lost all of their teeth, and a horny covering on the jaw margins replaces them. Turtles differ from other vertebrates in having a short trunk that is encased in a bony shell covered by horny plates (Fig. 3–20**B**). A carapace covers the trunk dorsally; a plastron covers the ventral surface. Most turtles live on land or in fresh water; but two contemporary families have returned to the sea. Sea turtles swim by means of large pectoral flippers that move up and down in unison (Fig. 3–20**C**). Females return to the beaches to lay their eggs.

Lepidosaurs.   Other sauropsids developed two windows, or fenestrae, in the temporal region of the skull in areas of reduced stress where bone is not needed. This is the diapsid skull. It became modified in different ways in different lineages. The more primitive diapsid reptiles are placed in the subclass **Lepidosauria.** Lepidosaurs are further characterized by loss of the lacrimal bone near the nose so the maxilla and nasal bones make contact with each other. They also share a specialized, thin, rodlike stapes, an ear ossicle capable of responding to airborne vibrations (Fig. 3–21). The most primitive lepidosaurs remain quadrupeds and retain

FIGURE 3–22 Lepidosaurs: **A**, The tuatara *Sphenodon* of New Zealand. **B**, A lizard. **C**, An amphisbaenian. **D**, A snake. (**A**, after Gadow. **C**, after Gans.)

a typical diapsid skull, palatal teeth, and a median eye. The tuatara, *Sphenodon*, which is now found only on a few islands off the New Zealand coast, is a living example of this stage of evolution. It is placed in the order **Sphenodonta** (Fig. 3–22**A**). Other contemporary lepidosaurs are the lizards, wormlike amphisbaenians, and snakes of the order **Squamata** (Fig. 3–22**B** to **D**). They are by far the most diverse of modern reptiles, including almost 6000 species. Some retain palatal teeth, a median eye, and other primitive features, but their diapsid skull is highly modified. Lizards and amphisbaenians have lost one of the bars of bone that border the temporal fenestrae; snakes have lost both. Consequently, jaw mechanics are very flexible. The jaws can be opened very wide, and snakes can swallow prey larger than the diameter of their heads. There is also a tendency for trunk elongation and limb reduction. Some lizards, most amphisbaenians, and all snakes lack legs. The long trunk and absence of legs may have evolved as adaptations for burrowing. Amphisbaenians and primitive snakes still burrow, but most extant snakes now live above ground.

Archosaurs.   Archosaurs (Fig. 3–23) retained an unmodified diapsid skull but tended to rear up on their hind legs, at least when they moved rapidly. Some became bipeds. Adaptations for bipedalism include an elongation of the ankle and other parts of the hind leg, changes in the ankle that allow the foot to be directed forward (Fig. 3–21), a reduction in the number of toes, and the evolution of a heavy tail that acted as a counterweight to the front of the body when the animal reared up on its hind legs. An unusual feature of the archosaurs is a perforation in the pelvic girdle in the base of the socket (acetabulum) that articulates with the hind leg. Archosaurs also have a skull fenestra either in front of the orbit or in the lower jaw or both. Archosaurs were the dominant reptiles during the Mesozoic. They include two orders of dinosaurs (orders **Saurischia** and **Ornithischia**), the flying reptiles (order **Pterosauria**), and the crocodiles (order **Crocodylia**). Many early dinosaurs were the size of chickens, but the group is best known for the large size of later species. Some archosaurs remained quadruped or reverted to a quadrupedal gait, but their enlarged hind legs and other features indicate bipedal tendencies in their ancestry. Of this large and diverse subclass, only the crocodiles and the birds that diverged from early archosaurs have survived.

Marine Reptiles.   Two groups of reptiles diverged early from other sauropsids and adapted independently and in different ways to the marine environment. Both lineages are characterized by a single temporal opening high up on the skull. Their phylogeny is uncertain, but it is possible that they diverged from early diapsids by the loss of the lower temporal opening and the widening of the arch of bone on the ventral border of the upper temporal opening (Carroll, 1988). These two groups were otherwise different, however, so their association in the subclass **Euryapsida** may not be a natural grouping (Fig. 3–17). One evolutionary line culminated in the ichthyosaurs (order **Ichthyosauria**). These reptiles looked like fish and swam by lateral undulations of the trunk and large caudal fin. Their paired appendages were reduced to small, finlike structures (Fig. 3–24**A**). The other line culminated in the marine plesiosaurs (order **Sauropterygia**). Plesiosaurs had short, broad trunks, long necks, and long, paddle-like paired appendages (Fig. 3–24**B**). Some were enormous, attaining a length of 15 meters.

Synapsids.   The synapsid line of tetrapod evolution appeared in the late Carboniferous and early Permian periods. Either synapsids diverged from the earliest stem reptiles (Fig. 3–17), or possibly stem reptiles and synapsids had a common ancestor. We treat them here, as do most authors, as a subclass of reptiles, the **Synapsida,** but some investigators regard them as a group that is independent of the reptiles.

A

B

C

D

FIGURE 3–23   Archosaurs: **A**, An early, bipedal saurischian dinosaur, *Coelophysis*. **B**, The ornithischian dinosaur *Triceratops*. **C**, A flying reptile, *Pteranodon*. **D**, A crocodile.

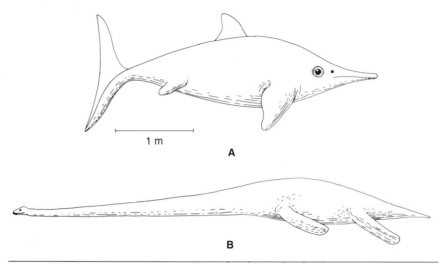

1 m

A

B

FIGURE 3–24   Mesozoic marine reptiles: **A**, The ichthyosaur *Ophthalmosaurus*. **B**, The plesiosaur *Elasmosaurus*. (After Watson.)

A single, laterally placed, temporal fenestra characterizes the skull roof. A similar temporal roof is found in mammals. The most primitive synapsids (order **Pelycosauria**) were very similar to captorhinids except for the small synapsid-type fenestra (Fig. 3–25**A**). The fenestra enlarged in later synapsids (order **Therapsida**), and the structure of their limbs and other parts of the postcranial skeleton suggests that they were becoming active and agile animals (Fig. 3–25**B**). They evolved many osteological features that foreshadowed the mammalian condition. Synapsids have left an excellent fossil record, and the transition through them to mammals is well documented.

Class Aves

Endothermy

Most birds and mammals have a level of metabolism three or four times greater than that of reptiles of comparable size and activity. The dominant aspect of the heat balance of a bird or mammal is the internal production of heat by its high metabolic level. These are **endothermic** vertebrates. "Warm blooded" is a loose synonym, but many ectotherms that live in warm habitats also have high body temperatures. Although birds and mammals are both endothermic, they descended from different groups of reptiles and evolved endothermy independently as they adapted to active modes of life.

Birds and mammals are able to maintain a high and constant body temperature both day and night over a wide range of ambient temperatures. Body temperature varies with the species but is on the order of 35°C to 40°C. Birds and mammals can be active over a far greater range of conditions than reptiles only at a high metabolic cost. Heat is produced by the activity of all parts of the body,

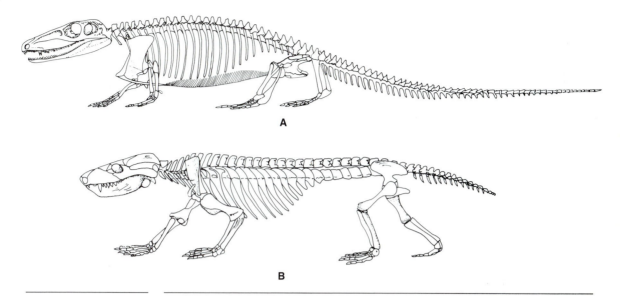

FIGURE 3–25   Synapsids: **A**, An early pelycosaur, *Varanosaurus*. **B**, A later therapsid, *Thrinaxodon*. (**A**, after Romer and Price. **B**, after Brink.)

but the visceral organs contribute a disproportionately large share. Usually, body core temperature is higher than environmental temperature, so heat flows from the core to the body surface to the environment. Temperature regulation is accomplished primarily by controlling the amount of heat lost at the body surface. Birds and mammals have evolved insulating layers of subcutaneous fat and surface feathers or hair that conserve heat. Both feathers and hair entrap a still layer of air next to the skin. Since the coefficient of thermal conductivity of air is only 0.00006 (that of water is 0.001), heat loss to the environment is minimized. The degree to which the feathers or hairs are elevated above the surface can vary the thickness of the layer of still air and hence the rate of heat loss. Aquatic birds and mammals have very oily feathers or hair that prevent water from reaching the body surface, and they also have thick layers of subcutaneous fat. The amount and rate of blood flow through the skin also can be controlled by the degree of vasoconstriction of cutaneous vessels and by the heart rate. Flow is extensive and rapid when heat needs to be lost; opposite changes occur when heat needs to be conserved. Finally, birds and mammals can regulate heat loss by controlling the amount of water evaporation. Birds and many mammals pant; some mammals also can sweat.

Controls of heat loss at the surface do not require the expenditure of much additional energy. A high and constant body temperature can be maintained at rest over a fluctuation in environmental temperatures of 5°C to 10°C with little extra metabolic work. This range is called the **thermal neutral zone,** and most birds and mammals spend much of their time in environments that fall within it. The control of body temperature when environmental temperatures exceed this range requires significant increases in metabolism. More heat is produced internally in cold weather. Heat loss at the surface is increased in hot weather by rapid blood circulation through the skin and by rapid panting or sweating.

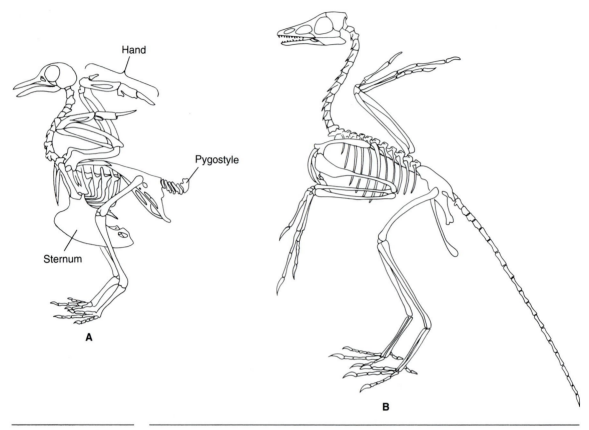

FIGURE 3–26 The skeleton of a modern bird (**A**) compared with a reconstruction of the skeleton of *Archaeopteryx* (**B**). (After Heilmann.)

Characteristics of Birds

Birds (class **Aves,** L. *aves* = birds) are characterized by their particular adaptations for endothermy and by their ability to fly. Feathers and subcutaneous fat provide insulation, and excess heat can be lost by panting. Flight requires wings, a light-weight body, and a high energy output. The pectoral appendages are modified as wings (Fig. 3–26A). The flying surfaces of the wings are composed of large, strong, and lightweight flight feathers that extend caudally from the arm and from three partially united fingers. Other flight feathers fan out from a group of fused caudal vertebrae known as the pygostyle. The broad, keeled sternum provides a large area for the origin of flight muscles. All the bones are exceptionally strong. Many contain air sacs that extend from the lungs into the space that is filled with bone marrow in mammals. A bird's hind legs act as shock absorbers when a bird lands, and when not flying, birds are bipeds or swimmers. Their archosaur-like pelvis is strong, and considerable fusion has occurred between the sacral and trunk ver-tebrae, pelvic bones, and bones of the lower leg and foot. The digestive, respira-tory, and circulatory systems are adapted to sustain a high level of metabolism. A horny bill replaces teeth in contemporary species.

The Origin and Adaptive Radiation of Birds

Archaeornithes.   The structure of modern birds points to an affinity with archosaurs, but the resemblance to archosaurs is even more striking when one examines the skeleton of *Archaeopteryx,* the best known bird from Jurassic deposits (**Fig. 3–26B**). *Archaeopteryx* was a crow-sized creature. It retained a long tail and an archosaur-like pelvis and hind legs. There were only four toes in the feet, as in archosaurs, and the metatarsals and ankle bones were not fused. Although partly crushed in the known fossils, the skull appears to have been diapsid. Teeth were present. The three fingers in the wing were not united; each bore a claw. There is no indication of an ossified sternum. Were it not for the impression of feathers in the fossil, the animal probably would be regarded as an unusual archosaur. Details of feather structure indicate that *Archaeopteryx* was capable of flapping, powered flight, but without an ossified sternum it could not have been a strong flier. The presence of a clavicle (the wishbone), which most archosaurs lack, and other technical features indicate that birds evolved from early, bipedal saurischian dinosaurs.

The closest living relatives of birds are the crocodiles. Because of the presence of so many primitive features, *Archaeopteryx* is placed in its own subclass, the **Archaeornithes** (Gr. *ornithes* = birds).

Neornithes.   The recent discovery of a 135-million-year-old, sparrow-sized *Sinornis* from the Lower Cretaceous in China shows the early evolution of avian flight and perching (**Fig. 3–27**). *Sinornis* had a broad sternum (breast bone) for the attachment of flight muscles and the most posterior vertebrae fused into a pygostyle. *Sinornis* was capable of modern sustained avian flight around inland lake habitats. In contrast to *Archeopteryx, Sinornis* must have had endothermic physiology to fuel the bulky flight muscles. *Sinornis* and some skeletal fragments from the upper Cretaceous period of Argentina may represent a unique subclass of birds (Walker, 1981), but possibly excepting these, all later birds are **Neornithes.** A member of this subclass has a short tail that usually bears a pygostyle. The fingers in the wing are partially united and lack claws. The sternum is well ossified, broad, and usually keeled. More fusion has occurred between vertebrae, among pelvic bones, and in foot bones than in *Archaeopteryx.* Cretaceous genera that retained teeth are placed in the superorder **Odontognathae**.

All other birds are toothless and they are sorted into two superorders, the **Paleognathae** and **Neognathae.** Paleognathous birds retain a relatively solid, archosaur-like bony palate. Neognathous birds have evolved a more mobile palate, which participates in the kinetic feeding movements of the skull. Most of the paleognathous birds evolved in areas where they were not subject to terrestrial predation, and their legs have become large and powerful, and wings are reduced. The broad, though usually unkeeled sternum and wing vestiges attest to their flying ancestry. Examples of terrestrial birds are the South American rhea, the African ostrich, the Australian emu, the New Guinea cassowary, and the New Zealand kiwi (**Fig. 3–28**).

Neognathous birds have undergone an extensive adaptive radiation, and approximately 8700 contemporary species are known. Ornithologists recognize 23

FIGURE 3–27   Reconstruction of *Sinornis* (From Sereno and Chenggang, 1992, permission of the American Association for the Advancement of Science.)

orders, all but one of which have living representatives (see synopsis). A phylogenetic tree can be constructed on the basis of distinctive, derived characters found in the groups, but the fossil record of birds is still incomplete.

A wealth of data on the configuration of bones and muscles (Cracraft, 1988) and DNA (Sibley and Ahlquist, 1990) has yielded two major evolutionary schemes of the radiation of neognathous birds. We adopt most components of the more conservative hypothesis proposed by Cracraft (Fig. 3–29) with the caveat that DNA evidence has led Sibley and Ahlquist to propose a significantly different hypothesis. For some groups we have deviated from Cracraft's hypothesis and adopted that of Sibley and Ahlquist. The differences between these two schemes reflect the major challenge for future biologists to reconstruct the history of life. The neognathous radiation has resulted in approximately 23 monophyletic lineages, each of which has, in turn, radiated to occupy various adaptive zones, especially among the songbirds (Passeriformes). A comparison of ducks, penguins, gulls, eagles, hummingbirds, and finches illustrates that as neognathous birds radiated, they adapted to different foods, habitats, and methods of locomotion. Birds, therefore, differ in bill structure, wing and tail size and form, and in the adaptations of their legs and feet. In some cases, such as on small islands, where birds have

FIGURE 3–28   Representative paleognathous birds: **A**, An ostrich. **B**, The kiwi.

not been subject to terrestrial predation, they have readapted to a completely terrestrial mode of life. It is beyond the scope of this book to characterize each order of birds, but a list of the orders is presented in the synopsis.

Class Mammalia

Characteristics of Mammals

Mammals are active, agile animals. They have an extraordinarily well-developed brain and a large behavioral repertoire. Mammals are endothermic, but they evolved endothermy independently of birds, for the synapsid line of evolution toward mammals diverged from stem reptiles very early. **Hair** rather than feathers supplements subcutaneous fat in retaining the heat produced internally by metabolic processes. Heat can be lost when needed by sweating in many species as well as by panting. As in birds, the digestive, respiratory, circulatory, and excretory systems are adapted to sustain a high level of metabolism, but there are many differences in the ways this is accomplished. The nasal cavities of most mammals are large, and their internal surface area is increased by scrolls of bone known as turbinate bones. This enables mammals to have a keen sense of smell, and the increased surface area also helps clean, moisten, and warm inspired air. Important changes occurred in feeding mechanisms during the evolution of synapsids. These culminated in significant changes in jaw structure and dentition. Mammalian teeth

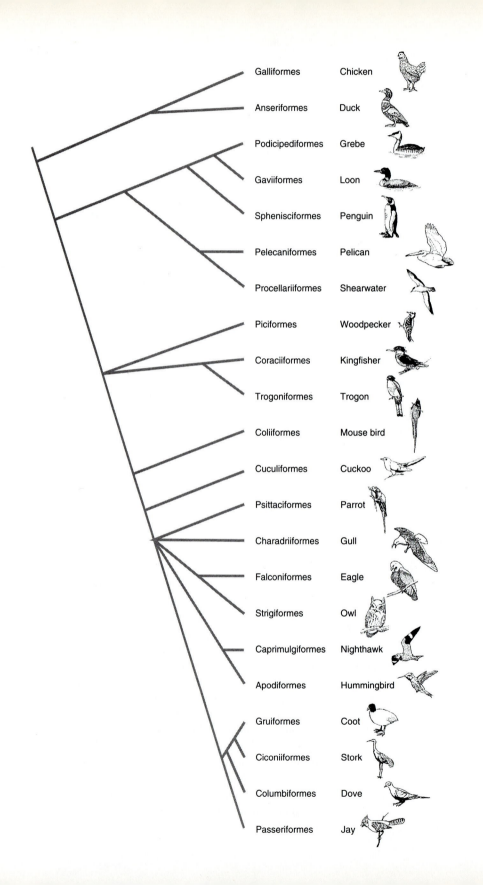

are specialized for different aspects of food capture and processing. Since mammals chew their food, precise occlusion is necessary. Tooth replacement is limited to a milk and a permanent set. A secondary palate evolved that separates the food from the respiratory passages in the mouth and much of the pharynx. This allows mammals to process food in their mouths and breathe at the same time. Having been freed from jaw functions, the quadrate and articular bones became specialized as two additional auditory ossicles. Mammals have three auditory ossicles rather than the single one usually present in other tetrapods. The cochlea, the sound-detecting part of the inner ear, is highly developed, giving mammals a keen sense of hearing.

The embryos of endothermic vertebrates must be reared in a warm environment when thermoregulatory mechanisms are developing. The egg-laying monotremes brood their eggs, as do birds. The embryos of other mammals undergo at least part of their development within a uterus. Newly hatched or born mammals are fed on milk secreted by the **mammary glands.** The term *mammal* derives from this feature (L. *mamma* = breast).

Primitive Mammals and the Origin of Mammalian Endothermy

Many, if not most, of the characteristics of mammals are associated with the evolution of endothermy. We cannot be certain when endothermy evolved. Some investigators believe it evolved gradually among the late synapsids as they became more active. The structure of their skeletons suggests that the therapsids were indeed active animals, but high activity and a high metabolism are not necessarily coupled. Crompton, Taylor, and Jagger (1978) have proposed an alternative hypothesis based on our knowledge of the metabolism of primitive living mammals and the probable mode of life of ancestral mammals.

The first mammals were small creatures that fed on insects, worms, and other invertebrates. Their eyes were small, but their auditory and olfactory organs were extraordinarily well developed. This sensory apparatus suggests that they were active at night and located their prey by smell and hearing. This notion is also consistent with ecological theory. To compete with reptiles, the early mammals must have occupied a different ecological niche, and the nocturnal, insectivorous niche would not have been available to most ectothermic reptiles. But to be active at night, the early mammals must have had some way of maintaining body temperature when the sun set. Probably they had an insulating layer of fat and fur. The first mammals may have lived like the tenrec of Madagascar. The tenrec avoids the heat of the day by burrowing into the ground, but it emerges at night to forage on insects, worms, and grubs that it finds by smell and hearing. It is endothermic but maintains a body temperature of about 28°C to 30°C, just a few degrees above the ambient nocturnal temperature. This is significantly lower than the body temperature of most mammals. The insulating layers of fat and fur enable the tenrec

◀ *FIGURE 3–29* A hypothesis of the adaptive radiation of the neognathous birds and the interrelationships of the monophyletic orders. (Modified from Cracraft.)

to maintain this body temperature at a relatively low energy cost. It has a reptilian level of metabolism; that is, it consumes no more energy than a reptile of the same size, temperature, and level of activity. "Setting" a body temperature slightly higher than the prevalent ambient temperature reduces the need to cool off by the evaporative loss of precious body water. This is particularly critical for a small animal, for when ambient temperatures exceed body temperature, its large surface-to-volume ratio would result in the rapid uptake of heat and the need to lose a great deal of water. Crompton and his associates believe that the tenrec and the related hedgehog reflect the adaptations of ancestral mammals and have remained in the ancestral niche.

When many reptile groups became extinct during the late Cretaceous, mammals could begin to exploit the diurnal resources that had previously been monopolized by the reptiles. For a small, primitive endotherm to become active in the daytime, body temperature should be about 10°C higher than that of a nocturnal species. This also would reduce the need for evaporative cooling during the daytime. Setting the body temperature higher than ambient daytime temperatures—and maintaining that setting at night—requires a much greater energy expenditure. It is probably in this context that mammals took the next step and evolved a much higher metabolism than reptiles. Endothermy, then, may have begun as an adaptation that enabled mammals to exploit a nocturnal niche and reached a higher level as mammals shifted to diurnal niches.

Some early mammals, including the egg-laying echidna, or spiny anteater, and the opossum do not easily fit this hypothesis because, although nocturnal, they have a body temperature and metabolism similar to that of diurnal mammals. This has led some investigators to question parts of the Crompton hypothesis, but Crompton and his associates argue that these mammals are only secondarily nocturnal.

Prototherians

Mesozoic mammalian fossils are fragmentary. Enough have been discovered, however, to show that mammals go back to the Triassic period, and that several radiations occurred during the Mesozoic (Fig. 3–30). Two subclasses, the **Prototheria** and **Theria** (Gr. *therion* = wild beast), can be distinguished on the basis of tooth structure and the extent to which certain bones contribute to the wall of the braincase. The earliest prototherians were the **Docodonta** and **Triconodonta.** The cusps of each of their molar teeth lay in a linear series, and the prootic bone (p. 219) contributed significantly to the braincase. These mammals appear to have been small, nocturnal insectivores with well-developed noses and ears but small eyes. The most abundant of later Mesozoic prototherians were rodent-like **multituberculates.** They had large, chisel-like front teeth and crushing multituberculate teeth in back. They are sometimes placed in a separate subclass, the **Allotheria.**

The only surviving prototherians are the platypus and two species of echidnas of Australia and New Guinea (Fig. 3–31). These mammals constitute the order **Monotremata.** Their fossil record is poor, and the adults lack teeth, so it is difficult to associate them with other groups of mammals. They are considered pro-

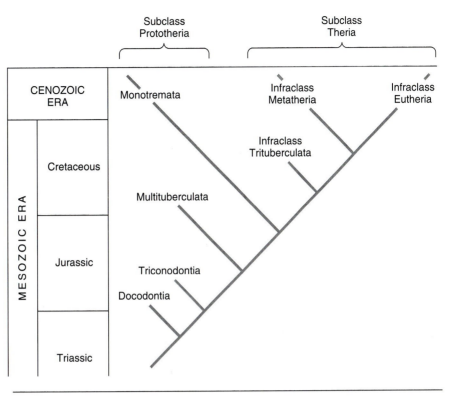

FIGURE 3–30   An overview of the phylogeny of mammals.

totherians because of the construction of the braincase. Monotremes are endotherms and have all of the distinctive characters of mammals, including mammary glands. Yet they continue to lay cleidoic eggs, and they have many skeletal and soft anatomy features that are similar to those of reptiles. For example, the cloaca has not been divided into separate digestive and urogenital passages, as in other mammals, so they retain a single opening for the discharge of fecal and urogenital products (Gr. *monos* = single + *trema* = hole). The platypus is a semiaquatic species that feeds in rivers with its ducklike bill. The nocturnal echidna gathers termites and ants with its long, horny bill and specialized tongue.

Early Therians

The earliest therians, the **Trituberculata,** probably diverged during the Jurassic from early triconodonts, but the three cusps of each molar tooth formed a triangle rather than being in one line (see Fig. 16–7). The alisphenoid bone rather than the prootic contributed to the lateral wall of the braincase. These conditions foreshadow the pattern seen in later mammals. We recognize two contemporary groups of therians, the infraclasses **Metatheria** (marsupials) and the **Eutheria,** which include the true placental mammals.

A   B

FIGURE 3–31   Contemporary prototherians: **A**, The platypus *Ornithorhynchus*. **B**, The Australian spiny anteater *Tachyglossus*.

Marsupials

Marsupials differ from eutherian mammals in many skeletal and dental details. Marsupials have a relatively smaller braincase than eutherians, an inturned, angular process on the lower jaw, and a more limited replacement of teeth, but they are distinguished primarily by their pattern of reproduction. Intrauterine life is very short. After birth the embryos attach themselves to the nipples, which often are located in a skin pouch called the **marsupium,** where they complete their development. We have sometimes regarded this reproductive pattern as inferior to the much longer intrauterine life that characterizes eutherian development, but it is well adapted to the unpredictable and often severe environments in which marsupials live (Chapter 21). Marsupials are different from eutherian mammals, but it does not follow that they are more primitive.

Marsupials appear in the fossil record during the early Cretaceous, when the large southern continent of Gondwanaland had begun to break up by continental drift. South America, Antarctica, and Australia were still connected, and a chain of islands lay between South and North America. Most investigators believe that the first marsupials evolved in North America, where the earliest fossils and the first radiation of marsupials have been found. Marsupials had spread to Europe and South America by the late Cretaceous. They became extinct early in the Tertiary in Europe and North America for reasons that are not completely understood. Other factors must have been involved in addition to competition with the newly arrived eutherians, since marsupials and eutherians coexist in some parts of the world today. Meanwhile, an extensive radiation of marsupials occurred in South America. When South America and North America again became connected in the early Tertiary, placental mammals moved into South America, and many marsupials became extinct. However, a variety of small opossum-like mammals survived in South and Central America, and the Virginia opossum invaded North America, where it has become well established.

A few marsupials had reached Australia by the Oligocene, presumably via a bridge of islands connecting Australia with South America by way of Antarctica (where marsupial fossils have recently been found). As Australia separated from Antarctica and drifted into lower latitudes, climatic conditions changed dramatically. Australia today has a great diversity of climates, ranging from tropical rain forests through temperate regions to deserts. Since only a few bats, rodents, and marine mammals also reached Australia, many ecological opportunities were avail-

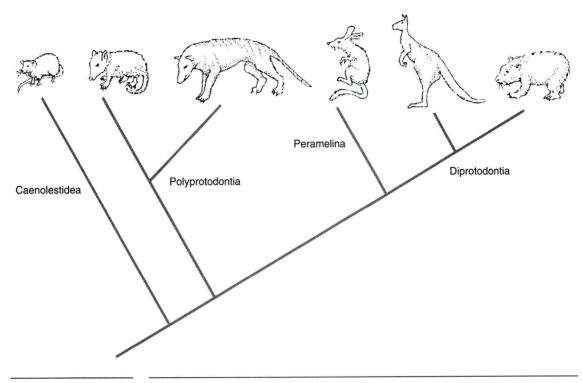

Caenolestidea

Polyprotodontia

Peramelina

Diprotodontia

FIGURE 3–32   The adaptive radiation of marsupials.

able. Marsupials underwent an extensive adaptive radiation and have occupied most of the insectivorous, carnivorous, and herbivorous niches that are occupied by eutherian mammals in other regions of the world (Fig. 3–32). Most of the 240 contemporary species of marsupials live in Australia. Although all these species were once placed in a single order, mammalogists now divide them into four orders. The caenolestids (order **Caenolestidea**), which occur primarily in South America, are rat-sized creatures that resemble ancestral mammals in their nocturnal habits and insectivorous and carnivorous diets. The polyprotodonts (order **Polyprotodontia**) include the opossums of North and South America, the native cats (dasyurids) of Australia, and the Tasmanian wolf. Most are small scavengers or predators that feed on invertebrates and small rodents. One group of dasyurids, the numbats or banded anteaters, have long tongues and other specializations that adapt them to eating termites. Bandicoots (order **Peramelina**) are hopping, rat-sized marsupials that forage in ground litter and dig among roots for food. The remaining Australian marsupials (order **Diprotodontia**) are specialized to feed on a variety of plants. The Australian possums are chiefly nocturnal, arboreal herbivores. Some have loose skin folds between their front and hind legs and glide gracefully from tree to tree as they forage. The diminutive honey possums have long, tubelike mouths with which they feed on nectar and pollen. The familiar koala is highly specialized to feed on the leaves of only a few species of eucalyptus trees. Wombats resemble large woodchucks. They shelter during the day in burrows that they have dug and emerge in the evening to feed. The dominant herbivores are the large kangaroos of open country and the smaller wallabies of more rocky and bushy terrain. They are terrestrial animals specialized for hopping. Despite their present diversity, studies of the structure of their blood serum proteins show that Australian marsupials are closely related and must have diverged from a common ancestral stock.

Eutherian Mammals

Eutherian mammals (infraclass **Eutheria**) date back to the Cretaceous and often are called placental mammals, but this term is not entirely appropriate because marsupials also have a placenta, although its structure is different (p. 773). Eutherian mammals have a longer gestation period than marsupials, and the young are born at a much more mature stage of development. There is no evidence that the eutherian method of reproduction is more efficient than the marsupial pattern.

Eutherians can be traced well back into the Cretaceous. Ancestral species probably continued to be small, insect-eating, nocturnal creatures, but they underwent an adaptive radiation very early in their history. The South American sloths, armadillos, and anteaters (order **Edentata**) and the scaly anteaters, or pangolins, of the Old World tropics (order **Pholidota**) were the first groups to diverge (Fig. 3–33). Some species remained generalized insectivores, but the anteaters became highly specialized for a diet of ants and termites.

Tenrecs, shrews and moles of the order **Insectivora** did not become as highly specialized as the edentates and pangolins and retained many primitive eutherian features. They are small, nocturnal insectivores with a well-developed nose and ears, but with small eyes. They have a full complement of teeth that are well adapted for piercing, killing, cutting, and crushing their food. Five clawed toes are present, and the first can oppose the others to some extent. The foot is not long and is placed flat on the ground, a posture termed **plantigrade.**

Animals with a primitive insectivorous dentition can easily adapt to feeding on other types of animals. Cats, weasels, and dogs (order **Carnivora**) are predaceous carnivores; but racoons, bears, and a few other species in this order have secondarily become omnivores. Seals and sea lions are marine carnivores.

Lemurs, monkeys, apes, and human beings belong to the order **Primates.** Ancestral primates are very similar to insectivores, differing primarily in features that adapt them to forage for insects, fruits, and soft plant food in the trees. Most primates are arboreal, but some, including human beings, have readapted to a terrestrial mode of life. The tree shrews of Southeast Asia (order **Scandentia**) resemble primitive primates in many ways, and some authors consider them to be primates. The flying lemurs of Southeast Asia (order **Dermoptera**) and bats (order **Chiroptera**) diverged early from the primate stock. Flying lemurs do not really fly, but glide from tree to tree as they forage. Loose skin folds between their front and hind legs are extended during a glide. The pectoral appendages and broad hand of bats form true wings. Most bats continue to feed on insects, but others have adapted to other food types.

FIGURE 3–33   A hypothesis of the phylogenetic relationships and adaptive radiation of the eutherian ▶ mammals. (Modified from Novacek.)

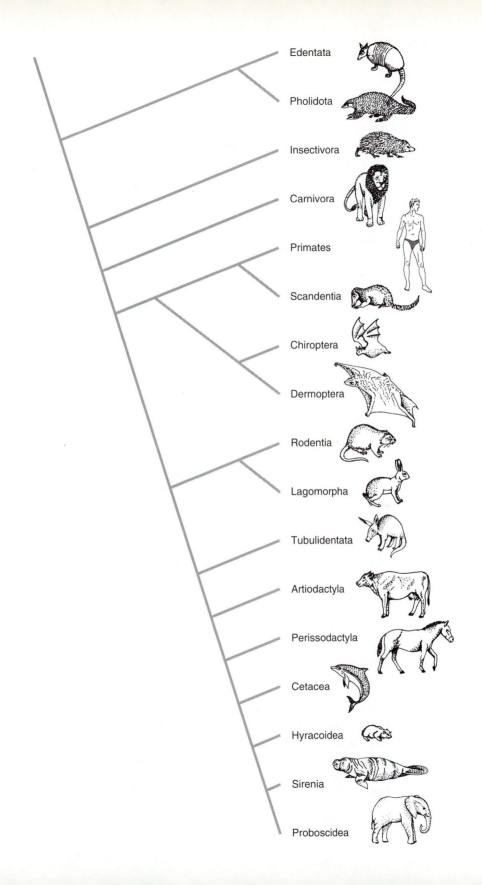

Edentata

Pholidota

Insectivora

Carnivora

Primates

Scandentia

Chiroptera

Dermoptera

Rodentia

Lagomorpha

Tubulidentata

Artiodactyla

Perissodactyla

Cetacea

Hyracoidea

Sirenia

Proboscidea

Most other mammals have adapted to eating plant food. Rabbits (order **Lago-morpha**) and rodents (order **Rodentia**) share enlarged front teeth and other adaptations for a gnawing, herbivorous mode of life, but these features evolved independently in the two groups.

Many groups of eutherians are adapted for browsing or grazing. Their jaws and tooth structure are specialized for feeding on plants, and the structure of their limbs enables them to run fast and escape their predators. Their limbs, and especially their feet, are very long, and they walk on the tips of their digits, each of which bears a hoof. Because they share hooves, these mammals are called **ungu-lates** (L. *ungula* = hoof), but their adaptive features evolved independently in several ungulate groups. Tapirs, rhinoceroses, and horses (order **Perissodactyla**) retain an odd number of toes, three or one. Pigs, giraffes, deer, sheep, and cattle (order **Artiodactyla**) retain an even number of toes, four or two. Elephants (order **Proboscidea**), conies (order **Hyracoidea**), sea cows (order **Sirenia**), and the African earth pig, or aardvark (order **Tubulidentata**), appear to have a common ancestor with the primitive ungulates. Porpoises and whales (order **Cetacea**) may have as well.

Vertebrate Evolution

In summary, the evolutionary history of the vertebrates can be depicted as a branching tree with new features characterizing each branch (Fig. 3–34). We can recognize seven major monophyletic lineages. Each newly evolved specialized feature (i.e., the emergence of jaws, lungs, paired limbs, amnion, and hair) seemed to have triggered new adaptive radiations and new ways of life. Such specialized characters are called **apomorphic** or **derived** characters that have evolved only once during vertebrate evolution. Vertebrates sharing such an apomorphy must therefore have shared a common ancestor. For example, all Sauropsida have lost the medial centrale bone element in the ankle (Fig. 3–34, character 1). Loss of this element reflects the monophyly of the Chelonia, Lepidosauria, Crocodilia, and Aves because they must have originated from a common ancestor in which the medial centrale was lost. Some zoologists therefore argue that birds should not be recognized as a separate class but should be linked with the crocodiles as well as dinosaurs.

This phylogenetic tree also reveals that some evolutionary lineages (e.g., the Agnatha) underwent very slow evolution or evolutionary **stasis,** whereas others (such as the Aves and some Osteichthyes) underwent very rapid rates of evolutionary transformations. Biologists try to explain these great differences in evolutionary rates. The successive emergence of the specialized features also illustrates one of the principles of Darwin's theory: descent with modification. Upon close examination, each new feature is composed of pre-existing building blocks, which either are rearranged or modified so new functions emerge. Thus, such modifications depend on historically (genetically) determined pre-existing structures and functions and the selection pressures prevailing in the environment. One of the major goals of zoologists is to understand to what extent and how fast these genetically determined pre-existing building blocks can change in response to various environmental factors.

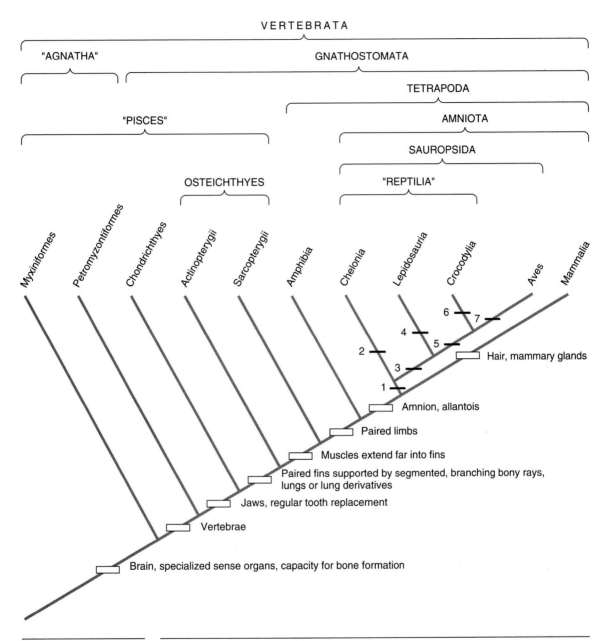

FIGURE 3–34   Phylogenetic interrelationships of the vertebrates. 1, Loss of the medial centrale of the ankle bones (see p. 282). 2, Carapace and plastron cover trunk. 3, Dorsal and lateral temporal openings (fenestrae). 4, Large sternum on which scapulocoracoid rotates (see p. 364). 5, Foot directed forward for much of the stride (see p. 361). 6, An akinetic skull (see p. 231). 7, Feathers. (Modified from Carroll.)

SUMMARY

1. Early vertebrates are ectothermic. They have no insulating layers at their body surface and cannot maintain an overall body temperature for an extended period that is different from that of the water in which they live. Because of the high thermal stability of water, this is not a problem for fishes.

2. The five orders of extinct ancestral vertebrates are commonly called ostracoderms. Early ostracoderms were marine, but most became freshwater. All were small, jawless creatures without well-developed paired appendages. Because of the absence of jaws, they are placed in the class Agnatha.

3. Modern cyclostomes, the lampreys and hagfishes, are the direct descendants of certain ostracoderms.

4. Fishes with jaws and paired appendages, known as gnathostomes, evolved in the late Silurian and early Devonian. These features allowed an extensive adaptive radiation of fishes into different methods of feeding and locomotion.

5. Placoderms are an extinct group of early jawed fishes.

6. Cartilaginous fishes (class Chondrichthyes) retain the embryonic cartilage as the adult skeletal material, have only small remnants of the ancestral dermal armor, and are heavy-bodied fishes without lungs or swim bladders.

7. Cartilaginous fishes include the living elasmobranchs (sharks, skates, and rays) and the odd-appearing holocephalans or chimaeras.

8. Bony fishes (class Osteichthyes) are characterized by the presence of at least some bone in the internal skeleton, a bony operculum that covers the gills, and lungs or swim bladders. Most species retain bony scales.

9. Acanthodians, or spiny sharks, are an extinct group of bony fishes characterized by extra pairs of spines between the paired appendages.

10. Most bony fishes are actinopterygians. They have fan-shaped paired fins supported by dermal rays. Early species had ganoid scales with many layers of enamel on the surface. Actinopterygians are represented today by the bichirs and sturgeons (chondrosteans) and a wide variety (20,000 species) of teleosts.

11. Sarcopterygians (lungfishes coelacanths, and rhipidistians) are bony fishes with lobate paired fins supported by a central axis of flesh and bone. Early members of the group had cosmoid scales, containing a great deal of dentine.

12. Although nearly extinct today, sarcopterygians are important because the rhipidistians gave rise to tetrapods.

13. The shift from aquatic to terrestrial life probably was triggered by increasing intraspecific and interspecific competition for food and mating resources in aquatic environments. At the same time, competition was less on land because prey and suitable mating habitats were abundant.

14. Because the physical conditions on land are so different, the transition from water to land was a major step in vertebrate evolution. Morphological and functional adaptations began in the rhipidistians, continued in the amphibians and reptiles, and reached a high degree of development in birds and mammals.

15. Amphibians typically have well-developed legs and other terrestrial adaptations, but they are restricted to moist and humid habitats by their inability to conserve body water, their need for water to reproduce, and their ectothermy. Some species are entirely aquatic.

16. Ancestral amphibians, the labyrinthodonts, are extinct. Amphibians are represented today by salamanders, frogs, and wormlike caecilians. All share a common tooth type and other features, so they are believed to be a monophyletic group, the lissamphibians.

17. Reptiles can exploit a wider range of terrestrial niches than amphibians because they have evolved horny scales and other features that reduce water loss, and the evolution of a cleidoic egg allows reproduction on land. Although still ectothermic, reptiles can sustain a high body temperature during warm and sunny periods by regulating their degree of absorption of solar radiation.

18. Reptiles radiated widely from the stem captorhinids and differentiated in their methods of feeding and locomotion. A synapsid line of evolution diverged very early and led to mammals. A sauropsid line led to other reptiles and to birds. Sauropsids were the dominant terrestrial reptiles during the Mesozoic era, but most groups became extinct. They are represented today by turtles, the tuatara of New Zealand, lizards, amphisbaenians, snakes, and crocodiles.

19. Endothermy evolved independently in birds and mammals. Insulating layers at the body surface, a high metabolism and core body temperature, and controls over heat loss at the body surface enable birds and mammals to be active over a wider range of environmental temperatures than other vertebrates.

20. Except for the presence of feathers and the modification of the pectoral appendage as a wing, ancestral birds resembled archosaurian reptiles.

21. Cretaceous birds retained teeth, but these have been lost in contemporary species. Contemporary species are grouped into paleognathous birds, which retain an archosaur-like palate, and neognathous birds with a more mobile palate that is involved in the feeding movement of the skull. The kiwi, ostrich, and most other paleognathous birds have become terrestrial and their wings are vestigial. Neognathous birds include the vast majority of living species.

22. Mammals are characterized by insulating hair, large nasal cavities with turbinate bones, a unique jaw joint, a dentition that allows them to masticate their food, a secondary palate, three auditory ossicles and a well-developed cochlea in the ear. Newborn young are nourished by milk secreted by the mammary glands.

23. The first steps in the evolution of endothermy in ancestral mammals may have been the development of insulating layers at the body surface and a slight elevation of body temperature. These adaptations enabled them to exploit a nocturnal, insectivorous niche not available to the diurnal reptiles. Subsequent steps evolved as mammals entered warmer diurnal niches, which became available with the extinction of many reptiles in the Cretaceous.

24. Prototherian mammals evolved from synapsids in the Triassic period. Although relatively abundant in the Mesozoic era, prototherian mammals are represented today only by the monotremes (platypus and spiny anteaters) of Australia and New Guinea.

25. Primitive therian mammals are found in the Jurassic. Marsupials and eutherians diverged from them during the Cretaceous period.

26. Marsupials have a short intrauterine life and complete their development attached to the mother's nipples. Opossums and a few other marsupials occur

in South and North America, but most contemporary species are confined to Australia. The wide variety of types includes native cats, banded anteaters, bandicoots, and many herbivorous species ranging from diminutive possums to large kangaroos.

27. Eutherian mammals are characterized by a much longer intrauterine life. They have radiated widely from stem insectivores and are now the dominant group of mammals in most parts of the world.

A Synopsis of Chordates

This synopsis includes the groups of recent and extinct chordates discussed in the text. Common names for representative animals in a group are given; generic names are given only when they have been used in the text.

Subphylum Tunicata
   Class Ascidiacea. Sea squirts
   Class Thaliacea
   Class Larvacea
Subphylum Cephalochordata. Amphioxus (*Branchiostoma*)
Subphylum Vertebrata
   Class Agnatha. Jawless vertebrates
      Subclass Monorhina
         °Order Osteostraci. Cephalaspids
         °Order Anaspida
         °Order Galeapsida
         Order Petromyzontiformes. Lampreys (*Petromyzon*)
         Order Myxiniformes. Hagfishes (*Bdellostoma, Myxine*)
      Subclass Diplorhina
         °Order Heterostraci
         °Order Coelolepida (Thelodontida)
   Class Placodermi
         °Order Antiarchi
         °Order Arthrodira. *Dunkleosteus*
   Class Chondrichthyes. Cartilaginous fishes
      Subclass Elasmobranchii
         Superorder Cladoselachimorphii
            Order Cladoselachii. *Cladoselache*
            °Order Pleuracanthodii
         Superorder Squalimorphii
            Order Squantiniformes. Angel sharks
            Order Squaliformes. Dogfishes (*Squalus*)
            Order Pristophoriformes. Saw sharks
         Superorder Galeomorphii
            Order Hexanchiformes
            Order Carchariniformes. Requiem sharks
            Order Lamniformes. Mako sharks

°Extinct group.

Order Orectolobiformes. Carpet sharks
Order Heterodontiformes. Bullhead sharks
Superorder Batoidimorphii (Batoidea). Skates, rays.
Subclass Holocephali
Order Chimaeriformes. Chimaeras
Class Osteichthyes. Bony fishes
°Subclass Acanthodii. Spiny sharks
Subclass Actinopterygii. Ray-finned fishes
Infraclass Chondrostei
Order Palaeoniscoidei. °Palaeoniscoids, bichirs (*Polypterus*), reed fish (*Calamoichthys*)
Order Acipenseroidea. Sturgeons (*Acipenser*), paddlefishes (*Polyodon*)
Infraclass Neopterygii
Division Ginglymodi
Order Lepisosteiformes, gar (*Lepisosteus*)
Division Halecomorphi
Order Amiiformes, Bowfin (*Amia*)
Division Teleostei
Superorder Osteoglossomorpha. Mormyrids, elephant snout fishes (*Mormyrus*), mooneye (*Hiodon*), arowana (*Osteoglossum*)
Superorder Elopomorpha. Tarpons (*Elops*), eels (*Anguilla*)
Superorder Clupeomorpha. Herrings (*Clupea*)
Superorder Euteleostei.
Order Salmoniformes. Trout, salmon (*Salmo*) pike (*Esox*)
Order Cypriniformes (Ostariophysi). Carps (*Cyprinus*), minnows, zebrafish, catfish, gymnotids, characins
Order Gadiformes. Cod (*Gadus*), whiting
Order Lophiiformes. Anglerfish (*Lophius*)
Order Atheriniformes. Flying fish (*Exocoetus*)
Order Cyprinidontiformes. Guppies (*Poecilia*)
Order Gasterosteiformes. Three spine stickleback (*Gasterosteus*)
Order Beryciformes. Squirrel fish (*Holocentrus*), orange roughy
Order Zeiformes. John Dories (*Zeus*)
Order Perciformes. Perch (*Perca*), bass (*Micropterus*) sunfish, angelfish (*Chaetodon*), wrasses, gouramis (*Trichogaster, Osphronemus*)
Order Scorpaeniformes. Lingcods, greenlings (*Hexagrammus*), sculpins
Order Tetraodontiformes. Pufferfish, boxfish, triggerfish (*Balistes*)
Order Pleuronectiformes. Flounder (*Pleuronectes*), Sole (*Solea*)
Order Channiformes. Snakeheads (*Channa*)
Order Dactylopteriformes
Order Pegasiformes

Order Lampridiformes

Order Synbranchiformes. Rice eels (*Synbranchus, Monopterus*)

Subclass Sarcopterygii. Fleshy-finned fishes

Order Dipnoi. Lungfishes (*Neoceratodus, Lepidosiren, Protopterus*)

°Order Rhipidistia, *Eusthenopteron, Panderichthys*

Order Coelacanthini. *Latimeria*

Class Amphibia

°Subclass Labyrinthodontia

°Order Ichthyostegalia. *Ichthyostega*

°Order Temnospondyli

°Order Anthracosauria

°Subclass Lepospondyli

Subclass Lissamphibia

Order Urodela (Caudata). Salamanders (*Ambystoma, Necturus*)

Order Anura (Salientia). Frogs (*Rana*), toads (*Bufo*)

Order Gymnophiona (Apoda). Caecilians

Class Reptilia

Subclass Anapsida

°Order Captorhinida. Captorhinids

Order Chelonia. Turtles

Subclass Lepidosauria. Primitive diapsids

°Order Eosuchia. Ancestral diapsids

Order Sphenodonta (Rhynchocephalia). Tuatara (*Sphenodon*)

Order Squamata

Suborder Sauria (Lacertilia). Lizards

Suborder Amphisbaenia. Amphisbaenians

Suborder Serpentes (Ophidia). Snakes

Subclass Archosauria. Advanced diapsids

°Order Thecodontia. Ancestral archosaurs

Order Crocodylia. Crocodiles, alligators

°Order Pterosauria. Flying reptiles

°Order Saurischia. Reptile-like dinosaurs

°Order Ornithischia. Bird-like dinosaurs

Subclass Euryapsida

°Order Ichthyosauria. Ichthyosaurs

°Order Sauropterygia. Plesiosaurs

Subclass Synapsida

°Order Pelycosauria. Primitive synapsids

°Order Therapsida. Advanced synapsids

Class Aves

Subclass Archaeornithes. *Archaeopteryx*

Subclass Neornithes

°Superorder Odontognathae. Cretaceous toothed birds (*Sinornis?*)

Superorder Paleognathae

Order Struthioniformes. Ostriches

Order Rheiformes. Rheas

Order Casuariformes. Emus, cassowaries

*Order Aepyornithiformes. Elephant birds
*Order Dinornithiformes. Moas
Order Apterygiformes. Kiwis
Order Tinamiformes. Tinamous
    Superorder Neognathae
        Order Galliformes. Fowl, pheasants, grouse
        Order Anseriformes. Geese, ducks, swans
        Order Podicipediformes. Grebes
        Order Gaviiformes. Loons
        Order Sphenisciformes. Penguins
        Order Pelecaniformes. Cormorants, pelicans, gannets
        Order Procellariiformes. Shearwater, albatrosses, petrels
        Order Piciformes. Woodpeckers, sapsuckers
        Order Coraciiformes. Kingfishers, hornbills
        Order Trogoniformes. Trogons
        Order Coliiformes. Mousebirds
        Order Cuculiformes. Cuckoos
        Order Psittaciformes. Parrots, cockatoos
        Order Charadriiformes. Gulls, oyster catchers, terns, sandpipers,
            auks
      *Order Diatrymiformes. *Diatryma*
        Order Falconiformes. Eagles, hawks, vultures
        Order Strigiformes. Owls
        Order Caprimulgiformes. Goatsuckers, nighthawks
        Order Apodiformes. Swifts, hummingbirds
        Order Gruiformes. Coots, rails, moorhens
        Order Ciconiiformes. Storks, flamingos
        Order Columbiformes. Pigeons, doves
        Order Passeriformes. Songbirds: crows, starlings, finches,
            warblers, blackbirds, robins, sparrows, wrens, swallows, jays
Class Mammalia
  Subclass Prototheria
      *Orders Docodonta and Triconodonta. Ancestral mammals of
        Triassic and Jurassic. *Megazostrodon*
      *Order Multituberculata. Rodent-like mammals of late Mesozoic
      Order Monotremata. Platypus *(Ornithorhynchus)*, spiny
        anteaters *(Tachyglossus, Zaglossus)*
  Subclass Theria
  Infraclass Trituberculata
      *Order Eupantotheria
      *Order Symmetrodonta
  Infraclass Metatheria (Marsupialia)
      Order Caenolestidea (Paucituberculata). Rat opossums
      Order Polyprotodontia (Marsupicarnivora). Opossums,
        Tasmanian devil, numbat
      Order Peramelina. Bandicoots
      Order Diprotodontia. Possums, gliders, koala, wombat,
        kangaroos, wallabies

Infraclass Eutheria
    Order Edentata. Sloths, anteater, armadillos
    Order Pholidota. Pangolin
    Order Insectivora. Tenrec, hedgehog, shrews, moles
    Order Carnivora. Carnivores
    Order Primates. Lemurs, monkeys, apes, human beings
    Order Scandentia. Tree shrews
    Order Chiroptera. Bats, flying fox
    Order Dermoptera. Flying lemur
    Order Rodentia. Rodents
    Order Lagomorpha. Rabbits, hares
    Order Tubulidentata. Aardvarks
    Order Artiodactyla. Pigs, camels, deer, giraffes, antelopes, cattle, sheep
    Order Perissodactyla. Tapir, rhinoceroses, horses
    Order Cetacea. Whales
    Order Hyracoidea. Conies
    Order Sirenia. Sea cows
    Order Proboscidea. Elephants

REFERENCES

Bakker, R. T., 1971: Dinosaur physiology and the origin of mammals. *Evolution,* 25: 636–658.

Bakker, R. T., 1972: Anatomical and ecological evidence of endothermy in dinosaurs. *Nature,* 238: 81–85.

Bartholomew, G. A., 1982: Physiological control of body temperature. *In* Gans, C., and Pough, F. H., editors: *Biology of the Reptilia,* vol. 12, New York, Academic Press.

Bemis, W. F., Burggren, W., and Kemp, N. E., 1986: *The Biology and Evolution of Lung-fishes.* New York, Alan R. Liss, Inc.

Benton M. J., editor, 1988: *The Phylogeny and Classification of the Tetrapods.* Vol. 1, *Amphibians, Reptiles and Birds.* Oxford, Clarendon Press.

Benton, M. J., editor, 1988: *The Phylogeny and Classification of the Tetrapods.* Vol. 2, *Mammals.* Oxford, Clarendon Press.

Bolt, J. R., 1977: Dissorophoid relationships and ontogeny, and the origin of the Lissamphibia. *Journal of Paleontology,* 51: 235–249.

Brink, A. S., 1957: Speculations on some advanced mammalian characteristics in the higher mammal–like reptiles. *Paleontol. Afr.,* 4:77–96.

Carey, F. G., Teal, J. M., Kanwisher, J. W., Lawson, K. D., and Beckett, J. S., 1971: Warm-bodied fish. *American Zoologist,* 11: 137–145.

Carroll, R. L., 1987: *Vertebrate Paleontology and Evolution.* New York, Freeman.

Colbert, E. H., and Morales, M., 1991: *Evolution of the Vertebrates,* 4th edition. New York, John Wiley and Sons.

Cracraft, J., 1988: The major clades of birds. *In* Benton, M. J., *The Phylogeny and Classification of the Tetrapods.* Vol. 1. Oxford, Clarendon Press.

Crompton, A. W., Taylor, C. R., and Jagger, J. A., 1978: Evolution of homeothermy in mammals. *Nature,* 272: 333–336.

Dean, B., 1909: Studies on fossil fishes. *Memoirs of the American Museum of Natural History,* 9: 211–287.

Duellman, W. E., and Trueb, L., 1986: *Biology of Amphibians.* New York, McGraw-Hill Book Company.

Eisenberg, J. F., 1981: *The Mammalian Radiations: An Analysis of Trends of Evolution, Adaptation, and Behavior.* Chicago, University of Chicago Press.

Feduccia, A., 1980: *The Age of Birds.* Cambridge, Harvard University Press.

Forey, P., and Janvier, P., 1993: Agnathans and the origin of jawed vertebrates. *Science,* 361: 129–134.

Gadow, H., 1901: *The Cambridge Natural History, Volume VII, Amphibians and Reptiles.* London, Macmillan & Company.

Gauthier, J. A., Kluge, A. G., and Rowe, T., 1988: The early evolution of the amniotes. *In* Benton, M. J., editor: *The Phylogeny and Classification of the Tetrapods.* Vol. 1. Oxford, Clarendon Press.

Graham-Smith, W., and Westoll, T. S., 1937: On a new long–headed dipnoan fish from the Upper Devonian of Scaumenac Bay, P. Q., Canada. *Transactions of the Royal Society of Edinburgh,* 59: 241–266.

Grande, L., and Bemis, W. E., 1991: Osteology and phylogenetic relationships of fossil and recent paddlefishes (Polyodontidae) with comments on the interrelationships of Acipenseriformes. *Journal of Vertebrate Paleontology,* II: supplement 1: 1–132.

Greenwood, P. H., Miles, R. S., and Patterson, C., editors, 1973: Interrelationship of fishes. *Zoological Journal of the Linnean Society of London,* 53: supplement 1.

Griffiths, M., 1978: *The Biology of Monotremes.* New York, Academic Press.

Heaton, M. J., 1980: The Cotylosauria: A reconsideration of a group of tetrapods. *In* Panchen, A. L., *q.v.*

Heilmann, G., 1927: *The Origin of Birds.* New York, D. Appleton and Company.

Heintz, A., 1931: Untersuchungen über den Bau der Arthrodira. *Acta Zoologica,* 12: 225–239.

Heintz, A., 1939: Cephalaspida from Downtonian of Norway. *Skr. norsk. vidensk.–Adak. i Oslo, mat.–nat. Kl.* No. 5: 119.

Jarvik, E., 1980: *Basic Structure and Function of Vertebrates.* London, Academic Press.

Johnson, G. D., and Anderson, W. D., Jr., editors, 1993: Proceedings of the symposium on phylogeny of the percomorpha. *Bulletin of Marine Science.* 52: 1–620.

Jordan, D. S., and Evermann, B. W., 1896–1900: The fishes of North and Middle America. *Bulletin of the U.S. National Museum,* 47 (parts 1–4): 1–3313.

Joysey, K. E., and Kemp, T. S., editors, 1972: *Studies in Vertebrate Evolution.* New York, Winchester Press.

Kemp, T. S., 1982: *Mammal-like Reptiles and the Origin of Mammals.* London and New York, Academic Press.

Kermack, D. M., and Kermack, K. A., editors, 1971: Early mammals. *Zoological Journal of the Linnean Society of London,* 50: supplement 1.

Lauder, G. V., and Liem, K. F., 1983: The evolution and interrelationships of the actinopterygian fishes. *Bulletin of the Museum of Comparative Zoology,* 150: 95–197.

Liem, K. F., 1987: Functional design of the air ventilation apparatus and overland excursions by teleosts. *Fieldiana: Zoology* 1379(37): 1–29.

Miles, R. S., and Westoll, T. S., 1968: The placoderm fish *Coccosteus cuspidatus* Miller ex Agassiz from the Middle Old Red Sandstone of Scotland. Part I. Descriptive morphology. *Transactions of the Royal Society of Edinburgh,* 67: 373–476.

Millot, J., 1955: The coelacanth. *Scientific American,* 193(6): 34–39.

Millot, J., and Anthony, J., 1958–1965: *Anatomie de Latimeria chalumnae.* Paris, Centre de la Recherche Scientifique.

Morell, V., 1993: *Archaeopteryx:* Early bird catches a can of worms. *Science,* 259: 764–765.

Moyle, P. B., and Cech, J. J., Jr., 1982: *Fishes, An Introduction to Ichthyology.* Englewood Cliffs, Prentice-Hall.

Moy-Thomas, J. A., 1971: *Palaeozoic Fishes,* 2nd edition. Revised by R. S. Miles. Philadelphia, W. B. Saunders Company.

Noble, G. K., 1931: *The Biology of the Amphibia.* New York, McGraw–Hill Book Company.

Novacek, M. J., Wyss, A. R., and McKenna, M. C., 1988: The major groups of eutherian mammals. *In* Benton, M. J., editor: *The Phylogeny and Classification of the Tetrapods.* Vol. 2. *Mammals.* Oxford, Clarendon Press.

Olson, E. C., 1971: *Vertebrate Paleozoology.* New York, Wiley-Interscience.

Ørvig, T., editor, 1968: *Current Problems of Lower Vertebrate Phylogeny.* New York, Interscience Publishers.

Panchen, A. L., editor, 1980: *The Terrestrial Environment and the Origin of Land Vertebrates.* London, Academic Press.

Regan, C. T., 1929: Fishes. *In Encyclopedia Britannica,* 14th edition, 9: 305–328.

Rieppel, O. C., 1988: *Fundamentals of Comparative Biology.* Basel, Birkhauser Verlag.

Ritchie, A., 1964: New Light on the morphology of the Norwegian Anapsida. *Skr. norsk. vidensk.–Akad. i Oslo. mat.–nat. KL. new series.* No. 14: 1–35.

Romer, A. S., and Price, L. I., 1940. Review of the Pelycosauria. *Geological Society of America, Special Papers,* No. 28: 1–538.

Rosen, D. E., P. L. Forey, B. G. Gardiner and C. Patterson, 1981: Lungfishes, tetrapods, paleontology and plesiomorphy. *Bulletin of the American Museum of Natural History,* 167: 159–276.

Schmalhausen, I. I., 1968: *The Origin of Terrestrial Vertebrates.* New York, Academic Press.

Schultze, H.-P., 1991: A comparison of controversial hypotheses on the origin of tetrapods. *In* Schultze, H.-P., and Trueb, L., editors: *Origins of the Higher Groups of Tetrapods. Controversy and Consensus.* Ithaca and London, Comstock Publishing Associates.

Schultze, H.-P., and Trueb, L., 1991: *Origins of the Higher Groups of Tetrapods. Controversy and Consensus.* Ithaca and London, Comstock Publishing Associates.

Sereno, P. C., and Chenggang, R., 1992: Early evolution of avian flight and perching: New evidence from the Lower Cretaceous of China. *Science,* 255: 845–848.

Sibley, C. G., and Ahlquist, J. E., 1990: *Phylogeny and Classification of Birds. A Study in Molecular Evolution.* New Haven, Yale University Press.

Stahl, B. J., 1974: *Vertebrate History: Problems in Evolution.* New York, McGraw-Hill Book Company.

Steen, M., 1938: On the fossil Amphibia from the Gas Coal of Nýřany and other deposits in Czechoslovakia. *Proceedings of the Zoological Society of London, Series* B, 108: 205–283.

Stensiö, E. A., 1927: The Downtonian and Devonian vertebrates of Spitzbergen. I. Family Cephalaspidae. *Skr. om Svalbard og Nordishavet.* No. 12: 1–391.

Stensiö, E. A., 1969: Elasmobranchiomorphi. Placodermata. Arthrodires. *In* Priveteau, J., *Traité de Paleontologie,* vol. 4, Part 2: 71–692. Paris, Masson S.A.

Tarsitano, S., 1991: Archeopteryx: Quo Vadis? *In* Schultze, H.-P., and Trueb, L., editors: *Origins of Higher Groups of Tetrapods. Controversy and Consensus.* Ithaca and London, Comstock Publishing Associates.

Thompson, K. S., 1971: The adaptation and evolution of early fishes. *Quarterly Review of Biology,* 46: 139–166.

Trueb, L., 1991: *Origins of the Higher Groups of Tetrapods. Controversy and Consensus.* Ithaca and London, Comstock Publishing Associates.

Trueb, L., and Cloutier, R., 1991: A phylogenetic investigation of the inter and intrarelationships of the Lissamphibia (Amphibia: Temnospondyli). *In* Schultze and Trueb, *q.v.*

Walker, C. A., 1981: New subclass of birds from the Cretaceous of South America. *Nature,* 292: 51–53.

Watson, D. M. S., 1937: The acanthodian fishes. *Philosophical Transactions of the Royal Society of London, Series* B, 228: 49–146.

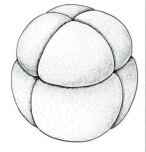

4 Early Development and Comparative Evolutionary Embryology

PRECIS

We will examine the development of chordates from the gametes to the establishment of the embryonic axis and the primordia of the organ systems. This will provide a basis for understanding the later development of the organ systems.

Outline

Ohe of the great marvels of life is the transformation of one seemingly simple cell, the fertilized egg, into a structurally and functionally complex organism. Descriptive embryologists have determined the succession of morphological changes by which this comes about, and developmental biologists are now studying how genes, and the proteins that they synthesize, control developmental processes. All cells of the body contain the same genes, but not all the genes are active or active to the same extent during the different periods of development or in the various tissues of the adult. We are just beginning to understand the regulation of gene activity during development. Seemingly different distributions of enzymes and other materials in the egg influence the way the genes contained in nuclei express themselves. As cells with different properties begin to emerge, their products influence the way adjacent tissues respond. This phenomenon has been called **induction**. Induction was discovered in the 1920s and 1930s by Spemann and other experimental embryologists, who observed the effects of transplanting bits of tissue from one embryo to another or from one part of an embryo to another. We now recognize that much of development is regulated by a succession of inductions, but tissues will respond to inductive influences only during a limited period, when they are said to be **competent** to respond. As tissues differentiate, they lose their competence to respond to some inductive influences, but not to others. The inducing materials are usually proteins that are produced by certain cells. These presumably bind with receptor molecules on the competent cells, enter the cells, and somehow affect the expression of their genes. Protein and peptide hormones appear to act in the same way (Chapter 14). Much research remains to be done, but the regulation of development may prove to be a special case of the broader field of regulation through which all parts of the organism respond to changes and function as an integrated whole.

We will examine early embryology in this chapter, from the gametes to the emergence of an embryonic axis and the primordia for the organ systems. This will provide a basis for understanding the later development, or **organogenesis**, of the organ systems. A knowledge of the development of organ systems will help us understand their structural and functional complexity and also help establish homologies and determine the course of evolution. Some of the variation in early development among different groups must be considered, but an extensive examination of this topic is beyond the scope of this book. We will also briefly discuss the roles of heterochrony and homeobox genes in major morphological transformations.

Gametes and Fertilization

Sperm

The reproductive cells, called **gametes**, develop in the gonads of sexually reproducing animals: sperm in the seminiferous tubules of the testis, eggs in the follicles of the ovary (Chapter 21). During gametogenesis the gamete-producing cells undergo two meiotic divisions, so the mature gametes are haploid; that is, each has only a single set of chromosomes rather than the double set present in other body cells. A mature **sperm** is a very small cell that has lost most of its cytoplasm

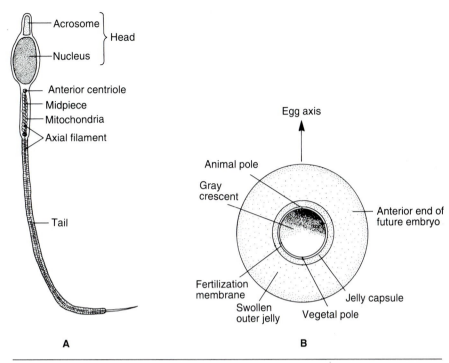

FIGURE 4–1 Gametes and zygote: **A**, A diagram of a mammalian spermatozoon. **B**, A fertilized frog egg. The frog egg is surrounded by jelly secreted by the oviduct, and its animal hemisphere is heavily pigmented. The pigment absorbs radiant energy, and the jelly insulates and protects the egg.

(Fig. 4–1**A**). It consists of a head, middle piece, and tail. The head contains the male nucleus and enzymes needed to penetrate the egg. Most of these enzymes are located in a cap, the **acrosome**, that covers the front of the nucleus. The middle piece contains coiled mitochondria that have the enzymes needed for the release of energy to the flagellum-like axial filament in the tail. Sperm of different species vary greatly in size and in the shape of their heads.

Eggs

A mature egg is a much larger cell that has accumulated enough energy reserves in its cytoplasm to initiate embryonic development. Its organization is surprisingly complex. It contains variable amounts of yolk, which is composed primarily of protein, phospholipids, and neutral fats, and it also contains enzymes, nucleic acids, and other materials needed to initiate development. The amount of yolk depends on how long the embryo will be nourished by food stored in the egg. There is little yolk in the **microlecithal** eggs (Gr. *mikros* = small + *lekithos* = yolk) of protochordates. These eggs hatch very soon into larvae that do not feed and metamorphose quickly into adults. Lampreys and amphibians have an intermediate amount of yolk in their **mesolecithal** eggs. Their larvae hatch at a later stage of development and then feed for themselves. A great deal of yolk is in the **macro-**

lecithal eggs of most fishes, whose free-living larvae, if present, are released late in development, and in the eggs of reptiles, birds, and monotremes, whose embryos develop into miniature adults before hatching or birth. The secondary reduction of yolk and other modifications in eutherian mammals that are related to development in a uterus will be considered later.

Materials in the egg are not distributed randomly but follow gradients. The small amount of yolk in the microlecithal eggs of protochordates has a rather even distribution. In vertebrate eggs the yolk is most concentrated toward one end, that known as the **vegetal pole**, and least concentrated at the other end, the **animal pole**. The gradient is particularly evident in macrolecithal eggs, where the egg nucleus and most of the cytoplasm are located at the animal pole. The future, or presumptive, anteroposterior axis of the embryo is not the same as the egg axis, but it has a relation to it and is usually determined by the point of entry of sperm. Often the anterior end of the embryo will lie between the animal pole and the equator of the egg (Fig. 4–1**B**).

At **ovulation**, when the egg is discharged from the follicle and ovary, it is primed and ready to develop into a new individual. Only a stimulus of some sort is needed to activate the egg and initiate continued development. Normally, the stimulus is sperm penetration, but other chemical or physical stimuli are effective in some species, in which case the egg develops **parthenogenetically** (i.e., without the contribution of the DNA of the sperm). If an egg is not activated within a few hours of ovulation, its delicately balanced internal organization breaks down, and the egg degenerates.

Fertilization

Fertilization involves sperm penetration, the combination of male and female nuclear material, and egg activation. Sperm penetration is a complex process because the egg is surrounded not only by its own plasma membrane but also by a **vitelline envelope** that is secreted by the ovary and oviduct. The sperm undergoes many changes as it approaches a conspecific egg. Membranes around its acrosome break down, sperm lysins are released, and a stiff acrosomal filament may form. The sperm is said to be capacitated when it is ready to penetrate the vitelline envelope and any other materials, such as adhering follicular cells, that may surround the egg.

Contact of the sperm head with the plasma membrane of the egg initiates a complex **cortical reaction** in the egg. This draws the sperm head into the egg and concurrently releases materials from small cortical granules at the egg surface that raise the vitelline envelope from the egg surface, preventing other sperm from entering. The vitelline envelope is now called the **fertilization membrane** (Fig. 4–1**B**). The ovulated egg usually is a secondary oocyte because its nucleus is arrested midway through the second meiotic division. Sperm entry triggers the completion of this division. One set of chromosomes is discarded in the second polar body, and then the haploid nucleus of the egg is ready to combine with the haploid sperm nucleus to form the diploid nucleus of the fertilized egg or **zygote** (Gr. *zygon* = yoke or union).

Finally, sperm penetration triggers a redistribution of materials within the egg cytoplasm that activates the egg and sets up a bilateral symmetry, if one was not already established in the unfertilized egg. Redistribution of some of the materials is evident in a frog egg, in which the animal hemisphere is heavily pigmented. Cortical pigments shift after sperm entry and leave a **gray crescent** on one margin of the equator of the egg (Fig. 4–1**B**). This is the presumptive posterodorsal part of the embryo. The first cell division of the zygote follows this redistribution of materials.

Cleavage

During **cleavage**, a period of rapid mitotic cell division, the unicellular zygote is converted into a multicellular embryo known as the **blastula** (Gr. *blastos* = germ or bud). No growth occurs during this period, so the cells, which are called **blastomeres** (Gr. *meros* = part), become smaller and smaller. The pattern of cleavage is correlated with the amount of distribution of yolk (Fig. 4–2). In microlecithal and mesolecithal eggs the entire zygote divides, so the cleavage is described as complete, or **holoblastic**. The first cleavage in chordates lies in the vertical plane, extends from the animal to the vegetal pole, and divides the embryo into prospective left and right sides. This cleavage bisects the gray crescent in amphibian zygotes. The second cleavage, also in the vertical plane, is at right angles to the first and results in the formation of four cells. The large amount of yolk in mesolecithal eggs slows the formation of cleavage furrows, so the first cleavage furrow may not have reached the vegetal pole before the second one begins at the animal pole. The third cleavage is in the horizontal plane and divides the embryo into eight cells. This cleavage lies near the equator in embryos that developed from microlecithal eggs, but it is displaced toward the animal pole in those that came from mesolecithal eggs. The resulting blastomeres are unequal in size. Cleavage continues in this fashion, tending to alternate between the vertical and horizontal planes. The cells that are formed remain close to the periphery, where gas and other exchanges with the environment occur. Since no growth occurs, and stored energy is used, the mass of the embryo decreases. A space that is known as the **blastocoele** appears within the embryo. The blastocoele is centrally located within blastulas that developed from microlecithal eggs, but it is displaced toward the animal pole in those that arose from mesolecithal eggs because of the large size of the yolk-filled cells in the vegetal hemisphere.

There is so much yolk in the macrolecithal eggs of most fishes, reptiles, and birds that cleavage is limited to a cytoplasmic disk at the animal pole. Cleavage is described as incomplete, or **meroblastic**, because the cleavage furrows do not extend into the large yolk mass. As a consequence, cleavage results in the formation of a skinlike disk of cells, the **blastoderm** (Gr. *derma* = skin), that lies on top of the yolk. A narrow cleft lies between the blastoderm and yolk. Part of this cleft becomes the blastocoele.

The significance of cleavage is that it converts a unicellular zygote into a multicellular embryo. The ratio of nucleus to cytoplasm, which is low at the onset of cleavage, increases to that of adult body cells. Experiments have demonstrated that all the nuclei of the blastomeres have the same genetic potential, but the

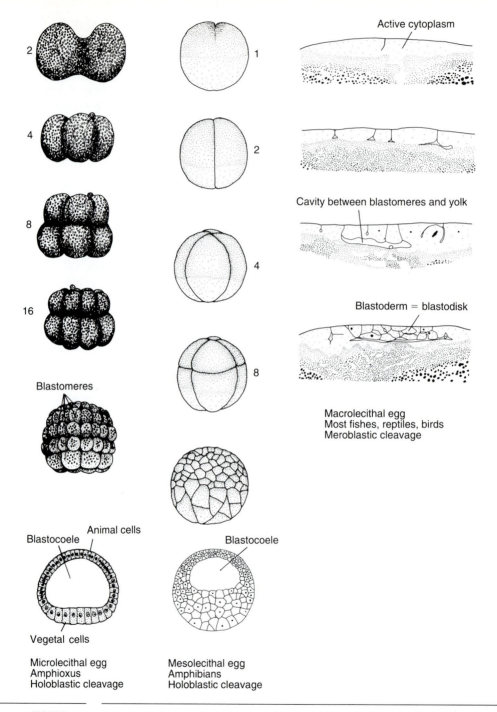

2

4

8

16

Blastomeres

1

2

4

8

Active cytoplasm

Cavity between blastomeres and yolk

Blastoderm = blastodisk

Macrolecithal egg
Most fishes, reptiles, birds
Meroblastic cleavage

Animal cells
Blastocoele

Vegetal cells

Blastocoele

Microlecithal egg
Amphioxus
Holoblastic cleavage

Mesolecithal egg
Amphibians
Holoblastic cleavage

FIGURE 4–2   Patterns of cleavage from the two-cell stage to the early blastula of amphioxus, an amphibian, and a bird. The blastulas of amphioxus and the amphibian embryo are shown both in lateral view and in sagittal section; for the bird, only vertical sections of cleaving cells on the yolk surface are shown. (After Balinsky and Fabian.)

nuclei lie in cytoplasm with different qualities because there was an unequal distribution of cytoplasmic materials in the zygote. The individual blastomeres differ in the amount of yolk, types of enzymes, and other substances that they contain. Certain of these differences influence the sequence in which nuclear genes are activated and how they express themselves. By staining parts of the surface of a blastula with vital dyes and following its development, it is possible to demonstrate that the presumptive fates of different regions of the embryo have already been established. Some regions will become gut, others neural tube, epidermis, notochord, head mesoderm, and so on.

Gastrulation, Mesoderm Formation and Early Neurulation

Cleavage is followed by **gastrulation**, when cells with different potentials come to the appropriate part of the embryo for further differentiation. These morphogenetic movements result from different rates of cell division and from changes in the size and shapes of cells. The single-layered blastula is converted to a **gastrula** (L. *gastrula* = little stomach) with well-defined layers of epithelium known as the **germ layers**. One germ layer, the **endoderm**, turns inward to form the primitive gut, or **archenteron**, which at this stage opens to the surface only by a **blastopore** at the posterior end of the gastrula (Fig. 4–3**D**). The endoderm will form the lining of most of the digestive tract and the glandular cells that develop from it. A second germ layer, the **ectoderm**, covers the surface of the embryo and will form the epidermis of the skin and the nervous system. Cells that will form the third germ layer, the **mesoderm**, move into the roof of the archenteron. The centrally positioned mesodermal cells that form the immediate roof of the archenteron are referred to as the **chordamesoderm** (Fig. 4–3**F**), which will give rise to the notochord. Separation of the mesoderm from the other germ layers may occur during or after gastrulation. The mesoderm will form skeletal structures, muscles, blood vessels, and other tissues of the body (Fig. 4–7). After completion of gastrulation, the ectoderm overlying the chordamesoderm is induced by the primordial notochord to thicken to become a **neural plate**, from which the brain, spinal cord, and neural crest will be derived (Fig. 4–3**E**). Gastrulation, mesoderm formation, and neurulation occur as three sequential and overlapping waves of development that generally occur cranial to caudal.

Amphioxus

The pattern of gastrulation is relatively simple in amphioxus. Its microlecithal egg develops into a blastula composed of small cells; those of the vegetal hemisphere are not much larger than other ones. The vegetal pole flattens and then folds inward by a process called **invagination** (Fig. 4–3**A** to **D**). As this process continues, the archenteron forms and the blastocoele becomes obliterated. Immediately following gastrulation, folds and creases begin to separate groups of epithelial cells as organ primordia (Fig. 4–3**E** to **H**). The prospective mesoderm forms as a series of **enterocoelic pouches** that bud off the dorsolateral walls of the archenteron. These pouches form a series of mesodermal segments, or **somites**. This process

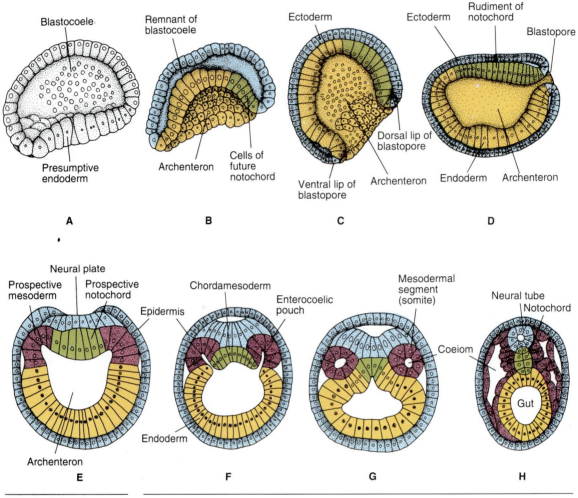

FIGURE 4–3 Gastrulation, mesoderm formation, and neurulation in amphioxus. Top row, sagittal sections of the formation of the gastrula. Bottom row, cross sections through late gastrulas and neurula to show the formation of the mesoderm, notochord, and neural tube. In all drawings ectoderm is shown in blue, endoderm in yellow, presumptive notochord in green, and mesoderm in red. (**A–C**, after Conklin. **E–H**, after Hatchek.)

occurs from anterior to posterior. A portion of the archenteric cavity is pinched off in the anterior somites as small cavities that will coalesce and become the **coelom**, but these spaces develop by cavitation within the more caudal somites and never connect with the lumen of the archenteron. The formation of mesoderm and coelom in amphioxus is similar to their development from enterocoelic pouches in echinoderms. As the somites are developing, the roof of the archenteron between them separates as the **notochord**, and a **neural tube** begins to form from the ectoderm overlying the notochord.

Amphibians

The larger, yolk-filled cells that form in the vegetal hemisphere during the cleavage of an amphibian's mesolecithal egg cannot move inward in the same way as in amphioxus. An **invagination** of a few cells on the gray crescent margin of the embryo forms a cleft that is the beginning of the archenteron (Fig. 4–4**A**). The dorsal margin of the cleft is the dorsal lip of the blastopore. This cleft lengthens as time goes on, grows laterally and ventrally, and eventually forms a circular blastopore. As the lips of the blastopore are developing, cells move from the dorsolateral surface of the embryo toward the blastopore, roll over the lips of the blastopore, and then continue to move forward beneath the ectoderm, thereby deepening and enlarging the archenteron (Fig. 4–4**B**). This process is known as **involution**. Involution begins at the dorsal lip of the blastopore, but soon cells begin to involute at the lateral lips, and eventually a few involute at the ventral lip. The blastocoele becomes obliterated as the archenteron enlarges. The movement of surface cells of the embryo toward the lips of the blastopore occurs faster than they are involuted. As a result, the prospective ectodermal cells, which originally were limited to the animal hemisphere, overgrow the yolky vegetal cells. This process is called **epiboly**. Finally, only a small plug of yolk-filled cells can be seen through the blastopore (Fig. 4–4**C**).

Cells that turn inward form the archenteron, the chordamesoderm, and the mesoderm. Prospective notochordal cells lie in the middorsal part of the archenteron roof; prospective mesoderm cells lie more laterally. Mesodermal cells move forward as a sheet between the endoderm and the ectoderm (Fig. 4–4**D**). The presumptive notochord separates from the mesoderm in the middorsal line, and endodermal folds meet beneath it to complete the roof of the archenteron (Fig. 4–4**D** and **E**). The ectoderm overlying the notochord thickens to form the **neural plate** (Fig. 4–4**D** and **E**).

Fishes, Reptiles, and Birds

The process of gastrulation is different in the macrolecithal eggs of fishes, reptiles, and birds. Since such eggs evolved independently in fishes and amniotes, it should not be surprising that their gastrulation processes differ, but the end result of gastrulation is essentially the same. We will confine ourselves to gastrulation in reptiles and birds. The central part of the blastoderm becomes clear and forms the **area pellucida**; the peripheral part remains as an **area opaca**. Cells separate, or delaminate, from the deep surface of the area pellucida at its prospective caudal end and migrate forward, forming a layer known as the **hypoblast** (Fig. 4–5**A**). The overlying cells now constitute the **epiblast**, and the space between these two layers is believed to correspond to the **blastocoele**. Cells then begin to move from the periphery of the epiblast toward its center, where they form a thickened, longitudinal ridge known as the **primitive streak** (Fig. 4–5**B**). As more cells move toward the primitive streak, cells already there turn inward and spread out laterally and anteriorly beneath the epiblast. The first cells to move inward along the primitive streak displace hypoblast cells medially and form endoderm. Cells that move inward later form mesoderm. When the process is complete, epiblast cells that

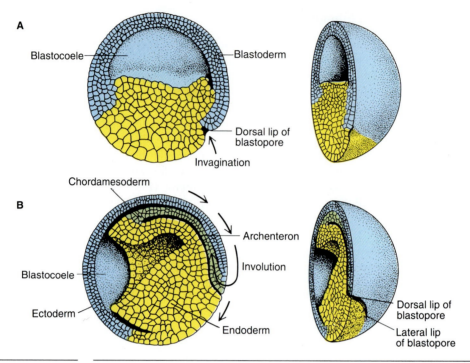

A

Blastocoele — Blastoderm

Dorsal lip of blastopore

Invagination

Chordamesoderm

B

Archenteron

Involution

Blastocoele —

Ectoderm —

Endoderm

Dorsal lip of blastopore

Lateral lip of blastopore

FIGURE 4–4   Gastrulation and mesoderm formation in amphibians: **A–C** are based on the frog; **D–E**, on a salamander. **A**, A sagittal section and end view of an early stage, as invagination is beginning. **B**, Similar views of a later stage, when involution and epiboly are occurring. **C**, The blastopore is now complete and a yolk plug protrudes through it. **D**, A sagittal section of a later gastrula of a salamander showing the spread of the mesoderm between the ectoderm and endoderm. **E**, A cross section of a later stage, when the notochord has formed, endodermal folds are completing the roof of the archenteron and the neural plate has been induced. (*A–C*, after Balinsky and Fabian. **D** and **E**, after Hamburger.)

remain on the surface are the ectoderm. All germ layers are derived from the primitive epiblast.

The inward movement of cells along the primitive streak has been called **ingression**, but it is functionally comparable to the involution of cells in amphibian embryos. Indeed, the primitive streak is considered the homolog of the blastopore. The process of ingression of cells along the primitive streak is completed first near the anterior end of the embryo, and the wave of completion spreads caudally. As ingression is completed, the primitive streak retreats caudally (from anterior to posterior), leaving the inturned mesoderm. Notochord and mesoderm separate, and the ectoderm overlying the notochord thickens as a **neural plate** (Fig. 4–5C). Mesoderm just lateral to the notochord and developing neural tube, the paraxial mesoderm, differentiates into a series of somitomeres and **somites** (Fig. 4–5D). These processes, too, progress from anterior to posterior. The coelom forms, as it does in amphibians, by a cavitation in the mesoderm lying lateral to the somites.

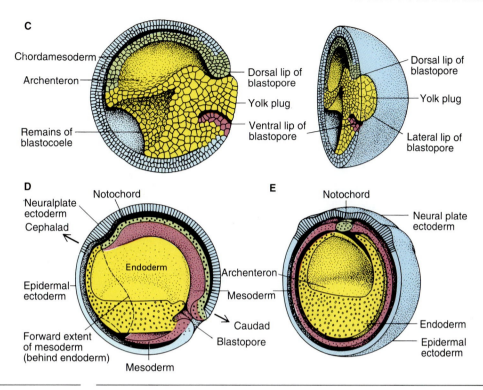

FIGURE 4–4   (*continued*)

Neurulation and the Neural Crest

As gastrulation is nearing an end and somites are forming, the chordamesoderm induces the overlying neural plate to differentiate into the neural tube characteristic of chordates (Figs. 2–7 and 4–5**C** and **D**). The ectoderm overlying the notochord first thickens and flattens to form a longitudinal **neural plate**. Then the margins of the neural plate elevate as a pair of **neural folds** with a **neural groove** between them. In vertebrates the folds are largest and most widely spaced anteriorly, in the region of the presumptive brain. When the neural folds meet dorsally, they unite to form the **neural tube**. The neural groove becomes the cavity that characterizes the central nervous system of chordates. Most of the nervous system will differentiate from the neural tube. The neural folds grow together above the neural tube, and separate from overlying ectoderm, which becomes the epidermis of the skin.

In vertebrates, but not in the protochordates, a column of ectodermal cells in each neural fold separates as the **neural crest** (Fig. 2–7). The neural crest comes to lie along the dorsolateral borders of the neural tube (Fig. 4–6**A**). Although derived from an epithelial layer, most of the neural crest cells transform into loosely packed, star-shaped cells known as **mesenchyme**. Mesenchyme is a widespread embryonic cell type. Most comes from mesoderm, but the neural crest contributes to it, especially in the head region. Mesenchyme of neural crest origin is called **ectomesenchyme**. Some neural crest cells remain close to their site of

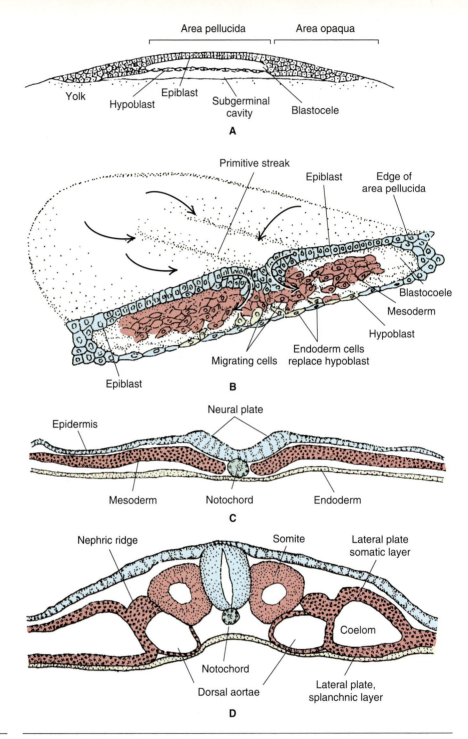

FIGURE 4–5 Gastrulation, mesoderm formation and neurulation in a bird: **A**, A longitudinal section of a blastoderm in which the hypoblast and epiblast have differentiated. **B**, A stereodiagram of the anterior half of the area pellucida; a primitive streak has formed, and endoderm and mesoderm cells are migrating inward. **C**, A cross section of an embryo at a later stage, when the notochord and neural plate have formed. **D**, A similar view of a later stage, when the neural tube is complete and the mesoderm is differentiating. (After Balinsky and Fabian.)

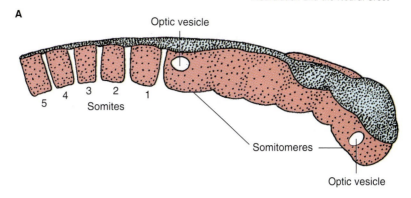

A

Optic vesicle

5 4 3 2 1
Somites

Somitomeres

Optic vesicle

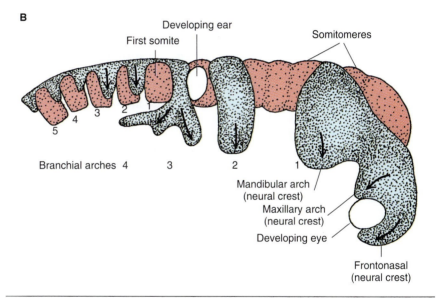

B

Developing ear

First somite

Somitomeres

5 4 3 2 1

Branchial arches 4   3   2   1

Mandibular arch
(neural crest)

Maxillary arch
(neural crest)

Developing eye

Frontonasal
(neural crest)

FIGURE 4–6   The neural crest of a generalized amniote vertebrate as seen in lateral view: **A**, An early stage. **B**, A later stage when neural crest cells are migrating around the eye, down the maxillary and mandibular arches, and between the branchial arches. The relationships between the neural crest cells and the mesodermal somitomeres in the head region and the somites in the body are indicated. (Modified from Noden.)

origin, but others spread, partly by ameboid action, throughout the embryo. Their movements appear to be genetically preprogrammed, even though they can be influenced by local environmental conditions within the embryo. They stream down around the developing eye and between the pharyngeal pouches (Fig. 4–6**B**), spread out between the surface ectoderm and the segmented somatic mesoderm, and extend between the neural tube and developing somites to the gut wall. Neural crest cells give rise to the cartilaginous visceral or branchial arches, including the upper and lower jaws, part of the front of the braincase, all the pigment cells of the body, parts of the teeth and bony scales, the sensory neurons of all the spinal nerves and many of the cranial nerves, the postganglionic autonomic neurons (p. 458), and the sheaths (Schwann cells) that invest most peripheral neurons. They also contribute to the connective tissue layers (meninges) that surround the brain and spinal cord. Neural crest cells not only contribute directly to many important structures but also appear to carry instructions that regulate many aspects of development. The neural crest is thus a key derived feature of vertebrates and is essential in organizing the vertebrate body plan.

As the neural tube forms, the embryo lengthens along its anteroposterior axis. Head and tail folds form in the embryos of most fishes, reptiles, and birds, and the embryo begins to separate from the large mass of yolk beneath it.

Mesoderm Differentiation and the Derivatives of the Germ Layers

Somites, their derivatives, and the coelom usually begin to differentiate in the mesoderm near the end of gastrulation, and the process continues through neurulation. By the time the neural tube is complete, three mesodermal areas can be recognized on each side of the embryo (Figs. 4–5**D** and 4–7). Thickened, segmental **somites**, or **paraxial mesoderm**, lie lateral to the neural tube and notochord (Fig. 4–6**B**). A broad, unsegmented **lateral plate**, or **lateral mesoderm**, extends laterally and ventrally between the archenteron and surface ectoderm. A **nephric ridge**, or **intermediate mesoderm**, which is segmented anteriorly, lies between the somites and lateral plate. The definitive coelom develops by a cavitation in the lateral plate. Portions of the coelom extend into the nephric ridge, and small, ephemeral spaces may occur within the somites.

As time goes on, each somite differentiates into three regions. The dorsolateral group of somite cells constitute the **dermatome**. Cells in the dermatome differentiate into mesenchymal cells that spread out beneath the surface ectoderm and form most of the dermis of the skin. Cells deep to the dermatome form the **myotome**, or embryonic muscle segment. The segmental myotomes extend ventrally between the ectoderm and lateral plate mesoderm and differentiate into muscle cells that form all (in anamniotes) or most (in amniotes) of the somatic muscles of the body wall and appendages. The most ventromedial part of the somite is the **sclerotome**. It differentiates into mesenchymal cells that migrate around the neural tube and notochord and form the vertebral column and occipital region of the skull. Other parts of the skull develop from mesenchymal cells derived from the cranial neural crest and from the dermatome.

The nephric ridge will differentiate into the kidney and the excretory and reproductive ducts. The lateral plate is split by the coelom into two layers. The medial **splanchnic layer** next to the endoderm will form the connective tissue and visceral muscles of the gut and heart walls. Part of it will form the mesenteries and the coelomic epithelium (visceral peritoneum) that covers the visceral organs. The lateral **somatic layer** of the lateral plate forms the coelomic epithelium bounding the coelom laterally (the parietal peritoneum), and it also may contribute to the somatic musculature and other tissue of the lateroventral body wall.

The organization of the neurula stage of the embryo foreshadows the structure of the adult. In Table 4–1 we summarize the developmental pathways leading to formation of the various organs in the adult. All pathways lead from one of the

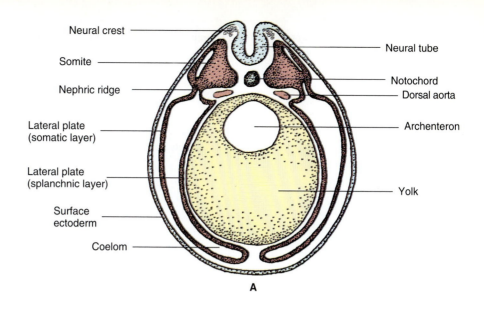

Neural crest

Somite

Nephric ridge

Lateral plate (somatic layer)

Lateral plate (splanchnic layer)

Surface ectoderm

Coelom

Neural tube

Notochord

Dorsal aorta

Archenteron

Yolk

A

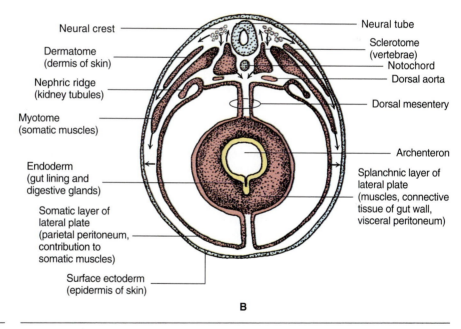

Neural crest

Dermatome (dermis of skin)

Nephric ridge (kidney tubules)

Myotome (somatic muscles)

Endoderm (gut lining and digestive glands)

Somatic layer of lateral plate (parietal peritoneum, contribution to somatic muscles)

Surface ectoderm (epidermis of skin)

Neural tube

Sclerotome (vertebrae)

Notochord

Dorsal aorta

Dorsal mesentery

Archenteron

Splanchnic layer of lateral plate (muscles, connective tissue of gut wall, visceral peritoneum)

B

FIGURE 4–7   Diagrams of the differentiation of the mesoderm in an embryo in which the moderate amount of yolk is incorporated in yolk-laden cells in the floor of the archenteron: **A**, A cross section showing the three primary mesodermal regions: somites, nephric ridge, and lateral plate. **B**, A later stage showing the structures that develop from each somite and each germ layer.

TABLE 4–1   **Differentiation of the Principal Organs of the Vertebrate Adult Body from the Three Germ Layers**

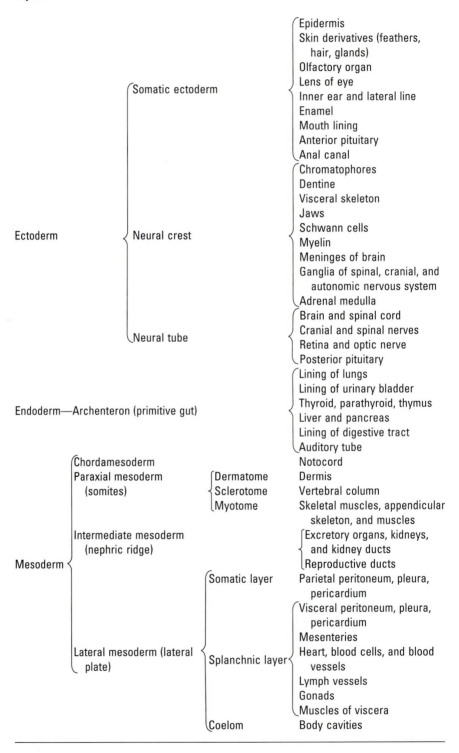

	Somatic ectoderm	Epidermis Skin derivatives (feathers, hair, glands) Olfactory organ Lens of eye Inner ear and lateral line Enamel Mouth lining Anterior pituitary Anal canal
Ectoderm	Neural crest	Chromatophores Dentine Visceral skeleton Jaws Schwann cells Myelin Meninges of brain Ganglia of spinal, cranial, and autonomic nervous system Adrenal medulla
	Neural tube	Brain and spinal cord Cranial and spinal nerves Retina and optic nerve Posterior pituitary
Endoderm—Archenteron (primitive gut)		Lining of lungs Lining of urinary bladder Thyroid, parathyroid, thymus Liver and pancreas Lining of digestive tract Auditory tube
	Chordamesoderm	Notocord
	Paraxial mesoderm (somites)	Dermatome — Dermis Sclerotome — Vertebral column Myotome — Skeletal muscles, appendicular skeleton, and muscles
Mesoderm	Intermediate mesoderm (nephric ridge)	Excretory organs, kidneys, and kidney ducts Reproductive ducts
	Lateral mesoderm (lateral plate)	Somatic layer — Parietal peritoneum, pleura, pericardium Splanchnic layer — Visceral peritoneum, pleura, pericardium / Mesenteries / Heart, blood cells, and blood vessels / Lymph vessels / Gonads / Muscles of viscera Coelom — Body cavities

three germ layers. Every organ is identified with a particular germ layer. This is a fundamental and common starting point for the development of all vertebrates. There is a high degree of homology between the three germ layers and the products derived from them. This is the primary evidence for common ancestry of all vertebrates.

The Extraembryonic Membranes of Reptiles and Birds

With the emergence of the cleidoic egg, amniotes can bypass the aquatic larval stage and reproduce on land. The cleidoic egg possesses extraembryonic structures that protect the embryo and sustain its metabolism. As the fertilized cleidoic egg passes down the oviduct and begins to develop, it is surrounded by materials secreted by the oviduct (Fig. 4–8**C**). The egg white that is laid down is at least 80 percent water; the rest is proteins, mostly **albumins**. A **shell membrane** of keratin fibers is secreted around the albumin, and a slightly porous, parchment-like or calcareous **shell** is laid down over the egg surface. The shell and shell membrane are protective layers, and the albumin and yolk provide the raw materials, water, and energy needed by the embryo during its development.

As the embryo is developing in the blastodisk on top of the large yolk mass, tissue layers extend from the embryonic body and form one or more **extraembryonic membranes**. First to develop is a yolk sac that forms by the spreading of tissue layers over the yolk. Since the yolk sac of a large-yolked fish embryo contains all three germ layers, it is described as a **trilaminar yolk sac** (Fig. 4–8**A**). An **extraembryonic coelom** extends from the embryo between the somatic and splanchnic layers of the part of the lateral plate that contributes to the yolk sac. As the yolk sac of a reptile or bird develops, the ectoderm and somatic layer of the lateral plate, which are part of the yolk sac in a fish, become elevated as **chorioamniotic folds** that grow over the embryo and meet above it (Fig. 4–8**B**). Only the endoderm and splanchnic mesoderm remain over the yolk in their **bilaminar yolk sac**. As the head and tail folds deepen and undercut the embryo, and as lateral body folds form, the embryo is raised off the yolk mass, but its archenteron remains connected to the yolk sac by a narrow **yolk stalk**.

Blood vessels and blood cells begin their development in the splanchnic mesoderm of the yolk sac and spread into the embryo. They convey materials from the yolk to the embryo. These vessels also bring in materials from the egg white, since the periphery of the yolk sac that is spreading over the yolk is in contact with the egg white for a long time.

Two additional extraembryonic membranes are formed in amniotes when the chorioamniotic folds meet above the embryo (Fig. 4–8**C**). The inner limbs of these folds form an **amnion** that surrounds the embryo; the outer limbs form the **chorion** that surrounds the amnion and eventually the entire yolk sac. **Amniotic fluid** accumulates in the amniotic cavity between the amnion and the embryo, so the embryo continues its development in a liquid environment. Although the cleidoic egg is laid on land, the embryo's immediate environment remains aquatic, like that of fish and amphibian larvae. The amniotic fluid also provides an important protective liquid cushion around the embryo. Additional protection is afforded by the chorion and by the egg white and shell.

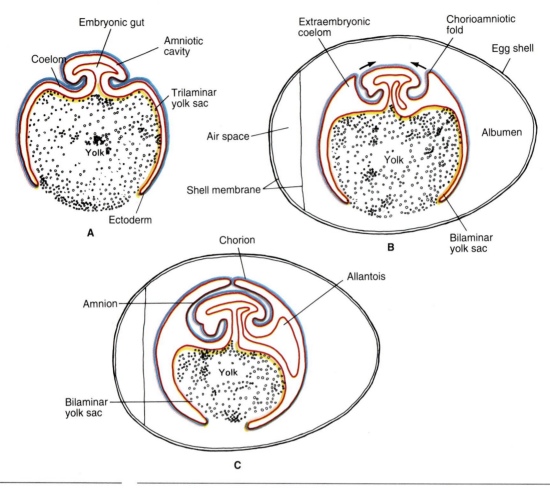

FIGURE 4–8 Diagrams of the extraembryonic membranes of vertebrates: **A,** The trilaminar yolk sac of a macrolecithal fish embryo. **B,** The bilaminar yolk sac of a reptile or bird, and the formation of the chorioamniotic folds. **C,** The extraembryonic membranes of a reptile or bird. Albumen, a shell membrane, and a shell, which are secreted by the oviduct, surround the embryos of reptiles and birds.

The third extraembryonic membrane of amniotes, the **allantois**, forms as an evagination of the posterior part of the archenteron, so its wall is composed of endoderm and vascularized splanchnic mesoderm (Fig. 4–8**B** and **C**). The allantois enlarges, nearly fills the extraembryonic coelom, and touches the chorion. Frequently, these two membranes fuse as a **chorioallantoic membrane**. The chorion and amnion are avascular membranes because they develop before blood vessels have entered the somatic mesoderm. The allantois vascularizes the chorion and brings embryonic vessels close to the inside of the porous egg membrane and shell. Embryonic gas exchange occurs by this route. Nitrogenous excretory products in the form of inert crystals of uric acid accumulate in the lumen of the allantois. In many species, after hatching or birth, the base of the allantois remains as the urinary bladder.

Developmental Modifications of Eutherian Mammals

The ancestors of therian mammals were reptiles and prototherians that laid macrolecithal cleidoic eggs, but therians evolved a pattern of reproduction in which the embryos are retained within the female uterus. This entailed many modifications, both in the female reproductive tract and in embryonic development. Here we will examine the condition in eutherians, or placental mammals; marsupial reproduction will be covered in Chapter 21.

Since the eutherian mother will provide the embryos with nutrients and other materials through the placenta, little maternal energy is used to deposit yolk in the eggs. Eutherian eggs have become secondarily microlecithal, and this simplifies cleavage. It is essential, however, that an implantation be established very early. Cleavage of the eutherian egg results in a ball of cells known as the **blastocyst** (Fig. 4–9**A**). The outermost cells of the blastocyst form a layer known as the **trophoblast** (Gr. *trophe* = nourishment + *blastos* = germ or bud), which is homologous to the chorionic ectoderm and forms the precursor of the fetal part of the placenta. The maternal part of the placenta is the vascularized and glandular uterine lining, or **endometrium** (Gr. *metra* = uterus). The trophoblast soon begins to penetrate the endometrium in human beings and other eutherians in which the fetal part of the placenta invades the uterine lining. The apposition or fusion of the extraembryonic membranes of the embryo with the endometrium of the mother's uterus is called the **placenta**, which is a structure for physiological exchange between fetus and mother. Penetration of the endometrium by the trophoblast leads to the implantation of the embryo. In ways not fully understood, the trophoblast also provides an immunological barrier that prevents the mother from developing antibodies against the embryo, which is foreign tissue because half its genes are paternal. The trophoblast thus prevents an immunological rejection of the embryo.

A sphere of cells known as the **inner cell mass** lies within the trophoblast, and some of its cells will form the embryo. The inner cell mass is attached to the trophoblast at the presumptive posterodorsal side of the embryo, but elsewhere it is separated from the trophoblast by a cavity. The subsequent differentiation of the embryo and its extraembryonic membranes differ considerably among eutherians; we will examine the human condition. Endodermal cells differentiate on the underside of the inner cell mass and spread laterally and ventrally within the cavity of the blastocyst to form the endodermal part of the yolk sac (Fig. 4–9**B**). There is no yolk in the yolk sac, but the dorsal part of it eventually will become separated by body folds from the rest of the yolk sac and form the archenteron. A cavity that appears among the cells in the dorsal part of the inner cell mass is the beginning of the amniotic cavity. Cells lining this cavity are regarded as ectoderm. The two-layered disk of cells (ectoderm and endoderm) between the cavities of the yolk sac and amnion is the **blastoderm** or **blastodisk**, and the embryo will develop from it.

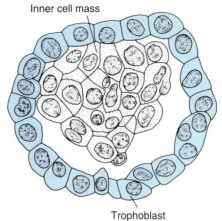

A

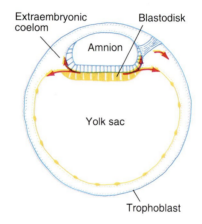

B

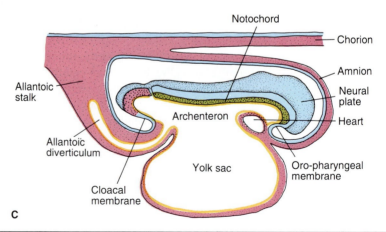

C

FIGURE 4-9   The early development of a eutherian mammal: **A**, Early blastocyst. **B**, The inner cell mass has differentiated into the amnion, blastodisk, and yolk sac. Mesoderm ingresses along the primitive streak and spreads as indicated by the arrows. **C**, The blastodisk differentiates into the embryo. (After Balinsky and Fabian.)

As in most fishes, reptiles, and birds, morphogenetic movements carry prospective mesoderm cells toward a longitudinal primitive streak, through which they ingress and spread out between ectoderm and endoderm. Neurulation, mesoderm differentiation, and the separation of the embryo from the yolk sac by body folds resemble the processes in fishes, reptiles, and birds. Mesoderm also spreads over the surfaces of the amnion and yolk sac, so these become typical, two-layered membranes (Fig. 4–9**C**). A stalk of tissue persists near the posterior end of the embryo and connects it with the ectodermal trophoblast. Mesoderm spreads via this stalk over the underside of the trophoblast and converts it to a more typical chorion. An allantois grows from the posterior part of the archenteron into this stalk. The cavity of the allantois does not become large in human embryos, but its wall enables blood vessels from the embryo to reach the chorion and vascularize the fetal part of the placenta. This part of the eutherian placenta is homologous to the chorioallantoic membrane of reptiles and birds. As the embryo enlarges, it is raised farther off the yolk sac and allantois. Thus the embryo is connected to the placenta by means of a cordlike stalk that also contains the allantois and yolk sac. This cordlike structure is the **umbilical cord**. The development of the mammalian embryo resembles that of reptiles and birds except for the reduction of yolk, simplification of cleavage, and the precocious development of those extraembryonic parts that contribute to the placenta and the umbilical cord.

Evolutionary Embryology

The connection between development and evolution has drawn the attention of biologists and philosophers for over a century. We have learned about these connections:

1. The extraembryonic membranes in the cleidoic egg are key adaptations for terrestrial vertebrate life, leading to the development of a placenta in eutherian mammals.
2. Active vertebrate life is the consequence of an embryonic key innovation, the neural crest cells and their derivatives, which gave rise to jaws, teeth, etc.
3. In Chapter 1, we discussed paedomorphosis as a way by which new forms can evolve.

Evolutionary developmental biology examines the connection between embryology and evolution. We will limit our discussion to two major concepts: heterochrony and the homeobox genes and their role in the origin of the vertebrate body plan.

Heterochrony

Heterochrony (Gr. *heteros* = other, different; *chronos* = time) simply means that the timing of development of a part of an embryo has changed with respect to its time of development in the ancestor of the embryo. Thus, it is a change in the relative time of appearance or rate of development of a structure. Heterochrony can produce two major developmental phenomena: paedomorphosis and peramorphosis, each of which can result in radical evolutionary transformations of parts of organisms or whole organisms.

Paedomorphosis and its two subdivisions, progenesis and neoteny, were discussed in Chapter 1. As in paedomorphosis, we can also recognize two forms of **peramorphosis** (Gr. *pere* = parent + *morphosis* = form): acceleration and hypermorphosis. Peramorphosis is the opposite of paedomorphosis and means "old looking." In **acceleration** a particular structure makes an early appearance and undergoes accelerated development so that it is overdeveloped by the time the animal matures. The extraordinarily large antlers in the Pleistocene Irish Elk may be an example, but this is still being debated among zoologists. **Hypermorphosis** (also known as **recapitulation**) also produces an overdeveloped structure. The ontogenetic trajectory is extended. A structure continues its development beyond the ordinary time of cessation when the animal matures. This phenomenon is common among invertebrates, but we know of no vertebrate examples. Thus, heterochrony can produce rather dramatic transformations in the anatomy of an animal without changing major genetic and developmental programs. Evolutionary transformations of structures are simply produced by changes in the timing of onset of a structure's development, in the rate of development, and in the timing of cessation of development.

The Homeobox and the Vertebrate Body Plan

All vertebrates are segmented during their development. Segmentation is most obvious in embryonic stages, but becomes masked in adults. A segment is a set of body parts that are present repeatedly in an embryo or adult. We have seen that somites are segments in the mesoderm that differentiate sequentially. Branchial or gill arches in the head also represent segments. In all vertebrates the hindbrain, an axial structure, is divided into a series of **rhombomeres** (Fig. 4–10). The paraxial mesoderm lying to each side of the developing brain is divided into **somitomeres** in the rostral part of the head and into **somites** in the caudal part of the head and in the trunk. The neural crest, which originates in the neural tube, gives rise to neural crest cells that actively migrate into ventral lateral positions, where they surround the forebrain (prosencephalon) and the pharynx to form the jaws, branchial arches, and cranial sensory neurons (ganglia) in a segmented pattern (Fig. 4–6). This segmental organization is fundamental in the vertebrate body plan.

In no other biological field is the connection between development and evolution better demonstrated than by the role of a sequence of homeotic genes in regulating segmented structures in both vertebrates and invertebrates.

Homeotic or **Hox** genes are composed of short segments of DNA (about 180 base pairs long). A cluster of homeotic genes is called a **homeobox**. Homeoboxes have been found in many invertebrate and vertebrate species. Many of the 180 bases found in the homeoboxes of widely differing species are identical. Fifty-nine of the sixty protein products (amino acid residues) of the Hox genes in frogs and the fruit fly, *Drosophila*, are identical. The similarity of products of Hox genes between mammals and *Drosophila* is about 70 percent.

The major function of Hox genes is to regulate the expression of other genes, which, in turn, determine the features characteristic of each body segment. This is the result of unique expression domains or combinations of overlapping expression domains. It has been shown that the protein encoded by the homeobox

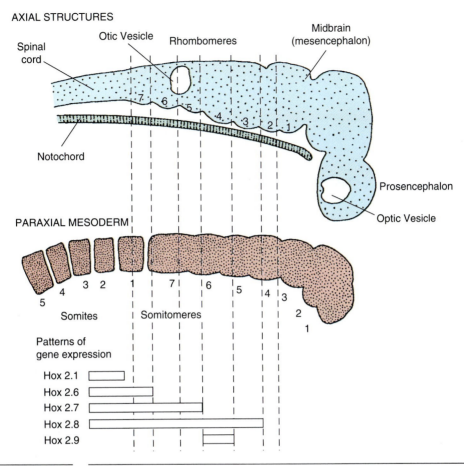

AXIAL STRUCTURES

FIGURE 4–10   Diagram of the lateral aspect of an amniote embryo with the ectoderm, the neural crest, and the circulatory and digestive systems removed. Homeobox-containing genes (hox) are shown with their respective anterior boundaries of gene expression in the hindbrain. (Modified from Noden.)

enables a number of gene regulators to recognize and bind to the gene under their control. Proteins made by a Hox gene are found in specific segments of the body.

The proteins produced by a specific Hox gene (for example, XlHBOX 1) are expressed in segments that are clearly aligned in the tissues along an anterior-posterior axis with sharply demarcated anterior and posterior boundaries. The vertebrate embryo is subdivided into anteroposterior fields of cells with different developmental capacities. This subdivision is based on Hox gene expression patterns and precedes the differentiation of specific organs. Because we know that chordamesoderm induces neural tissue, it is probable that the chordamesoderm expressing XlHBOX 1 also induces cells in the overlying neural plate to express the gene as well. Each Hox gene is active in a specific segment or segments of the body. All vertebrates have four homeobox complexes, each located on a separate

FOCUS 4–1   # The Homeobox and the Vertebrate Limb

As we have learned, the concept of the homeobox is rapidly becoming the cornerstone of developmental biology. Homeobox gene activity may shed light on the genetic and developmental mechanisms underlying evolutionary transformations of anatomical structures. Zoologists have long debated the evolutionary origin of the tetrapod limb from piscine fins. The expression of Hox genes may reveal how this transformation occurred. It has been shown that the XlHBOX 1 gene is first expressed in a circular region of the limb bud mesoderm, which is the precursor of the pectoral fin in the teleostean zebrafish. As the cells proliferate, XlHBOX 1 protein forms a steep gradient in the pectoral fin bud. Identical XlHBOX 1 protein gradients occur in the developing forelimbs of the frog, chicken, and mouse, indicating that XlHBOX 1 is an ancient gene whose function in the limb gradient field predates the appearance of tetrapod structures. A recent study by Morgan et al. (1992) shows that four members of a Hox gene cluster (Hoxd-10, -11, -12, and -13; see the figure) are expressed in the hind limb bud of a chick embryo. This Hox gene complex specifies the fate of cells along the anteroposterior axis of the limb. Hoxd-10 specifies the most anterior digit (I). Hoxd-11 is subsequently expressed and together with Hoxd-10 specifies the next digit (II). Digit III corresponds to an expression of a combination of Hoxd-10, -11, and -12, and digit IV to a combination of Hoxd-10, -11, -12, and -13, which are expressed in overlapping domains. It is the unique combination of Hox genes in a particular domain that controls the development of the digits. These Hox genes also specify positional identity along the anteroposterior axis of the leg. The skeleton of the leg shows that the number of bones in each toe, including the element that bears the terminal claw, reflects its designation (i.e., digit I has two phalanges, digit II has three phalanges, digit III has four, and digit IV has five; see the figure). Thus, the first domain lacks the Hoxd-11 gene, while the second domain possesses Hoxd-11, which specifies the development of digit II. In an ingenious experiment, Morgan et al. (1992) infected the cells of the developing limb bud of a chick with a virus carrying the Hoxd-11 gene from a mouse. The virus infected all cells of the developing limb bud with the Hoxd-11 gene (see the figure). As a result, the combination of Hox genes in the first domain was altered from Hoxd-10 to Hoxd-10 + 11. The combinations of Hox genes in the remaining domains remained unaltered. The altered combination of the first domain by the addition of Hoxd-11 resembled that of the second domain. The altered first domain gave rise to a structurally transformed digit (II' in the figure). The original digit I was transformed into a digit II with three phalanges. Thus, localized ectopic expression of the Hoxd-11 gene in the limb of the developing chick results in a transformation of digit I into digit II. This is a direct demonstration that homeobox genes control the patterning of the vertebrate limb. These results, and those involving the determination of head segmentation, show that the homeobox genes play a central role in the evolution of the vertebrate body.

chromosome. Hox genes are arranged in a precise order on the linear DNA molecule. The order of the Hox genes in a sequence or complex corresponds directly to where the genes are expressed in the more posterior parts of the body. Genes to the right are expressed in the same order that they are aligned in the DNA molecule. Thus, Hox genes are expressed linearly, thereby establishing a key characteristic of the vertebrate body plan (i.e., segmentation along an anteroposterior axis).

FOCUS 4–1   *continued*

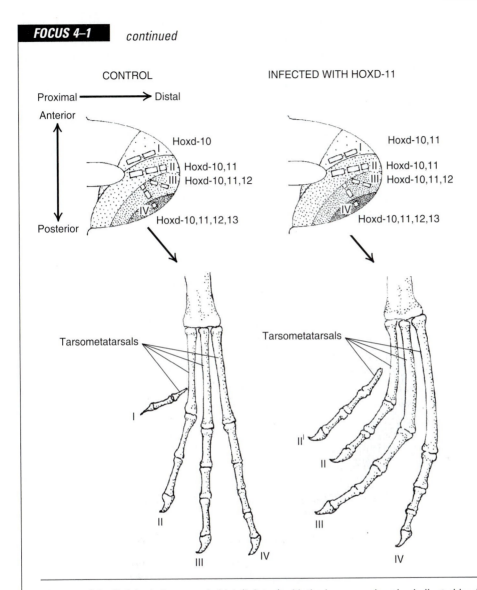

CONTROL

INFECTED WITH HOXD-11

Proximal ——→ Distal

Anterior

Hoxd-10

I

II Hoxd-10,11
III Hoxd-10,11,12

IV

Posterior

Hoxd-10,11,12,13

Hoxd-10,11

I

II Hoxd-10,11
III Hoxd-10,11,12

IV

Hoxd-10,11,12,13

Tarsometatarsals

Tarsometatarsals

I

II

III

IV

II'

II

III

IV

Diagram of the limb bud of a normal chick (left top) with the hox gene domains indicated by different shades. The skeleton of the normal foot is shown (left bottom). In the limb bud on the right top, the cells were infected with the hoxd-II gene, resulting in the total transformation of digit I into digit II' (right bottom). (Modified after Morgan et al.)

During early development, the vertebrate head is subdivided into a series of anteroposterior segments (Fig. 4–10). Hox genes 2.1, 2.6, 2.7, 2.8, and 2.9 are arranged sequentially on the DNA molecule in an order that corresponds to where the genes are expressed in the hindbrain, with the boundaries of their expression corresponding to boundaries between rhombomeres (Fig. 4–10). Hox gene 2.6 is expressed in the most posterior rhombomere, number 7, while Hox 2.7 is expressed in the next anterior rhombomere, number 6. This illustrates the unifying principle of homeobox clusters: genes expressed posteriorly are located at the left of the sequence (e.g., Hox 2.6), and those expressed anteriorly are at the right (e.g., Hox 2.8 and 2.9). Perturbation experiments involving the inactivation of a Hox gene result in segmentation defects, while addition of a Hox gene leads to the formation of an additional segmental structure (Focus 4–1).

The occurrence of Hox genes in many invertebrate taxa supports the notion that the first Hox genes must have evolved approximately 600 million years ago in flatworms or other primitive metazoans with anteroposterior polarities. This truly remarkable constancy suggests that once the molecular system for determining the anteroposterior axis of animals was established, it was retained throughout evolutionary history and served as the basis for the establishment of diverse body plans in animal phyla.

SUMMARY

1. Gametes are haploid cells that contain only a single set of chromosomes. Sperm are small, motile gametes that are capable of reaching and penetrating an egg. Eggs are much larger, nonmotile gametes that contain enough food reserves to initiate development. The amount of yolk in the egg correlates with the future pattern of nutrition of the embryo or larva.

2. Fertilization is the union of a haploid sperm with a haploid egg to form a diploid zygote. It involves several events: sperm penetration of the egg, the combination of male and female nuclear material, and the activation of the egg.

3. During cleavage the single-celled zygote is converted to a multicelled embryo called a blastula, but no growth occurs. Cells in different parts of the blastula contain different enzymes and other components and so differ in their presumptive fates.

4. During gastrulation the single-layered blastula is converted to a two-layered gastrula. The gastrula is covered by ectoderm and has a simple gut cavity, the archenteron, that is lined by endoderm and opens to the surface posteriorly at the blastopore. Mesoderm moves into the roof of the archenteron at the dorsal lip of the blastopore or its equivalent (the primitive streak) and spreads between the ectoderm and endoderm. Patterns of gastrulation and mesoderm formation are greatly influenced by the amount of yolk.

5. Formation of the ectodermal neural tube is induced by the underlying chordamesoderm. The central nervous system develops from the neural tube.

6. Ectodermal neural crest cells separate during the formation of the neural tube and spread throughout the embryo. They give rise to many structures (branchial skeleton, other cranial bones, pigment cells, many peripheral neurons, parts of teeth, and bony scales) and also help regulate development and the emergence of the vertebrate body plan by patterning mesoderm and ectoderm.

7. The mesoderm differentiates into a series of segmental somites that lie lateral to the neural tube and notochord, a nephric ridge, and a lateral plate that spreads around the archenteron and yolk. The coelom develops by cavitation within the lateral plate, but it may enter the nephric ridge and somites.

8. Each somite breaks up into three regions. A lateral dermatome of mesenchy-

mal cells spreads out beneath the ectoderm and forms the dermis of the skin. A myotome gives rise to all or most of the somatic muscles of the body. A medial sclerotome of mesenchymal cells migrates around the neural tube and notochord and forms most of the axial skeleton.

9. The nephric ridge gives rise to the kidney tubules and the urinary and genital ducts.

10. The lateral plate mesoderm is divided by the coelom into an outer somatic and an inner splanchnic layer. The somatic layer forms the parietal peritoneum and contributes to the somatic muscles in many species. The splanchnic layer forms the connective tissue and visceral muscles of the gut and the heart walls and visceral peritoneum.

11. The yolk of macrolecithal eggs becomes suspended from the embryo in an extraembryonic membrane known as the yolk sac. The yolk sac is trilaminar in fishes, including all three germ layers, but is bilaminar in reptiles and birds, consisting only of the endoderm and splanchnic mesoderm.

12. The ectoderm and somatic mesoderm rise off the yolk in reptiles and birds to form additional extraembryonic membranes. The amnion protects the embryo in a cushion of water, and a protective chorion surrounds the amnion, embryo, and yolk sac. Albumin, a shell membrane, and a shell, all of which are secreted by the oviduct, lie peripheral to the chorion.

13. The last extraembryonic membrane of amniotes, the allantois, extends into the extraembryonic coelom. It functions in gas exchange and serves as a site for the accumulation of excretory products. Its base may remain as the urinary bladder.

14. The development of eutherian mammals is similar to the pattern seen in reptiles and birds except for the secondary reduction of yolk, the simplification of cleavage, and the precocious development of those extraembryonic membranes that contribute to the placenta. The placenta is composed of the embryonic chorioallantoic membrane and part of the maternal endometrium.

15. Evolutionary changes are closely linked to certain changes in patterns of development. Radical evolutionary transformations of parts of organisms or whole organisms may result from paedomorphosis and peramorphosis. In paedomorphosis, a change occurs in the rate of development of reproductive organs relative to somatic ones. In peramorphosis, a particular structure undergoes accelerated development or continues to develop for a longer time than in its ancestor.

16. The expression of segmental structures in the hindbrain and limb in vertebrates is regulated by groups of homeotic or Hox genes. These genes also occur in many invertebrates, and their products (amino acid sequences) show a 70 percent similarity in such diverse animals as the fruit fly *Drosophila* and mammals. Many of these genes must have evolved 600 million years ago before arthropods and vertebrates diverged.

REFERENCES

Arey, L. B., 1974: *Developmental Anatomy*, 7th edition. Philadelphia, W. B. Saunders Company.

Balinsky, B. J., assisted by Fabian, B. C., 1981: *An Introduction to Embryology*, 5th edition. Philadelphia, Saunders College Publishing.

Conklin, E. G., 1932: The embryology of *Amphioxus. Journal of Morphology*, 54:69–118.

Corliss, C. E., 1976: *Patten's Human Embryology*. New York, McGraw-Hill Book Company.

De Robertis, E., Oliver, G., and Wright, C. V. E., 1990: Homeobox genes and the vertebrate body plan. *Scientific American*, 263:46–52.

Feduccia, A., and McCrady, E., 1990: *Torrey's Morphogenesis of the Vertebrates*, 3rd edition. New York, John Wiley.

Gilbert, S. F., 1991: *Developmental Biology*, 3rd edition. Sunderland, Mass., Sinauer Associates Inc.

Hall, B. K., 1992: *Evolutionary Developmental Biology*. London, Chapman and Hall.

Hall, B. K., and Hörstadius, S., 1988: *The Neural Crest*. Oxford, Oxford University Press.

Hamilton, W. J., Boyd, J. D., and Mossman, H. W., 1972: *Human Embryology*, 4th edition. Baltimore, Williams and Wilkins.

Hatschek, B., 1883: The amphioxus and its development. London, Sonnenschein.

Kessel, M., Balling, R., and Grass, P., 1990: Variations of cervical vertebrae after expression of a Hox-1.1 transgene in mice. *Cell*, 61:301–308.

Lillegraven, J. A., 1975: Biological considerations of the marsupial-placental dichotomy. *Evolution*, 29:707–722.

Lillegraven, J. A., 1985: Use of the term "trophoblast" for tissues in therian mammals. *Journal of Morphology*, 183:293–299.

Maderson, P. F. A., 1975: Embryonic induction and evolution. *American Zoologist*, 15:315–327.

McKinney, M. L., 1988: *Heterochrony in Evolution: A Multidisciplinary Approach*. New York, Plenum Press.

Morgan, B. A., Izpisua-Belmonte, J.-C., Duboule, D., and Tabin, C. J., 1992: Targeted misexpression of Hox-4.6 in the avian limb bud causes apparent homeotic transformations. *Nature*, 358:236–239.

Noden, D. M., 1991: Vertebrate craniofacial development: The relation between ontogenetic process and morphological outcome. *Brain, Behavior and Evolution*, 38:190–225.

Northcutt, R. G., and Gans, C., 1983: The genesis of neural crest and epidermal placodes: A reinterpretation of vertebrate origins. *Quarterly Review of Biology*, 58:1–28.

Raff, R. A., Kaufmann, P. C., 1983: *Embryos, Genes and Evolution*. New York, The Macmillan Company.

Thompson, K. S., 1983: Reflections on the neural crest. *American Scientist*, 71:72–74.

5 | The Integument

PRECIS

The skin, or integument, is the interface between the external and the internal environments and therefore plays a vital role in protecting the body and maintaining the integrity of the internal environment. Its structure and the nature of its derivatives (scales of many types, glands, feathers, hair) are closely correlated with the vertebrate's environment and mode of life.

Outline

The skin or **integument** (L. *integumentum* = covering) is one of the boundary layers of the body that forms the interface between the animal's internal environment and the outside world. Other important boundary layers are the linings of the digestive and respiratory tracts. As the most exposed boundary layer, the skin is a remarkably complex organ. It is composed of many tissues including epithelium, connective tissue, fat, and smooth muscle, and it contains blood vessels, glands, sensory receptors, nerves, and other structures. Since vertebrates maintain an internal environment distinctly different from their external environment, the skin has many important functions. It protects the body against abrasion, undue exchanges of water and salts, ultraviolet radiation, and other assaults of the external world. But it does not isolate the body. Many sensory signals are received through the skin, and exchanges of water, ions, and gases occur through it in some species. Secretions of its glands may be protective, serve as chemical signals in communication, and have a nutritive function. The skin is involved in vitamin D synthesis and the regulation of blood pressure. Coloration of the skin frequently camouflages an animal, warns potential predators, and plays a role in many other aspects of behavior. In birds and mammals, and to a lesser extent in reptiles, the skin helps regulate body temperature. Finally, the skin also is a complex ecosystem containing a characteristic flora and fauna of viruses, bacteria, fungi, yeasts, and mites. It is truly a jack of all trades.

General Structure and Development of the Skin

Although the integument varies considerably among species and even between parts of the body in an individual, its basic structure and development are much the same in all vertebrates (Fig. 5–1). An outer **epidermis** (Gr. *epi* = upon + *derma* = skin) that is composed of a stratified, squamous epithelium develops embryonically from the ectoderm. It rests on a basement membrane of delicate fibrils and overlies a dermis of fibrous connective tissue. The **dermis** develops

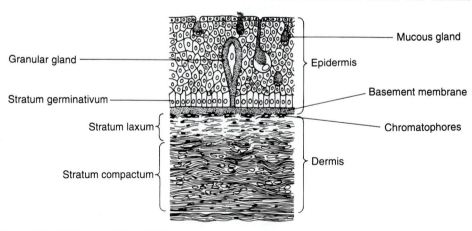

FIGURE 5–1   The structure of fish skin. This drawing is based on a species without bony scales, so the cellular layers of the skin are clear. (After Portmann.)

from mesenchymal cells, most of which are derived from the mesodermal dermatomes of the somites.

One or two cell layers of the epidermis, which are located just above the basement membrane, constitute the **stratum germinativum**. These cuboidal cells multiply mitotically. Cells that do not remain in the stratum germinativum move toward the body surface, differentiate, flatten, and eventually slough off either individually or in sheets. Epidermal cells have the capacity to synthesize **keratin** (Gr. *kerat-* = horn), a water-insoluble, horny protein that may fill the cells and replace other organelles. This capacity is best developed in terrestrial vertebrates, where dead, keratin-filled cells form a **stratum corneum** (L. *cornu* = horn) on the skin surface. The epidermis is only a few cells deep in fishes and amphibians but becomes thick in reptiles and mammals.

The dermis is the distinctive part of vertebrate skin; the invertebrate integument lacks such a layer. The dermis is a fibrous connective tissue composed primarily of an extracellular matrix of **collagen** and **elastic fibers** embedded in a ground substance of **proteoglycans** and other macromolecules, many of which bind with water. The primary cellular elements of the dermis are elongated **fibroblasts**, which synthesize and extrude the fibers. Other cell types include pigment cells, fat cells, smooth muscle fibers, scattered white blood cells, and **macrophages**. Macrophages are phagocytic cells that help protect the body against foreign micro-organisms by engulfing and degrading them. Fibroblasts, collagen fibers, and other dermal structures differentiate embryonically just beneath the basement membrane of the epidermis and sink inward.

In mature skin the dermis is thicker than the epidermis and typically consists of two layers. Collagen fibers are irregularly arranged in the superficial **stratum laxum** (L. *laxus* = loose) and more tightly packed in the deeper **stratum compactum**. Nerves and sense organs occur in the dermis, and free nerve endings penetrate the epidermis in some species. Fat and muscle may be present in deeper parts. Many blood vessels travel through the dermis. Papillae of the dermis sometimes protrude into the epidermis and bring capillaries close to the body surface, but capillaries normally do not enter the epidermis. Capillaries do enter the epidermis in some salamanders that depend on cutaneous respiration (p. 653).

Teeth, bony scales, horny scales, feathers, hair, and other skin derivatives develop as a result of intricate interactions between the dermis and the overlying epidermis. The cells that form these structures must be competent to respond to inductive influences of adjacent cells and to environmental influences. Neural crest cells that migrate between the epidermis and developing dermis have been implicated in the formation of many of these structures. Feathers, hair, and other keratinized structures, as well as glands, are composed of epidermal cells. Parts of these structures grow down into the dermis during development. Superficial bones are dermal derivatives. Teeth and the bony scales of many fishes are composed of both epidermal and dermal products. Many of these structures are hard and perform the supportive and protective functions we associate with skeletons, so they are appropriately called the **integumentary skeleton**. The integumentary skeleton develops within or just beneath the skin and should not be confused with the deeper cartilage and bones of the **endoskeleton** or with the **exoskeleton** of arthropods, mollusks, and many other invertebrates, which is a secretion on the

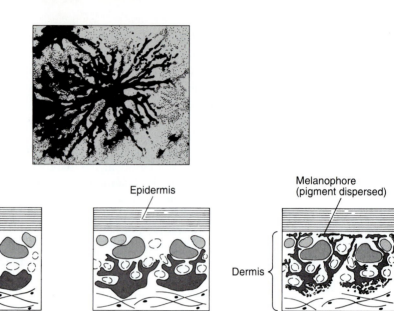

A

B

Xanthophore
Erythrophore
Iridophores
Melanophore
(pigment withdrawn)

Epidermis

Melanophore
(pigment dispersed)

Dermis

FIGURE 5–2 Pigment cells: **A**, A surface view of a melanophore in which the pigment is dispersed throughout the cell. **B**, Vertical sections through the skin of the rainbow lizard showing two dermal chromatophore units. Pigment is withdrawn from melanophore processes in the left figure and the lizard is orange; in the middle figure a partial dispersal of pigment produces a terracotta color; on the right full dispersal turns the animal a chocolate brown. (After Harris.)

body surface. Glands, bony scales, and keratinized structures differ greatly among the different groups of vertebrates, according to their environments.

Skin Coloration and Its Function

The skin of all vertebrates, except albinos, contains pigments of various types within cells collectively known as chromatophores (Gr. *chroma* = color + *phoros* = bearing, from *pherein* = to bear). **Chromatophores** are of neural crest origin. They are in the upper part of the dermis in fishes, amphibians, and reptiles, but they penetrate or are located in the epidermis in birds and mammals. When the pigment is a dark melanin, the cells are called **melanophores** (Gr. *melania* = blackness). The melanin may be brown or black or more yellow and reddish. Melanophores are star-shaped cells with long, branching processes (Fig. 5–2A). The pigment itself is synthesized and contained within cellular organelles, the **melanosomes**. In anamniotes and reptiles the melanosomes may migrate into the processes of the melanophores, which maximizes the color, or concentrate near the center of the cell. Pigment is synthesized within the melanophores of birds and mammals, but most of it is transferred to feather, hair, and other epidermal cells.

Most mammals only have melanophores, but the other vertebrates also have brighter colored pigment cells. **Iridophores** (Gr. *iris* = rainbow) contain reflective platelets of purine, usually guanine, that may give them a silvery appearance. **Xanthophores** (Gr. *xanthos* = yellow) and **erythrophores** (Gr. *erythros* = red) contain yellowish or reddish pigments, respectively, that are composed of pteridines and carotenoids. Frequently, three or more chromatophores are organized into a **dermal chromatophore unit** (Fig. 5–2**B**). Iridophores in the center of the unit are surrounded by processes of a deeper melanophore and are overlain by xanthophores or erythrophores. Different colors result from different combinations of these cells and patterns of pigment dispersal within the melanophores. If melanin is dispersed in the processes of the melanophore overlying the iridophore, the body appears dark, or perhaps yellowish if pigment is well dispersed in the xanthophore. If pigment is withdrawn from the overlying melanophore processes, light reaches the iridophores and is dispersed. Short wavelengths (blue) are reflected through the skin; and longer ones are absorbed by the underlying melanophore. When the blue light comes back through a filter of yellow xanthophores, the color is green. An African bird, the touraco, is the only vertebrate in which a green pigment is known.

In some species minute ridges on surface cells refract light and produce iridescent colors that change with the angle of observation. Hummingbirds have such colors. In a few vertebrates the degree of vascularity of the skin produces color. Blushing is a familiar example.

Pigmentary colors are subject to change. More or less pigment may be synthesized as an animal slowly adjusts to its background, to different seasons, or to the degree of solar radiation (as in a suntan). Slow changes of these types are referred to as morphological color changes. Fishes, amphibians, and reptiles can also change color physiologically. Rapid changes occur by a migration of pigment within the chromatophores. Migration of pigment within the melanophores of many fishes is controlled both by sympathetic stimulation, which leads to an aggregation of melanosomes, and by melanophore-stimulating hormone produced by the pituitary gland, which causes a dispersal of melanosomes. Control in amphibians and reptiles appears to be primarily or exclusively hormonal. Melanophore-stimulating hormone promotes pigment dispersal; norepinephrine from the adrenal gland causes pigment aggregation.

Skin color plays an important role in the life of vertebrates. It may be concealing, or **cryptic** (Gr. *cryptos* = hidden), and help hide an organism from its predators or enable a predator to stalk its prey undetected. (See Focus 5–1.) It may be **aposematic** (Gr. *apo* = away + *semat-* = signal) and advertise the presence of dangerous, venomous, or distasteful species that profit from making known their presence. After a few unpleasant experiences, in which a few individuals may be injured or killed, predators learn to avoid members of these species. Skin coloration is also important in species recognition, establishment of territories, courtship, and other types of communication. Color changes help many reptiles thermoregulate (p. 84).

Some degree of skin pigmentation protects deeper body tissues from injury by ultraviolet radiation. In some small fish and other vertebrates, especially those active in the daytime, melanophores also occur in membranes around the nervous

---

FOCUS 5–1   **Adaptive Camouflage in Flatfishes**

The ability of organisms to change appearance according to their immediate local surroundings is called adaptive camouflage. This ability reaches a pinnacle in flatfishes (flounders, sole). This mechanism is adaptive, and concealed individuals survive predation significantly better than less cryptic ones. The disguise evolved by flatfishes is so good that a viewer is unlikely to distinguish between the fish and its surroundings. The disguise functions like a malleable mirror having a reflection that is interpreted only in relation to the immediate local surroundings (see the figure).

Animals have only a limited number of designs to minimize their visibility. These designs are based on four principles: (1) color resemblance; (2) obliterative shading (a graded lightening and darkening on the surface that diminish the appearance of relief); (3) disruptive coloration (a pattern of colors and tones that serves to divide the continuous body outline and body surface into sections); and (4) shadow elimination. Camouflage in flatfishes is primarily based on disruptive coloration (Saidel, 1988) involving adaptive changes in skin reflectance

and contrast between discrete skin areas (see the figure). These changes occur on a mural provided by the distribution of different chromatophores and dermal reflecting cells (iridophores). By modulating the various dermal chromatophore units via sympathetic nerve stimulation, melanophore-stimulating hormone, and the newly discovered melanophore-concentrating hormone (Kawauchi et al., 1983), the flatfish can make rapid changes of its skin appearance, blending it with its local surroundings. The entire control system is superimposed by the visual system. Blinded flatfishes darken and are unable to engage the adaptive camouflage mechanism when placed on light or contrasting backgrounds. At first it was thought that a flatfish can reproduce its background very faithfully. However, there is only a limited range of preset patterns of stimulation of chromatophore units that, in turn, have fixed patterns of distribution. Thus the fish has only a restricted range of options (see the figure) by which it can remain unseen by the viewer. The flatfish's camouflage strategy is to condition the viewer's eyes for certain characteristics of the local surroundings and to elim-

---

system and even in the peritoneum lining the body cavity. The pigmentation of human skin also affords some protection against ultraviolet radiation.

The Skin of Fishes

Structure

Integumentary structure is correlated closely with a vertebrate's mode of life and environment. All fishes are aquatic and encounter similar environmental problems, and therefore the basic structure of the skin is similar in most groups, apart from the nature of their bony scales. The epidermis is relatively thin, and most of its cells are living (Fig. 5–1). The surface cells frequently are covered with complex patterns of microridges visible only with electron microscopy. In some cases the microridges increase the surface area available for exchanges between the animal and its environment, but their function is uncertain in most cases. Keratin may be deposited in limited areas, as in the horny teeth of cyclostomes and the nuptial

FOCUS 5-1 *continued*

inate the sharp boundaries of its body by means of transparent fins along the entire perimeter of its body. With the transparency of the fins, the viewer sees a combination of the substratum and skin textures and no sharp boundary. At the same time, the fish adapts its skin to its local surroundings by producing an average reflectance of its background (see the figure). In its natural habitat this strategy will result in the fish being unseen not only by the human, but also by the predator's eye.

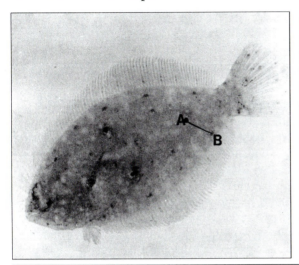 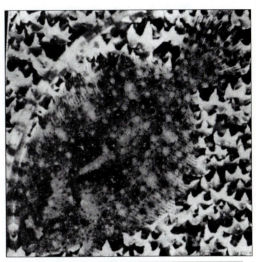

Adaptive changes in the flounder *Paralichthys lethositma* to a white (left) and patterned (right) background. The spots identified by **A** and **B** are maintained on both backgrounds. (From Saidel.)

tubercles that develop in many male fishes during the breeding season, but most epidermal cells generally do not synthesize keratin.

Unicellular glands, most of which produce mucus, are abundant in the epidermis. Their secretions, together with secretions of the surface cells, form a mucous cuticle and a generally slimy surface. Surface mucus is believed to reduce water exchanges between the fish and its environment and so help the excretory organs maintain a stable internal environment (Chapter 20). Mucus also protects the body from bacterial invasion and the attachment of ectoparasites. It may also reduce friction, especially during sudden bursts of speed, as when a fish tries to escape a predator. Granular unicellular glands are also present, but the functions of many are not known. In minnows, catfishes, and other ostariophysans, some of these glandular cells contain an alarm substance that is released when the fish is injured and the skin ruptured. This acts as a chemical messenger, or **pheromone** (Gr. *pherein* = to bear + *hormaein* = to excite), and triggers a fright reaction in nearby members of the species. **Multicellular glands** that grow down into the dermis are rare in fishes. Hagfishes have large slime glands; some fishes have poison glands that produce toxic materials and are often associated with fin spines;

FIGURE 5–3   The photophore beneath the eye of the lantern fish *Anomalops* may be exposed or covered by a skin flap shade.

some deep-sea species have light-generating glands, or **photophores**. In some cases the secretions of the photophores nourish a bacterial colony that generates light; in others chemical reactions within the gland produce light. The light organs beneath the eyes of the nocturnally feeding lantern fish function as head lights and appear to help the fish find prey (Fig. 5–3). In other cases the patterns of light are of value in species recognition. In many deep-sea angler fishes a light organ dangled over the mouth on a modified dorsal fin spine acts as a lure.

The collagen fibers of fish dermis frequently are more regularly arranged than in other vertebrates. They develop in layers that spiral around the body at an approximately 45-degree angle to the longitudinal axis. Adjacent layers of fibers are nearly perpendicular to each other. As in plywood, this arrangement strengthens the skin so that body shape is maintained during undulatory swimming, and the skin itself acts as an exotendon transmitting muscular forces (p. 348).

Bony Scales

Bony scales more than any other single feature characterize the skin of most fishes. The tissues that may contribute to them are bone, dentine, and enamel. **Bone** consists of an extracellular matrix of collagen fibers embedded in an organized ground substance of protein polysaccharides and calcium phosphate. This matrix is laid down by bone-forming cells, known as **osteoblasts** (Gr. *osteon* = bone + *blastos* = bud) that have differentiated from mesenchymal cells of the dermis. Calcium phosphate crystals known as **hydroxyapatite**, which are synthesized by the osteoblasts, bind to the collagen fibers and constitute 70 percent of the bone by weight. As the bone develops, the bone-forming cells, now called **osteocytes**, usually become entrapped in the matrix (Fig. 5–4). They lie in small cavities, the **lacunae**, that are interconnected by minute canals, the **canaliculi**. Processes of the osteocytes extend through the canaliculae. This type of bone is called cellular bone. In some cases, notably in the scales of teleost fishes, the bone is acellular. Bone-forming cells are found on the periphery of the developing scale. As the bone matrix is formed, the cells move centrifugally, away from the center of the scale. No bone cells or processes are left behind.

Dentine and enamel (or enamel-like) layers may be deposited over the bone on the surface of the scales. The structure of the scales and formation of these layers vary considerably in different fish groups. Investigators are not in complete

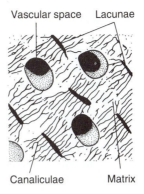

Vascular space   Lacunae

Canaliculae   Matrix

FIGURE 5–4   A microscopic section of bone.

agreement on scale structure. Often scales develop embryonically in a manner similar to teeth (Fig. 5–5). First mesenchymal cells aggregate in small papillae just beneath the basement membrane of the epidermis. Although located in the dermis, many of these cells are believed to be of neural crest origin. The basal cells of the overlying epidermis respond by differentiating into **ameloblasts** (Middle English *amel* = enamel + Gr. *blastos* = bud) that collectively form an **enamel organ**. Under the inductive influence of the enamel organ, underlying dermal cells differentiate into **odontoblasts** (Gr. *odont-* = tooth) that begin to lay down dentine between themselves and the enamel organ. The enamel organ then produces enamel on top of the dentine. This process continues, with the enamel organ retreating in one direction and the odontoblasts in the other.

Enamel, which is of epidermal origin, is the hardest tissue in the body: approximately 96 percent by weight is crystals of hydroxyapatite. Its matrix includes distinctive proteins called **amelogenins**. As enamel is laid down and the enamel organ retreats, no cells or cell processes are left behind in the enamel. In some cases the enamel is deposited in successive waves and has a lamellar appearance. This type of enamel, which occurs on many scales, was at first thought to be different from the enamel on teeth and was called **ganoine** (Gr. *ganos* = sheen). It is now believed that the enamel of teeth differs from that of ganoine only in

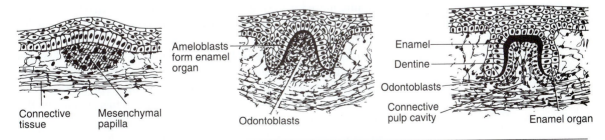

Connective tissue   Mesenchymal papilla

Ameloblasts form enamel organ

Odontoblasts

Enamel
Dentine
Odontoblasts
Connective pulp cavity
Enamel organ

FIGURE 5–5   Three vertical sections through the skin of a fish to show the interaction between a mesenchymal papilla in the dermis and the epidermis in forming enamel and dentine. Enamel and dentine are added to the surface of the bony scales of many fishes.

the pattern of deposition. They are the same material, and both are derivatives of the ectodermal epidermis.

Dentine, which is also produced by cells of neural crest origin, is similar to bone and differs primarily in the arrangement of the cells and matrix. As the odontoblasts form dentine and retreat, they leave long, cytoplasmic processes that become lodged in dentine tubules. **Cosmine** (Gr. *kosmios* = well ordered) is a variety of dentine in which the dentine tubules are grouped into radiating tufts.

The primary dermal armor of the ancestral ostracoderms consisted of thick **cosmoid plates** containing all of these hard materials (Fig. 5–6**A**). The deepest part of the plate was made up of layers of bone, usually cellular. Spongy bone lay superficial to the lamellar bone. Many of the spaces in the scale accommodated blood vessels that nourished the developing scale (Bemis and Northcutt, 1992). Others contained mucous glands and parts of the aquatic lateral line and electro-receptive systems (p. 413) that enabled these fishes to detect water movements of various kinds and weak electric currents. The surface of the plate was composed of a series of denticles. Each denticle contained dentine of the cosmoid type and a thin cap of enamel-like material. Since this enamel-like material may be hard cosmine that developed from the dermis (unlike enamel, which is always of epidermal origin), it is often called **enameloid**. Most of these early ostracoderms were small, fresh-water fishes that moved slowly as they fed along the bottoms of ponds and streams. The heavy armor may have helped keep them close to the bottom and protected them from predators, such as aquatic scorpions, or euryp-terids. The bone of the scales also may have provided a reservoir of needed calcium and phosphate ions. This may have been a prerequisite for early vertebrates to penetrate fresh water, where these ions are in short supply. Finally, the scales may have helped a fish in fresh water solve its water balance problem. Because the concentration of salts within its body fluids was greater than that in the surrounding fresh water, large amounts of water would diffuse into the body by osmosis through gills and any other permeable surface. The enamel or enameloid on the scales may have reduced this influx of water over the general body surface.

As fishes became more active creatures, there would be advantages to a less cumbersome armor. The thick, bony plates became reduced in different ways in different lines of fish evolution. Living cyclostomes have no scales at all. Presumably, the ancestral armor has been lost, but it is possible that they evolved from some unknown, unarmored ostracoderm. Placoderms lost the armor on most of the trunk and tail but retained heavy denticulate cosmoid scales and plates, similar to the ancestral armor, on the head and thorax. Most of the bony material has disappeared in cartilaginous fish. The spiny **dermal denticles** or **placoid scales** of sharks and other cartilaginous fishes are composed of dentine surrounding a vascular pulp cavity and capped by a hard material (Fig. 5–6**B**). The nature of this material has long been a puzzle. Recent studies have shown that it contains the amelogenin proteins found in enamel, but it also contains some fibrous material of dermal origin. Many investigators now call this superficial layer enamel; others hedge and call it enameloid. A thin layer of what appears to be acellular bone underlies the dentine. In a sense, the dermal denticles represent isolated denticles of the ancestral armor.

Early bony fishes retained heavy, bony scales. Ancestral sarcopterygians had rhomboid **cosmoid scales** without denticles and similar in composition to the ostracoderm plates. Ancestral actinopterygians and acanthodians had rhomboid **ganoid scales** (Fig. 5–6**C**) composed of lamellar bone overlain by many layers of enamel (ganoine). Some ganoid scales contain small amounts of vascular spongy bone and dentine; others do not. Ganoid scales occur today only in *Polypterus* and in the garpikes. Other actinopterygians have very thin scales that develop in overlapping skin folds, so the scales themselves overlap. The scale is composed primarily of what appears to be acellular bone underlain by dense, fibrous material (Fig. 5–6**D**). This gives the scale great flexibility. The surface of **cycloid scales** is sculptured by a pattern of growth rings (Fig. 5–6**E**). **Ctenoid scales** have in addition a series of comblike projections, ctenii, on the posterior part of the scale nearest the skin surface (Fig. 5–6**F**). The origin of the sculpturing on the most superficial part of the scale is uncertain, but some investigators believe that it may represent a thin layer of dentine or enamel. Contemporary sarcopterygian fishes also have cycloid scales, which evolved independently.

The Skin of Amphibians

The aquatic larvae of amphibians have skin similar to that of fishes. After metamorphosis, the epidermis remains relatively thin, but keratin is synthesized by the epidermal cells. As it accumulates, the cells die and form a horny layer, the **stratum corneum**, on the skin surface (Fig. 5–7). The stratum corneum is thin in amphibians, seldom more than one or two layers of cells, so cutaneous respiration is made possible while simultaneously providing protection against desiccation and abrasion, which confront all land vertebrates. It is sloughed off (desquamation) periodically in large sheets only one or two cells thick. The discarded tissue is usually eaten. Desquamation is hormonally controlled. Evidence for this is that hypophysectomized toads do not slough their keratinized layer, but instead keratinized cells merely continue to pile up.

Unicellular glands, so prevalent in fishes, are absent, but multicellular, alveolus-shaped glands grow down into the dermis from the stratum germinativum. Most are **mucous glands** whose secretions protect the skin surface, reduce the loss of body water to a limited extent, and keep the surface moist. In contemporary amphibians considerable gas and ionic exchanges occur across the skin—hence the prolific dermal vascularization. Other **granular glands** secrete toxic substances. The **parotid gland** (Gr. *para* = beside + *ot-* = ear) of toads that forms a bulge behind the ear is an aggregation of granular glands. Its irritating secretions discourage experienced snakes, racoons, and other predators. Some tropical frogs have highly toxic skin glands. Their presence is often coupled with a brilliant red or yellow aposematic coloration. Some Amazonian Indians use the toxin on their arrow tips. Glands on the back of the Surinam toad, *Pipa pipa*, help nourish the tadpoles that develop in pits in the skin on the mother's back.

Amphibian coloration is as highly developed as in fishes. Chromatophores include melanophores in both epidermis and dermis, and numerous xanthophores and iridophores in the dermis.

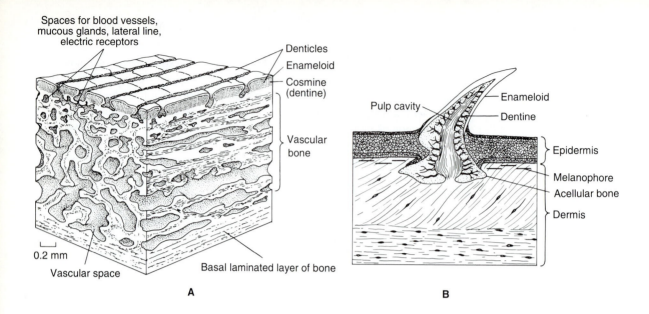

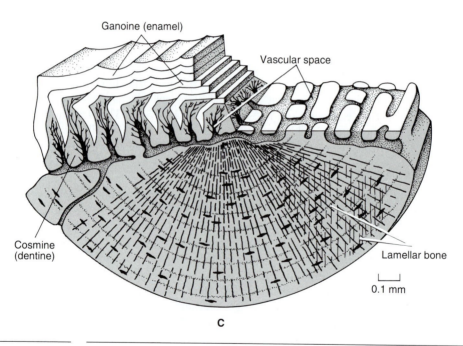

FIGURE 5–6 Some bony scales of fishes: **A**, A vertical section through a cosmoid plate of an ostracoderm (a heterostracan). **B**, The dermal denticle of a cartilaginous fish. **C**, A vertical section through the ganoid scale of a primitive actinopterygian (a palaeoniscoid). (**A**, after Kiaer. **C**, after Ørvig.)

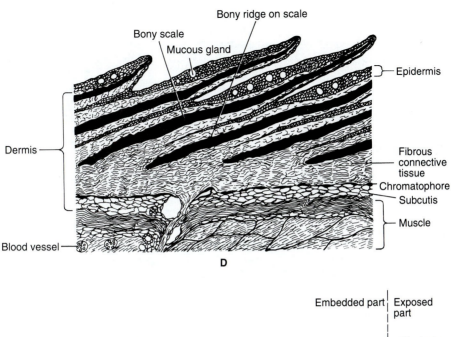

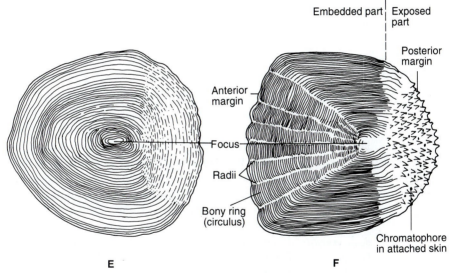

FIGURE 5–6 *continued*
D, A vertical section through the skin and cycloid scale of a teleost. **E**, Surface view of a cycloid scale. **F**, Surface view of a ctenoid scale.

Small bony scales were present in the dermis of early labyrinthodonts, but they have been lost in most other amphibians. Caecilians and some anurans retain a few bony nodules, called **osteoderms**, deep in the grooves in their skin. They are homologues of dermal scales and are remnants of osteoderms in primitive, extinct amphibians. In primitive amphibians one large osteoderm became incorporated into the pectoral girdle as the **interclavicle**.

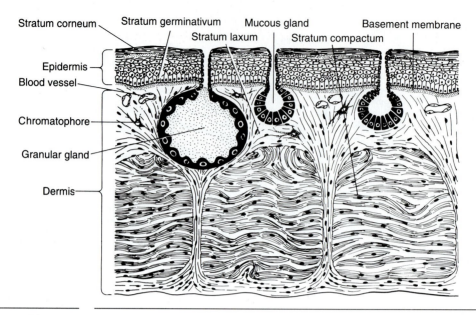

FIGURE 5–7   A vertical section through the skin of a frog. (After Portmann.)

The Skin of Reptiles

As vertebrates adapted more completely to the terrestrial environment, the skin became more important in protecting them against abrasion and loss of body water. The epidermis of reptiles has a thick stratum corneum composed of many layers of dead, keratin-filled cells. These are organized as horny plates in turtles and as **horny scales** in lizards and snakes (Fig. 5–8A). Reptiles lose less water by evaporation through the epidermis than amphibians. The large amount of keratin is partly responsible, but phospholipids bound to the keratin are probably more important in reducing water loss than the keratin itself.

The epidermis of lizards and snakes is particularly complex. The functional scale consists of many layers of cells. The outer layers are heavily keratinized, but the deepest several layers are living. All these cells constitute a mature **outer epidermal generation** (Fig. 5–8B). An immature **inner epidermal generation** of cells, which has been produced more recently by the stratum germinativum, lies beneath. The same keratinized and living layers that characterize the outer generation begin to differentiate in the inner generation prior to skin shedding. A new layer of scales is ready to take over for the old layer before the old layer is lost. During molting a deep, unkeratinized layer of cells of the outer epithelial generation undergoes autolysis and breaks down. A fission zone develops, and the outer epidermal generation is shed as a unit. The horny plates of the shell of most turtles are not shed but rather wear away at the surface. As the turtle grows, newly formed horn protrudes beyond the margin of the old plates. This results in a series of growth rings around the margin of each plate.

The terminal phalanx of each digit is encased in a protective claw. Claws also help the animal grip the substratum during locomotion. Although claws are present

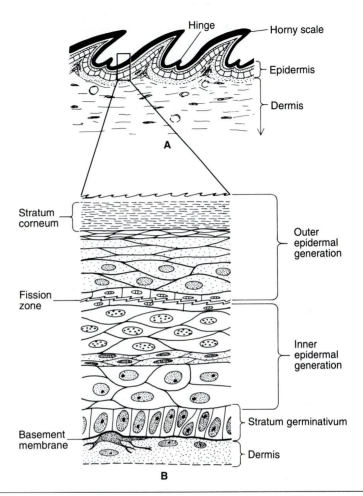

FIGURE 5–8   Reptile skin: **A**, A vertical section through a horny scale and the skin of a squamate reptile. **B**, An enlargement of a section of the scale. (After Maderson.)

in a few amphibians—the South African clawed frog (*Xenopus laevis*) is a familiar example—all reptiles with limbs have them. Claws, like horny scales, are derivatives of the stratum germinativum. Their hardness derives from the incorporation of calcium salts along with keratin.

Some epidermal scales are modified for particular functions (e.g., over the tips of digits they act as suction pads enabling, for example, geckos to crawl up vertical walls or move upside down on the ceiling). Modified ventral scales of snakes play an important role in locomotion. Except for **apical pits**, reptilian skin is generally devoid of sense organs. Apical pits are situated one to seven in a row, at or near the posterior margin of the epidermal scales. A tiny hairlike filament protrudes from each pit and has a tactile function.

The reptile dermis contains no mucous glands. The few glands that are present are **scent glands**, producing musk and other secretions used in courtship and mating behavior. These occur in specific places in different species: under the thigh in some lizards, on the underside of the lower jaw in crocodiles, and near the cloaca in some turtles.

Some reptiles (such as the skinks among lizards) retain small, thin osteoderms in their dermis. In crocodilians, some lizards, and *Sphenodon*, osteoderms form a series of riblike structures called **gastralia** in the ventrolateral abdominal wall. Their function is still unknown, even though it is hypothesized that they play a role in stiffening the pleuroperitoneal cavity as part of the mechanism for ventilating the lungs.

The Skin of Birds

The epidermis of birds is relatively thin and is not heavily cornified over most of the body surface. Horny scales do develop on the legs and feet of most birds. They are not shed. Claws are present, and the margins of the toothless jaws are covered by a horny beak.

The most conspicuous derivatives of the epidermis are the feathers. In addition to forming the flying surfaces of the wings and tail, feathers entrap air and reduce airflow next to the skin. This reduces evaporative loss of body water and also forms an insulating layer that enables birds to maintain a high and constant body temperature. These functions of feathers may have evolved before their role in flight. Those that cover the body surface are the **contour feathers**. The larger and stiffer contour feathers that form the flying surfaces are **flight feathers**. A typical contour feather (Fig. 5–9) has a central axis whose base, the quill or **calamus** (L. *calamus* = reed), is lodged in a **feather follicle** in the dermis. The distal part, known as the shaft or **rachis** (Gr. *rhachis* = spine), supports a **vane**. During development dermal tissue and blood vessels enter the proximal end of the quill through an opening, the **inferior umbilicus**. The vane is composed of many **barbs** that branch obliquely from each side of the rachis. The barbs in turn bear many **barbules**, which have minute hooklets that interlock with the hooklets of adjacent barbules, so the vane is a coherent, strong, and flexible structure. If barbules become separated, the bird can interlock them again by preening its feathers with its bill. Because hooklets are absent from the barbules of ostriches and many other flightless species, the feathers are fluffy. The contour feathers of some primitive birds are double, for a small second vane, known as the **afterfeather**, arises from a **superior umbilicus** located at the distal end of the calamus. Traces of after-feathers are sometimes represented in more advanced species by a small tuft of barbules.

Other types of feathers include down and bristles (Fig. 5–10). **Down**, which is found beneath the contour feathers in some species, consists of very fluffy barbs that arise from the distal end of the quill. Down is excellent insulation. It is particularly important in young birds, whose small body size gives them a large heat-losing surface relative to their heat-producing mass. Bristles, or **filoplumes** (L. *filum* = thread), are short, stiff feathers in which the barbs are reduced or lost. Those around the nostrils and eyes of some species keep out dirt. Longer ones around the mouths of nighthawks and flycatchers help the birds net insects.

Colors in feathers are the product of two mechanisms: physical structure and

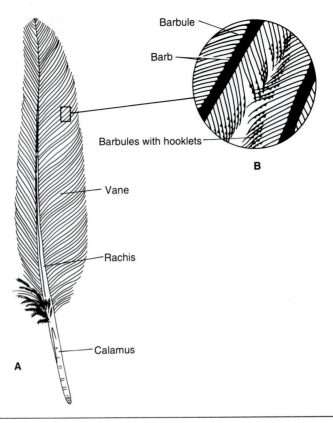

Barbule

Barb

Barbules with hooklets

B

Vane

Rachis

Calamus

A

FIGURE 5–9   The structure of a contour feather of a bird: **A**, Entire feather. **B**, An enlargement of a small part of the vane.

the presence of chemical pigments. Red, yellow, brown, and black are of pigment origin, whereas physical structure and the associated light scattering produce white, blue, and iridescent colors. Brown and black are caused by melanin taken up from the germinal region of the follicle. Red, orange, and the pink of the flamingo are caused by carotinoids in the keratinized cells of the feather and are of dietary origin. White is produced by air in cells and by the polygonal shape of the barbule cells, which breaks up the light and reflects and refracts all wavelengths. Blue is caused by the scattering action of particles in cells located beneath the outer layer of the feather barbs, resulting in the reflection of blue, while the longer wavelengths are transmitted. The combination of structural blue with yellow pigment results in green. This effect can be seen in hummingbirds.

The dermis from various regions in the body determines whether overlying epidermis remains flat and devoid of feathers, forms scales, or gives rise to different types of feathers.

Feathers develop by an inductive interaction between the epidermis and the dermal mesenchyme that leads first to the formation of a cone-shaped dermal papilla that pushes up the overlying epidermis (Fig. 5–11). Mitotic divisions in a collar-like zone of the stratum germinativum near the base of the papilla form a

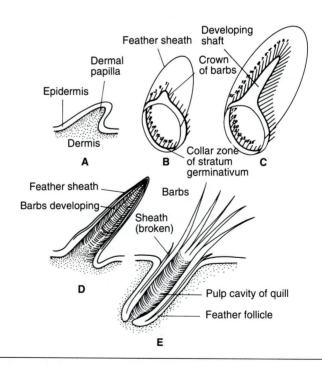

FIGURE 5–10   Some feather types: **A**, Down. **B**, Filoplumes.

FIGURE 5–11   Five stages in the development of a feather. (After Lillie.)

crown of barbs. These are covered by a horny sheath of epidermis. As development proceeds, cells on one side of the collar divide more rapidly than others and form a shaft that extends upward, carrying with it the barbs formed in the collar. Meanwhile, the base of the feather recedes into the skin, accompanied by layers of epithelial cells that form the feather follicle. Except in a few primitive species of birds, feathers do not develop uniformly over the body surface but fan out from distinct feather tracts.

Horny scales also begin when an underlying dermal papilla pushes up the overlying epidermal cells. This suggests that feathers and scales are homologous structures and that feathers may have evolved from the horny scales of reptiles. The feathers on the legs of owls and some other birds develop at the tips of horny scales, which reinforces the notion of a homology between these structures.

Feathers wear out and are replaced. Most species have an annual molt in the summer after the breeding season, but much variation occurs among species. Many songbirds have two molts a year and distinct winter and breeding plumages. Usually, only a few feathers are lost at a time, so the molting period extends over several weeks and the bird remains covered and able to fly.

Avian dermis lacks ossifications. It is a thin, well-vascularized layer, with interlacing collagenous fibers and several types of sensory receptors. Many smooth muscles are associated with the feathers. Feather position is crucial for flight, behavior, and thermoregulation. The position of a feather can be regulated by a group of muscles. **Erectors** lift feathers, **depressors** lower them, **retractors** pull them inward, and **rotators** turn them (Fig. 5–12). Because of the complex configuration, each of these muscles may affect adjacent feathers (e.g., one muscle may depress one feather and elevate an adjacent one).

The evolution of feathers and skin structure is closely correlated with the evolution of endothermy. Air can be trapped by the feathers and held close to the skin as an insulating layer. The thickness of the layer of air is controlled by the dermal smooth muscles, which determine the positions of the feathers. The skin is also very vascular. Smooth muscles in the walls of the blood vessels regulate the blood flow through the skin and the degree to which body heat may be lost or conserved.

Although not organized as distinct glands, many of the epidermal cells of birds secrete lipids which help protect and waterproof the skin (Merton, 1984). The most conspicuous integumentary gland of birds is a single, branched, alveolar **uropygial gland** (Gr. *oura* = tail + *pyge* = rump). This is located above the base of the tail and produces fatty and waxy secretions that a bird spreads over its feathers as it preens. The gland is largest in aquatic species, which use its secretions to waterproof the feathers. It is absent from ostriches and a few other species. Beyond this, a few wax glands occur in the external ear canal.

The Skin of Mammals

The Epidermis and Its Derivatives

Mammals have a thick epidermis with a many-layered stratum corneum of dead and flattened cells filled with keratin bound to phospholipids (Fig. 5–13). It is

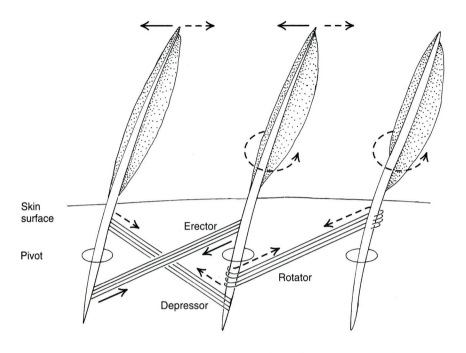

Skin surface

Erector

Pivot

Rotator

Depressor

FIGURE 5–12   Simplified diagram of feather muscles showing the erector running between the posterior margin below the pivot of the shaft of the feather and the anterior margin above the pivot of the next feather. The depressor runs between the posterior margin above the pivot to the posterior margin below the pivot of the next feather. The rotator wraps around the shaft of two successive feathers attaching below the pivot of one feather and above the pivot of the adjacent feather. Combined actions of these muscles can position the feathers in many ways.

effective in reducing water loss and other exchanges between the body and the external environment. Various transitional layers can be recognized between the stratum germinativum and stratum corneum as newly formed cells differentiate, accumulate keratin, and die. The stratum corneum is exceptionally thick and forms footpads on the soles of the feet and toes of many mammals.

Hair and Other Keratin Derivatives.   **Hair** is the distinctive keratinized derivative of mammalian skin. In furred species it affords mechanical protection and entraps a layer of still air that provides excellent insulation. Human body hairs are much reduced and are primarily tactile. A representative hair consists of a **shaft** of dead, cornified cells, the base of which is embedded in a **hair follicle** in the dermis (Fig. 5–13). Most of the follicle wall is composed of epidermal cells that grow down into the dermis with the developing hair. The epidermal part of the follicle is surrounded by fibrous dermal tissue containing nerve endings. The base of the follicle is enlarged, forming a **root** into which a conical **hair papilla** protrudes. The papilla contains nerve endings and many capillaries that nourish the devel-

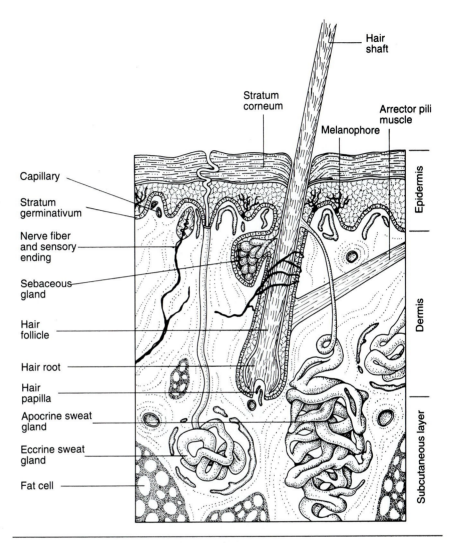

FIGURE 5–13   A diagram of a vertical section through human skin.

oping hair. As new cells are formed by mitosis just above the hair papilla, they elongate and are added to the base of the hair shaft. A hair shaft usually has a medulla of shrunken, dead cells, a thickened cortex that imparts most of the strength, and an outer cuticle of overlapping, scaly cells (Fig. 5–14). The pattern of the cuticular scales differs enough among mammals that it can be used to distinguish many species.

Hair color is the result of pigment within the cells of the cortex. White is caused by the absence of pigment together with the refraction of light from air spaces between the cells of the medulla. The amount of pigment determines the shade of hair color.

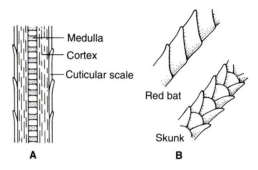

Red bat

Skunk

A   **B**

FIGURE 5–14   The structure of hair: **A**, A longitudinal section of a hair. **B**, Surface views of variation in cuticular scales. (After Storer.)

Each follicle and hair shaft usually lie somewhat oblique to the skin surface. A group of smooth muscle fibers of dermal origin form an **arrector pili muscle** (L. *arrectus* = set upright + *pilus* = hair) that originates from the epidermis and attaches to one side of the follicle in such a way that its contraction can pull the hair upright. These muscles are under the control of the sympathetic nervous system and often contract under stress. Human goose-bumps and the hackles that rise along the back of the neck when two hostile dogs confront each other are familiar examples. As temperature in the external environment drops, the change is detected by peripheral sensory receptors in the skin (see p. 441), and messages are relayed to the central centers in the spinal cord and brain. Responses are then sent via the sympathetic nerves to the arrector pili muscles, raising the hairs and thereby increasing the thickness of the layer of insulating air trapped near the skin surface. Regulating the thickness of the insulating air layer helps mammals maintain a relatively constant internal temperature through a wide range of temperature changes. The degree of blood flow through the skin is also a factor (p. 170).

Hairs vary greatly on different parts of the body. Our fine body hairs lack a medulla and grow very little after they are formed, whereas scalp hair grows continuously. Hair life ranges from a few months for hair in our armpits to nearly four years for scalp hair. Growth stops when mitosis ceases in the root. After a period of quiescence, during which the base of the old hair shaft separates from the root, mitosis resumes and the newly forming shaft pushes out the old hair.

Hairs also differ greatly among species. Most furred mammals have long **guard hairs** that protect a denser coat of soft **underhair**. Guard hairs are modified as **quills** in echidnas, hedgehogs, and porcupines. Long, tactile whiskers, **vibrissae**, occur on the snouts of cats, dogs, rats, and many other mammals (p. 407). Patterns of hair replacement vary. Kangaroos, foxes, and some other species molt once a year. Mink and weasels molt twice a year and have different colored winter and summer pelages.

Hair is nearly absent in cetaceans, where its insulating function is performed by a layer of subcutaneous blubber (p. 171). It is also reduced in elephants, rhinoceroses, hippopotamuses, and some other large mammals living in hot climates.

Zoologists believe that hair is a unique derivative of the epidermis and did not evolve from horny scales. Unlike horny scales and feathers, hair develops from an

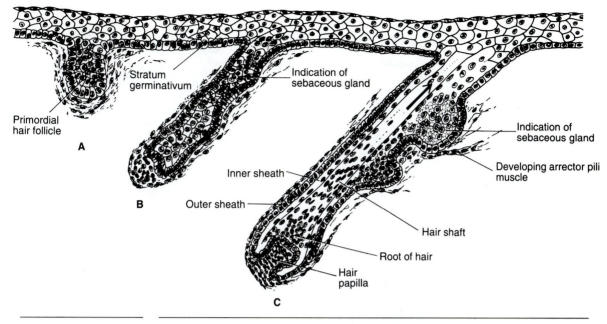

Stratum germinativum

Indication of sebaceous gland

Primordial hair follicle

A

B

Inner sheath

Outer sheath

Indication of sebaceous gland

Developing arrector pili muscle

Hair shaft

Root of hair

Hair papilla

C

FIGURE 5–15   Diagrams of hair development. (After Corliss.)

ingrowth of the epidermis into the dermis (Fig. 5–15). Both hairs and scales are present on the tails of some rodents. The hairs emerge in groups of three to five between the scales. This pattern of hair clustering can sometimes be seen on parts of the body where there are no scales. Maderson (1972) has proposed that hairs evolved as tiny outgrowths between the scales that may at first have had a tactile function. As these processes became more numerous, they began to serve as insulation.

Claws are present in most mammals. Claws are modified as nails in primates and as hooves in ungulates (Fig. 5–16).

Some mammals have other cornified derivatives of the skin. The **horns** of sheep, cattle, and African antelopes, which are used in defense and courtship, consist of a core of dermal bone covered by a thick layer of keratinized cells that develop from the stratum germinativum (Fig. 5–17). Horns usually occur in both sexes, grow and wear away continually, and do not branch. They are not shed. The pronghorn of western North America has a unique type of horn with one branch and a covering that is shed annually. Horns should not be confused with **antlers** that perform similar behavioral functions. The branched antlers of deer and related mammals are bony outgrowths of the skull that are covered by skin, the velvet, only during their growth. As the antlers mature, the velvet is sloughed off. Antlers are usually found only on males; caribou are an exception. After the mating season, bone is reabsorbed at the base of the antlers, and they fall off. The horn of the rhinoceros is yet another type. It consists of a solid mass of hairlike, keratin fibers cemented together. It is tragic that its alleged magical and aphrodisiac properties have encouraged poachers and placed rhinoceroses on the endangered species list.

The large, toothless whales have plates of **baleen** that hang down from each side of the palate and filter plankton from mouthfuls of water (Fig. 5–18). Each

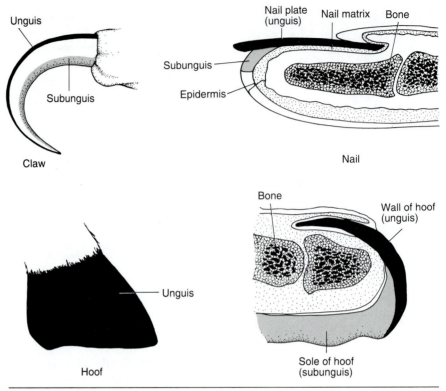

FIGURE 5–16   Surface view of a claw, sagittal section of a fingernail, surface view and sagittal section of a hoof. (After Waterman.)

plate is composed of hairlike keratin fibers that are cemented together at the base of the plate but free distally. In the early 20th century, baleen plates were used as corset stays.

Glands.   Multicellular cutaneous glands became abundant during the evolution of mammals. **Sebaceous glands** (L. *sebum* = tallow) are branched alveolar glands (Fig. 5–13) that produce oily and waxy secretions, which are discharged when the cells producing the materials break down. Sebaceous glands usually open into the hair follicles, lubricating and waterproofing the hairs, but those on the nipples and some other parts of the body open directly to the surface. Long sebaceous glands, the **tarsal glands**, open on the edge of the eyelids. Their oily secretion helps protect the surface of the eyeball and coats the rim of the eyelid so that tears do not overflow.

Sweat glands occur in most mammals, but they usually are not so widely distributed on the skin surface as in human beings. Often they are limited to the snout, the tail base, or the soles of the paws. Cetaceans, sea cows, and a few other mammals have none. Sweat glands are coiled tubular glands (Fig. 5–13) whose secretions are released without cell destruction. The contraction of **myoepithelial**

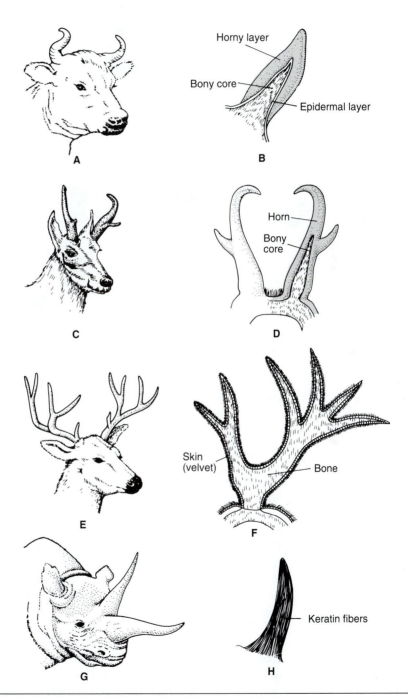

FIGURE 5–17 Types of horns and antlers as seen in surface view and longitudinal section. **A** and **B**, cow. **C** and **D**, pronghorn. **E** and **F**, deer. **G** and **H**, rhinoceros.

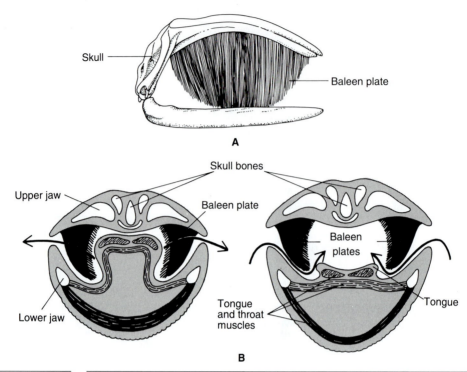

FIGURE 5–18 The baleen plates of a toothless whale: **A,** Lateral view of the skull and plates. **B,** Transverse sections through the mouth and plates with the tongue elevated as water is expelled,
and depressed as water and food are drawn in. Arrows indicate the feeding current. (**A,**
after Vaughan. **B,** after Portmann.)

cells around the secretory cells helps discharge the sweat. **Eccrine sweat glands**
secrete a watery solution that is discharged directly to the skin surface. Although
sweat may contain some urea and salts, its primary function is evaporative cooling
of the surface. **Apocrine sweat glands** are similar morphologically but discharge
a thicker and frequently odoriferous secretion into the hair follicles. Most are
limited to the armpits and genital area in humans. The **scent glands** of other
mammals, which may occur on any part of the body, are modified apocrine glands.
Their secretions are pheromones used in marking trails and territories, in courtship
behavior, and sometimes in defense. Some scent glands discharge into the anal or
urinary areas, making the pheromones a component of the feces or urine. We have
all seen dogs leaving identifying markers on hydrants and trees. The **wax glands**
of our external ears are also modified apocrine glands.

Mammary glands probably evolved from sweat or sebaceous glands, but it
is uncertain from which. Mammary glands resemble sebaceous glands in being
compound alveolar glands, and they resemble sweat glands in having myoepithelial
cells. They develop embryologically from a pair of continuous epithelial milk ridges
that grow down into the dermis and extend from the armpit to the groin (Fig. 5–
19**A** and **B**). The definitive glands may develop along most of the ridge, as in the
long series of mammary glands in pigs, or from limited parts of it. Glands develop

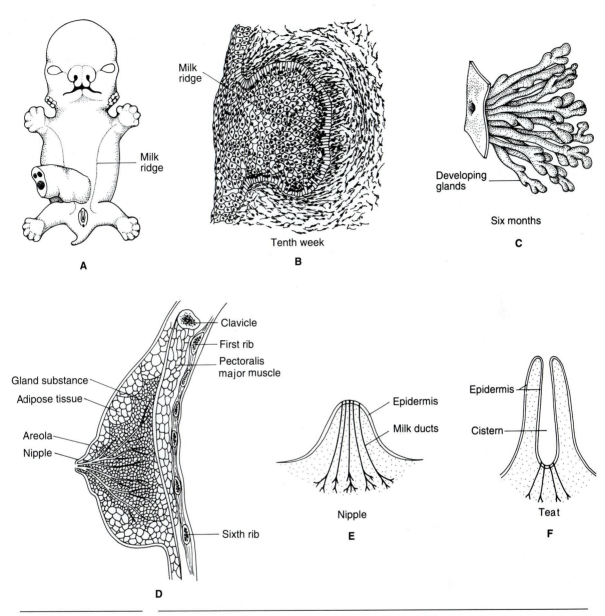

FIGURE 5–19   Mammary glands: **A–C**, The development of a mammary gland from the milk ridge. **D**, A mature human mammary gland. **E**, A vertical section through a nipple. **F**, A vertical section through a teat. (**A–C**, after Corliss. **D**, after Kluge. **E** and **F**, after Rand.)

near the groin in ungulates; primates have a single pair of pectoral glands. Extra patches of glandular material may develop abnormally along parts of the milk ridge that usually atrophy.

The mammary glands contain considerable adipose tissue along with secretory cells. They secrete milk under hormonal influence only during lactation (p. 544).

Milk contains water, carbohydrates, fat, protein, various minerals, and antibodies. Parts of the duct system are enlarged and store milk until the young are fed. The ducts open to the surface in different ways. In monotremes the ducts discharge into a small depression on the belly. The parent rolls over and, since monotremes have horny beaks rather than fleshy lips and cheeks and cannot suck efficiently, the young lap the milk from the hairs. Both male and female monotremes have functional glands. In other mammals the functional glands are restricted to females, and the ducts open onto nipples or into teats (Fig. 5–19**E** and **F**). Many ducts reach the surface of **nipples**. A **teat**, as in a cow, is formed by the elevation of a collar of skin around a large cistern into which the ducts discharge. The teat itself has a single opening from the cistern. Mammary glands are a particularly important derived feature of mammals, for they necessitate a close association between the young and mother. This association provides young mammals more time to learn to cope with the rigors of their world, and it plays a key role in the evolution of social structure in mammals.

Vitamin D Synthesis.   Mammalian epidermis is also the site for the synthesis of vitamin D, the "sunshine" vitamin. Under the influence of ultraviolet rays, provitamin **7-dehydrocholesterol** in epidermal cells is converted to **vitamin D**. Vitamin D plays a key role in the absorption of calcium from the intestine and the deposition of calcium and other minerals in bone. The rate of vitamin D synthesis depends on the degree of skin pigmentation, for heavy pigmentation limits the amount of ultraviolet radiation that penetrates the skin. Many investigators believe that different degrees of pigmentation among people living in different parts of the world are adapted to the amount of ultraviolet radiation. This helps explain the marked correlation between the heavy pigmentation of people living at equatorial latitudes, and the lighter pigmentation of people living at higher latitudes. Normal suntanning in the summer in lightly pigmented people may be a mechanism to maintain a constant rate of vitamin D synthesis despite the seasonal increase in ultraviolet radiation.

The Dermis and Subcutaneous Tissue

The dermis of mammals is essentially the same as that of other vertebrates (p. 145). Except for the dermal bony elements of the skull and pectoral girdle, dermal bone is not common in mammalian skin. Bone is found under the keratinized epidermal plate of armadillos. In the related, extinct glyptodonts, most of the body was sheathed by dermal plates that formed a thick shield several centimeters thick.

Mammalian dermis has a particularly extensive vasculature (Fig. 5–20). The networks of arterioles, capillaries, and venules play important roles in thermoregulation and blood pressure control. Precapillary sphincters in arterioles, and postcapillary sphincters in the venules, regulate capillary blood flow. A free flow of blood in the dermal capillaries, which possess a very large surface area, allows heat to radiate from the body, thereby cooling it. When heat needs to be conserved, the sphincters constrict and blood flow through the capillaries is reduced greatly. The degree to which the hairs are elevated also affects heat loss from the surface (p. 164).

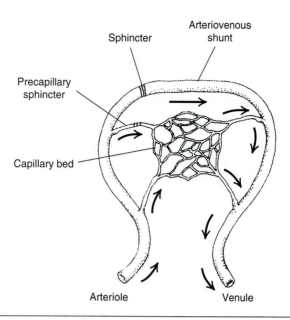

FIGURE 5–20   A simplified scheme of the dermal vascular arteriovenous shunt and capillary beds with the various sphincters.

The dermal vasculature helps regulate blood pressure in the body as a whole because various shunts and ateriovenous anastomoses in the skin affect blood volume in other parts of the body. If body blood pressure rises, arteriovenous shunts close and precapillary sphincters in the dermis open, allowing a greater volume of blood to flow through the extensive capillary beds. This reduces the volume of blood flowing through other parts of the body, which results in a general lowering of blood pressure. When blood pressure in the body drops too low, the arteriovenous shunts in the dermis open. Blood is shunted away from the skin, increasing blood volume in the rest of the body, and blood pressure rises.

A layer of subcutaneous tissue, which is composed of bundles of connective tissue fibers interspersed with fat, lies between the dermis and the muscles of the body wall. This layer is far thicker in birds and mammals than in other vertebrates and helps insulate the body. It forms an exceptionally thick layer of **blubber** in cetaceans and other marine mammals. The blubber represents 40 percent or more of body weight in some of the larger whales, and its fat content is 40 to 60 percent. The blubber of whales helps streamline body shape, provides buoyancy, and is an important fuel reserve. Migrating whales do most of their feeding in feeding grounds close to the polar regions, where cold water and upwelling of minerals make plankton and other food abundant. Blubber thickness increases greatly during this period. Whales feed less frequently, or not at all, as they migrate to their temperate breeding grounds. They utilize the fuel in blubber, and the thickness of this layer decreases significantly. Blubber is important insulation for these animals in their cold feeding grounds. A thinning of the blubber when the whales enter warmer waters allows more body heat to be lost and prevents heatstroke.

Heat loss is a problem for whales because their skin is far less vascular than that of other mammals, and they do not sweat or pant.

The Evolution of Vertebrate Skin

As we have seen, the earliest known vertebrates—the ostracoderms—had a dermal armor of bony plates. This primary dermal armor underwent radical changes during the evolution of the vertebrates. The first transformation occurred in the placoderms, in which the armor broke up into smaller units, the denticulated cosmoid scales (Fig. 5–21). With the evolution of teeth from scales in the mouth, the placoderms could perform new predatory functions, occupy new feeding niches, and assume new habits. The selective advantage of teeth was so great that they were retained throughout the subsequent adaptive radiation of vertebrates, with the notable exception of modern birds. In the Chondrichthyes the dermal denticles of the denticulate cosmoid scales became transformed into placoid scales.

The denticulate cosmoid scales lost the superficial dermal denticles and became adenticulate in ancestral bony fishes. The adenticulate cosmoid scales of the sarcopterygians gave rise to the dermal pectoral girdle and dermatocranium, both of which are retained but modified in the subsequent evolution of the vertebrates. Adenticulate cosmoid scales also gave rise to the osteoderms and gastralia in some reptiles. Within the actinopterygian fishes, the adenticulate cosmoid scales became transformed into ganoid scales, which, in turn, became modified successively into cycloid and ctenoid scales. The evolutionary history of the dermal skeleton is that of gradual morphological transformations accompanied by drastic functional changes. Thus the original protective function of the adenticulate cosmoid armor was changed into a locomotory function with the establishment of the dermal pectoral girdle, which resulted in a shift of the sarcopterygians into an entirely new adaptive zone with an array of new selection pressures. Once the pectoral girdle had emerged, it led to an avalanche of evolutionary diversifications in the locomotory apparatus of the vertebrates.

Feathers and hair are often considered key evolutionary innovations, which can be defined as any changes in structure and function of an organ that permit organisms to assume radically new ways of life. Feathers and hair are the essence of endothermy of birds and mammals, respectively. Endothermy is expressed in a high body temperature that is maintained at nearly a constant level. This made possible a higher level of metabolic activity, which is a prerequisite for flight in birds, and enabled mammals to be nocturnally active in a world dominated by diurnal reptiles. Constant high metabolic rates enable the nervous system to function more rapidly, enhancing the rate of information processing and, thereby, prey capture and predator avoidance. Endothermy conveys significantly greater stamina and capacity to produce continuous muscle contractions.

Feathers are unquestionably derived from reptilian scales. It is reasonable to accept the hypothesis that reptiles ancestral to the birds already had feathers that functioned originally in thermoregulation or in a way not connected with flight. These ancestral feathers represented one of the building blocks that led to the new function: flight. This example explains how an incipient structure could be

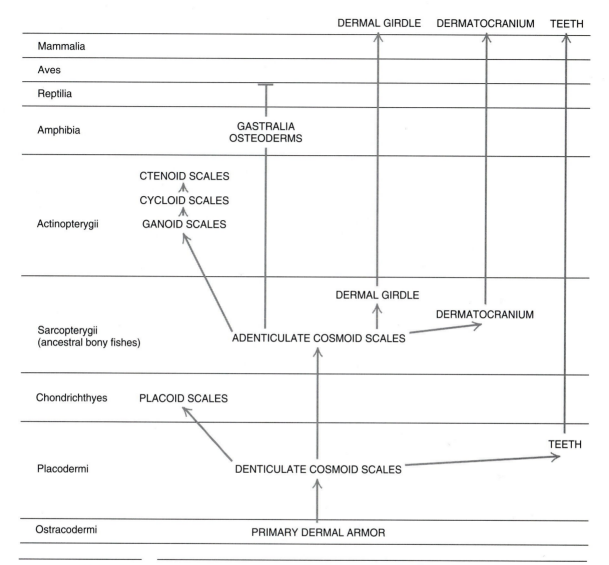

FIGURE 5–21   A simplified evolutionary history of the dermal armor and its derivatives in the major verte-brate groups. (Modified from H. M. Smith.)

favored by natural selection before reaching the elaboration for a radically new role. Similarly, it can be argued that mammalian hair originally functioned as tactile structures. These tactile structures became converted into devices (hairs) for endo-thermy. Hair does not fossilize, so we can only speculate on early stages of its evolution. Hair probably conveyed such a high selective advantage that it evolved very rapidly without intermediate stages. Thus feathers and hair may originally have functioned in a different way than their current primary function. Feathers, originally of value in thermoregulation, were also **preadapted** for flight; and hair, originally tactile, was also preadapted for thermoregulation. An organ is thought to be preadaptive when an adaptive feature originated in a context different from its eventual function. The organ was then redeployed by evolution as a result of having been fortuitously suited to the new function (Skelton, 1985).

SUMMARY

1. The skin forms the interface between a vertebrate and its outside world. It protects the body in many ways, contains receptors, participates in some species in gas, water, and ion exchanges, sometimes helps nourish young, has a thermoregulatory role in birds and mammals, and participates in vitamin D synthesis and blood pressure regulation in mammals.

2. The skin consists of an epithelial epidermis of ectodermal origin and a connective tissue dermis derived from the dermatomes of the somites. Neural crest cells migrate between these layers, form chromatophores, and participate in many of the inductive interactions between the dermis and the epidermis that lead to the formation of skin derivatives: scales of many types, feathers, hair, glands.

3. Skin color results primarily from the combination of chromatophore types and the degree of dispersal of pigment within them. Color can change morphologically, depending on the amount of pigment synthesized, and physiologically (in anamniotes and reptiles) as pigment migrates within the melanophores.

4. Skin color is important in the life of most vertebrates. Color may help conceal an individual, advertise an individual's presence, serve for species recognition, and send signals for courtship and establishing territories. Color also protects deeper body tissues from ultraviolet radiation, and it has a thermoregulatory role in some reptiles.

5. Fishes live in a more stable environment than terrestrial vertebrates and have a simpler skin structure. Typically, no cornification occurs, and most glands are unicellular mucous glands.

6. Most fishes have bony scales that develop in the skin from dermal-epidermal interactions. The scales of ancestral vertebrates were thick structures overlain by dentine and enamel-like materials. These scales may have offered mechanical protection and served as a reservoir for calcium and phosphate ions, and the enamel-like covering may have reduced water exchanges between the animal and its environment.

7. Scales became much thinner and lighter in later fishes and were lost in some groups. Cartilaginous fishes have only dermal denticles. Early actinopterygians had ganoid scales, and early sarcopterygians had cosmoid scales, but these are reduced to thin disks of bone (cycloid or ctenoid scales) in contemporary species.

8. Some cornification occurs in amphibian skin, and their skin glands are multicellular mucous or poison glands. Bony scales have been lost in most contemporary groups.

9. Heavily cornified skin and horny scales evolved in reptiles. Most species shed

their scales periodically. Keratin and phospholipids in the cells of the scales reduce evaporative water loss through the skin. Osteoderms, some of which form riblike gastralia, are present in many reptiles. Apart from scent glands in some species, reptile skin is aglandular.

10. Bird skin is characterized by the presence of feathers, which are homologous to horny scales. Feathers protect the body surface, reduce evaporative water loss, and insulate the body. The large flight feathers on the tail and wings form the flying surfaces. No dermal ossifications are present in bird skin. Distinct skin glands are generally absent except for a uropygial oil gland above the tail base and wax glands in the external ear canal.

11. Hair is the most obvious derived feature of mammalian skin. Hair protects the body surface and, with subcutaneous fat, is a thermal insulating layer. Hair distribution over the body and its modifications vary greatly among mammals.

12. Claws are retained in most mammals but are modified as nails in primates and as hoofs in ungulates.

13. Other keratinized derivatives of the skin are horns of various types (not to be confused with bony antlers) and the baleen plates of the toothless whales.

14. Numerous glands evolved in mammal skin. These include sebaceous glands, sweat glands, and mammary glands, by which females nourish their infants. Nutrition of the young by mammary glands requires a longer association between mother and offspring than occurs in most other vertebrates.

15. Mammalian epidermis is the site for the ultraviolet-light-mediated conversion of a provitamin into vitamin D.

16. The dermis of mammals has an extensive and complex vasculature. The degree to which blood flows through the dermal capillaries, or bypasses them through arteriovenous shunts, affects thermoregulation and general body blood pressure.

17. During the evolution of vertebrate skin, the primitive dermal armor of ostracoderms broke up into smaller denticulated cosmoid scales in placoderms. Certain of these scales became teeth, and teeth enabled vertebrates to occupy new feeding niches and habitats. The superficial parts of the denticulated scales became the placoid scales of cartilaginous fishes. The ancestral, denticulated scales evolved into a wide variety of scales in bony fishes and formed the dermal parts of the pectoral girdle, the dermal parts of the skull, and the osteoderms of early tetrapods.

18. Feathers and hair are key evolutionary innovations that enabled birds and mammals to become endothermic. Endothermy opened new niches to birds and mammals and made possible many changes in their mode of life.

REFERENCES

Bagnara, J. T., and Hadley, M. E., 1973: *Chromatophores and Color Change.* Englewood Cliffs, N.J., Prentice-Hall.

Bemis, W. E., and Northcutt, R. G., 1992: Skin and blood vessels of the snout of the Australian lungfish, *Neoceratodus forsteri*, and their significance for interpreting the cosmine of Devonian lungfishes. *Acta Zoologica* (Stockholm), 73:115–139.

Cott, H. B., 1957: *Adaptive Coloration in Animals*, 2nd edition. London, Methuen and Company.

Kawauchi, H., Kawazoe, I., Tsubokawa, M., Kishida, M., and Baker, B. I., 1983: Characterization of melanin-concentrating hormone in chum salmon pituitaries. *Nature*, 305: 321–323.

Kiaer, J., 1928: The structure of the mouth of the oldest known vertebrates, pteraspids and cephalaspids. *Paleobiologica*, 1:117–134.

Kemp, N. E., 1984: Organic matrices and mineral crystallites in vertebrate scales, teeth and skeletons. *American Zoologist*, 24:965–976.

Lillie, F. R., 1942: On the development of feathers. *Biological Reviews of the Cambridge Philosophical Society*, 17:247–266.

Maderson, P. F. A., symposium organizer, 1972: The vertebrate integument. *American Zoologist*, 12:12–171.

Merton, G. K., 1984: *Glandular functions of avian skin; An Overview*. Journal of the Yamashing Institute of Ornithology, 16:1–12.

Montagna, W., and Parakkal, P. F., 1972: *The Structure and Function of the Skin*, 3rd edition. New York, Academic Press.

Ørvig, T., 1957: Paleohistological notes. I. On the structure of the bone tissue in certain Palaeonisciformes. *Ark. Zool.*, 10:481–490.

Parakkal, P. F., and Alexander, N. J., 1972: *Keratinization: A Survey of Vertebrate Epithelia*. New York, Academic Press.

Rao, K. R., and Fingerman, M., symposium organizers, 1983: Chromatophores and color changes. *American Zoologist*, 23:461–592.

Saidel, W. M., 1988: How to be unseen: An essay in obscurity. *In* Atema, J., Fay, R. R., Popper, A. N., and Tavolga, W. N., editors: *Sensory Biology of Aquatic Animals*. New York, Springer-Verlag.

Skelton, P. W., 1985: Preadaptation and evolution in rudist bivalves. *Paleontology*, 33:159–173.

Storer, T. I., 1943: *General Zoology*. New York, McGraw-Hill Book Company.

6 Properties and Mechanics of Structural Materials

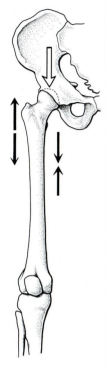

PRECIS

We introduce the skeletal system in this chapter by exploring the properties of cartilage, bone, and other structural materials, and by considering the mechanical principles that govern their interaction and distribution. After discussing the types of stresses that structural materials must meet, and the resulting strains, we describe the properties and growth of structural materials. The individual cartilages and bones are united by joints and form lever systems that transmit muscle forces to some point where the force generated by muscles is applied: another skeletal element, teeth, or the feet. Since these forces are vector quantities, they can be resolved into components acting in different directions. Conversely, the components can be combined to give a resultant force. The properties of structural materials, and the mechanical principles we discuss, affect the distribution of structural materials in the body.

Outline

Forces Acting on the Body
Properties and Growth of Structural Materials
    Hydroskeletons, the Notochord
    Dense Connective Tissues
    Cartilage
    Bone
    Bone Growth
    Other Functions of Bones
Joints
Lever Systems
Components and Resultants of Forces
Distribution of Materials

Bone, cartilage, and other structural materials form a skeletal framework that maintains body shape, supports the body against the pull of gravity, resists shortening when muscles contract, and transfers muscle forces to the appendages, jaws, and other points where the forces are applied. Movement requires that the structural materials either bend or form discrete units or elements—the individual bones and cartilages—that are linked together at joints. Strong connective tissue ligaments bind skeletal elements together at joints, and connective tissue tendons attach the muscles to the bones and cartilages. In this chapter we will explore the properties of these structural materials and the mechanical principles that govern their interactions and distribution. This will serve to introduce the skeletal and muscular systems, which are responsible for the support and movement of the body.

Forces Acting on the Body

To analyze the nature and distribution of structural materials, we first need to understand the nature of the forces that act on and within the body. Physicists define **force** as mass multiplied by acceleration. The push or pull needed to accelerate 1 kg 1 meter per second per second is known as 1 **newton.** In skeletal systems, forces are represented by **loads** such as the weight of the body pushing down on joint surfaces or feet on the ground, the pull of muscles on their attachments, or the bite force between teeth of the upper and lower jaws. Since these and similar forces are concentrated at surfaces, it is convenient to express them as force per unit area (e.g., grams per square centimeter, or g/cm^2). The force per unit area is known as the **pressure** or **stress.** When, for example, the limb bones support the weight of the body and at the same time resist the opposite but equal force that the ground exerts on the legs, the bones are stressed and tend to be deformed. The deformation resulting from stress is known as **strain.**

The major stresses and the strains that they generate in structural materials are shown in Figure 6–1. Two parallel forces moving directly toward each other subject the material to **compression** stress. The resulting strain is expressed by a shortening and widening of the material. Two parallel forces pulling directly away from each other subject the material to **tension** stress. Strain resulting from tension is represented by the lengthening and narrowing of the material. Two parallel forces moving toward each other, but not directly opposite each other, subject the material to **shear.** This causes one part of the material to slide relative to another part. Rotational forces applied in opposite directions induce **torsion** and cause the material to twist. Although one type of stress may predominate, many of the structural elements of the body are subject to a complex combination of these forces as they support the weight of the body and resist the pull of muscles that attach to the bones. Bending, for example, results from the interaction of compression and tension. These forces, as well as genetic factors, affect the type of structural material utilized (e.g., ligament, cartilage, bone), the shape of the structural element, and the distribution of materials within it.

The deformation resulting from stress can be expressed graphically in a **stress-strain curve** (Fig. 6–2). In a rigid material like bone, strain, or deformation, at

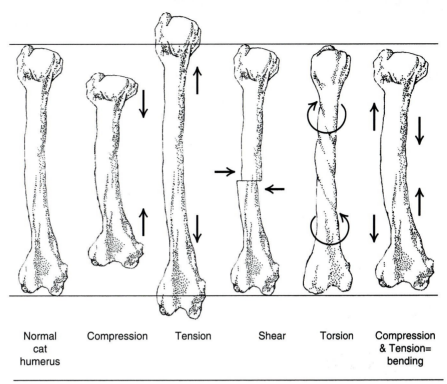

| Normal cat humerus | Compression | Tension | Shear | Torsion | Compression & Tension= bending |

FIGURE 6–1   The major stresses to which materials are subjected, and the resulting deformations, as seen in the humerus of a cat. A normal humerus is shown at the left, and those deformed by the stresses indicated by arrows are shown to the right. The deformations have been exaggerated.

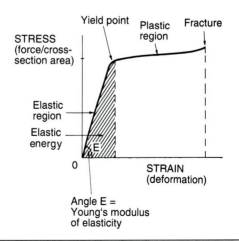

FIGURE 6–2   A stress-strain curve for bone under tension or compression stress. (After Currey.)

first increases only slightly as stress increases greatly. This portion of the curve is known as the elastic region because the bone will return to its original shape when the stress is released; that is, both stress and strain return to zero. The area under this portion of the curve represents the elastic energy that caused the deformation, and most of this energy is recovered when the stress is released. When stress reaches the yield point of the material, a further increase in stress causes considerable deformation or strain, and the bone behaves plastically. When stress is released and falls to zero, the strain does not return to zero; that is, the bone is permanently deformed. Forces in bones normally lie within the elastic region of the curve, but bones behave plastically over a considerable strain before they reach the fracture point. The slope of the curve in the elastic region represents **Young's modulus of elasticity,** which also can be defined as stress divided by strain. Rubber, which deforms greatly under low stress, has a low modulus, whereas bone, which deforms only slightly under high stress, has a high modulus. The modulus of elasticity applies only to stresses up to the yield point of the material.

Properties and Growth of Structural Materials

The structural materials of the body differ with respect to the stress or combination of stresses they are adapted to meet, and this is an important factor that determines their distribution and use in the body.

Hydroskeletons, the Notochord

Water, when it is limited to a confined space that is surrounded by firm tissues, is a structural material in the sense that it forms a firm but flexible **hydroskeleton.** The proboscis and collar coelom of enteropneust hemichordates can be filled with sea water and act as a hydroskeleton during burrowing (p. 27). The **notochord** of protochordates and embryonic vertebrates, which is a longitudinal, dorsal rod of turgid cells, is their primary structural element. The notochord is a hydroskeleton that prevents the body from being shortened or compressed by the contraction of the myomeres, yet allows the body to bend from side to side and allows power swimming movements (p. 34). The resistance of the notochord to compression derives from the incompressibility of its liquid-filled cells, which are surrounded by a firm connective tissue sheath. As development proceeds, the vertebral bodies develop around the notochord, and the notochord usually is greatly reduced or disappears. It persists as a well-formed structure in the adults of hagfishes, lampreys, and a few other fishes and early tetrapods. In fossil lungfishes, rhipidistians and early labyrinthodonts, for example, the large space extending through the centra of the vertebral column probably lodged a notochord.

Dense Connective Tissues

Other structural materials are specialized forms of connective tissue. In its most common form, connective tissue is a loose web of cells and extracellular fibers that permeates all of the organs of the body, binding other tissues together. We find it beneath the skin, in muscles, and in the walls of blood vessels and digestive

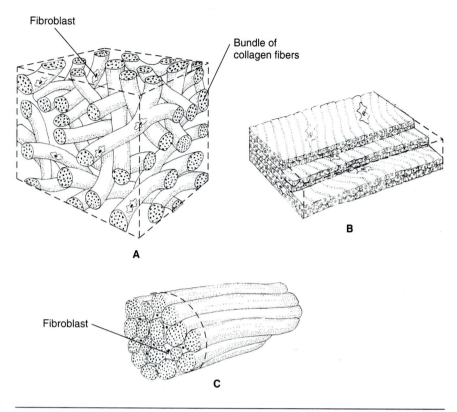

FIGURE 6–3   Arrangement of bundles of collagen fibers in some dense connective tissues: **A**, dense irregular connective tissue; **B**, a ligament; **C**, a tendon. (After Williams et al.)

organs. **Loose connective tissue** is produced by cells known as **fibroblasts** that extrude around themselves an extensive **matrix.** The matrix consists of a viscous **ground substance** interspersed with numerous **collagen fibers.** Collagen is a protein consisting of filamentous molecules of **tropocollagen** that are organized first into collagen microfibrils, which are visible only with the electron microscope, and finally into the collagen fibers visible by normal microscopy. Cross linkages between tropocollagen molecules give collagen a very high tensile strength, yet it is flexible and bends easily. Collagen itself has a relatively low elasticity, but some connective tissue incorporates fibers of another protein, **elastin,** that can stretch and recoil to its normal length.

Fibroblasts are less numerous in the dense connective tissues than in loose connective tissues, and the collagen fibers form densely packed bundles (Fig. 6–3). **Dense irregular connective tissue,** which forms capsules around the kidney and many other organs, consists of an irregular array of collagen bundles. Ligaments and tendons are composed of very tightly packed bundles of collagen fibers. Collagen fibers in **ligaments** often form layers, with fibers going in different directions. Ligaments unite bones at joints, holding them together yet permitting movement. In **tendons** the bundles have a more cable-like orientation. Tendons unite the muscles to bones and cartilages. Because of the high tensile strength of collagen, ligaments and tendons resist tension and twist very well, but they bend

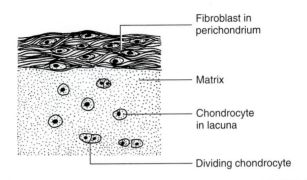

Fibroblast in
perichondrium

Matrix

Chondrocyte
in lacuna

Dividing chondrocyte

FIGURE 6–4   Hyaline cartilage.

easily and do not resist compression. Because of this, ligaments and tendons are binding, not supporting structures.

Cartilage

Both cartilage and bone are dense connective tissues whose matrix contains, in addition to collagen fibers, special molecules that give them rigidity. Their rigidity enables cartilage and bone to resist compression well and hence provide support. Cartilage is nearly as strong as bone in resisting compression, but it does not resist tension and shear as well. It has a lower modulus of elasticity than bone, and so is more flexible and elastic. The cellular elements of cartilage, known as **chondrocytes** (Gr. *chondros* = cartilage + *kytos* = cell), lie in small spaces, the **lacunae** (L. *lacuna* = pit), within a firm and extensive extracellular matrix that the chondrocytes have extruded (Fig. 6–4). The matrix consists of collagen and elastic fibers embedded in a ground substance that contains a great deal of water. Some of the water is free, but much is combined in a unique way with giant molecules known as **proteoglycans.** The resistance of cartilage to compression, its flexibility, and its smoothness derive from the water within it. When the cartilage is under compression, some of the free water is expelled and contributes to the lubrication at the joint surfaces. The cartilage becomes thinner. When loads are released, the osmotic pressure of molecules within the matrix draws water back in, and the cartilage returns to its original thickness. Because of these changes in the cartilage, humans who spend a great deal of time standing or walking are approximately 1 cm shorter at the end of the day than at the beginning. Unlike other connective tissues, cartilage seldom has a vascular or nerve supply. The metabolic rate of cartilage is low, and most cartilaginous elements are small enough to be supplied by diffusion from adjacent tissues. Cartilage canals containing blood vessels enter the larger cartilages.

Because their matrix contains several types of materials and spaces that lodge the cellular elements, cartilage and bone are **composite materials.** Composite materials are much stronger than homogeneous materials such as glass. A crack can propagate rapidly through a homogeneous material, but the force at the apex of a crack spreading through a composite material becomes blunted and dissipates

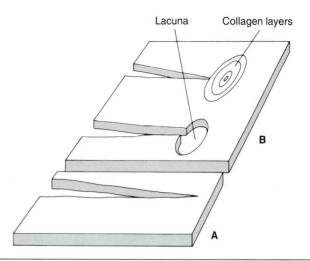

Lacuna   Collagen layers

B

A

FIGURE 6–5   The propagation of a crack through a homogenous material (**A**) and its interruption by a space or a change of material in a composite material (**B**). (After Lanyon and Rubin.)

when it impinges on a different material, such as a space containing a cartilage or bone cell (Fig. 6–5). Fiberglass, which consists of minute glass fibers embedded in resin, is a familiar example of a composite material.

The most common type of cartilage, **hyaline cartilage** (Gr. *hyalos* = glass), occurs on the ends of the limb bones, on the ventral ends of the ribs, and forms the cartilages of the larynx and tracheal rings. The matrix is translucent. Other types of cartilage are distinguished from hyaline cartilage primarily by the type and amount of the fibrous component of the matrix. **Elastic cartilage** of the mammalian external ear and epiglottis contains a particularly dense network of branching elastin fibers. Elastic cartilage is yellowish in fresh preparations. **Fibrocartilage** contains a great many collagen fibers and often grades into the dense connective tissues of tendons and ligaments. It is found in the intervertebral disks as well as the pubic symphysis, and often is associated with the attachment of tendons and ligaments.

Calcium salts are often deposited in cartilage, forming **calcified cartilage.** Although calcified cartilage has some of the rigidity of bone, it is quite different from bone in its histological structure. Calcification of cartilage is one process that occurs during the replacement of cartilage by bone (p. 185). Calcified cartilage also occurs in parts of the vertebrae of many cartilaginous fishes, where it adds strength to the cartilage.

Cartilage grows from the recruitment of fibroblasts from the connective tissue **perichondrium** (Gr. *peri* = around) that covers the cartilage, and internally by the mitotic division of chondrocytes. Daughter chondrocytes separate and synthesize more matrix. Because of the ease and speed with which cartilage grows without requiring complex remodeling, it is an excellent embryonic skeletal material. Most of the internal skeleton of a vertebrate embryo is cartilage. Later in the development of the majority of species, nearly all of the cartilage is replaced by bone; but some cartilage persists where its ease of growth, smoothness, or elasticity

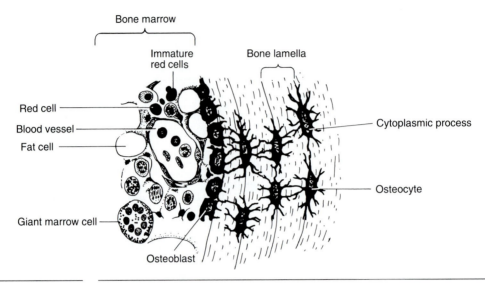

FIGURE 6–6   A microscopic section of bone to show its formation and histological structure. This section of bone lies beside the marrow cavity. (After Corliss.)

are particularly important: at or near the ends of limb bones, on the articular surfaces of bones at movable joints, at the distal ends of the ribs, and so forth. The internal skeleton of adult chondrichthyan fishes is entirely cartilaginous, and considerable cartilage persisted in the skeletons of early bony fishes.

Cartilaginous fishes lack the hydrostatic swim bladder that allows bony fishes to float at any depth in the water with little muscular effort. They are relatively dense fishes that tend to sink. A skeleton of cartilage, which is less dense than bone, makes them slightly more buoyant.

Bone

Bone is a highly vascular, mineralized, dense connective tissue that is hard, resilient, and capable of slowly changing as forces on the body change during an individual's life. Bone is particularly strong because its mineral component imposes rigidity and resists compression, and its fibrous component provides some flexibility and resists tension and torsion.

Bone is the primary skeletal tissue of most adult vertebrates. Its histological structure is the same as in many of the bony scales of fishes (see Fig. 5–4). Fibroblasts transform into **osteoblasts** (Gr. *osteon* = bone + *blastos* = bud) that produce a characteristic matrix of collagen fibers and protein polysaccharides (Fig. 6–6). The first-formed matrix, known as **osteoid,** becomes calcified by the binding of calcium phosphate crystals to the collagen fibers. The crystals are in the form of **hydroxyapatite.** As matrix continues to be produced, some bone-forming cells become trapped as **osteocytes** within spaces in the matrix, the lacunae. Osteocytes are star-shaped cells whose radiating processes lie within minute canaliculi that permeate the matrix. The collagen fibers, which comprise about 40 percent of the dry weight of bone, form randomly organized bundles in the **woven-fibered bone** of young individuals. As an individual matures, the collagen bundles become organized into parallel sheets, or lamellae, with rows of osteocytes between the layers. Layers in **lamellar bone** commonly have different fiber directions, which adds to the strength of the bone.

Bones differ in their degree of calcification according to their functions. The mammalian femur, which is a load-bearing bone, has a high mineral content (67 percent) and is stiff. An antler, which does not support weight but must resist impact and shear when males butt and lock heads, is only 59 percent mineral. Because so many variables are involved (location, function, sex, age, and species), it is difficult to generalize about the strength of bone, but it is clear that bone is a strong material. Most bone can resist compression of 2000 kg/cm²; this is about four times the compressive strength of concrete. Compact bone resists tension of about 100 kg/cm². This is about the same as tendons and ligaments. However, tendons and ligaments have an advantage when tension is the only stress because they are more flexible and lighter. Compact bone has a greater resistance to shear than tendons or ligaments, about 800 kg/cm².

Bone Growth

If bone develops directly within connective tissue, it is called **membrane or dermal bone.** Many of the superficial skull bones and some of the pectoral girdle bones are membrane bone. Membrane bone begins to form as small rods, or **trabeculae** (L. *trabecula* = small beam), that coalesce when they meet (Fig. 6–7). As development continues, layers of bone are deposited around the periphery of the latticework of trabeculae.

Bone that forms around and within the cartilages of the embryonic endoskeleton, largely replacing them, is called **cartilage replacement bone.** This bone is composed of the same types of cells and matrix as dermal bone, but its architecture is frequently more complex, for it resists a more complex set of stresses. An endoskeletal bone appears first in the embryo as a condensation of mesenchyme. The mesenchyme chondrifies and forms a small, cartilaginous replica of the adult bone (Fig. 6–8). As development proceeds, the cartilage continues to grow, and bone also begins to form both within the cartilage (**endochondral bone)** and around its periphery (**perichondral bone).** Details of ossification patterns vary somewhat among vertebrates (Shapiro, 1992). The first ossification in the long bone of a mammal is the formation of a collar of perichondral bone around the middle of the shaft, or **diaphysis.** The bone-forming osteoblasts are derived from cells in the perichondrium, which now becomes the **periosteum.** Next, chondrocytes within the diaphysis enlarge and line up in rows. Some of the matrix is reabsorbed, but calcium is deposited in the remaining matrix near the enlarged chondrocytes. The chondrocytes break down, and the cartilage becomes honeycombed. Vascular connective tissue invades the area from the periosteum. Some of the fibroblast cells differentiate into osteoblasts, line up along the remnants of the matrix, and begin to form interlacing bony trabeculae. The spaces between the trabeculae are filled with bone marrow (p. 189).

Endochondral ossification proceeds from the center of the diaphysis toward each end. This process would soon replace the original cartilaginous primordium

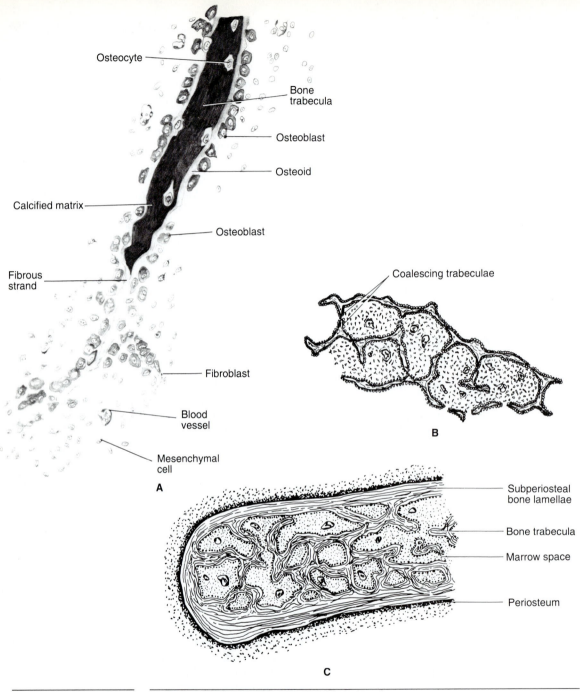

Osteocyte

Bone trabecula

Osteoblast

Osteoid

Calcified matrix

Osteoblast

Fibrous strand

Fibroblast

Blood vessel

Mesenchymal cell

A

Coalescing trabeculae

B

Subperiosteal bone lamellae

Bone trabecula

Marrow space

Periosteum

C

FIGURE 6–7   Three stages in the formation of membrane or dermal bone: **A**, Osteoblasts are lined up on a group of collagen fibers and are forming a bony trabecula. **B**, Trabeculae coalescing. **C**, Layers of bone are deposited around the periphery of a latticework of trabeculae to form a flat bone. (After Corliss.)

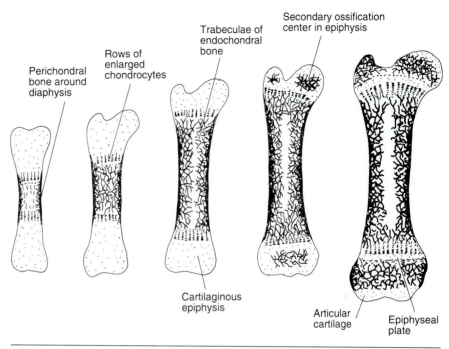

Perichondral
bone around
diaphysis

Rows of
enlarged
chondrocytes

Trabeculae of
endochondral
bone

Secondary ossification
center in epiphysis

Cartilaginous
epiphysis

Articular
cartilage

Epiphyseal
plate

FIGURE 6–8   The formation of cartilage replacement bone around (perichondral bone) and within (endochondral bone) the femur of a mammal. Cartilage is lightly stippled; bone is shown in black.
(After Corliss.)

of the bone except that the cartilage at each end, which is called the **epiphysis,**
continues to grow as bone forms in the diaphysis. In this way the entire element
increases in length. In anamniotes, reptiles, and birds the epiphyses remain cartilaginous until adult stature is attained. Secondary centers of ossification develop
in the epiphyses of mammals. However, cartilaginous **epiphyseal plates** remain
between the epiphyses and the diaphysis, and cartilage continues to grow in these
plates until the adult stage, when the ossification centers in the epiphyses and
diaphysis unite. Union occurs at different ages in different bones, and these differences can help determine the age of a skeleton. The epiphyseal plates of mammals, together with the secondary ossification centers in the epiphyses, allow for
both growth in length and strength at the ends of the bone next to the joint.

Growth in girth occurs by the continued formation of perichondral bone
around the periphery of the cartilaginous primordium from cells recruited from
the periosteum. As the bone grows in length and diameter, it must be continually
remodeled because the rigidity of mineralized bone precludes it from expanding
from within as cartilage does. Bone is removed by the action of **osteoclasts** (Gr.
osteon = bone + *klastos* = broken) that differentiate by the fusion of precursor
cells from the bone marrow. Osteoclasts secrete acid (hydrogen ions) that dissolves
the mineral component of bone and enzymes that digest the collagen. Osteoclasts
in effect tunnel into old bone (Fig. 6–9). Blood vessels invade the tunnel and
osteoblasts line up on the tunnel's inner wall. The osteoblasts deposit concentric

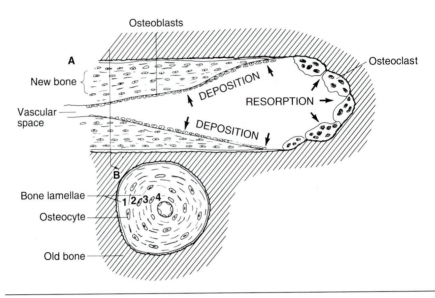

FIGURE 6–9 The formation of an osteon or haversian system within old bone. **A**, A longitudinal section. **B**, A transverse section at the level indicated. Numbers indicate the sequence in which lamellae are formed. (After Lanyon and Rubin.)

layers of bone to form a columnar **osteon,** or **haversian system.** The most recently formed layer of bone in an osteon is next to the central vascular space. As the bone continues to be remodeled, newly formed osteons replace the old ones, but fragments of former osteons often lie between the current ones.

The marrow cavity also enlarges as the bone grows, so most of the bone is located peripherally. Bone tissue on the surface is deposited in layers that go around the entire element. More deeply, the bone tissue is deposited as osteons, most of which extend vertically through the bone (Fig. 6–10). The collagen fibers in adjacent lamellae of an osteon spiral around the central vascular space at different pitches, and this, together with the columnar shape of an osteon, gives bone great strength. The dense, peripheral bone tissue is known as **compact bone.** Bone tissue near the ends of the bone and adjacent to the marrow cavity tends to retain its trabecular structure and is known as **cancellous bone.**

Other Functions of Bones

Bones are structural elements, but they also have other vital functions. Bone tissue is an important reservoir of calcium and phosphate ions, which are essential in many biochemical processes including muscle contraction and the storage and release of energy. Calcium and phosphate ions continually move between the bones and body fluids. The early evolution of bone among ancestral vertebrates may have made it possible for them to move from a marine environment into a freshwater one where these ions are in short supply.

The cylindrical cavities that develop in the long bones of the appendages, as well as the spaces between the trabeculae of all bones, are filled with bone marrow.

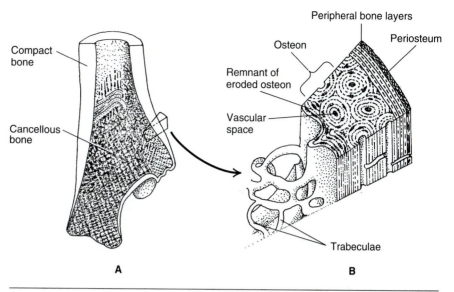

Compact bone

Cancellous bone

Peripheral bone layers

Osteon

Periosteum

Remnant of eroded osteon

Vascular space

Trabeculae

A

B

FIGURE 6–10   Mature bone structure: **A**, Section of a long bone. **B**, An enlargement of a portion of it. (After Krejsa.)

During fetal and much of juvenile life, all of the marrow is **red bone marrow.** This is a highly vascular tissue containing stem cells that differentiate into both red blood cells and many types of white blood cells. Red blood cells play a crucial role in gas transport in the blood, and white blood cells are an essential part of the body's defense and immune systems. As an individual ages, much of the red marrow in the long bones is replaced by fat cells, forming **yellow bone marrow.** Red bone marrow persists in many other bones including the vertebrae and skull bones.

Joints

The individual cartilages and bones of the skeleton are connected to one another at joints or **articulations.** The structure of joints is determined by the degree and direction of movement needed, and by the forces acting on the joint. We normally think of joints as allowing free movement between skeletal elements, but movement is restricted at many joints. Such joints are known as **synarthroses** (Gr. *syn* = together, joined + *arthrosis* = articulation). In addition to providing great strength, synarthroses allow for growth and often for limited movement. Common synarthroses are sutures, synchondroses, and symphyses. **Sutures** can be seen in the skull, where two dermal bones are separated yet united by a **sutural ligament** of connective tissue (Fig. 6–11**A**). The connective tissue periosteum, which covers the bones, continues across the joint and helps bind the bones together. The sutural ligament is a remnant of the sheet of connective tissue in which the bones developed, and in old age may become ossified. **Synchondroses,** which also can be seen in the parts of the skull, occur where two bones ossify in a continuous

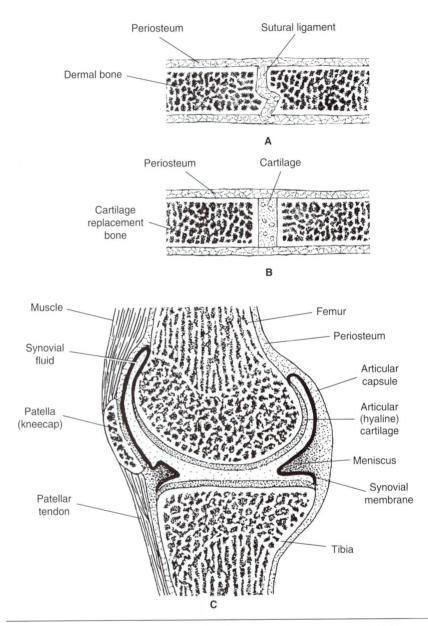

FIGURE 6–11   Diagrams of representative joints. **A**, A suture; **B**, a synchondrosis; **C**, a synovial joint, based on the human knee. (After Williams et al.)

sheet of cartilage and remain separated by a thin plate of cartilage (Fig. 6–11**B**). Synchondroses also occur during development of the long bones of the appendages, where the epiphysis and diaphysis are separated by an epiphyseal plate of cartilage. Examples of **symphyses** are the mental symphysis at the chin between the two halves of the lower jaw, the intervertebral symphyses between the centra of the vertebral column, and the pelvic symphysis between the two halves of the pelvic girdle. All lie in the median plane of the body. Usually a deformable disk of fibrocartilage separates the bones and permits some motion. Ligaments, in addition to the periosteum, cross the joint (see Fig. 10–8, p. 355).

Most joints allow considerable movement, and these are known as **diarthroses** (Gr. *di* = two + *arthrosis* = articulation), or **synovial joints.** Bones at a synovial joint are capped by smooth **articular cartilages,** and are encased by a strong, fibrous **articular capsule** (Fig. 6–11**C**). The enclosed articular cavity is lined by a **synovial membrane,** but the membrane does not cover the cartilages. A small amount of a viscous **synovial fluid,** which is secreted by the synovial membrane, acts as a lubricant and helps nourish the cells in the articular cartilages. The coefficient of friction on the articular surfaces is very low, about comparable to that of ice on ice. Thickened bands of fibers within the articular capsule sometime form intrinsic ligaments. Other ligaments usually cross the joint, strengthening it further and sometimes restricting movement in certain directions. Muscles that span the joint also reinforce it.

Some synovial joints contain a complete **disk** of fibrocartilage, or a partial disk known as a **meniscus** (Fig. 6–11**C**). The disk or meniscus often improves the fit or congruity between the bones and may act as a shock absorber. Joints containing a disk or meniscus often have a sliding or translational movement between the bones in addition to hinge, rotary, or other motions.

Lever Systems

The skeletal and muscular elements of the body are in effect machines that transmit forces from one point to another, usually changing the magnitude and direction of the force in the process. More specifically, many skeletal elements form rigid bars or **levers** that rotate or pivot about a fixed point, known as the **fulcrum.** A muscular force, called the **in-force,** is applied to one part of the lever, and the lever transmits this force to a point where the force is applied (e.g., another skeletal element, the jaws, the feet). The applied force is called the **out-force.** The relative magnitudes of the in-force and out-force, and the speeds with which the ends of the levers move, are determined by the points at which the forces are applied relative to the fulcrum.

A seesaw (Fig. 6–12**A**) is a familiar example of a lever system. A force applied to one end of the plank on which one sits (the in-force) will raise a load (the out-force) at the other end. We learned as children that if the plank was placed eccentrically on the fulcrum, a small child at the long end of the plank could raise a heavy person at the other. This is because the turning forces, called **moments** or **torques,** are the products of the forces times the length of their lever arms. In this case the in-torque has a long lever arm so the in-torque can counter a heavy load at the out-torque end. When the seesaw is in balance, the in- and out-torques are in equilibrium:

$$F_i \times L_i = F_o \times L_o$$

It is important to realize in making the calculations that the effective lengths of the lever arms are the perpendicular distances from the line of action of the forces to the fulcrum (Fig. 6–12**A**). Since the forces (in this case the pull of gravity) act

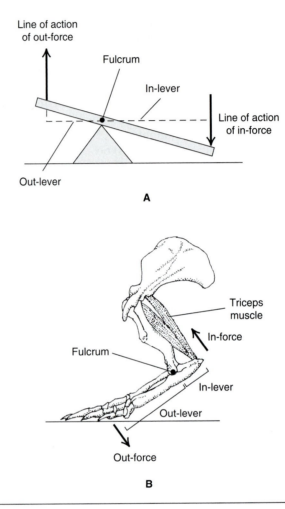

Line of action
of out-force

Fulcrum

In-lever

Line of action
of in-force

Out-lever

A

Triceps
muscle

In-force

Fulcrum

In-lever

Out-lever

Out-force

B

FIGURE 6–12   First-order levers. **A**, A seesaw; **B**, the extension of the forearm and hand of an armadillo.

vertically, the length of the lever arms would correspond to the length of the plank from the forces to the fulcrum only when the plank is horizontal.

The extension of the forearm at the elbow is an anatomical example of a seesaw-type lever (Fig. 6–12**B**). The elbow joint is the fulcrum, the in-force is delivered by the triceps muscle to the olecranon at the end of the ulna, and the in-lever arm is the perpendicular distance from the line of action of this force to the elbow joint. The out-force pushes down and backward on the ground, and its lever arm is the perpendicular distance from its line of action to the elbow joint.

Engineers call lever systems of the type we have described, in which the in-force is applied on one side of the fulcrum and the out-force is delivered on the other side of the fulcrum, **first-order levers**. In **second-order levers** both forces lie on the same side of the fulcrum. The in-force is applied to the end of the lever farthest from the fulcrum, and the out-force is delivered between it and the fulcrum. A wheelbarrow is a familiar example (Fig. 6–13**A**). An anatomical example is the extension of the foot by the gastrocnemius and soleus muscles when one

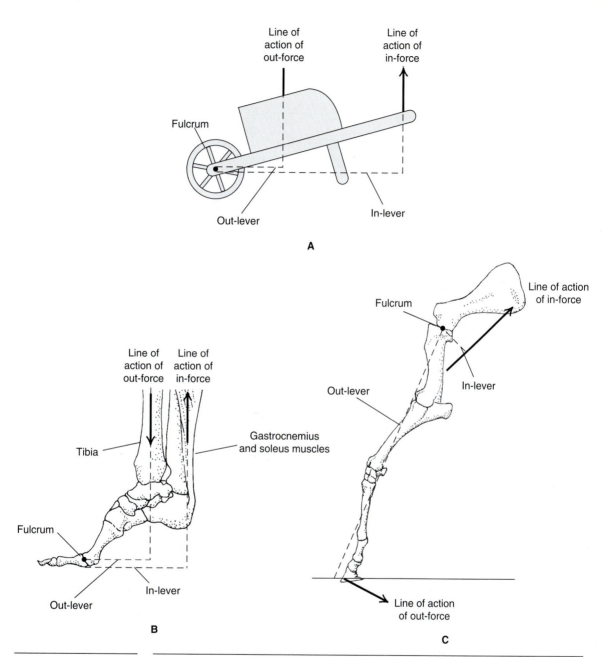

FIGURE 6–13   Lever systems. **A**, A wheelbarrow, a second-order lever. **B**, Action of the gastrocnemius muscle in raising the body on the toes, a second-order lever. **C**, Retraction of the front leg of a horse at the shoulder, a third-order lever.

raises the body on the toes (Fig. 6–13**B**). In **third-order levers** both forces are also on the same side of the fulcrum, but the in-force is applied closer to the fulcrum. An example is the retraction of the entire front leg at the shoulder joint (Fig. 6–13**C**). The fulcrum is the shoulder joint, the in-force is applied by retractor

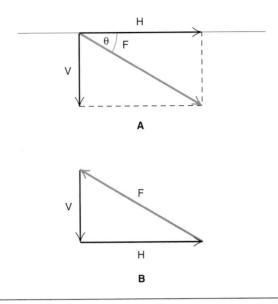

FIGURE 6–14   **A**, A parallelogram of force resolving the vector F into its vertical (V) and horizontal (H) components. **B**, The construction of a triangle of force, summing the vectors H and V to derive the resultant F.

muscles (such as the teres major) that attach near the proximal end of the humerus, and the out-force is delivered by the distal end of the limb to the ground.

It can be seen from the diagrams that lever systems in the body are often mechanically inefficient with respect to the magnitude of the forces, because the in-lever arms often are shorter than the out-lever arms. A greater muscular force must be generated on the in-lever than is delivered by the out-lever. These lever systems are advantageous, however, in that muscles are kept close to the fulcrums so the limbs are not bulky, and a short contraction of the muscles can cause an extensive movement at the out-end of the levers.

Components and Resultants of Forces

The forces acting on lever systems are **vector quantities** because they have both a magnitude and a direction. In analyzing forces, the forces are shown as arrows. The direction of an arrow is in line with the direction of the force, and its length represents the magnitude of the force on some convenient scale. Sometimes it is desirable to resolve a single vector into components acting in different directions. Consider, for example, the vector delivered to the ground by the foot in Figure 6–12**B**). The question may be asked, How much of the force (F) of this vector acts vertically (V) to support the body and how much acts horizontally (H) to propel the animal forward? The problem can be solved geometrically by constructing a parallelogram of forces in which F is the diagonal of the parallelogram, V one side, and H one end (Fig. 6–14**A**). The lengths of V and H represent the magnitude of their forces on the scale to which F was drawn. The problem can also be solved trigonometrically provided that the magnitude of F and the angle θ are known: V = F sin θ, H = F cos θ.

Conversely, two or more components of a force can be added to give a single resultant vector. This can be done by constructing a parallelogram whose side equals V and whose length equals H. The diagonal of the parallelogram is the resultant vector F. Stated another way, the combined action of V and H on a single point is comparable to the vector F. Alternatively, the resultant F can be determined by drawing vector V and placing the tail of vector H at the head of V (Fig. 6–14**B**). The angle between H and V in this diagram must be the same as in the parallelogram of force diagram. Drawing a line from the head of H to the tail of V produces the resultant F. Such a diagram, which represents one half of a parallelogram, is often called a **triangle of force.**

Distribution of Materials

The properties of the structural materials, and the mechanical principles we have discussed, affect the distribution of materials in the body. When an engineer designs a bridge, he or she carefully analyzes the nature and magnitude of the stresses that must be resisted and their distribution through the structure. Suitable materials are placed along the lines, or trajectories, of those stresses. It would be dangerous to use a material designed to meet tension, such as a cable, when compression must be resisted, and it would be uneconomical and inefficient to place materials where there are no stress trajectories. So, too, biological materials are not organized in a random way. Bone is present where compression or a combination of compression, tension, and shear must be met. Ligaments occur where only tension must be resisted.

The size and shape of bones, and the distribution of materials within them, are determined by genetic factors and by the combination of forces that act on them. The basic shape of each bone is determined genetically and will develop even when a limb is paralyzed and not subject to any stress. But the distribution of materials within a bone, its cross-sectional shape and thickness, the size and shape of the processes to which muscles attach, and the size and shape of its articular surfaces require stressing to develop normally. The trabeculae within a developing bone have a random architecture at first. As stresses are applied, bone is remodeled and the trabeculae form tracts that follow the stress lines. Stress lines in an animal change during growth and must meet the needs of a moving animal. They are far more complex and changeable than in a static bridge.

If a bone is simply compressed by forces distributed evenly at each end, bone trabeculae tend to run longitudinally parallel to the applied stress. This is the case in many human vertebral centra (Fig. 6–15).

If the weight that a bone must support is applied to one side of the bone, as when the weight of the body falls on the medially projecting head of the mammalian femur, the bone tends to bend. Bone on the medial side of the shaft is compressed, whereas bone on the lateral side is under tension. The situation is analogous to the distribution of stress lines in a simple crane (Fig. 6–16). Measurements reveal that the magnitude of the compression and tension is greatest

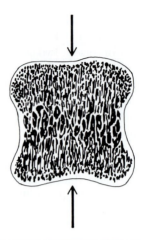

FIGURE 6–15   A longitudinal section through the vertebral body or centrum of a human vertebra. Trabec-ulae are parallel to the compression forces (shown by arrows) that act on it.

at the periphery of the shaft and drops to zero near the center. For this reason a long limb bone can be a hollow cylinder with compact bone on the periphery and a marrow cavity in the center. The bone must be a cylinder because, as the bone changes position during locomotion, stresses develop on its anterior and posterior surfaces as well as on its medial and lateral surfaces. At the top of the crane, or in the neck and head of the femur, compression and tension lines cross each other at right angles. The stresses are met in the femur by crossing bone trabeculae. The situation is more complex in the femur because of the attachment and pull of muscles on bone processes.

Muscle forces acting on a bone sometimes reduce the stresses within it. For example, certain muscles on the lateral side of the femur that extend from the hip to the lower end of the bone exert a pull that subjects the lateral side to compression. This reduces the tension stress that supporting body weight generates on this side. The pull of lateral muscles also subjects the medial side to tension and reduces the compression stresses normally there.

If a bone need resist bending in only one direction, then the engineer's I beam or the carpenter's joist illustrates a more economical use of materials than a cylinder, for the beam concentrates materials on the top and bottom, where the tension and compression stresses are located (Fig. 6–17). Separating the top and bottom surfaces by a vertical section of the I beam is important. The formula for resisting bending in this situation is

$$\text{resistance to bending} =$$
a constant determined by the nature of the material
$\times$ the width of the beam $\times$ the height^2 of the vertical segment.

Height, being squared, is particularly important. Biological examples of these principles can be seen in sections through many bones: the ilium of the pelvic girdle, a neural spine of a vertebra, and the lower jaw.

It is clear from the pattern of bone deposition that bone develops where

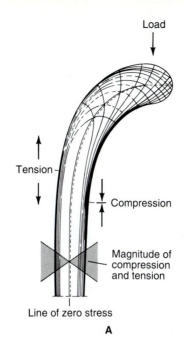

Load

Tension

Compression

Magnitude of compression and tension

Line of zero stress

A

Load

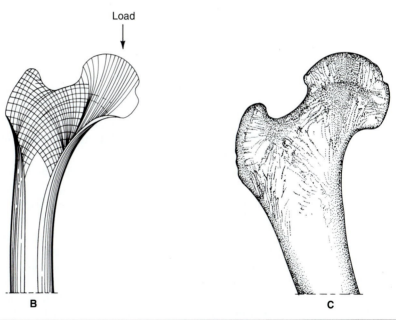

B

C

FIGURE 6–16 Stress trajectories in a simple crane (**A**) and the upper part of a mammalian femur (**B**) in response to a load applied on the medial side. Material on the concave side of the crane and the medial side of the femur is under compression; that on the opposite side is under tension. The magnitude of both stresses decreases to zero in the center. **C**, A drawing of a radiograph of trabeculae within the head of a human femur. (**A**, after Murray; **B** and **C** after Lanyon and Rubin.)

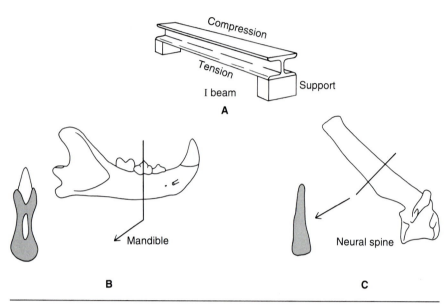

FIGURE 6–17   **A,** The distribution of stresses in an I beam that is supported at each end. Biological examples of this principle can be seen in the mandible of a mammal **(B)** and the neural spine of a mammalian vertebra **(C).**

stresses must be met. This has also been demonstrated experimentally. In one study, Lanyon and his associates (1982) removed a portion of the ulna shaft from a sheep's limb (Fig. 6–18), thus increasing the load-bearing stresses on the radius. The animal was walked for an hour each day. After a year the radius had increased in thickness over control animals by the equivalent of the amount of bone removed from the ulna.

Much research is being done to understand the factors that control the remodeling of bones to meet the stresses upon them (Currey, 1984b). As we have seen, an interchange of calcium and phosphate ions between the blood and bone continually occurs. Estimates in human beings are that nearly one quarter of the blood calcium interchanges with bone calcium every minute, yet blood levels of calcium remain remarkably constant. Interchanges are regulated by many hormones, including growth hormone, parathyroid hormone, and calcitonin; by certain vitamins (e.g., vitamin D); and by several factors derived from bone marrow cells. But it is difficult to see how this generalized chemical environment can target bone reabsorption and deposition in a way that leads to remodeling to meet stresses. Intermittent and variable stress appears to be necessary for bone deposition. Bone is reabsorbed when gravitational stresses are reduced, as in bed-ridden patients and astronauts subject to zero gravity. A constant pressure also promotes reabsorption. Orthodontists move displaced teeth through the jaw bone by using braces that apply a constant pressure.

The selective reabsorption and deposition of bone in response to changing stresses requires that a strain generates some signal that can selectively activate or inhibit osteoblasts and osteoclasts. Stressed bone does generate an electric

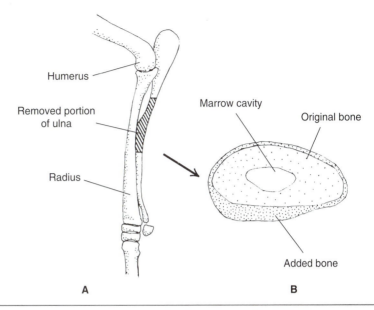

FIGURE 6–18   **A**, The forearm of a sheep, showing the portion of the ulna that was experimentally removed. **B**, A cross section of the radius at the level indicated after one year. (After Lanyon et al.)

potential, perhaps by the piezoelectric effect of either the mineral or collagen component of bone, but the origin of the electric potential is not certain. Many investigators believe that the electric potential is the signal, but others are doubtful. Much remains to be learned about bone remodeling.

SUMMARY

1. Skeletons maintain body shape, support the body against gravity, and transfer muscle forces to points of application.
2. The force per unit area applied to skeletons is called stress; the deformation that stress produces is called strain. The major stresses that skeletons resist are compression, tension, shear, and torsion. The relationship between stress and strain can be expressed in a stress-strain curve.
3. The skeletal materials of vertebrates are the notochord (a hydroskeleton) and dense connective tissues. Dense connective tissues contain many bundles of collagen fibers. They form ligaments and tendons and resist tension very well. Cartilage and bone are dense connective tissues whose matrix also contains molecules that provide rigidity and resistance to compression.
4. Cartilage is lighter than bone and resists compression but not the other stresses. It is very flexible and can grow from within without complex remodeling. Its distribution in embryonic and adult vertebrates correlates with these properties.
5. Bone, the primary skeletal tissue of most adult vertebrates, resists all stresses well. Osteoblasts produce a matrix that contains many collagen fibers to which calcium phosphate salts bind. Osteoblasts that become trapped in the matrix

are known as osteocytes. Since bone and cartilage are composite material, they have great strength.

6. Dermal or membrane bone develops by the direct deposition of bone in connective tissue; cartilage replacement bone forms within cartilaginous primordia (endochondral bone) and around their periphery (perichondral bone). As bone grows, it is continuously remodeled. Osteoclasts burrow tunnels in old bone, and osteoblasts lay down concentric layers of new bone within the tunnels to form haversian systems.

7. The individual cartilages and bones are connected to one another by joints or articulations. Synarthroses (sutures, synchondroses, and symphyses) are joints that provide strength and allow for growth, but movement is limited. Diarthroses or synovial joints have a more complex structure and allow for considerable movement.

8. Many skeletal elements are levers that rotate or pivot at a fixed point known as the fulcrum. Muscles apply an in-force to one part of the lever, and the lever delivers an out-force to a point where the force is to be applied. The turning forces, called moments or torques, are the products of the forces times the length of their lever arms. The relative magnitude of the in- and out-forces, and the speed with which the ends of a lever move, are determined by the points at which the forces are applied relative to the fulcrum. Three types, or orders, of levers are recognized.

9. The forces acting on levers are vector quantities having both a magnitude and a direction. A vector can be resolved into component forces acting in different directions. Conversely, component forces can be combined to give a resultant force.

10. The size and shapes of bones, and the distribution of materials within them, are determined by genetic factors and by the stresses that are applied throughout life. Bone is deposited where stresses are present. Bone is compact around the periphery of limb bones because the stresses are greatest here. It is arranged in a complex trabecular pattern in the heads of many limb bones, where compression and tension trajectories cross each other. If bone needs to resist bending in only one plane, it usually takes the form of an I beam or a carpenter's joist.

REFERENCES

Alexander, R. McN., 1983: *Animal Mechanics*, 2nd edition. Oxford, Blackwell Scientific Publications.

Corliss, C. E., 1976: *Patten's Human Embryology*. New York, McGraw-Hill.

Currey, J., 1984a: Comparative mechanical properties and histology of bone. *American Zoologist*, 24:5–12.

Currey, J., 1984b: *The Mechanical Adaptations of Bones*. Princeton, Princeton University Press.

Fawcett, D. D., 1986: *Bloom and Fawcett, A Textbook of Histology*, 11th edition. Philadelphia, W. B. Saunders Company.

Frost, H. M., 1967: *An Introduction to Biomechanics*. Springfield, Ill., Charles C. Thomas.

Hall, M. C., 1966: *The Architecture of Bones*. Springfield, Ill., Charles C. Thomas.

Halstead, L. B., 1974: *Vertebrate Hard Tissues*. London, Wykeham Publications.

Lanyon, L. E., Goodship, H. E., Pye, C., and McFie, H., 1982: Mechanically adaptive bone remodeling: A qualitative study on functional adaptation in the radius following ulna osteotomy in the sheep. *Journal of Biomechanics,* 15:141–154.

Lanyon, L. E., and Rubin, C. T., 1985: Functional adaptations in skeletal structures. *In* Hildebrand, M., Bramble, D. M., Liem, K. F., and Wake, D. B., editors: *Functional Vertebrate Morphology.* Cambridge, Harvard University Press.

McLean, F. C., and Urist, M. R., 1968: *Bone.* Chicago, University of Chicago Press.

Murray, P. D. F., 1936: *Bones.* Cambridge, Cambridge University Press.

Shapiro, F., 1992: Vertebral development in the chick embryo during days 3–9 of incubation. *Journal of Morphology,* 213:317–333.

Wainwright, S. A., 1988: *Axis and Circumference, The Cylindrical Shape of Plants and Animals.* Cambridge, Harvard University Press.

Williams, P. L., Warwick, R., Dyson, M., and Bannister, L. H., 1989, editors: *Gray's Anatomy,* 37th British edition. Edinburgh, Churchill Livingstone.

7 Evolution of the Head Skeleton

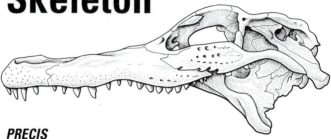

PRECIS

In this and the succeeding two chapters, we will examine the structure and evolution of the skeletal and muscular systems. It is necessary to understand their structure before we can explore many aspects of their functional anatomy. We will divide the skeleton into the head and postcranial skeletons. The head skeleton forms a natural group of functionally interrelated parts. It includes the chondrocranium, which is a part of the axial somatic skeleton; the visceral skeleton; and the dermal bones of the integumentary skeleton that usually sheathe the other components. Examining the head skeleton in fishes will lay the foundation for considering the changes that occurred during the evolution of amphibians and amniotes.

Having examined the properties and mechanics of the structural materials, we can now consider how they are assembled into the skeletal framework that supports the body, protects the brain and some other internal organs, and may house blood-forming tissues in the marrow cavities of the bones. In this and the next two chapters, we will deal primarily with the structure and evolution of the skeletal and muscular systems.

We often think of skeletons as dead and static, but they are dynamic, living tissues capable of growth, repair, and changing in composition and configuration as an animal grows and adapts to changing forces acting on its body. Bone, for example, is an important reservoir for calcium and phosphate ions, which are continuously interchanged with the ions in body fluids. At times the degree of mineralization of bones may become less, as when calcium and phosphate ions are needed for other purposes during pregnancy, lactation, or the formation of new skeletal materials such as antlers and teeth. At other times bones store these ions and mineralization increases. We often take for granted the properties of growth, repair, and change in skeletal materials, but these properties are unique to living organisms. They are not found in lumber, steel, concrete, and other materials builders use in construction.

So much can be learned from the skeleton that vertebrate morphologists often put more emphasis on this organ system than on others. The skeleton, including bony scales and teeth, fossilizes better than any other tissues of vertebrates, so much of what we know about the course of vertebrate evolution has come from analyses of skeletal remains. Seldom is an entire skeleton preserved, but paleontologists can make reconstructions by combining the known parts of several skeletons, provided that the separate fossils have overlapping parts that confirm they are the same species. Often missing parts can be reconstructed from adjacent parts with which they articulate, or by comparison with closely related contemporary species. Given a good reconstruction of the skeleton, much can be deduced about the soft parts and mode of life of the species. Muscles frequently attach on skeletal protuberances, or leave scars on the surface of bones. The shapes of joint surfaces of limb bones, combined with an analysis of the proportion of limb segments, can tell us a great deal about the posture of the animal and its mode of locomotion. Jaw proportions and the structure of the teeth indicate feeding habits. Sensory canals on the head leave grooves or pits in the skull. Cranial nerves and blood vessels pass through foramina in the skull. Endocranial casts often show the size and shape of parts of the brain. Canals in the skull that lodge the semicircular ducts of the inner ear tell us the position in which the head was held.

Divisions of the Skeleton

As we discussed in the chapter on the integument (Chapter 5), vertebrates have an **integumentary** or **dermal skeleton** represented by bony scales and plates that develop in or just beneath the skin. The main part of the vertebrate skeleton, however, is a deeper **endoskeleton** of cartilage or cartilage replacement bone. It is to this part of the skeleton, and the dermal bones that become associated with parts of it, that we now turn. Romer (1972) proposed that the vertebrate body is

TABLE 7–1	**Division of the Skeleton**
Integumentary Skeleton (Dermal Bone)	**Endoskeleton (Cartilage Replacement Bone)**

Bony scales
Dermal plates and bones
Teeth
	Somatic skeleton
	    Axial skeleton (chondrocranium, vertebrae, ribs, sternum)
	    Appendicular skeleton (girdles, bone of appendages)
	Visceral skeleton
	    Mandibular arch
	    Hyoid arch
	    Five branchial arches

organized as a tube within a tube. The "outer" body wall tube is somatic and the "inner" gut tube is visceral. Skeletal elements, muscles, nerves, and blood vessels associated with these two parts of the body can be called **somatic** and **visceral.** We will utilize this division because it is helpful; however, anatomists now recognize that some embryonic tissues move between these compartments, and the concept breaks down to some extent, especially in the head. Thus we divide the skeleton into a **somatic** and **visceral skeleton** that are located, respectively, in the "outer" and "inner" tubes of the body. The somatic skeleton can be subdivided into axial and appendicular parts. The chondrocranium of the skull, which helps encase the brain and certain sense organs; the vertebral column and ribs; and the skeleton of the median fins and sternum, when present, all lie in the longitudinal axis of the body and belong to the **axial skeleton.** The skeleton of the paired appendages and girdles lies more laterally in the body wall and constitutes the **appendicular skeleton.** The visceral skeleton in the gut wall forms skeletal arches that are associated with the pharyngeal pouches. It can be subdivided into specific visceral arches. In gnathostomes a **mandibular arch** (visceral arch 1) lies anterior to the first pharyngeal pouch; a **hyoid arch** (visceral arch 2) lies between the first and second pouches; and five **branchial arches** (visceral arches 3 to 7) lie between the remaining pouches and behind the last one. A summary of the divisions of the skeleton is shown in Table 7–1. In tracing the evolution of the skeleton, we will use these divisions to identify homologous structures, but parts of these divisions combine to form functional units that are conveniently described together: the head skeleton, trunk skeleton, and appendicular skeleton.

Composition of the Head Skeleton of Fishes

Vertebrates are distinguished from protochordates by a high degree of **cephalization.** As the ancestors of vertebrates increased in activity, the front of the body first encountered new environments. Sense organs became concentrated here, and the associated nervous tissue formed a brain. The mouth, too, was located at the front of the body, and gills became localized here. In short, a head evolved. Skeletal structures and somatic muscles in the portion of the head caudal to the inner ear develop from mesodermal somites, as these tissues do in most of the body. But the mesoderm that lies on each side of the neural tube (the **paraxial mesoderm**) rostral to the ear is organized, at least in amniotes, as a series of bumps and grooves that form **somitomeres** rather than sharply defined somites. The somitomeres apparently lack some of the developmental potential of typical somites because many tissues and structures in the front of the head develop from mesenchyme that migrates from the ectodermal neural crest rather than from the somitomeres. Neural crest contributes to the dermis of the skin in the head, to some muscles, and forms cartilages and membrane bones.

The head skeleton houses the brain and special sense organs and includes the visceral arches in the pharyngeal wall. Although originally associated with the gill apparatus, anterior arches became modified for feeding and contributed to the jaws. Others remain as gill arches in fishes. In tetrapods the visceral arches have undergone many transformations related to feeding, swallowing, breathing, and hearing. For purposes of analysis, it is convenient to divide the head skeleton into three major components:

1. The **chondrocranium** is the anterior part of the axial skeleton. It encases the special sense organs and contributes to the skeletal elements encasing the brain (the braincase or cranium).
2. The **visceral arches,** or **splanchnocranium,** support and move the gills and contribute to the jaws in gnathostomes.
3. A **dermatocranium** of dermal bones usually encases the other two components and contributes to the braincase and jaws. Teeth, which evolved from part of the integumentary skeleton, attach to the jaws and sometimes to other skeletal elements in the mouth and pharynx.

The three components of the head skeleton are distinct in many early fishes, but become confusingly mixed in more advanced vertebrates. The entire head skeleton may loosely be called the **skull,** but the term *skull* is usually limited to those parts of the head skeleton that become united in encasing the brain and forming the face and upper jaw.

The Chondrocranium

The **chondrocranium** (Gr. *chondros* = cartilage + *kranion* = skull), sometimes called the neurocranium, usually is a trough of cartilage or cartilage replacement bone that covers the ventral, lateral, and posterior surfaces of the brain and encapsulates the ear and nose. The caudal portion of the chondrocranium adjacent to the notochord develops embryonically from the mesodermal sclerotomes of the anterior somites, but much of its rostral portion arises from neural crest cells.

The composition of the chondrocranium can be understood by examining its embryonic development. The notochord extends forward beneath the brain nearly as far as the pituitary gland. A pair of cartilaginous rods, the **parachordals,** develop on either side of the front of the notochord (Fig. 7–1). As they enlarge, they unite and form a **basal plate,** which usually is perforated anteriorly by a basicranial fenestra through which arteries enter the cranial cavity. A pair of

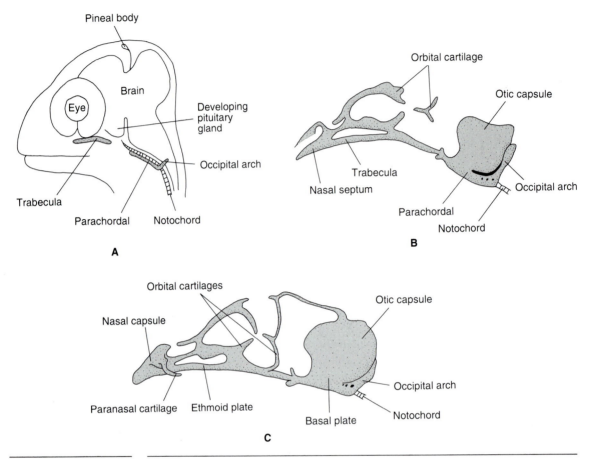

Lateral views of three stages in the development of the chondrocranium of a lizard. (The term *chondrocranium* as used in this book and many others is limited to the cartilages shown that develop around the brain and special sense organs. Some authors include the visceral skeleton in the chondrocranium.) (After DeBeer.)

trabeculae extend forward beside and in front of the pituitary gland. The trabeculae form the floor of the rostral part of the braincase, leaving a space between them for the pituitary gland. Their anterior ends may unite to form an **ethmoid plate** that lies between the pair of developing nasal sacs, and then the trabeculae may continue forward to support the rostrum.

Posteriorly, a variable number of occipital elements, which are serially homologous to developing vertebrae, surround the notochord and the back of the brain. Their union forms the **occipital arch,** which protects the back of the brain. The **foramen magnum** for the passage of the spinal cord perforates the occipital arch, and one or a pair of **occipital condyles** that articulate with the first vertebra develop ventral or lateral to the foramen magnum.

Nasal capsules develop around the nasal sacs and unite with the ethmoid plate. **Otic capsules** (Gr. *ot-* = ear) develop about the parts of the ear that lie

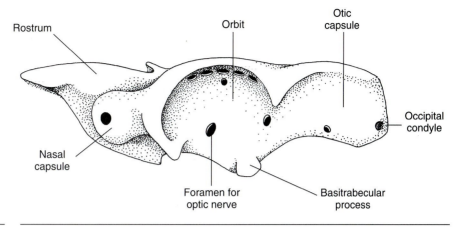

Rostrum

Orbit

Otic
capsule

Occipital
condyle

Nasal
capsule

Foramen for
optic nerve

Basitrabecular
process

FIGURE 7–2   A lateral view of the chondrocranium of a cartilaginous fish, based on *Squalus*. The small dark openings are foramina for cranial nerves and blood vessels.

within the braincase and unite with the basal plate and occipital arch. This part of the ear, known as the inner ear, is composed of the semicircular ducts and associated sacs that contain the receptive cells for equilibrium and hearing. Most tetrapods also have external and middle ears that receive and transmit ground or airborne sound waves to the inner ears; these are not present in fishes. **Optic capsules** begin to form around the eyeballs but do not unite with the rest of the chondrocranium because the eyeballs must be free to move. The optic capsules contribute to the fibrous tunic in the walls of the eyeballs and sometimes ossify as a ring of **sclerotic bones** (Gr. *skleros* = hard).

The lateral walls of the chondrocranium between the eyes develop from a complex set of rods and pillars known as the **orbital cartilages.** These cartilages coalesce with each other and adjacent parts of the chondrocranium, leaving foramina for the optic and other nerves entering the orbit.

In the fully formed chondrocranium of many fishes only two rods of cartilage cross the top of the brain, the occipital arch and a **synotic tectum** (L. *tectum* = roof) that extends between the otic capsules. The brain is covered dorsally by dermal bones. In cyclostomes and the chondrichthyan fishes that have no dermal bones, cartilage grows over the top of the brain (Fig. 7–2). As development proceeds in most species of bony fishes, cartilage replacement bones ossify within the chondrocranium and replace much of the cartilage. Their pattern will be considered later.

The Visceral Skeleton

The visceral skeleton, or **splanchnocranium** (Gr. *splanchnon* = gut) consists of a series of arches of cartilage or cartilage replacement bone that lie between the pharyngeal pouches. Since the first arch lies in front of the first pouch and the last one behind the last pouch, there is one more arch than pouches. The arches support the gills, but more importantly they add flexibility and elasticity to the pharynx. Muscle contraction compresses the arches, pharynx, and pouches, and the elastic recoil of the arches contributes to the expansion of these structures. The expansion and contraction of the pharynx are important movements in feeding and respiration of early vertebrates. Unlike the caudal part of the chondrocranium and the vertebral column, the visceral arches arise from the migration of neural crest cells into the region and not from the somites. Muscles associated with the visceral arches lie in the wall of the pharynx.

Ancestral vertebrates probably were scavengers and filter feeders. A typical ostracoderm had a jawless mouth, a large pharynx with a floor that could move up and down like a pump, and numerous gill slits through which a large volume of water could escape (see Fig. 3–3**B**). The visceral arches, numbering at least 16 in some species, lay between the gill slits and were attached to the integumentary skeleton. Contemporary adult cyclostomes are scavengers and bloodsuckers. Hagfishes can grasp and tear soft tissue, but lampreys rasp the surface of their prey with a horny tongue, and suck blood. Cyclostomes have fewer visceral arches than ostracoderms, but more than most other fishes. The lamprey has seven gill pouches and eight visceral arches. These are united to form a complex **branchial basket** from which extensions go into the tongue and around the pericardial cavity (Fig. 7–3**A**).

Jaws that move up and down probably evolved when certain ostracoderms began to feed on larger prey. The first visceral arch near the mouth opening was in a strategic position to help the fish open its mouth widely and engulf and seize food. Natural selection must have favored such a mechanism because a visceral arch became part of the jaw mechanism. Zoologists are not certain whether the jaw-forming arch was the first in the series, or whether it lay caudal to a premandibular arch that was later lost. The paired trabeculae of the chondrocranium resemble part of a premandibular arch in some ways. Moreover, a distinct cranial nerve is associated with each visceral arch, and early vertebrates have a cranial nerve (the profundus, p. 477) that may have been associated with a premandibular arch. This nerve becomes a branch of another nerve in most vertebrates.

When present, dermal bones contribute to the jaws, but in cartilaginous fishes, which lack these bones, the jaws are formed entirely by the first visceral arch. This is known in all gnathostomes as the **mandibular arch** (Fig. 7–3**B**). It consists of a dorsal **palatoquadrate cartilage** (the upper jaw of sharks) and a lower **mandibular cartilage** (the lower jaw). The second visceral arch, known as the **hyoid arch,** lies close behind the mandibular arch. Its dorsal segment, the **hyomandibular cartilage,** suspends the palatoquadrate in many fishes because it extends as a prop from the otic capsule of the chondrocranium to the posterior end of the palatoquadrate. From here the hyoid arch continues ventrally into the floor of the mouth and pharynx. There is not enough space for a complete gill pouch between the mandibular and hyoid arches. The pouch that one would expect to find between these arches is reduced to a **spiracle** or lost entirely.

The remaining visceral arches are associated with the gill apparatus and are known as **branchial arches** (Gr. *branchia* = gills). Nearly all jawed fishes have five branchial arches (visceral arches 3 to 7 or branchial arches 1 to 5), but a few unusual sharks have more. The first four branchial arches lie in the base of the

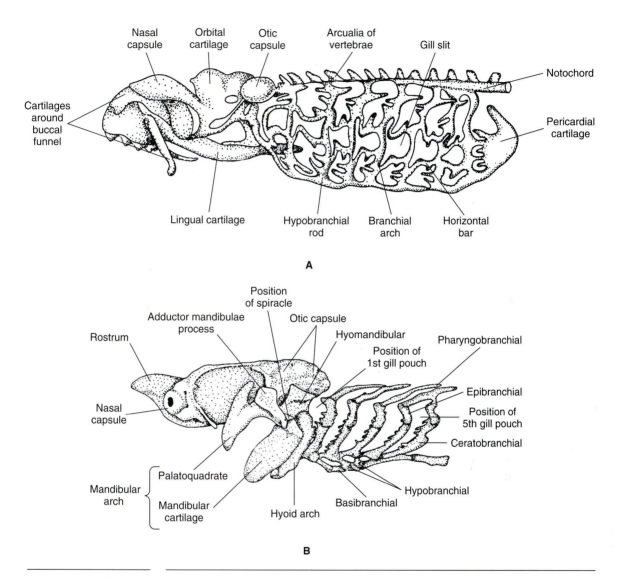

Nasal capsule   Orbital cartilage   Otic capsule   Arcualia of vertebrae   Gill slit

Notochord

Cartilages around buccal funnel

Pericardial cartilage

Lingual cartilage   Hypobranchial rod   Branchial arch   Horizontal bar

A

Position of spiracle

Adductor mandibulae process   Otic capsule

Rostrum   Hyomandibular   Pharyngobranchial

Position of 1st gill pouch

Epibranchial

Nasal capsule   Position of 5th gill pouch

Ceratobranchial

Palatoquadrate   Hypobranchial

Mandibular arch   Basibranchial

Mandibular cartilage   Hyoid arch

B

FIGURE 7–3   Lateral views of the chondrocranium and visceral skeleton of three fishes: **A,** The chondro-cranium and branchial basket of the lamprey. **B,** A cartilaginous fish, *Squalus.*

gills. The last arch lies posterior to the last gill pouch and sometimes is very small. A representative branchial arch is a >-shaped structure consisting of a dorsal **epibranchial cartilage,** which is serially homologous to the hyomandibular and palatoquadrate, and a ventral **ceratobranchial cartilage.** The ceratobranchials are connected by short **hypobranchial cartilages** to **basibranchial cartilages** that lie in the floor of the pharynx. The basibranchials frequently fuse. In cartilaginous fishes **pharyngobranchial cartilages** extend from the tops of the arches into the roof of the pharynx.

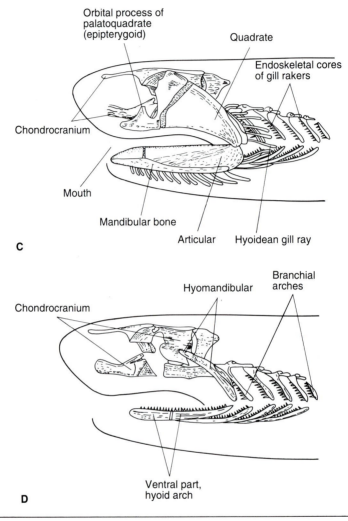

C

D

FIGURE 7–3   *continued*
C and **D**, Reconstructions of an early bony fish, the acanthodian *Acanthodes*. Overlying dermal bones have been removed in *Acanthodes* to show deeper parts of the head skeleton. The mandibular arch is in place in **C** but has been removed in **D** to show the remaining visceral arches. (**A**, after Young. **B**, from Walker and Homberger. **C** and **D**, after Jarvik.)

In bony fishes parts of the visceral skeleton often ossify (Fig. 7–3**C** and **D**). The most important bones in the mandibular arch are a **quadrate** that develops in the posterior end of the palatoquadrate cartilage, an **epipterygoid** that ossifies anterior to the quadrate, and an **articular** that ossifies in the posterior end of the mandibular cartilage. These are, of course, cartilage replacement bones. The jaw joint of all jawed vertebrates, other than mammals, lies between the quadrate and articular, or the posterior ends of the palatoquadrate and mandibular cartilages. The hyoid arch also usually ossifies, as do parts of the branchial arches.

The Evolution of Jaw Suspension in Fishes

When the first visceral arch became or contributed to the jaws, the upper part of the jaw, the palatoquadrate, needed to be anchored in some way. Its posterior end could then serve as a fulcrum for the movement of the lower jaw. The palatoquadrate is suspended in different ways among the many groups of jawed fishes. Zoologists are not certain how this was done among placoderms. In early cartilaginous and in early bony fishes, the palatoquadrate had one or more movable articulations with the chondrocranium, and the hyomandibular extended as a prop from the otic capsule of the chondrocranium to the posterior end of the palatoquadrate (Fig. 7–4). This type of jaw suspension is called **amphistylic** (Gr. *amphi* = both + *stylos* = pillar). An amphistylic suspension is also found in the rhipidistian ancestors of terrestrial vertebrates.

Advanced cartilaginous and bony fishes have a more flexible jaw mechanism that allows the jaws to be protruded downward or forward during feeding, as we will discuss in Chapter 16. A more flexible jaw evolved when the palatoquadrate lost its primitive connection to the chondrocranium. The palatoquadrate is stabilized only by the hyomandibular. This type of suspension, called **hyostylic,** appears to be the most advanced type in fishes (Fig. 7–4). It evolved independently, and in somewhat different ways, in the more advanced cartilaginous and bony fishes as their methods of feeding changed. Although the palatoquadrate of contemporary sharks is not fused to the chondrocranium, the palatoquadrate does have an

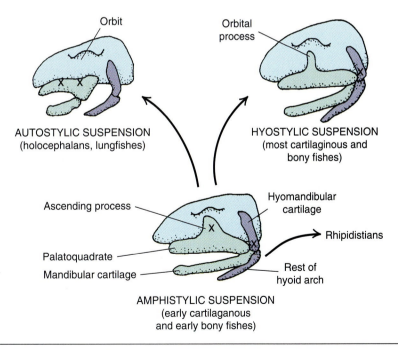

AUTOSTYLIC SUSPENSION
(holocephalans, lungfishes)

Orbit

Orbital process

HYOSTYLIC SUSPENSION
(most cartilaginous and bony fishes)

Ascending process

Hyomandibular cartilage

Rhipidistians

Palatoquadrate

Mandibular cartilage

Rest of hyoid arch

AMPHISTYLIC SUSPENSION
(early cartilaganous and early bony fishes)

FIGURE 7–4 The probable evolution of the suspension of the palatoquadrate cartilage in fishes. The embryonic cartilaginous elements are shown, but many of these become ossified in bony fishes. X = points of palatoquadrate suspension. See Fig. 7–5 for color code.

orbital process that extends dorsally beside the chondrocranium just in front of the basitrabecular processes. The orbital processes act as "guide rails," permitting the jaws to protrude but not to move from side to side.

Holocephalans, lungfishes, and other species that feed on shellfishes and other hard food have evolved an **autostylic** (Gr. *autos* = self) suspension in which the palatoquadrate is fused or firmly articulated to the chondrocranium and the hyomandibular is not involved. An autostylic suspension also evolved in early tetrapods in which the hyomandibular assumed other functions.

The Dermatocranium

The most conspicuous parts of the head skeleton of ostracoderms and most other early fishes are the large dermal plates that cover the head and help support and protect the brain, gills, and other soft structures. These dermal plates, which collectively form the **dermatocranium,** probably evolved in close association with the lateral line sensory system and electroreceptors that lie within canals and pits in the bones. Contemporary cyclostomes and cartilaginous fishes lack a dermal skeleton, but we are uncertain whether the dermal bones have been lost or were never present in their ancestors. The earliest known representatives of these groups have no trace of dermal bone other than the tiny placoid scales in the skin of chondrichthyans. Other fishes have a confusing array of dermal elements, although homologies can be established by using sensory canals, nerve foramina, and other openings as landmarks. We will not study the dermal bones in detail until they become stabilized in rhipidistian fishes and ancestral tetrapods.

For descriptive purposes it is convenient to group the dermal bones into six series (Fig. 7–5). (1) A **dermal roof** covers the top and sides of the head. The chondrocranium lies beneath it. Jaw muscles, the eyeball, and the palatoquadrate cartilage lie between it and the chondrocranium. The rostroventral border of the dermal roof contributes to the upper jaw. Caudally, the roof unites with the quadrate bone and frequently with the hyomandibular. (2) A **palatal series** of dermal bones develops in the roof of the mouth and covers most of the ventral surface of the palatoquadrate cartilage, leaving at least one large opening, the subtemporal fenestra, for the passage of jaw muscles to the lower jaw. (3) A slender **parasphenoid** lies on the ventral surface of the chondrocranium. (4) A **lower jaw series** of dermal bones nearly completely encases the mandibular cartilage and unites with the articular bone. Finally, (5) an **opercular series** covers the branchial region laterally, and (6) a **gular series** covers it ventrally. The pattern of the bones in the head skeleton of an early actinopterygian fish is shown in Figure 7–6.

Teeth

Most jawed fishes have teeth attached to the jaws and frequently to the palatal bones and parasphenoid. Teeth may also attach to the branchial arches, especially in teleost fishes, which often lack teeth on the jaws. Teeth are functionally a part of an animal's food-gathering apparatus, and their distribution, structure, and pattern of replacement are related closely to feeding habits. We will consider them when we discuss feeding mechanisms (Chapter 16).

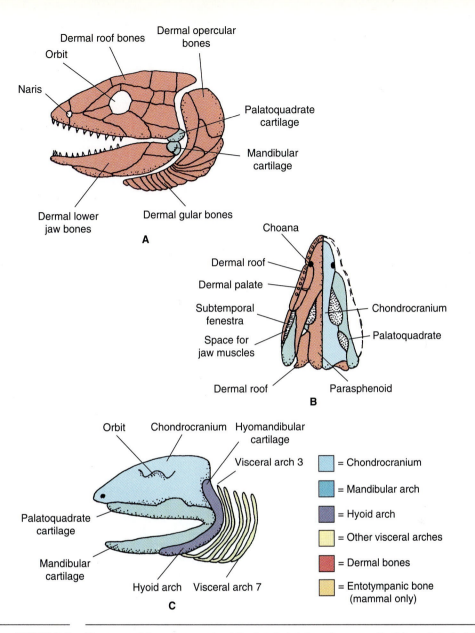

FIGURE 7–5 Diagrams of the components of the head skeleton of a generalized bony fish: **A**, A lateral view showing the superficial dermal bones that cover most of the other components. **B**, A ventral view of the skull with the dermal bones removed from the right side of the drawing. **C**, A lateral view after the removal of all of the dermal bones, leaving the chondrocranium and visceral arches.

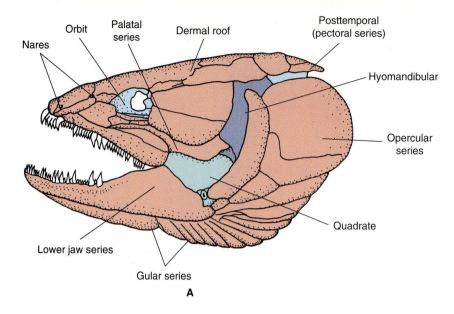

A

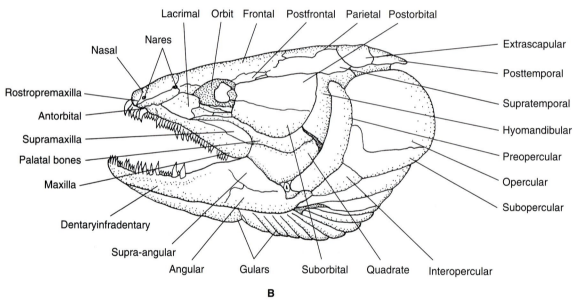

B

FIGURE 7–6 Lateral views of the head skeleton of *Amia*, a surviving early neopterygian. **A**, Components of the head skeleton are color coded as in Figure 7–5. **B**, Individual bones are identified. (From Walker and Homberger.)

The Head Skeleton of Amphibians

Bony fishes, especially the teleosts, radiated widely as they adapted to different modes of life and feeding patterns. Their skulls vary more widely than the skulls of all other vertebrates combined. However, we need not examine this diversity to understand the evolution of the head skeleton of terrestrial vertebrates. Ancestral terrestrial vertebrates, the labyrinthodonts, evolved from rhipidistian fishes, and, for the most part, retained the basic skull form found in this group of fishes. The labyrinthodont skull was inherited with only a few modifications by ancestral amniotes. Most contemporary reptiles, birds, and mammals retain the essential features of this skull. Contemporary amphibians, on the other hand, have highly modified skulls in which many of the ancestral elements have been lost. The labyrinthodont skull, therefore, is a far better starting point for understanding the evolution of the tetrapod skull than is the skull of a frog or salamander.

Labyrinthodonts

General Features of the Skull.   The labyrinthodont skull has the same components that we have described for the head skeleton of fishes: chondrocranium, dermatocranium, and visceral arches. But changes in feeding, gas exchange, and sensory systems that accompanied the transition of vertebrates from water to land did affect skull morphology.

A major change occurred in skull proportions (Fig. 7–7**A** and **B**). The facial portion of the skull, from the eyes to the snout, is short in a rhipidistian but became much longer in a labyrinthodont. The resulting longer jaws are probably correlated with differences in feeding. Feeding changes appear to have taken place because the vertical joint near the center of the chondrocranium of rhipidistians, which would have allowed the upper jaw and front of the skull to move up and down slightly relative to the back of the skull, is lost. A prominent immovable joint remained at this site in early labyrinthodonts.

Both rhipidistians and labyrinthodonts had a pair of internal nostrils, or **choanae** (Gr. *choane* = funnel), that opened near the front of the roof of the mouth (Fig. 7–7**C** and **D**). Choanae are part of the air passages leading to lungs in living vertebrates, and their presence in these extinct species indicates that lungs were probably present. Rhipidistians also breathed with gills, for they had a complete set of branchial arches that were covered by the opercular and gular bones. Gills were lost in the transition from water to land, and with them the opercular and gular bones that covered the branchial region. The branchial arches themselves became reduced in terrestrial vertebrates, but the ventral portions of the hyoid arch and several branchial arches form a **hyobranchial apparatus** that supports the newly evolved tongue and larynx. Parts of the branchial arches contribute to the cartilages of the larynx.

The loss in labyrinthodonts of much of the branchial apparatus, together with the loss of dermal bones that connected the pectoral girdle to the back of the skull in rhipidistians, enabled the head to move independently of the trunk. A short neck region began to develop.

Rhipidistians had the primitive **parietal foramen** on the top of the skull for

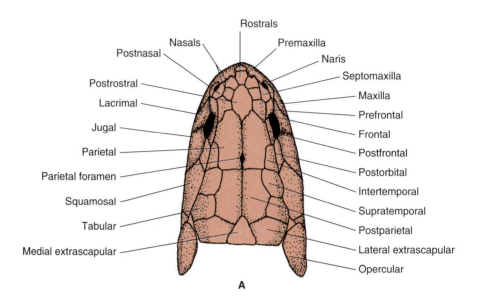

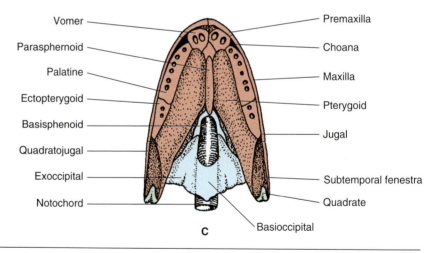

FIGURE 7–7 Evolution of the tetrapod skull: A comparison of the skull of a rhipidistian fish, *Eusthenopteron* (**A** and **C**), and a labyrinthodont amphibian, *Palaeogyrinus* (**B** and **D**), as seen in dorsal and palatal views. (**A** and **C**, from Romer and Parsons, after Jarvik; **B** and **D** after Romer and Parsons.)

a median eye, and this was retained in labyrinthodonts (Fig. 7–7**A** and **B**). Changes occurred in other sense organs as vertebrates moved from water to land, but few affected the skull. The earliest known labyrinthodont, *Ichthyostega*, retained **lateral line** canals on its skull, but this sensory system, which detects water movements, was lost in most labyrinthodonts.

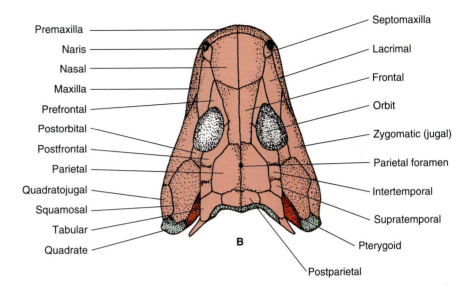

Premaxilla
Naris
Nasal
Maxilla
Prefrontal
Postorbital
Postfrontal
Parietal
Quadratojugal
Squamosal
Tabular
Quadrate

Septomaxilla
Lacrimal
Frontal
Orbit
Zygomatic (jugal)
Parietal foramen
Intertemporal
Supratemporal
Pterygoid

B

Postparietal

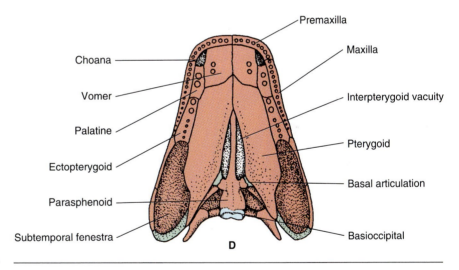

Choana
Vomer
Palatine
Ectopterygoid
Parasphenoid
Subtemporal fenestra

Premaxilla
Maxilla
Interpterygoid vacuity
Pterygoid
Basal articulation
Basioccipital

D

FIGURE 7–7 *continued*

Although the hyomandibular may have braced the palatoquadrate in very early labyrinthodonts, the suspension of the palatoquadrate soon became autostylic. The freed hyomandibular became the **stapes** in frogs, an auditory ossicle that transmits high-frequency, airborne vibrations from the tympanic membrane to the fenestra ovalis on the side of the otic capsule. We are uncertain of the function of the stapes in labyrinthodonts. Labyrinthodonts had an "otic notch" (Fig. 7–8**C**) high on the dermal roof that investigators once believed lodged a tympanic membrane, but this is now doubtful. Their stapes was too large and massive a bone to have responded to the high-frequency vibrations of the sort detected by a tympanic ear.

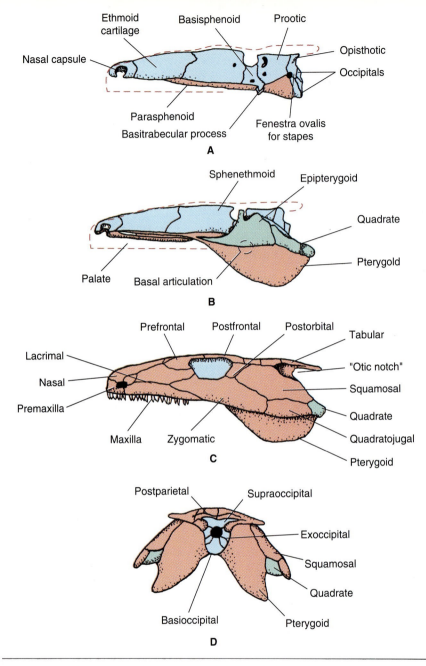

FIGURE 7–8 The skull of a labyrinthodont, *Palaeogyrinus*. **A**, A lateral view of the chondrocranium. **B**, A similar view with the addition of the palatoquadrate and part of the dermal palate (pterygoid). **C**, A lateral view of the skull with the addition of the dermal roof. **D**, A posterior view of the skull. (After Romer and Parsons.)

The stapes may have transmitted lower frequency vibrations of sufficient intensity by bone conduction from the jaw joint and cheek region to the otic capsule, (Clack, 1989), or it simply may have braced the braincase against the cheek. It is clear from fundamental differences in the tympanic ear of frogs, certain reptiles, and mammals that a tympanic ear evolved independently among terrestrial vertebrates several times, and was not inherited from a common type present in ancestral terrestrial vertebrates.

The Chondrocranium.   In examining the structure of the labyrinthodont skull, we will begin with the deepest part and work our way outward. Most of the chondrocranium was ossified (Fig. 7–8**A** and **D** and Table 7–2). Four bones ossified in the occipital arch around the foramen magnum: a **supraoccipital** dorsally, a **basioccipital** ventrally, and a pair of **exoccipitals** laterally. A single **occipital condyle** was present ventral to the foramen magnum. Two ossifications developed in the otic capsule: an **opisthotic** posteriorly and a **prootic** anteriorly. A fenestra ovalis, to which the stapes articulated, lay between them. The floor of the chondrocranium between the otic capsules was formed by a **basisphenoid** that bore the **basitrabecular processes** with which the palatoquadrate and palate articulated. A slight movement was possible at this joint. A long, trough-shaped **sphenethmoid** formed the floor and lateral walls of the chondrocranium between the eyes. The nasal capsules and ethmoid region between them were unossified. A dermal **parasphenoid** developed in the roof of the mouth and became attached to the underside of the chondrocranium.

The Palatoquadrate.   The front part of the palatoquadrate remained unossified and usually atrophied. Its posterior end was ossified in labyrinthodonts as a **quadrate,** which became united with adjacent parts of the skull: otic capsule, dermal roof, and palate (Fig. 7–8**B**). The middle part ossified as an **epipterygoid** bone that had a basal articulation with the basitrabecular processes on the chondrocranium, and also had an ascending process that extended dorsally beside a gap in the lateral wall of the chondrocranium.

The Dermal Roof.   The number of bones forming the dermal roof decreased in the evolutionary transition from rhipidistians to labyrinthodonts, but the pattern of bones remained much the same. For purposes of description it is convenient to sort these bones into groups (Table 7–2 and Figs. 7–7**B** and 7–8**C**). (1) One group formed the margin of the upper jaw: **premaxilla** (sometimes called the incisive bone in mammals) and **maxilla.** (2) A second group lay along the top of the skull just lateral to the middorsal line: **nasal, frontal, parietal,** and **postparietal.** (3) A third group surrounded the orbit: **lacrimal, prefrontal, postfrontal, postorbital,** and **jugal** (the zygomatic bone of a mammal). The lacrimal usually reached the external nostril but in some species a small **septomaxilla** lay rostral to the nasal. (4) A temporal group was located lateral to the middorsal bones and posterior to the circumorbital group: **intertemporal, supratemporal,** and **tabular.** (5) Finally, a group occupied the cheek region: **squamosal** and **quadratojugal.** The last bone extended between the jugal and the quadrate along the margin of the skull.

(Text continues on page 223)

TABLE 7–2 Components of the Tetrapod Skull and Lower Jaw

Labyrinthodont and Captorhinid	Necturus	Turtle	Alligator	Eutherian Mammal
Skull				
Chondrocranium (Cartilage replacement bone)				
Basioccipital (1)	Unos.	X (1)	X (1)	X (1) ⎫
Exoccipital	X	X	X Fused with opisthotic	X ⎬ Occipital
Supraoccipital (1)	Unos.	X (1)	X (1)	X (1) ⎧
Opisthotic	X Operculum	X	X	X ⎬ Petrosal part of temporal
Prootic	X	X	X	X ⎭
Basisphenoid (1)	Unos.	X (1)	X (1)	X (1) Body of basisphenoid
Sphenethmoid (1)	Unos.	Unos.	X Laterosphenoid	X (1) Presphenoid
Unossified ethmoid region	Unos.	Unos.	Unos.	X Ethmoid
Unossified nasal capsule	Unos.	Unos.	Unos.	X Turbinates
Visceral arches (Cartilage replacement bone)				
Palatoquadrate				
Quadrate	X	X	X	X Incus
Epipterygoid	0	X	0	X Wing of basisphenoid
Hyomandibular				
Stapes	X	X	X	X
Dermal bones				
Roof				
Tooth-bearing marginal bones				
Premaxilla (incisive)	X	X	X	X
Maxilla	0	X	X	X

Median series			
Nasal	O	O	X
Frontal	X	X	X
Parietal	X	X	X
Postparietal	O	O	X Interparietal, often a part of occipital
Circumorbital series			
Lacrimal	O	O	X
Prefrontal	O	X	O
Postfrontal	O	O	O
Postorbital	O	X	O
Jugal (zygomatic)	O	X	X
Temporal series			
Intertemporal	O	O	O
Supratemporal	O	O	O
Tabular	O	O	X ? Part of occipital
Cheek bones			
Squamosal	X	X	X Squamous part of temporal
Quadratojugal	O	X	O
Palate and underside of chondrocranium			
Parasphenoid	X Fused with basisphenoid	X Reduced	X ? Part of basisphenoid
Vomer	X (1)	X	X (1)
Palatine	X	X	X
Ectopterygoid	O	X	X ? Part of basisphenoid
Pterygoid	X	X	X Pterygoid process of basisphenoid

TABLE 7–2 *continued*

Labyrinthodont and Captorhinid	Necturus	Turtle	Alligator	Eutherian Mammal
Lower jaw				
Visceral arches				
(Cartilage replacement bone)				
Mandibular cartilage				
Articular	Unos.	X	X	X Malleus
Dermal bones				
Lateral series				
Dentary	X	X	X	X
Splenials (2)	X	0	X	0
Surangular	0	X	X	0
Angular	X	X	X	X Tympanic part of temporal (entotympanic; a new cartilage replacement bone)
Medial series				
Coronoids (3)	0	X	X	0
Prearticular	0	X	X Fused with articular	X Anterior process of malleus

The components of the skull and lower jaw of a labyrinthodont and captorhinid, together with the part of the skeleton to which they belong, are shown in the left hand column. The homologies between these elements and those of certain other tetrapods are shown in the right hand columns. An X indicates that an element is present; 0, that it is absent; Unos., that the region is unossified. All the elements are paired unless indicated to the contrary by a number in parentheses: (1) indicates a median element; (2) or (3) that two or three of these are present on each side.

The area beneath the temporal and cheek bones and lateral to the chondrocranium is known as the **temporal fossa.** It was largely filled by a jaw-closing, adductor mandibulae muscle that arose from the deep surface of the roofing bones and the side of the chondrocranium and passed through the **subtemporal fenestra** to the lower jaw (Fig. 7–7**D**). The complete temporal roof characteristic of rhipidistians and labyrinthodonts is called an **anapsid** roof.

The Palatal Bones.   A group of paired dermal, palatal bones ossified in the roof of the mouth: **vomer, pterygoid, palatine,** and **ectopterygoid** (Fig. 7–7**D**). The **epipterygoid** bone, an ossification of the palatoquadrate cartilage, lay dorsal to the pterygoid and gave the palate a movable articulation with the braincase. Spaces known as **interpterygoid vacuities,** which lay between the pterygoid bones and the chondrocranium, permitted the upper jaws and palate to move on the chondrocranium.

The Lower Jaw.   The lower jaw of a labyrinthodont was also similar to that of a rhipidistian (Fig. 7–9). The posterior end of the mandibular cartilage was ossified as an **articular** bone, and the rest of the cartilage was nearly completely sheathed by dermal bone. A **dentary** formed much of the front half of the jaw. Two **splenials** lay ventral to the dentary; two or three **coronoid** bones were medial and posterior to it. An **angular** and **surangular** lay posterior to the dentary on the lateral surface of the jaw, and a **prearticular** extended between the dentary and articular on the medial side. A **prearticular fossa,** anterior to the articular, lay between the dermal bones on the medial surfaces of the lower jaw. The temporal muscle emerging through the subtemporal fenestra attached along the margins of the fossa, and blood vessels and nerves entered the lower jaw through it.

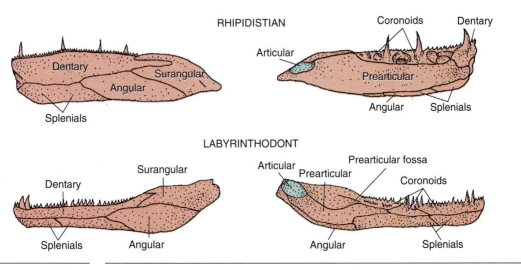

FIGURE 7–9   The lower jaws of rhipidistians (top row) and labyrinthodonts (bottom row) as seen in lateral views (left) and medial views (right). (After Romer and Parsons.)

Teeth.   Labyrinthodonts had fishlike, conical teeth loosely attached to the jaws (Figs. 7–7**C** and **D** and 7–9). The teeth of most fishes and early tetrapods are used simply to grasp food, not to masticate it. An unusual feature, also seen in some rhipidistian teeth, was an infolding of the enamel into the dentine, giving the teeth a labyrinthine appearance in cross section. The term *labyrinthodont* derives from this feature. Teeth were numerous. They occurred on the premaxilla and maxilla of the upper jaw, on many of the bones of the palate, and often on the parasphenoid. The dentary was the primary tooth-bearing bone of the lower jaw, but a few teeth occurred on the coronoids.

Contemporary Amphibians

In the evolution of contemporary amphibians, the skull became broad and flat (Fig. 7–10**A** to **C**). Most of the chondrocranium is unossified (Table 7–2). The basioccipital region is absent, and the paired occipital condyles are borne on the exoccipitals. There are fewer dermal bones, especially in the temporal region of the skull. The frontal and parietal bones are fused in frogs but remain distinct in salamanders. A frog has a well-developed stapes or columella that extends from the tympanic membrane to the otic capsule, but a salamander has no tympanic membrane and only a small stapes. We will consider the unusual hearing mechanisms of contemporary amphibians in Chapter 11.

The mandibular cartilage usually remains unossified, but its caudal end (the articular region) articulates with the quadrate. Many of the dermal lower jawbones are lost, but the dentary, angular, and sometimes the splenials remain (Fig. 7–10**D**). Salamanders retain teeth on the lower jaw, but frogs have lost them.

Amphibians retain gills only as larvae. The branchial arches are much reduced compared with those of fishes, but the ventral parts of many of these arches unite with the ventral part of the hyoid arch to form the **hyobranchial apparatus.** Much of the **hyobranchial** apparatus is not ossified, so it is not well known in fossils. In living amphibians it is embedded in the base of the tongue. Ligaments extend from the ventral part of the hyoid arch to the skull base, attaching near the stapes. The hyobranchial apparatus forms a sling for the support of the tongue and larynx, and for the attachment of muscles. Hyobranchial movements are used in food capture, swallowing, and gas exchange. The ventral parts of the hyoid arch and of the first three branchial arches (visceral arches 3, 4, and 5) contribute to it (Fig. 7–10**D**). The sixth visceral arch forms the **lateral laryngeal cartilages** of the larynx. The seventh visceral arch probably was lost completely in terrestrial vertebrates.

The Head Skeleton of Reptiles and Birds

The skull of the stem reptiles, the captorhinids, was similar in most respects to that of labyrinthodonts, but the skull was higher, somewhat narrower, and lacked an otic notch (Fig. 7–11**A**). Captorhinids retained the same complement of dermal bones as the labyrinthodonts, and the temporal region of the skull remained complete, or anapsid. The parietal foramen for a median eye was retained. The

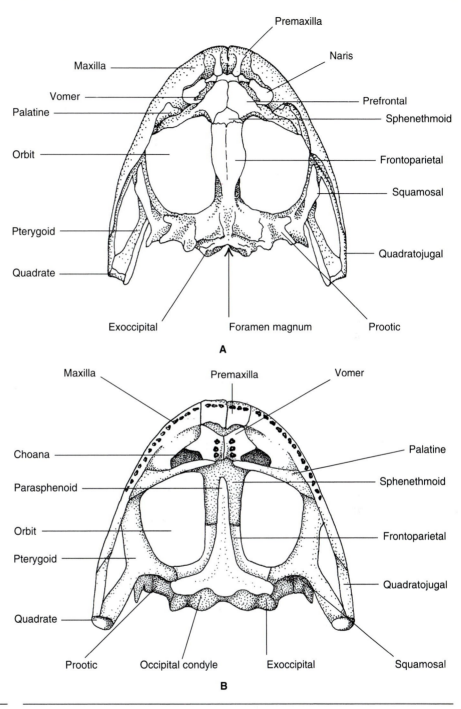

FIGURE 7–10 The skull and hyoid apparatus of contemporary amphibians: **A** and **B**, Dorsal and ventral views of the skull of a bullfrog, *Rana*.

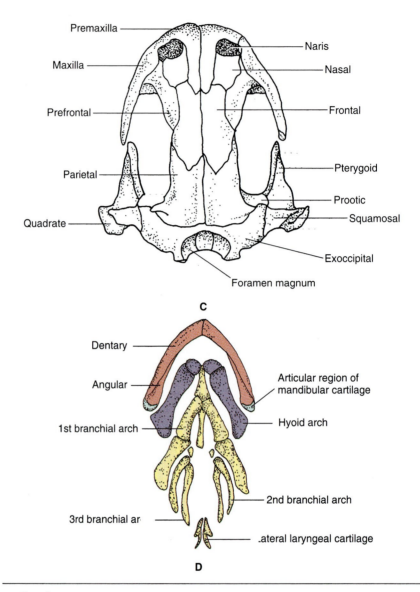

C, Dorsal view of the skull of a salamander, *Ambystoma*. **D**, A ventral view of the lower jaw, hyoid apparatus, and larynx of the salamander *Necturus*. (**C**, after Romer and Parsons. **D**, from Walker and Homberger.)

FIGURE 7–10 *continued*

occipital and otic region of the chondrocranium was ossified, but the rostral part was largely cartilaginous. A single occipital condyle was present. The chondrocranium articulated with the dermal roof and palate but was not firmly united with them. A large stapes butressed the chondrocranium by extending from the otic capsule to the quadrate. The medial end of the stapes fitted into a fenestra ovalis in the otic capsule, but hearing must have been limited to detecting low-frequency,

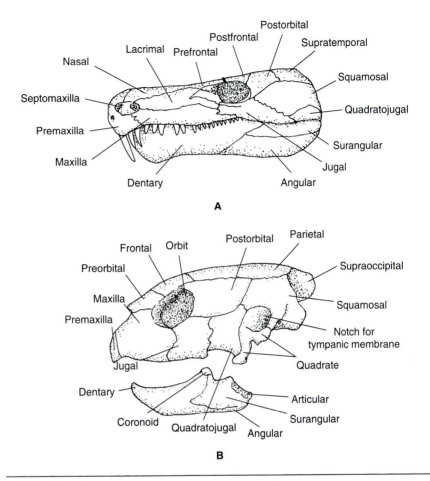

FIGURE 7–11   The skull of anapsid reptiles: **A,** A lateral view of the skull of a primitive captorhinid reptile, *Limnoscelis*. **B,** A lateral view of the skull of a sea turtle, *Caretta*. (**A,** after Romer and Parsons.)

high-intensity vibrations through bone conduction. None of the bones in the ear region was shaped in such a way that it could hold a tympanic membrane. Teeth were numerous but smaller than in labyrinthodonts. The small captorhinids were probably insectivores.

Turtles represent one line of descent from the captorhinids, and they retain the primitive anapsid temporal roof (Fig. 7–11**B**). This is very evident in sea turtles, but in other species the temporal roof has been partly lost by the development of a deep notch that extends rostrally from its posterior edge. Turtles have lost many of the dermal bones present in labyrinthodonts and captorhinids (Table 7–2). The snout is short. Teeth have been lost in contemporary species and are functionally replaced by horny sheaths over the jaw margins, but palatal teeth were retained in Triassic species of *Triassochelys*. The parietal foramen has also been lost. Turtles evolved a tympanic ear well adapted to detect airborne sound waves. The large quadrate bears a notch that supports a tympanic membrane, and the stapes is a slender rod that extends from the membrane to the fenestra ovalis.

Temporal fenestrae developed independently in the temporal region of the roof in other lines of reptile evolution (Fig. 7–12). The temporal fenestrae appear to have developed as early amniotes radiated and adapted to many different life styles and modes of feeding. Jaws and teeth changed in size and shape, as did the size and distribution of the adductor or jaw-closing muscles. Forces acting on the temporal roof from which some of these muscles arose changed. As we have seen, bone tends to form and thicken in stressed areas and to thin out and disappear where stresses are absent. Sutures where several bones meet in the temporal region would be potential weak points, and a redistribution of forces in the temporal roof avoided stressing them. Bone appears to have *initially* disappeared in such locations, and the jaw-closing muscles may have arisen from the strong periphery of developing fenestrae. As the fenestrae enlarged, they also provided an area into which underlying jaw muscles could bulge when they contracted and shortened. It is unlikely, however, that this bulging could have been the initial cause of fenestration, for the fenestrae would have been too small at first to serve this function.

Different patterns of fenestration developed in different lines of reptile evolution (Fig. 7–12). The earliest line to diverge were the synapsid, or mammal-like reptiles. They are characterized by having a **synapsid** skull (Gr. *syn* = together + *aspis* = arch) with a single temporal fenestra located low on each side of the skull, and bounded dorsally by the squamosal and postorbital bones. Mammals retain this type of skull, although the postorbital bone has been lost, and the fenestra is much larger than in synapsids and often merges with the orbit.

In other lines of reptile evolution both an upper and a lower temporal opening developed. This is the **diapsid** skull (Gr. *di-* = two). It is retained in the tuatara, *Sphenodon*, of New Zealand (Figs. 7–12 and 7–13**A**), in crocodilians and other archosaurs (Figs. 7–12 and 7–13**B**), and in modified form in lizards and snakes (Fig. 7–12). Lizards have lost the arch of bone ventral to the lower opening, and snakes, in addition, have lost the squamosal-postorbital arch between the two openings. These modifications increase the mouth gape. The quadrate, and in snakes the squamosal as well, is free to participate in jaw movements in addition to the usual movement between the quadrate and articular bones. This design enables that snakes can swallow prey larger than the diameter of their bodies (p. 587).

Two groups of extinct marine reptiles, the plesiosaurs and ichthyosaurs, have a single temporal fenestra located high on the skull dorsal to the postorbital and squamosal. This is called a **euryapsid** skull (Gr. *eurys* = wide). Intermediate fossils suggest that the plesiosaur skull evolved from a diapsid type by the loss of the lower temporal bar and the broadening of the remaining one. This may have been true for ichthyosaurs too, but the evidence is less clear.

Contemporary lepidosaurs, represented by *Sphenodon*, and archosaurian reptiles, represented by alligators and crocodiles, have lost many of the small dermal bones that were present in the temporal region of the skulls of captorhinids, but they retain most of the others (Fig. 7–13). More dermal bones are present than in turtles. *Sphenodon* and many lizards retain the parietal foramen for a median eye, but this is lost in snakes and most archosaurs. The part of the chondrocranium

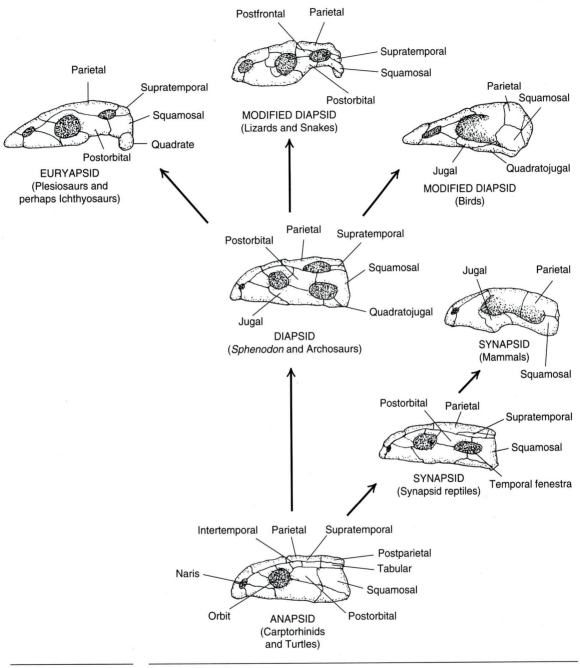

FIGURE 7–12 The evolution of temporal fenestration in reptiles, birds, and mammals.

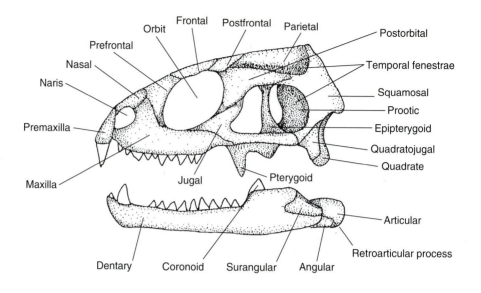

A

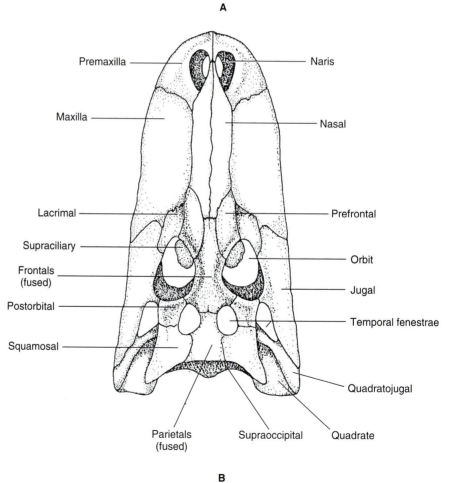

B

FIGURE 7–13 **A**, A lateral view of the skull and lower jaw of *Sphenodon*, a primitive diapsid reptile.

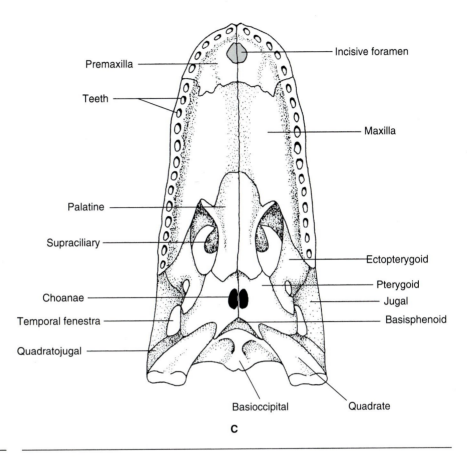

Premaxilla

Teeth

Palatine

Supraciliary

Choanae

Temporal fenestra

Quadratojugal

Incisive foramen

Maxilla

Ectopterygoid

Pterygoid

Jugal

Basisphenoid

Basioccipital   Quadrate

C

FIGURE 7–13 *continued*
B and **C**, Dorsal and palatal views of the skull of an alligator, an archosaur.

between the eyes is frequently unossified. In this region the brain is encased by plates of cartilage, but these are seldom seen in dried skulls. The dentary bone expands in the lower jaw and bears all of the teeth. Splenials and one of the coronoids present in stem reptiles are lost, but the other lower jawbones remain (Fig. 7–13**A**). The articular bone bears a prominent **retroarticular process** to which a muscle that opens the jaws attaches.

Sphenodon, lizards, and snakes retain a primitive palate capable of moving on the braincase, but in most other groups the primitive interpterygoid vacuities have been lost, so the dermal palatal bones are firmly united to the braincase. In crocodilians (Fig. 7–13**C**) shelflike extensions of the premaxillary, maxillary, palatine, and pterygoid bones have grown ventrally and medially and unite ventral to the original roof of the mouth, forming a secondary palate. The choanae now open far back in the mouth cavity. A flap of flesh completes the separation of the mouth cavity and respiratory passages. The secondary palate in this case is an adaptation for aquatic life. A crocodile can seize and manipulate prey beneath the water while continuing to breathe, provided the tip of its snout is out of the water.

Birds evolved from a group of early archosaurian reptiles, the saurischian order of dinosaurs. The skull of the earliest known bird, *Archaeopteryx*, is badly crushed, but it is clear that teeth were present and the braincase was modest in size; the temporal region of the skull appears to have been diapsid. In modern birds, the bar between the two temporal fenestrae has disappeared (Fig. 7–14). The fenestrae have enlarged and united with an expanded orbit. Only a slender bar of bone, composed of the jugal and quadratojugal, bounds these openings ventrally. Considerable brain enlargement also occurred during the evolution of modern birds. The bird braincase, composed of the chondrocranium and dermal roofing elements, is much larger than in reptiles, globular in shape, and completely ossified. The jaws form a long, narrow beak. Teeth have been lost in living species, and the beak is covered by a horny sheath. The palate is movably articulated with the braincase. This, together with a joint near the front of the skull roof, allows the front of the upper bill to be raised and lowered slightly during feeding (p. 587). Sight is particularly important for birds, and the eyes and orbits are exceptionally large. A ring of **sclerotic bones** has developed in the wall of the eyeball.

The larynx of a reptile or bird is larger and more complex than that of an amphibian (Fig. 7–15C). A ring-shaped **cricoid cartilage** has been added to the lateral laryngeal cartilages present in amphibians (Fig. 7–10D). The lateral laryngeal cartilages are derivatives of the fourth branchial arch (visceral arch 6), and this arch and possibly also the third branchial arch contributes to the cricoid cartilage. The third branchial arch is no longer a part of the hyobranchial apparatus of reptiles and birds, for this apparatus consists only of derivatives of the hyoid and first two branchial arches (Fig. 7–15A and B).

The Evolution of the Mammalian Head Skeleton

Mammals evolved from captorhinids through the two orders of synapsids, the early pelycosaurs and the later therapsids. This line of evolution is characterized by improved adaptations to terrestrial life. Many sense organs increased in size and sensitivity. The brain enlarged and became more complex, and the range and complexity of behavioral responses increased. Mammal-like reptiles became more active and mammals endothermic. Increased activity required a greater food intake and an increased volume of gas exchange. All of these changes affected the skull and lower jaw, both of which enlarged and became stronger. Increased strength was attained partly by reducing the number of bones and joints between them. Many bones became progressively smaller and eventually were lost as others expanded to occupy the space left. Except for the jugal and lacrimal, the circumorbital bones were lost, as were many in the temporal region of the skull: tabular, intertemporal, supratemporal, and quadratojugal bones. Only the dentary remains in the lower jaw of mammals (Table 7–2). Some other bones fused together to form complex bones that develop embryonically from several centers of ossification, each center representing an originally separate element. Among these complex bones are the occipital, temporal, and sphenoid (Table 7–2).

Many other changes also occurred in the skull and lower jaw, for they form an integrated unit so changes in one area affect stresses and the shape of elements in other areas. It is difficult, however, to discuss the skull as a whole. For purposes of analysis, we will break it up into smaller units related to particular functions.

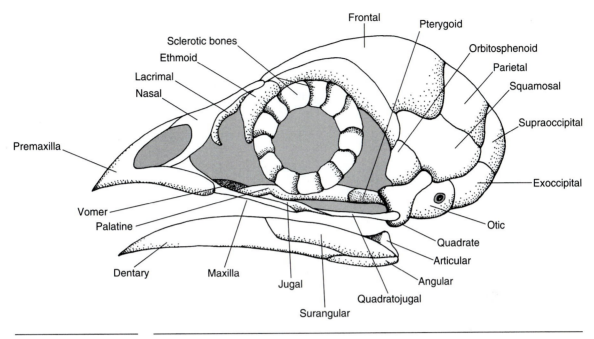

FIGURE 7–14   A lateral view of the skull and lower jaw of a modern bird.

We will first look at those parts of the skull that house major sense organs, go on to consider how the brain is encased, and then examine the changes that are correlated with changes in feeding mechanisms. Changes in feeding affected the palate and breathing, jaw muscles and temporal fenestration, lower jaw and middle ear structure, and teeth.

Skull Changes Correlated with Changes in Sense Organs

Evolutionary changes in the ears, nose, and median eye affected several aspects of the skull. The median eye of ancestral terrestrial vertebrates remained in pelycosaurs but became very small in late therapsids. The median eye was transformed in mammals into the pineal gland, an endocrine gland attached to the top of the diencephalon of the brain and usually covered by the cerebral hemispheres. The retreat of the pineal complex from the surface of the skull was accompanied by the loss of the parietal or pineal foramen (Fig. 7–17**C**).

The senses of smell and hearing became particularly important in ancestral mammals. The otic capsule of the chondrocranium was large, and spaces within it indicate that a cochlea had evolved. This is the part of the inner ear that gives mammals their keen sense of hearing. Other changes that accompanied jaw changes increased the sensitivity of the external and middle ears (p. 241). The ear of mammals is very sensitive to high-frequency airborne vibrations. The nasal cavities also enlarged, and parts of the nasal capsule ossified as scroll-shaped **turbinate** bones that greatly increased the surface area for olfactory receptors, and for cleaning, moistening, and warming inspired air. Conditioning inspired air is

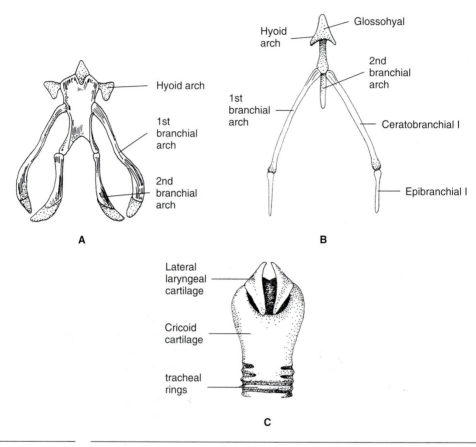

FIGURE 7–15   The hyobranchial apparatus and larnyx of reptiles and birds: **A**, A dorsal view of the hyobranchial apparatus of a turtle. **B**, A comparable view of the hyobranchial apparatus of a bird. **C**, A dorsal view of the larynx of a turtle. (**A** and **C**, after Romer and Parsons.)

important for an animal that is becoming more active and endothermic. Ancestral mammals apparently had a keen sense of smell and hearing, as do contemporary nocturnal insectivores. It is likely that early mammals occupied a similar ecological niche. This niche was not exploited by the larger, diurnal, and ectothermic reptiles. The type of dentition found in ancestral mammals (p. 242) reinforces this hypothesis.

Braincase Evolution

The brain increased greatly in size in the line of evolution toward mammals as the cerebrum and cerebellum enlarged and increased in complexity. Enlargement of the chondrocranium did not keep pace with the increase in brain size, so the mammalian chondrocranium forms only the rear, floor, and front of the braincase (Figs. 7–16, 7–17, and 7–18). Other bones encase the sides and top of the brain. The chondrocranium becomes completely ossified. The four occipital bones of

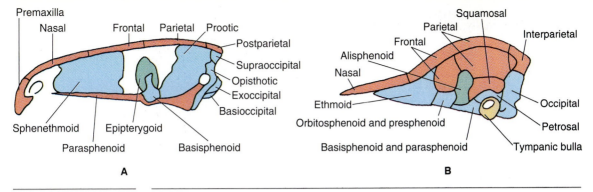

A

B

FIGURE 7–16 Sagittal sections through the skull to show the groups of bones that form the braincase: **A**, An ancestral captorhinid. **B**, A mammal. Color code as in Figure 7–5. (After Romer and Parsons.)

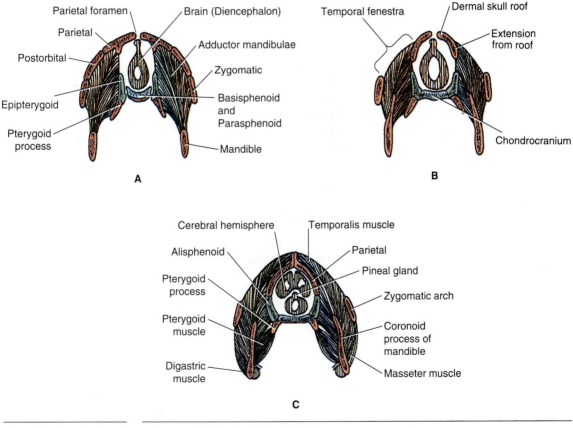

A

B

C

FIGURE 7–17 Evolution of the mammalian skull as seen in cross sections through the temporal roof: **A**, An ancestral reptile (a captorhinid). **B**, A synapsid. **C**, A mammal. (From Walker and Homberger.)

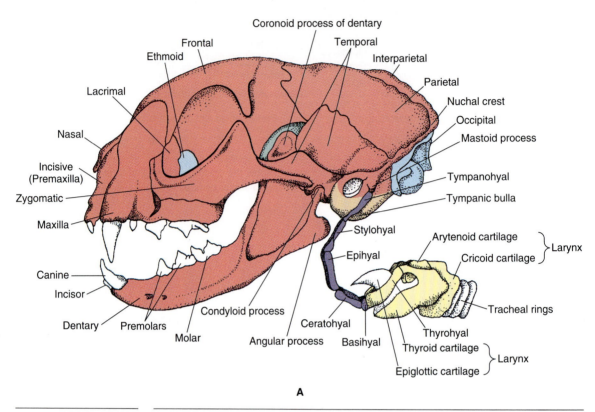

FIGURE 7–18 Composition of the mammalian head skeleton as seen in a cat; color coded as in Figure 7–5. **A**, Lateral view of the skull, lower jaw, hyoid apparatus, and larynx.

ancestral terrestrial vertebrates are represented by distinct centers of ossification, but all unite in many adult mammals to form a single occipital bone (Table 7–2). The originally single occipital condyle, which was located ventral to the foramen magnum and borne primarily by the basioccipital bone, became double as it shifted dorsally and laterally to the exoccipital bones. Mammals have a pair of occipital condyles located lateral to the foramen magnum. Vertical movements of the head occur between them and the first cervical vertebra, the atlas. Rotation occurs at a specialized joint that evolved between the first two cervical vertebrae (p. 272). These joints give mammals great flexibility in their head movements. Flexibility of head movements and a longer neck enable mammals to turn their head in many directions to explore their environment and reach up and down to feed.

The two otic bones united to form a hard **petrosal** bone (Gr. *petros* = stone) that encases the inner ear. The petrosal united with the squamosal of the dermal roof to form most of the mammalian **temporal** bone.

The basisphenoid is represented in mammals by one center of ossification, and the primitive sphenethmoid by three: a pair of orbitosphenoids that form part of the medial walls of the orbits, and a presphenoid under the brain between them. These bones remain separate in some mammals, but in many species all

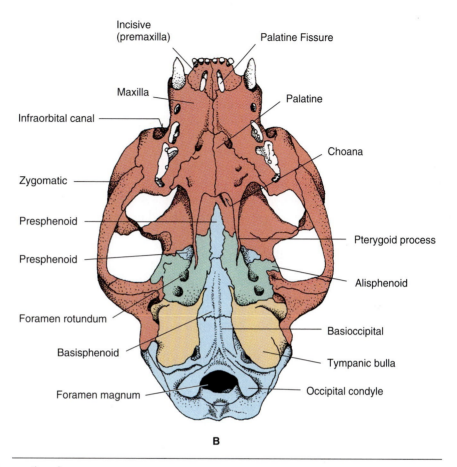

Incisive (premaxilla)
Palatine Fissure
Maxilla
Palatine
Infraorbital canal
Zygomatic
Choana
Presphenoid
Presphenoid
Pterygoid process
Alisphenoid
Foramen rotundum
Basisphenoid
Basioccipital
Foramen magnum
Tympanic bulla
Occipital condyle

B

FIGURE 7–18 *continued*
B, Ventral view of the skull. (From Walker and Homberger.)

unite to form the greater part of the wedge-shaped **sphenoid** bone (Gr. *sphen* = wedge + *eidos* = form). The dermal parasphenoid united with them.

The ethmoid region of the chondrocranium and the nasal capsules, which were largely unossified in labyrinthodonts and captorhinids, are represented in mammals by the ethmoid and turbinate bones respectively. The **ethmoid** bone (Gr. *ethmos* = sieve) forms the front of the braincase adjacent to the olfactory bulbs of the brain. It is perforated by many **cribriform foramina** (L. *cribum* = sieve + *forma* = shape) through which processes of olfactory cells pass from the nose to the olfactory bulbs. Part of the ethmoid bone extends rostrally to form most of the vertical nasal septum between the nasal cavities.

The epipterygoid, which formed a movable articulation between the palate and the chondrocranium in labyrinthodonts and captorhinids, enlarged greatly and united with the basisphenoid. In mammals it is known as the **alisphenoid** (L. *ala* = wing), for it forms the great wing of the sphenoid that contributes to the median wall of the orbit. The alisphenoid is a cartilage replacement bone, but its phylogenetic origin is the palatoquadrate cartilage of the visceral skeleton and not the chondrocranium.

The paired frontals, parietals, and postparietals, all dermal roofing bones, continue to cover the brain dorsally. The primitive paired postparietals have united to form a median **interparietal** bone, which frequently fuses with and becomes a part of the occipital bone. Sheetlike extensions of the frontals, parietals, and squamosals have extended from the dermal roof inward, passing medial to the temporal muscles, and united with the alisphenoid and chondrocranial elements (Fig. 7–17). These dermal extensions form much of the lateral walls of the mammalian braincase. Although derived from the dermal roof, the dermal extensions are not part of the original roof itself, which lay lateral to the jaw muscles. In the temporal region much of the original roof disappeared as the synapsid temporal fenestra evolved and enlarged in synapsids and early mammals (Fig. 7–20).

Palatal Evolution

Increasing levels of activity and eventually endothermy were made possible by changes in feeding mechanisms that enabled late therapsids and mammals to gather more food and to cut up and chew the food in the mouth before swallowing it. Jaw muscles became more powerful and complex and the jaws and bite force stronger. Correlated with these changes, the palate lost its primitive flexibility and became firmly united with the braincase. The dermal pterygoid bones were reduced to winglike, pterygoid processes on the ventral surface of the sphenoid (Figs. 7–17**C** and 7–18**B**). The names of these bones come from their appearance in mammals (Gr. *pteryg-* = wing or fin + *eidos* = form). The ectopterygoids appear to be incorporated in these processes. Pterygoid muscles extend from them to the medial side of the lower jaw.

The choanae of pelycosaurs and early therapsids continued to open into the mouth cavity through the anterior part of the palate (Fig. 7–19**A**), but in late therapsids shelflike extensions of the premaxillary, maxillary, and palatine bones grew ventrally and medially and united to form a small **hard palate** that lay ventral to the original primary palate, or roof of the mouth (Fig. 7–19**B**). As the hard palate evolved and extended caudally, the anterior part of the primary palate regressed and provided space for the enlarging nasal cavities. In mammals the originally paired vomer bones became a single, median element that lies dorsal to the hard palate and forms the ventral part of the nasal septum (Fig. 7–19**C**). The choanae of mammals open far back in the mouth cavity near the caudal border of the hard palate. A fleshy **soft palate** continues caudally from the hard palate, attaching along the ventral border of the pterygoid processes. The soft palate separates the nasopharynx, into which the choanae lead, from the oropharynx, into which the mouth cavity leads. Air and food passages are separated until they come together in the more caudal, laryngeal portion of the pharynx.

The evolution of the hard and soft palate, together known as the **secondary palate,** allows mammals to manipulate and chew food while still breathing. Breathing need be stopped only momentarily when food is swallowed and passes through the laryngopharynx to the esophagus. An increased intake of food and a large volume of gas exchange are important for an endotherm, and the evolution

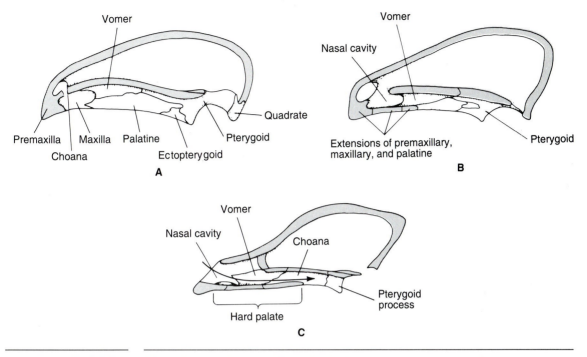

FIGURE 7–19   The evolution of the mammalian palate: **A**, A primitive therapsid retains the ancestral tetrapod condition. **B**, An advanced therapsid showing an early stage in the evolution of the hard palate. **C**, A mammal with a well-developed hard palate. (Modified after Romer and Parsons.)

of the secondary palate is correlated, at least partly, with the evolution of endothermy. But mammals also chew their food, and chewing requires strong upper and lower jaws. Thomason and Russell (1986) have shown that the secondary palate greatly strengthens the upper jaw and snout of mammals, especially against torsion and lateromedial bending. It is quite possible that this factor contributed to the evolution of the hard palate, especially in early stages when endothermy may not have been a factor.

Temporal Fenestration and Jaw Muscles

A small, synapsid type of temporal fenestra appeared in early pelycosaurs. The primary jaw-closing muscle, the **adductor mandibulae,** arose from the periphery of the fenestra and from a tendinous sheet that covered it (Fig. 7–20**A**). The adductor mandibulae muscle inserted on a small, dorsally projecting coronoid eminence on the lower jaw. A **pterygoideus** muscle arose from the lateral side of the palate and inserted on the medial side of the lower jaw (Fig. 7–20**C**).

The adductor mandibulae muscle became much larger and more powerful in therapsids and differentiated into a **temporalis** and **masseter** muscle. The masseter later subdivided into superficial and deep portions (Fig. 7–20**B**). The temporal fenestra enlarged considerably, thereby accommodating the increased

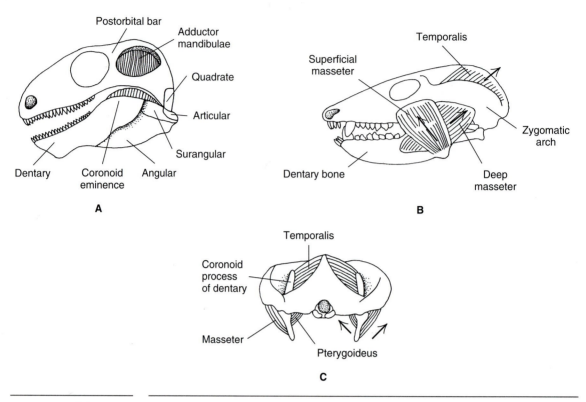

FIGURE 7–20   Jaw muscles in mammal-like reptiles. **A,** Lateral view of the skull and lower jaw muscles of a pelycosaur, based on *Dimetrodon*. **B,** A similar view of the jaw muscles of a late therapsid, based on *Probainognathus*. **C,** A posterior view of the skull, lower jaw, and jaw muscles of a late therapsid. Arrows indicate the direction of pull (line of action) of the jaw muscles. (**A,** after Kemp. **B** and **C,** after Carroll.)

muscle mass. The only parts of the original dermal roof remaining in this region are the postorbital bar, a bar of bone (now called the **zygomatic arch**) that bounded the fenestra and orbit ventrally, and a narrow, middorsal strip of bone. The temporalis continued to arise from the margins of the fenestra, but its insertion on the dentary was now on a large coronoid process of the dentary bone that evolved from the smaller coronoid eminence. The temporalis attached on the dorsal border and medial side of this process (Fig. 7–20**C**). The masseter muscle took its origin from the zygomatic arch. An outward-bowing zygomatic arch allowed the deep masseter to insert on the lateral surface of the coronoid process. The superficial masseter inserted near the posterior angle of the lower jaw. The pterygoideus muscle remained essentially unchanged.

These muscle changes not only made for a strong bite, but gave greater control over the movements of the lower jaw (Fig. 7–20**B** and **C**). Temporalis and deep masseter muscles pull upward and caudally; the anterior masseter, upward and rostrally. The masseter complex as a whole also pulls somewhat laterally, thereby balancing the medial pull of the pterygoideus.

Jaw and Ear Changes

Changes in jaw muscles were accompanied by changes in the jaws, which culminated in a new jaw joint and additional auditory ossicles in the middle ear. In pelycosaurs, as in other reptiles, the jaw joint lay between a large quadrate bone in the upper jaw and a large articular bone in the lower jaw (Fig. 7–21A). Both of

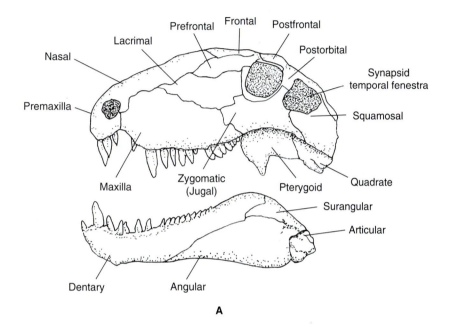

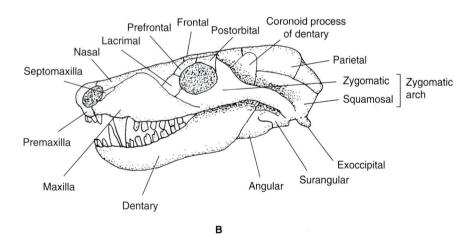

FIGURE 7–21   The evolution of the temporal fenestra and lower jaw in synapsids: **A**, A lateral view of the skull and lower jaw of a pelycosaur, *Dimetrodon*. **B**, Similar view of the skull and lower jaw of an advanced therapsid, *Thrinaxodon*. (**A**, after Kemp. **B**, after Romer and Parsons.)

these are ossifications in the mandibular arch and hence cartilage replacement bones. All other jaw bones are dermal. There is no indication that a tympanic membrane lay behind the quadrate as it does in turtles and some lizards. The stapes was large and extended from the otic capsule to the quadrate. It was too massive to have responded to high-frequency air vibrations. Lower frequency vibrations that were strong enough probably could be transmitted from the lower jaw through the jaw joint (articular and quadrate) to the stapes and otic capsule. The dentary was the largest bone of the lower jaw and the primary tooth-bearing element. Postdentary bones (angular, surangular, articular) were relatively large and firmly united through sutures with the dentary.

The dentary bone enlarged in therapsids as muscle forces acting on it became stronger and more complex. Postdentary bones became smaller and less firmly articulated with the dentary (Fig. 7–21**B**). The quadrate too became smaller and loosely articulated with the squamosal. This poses a paradox. If the jaws and bite were becoming stronger, how was it possible for the ones bearing the jaw joint (articular and quadrate) to become smaller and loosely articulated with their neighbors? Part of the answer appears to be a redistribution of forces in the jaw such that, although the bite was stronger, there was little force at the joint (Focus 7–1). Another part of the answer was the evolution at the same time of a tympanic ear sensitive to high-frequency airborne vibrations. We will consider the details of ear evolution later (p. 424). Briefly, a reflected process of the angular bone in late therapsids was shaped to hold a tympanic membrane, and the loosening of the joint-bearing bones (articular and quadrate) from the rest of the jaw made them more responsive to vibrations received by the tympanic membrane.

As these trends continued in very late therapsid reptiles, a part of the dentary bone (the condyloid process) made contact with the enlarging squamosal bone lateral to the articular and quadrate (Fig. 7–22). Four bones now participated in the jaw joint. The rounded articular condyle of the lower jaw was composed of the articular and dentary bone. The mandibular fossa in the skull that received the condyle was formed by the quadrate and squamosal. At the reptile-mammal transition, the dentary-squamosal component of the joint became larger and soon the only part of the joint. The articular and quadrate, already partly free from surrounding bones, became further reduced in size and detached from the joint. These bones then became incorporated into an expanding middle ear cavity as the additional auditory ossicles characteristic of mammals: articular = **malleus,** quadrate = **incus.** The small prearticular bone of the lower jaw formed a process on the malleus. The angular bone and its reflected lamina became the **tympanic** bone that holds the tympanic membrane and partly encases the middle ear cavity. The encasement of this cavity is completed in eutherian mammals by the **entotympanic,** a new cartilage replacement bone without a homologue in reptiles. The tympanic and entotympanic bones unite with the squamosal and otic capsule to form the mammalian temporal bone (Table 7–2). Other postdentary bones were lost, so the mammalian lower jaw consists of only the dentary bone. A single lower jaw bone, a dentary-squamosal jaw joint, and three auditory ossicles (malleus, incus, and stapes) are important derived features that define mammals.

Teeth   Changes in feeding mechanisms that affected other parts of the skull and lower jaw also affected the teeth. Ancestral pelycosaurs had conical teeth on their

FOCUS 7–1   Changes in Forces Acting on the Lower Jaw

To analyze the forces acting on the lower jaw at different stages in the evolution of mammals, investigators have applied the principles of levers to the jaw. These principles can be appreciated by looking first at the lower jaw of a cat when it is biting something with its front teeth (see figure **A,** top). The primary fulcrum is, of course, the **jaw joint fulcrum** (J). Teeth are engaging food at the front of the jaw (O). To simplify this example, we will look only at the effects of the temporalis muscle. The temporalis muscle exerts a force at its insertion on the top of the coronoid process of the dentary bone. This force has the line of action F, which can be resolved into horizontal and vertical components. The horizontal component is a sliding or

translational force that acts along the line OJ and pulls the lower jaw into the jaw joint. The vertical component is of more interest to us. It is a rotational force or moment that generates a bite force (B) at the front of the jaw (see figure **A,** middle). The magnitude of this moment is the force (F) times its moment arm (M). Recall (p. 191) that the moment arm in any lever system is the perpendicular distance from the line of action of the force to the fulcrum, in this case the jaw joint. At static equilibrium the moment generated by the temporalis will be opposed by an equal but opposite moment, which is the bite force at the front of the jaw times the length of its lever arm to the jaw joint: $FM = B\ (OJ)$. The bite force itself can be derived from this formula and is $B = FM/(OJ)$.

Bramble (1978) has taken this analysis a step further and points out that the point where the teeth engage food acts as a secondary or **occlusal fulcrum** (O) (see figure **A,** bottom). When the line of action of the temporalis muscle (F) is projected forward, it falls below the tooth row (line OJ) and its moment arm to the occlusal fulcrum is m. The moment of the temporalis acting through O is Fm and it will result in an upward or positive force (r) acting on the jaw joint: $Fm = r(OJ)$, $r = Fm/(OJ)$.

As shown in Figure **B,** the line of action of the temporalis muscle and the height of the coronoid process have profound effects on force r acting vertically at the jaw joint. When the force of the temporalis has the alignment F_1, as in the previous example, its moment arm (m_1) falls below the line OJ, the jaw rotates counterclockwise at O, and a positive vertical force ($r+$) is exerted at the jaw joint. When the force of the temporalis has the alignment F_2, its moment arm (m_2) falls above the line OJ, the jaw rotates clockwise at O, and the vertical force at the jaw joint is negative ($r-$). In principle, the occlusal fulcrum can be at any point along the tooth row of the lower jaw (see figure **C,** 01 to 07). When

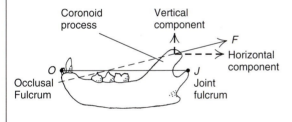

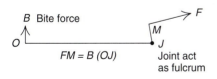

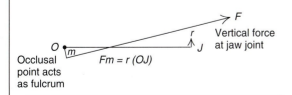

A, The lower jaw of a cat and the lever systems that act on it through the joint and occlusal fulcra.

continued

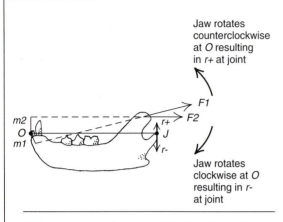

B, The lower jaw of a mammal showing the resultant at the jaw joint of differences in the line of action of the temporalis muscle acting through the occlusal fulcrum.

the occlusal fulcrum is far forward (01), force *r* at the jaw joint will be positive; when the occlusal fulcrum is near the back of the tooth row (07), the force at the jaw joint will be negative.

We can now apply these principles to the lower jaw in a series of mammal-like reptiles (see figure **D**). We will assume an occlusal fulcrum at the cheek teeth near the back of the jaw. The adductor muscle is undivided in primitive species, but becomes differentiated into the temporalis muscle and masseter complex of muscles in later species. When the adductor is divided, we will consider only the effect of the temporalis muscle. In pelycosaurs (see figure **D**, bottom), the line of action of the adductor muscle passed posterior to point *O*, so there was a strong positive vertical force at the jaw joint. Postdentary bones were an important component of the jaw

joint; they were large and firmly articulated with the dentary bone. As one progresses through the series of mammal-like reptiles, a coronoid process developed on the dentary bone, and the line of action of the main adductor muscle, now the temporalis muscle, became less vertical. As a consequence of these factors, the line of action of the temporalis muscle moved forward relative to point *O*. In an intermediate therapsid (see figure **D**, middle), the line of action of the temporalis muscle passed directly through the occlusal fulcrum. Vertical forces at the jaw joint would be zero or neutral. Postdentary bones became smaller and were less firmly articulated to the dentary bone. By late therapsids (see figure **D**, top), the line of action of the temporalis muscle was clearly such that vertical forces at the jaw joint would be strongly negative.

The deep and superficial masseter muscles began to differentiate in intermediate and late therapsids. Bramble has analyzed the forces that

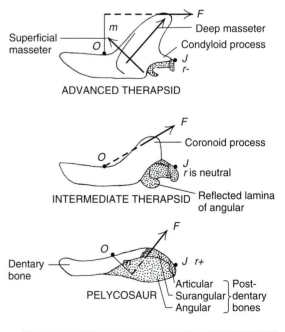

ADVANCED THERAPSID

INTERMEDIATE THERAPSID

PELYCOSAUR

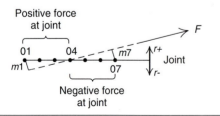

C, Diagram showing the effect on the jaw joint of different locations of the occlusal fulcrum.

D, Changing effects on the jaw joint of shifts in the line of action of the temporalis muscle during the evolution of synapsids. (*Redrawn from Bramble.*)

the masseter complex generated and their interaction with the forces of the temporalis muscle.
We need not go into details but simply point out
that the masseter complex of muscles added significantly to the bite force and to the control of
jaw movements. The masseters also added a positive vertical force at the jaw joint, but, since the
vertical force that the temporalis muscle generated at the joint was becoming strongly negative,
the combined forces at the jaw joint were close
to zero. Under these circumstances, the post
dentary bones could become very small and
loosely articulated to the dentary bone. Postdentary bones now were probably being affected

more by selective forces favoring the evolution
of an ear sensitive to airborne vibrations than
providing strength to the jaw joint. It was soon
after this stage that a new dentary-squamosal jaw
joint began to evolve beside the former articular-
quadrate joint. Thus we see that the development of a coronoid process, the differentiation
of the adductor muscles, and the changing alignment of the temporalis muscle resulted in a
strong bite force but kept forces at the jaw joint
of late therapsids close to neutral. Neutral forces
at the jaw joint were a prerequisite for the
change in the jaw joint and the evolution of the
mammalian middle ear.

palate and jaws adapted for capturing and holding small prey, which they swallowed whole or in large chunks (Fig. 7–21A). With the evolution of a hard palate
in therapsids, palatal teeth were lost and other teeth limited to the bones forming
the jaw margins, specifically to the premaxilla and maxilla in the upper jaw and to
the dentary in the lower jaw. In late therapsids, the teeth became firmly set in
sockets, a condition described as **thecodont.** The teeth also differentiated as they
performed different functions, a condition described as **heterodont.** Teeth at the

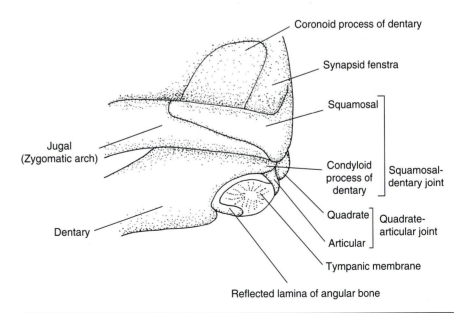

FIGURE 7–22   A lateral view of the jaw joint region of the skull and lower jaw of an advanced cynodont
therapsid that was close to the reptile-mammal transition. (After Allin.)

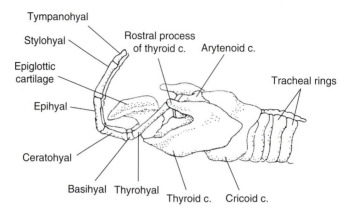

FIGURE 7–23   A lateral view of the hyoid apparatus and larynx of a cat. (From Walker and Homberger, after Nickel, Schummer, and Seiferle.)

front of the jaws became small, nipping **incisors,** a pair of large **canine** teeth followed them, and the **cheek teeth** were adapted for cutting and tearing apart prey (Fig. 7–21**B**).

In the earliest mammals of the Triassic period, the cheek teeth had differentiated into **premolars** and **molars.** Each molar had three cones, or cusps, in a longitudinal, linear series; hence these mammals are known as **triconodonts.** The shapes of the premolars and molars suggest that these creatures fed on insects and small arthropods. The teeth did not emerge from the gums in very young triconodonts. The absence of teeth in young individuals is evidence that they fed on milk. A young reptile, in contrast, hatches with a full complement of teeth and captures prey for itself.

When teeth do appear during the postnatal development of mammals, the control of jaw movements made possible by the more complex jaw musculature allows the teeth of upper and lower jaws to mesh, or occlude, very precisely. Specialized teeth, firmly set in sockets and with a good occlusion, enable mammals to cut and chew food and process the large amount of food needed to sustain an endothermic metabolism. Good occlusion is maintained during growth by a limited tooth replacement. As the jaw grows larger, the first, or milk set of incisors, canines, and premolars are replaced by larger permanent teeth. Molars are not replaced, but appear sequentially as the jaw enlarges. Teeth are considered further in Chapter 16.

The Mammalian Hyoid Apparatus and Larynx

The hyobranchial apparatus of mammals, usually called the hyoid apparatus, is an integral part of the feeding mechanism, for it and its muscles participate in tongue movements, opening and closing the jaws, and in swallowing (p. 589). The hyoid apparatus consists of a transverse body, or **basihyal,** a lesser horn, or **ceratohyal,** and a greater horn, or **thyrohyal** (Figs. 7–18**A** and 7–23). All are embedded in the musculature in the base of the tongue. Either a ligament or a chain of small bones (**epihyal, stylohyal, tympanohyal**) extends from each lesser horn to the tympanic region of the skull base. In some mammals, including humans, the dorsal part of this chain of bones has fused to the temporal bone forming its **styloid process** (Gr. *stylos* = pillar + *eidos* = form). The thyrohyal attaches to the thyroid cartilage of the larynx. The thyrohyal is a derivative of the third visceral arch; the rest of the hyoid apparatus derives from the ventral half of the hyoid arch.

The fourth and fifth visceral arches, which are incorporated in the hyobranchial apparatus of amphibians (Fig. 7–10**D**), form the **thyroid cartilage** of the larynx of eutherian mammals (Gr. *thyreos* = door-shaped shield + *eidos* = form). The sixth arch, which forms small lateral laryngeal cartilages in amphibians, differentiates into the **cricoid cartilage** (Gr. *krikos* = ring) and a pair of **arytenoid cartilages** (Gr. *arytaina* = ladle). The arytenoids extend into the vocal cords and can move them together or apart. The evolution of a more complex larynx enables mammals to have a larger repertoire of sounds than reptiles, and the resulting improved communication has been an important feature of mammalian evolution. The mammalian **epiglottis,** which helps deflect food around the entrance to the larynx and into the esophagus, is supported by a newly evolved **epiglottic cartilage.**

SUMMARY

1. The proper reconstruction and analysis of fossil skeletons can provide us with a great deal of information about the course of vertebrate evolution and the mode of life of extinct species.
2. The skeleton consists of a superficial integumentary skeleton that includes dermal elements, and a deeper endoskeleton of cartilage or cartilage replacement bone. The endoskeleton is divided into a somatic skeleton located in the body wall and a visceral skeleton in the wall of the pharynx. Parts of the integumentary skeleton become associated with the endoskeleton in the head and pectoral region.
3. The head skeleton consists of the chondrocranium (an axial part of the somatic skeleton), the visceral skeleton (also called the splanchnocranium), and usually dermal bones of the integumentary skeleton that encase the other parts (dermatocranium).
4. The chondrocranium originates from mesenchyme cells that are derived partly from the mesodermal sclerotome and partly from the ectodermal neural crest. These cells first form cartilaginous rods and capsules that then fuse together to form the chondrocranium.
5. The visceral skeleton forms an important part of the head skeleton of fishes. It consists of a mandibular arch that forms or contributes to the jaws, a hyoid arch that often helps to suspend the mandibular arch from the chondrocranium, and a series of five branchial arches that are part of the gill apparatus.
6. The dermal bones that ensheathe the other parts of the head skeleton can be grouped into a dermal roof, a palatal series, a parasphenoid under the chondrocranium, a lower jaw series, and an opercular and gular series that cover the branchial arches.

7. Teeth are numerous in fishes and may occur on the jaws, palate, parasphenoid, and branchial arches. They are used primarily to grasp food.

8. The earliest terrestrial vertebrates, the labyrinthodonts, had head skeletons essentially similar to those of their rhipidistian ancestors. Feeding changes presumably led to an increase in snout and jaw length and to the loss of a joint near the middle of the chondrocranium. Lateral line canals were lost. Except for the anterior end and part of the lateral walls, the chondrocranium was ossified. The palatoquadrate had an autostylic suspension, and the hyomandibular became an auditory ossicle, the stapes, in many groups. The ventral part of the hyoid arch and the remains of the branchial arches formed a hyobranchial apparatus, to which tongue muscles attach, and a small larynx. Dermal opercular and gular bones were lost with the gills, and the pectoral girdle lost its connection with the skull. A short neck began to develop.

9. Contemporary amphibians have exceptionally broad and flat skulls, many dermal bones have been lost, and most of the chondrocranium is unossified.

10. The skull of a stem captorhinid reptile was higher and narrower than in labyrinthodonts. The temporal roof was solid, or anapsid.

11. Turtles retain the anapsid temporal roof but have exceptionally short snouts, and they have lost many of the dermal bones. Horny sheaths replace teeth on the jaw margins.

12. Temporal fenestrae developed in the dermal roof in other lines of reptile evolution at areas of reduced stress where several bones come together. Their sharp margins provide a firm attachment for jaw-closing muscles, and when the fenestrae are large enough, they provide an area into which jaw muscles can bulge when they contract.

13. The position of the temporal fenestrae is quite distinctive in the various lines of amniote evolution, and presumably reflects differences in jaw musculature and stresses in the skull. The line of evolution to mammals had a single synapsid-type fenestra low on the side of the skull. Other reptile lines had a pair of temporal fenestrae (diapsid condition), or some modification of this pattern.

14. The line of evolution through synapsids to mammals is characterized by an increased level of activity and more complex behavioral responses. These affect skull evolution by their effects on certain sense organs, brain size, and breathing and feeding mechanisms.

15. A reduction in the number of bones by loss and fusion made the mammalian skull and lower jaw stronger than that of a reptile.

16. The conversion of the median eye complex to a pineal gland led to a loss of the parietal foramen. The otic capsule enlarged as the inner ear evolved a cochlea and became more sensitive to sound waves. An expansion and partial ossification of the nasal capsules as turbinate bones correlates with an increased importance of the sense of smell and with the conditioning of inspired air.

17. The mammalian braincase became completely ossified. Extensions of dermal roofing bones and a large epipterygoid cover parts of the enlarged brain that are not covered by chondrocranial elements.

18. The evolution of a hard palate strengthened the upper jaw, allowed for the expansion of the nasal cavities, and allowed for a more complete separation

of food and air passages. The simultaneous eating and breathing that the hard palate permits is needed to sustain the high metabolism of mammals.

19. The enlargement and differentiation of the adductor jaw musculature led to an enlargement of the temporal fenestra. These changes in the jaw muscles and in their alignment resulted in a stronger bite force and the precise control over jaw movements needed for chewing, and at the same time led to a reduction of vertical forces acting at the jaw joint. Force changes at the jaw joint led to a reduction in size of the postdentary bones, their loose articulation with the dentary, and a new squamosal-dentary jaw joint. The former jaw joint bones (quadrate and articular) became incorporated in the middle ear as additional auditory ossicles.

20. Teeth became firmly set in sockets and differentiated for different functions. A limited replacement of teeth ensures the good occlusion that is needed to chew food.

21. The mammalian hyoid apparatus supports the tongue and provides attachment for the muscles used in positioning the food between the teeth and in swallowing. It is composed of derivatives of the ventral part of the hyoid and first branchial arch. The second and third branchial arches contribute to an expanded larynx.

REFERENCES

Allin, E. F., 1975: Evolution of the mammalian ear. *Journal of Morphology*, 147:403–438.

Barghusen, H. R., 1972: The origin of the mammalian jaw apparatus. *In* Schumacher, G. H., editor: *Morphology of the Maxillo-Mandibular Apparatus*. Leipzig, Georg Thieme.

Bramble, D. M., 1978: Origin of the mammalian feeding complex, models and mechanisms. *Paleobiology*, 4:271–301.

Clack, J. A., 1989: Discovery of the earliest known tetrapod stapes. *Nature*, 342:425–427.

Crompton, A. W., 1963: The evolution of the mammalian jaw. *Evolution*, 17:431–439.

Crompton, A. W., and Parker, P., 1978: Evolution of the mammalian masticatory apparatus. *American Scientist*, 66:192–201.

DeBeer, G, R., 1937: *The Development of the Vertebrate Skull*. London, University of Oxford Press.

Frazzetta, T. H., 1968: Adaptive problems and possibilities in the temporal fenestration of tetrapod skulls. *Journal of Morphology*, 125:145–158.

Hall, B., and Hanken, J., editors, 1993: *The Vertebrate Skull*. Chicago, University of Chicago Press.

Hopson, J. A., 1966: The origin of the mammalian middle ear. *American Zoologist*, 6:437–450.

Moore, W. J., 1981: *The Mammalian Skull*. Cambridge, Cambridge University Press.

Nickel, R., Schummer, A., and Seiferle, E., 1986, 1979, 1981: *The Anatomy of Domestic Animals*, Volumes 1, 2, and 3. New York, Springer Verlag.

Romer, A. S., 1972: The vertebrate as a dual animal—somatic and visceral. *In* Dobzhansky, T., Hecht, M. K., and Steere, W. C., editors: *Evolutionary Biology*, 6:121–156. New York, Appelton-Century-Crofts.

Thomason, J. J., and Russel, A. P., 1986: Mechanical factors in the evolution of the mammalian secondary palate, a theoretical analysis. *Journal of Morphology*, 189:199–213.

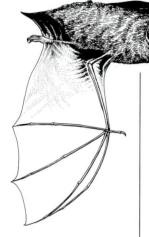

8 | Evolution of the Postcranial Skeleton

PRECIS

The postcranial skeleton can be divided into the trunk and appendicular skeletons. The trunk skeleton includes the vertebral column, median fins, ribs, and sternum, all of which belong to the somatic axial skeleton. The appendicular skeleton, also a part of the somatic skeleton, consists of the skeleton of the paired fins in fishes and limbs in tetrapods, and the girdles in the body wall to which the appendages attach. Both trunk and appendicular skeletons form the basic framework for the body and participate in body support and locomotion. We will consider their structure and evolution in this chapter, but defer most aspects of their functional anatomy until we have examined the muscles that act on them.

Some of the somatic skeleton contributes to the skull and lower jaw, as we have seen, but most of it forms the postcranial skeleton, which includes the trunk portion of the axial skeleton (vertebral column, median fins, ribs, and sternum) and the appendicular skeleton (paired fins or limbs and the girdles to which they attach). Trunk and appendicular skeletons provide the basic framework for the body and are actively involved in support and locomotion. Since fishes are buoyed by the water in which they live, their vertebral axis is primarily a compression strut. It prevents the body from shortening and converts trunk muscle contractions into lateral undulations that power swimming movements. Although the vertebral axis of terrestrial vertebrates participates in locomotion in some species, it acts primarily as a beam that transfers body weight to the girdles and appendages. Thrusts of the appendages on the ground propel most tetrapods. The ribs are thought to be part of the locomotor apparatus of fishes. In terrestrial vertebrates, the ribs help encase the lungs and abdominal viscera and participate in amniotes in ventilating the lungs. We will examine the structure and evolution of the trunk and appendicular skeletons, beginning with the trunk skeleton.

The Structure and Development of Vertebrae and Ribs
Vertebral Structure

Although vertebrae vary considerably among species, and also among regions of the body of an individual, each vertebra typically has a **vertebral, or neural, arch** that extends dorsally around the spinal cord (Fig. 8–1). Intervertebral foramina, through which spinal nerves pass, usually lie between the bases of successive neural arches, but spinal nerve foramina perforate the neural arches in a few species of fishes. Each vertebra in the tail region of fishes and many tetrapods also has a **hemal arch** that extends ventrally around the caudal artery and vein. Neural and hemal arches not only protect the spinal cord and caudal blood vessels, but also serve for the attachment of muscles used in locomotion and body support. Their surface area and mechanical advantage as lever arms frequently are increased by the development of spinelike processes that extend from the apices of the arches. Since these arches always lie in an intersegmental position between the muscle segments of the body, each is acted on by muscles that develop from adjacent segments. The vertebral column as a whole can be bent by the contraction of trunk and tail muscles.

The notochord lies between the developing neural and hemal arches in the embryos of all vertebrates. It expands and persists in the adults of many early fish groups, and the arches rest directly on it or on ossifications (arch bases) that develop on its surface. In the adults of other fishes and tetrapods, **vertebral bodies, or centra,** largely or completely replace the notochord, and the arches unite with them (Fig. 8–1). The centra usually are also intersegmental, but this is not always the case. In *Amia*, for example, the myosepta that separate the muscle segments of the body pass between the vertebral centra, so the centra are segmental and not intersegmental. The arches of *Amia*, however, are intersegmental. In part of the vertebral column of some fishes and early tetrapods, two centra are present per body segment, a condition referred to as **diplospondyly** (Gr. *diploos*

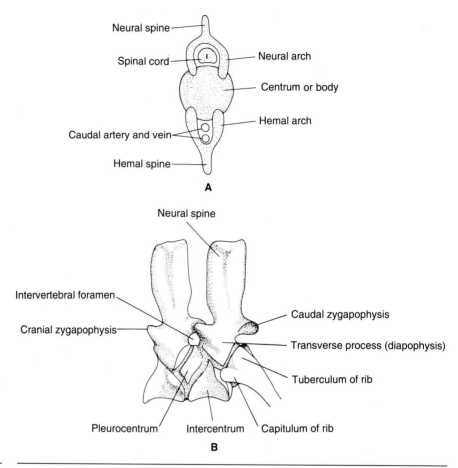

FIGURE 8–1   Vertebral structure: **A**, An end view of a caudal vertebra of a fish to show the neural and hemal arches. **B**, Lateral views of two trunk vertebrae of an early labyrinthodont; a rib is attached to one. Note the zygapophyses that unite successive vertebrae and the transverse processes to which the tuberculum of a rib attaches. Each centrum in this case consists of an anteromedian intercentrum and a posterodorsal pleurocentrum on the right and left sides. (**B**, after Schmalhausen.)

= double + *spondylos* = vertebra). Diplospondyly increases vertebral flexibility during locomotion. **Intervertebral pads,** or **disks,** which are composed of remnants of the notochord, connective tissue, and sometimes fibrocartilage, lie between and link adjacent centra.

Both the anterior and the posterior surfaces of the centra are concave in many fishes and some early tetrapods, a shape termed **amphicoelous** (Gr. *amphi* = both + *koilos* = hollow) (Fig. 8–2). The intervertebral pads occupy the concavities. Considerable notochordal tissue is present, and a thin strand of notochord may perforate each centrum. Many vertebral shapes occur in terrestrial vertebrates because the degree and direction of vertebral column movement and the amount of support needed varies among groups. Strength can be increased when one

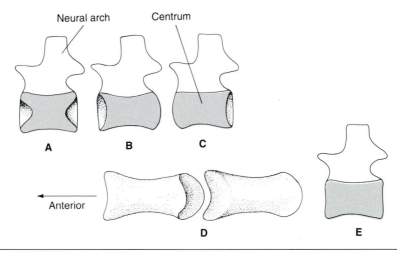

FIGURE 8–2   The shapes of centra: **A**, Amphicoelous. **B**, Procoelous. **C**, Opisthocoelous. **D**, Heterocoelous. **E**, Acoelous. A ventral view of two successive centra is shown in **D**; the others are shown in sagittal section. Anterior is toward the left. (**D**, after Wake.)

surface of a centrum is concave and the other convex; the convex surface of one centrum fits into the concavity of the adjacent one. If the concavity is on the anterior surface, the shape is **procoelous** (Gr. *pro* = before); if on the posterior surface, **opisthocoelous** (Gr. *opisthen* = behind). In **heterocoelous** vertebrae, one surface of a centrum is saddle shaped in the horizontal plane, and it articulates with a reciprocal, vertical, saddle-shaped surface on the centrum of the adjacent vertebra. This arrangement, which is found in the neck vertebrae of birds, combines considerable strength with great freedom of movement. Both central surfaces of mammals and some other vertebrates are nearly flat, a shape termed **acoelous.**

Some teleosts and nearly all terrestrial vertebrates have paired articular processes, called **zygapophyses** (Gr. *zygon* = yoke + *apo* = away from + *physis* = growth), that extend forward and backward from the neural arch (Fig. 8–1**B**). The articular surfaces of the caudal zygapophyses face ventrally (and sometimes slightly laterally) and overlap the upward, medially facing articular surfaces of the cranial zygapophyses of the next posterior vertebra. Zygapophyses strengthen the union between vertebrae and restrict motion in certain directions. Paired **transverse processes** of several types usually extend laterally from the centrum or neural arch. Since they usually serve for rib attachment, we will consider them when we discuss ribs.

Vertebral Development

Vertebrae develop embryonically near the intersections of the connective tissue septa of the body. Since cartilage and bone frequently form in these septa, they are collectively called skeletogenous septa (Fig. 8–3). The embryonic **myosepta** lie in the transverse plane between developing muscle segments, or myotomes.

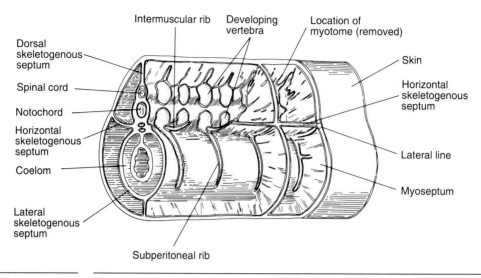

Intermuscular rib   Developing vertebra   Location of myotome (removed)

Dorsal skeletogenous septum

Spinal cord

Notochord

Horizontal skeletogenous septum

Coelom

Lateral skeletogenous septum

Skin

Horizontal skeletogenous septum

Lateral line

Myoseptum

Subperitoneal rib

FIGURE 8–3   A stereodiagram of the skeletogenous septa of a jawed fish. Vertebral elements and ribs develop at the intersections of these septa. (After Goodrich.)

Myosepta extend inward from the skin to the spinal cord and notochord, where they intersect with longitudinal, vertical septa lying in the sagittal plane. A **dorsal skeletogenous** septum extends from the spinal cord to the middorsal line, and in the tail, a **ventral skeletogenous** septum extends from the notochord to the midventral line. In the trunk, the ventral skeletogenous septum splits adjacent to the body cavity and passes ventrally on each side of the coelom as **lateral skele-togenous** septa. Finally, gnathostomes have a pair of **horizontal skeletogenous** septa that lie close to the frontal plane. In fishes they extend inward from the skin at the level of the lateral line to the notochord and spinal cord.

The components of the vertebrae develop in or near intersections of these septa from concentrations of mesenchymal cells, most of which are derived from the sclerotomal portion of the embryonic somites. The condensed mesenchymal cells transform into cartilage-forming cells, and in most species the cartilage is later replaced by bone. Membrane bone may be added peripherally to the cartilage replacement bone in some cases.

Since the sclerotomes are a part of the somites, they are at first segmental in position and lie adjacent to the myotomes. Some migration of sclerotomal tissue must occur because most or all of the vertebral components become interseg-mental and lie between the myotomes. How this shift in position comes about, particularly with respect to the centra, is not entirely clear, and it varies among vertebrate groups. The first indication of vertebral formation is the condensation of mesenchymal cells lateral to the spinal cord in an intervertebral position between the developing spinal nerves. These condensations will form the neural arches. Hemal arches form intersegmentally in the caudal region between branches of the caudal artery and vein. As the neural and hemal arches differentiate, they first form paired blocks of cartilage collectively called **arcualia** (Fig. 8–4A). The arcualia generally ossify later in development. Usually, only a single pair

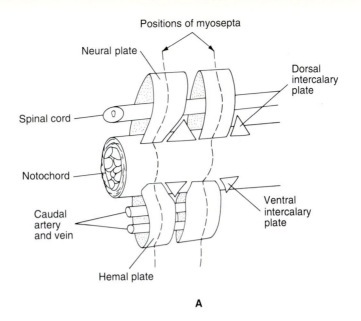

A

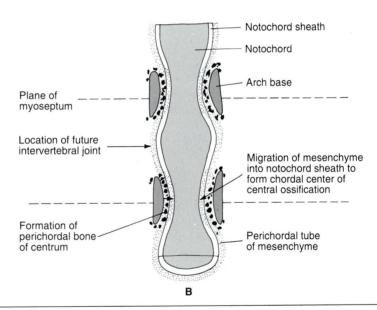

B

FIGURE 8–4 Vertebral development in anamniotes: **A**, A lateral view of the formation of the neural and hemal arches of a fish. **B**, A frontal section through the notochord and perichordal tube to show the elements that may contribute to the centra in a fish.

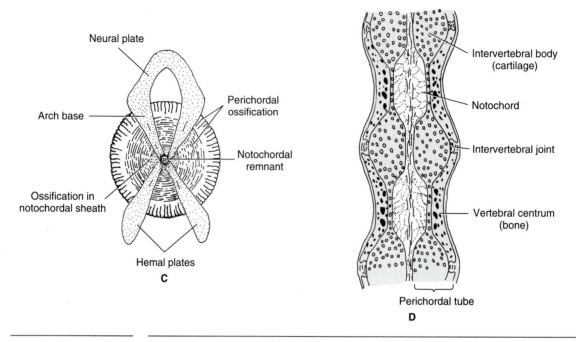

FIGURE 8–4 *continued*

C, A transverse section through the vertebra of a bony fish in which most of the centrum forms as a perichordal ossification, but the arch bases and some ossification within the notochordal sheath make a contribution. **D**, A frontal section of the development of the centra in contemporary amphibians. (*C*, after Wake. *D*, after Jollie.)

of arcualia, **neural plates,** develops around the spinal cord for each vertebra, but a small second pair, called **dorsal intercalary plates,** forms between the neural plates in cartilaginous and some early bony fishes. Well-developed **hemal plates** develop around the caudal artery and vein and protect them from various forces; in particular, they prevent the vessels from collapsing when waves of muscle contraction sweep down the tail during the locomotion of fishes. Small **ventral arch bases,** which are serially homologous to the hemal plates, frequently form beneath the developing trunk vertebrae and usually extend a short distance into the lateral skeletogenous septa. These extensions form the basapophyses onto which certain ribs of fishes attach (p. 257). Hemal arches are not needed in the trunk because the major artery and veins lie between the vertebral column and the body cavity and are not surrounded by muscles. Cartilaginous and early bony fishes also may have small **ventral intercalary plates** lying between the bases of the hemal arches.

In most fishes, other mesenchymal cells condense to form a continuous **perichordal tube** around the notochord and its sheath. The centra of a cartilaginous fish develop by the chondrification of sclerotomal cells that invade the notochordal sheath at intersegmental intervals. These are **chordal** centra. The centra of a bony fish develop primarily from the direct intersegmental deposition of bone by sclerotomal cells that form the continuous perichordal tube (Fig. 8–4**B**). These are primarily **perichordal** centra, but some cells may invade the notochordal sheath and ossify there. As the ossifying centra enlarge, they may spread over the bases of the dorsal and ventral arches and incorporate them into the centra (Fig. 8–4**C**).

Centrum formation is somewhat similar in contemporary amphibians (Fig. 8–4**D**). Cartilage and bone begin to form in a continuous perichordal tube and gradually thicken. Cartilage becomes particularly thick in the intervertebral areas, where it forms a series of **intervertebral bodies.** In many salamanders each intervertebral body remains as a distinct cartilaginous body between ossified amphicoelous centra. In frogs each intervertebral body usually ossifies and fuses with either the front of the centrum (opisthocoelous vertebra) or the posterior surface (procoelous vertebra).

Centrum formation is different in amniotes, for a continuous perichordal sheath never develops. Mesenchymal cells become more concentrated or densely packed during early development in the caudal half of each sclerotome than in the anterior portion (Fig. 8–5**A** and **B**). The cells in the caudal part of the sclerotomes further concentrate around the notochord in an intersegmental position (Fig. 8–5**C**). The vertebral centra develop here primarily from cells that are derived from the caudal parts of the sclerotomes. The fate of the more scattered cells in the anterior portions of the sclerotome is less certain. Anatomists have long believed that they migrated forward to join the posterior half of the next anterior segment. According to this view, each vertebral centrum is derived from sclerotomal cells from two adjacent segments. More recently investigators have questioned this notion and contend that the cells in the anterior parts of the sclerotomes dissipate and make no contribution to the centra.

Ribs and Transverse Processes

Ribs develop in the myosepta of the trunk and tail. At first they are cartilaginous, but they ossify later in development in most vertebrate groups. Ribs strengthen the myosepta and body wall and provide attachment for many trunk and tail muscles. Several types of ribs are found in fishes (Fig. 8–3). **Intermuscular ribs** develop in the myosepta at their intersection with the horizontal skeletogenous septa. They extend laterally between the epaxial muscles lying above the horizontal septum and the hypaxial muscles beneath it. They attach by single heads to the lateral surface of the centra. **Subperitoneal ribs** develop in the myosepta at their intersection with the lateral skeletogenous septa. They attach by single heads to lateral extensions of the ventral arch bases that are known as the **basapophyses.** Basapophyses are transverse processes that are serially homologous to the bases of the hemal arches. Subperitoneal ribs extend ventrally in the body wall next to the coelom.

Some fishes have intermuscular ribs; some have subperitoneal ribs; some have both types; and some have ribs that do not fit easily into either category. Chondrichthyan fishes, for example, have short ribs in an intermuscular position, but these attach to the centra far ventrally on basapophyses, an attachment usually occupied by subperitoneal ribs. The potential for rib development exists in the myosepta where they intersect with other skeletogenous septa. Ribs in one group are homologous to the ribs of another group in a general sense, but probably ribs

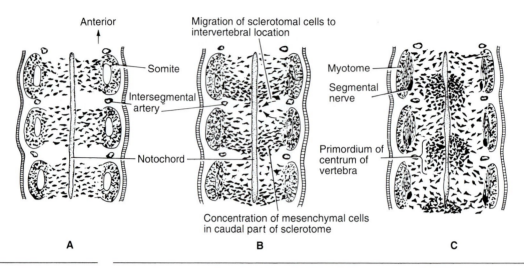

A   **B**   **C**

FIGURE 8–5   Three frontal sections to show the intervertebral development of the amniote centrum from the segmental sclerotomes. (After Corliss.)

have developed independently and in somewhat different sites in the various fish groups. Precise homologies are not always possible.

The ribs of tetrapods also serve for the attachment of trunk muscles, and their movements in most amniotes are an important factor in ventilating the lungs. Their formation is somewhat different from that seen in fishes. They are subperitoneal in position but attach to the vertebrae farther dorsally than either the subperitoneal or the intermuscular ribs of fishes. Typically, each tetrapod rib has two points of articulation with a vertebra (Figs. 8–1**B** and 8–6). A terminal **capitulum** articulates

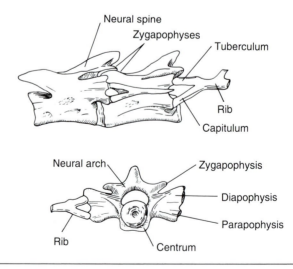

FIGURE 8–6   A lateral and an end view of the trunk vertebra of a salamander, *Necturus*, to show rib attachments. (After Goodrich.)

on the lateral surface of a centrum, sometimes on a small transverse process known as a **parapophysis**. The **tuberculum** of a rib, which is slightly distal to the capitulum, articulates on a transverse process that extends laterally from the base of the neural arch. This type of transverse process is known as a **diapophysis** (Gr. *dia* = through, across). The tuberculum of a rib is not always present. A pair of embryonic ribs develop on all the tetrapod vertebrae except the more distal, caudal ones. These rib primordia may form distinct ribs in adults, or they may become inseparably fused to the diapophysis and parapophysis to form an enlarged transverse process called a **pleurapophysis** (Gr. *pleura* = side, rib).*

The Evolution of the Trunk Skeleton

Fishes

The Vertebral Axis.   The evolution of the trunk skeleton and its degree of regional differentiation correlate closely with the environment in which a species lives (water or land), its method of locomotion, and the forces acting on its body. Water is a relatively dense and viscous medium when compared to air; it provides a great deal of buoyancy but offers resistance to motion. The trunk skeleton of fishes, therefore, plays only a small role in supporting them against gravity, but it is central to locomotion. The vertebral column acts primarily as a compression strut that enables a fish to push through the water. It also resists telescoping of the body when longitudinal trunk and tail muscles contract. Since telescoping cannot occur, and the vertebral axis is jointed, muscle contraction is converted to a series of lateral undulations that sweep down the body and provide the propulsive thrust for most fishes (p. 346).

The first trunk vertebra articulates with the skull and is modified slightly in this connection. The remaining **trunk vertebrae** are essentially alike, differing slightly among species with respect to the amount of the trunk that undulates. **Caudal vertebrae** differ in having hemal arches that protect the caudal artery and vein from forces developed by the surrounding tail muscles. Hemal arches play an important role in the hemodynamics of blood circulation in fishes (p. 682).

Neural arches are poorly developed or absent in hagfishes, but in other groups they provide firm attachment for dorsal epaxial muscles. Vertebral centra are absent, and an expanded notochord persists in the adults of agnathans, early chondrichthyans, lungfishes, the living coelacanth (*Latimeria*), and chondrosteans (Fig. 8–7**A**). The vertebral axis need not be strong because none of these fishes push rapidly through the water. The vertebral axis is stronger in other groups. Advanced cartilaginous fishes have chordal centra that chondrify within the sheath of the notochord. Frequently, the centra are strengthened by the deposition of calcium salts deep within the cartilage next to a small, persistent notochord. The spinal cord of cartilaginous fishes is also completely ensheathed by plates of cartilage

---

*Technically, a pleurapophysis is the rib component of the transverse process, but since it is indistinguishable from the other components in an adult, the term is usually applied to the entire transverse process.

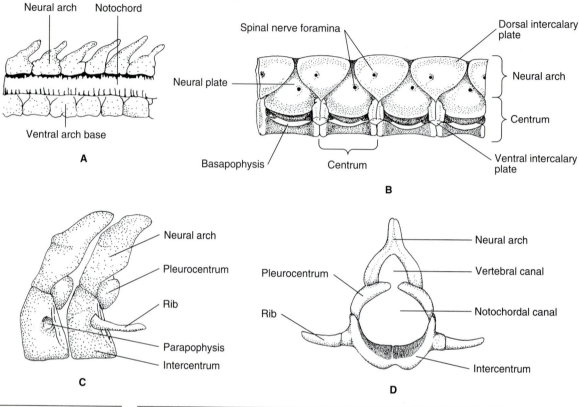

FIGURE 8–7 Some vertebrae of fishes. **A**, A lateral view of the vertebral axis of a sturgeon in which the notochord persists throughout life. **B**, A lateral view of the trunk vertebrae of a dogfish. **C**, A lateral view of the trunk vertebrae of a rhipidistian. **D**, An end view of a trunk vertebra of a rhipidistian. (**A**, After Goodrich. **B**, from Walker and Homberger. **C** and **D**, after Jarvik.)

rather than by distinct neural arches that are interrupted at segmental intervals (Fig. 8–7**B**). Their continuous neural arch does not unduly restrict lateral undulations because cartilage is flexible. Advanced bony fishes have perichordal centra that ossify primarily by the deposition of bone around the notochord. The rhipidistian ancestral to terrestrial vertebrates had unusual centra (Fig. 8–7**C** and **D**). The notochord persisted but was flanked by small ossifications that strengthened the vertebral axis. Pairs of small **pleurocentra** lay along the dorsolateral surface of the notochord just caudal to the bases of the neural arches, and U-shaped **intercentra** lay along the ventrolateral surface of the notochord just cranial to the pleurocentra. In certain regions of the vertebral column, it is evident that the intercentra developed by the fusion of a pair of lateral halves. Each centrum is thus composed of three blocks of bone—a pair of pleurocentra and an intercentrum.

Ribs are absent in lampreys and hagfishes and poorly developed in chondrichthyan fishes, but they are well developed and often numerous in bony fishes. Many bony fishes have both intermuscular and subperitoneal ribs, and some also have accessory ribs that develop in other parts of the myosepta. Since fishes exchange gases with the environment through their gills, their ribs have no respiratory function. Rather, the ribs are part of the locomotor apparatus for they strengthen the myosepta, to which the longitudinal muscle fibers of the trunk and tail attach, and they help transfer forces to the vertebral axis.

The vertebral axis ends in an expanded caudal fin that delivers a strong thrust to the water. Caudal fins, along with all other fins, are supported distally by slender fin rays that develop in the skin on each side of the fin. In chondrichthyan fishes the fin rays are unsegmented, horny **ceratotrichia** (Gr. *kerat-* = horn + *trich-* = hair). Osteichthyans have **lepidotrichia** (Gr. *lepid-* = scale) composed of rows of small bony segments that evolved from rows of bony scales (Fig. 8–8). Lepidotrichia usually branch distally. The shape of the caudal fin differs among groups of fishes in relation to their methods of swimming and buoyancy control (p. 350). Osteostracans (among the ostracoderms), most chondrichthyans, and primitive osteichthyans have a **heterocercal** tail (Gr. *heteros* = other + *kerkos* = tail) in which the vertebral axis turns upward into an expanded dorsal lobe, which it stiffens (Fig. 8–8). This appears to have been the ancestral type of caudal fin.

An early variation of the heterocercal tail was the reverse heterocercal, or **hypocercal,** tail found in some ostracoderms (heterostracans and anapsids). The vertebral axis turned downward into an expanded ventral lobe. This tail appears to be correlated with a unique pattern of feeding in which the head pointed downward (Fig. 3–3**C** and **D**).

The caudal fins of more advanced bony fishes have become symmetrical in one of two ways. In the line of evolution toward teleosts the caudal fin became symmetrical only externally. Internally, a reduced vertebral axis tips sharply upward. Most of the caudal fin is comparable to only the ventral lobe of the heterocercal tail and is supported by enlarged hemal spines known as **hypural bones** (Gr. *hypo* = under + *oura* = tail). Only a small part of the fin is supported by neural spines, or **epural bones.** This type of tail is known as **homocercal.** Early neopterygians have an abbreviated heterocercal tail that is intermediate between the heterocercal and homocercal types.

In the line of evolution toward contemporary lungfishes and coelacanths, the tail became symmetrical in another way. The vertical axis straightened out, and the dorsal and ventral lobes of the fin became equal in size. This type of caudal fin is known as a **diphycercal** tail (Gr. *diphyes* = twofold).

Median Fins.   Most fish have one or two **dorsal** fins, and many species have an **anal** fin on the tail located just caudal to the cloacal aperture. Dorsal and anal fins help stabilize the fish by reducing the tendency to roll from side to side and to yaw from left to right as the animal moves through the water. Some species move slowly and maneuver by undulating these fins. The external parts of the dorsal and anal fins, like the caudal fins, are supported by ceratotrichia or lepidotrichia (Fig. 8–9). Frequently, one or more spines form a cutwater at the leading edge of a fin. Usually, each spine has evolved by the elongation of a single scale, but some have evolved from the fusion of paired lepidotrichia. The large spine at the leading edge of the dorsal fin of a goldfish or carp is an example. Like lepidotrichia, this spine can be divided into left and right components, for which reason it is sometimes

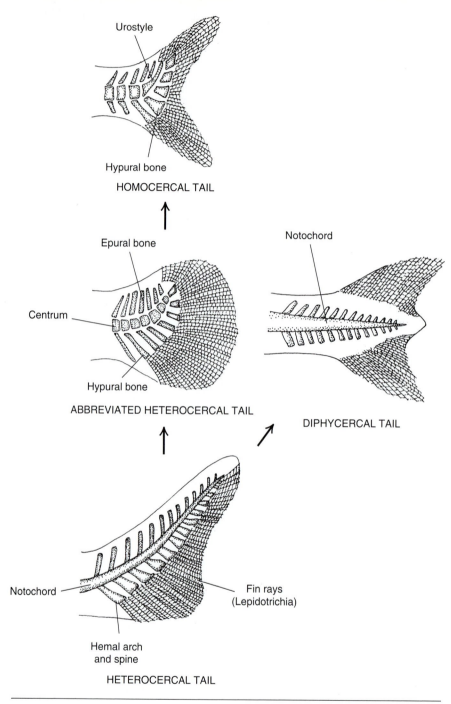

Urostyle

Hypural bone

HOMOCERCAL TAIL

Epural bone

Centrum

Hypural bone

ABBREVIATED HETEROCERCAL TAIL

Notochord

DIPHYCERCAL TAIL

Notochord

Fin rays
(Lepidotrichia)

Hemal arch
and spine

HETEROCERCAL TAIL

FIGURE 8–8   A diagram of the evolution of major caudal fin types in fishes.

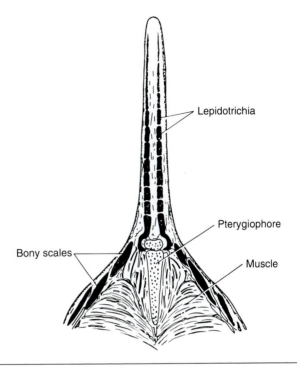

Lepidotrichia

Pterygiophore

Bony scales

Muscle

FIGURE 8–9   A transverse section through the dorsal fin of a bony fish. (After Goodrich.)

called a spinous ray. Poison glands are associated with the spines in some species. The fin rays of the dorsal and anal fins attach to deeper, internal rods of cartilage or bone that collectively are called **pterygiophores** (Gr. *pterygion* = wing or fin + *phoros* = bearing). Pterygiophores, in turn, may attach to the neural or hemal spines.

Amphibians

The trunk skeleton was greatly affected by the transition of vertebrates from water to land. Air, which is less dense than water, provides little support against gravity, but on the other hand it offers little resistance to movement. Thrusts of the appendages on the ground, rather than of the trunk and tail against the water, provide the major propulsive forces for locomotion, but lateral undulations of the trunk and tail assist the appendages in many early amphibians and early reptiles, and in such contemporary species as salamanders and snakes. The vertebral column acts primarily as a supporting beam that transfers the weight of the body to the girdles and appendages when the animal walks. It must resist bending in the vertical plane so these low-slung animals do not drag along the ground. The anterior part of the vertebral column is specialized to allow for the mobility of the head that characterizes terrestrial vertebrates. Head and trunk no longer move as a unit, as they must in fishes pushing through the denser water.

The individual vertebrae of the earliest amphibians were stronger than those of their rhipidistian ancestors, but apart from this, there is a remarkable resemblance in vertebral morphology. Early labyrinthodonts had **rhachitomous** vertebrae (Gr. *rhachis* = spine + *tomos* = cut) in which each centrum was "cut" into a large intercentrum and two smaller pleurocentra (Fig. 8–10). This closely resembled the diplospondylous vertebrae of the rhipidistians (Fig. 8–7C and **D**). The notochord persisted but was more constricted than in rhipidistians. The intercentrum was a U-shaped or sometimes circular element, and the pleurocentra were more massive than in rhipidistians. Neural arches were well developed and bore strong zygapophyses, but they articulated with the pieces of the centrum rather than being fused with them.

Vertebrae became stronger in later labyrinthodonts in different ways in different evolutionary lineages (Fig. 8–10). The pleurocentra became reduced in some labyrinthodonts, while the intercentrum expanded and eventually became the sole component of the centrum. These vertebrae are described as **stereospondylous** (Gr. *stereos* = solid + *spondyle* = vertebra). Contemporary amphibians have vertebrae with a single centrum, but we do not know whether it represents an intercentrum, the pleurocentra, or a union of these elements. Embryology offers no clues, and the origin of modern groups from the labyrinthodonts is not well represented in the fossil record. The neural arch of contemporary species is fused to the centrum. A rhachitomous vertebra persisted for a long time in the line of labyrinthodont evolution toward reptiles, but the paired pleurocentra gradually enlarged and fused to become the dominant part of the centrum. The intercentrum was reduced to a small piece located ventrally between the definitive centra. In one branch of labyrinthodont evolution, both the pleurocentra and intercentra became large, disk-shaped elements. These are described as **embolomerous** vertebrae.

Labyrinthodonts had ribs on most of their vertebrae. Those on the trunk were strong and well developed, and they curved ventrally. It is unknown whether the anterior trunk ribs articulated with a sternum ventrally. No sternum is present in the fossils, although an unossified plate of cartilage may have been present. Each rib was two headed (Fig. 8–10). The **capitulum** of a rib articulated on the intercentrum, and its **tuberculum** articulated on the diapophysis of the neural arch. Since some lateral undulations of the trunk and tail probably occurred during locomotion, the ribs continued to provide the attachment for locomotor muscles, but an important function probably was to strengthen the body wall of these relatively large animals. Ribs may also have prevented the weight of the body from collapsing the lungs and abdominal viscera when these animals lay on the ground. We do not know whether rib movements were used in ventilating the lungs. Ribs are relatively insignificant in modern amphibians, whose body weight is small. Salamanders, caecilians, and very primitive frogs have short ribs (Fig. 8–11). In other frogs short ribs have fused to the sides of the vertebrae to form pleurapophyses. Frogs and most salamanders have a small **sternum** associated with the ventral part of the pectoral girdle.

The trunk skeleton of an amphibian has more regional differentiation than that of a fish because the functions it performs and the stresses it resists are more varied (Fig. 8–11). A single **cervical vertebra** (L. *cervix* = neck), the **atlas,** allows the head some mobility. The trunk vertebrae are essentially alike. A single **sacral**

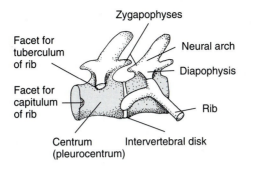

Zygapophyses

Facet for
tuberculum
of rib

Neural arch

Diapophysis

Facet for
capitulum
of rib

Rib

Centrum
(pleurocentrum)

Intervertebral disk

LATE REPTILE, MAMMAL

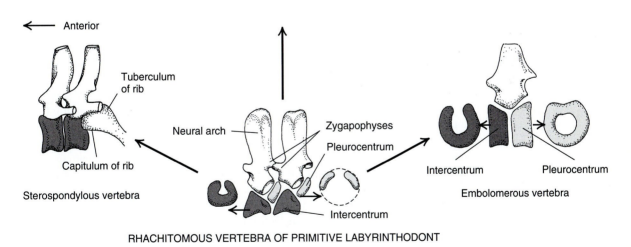

Neural arch

Tuberculum of rib
attaches to diapophysis

Pleurocentrum
or centrum

Intercentrum

Capitulum of rib attaches
to intercentrum

EARLY REPTILE

Anterior

Tuberculum
of rib

Neural arch

Zygapophyses

Pleurocentrum

Capitulum of rib

Intercentrum

Sterospondylous vertebra

Intercentrum

Pleurocentrum

Embolomerous vertebra

RHACHITOMOUS VERTEBRA OF PRIMITIVE LABYRINTHODONT

FIGURE 8–10   A lateral view of the evolution of trunk vertebrae in terrestrial vertebrates.

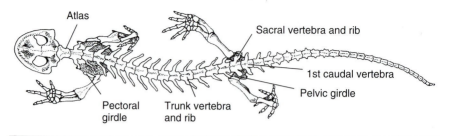

FIGURE 8–11   Dorsal view of the skeleton of a salamander showing regional specialization of the vertebral column. (From Romer and Parsons, after Schaeffer.)

vertebra (L. *sacr-* = sacred, so called because this part of the body of domestic animals was offered in sacrifices) and pair of ribs articulate with an expanded pelvic girdle and transfer body weight to the hind legs. Weight transfer to the pectoral girdle of nearly all tetrapods is not by bone but by muscular connections between the trunk skeleton and the scapula. **Caudal vertebrae** follow the sacral vertebra. Hemal arches are absent from the first several caudal vertebrae but are present on the caudal vertebrae that lie behind the cloaca. Ribs and zygapophyses are usually absent. The caudal vertebrae become progressively smaller toward the tail tip, and the hemal and neural arches gradually disappear.

Trunk length varies greatly among amphibians. Some caecilians have 285 vertebrae; most frogs only have 8 plus a long, terminal **urostyle** (Gr. *oura* = tail + *stylos* = pillar) that evolved from several fused caudal vertebrae. The short, strong, vertebral column of a frog is adapted for jumping and swimming by powerful thrusts of the hind legs (p. 379).

Reptiles

Correlated with their increased activity and greater penetration of the terrestrial environment, reptiles have evolved much stronger trunk skeletons than amphibians. Intercentra generally are lost in contemporary species (Fig. 8–10), although they persist in parts of the vertebral column in some groups. (The term *intercentrum* derives from its position in ancestral reptiles.) Intercentra may remain in the tail as points of attachment of the hemal arches, now called **chevron bones,** or they may become incorporated in the chevron bones. *Sphenodon* retains intercentra in the trunk region, and many lizards retain them in the neck. The strong, definitive centrum of reptiles has evolved from the fusion and expansion of the pleurocentra. The neural arch has fused to it. Remnants of the notochord perforate the centra of *Sphenodon*, but otherwise notochord remnants, if present, are limited to the intervertebral disks.

Ribs are present on most of the vertebrae extending from the atlas through the anterior caudal vertebrae. Those in the anterior part of the trunk are long and articulate by costal cartilages with a sternum (Fig. 8–12). The sternum also unites with the pectoral girdle. With the loss of the intercentrum to which it originally attached, the capitulum of a rib usually has an intervertebral articulation, part of it attaching to the posterior margin of one centrum and part to the anterior margin

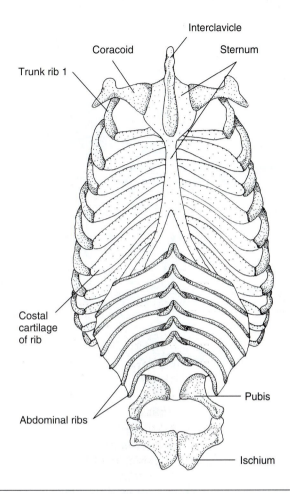

FIGURE 8–12   A ventral view of the ribs and sternum of a crocodile. The ventral parts of the pectoral and pelvic girdles are also shown. (After Romer and Parsons.)

of the next caudal centrum (Fig. 8–10). Crocodiles retain primitive, two-headed ribs, but the tuberculum tends to be lost from the ribs in other contemporary reptiles.

Regional differentiation of the vertebral axis is more extensive than in amphibians. All reptiles have more cervical vertebrae, which permit a freer movement of the head. The atlas and the second cervical vertebra, which is called the **axis,** begin to show some of the modifications for head movements that reach a higher degree of specialization in mammals (Fig. 8–16). The remaining cervical vertebrae are characterized by having relatively short ribs. Turtles and many lizards have eight cervical vertebrae; crocodilians, nine; and numerous extinct groups had more. Often, many of the cervical vertebrae and the anterior trunk vertebrae have a ventral projection from the centrum known as a **hypapophysis** (Gr. *hypo* = under). Hypapophyses are points of attachment for muscles and ligaments that help move the head (Fig. 8–13**B** and **C**).

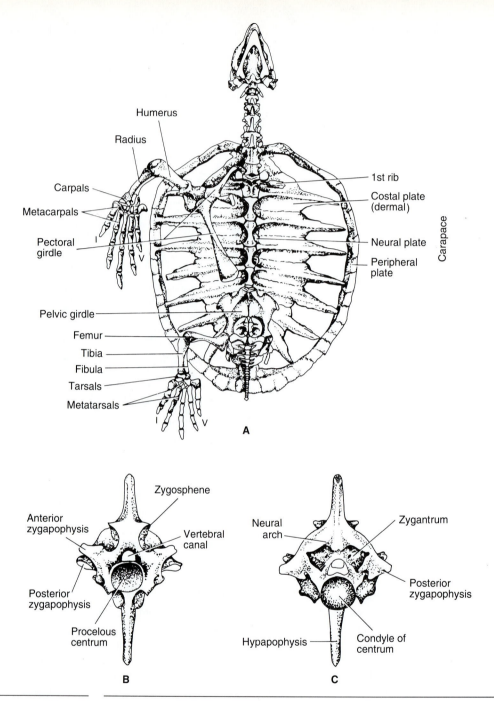

Humerus

Radius

Carpals

Metacarpals

Pectoral
girdle

1st rib

Costal plate
(dermal)

Neural plate

Peripheral
plate

Carapace

Pelvic girdle

Femur

Tibia

Fibula

Tarsals

Metatarsals

A

Zygosphene

Anterior
zygapophysis

Vertebral
canal

Posterior
zygapophysis

Procelous
centrum

B

Neural
arch

Zygantrum

Posterior
zygapophysis

Hypapophysis

Condyle of
centrum

C

FIGURE 8–13 Some modifications of trunk vertebrae in reptiles: **A**, A ventral view of the skeleton of a sea turtle, *Eretmochelys*. Trunk vertebrae and ribs are united with the carapace. **B** and **C**, Anterior and posterior views, respectively, of two adjacent procoelous vertebrae from the neck region of a boa constrictor. (After Bellairs.)

The first trunk vertebra bears a pair of ribs that articulate by **costal cartilages** with the sternum (Fig. 8–12). Only two additional ribs reach the sternum in *Sphenodon,* but eight or nine reach it in crocodilians. Trunk ribs help support the body and protect the lungs from excess pressure when the animal is lying on the ground. Movements of the trunk ribs also are the major mechanism for ventilating the lungs in most reptiles (p. 655). Ribs on the remaining trunk vertebrae decrease in size toward the sacrum. The most posterior trunk ribs of lizards fuse with the vertebrae to form pleurapophyses.

Reptiles have a stronger sacral region than amphibians, for most have two sacral vertebrae and ribs. Some groups have more. The sacral ribs are distinct elements in most contemporary reptiles, but they are represented by pleurapophyses in lizards.

Tail length and the number of caudal vertebrae vary considerably. Chevron bones are present on a few caudal vertebrae posterior to the cloaca. Distinct ribs are present on the proximal caudal vertebrae of many reptiles but have become pleurapophyses in lizards.

Most lizards can spontaneously drop off, or **autotomize,** much of the tail when attacked by a predator. The tail writhes on the ground and attracts the attention of the predator as the rest of the animal slinks away and eventually regenerates a new tail. One or more caudal vertebrae have an autotomy septum that extends through the vertebra, and cleavage of the tail occurs at these points. A few salamanders can also autotomize their tails, but separation occurs between vertebrae.

Many reptiles and a few fossil amphibians have **abdominal ribs** embedded in the ventral abdominal wall (Fig. 8–12). Presumably, they help support and protect this part of the body. Some abdominal ribs, known as **gastralia,** are dermal and may have evolved from rows of bony scales. Others, called **parasternalia,** are cartilage or cartilage replacement bone. The sternum may have evolved from parasternalia.

Reptiles have undergone such an extensive adaptive radiation that a book of this scope can consider only a few of the specializations of their trunk skeletons. The trunk region of a turtle is short and includes only ten vertebrae and ribs. Most of these ribs have united with dermal ossifications to form the **carapace,** or dorsal part of the shell (Fig. 8–13**A**). The **plastron,** or ventral part of the shell, is composed of dermal ossifications and also incorporates gastralia and the ventral dermal elements of the pectoral girdle (p. 285). The elongated body of a snake may contain 200 or more vertebrae. A snake moves primarily by lateral undulations of the trunk and tail. Not surprisingly, extra intervertebral joints help articulate such a long and flexible body. A pair of processes called **zygosphenes** (Gr. *zygon* = yoke + *sphen* = wedge), which are located on the anterior surface of a neural arch dorsal to the zygapophyses, project into a pair of sockets, the **zygantra** (Gr. *antron* = cave) on the posterior surface of the next anterior neural arch (Fig. 8–13**B** and **C**). Some lizards also have these extra articulations. Since a snake lacks legs and a sternum, distinct vertebral regions, apart from the atlas-axis complex and tail, cannot be recognized. All the vertebrae, except for the caudal ones, bear ribs that curve ventrally and may attach to the large ventral scales. In addition to supporting the body and ventilating the lungs, rib movements are important in locomotion in some species of snakes.

Birds

Modifications of the trunk skeleton of birds are correlated with their unique patterns of locomotion—flying and bipedal walking. Since the pectoral appendages are so highly specialized as wings, they cannot be used in another way. The head and bill perform all the feeding, nest-building, and other manipulative functions that the pectoral appendages often have in other terrestrial vertebrates. An exceedingly long and flexible neck allows for the great mobility of the head (Fig. 8–14**A**). A bird can turn its head 180 degrees to the left or right. Contemporary species have 11 to 25 cervical vertebrae with heterocoelous centra (Fig. 8–2**D**). Since the trunk vertebrae form the fulcrum on which the wings move up and down, the back must be short and rigid. It contains only three to ten vertebrae, most of which are partially fused. Bipedal locomotion, and the action of the hind legs as shock absorbers when the bird alights, necessitates a strong sacrum. Adjacent trunk and tail vertebrae fuse with the two sacral vertebrae of ancestral reptiles to form a **synsacrum** that includes 10 to 23 vertebrae (Fig. 8–14**B**). The synsacrum, in turn, has fused with the pelvic girdle. Ancestral birds had a long, reptilian tail, but in contemporary species the terminal four to seven caudal vertebrae have united to form a large element, the **pygostyle** (Gr. *pyge* = rump + *stylos* = pillar), that supports the large tail feathers. Six or seven more proximal caudal vertebrae are unfused and allow for the changes in tail position needed in flight and maneuvering. Movement within the vertebral column is limited to the cervical region, to the region between the trunk vertebrae and synsacrum, and to the proximal caudal vertebrae.

Short cervical ribs are present on most of the cervical vertebrae. The trunk vertebrae bear ribs that unite with an expanded and usually keeled sternum from which powerful flight muscles arise (Fig. 8–14**A**). The ventral, or sternal, portions of the ribs, which are represented by costal cartilages in most terrestrial vertebrates, are ossified. Dorsal and ventral parts of the ribs are united by movable joints. The rib cage is strong yet flexible enough to ventilate the lungs. Most of the dorsal rib segments bear **uncinate processes** (L. *uncinus* = hook) that overlap the next posterior ribs. These processes further strengthen the rib cage and serve as lever arms onto which certain respiratory muscles attach.

Mammals

Mammals are active, agile animals with strong yet flexible trunk skeletons that allow for free head movements in most species, transfer body weight to the girdles and appendages, may participate in locomotor movements, and have an important role in respiration. Functions and stresses vary all along the vertebral column, so no two vertebrae are exactly alike. Nevertheless, the vertebrae can be sorted into five groups (Fig. 8–15).

With few exceptions, all mammals have seven **cervical vertebrae** regardless of neck length. The seven cervical vertebrae in the giraffe's neck are very long,

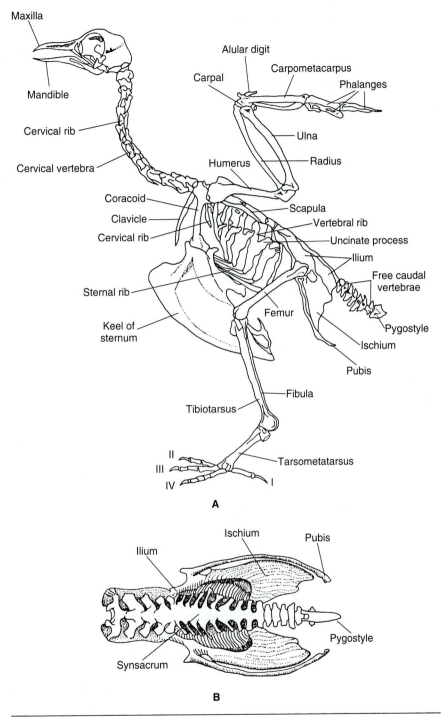

FIGURE 8–14 The skeleton of modern birds: **A**, A lateral view of the skeleton of a pigeon. **B**, A ventral view of the synsacrum and adjacent elements of a chicken. (**A**, after Pettingill. **B**, after Welty and Baptista.)

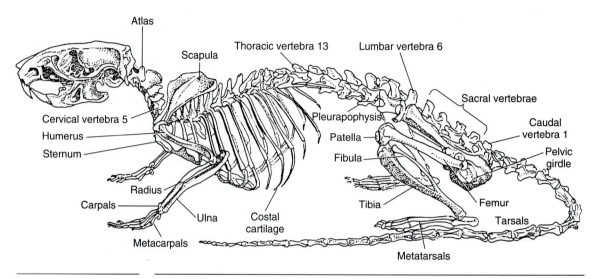

FIGURE 8–15 A lateral view of the skeleton of a rat. (After Hebel and Stromberg.)

whereas those in a whale are short and fused together in some species. The head and trunk of a whale move together as the animal pushes through the water. Fusion of some cervical vertebrae also occurs in certain terrestrial mammals as adaptations to unique modes of life. Fusion of posterior cervical vertebrae in jumping rodents and kangaroos prevents head bobbing. Fusion of four cervical vertebrae in armadillos stiffens the neck in this burrowing species (Fig. 8–17**A**).

Cervical vertebrae are characterized by having pleurapophyses, each of which is usually perforated by a **transverse foramen** through which vertebral blood vessels pass on their way to the brain. The transverse foramen lies between the capitulum and tuberculum of the ancestral cervical rib.

Since one or more vertebral elements usually contribute to the back of the skull during embryonic development (p. 206), the craniovertebral joint of most vertebrates is a modified intervertebral joint. In all terrestrial vertebrates, the first cervical vertebra, the **atlas,** is specialized to permit the head to move independently of the trunk. The second cervical vertebra, the **axis,** contributes to this movement in some reptiles. The atlas and axis become highly specialized in the evolution from late synapsid reptiles to mammals and, together with changes in the occipital condyle on the back of the skull, form a universal joint that allows exceptional freedom of movement of the head. The occipital condyle of reptiles is a hemispherical knob ventral to the foramen magnum that articulates with a concavity on the centrum of the atlas. Movement occurs here but rotation and dorsoventral flexion of the head is limited by a proatlas that lies dorsally between the atlas and skull (Fig. 8–16**A**). The **proatlas** represents a neural arch of a vertebra, the rest of which is incorporated in the skull. In late synapsids, the single occipital condyle divided into a pair that shifted dorsally and came to lie lateral to the foramen magnum. Complementary articular facets evolved on the neural arch of the atlas, and the centrum of the atlas is represented only by its small intercentrum (Fig. 8–17**B**). The proatlas was lost as was the neural spine of the atlas. These

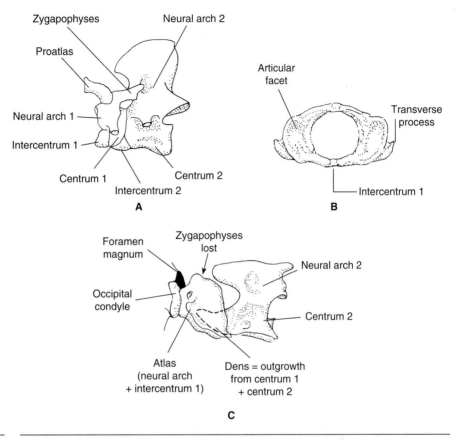

FIGURE 8–16   Evolution of the atlas and axis. **A**, A lateral view of the first two cervical vertebrae of an early synapsid, *Dimetrodon*. **B**, An anterior view of the atlas of a mammal. **C**, A lateral view of the atlas and axis of a mammal. (**A**, after Jollie. **B**, after Romer and Parsons.)

changes allowed considerable up and down movement of the head; but excessive dorsoventral flexion, which would stress the spinal cord, was prevented by changes in the axis. The centrum of the axis evolved a long, toothlike process, the **dens,** that protrudes into the vertebral canal of the atlas (Fig. 8–16**C**). A **transverse ligament** within the vertebral canal of the atlas crosses the dorsal surface of the dens and prevents hyperflexion. The dens evolved as an outgrowth of the centrum of the axis, but it probably incorporates the intercentrum of the axis and the missing pleurocentrum of the atlas. Loss of the zygapophyses between the axis and atlas, and the evolution of new articular surfaces between the centrum of the axis and the neural arch of the atlas, allows rotational movements to occur at the axis-atlas joint.

The trunk of a mammal has differentiated into **thoracic** (Gr. *thorac-* = chest) and **lumbar** (L. *lumbus* = loin) regions (Fig. 8–15). Thoracic vertebrae bear ribs, most of which connect by costal cartilages with the sternum. The short, embryonic ribs of the lumbar vertebrae fuse to the sides of the vertebrae and form conspicuous pleurapophyses. Differentiation of the trunk into the thoracic and lumbar

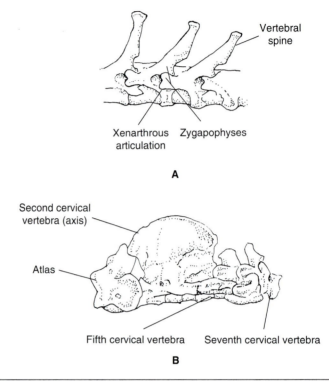

Vertebral spine

Xenarthrous articulation   Zygapophyses

A

Second cervical vertebra (axis)

Atlas

Fifth cervical vertebra   Seventh cervical vertebra

B

FIGURE 8–17   Some vertebral specializations in armadillos. **A**, A lateral view of three lumbar vertebrae showing the additional xenarthrous articulations. **B**, A lateral view of the cervical vertebrae showing fusion of the second through fifth vertebrae. (After Vaughan.)

regions is correlated with the evolution of a diaphragm that attaches to the caudal ribs and separates the pleural cavities, containing the lungs, from the peritoneal cavity. The ribs maintain the integrity of the pleural cavities as the diaphragm changes their volume and ventilates the lungs. Rib movements participate in strong ventilatory movements. Ribs also transfer the weight of the anterior portion of the trunk to the pectoral girdle via a muscular sling that extends to the scapula (see Fig. 10–10).

Most mammals have 12 to 15 thoracic vertebrae and 4 to 7 lumbar ones (Fig. 8–15). In many species extra processes on the lateral surface of thoracic and lumbar vertebrae are points of attachment for muscles and ligaments and reinforce articulations between vertebrae. For example, edentates have **xenarthrous** (Gr. *exenos* = stranger + *arthron* = joint) articulations on the lateral surface of their lumbar vertebrae ventral to the zygapophyses (Fig. 8–17**B**). These articulations help brace the trunk when armadillos burrow (Gaudin and Biewener, 1992), and probably enable anteaters to rear upright and defend themselves with their powerful front legs and claws, and help arboreal sloths support a relatively heavy body as they reach out to grasp another tree limb.

Cetaceans have only 9 thoracic and as many as 21 lumbar vertebrae. A long, flexible lumbar region participates in the vertical oscillations of the tail that are used in swimming. Vertical movements are facilitated by the loss of zygapophyses in this region. The reduction in body support that accompanies their loss is not a problem in water, but stranded whales are unable to support body weight, cannot ventilate their lungs normally, and suffocate.

The weight of the posterior part of the body of a terrestrial mammal is transferred to the pelvic girdle through a strong **sacrum,** containing at least three sacral vertebrae and their ribs, which are usually fused. More sacral vertebrae are present in species in which stresses on the sacrum are greater. All the body weight of a human being, a biped, passes through the pelvic girdle, and five sacral vertebrae and ribs are incorporated in the sacrum.

Tail functions and length vary considerably among mammals. Cetaceans have powerful tails whose distal ends bear the horizontal flukes used in locomotion. The flukes are stiffened by dense connective tissue, not by skeletal elements. Only three to five reduced caudal vertebrae are present internally in human beings. They usually are fused to form a **coccyx** (Gr. *kokkyx* = cuckoo, because of its resemblance to a cuckoo's bill), to which certain anal and perineal muscles attach. Most mammals have modest tails. They may be used in locomotion (the prehensile tails of New World monkeys), in balance (arboreal species), or as part of a mammal's behavioral repertoire to express emotions (a dog's wagging tail) or warning (white-tailed deer). Proximal caudal vertebrae may bear small chevron bones, but these are lost distally, as are most other parts of the vertebrae. Terminal vertebrae consist of only reduced centra.

The Origin of the Appendicular Skeleton

Most fishes swim by lateral undulations of their trunk and tail, and their paired fins are used to help stabilize the body and to maneuver. Most ostracoderms, however, lay on or swam slowly along the bottom and did not have the type of stabilizing mechanism provided by paired fins. Evidence for bottom dwelling is their heavy body armor, a somewhat flattened body shape, and dorsally directed eyes. Only some pterapsids among the ostracoderms had paired pectoral flaps just behind the lateral horns of their head shield. Although these flaps had no internal skeleton, we believe that muscles entered them and they served as primitive fins.

Ancestral jawed fishes became much more active animals with a streamlined body form and some reduction of the heavy, ancestral armor. They probably swam as contemporary fishes do (p. 343). As fishes became more active and adapted to more diverse habitats and ecological niches, natural selection would have favored the enlargement of any lateral protuberances or flaps that assisted the median fins in reducing roll and pitch. Large pectoral fins attached low near the front of the body would also have acted as hydroplanes and raised a heavy-bodied fish off of the bottom as it swam. Paired fins probably evolved in this context, but early stages in their evolution are not known.

A classic theory of the origin of paired limbs postulates that ancestral fishes had a continuous, longitudinal **fin fold** on each side of the trunk resembling somewhat the metapleural folds of amphioxus (Balfour, 1876; Thacher, 1877). Probably under the influence of hox genes, parts of this fold regressed and other parts enlarged to form the two paired fins or limbs of later vertebrates (Ahlberg, 1992; Tabin and Laufer, 1993). Many studies on chick embryos, most recently by Stephens et al. (1992), support the notion that a continuous, longitudinal **limb-forming zone** of opaque cells exists at an early stage in the lateral plate mesoderm. This is consistent with the classic fin fold hypothesis of limb origin, although it is unlikely that adult vertebrates ever had a continuous fin fold. The closest paleontologists have come to finding a fin fold is a small, lateral keel of enlarged bony plates in some cephalaspid and anaspid ostracoderms. Paired fins may have evolved by the enlargement of certain parts of such a keel. As fins became larger, inward extensions of cartilage or replacement bone from their bases probably coalesced to form the endoskeletal part of the girdle. Overlying dermal plates in the pectoral region formed the dermal part of the girdle and helped support the girdle by connecting it to the back of the skull. The pelvic girdle never has a dermal component. In whatever way fins and girdles formed, it is likely that they evolved independently and in somewhat different ways in different lineages. Early groups of jawed fishes are characterized by distinct types of paired fins.

The Appendicular Skeleton of Jawed Fishes

In one group of placoderms, the arthrodires, well-developed dermal plates covered the front of the trunk, and the skull had a distinctive and movable articulation with them (see Fig. 3–5**B**). Many of these thoracic plates appear comparable to the **dermal elements** of a pectoral girdle (Fig. 8–18**A**). A small **scapulocoracoid** cartilage, which formed the endoskeletal part of the pectoral girdle, lay beneath the ventral part of the thoracic armor and bore an articular surface for a pectoral fin whose structure is not well known. Another group of placoderms, the antiarchs, had a jointed pectoral appendage that was covered with dermal plates and resembled an arthropod's limb (see Fig. 3–5**C**). It may have been used to crawl or hold the front of the body elevated. Small pelvic fins were also present.

Early chondrichthyan fishes had large pectoral fins and smaller pelvic ones that had long attachments to the trunk (see Fig. 3–6**A**). These **broad-based fins** were supported by many cartilaginous pterygiophores that attached to deeper cartilages and small girdles in the body wall (Fig. 8–18**B**). They probably were effective stabilizing organs, and the pectoral ones were large enough, and attached low enough on the body wall, to have acted as hydroplanes. The fin base became constricted in later sharks. Most contemporary species have only three **basal pterygiophores** in the pectoral fin and two in the pelvic fin (Fig. 8–18**C** and **D**). **Radial pterygiophores** extend from the basal pterygiophores and support the horny, distal ceratotrichia, which develop in the skin and are similar to those in the median fins (p. 261). Such **narrow-based fins** are not merely stabilizers; because they can be rotated at their bases, they can also be used in turning, braking, and other maneuvers. In the absence of dermal bone in the pectoral girdle, the scapulocoracoid of a cartilaginous fish is larger than that of other groups and provides a sufficient area for the attachment of muscles. Part of it extends across the midventral line and interconnects the left and right girdles. The pelvic

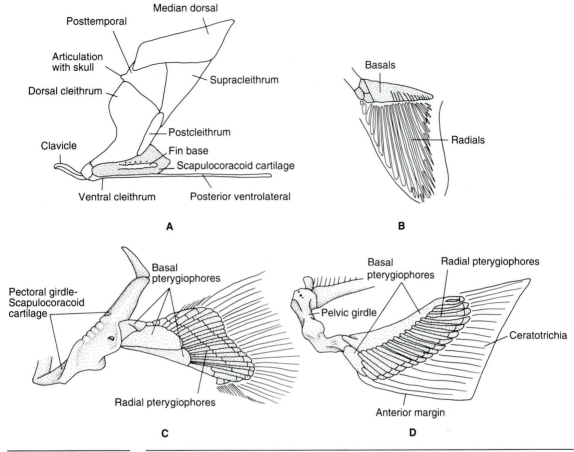

Median dorsal
Posttemporal
Articulation with skull
Dorsal cleithrum
Supracleithrum
Postcleithrum
Clavicle
Fin base
Scapulocoracoid cartilage
Ventral cleithrum
Posterior ventrolateral

A

Basals
Radials

B

Pectoral girdle-Scapulocoracoid cartilage
Basal pterygiophores
Radial pterygiophores

C

Basal pterygiophores
Radial pterygiophores
Pelvic girdle
Ceratotrichia
Anterior margin

D

FIGURE 8–18   The appendicular skeletons of arthrodires and chondrichthyan fishes: **A**, A lateral view of the thoracic dermal armor of the arthrodire *Dunkleosteus*. Many of these plates appear comparable to the dermal portion of the pectoral girdle of other fishes. **B**, A dorsal view of the broad-based pectoral fin of the Devonian shark, *Cladoselache*. **C** and **D**, Laterodorsal views of the narrow-based pectoral and pelvic fins, and their girdles, of a contemporary small shark, *Squalus*. (**A**, after Westol. **B**, after Moy-Thomas and Miles. **C** and **D**, after Jollie.)

girdle is a simple transverse rod of cartilage in the ventral body wall just anterior to the cloaca.

Among the early bony fishes, acanthodians had conspicuous pectoral and pelvic spines, which apparently functioned as cutwaters, on the leading edge of pectoral and pelvic fins (Figs. 3–8**A** and 8–19**A**). Basal and radial elements that attached to a scapulocoracoid have been found. A series of paired spines lay between the pectoral and pelvic spines, but they did not support fins.

Throughout their evolution, actinopterygian fishes have been characterized by their narrow-based, fan-shaped paired fins. We do not know whether there was a broad-based stage in the evolution of these fins, as there appears to have been

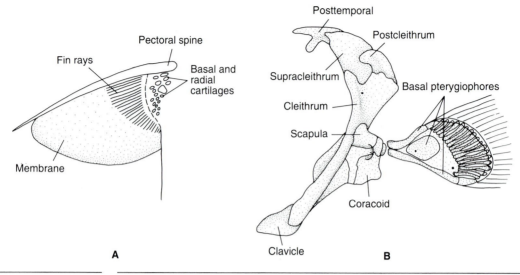

FIGURE 8–19   Pectoral fins and girdles of acanthodians and actinopterygians: **A**, A reconstruction of a dorsal view of the pectoral fin of the acanthodian *Acanthodes*. **B**, A lateral view of the pectoral fin and girdle of the primitive actinopterygian *Polypterus*. (**A**, after Stahl. **B**, partly after Jarvik.)

in the evolution of the narrow-based fins of contemporary cartilaginous fishes. The pectoral fin of the chondrostean *Polypterus* is representative. It has three basal pterygiophores to which many radials attach (Fig. 8–19**B**). Most of the fin is supported, as are the median fins, by lepidotrichia, which evolved from rows of bony scales in the skin. A scapulocoracoid forms the endoskeletal part of the pectoral girdle. A distinct **scapula** and distinct **coracoid** ossify within it in *Polypterus,* and in many other actinopterygians. A large **cleithrum,** to which the scapulocoracoid attaches, forms the major part of the dermal portion of the pectoral girdle. A small **clavicle** extends ventrally from the cleithrum in early actinopterygians, but it is lost in more advanced actinopterygians. Many species have one or more **supracleithral** and **postcleithral** elements. A **posttemporal** bone usually anchors the pectoral girdle to the back of the skull. Recall that head and trunk move as a unit in fishes. The structure of the pelvic fin is similar to that of the pectoral fin, but the pelvic girdle consists only of a pair of small ventral plates of cartilage or cartilage replacement bone. The pelvic girdle does not articulate with the vertebral column in fishes because the body is supported by water, not by limbs.

Sarcopterygians have only a single basal pterygiophore in their lobate, fleshy fins. The monobasic fin of dipnoans is biserial, for it has a long, central axis with many short, radial pterygiophores extending from it to both margins of the fin (Fig. 8–20**A**).This type of fin was called an **archipterygium** by the late 19th-century German anatomist, Gegenbaur, because he regarded it as ancestral. Gegenbaur believed that girdles were displaced visceral arches and that fins were enlarged branchial rays. This hypothesis became untenable when anatomists recognized the many fundamental differences between the visceral skeleton, to which gill arches belong, and the somatic skeleton, to which the appendicular skeleton

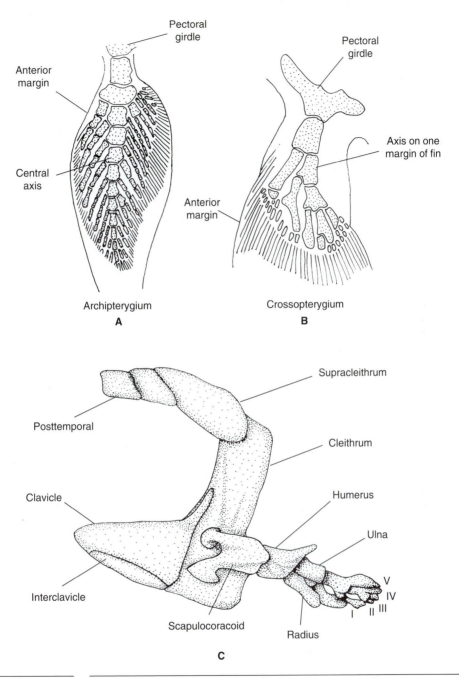

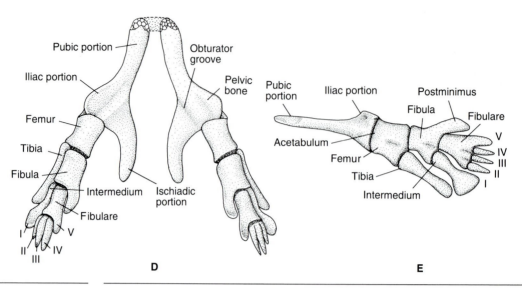

FIGURE 8–20 *continued*

D and **E**, Dorsal and lateral views, respectively, of the pelvic girdle and fin of *Eusthenopteron*. (**C–E**, after Jarvik.)

belongs. Rhipidistians and coelacanths have a uniserial, monobasic fin in which the fin axis extends along one margin and radial elements extend from the axis to the other margin (Fig. 8–20**B**). This type of fin is called a **crossopterygium** (Gr. *krossoi* = tassels + *pteryg-* = wing or fin).

The pectoral girdle of rhipidistians is similar to that of primitive actinopterygians except for the presence of an interclavicle that connects the two sides ventrally (Fig. 8–20**C**). Some dipnoans also have an interclavicle, but their pectoral girdle as a whole is much reduced. The partial support of the scapulocoracoid by an enlarged first rib is a unique feature of dipnoans. The pelvic girdle of sarcopterygians consists of a pair of ventral cartilages or bones (Fig. 8–20**D** and **E**). Each has a large, anteroventral pubic process, a smaller, medioventral ischiadic process, and a small, dorsally projecting iliac process. The iliac process does not reach the vertebral column and ribs.

The Appendicular Skeleton of Ancestral Tetrapods and its Origin

The Limbs of Ancestral Terrestrial Vertebrates

The appendicular skeleton underwent many changes during the transition from water to land, where an animal needs more support. The paired fins became limbs and the girdles strengthened. Body weight is transferred from the trunk skeleton to the ground through the girdles and limbs. Articulations between the girdles and limbs, and between limb segments, give the limb great flexibility. After a foot is moved forward and placed on the ground, retractor muscles acting on the limb advance the body relative to the foot (p. 361). Undulations of the trunk also help advance the body.

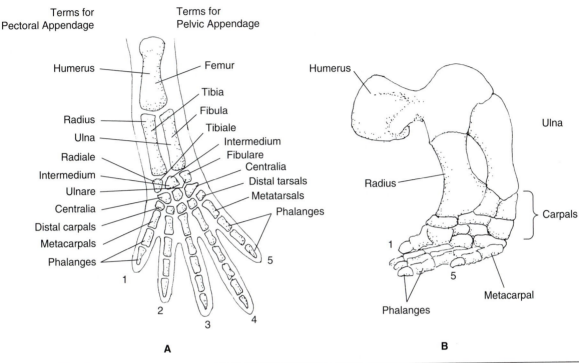

FIGURE 8–21   The limbs of ancestral terrestrial vertebrates. **A**, A generalized diagram of the cheiropterygium of a terrestrial vertebrate, bones of the pectoral appendage are labeled on the left; those of a pelvic appendage are on the right. **B**, The pectoral appendage of a labyrinthodont in a standing position. Note the horizontal posture of the humerus. (**B**, partly after Gregory.)

Early labyrinthodonts had the typical **cheiropterygium** (Gr. *cheir* = hand) with the three distinct segments found in both the pectoral and the pelvic appendages of terrestrial vertebrates (Fig. 8–21A). The proximal segment, the **brachium** in the front leg or the **thigh** in the hind leg, contained a single element, the **humerus** or **femur,** respectively. The forearm, or **antebrachium,** contained a **radius** and **ulna;** the shank, or **crus,** had the **tibia** and **fibula.** The radius and tibia lay on the medial border of the limb. The third segment, the hand (**manus**) or foot (**pes**), consisted of a series of small, podial elements that are called **carpals** in the wrist and **tarsals** in the ankle. The **digits** extended distally from the podials. The proximal part of each digit, which lay in the palm of the hand or the sole of the foot, was formed by elongated metapodials (**metacarpals** and **metatarsals**); the free part of each digit was formed by a series of **phalanges** (Gr. *phalanx* = soldiers in a battle line). Most terrestrial vertebrates have five digits, and we often assume that this was the primitive number. Recently discovered specimens show that the number of digits in early amphibians ranged from four to eight. The number five appears to be a secondary stabilization. The most medial digit is called the first one.

The numerous podial elements were arranged in a proximal and distal series with several central elements, known as **centralia,** between them (Fig. 8–21**A**). The three proximal tarsals and carpals are named according to their position relative to the adjacent antebrachial or shin bones: **radiale, intermedium,** and **ulnare** in the manus, and **tibiale, intermedium,** and **fibulare** in the pes. Most early tetrapods had five **distal carpals** and **distal tarsals,** one for each digit.

The number of phalanges varies with digit length. The pattern of locomotion of primitive tetrapods requires an increase in digit length from the medial toward the lateral side of the foot. The last digit decreases abruptly in length. Early reptiles and some of the labyrinthodonts had two phalanges in the first digit, three in the second, four in the third, five in the fourth, and then a decrease to three or four in the fifth. This is expressed as a **phalangeal formula** of 2–3–4–5–3 or 4.

The humerus and femur of a labyrinthodont were short and stocky bones because the animal did not raise itself far from the ground and its limbs were splayed (Fig. 8–20**B**). That is, when the animal was standing, its humerus and femur extended laterally from the girdles. The bones of the antebrachium and shin were flexed and extended vertically to the ground. A sharp bend or extension at the wrist and ankle placed the palm of the hand and sole of the foot flat on the ground. The head of the humerus and femur were on the proximal end of the bone. Large processes for the attachment of muscles usually lay beside the heads. Prominent ridges for the attachment of other muscles often extended down the shaft. The radius and ulna of the forearm, and the tibia and fibula of the shin, were also short and stocky, and they were nearly the same size because they shared equally in the transfer of body weight.

The Girdles of Ancestral Terrestrial Vertebrates

Because the girdles of terrestrial vertebrates transfer body weight to the appendages and receive the thrusts of the legs, they must be strong. The endoskeletal part of the pectoral girdle of labyrinthodonts had expanded and had become the major part of the girdle. It usually ossified as a single scapulocoracoid element (Fig. 8–22**A**). A **glenoid fossa** received the humeral head. The ventral coracoid region was broad and extended toward the midventral line. The dermal girdle, in contrast, was reduced. The cleithrum remained, but the postcleithral and supracleithral elements and the posttemporal bone were lost. This freed the pectoral girdle from the skull, and head and trunk could now move independently. The clavicle remained. A large interclavicle connected the left and right sides of the girdle ventrally, thereby strengthening the pectoral arch. A terrestrial vertebrate does not have a direct skeletal attachment between the pectoral girdle and the vertebral column. Weight is transferred from the vertebral column to the scapula by a muscular sling (p. 358). Ventrally, the girdle of tetrapods usually connects with the sternum.

The pelvic girdle of a labyrinthodont had a prominent **ilium** that extended dorsally to articulate directly with the one pair of sacral ribs (Fig. 8–22**B**). Part of the ilium also extended caudally and served for the attachment of powerful tail muscles. The ventral part of the pelvic girdle was greatly expanded and ossified as an anterior **pubis** and posterior **ischium.** An **obturator foramen** for the obtu-

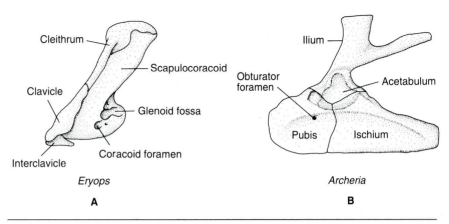

FIGURE 8–22   Lateral views of the girdles of a labyrinthodont. **A**, Pectoral girdle. **B**, Pelvic girdle. (After Romer and Parsons.)

rator nerve perforated the pubis. The girdle was further strengthened by the union of pubis and ischium of opposite sides at a midventral **pelvic symphysis.** The head of the femur articulated with a socket known as the **acetabulum,** located at the point where all the pelvic bones united.

Origin of the Tetrapod Limb and Girdle

By comparing Figures 8–20**B** and 8–21**A,** it is easy to visualize the morphological transformation of the rhipidistian paired fin into the limb of a labyrinthodont. The single proximal element of the rhipidistian is homologous to the humerus or femur of the cheiropterygium, and the second axial element of the fin and the first radial are homologous to the radius and ulna or tibia and fibula. Beyond that, homologies are less certain, but the distal proliferation of elements in the crossopterygium is comparable in a general way to the proliferation of elements in the hand or foot.

   The pectoral girdles of a rhipidistian and a labyrinthodont are similar (cf. Figs. 8–20**C** and 8–22**A**). Only the loss of elements connecting the girdle to the skull and an expansion of the scapulocoracoid portion of the girdle are necessary for the conversion. The pelvic girdle of the labyrinthodont did enlarge considerably and acquire a connection to the sacrum, but iliac, pubic, and ischiadic processes were present in the rhipidistian pelvis. The similarities between the girdles and the strong resemblance between the crossopterygium and the cheiropterygium contribute to the evidence that labyrinthodonts evolved from rhipidistians rather than from dipnoans.

   These changes did not come about abruptly. The rhipidistians ancestral to terrestrial vertebrates probably lived in freshwater environments subject to stagnation and periodic drought (p. 75). Their sturdy, paired fins doubtless enabled them to push through shallow, vegetation-choked water, dig into mud at the bottoms of ponds, and briefly emerge on land to escape enemies or search for food. As these forays on land became more extensive, selection would have favored stronger limbs and girdles and their transformation into the ancestral tetrapod condition.

The Limbs and Girdles of Contemporary Amphibians and Reptiles

The limbs and girdles of other amphibians and of reptiles are variations on the pattern established in the ancient labyrinthodonts. Although most reptiles retain five digits, contemporary amphibians only have four fingers in their hands. Some reduction occurs in the number of tarsals and carpals through loss or fusion (Fig. 8–23**A** and **B**). The proximal tarsal and centralia frequently fuse in reptiles to form a single unit, and the functional ankle joint lies between it and the distal tarsals. Because the joint lies near the center of the ankle, it is called a **mesotarsal joint.** It probably increases the capacity for rotation within the joint. The number of phalanges is reduced in contemporary amphibians, and they typically have only one to three phalanges in their digits. Many reptiles retain the primitive phalangeal formula of 2–3–4–5–3 or 4.

Reptiles and mammals usually have a small **pisiform** on the ulnar side of the wrist (Fig. 8–23**A**). The pisiform is not a supporting part of the skeleton but rather a **sesamoid** bone (Gr. *sesamon* = sesame seed + *eidos* = form). Sesamoid bones

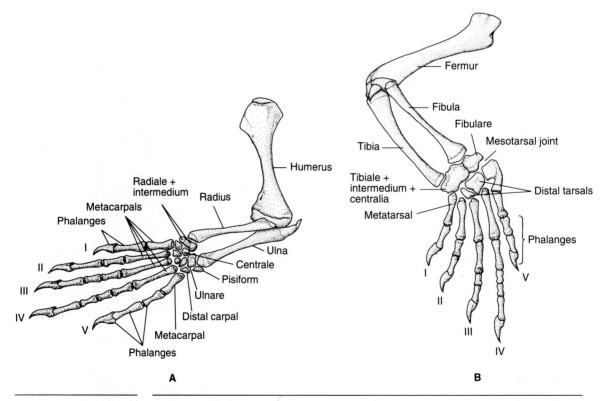

FIGURE 8–23 The limb skeleton of a modern lizard, *Varanus:* **A,** Pectoral appendage. **B,** Pelvic appendage. (After Bellairs.)

lie in the tendons of certain muscles and alter the direction of pull of the muscle, as the pisiform does, or facilitate the movement of a tendon across a joint.

The long bones of the limbs become more slender in contemporary amphibians and reptiles. The long bones of the antebrachium and crus sometimes fuse in species where extra rigidity is needed, as in the limbs of frogs (see Fig. 10–27).

The endoskeletal part of the pectoral girdle ossifies as a distinct scapula and coracoid in modern amphibians and in the stem reptiles and contemporary species. (Fig. 8–24**A** to **D**). The coracoid should be called an **anterior coracoid** to distinguish it from an additional posterior coracoid that appears in the line of evolution to mammals. The anterior coracoid can be distinguished by its having a small **coracoid foramen** for a pectoral nerve. The clavicle and interclavicle are usually retained, but the cleithrum is lost except in a few early frogs. The clavicles and interclavicles are incorporated in the front of the plastron in turtles; and a few reptiles, including crocodiles, lose the clavicle.

Most contemporary amphibians and reptiles also have a midventral **sternum** attached to the caudal part of the coracoid plate. It is cartilaginous in most species but partly calcified in diapsid reptiles. As we have seen, the anterior trunk ribs of reptiles also attach to the sternum, but the very short ribs of amphibians do not. There is no fossil evidence for a sternum in ancestral amphibians and reptiles, but a cartilaginous sternum, which would not have been fossilized, cannot be ruled out.

A peculiarity of the pelvic girdle of modern amphibians is the failure of the pubic region to ossify (Fig. 8–24**E**). Moreover, frogs have an exceptionally long ilium that is related to the development of powerful thrusts by the hind legs (p. 380). A puboischiadic fenestra develops between the pubis and ischium in contemporary reptiles (Fig. 8–24**F** and **G**).

As reptiles entered new habitats and ecological niches, their methods of locomotion and appendicular skeletons changed considerably. Extinct plesiosaurs evolved large, oarlike appendages as they adapted to the marine environment, but the appendages of the marine ichthyosaurs, which swam by fishlike undulations of the trunk and tail, became small, stabilizing keels (see Fig. 3–24). Snakes evolved long, undulating trunks and lost their appendicular skeletons. This probably was originally an adaptation for burrowing, but most snakes now live above ground.

The Appendicular Skeleton of Birds

Wings

One group of extinct reptiles (the pterosaurs), birds, and bats adapted for flight and evolved wings. In each group the wing is a modified pectoral appendage, but differences in wing construction indicate that the wings evolved independently (Fig. 8–25). The pterosaur wing was a large membrane that attached to the side of the body and was supported by the arm and a greatly elongated fourth finger. The fifth digit and the terminal, claw-bearing phalanx of the fourth digit were lost. The bat wing is also a membrane, but it is supported by the last four elongated fingers. The first digit forms a grasping hook and remains free of the wing.

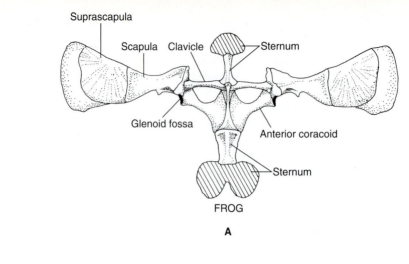

Suprascapula

Scapula · Clavicle · Sternum

Glenoid fossa

Anterior coracoid

Sternum

FROG

A

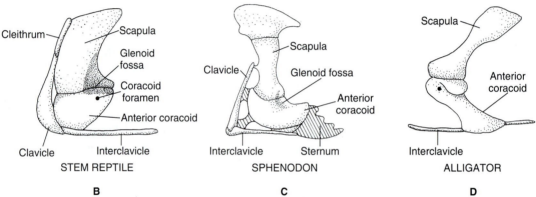

Cleithrum · Scapula

Glenoid fossa

Coracoid foramen

Anterior coracoid

Clavicle · Interclavicle

STEM REPTILE

B

Scapula

Clavicle · Glenoid fossa

Anterior coracoid

Interclavicle · Sternum

SPHENODON

C

Scapula

Anterior coracoid

Interclavicle

ALLIGATOR

D

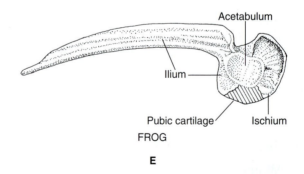

Acetabulum

Ilium

Pubic cartilage · Ischium

FROG

E

FIGURE 8–24 The girdles of some contemporary amphibians and reptiles. The pectoral girdle of the frog, **A**, is shown in ventral view; all others are in lateral view. **A–D**, Pectoral girdles. **E–G**, Pelvic girdles. Cartilaginous regions are hatched. (**A** and **E**, after Walker. Others, partly after Jollie.)

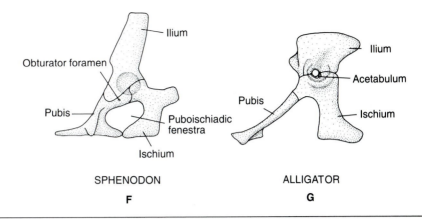

SPHENODON

F

ALLIGATOR

G

FIGURE 8–24   *continued*

The flying surface of a bird's wing is not a membrane; rather it is formed by long flight feathers that attach along the posterior border of the humerus, ulna, and hand (Figs. 8–14**A** and 8–25). Tail feathers arising from the pygostyle (p. 270) form another flying surface that is used primarily at low air speeds and in maneuvering. Feathers are modified horny scales (p. 161) and form a very lightweight and adjustable flying surface, which is easily repaired because worn feathers are replaced periodically. The bones of a bird's wing, like the rest of the skeleton, are exceptionally lightweight yet strong. Although the surface bone is thin, internal struts and ridges along stress lines provide strength where needed. An air sac from the lungs extends into the humerus through a foramen at its proximal end. No rotation occurs at the elbow, and the radius and ulna extend parallel to each other to the wrist.

The hand is greatly modified. Only three fingers are present (as was the case in the saurichian dinosaur ancestors of birds), and they are reduced. The phalangeal formula in specimens of *Archeopteryx,* the first known bird, is clearly 2–3–4. This suggests that the fingers are the first, second, and third ones. But the relationships of these fingers to carpals that appear, and disappear, during the embryonic development of the wing of modern birds have led some embryologists to interpret the fingers as being the second, third, and fourth. Only two distinct carpals remain in the adult wrist, but these allow the hand and the feathers it bears to rotate relative to the rest of the wing. The other carpals and the metacarpals of the three fingers have fused to form a long **carpometacarpus.** The most anterior digit (whether it is first or second) can move independently of the other two and bears a small, distinct tuft of feathers, known as the **alula** (L. *alula* = little feather). This tuft of feathers can be elevated from the rest of the wing to form an air slot that is important in increasing lift at certain stages of flight (p. 382).

The pectoral girdle and thoracic skeleton of birds form a strong support for the action of the wings (Fig. 8–14**A**). The scapula is a narrow blade, the anterior coracoid is a strong strut that braces the shoulder joint, and the two clavicles have fused together ventrally to form the wishbone or **furcula** (L. *furcula* = small fork). Major flight muscles arise from the large, keeled sternum (p. 388).

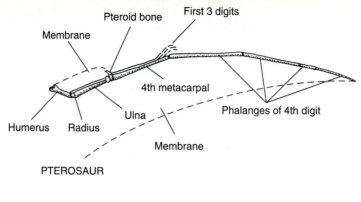

PTEROSAUR

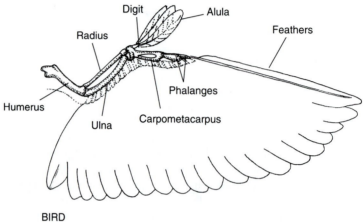

BIRD

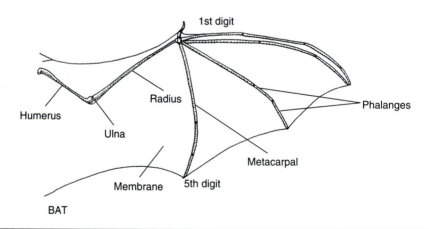

BAT

FIGURE 8–25 Vertebrate wings. (After Norberg.)

The Pelvic Girdle and Appendage

Since the pectoral appendages are wings, birds are bipeds. Indeed, bipedalism must have evolved before flight because the front legs could not have been modified as wings if they were needed in quadrupedal locomotion. The pelvic girdle and hind legs are exceptionally strong, not only supporting all the weight of the bird but also acting as shock absorbers when it lands. The pubis has turned caudal beside the ilium, and all the pelvic bones are firmly united (Fig. 8–14**A**). The long ilium is fused with the synsacrum. The midventral pelvic symphysis has been lost, so the pelvic canal between the two halves of the girdle is much larger than in other terrestrial vertebrates. This permits a caudal displacement of the viscera, which, together with a shortening of the trunk, brings the center of gravity of the bird over the hind legs. A large pelvic canal also allows for the passage of large eggs with fragile shells.

The hind legs have rotated from the primitive splayed position to a position beneath the body. The head of the femur necessarily shifts to the medial side of the proximal end of the bone. A bird, like human beings and many other bipeds, maintains balance on one leg when the other swings forward by adducting the leg (that is, pulling the supporting leg toward the midventral line and shifting body weight so that the foot is directly beneath the projection of the body's center of gravity).

The tibia is the primary supporting element of the shin, and the fibula is reduced to a thin splint needed for the attachment of certain muscles. The ankle joint, as in many reptiles, is a mesotarsal joint. Since the legs are directly beneath the body, the only movement at the ankle is a hinge action. The joint is strengthened by the fusion of the proximal tarsals with the tibia. Accordingly, the tibia is most accurately described as a **tibiotarsus.** The fifth toe has been lost, but the metatarsals of the remaining toes and the distal tarsals are fused to form a long **tarsometatarsus.** The remaining toes have the characteristic reptilian phalangeal formula of 2–3–4–5. The first toe usually is turned caudad; this increases stability.

The Mammalian Appendicular Skeleton and its Evolution

Early synapsids, the pelycosaurs, walked in the manner of labyrinthodonts and early reptiles, with whom they were contemporary. Their limbs were splayed and the humerus and femur moved back and forth close to the horizontal plane. Their limb bones were also short and chunky (Fig. 8–21**B**), and their girdles were greatly expanded ventrally to accommodate the powerful adductor musculature needed to raise them from the ground. A technical, but phylogenetically important, difference was the ossification of a second element, the **posterior coracoid,** in the coracoid plate (Fig. 8–26**A**).

Advanced synapsids, the therapsids, became more active animals and began to acquire many of the appendicular features characteristic of mammals. Mammals walk with their legs rotated under the body and moving back and forth close to the vertical plane (Fig. 8–27). We discuss the functional anatomy of locomotion later (p. 364), but it is evident that this limb position provides better mechanical

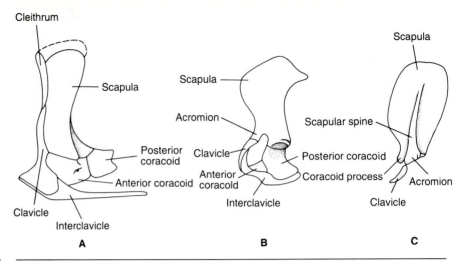

FIGURE 8–26 Lateral views of the evolution of the mammalian pectoral girdle: **A**, An early synapsid, *Dimetrodon*. **B**, The platypus *Ornithorhynchus*. C, An opossum, *Didelphis*. (**A** and **B**, after Jollie. **C**, after Romer and Parsons.)

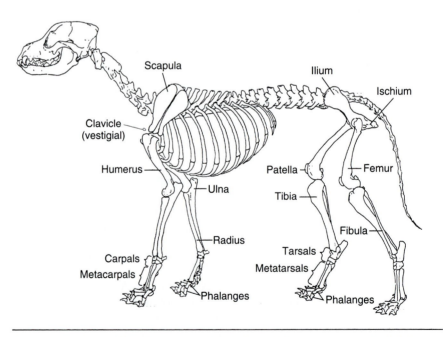

FIGURE 8–27 The skeleton of a dog. Note the posture of the limbs and the proportions of the limb bones. (After Miller, Christensen, and Evans.)

support and enables the limbs to swing through longer arcs, thus increasing the efficiency of locomotion.

Changes in girdle and limb structure are correlated with the changes in limb position and movements. A strong, ventral adductor musculature is no longer needed in the pectoral region to raise the body from the ground and support it. Dorsal musculature, on the other hand, expands because it becomes more important in bracing the shoulder joint and in the fore and aft swing of the foreleg at the shoulder. Indeed, some of the original ventral musculature migrates dorsally (Chapter 9). As ventral muscles become less important, the coracoid region becomes progressively smaller (Fig. 8–26**B** and **C**). Therapsids and monotreme mammals retain two small coracoids, but the anterior one becomes excluded from the glenoid fossa. The anterior coracoid is lost in therian mammals, and the posterior coracoid is reduced to a small **coracoid process** on the scapula beside the glenoid fossa. Loss of the coracoids allows the glenoid fossa itself to face ventrally directly over the head of the humerus. The scapula expands to accommodate the increased dorsal musculature. The original cranial border of the scapula, which can be identified by a process, the **acromion,** to which the lateral end of the clavicle attaches, rolls outward to form a ridge known as the **scapular spine.** The supraspinous portion of the scapula located cranial and dorsal to the spine represents a newly evolved shelf of bone.

Reduction continues in the dermal portion of the pectoral girdle (Fig. 8–26**B** and **C**). The cleithrum is lost in all mammals. Monotremes retain an interclavicle, but this is lost in therians; instead, the medial end of the clavicle articulates with the front of the sternum. The clavicle persists in many mammals as a brace that helps stabilize the position of the shoulder, but it is reduced in running species (Fig. 8–27).

Similar changes occur in the pelvic girdle. The large pubis and ischium in the ventral region of pelycosaurs become smaller in therapsids and mammals, but both elements persist (Fig. 8–28). The acetabulum continues to face laterally. The obturator foramen, through which the obturator nerve passes in early mammals, enlarges and also accommodates the bulging of certain pelvic muscles. With the reduction of heavy tail muscles, the ilium loses its caudally directed process and inclines cranially. All elements of the girdle are firmly fused in adult mammals. Monotremes and marsupials of both sexes have a pair of rod-shaped **epipubic bones** that extend forward from the pubis into the ventral abdominal wall. Their function is uncertain, but White (1989) has proposed that muscles associated with them help protract the pelvic limb during locomotion.

The humerus and femur and the bones of the forearm and shin become longer than in early tetrapods (Fig. 8–27), and this, too, increases step length. Muscular processes on the humerus and femur are present, but they are relatively small except in burrowing species that have exceptionally powerful limbs. Since the glenoid fossa faces ventrally, the head of the humerus remains at the proximal end of the bone, but the head of the femur shifts to the medial side of the proximal end because the acetabulum continues to be directed laterally.

Since the limbs are carried nearly beneath the body, the bones of the forearm and shin no longer share equally in weight transfer, as they did in early tetrapods. The radius is the major bone in transferring body weight to the wrist (Fig. 8–27).

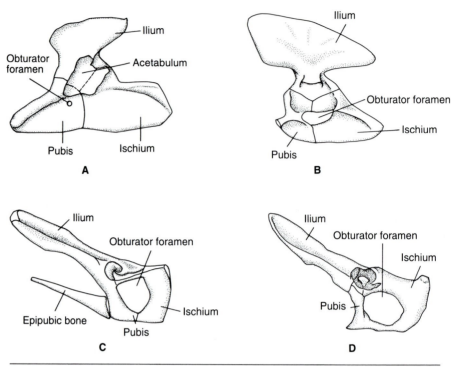

FIGURE 8–28   Lateral views of the evolution of the mammalian pelvic girdle: **A**, An early synapsid, *Dimetrodon*. **B**, A late synapsid, *Cynidiognathus*. **C**, The opossum *Didelphis*. **D**, An eutherian mammal. (After Romer and Parsons.)

The ulna contributes to the elbow joint and also is the axis about which the radius rotates. Since the elbow points caudally, a rotation of the radius across the front of the ulna is essential for the toes to point forward and for the sole of the front foot to be on the ground. The capacity of the radius to rotate around the ulna persists in many mammals, including human beings. We can direct the palm of the hand toward the ground in the **prone** position, or upward in the **supine** position.

Since both the knee and the hind foot point forward (Fig. 8–27), the major movement at the knee and ankle is a hinge action. Rotation is limited. A major muscle that extends the lower leg crosses the front of the knee joint, and a sesamoid bone called the **patella** evolves in its tendon. The patella facilitates the movement of this tendon as it slides across the knee during the movements of the lower leg. Tibia and fibula parallel each other. The tibia is the major weight transfer bone, and it is substantially larger than the fibula.

A hinge action is the primary movement at the wrist and ankle because the feet point forward at all times during a step cycle. A reduction in the number of carpals and tarsals occurs during the evolution of mammals (Fig. 8–29). Only one distinct centrale remains in the carpus and tarsus of most species, and humans do not normally have the one in the wrist. The tibiale, intermedium, and probably another centrale unite in the tarsus to form a large element called the **astragalus.**

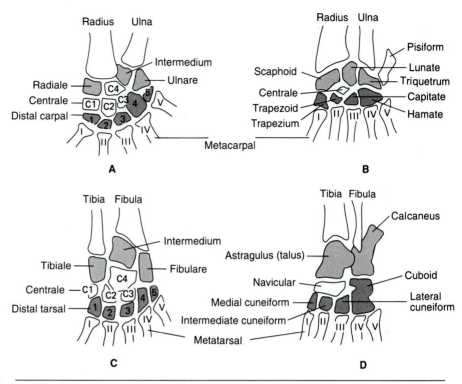

FIGURE 8–29   The evolution of the mammalian carpus (top row) and tarsus (bottom row). The wrist and ankle of an ancestral tetrapod are shown on the left (**A** and **C**), and those of a mammal on the right (**B** and **D**). (Modified after Romer and Parsons.)

The hinge action at the ankle occurs between the astragalus and tibia rather than at a mesotarsal joint as in many reptiles and birds. The fibulare, now called the **calcaneus,** develops a caudally directed process that acts as a lever arm for the attachment of shank muscles that extend the foot. Alternative names are available for other carpals and tarsals, based on their shapes in human beings (Fig. 8–29).

Toes are more equal in length in the forward-directed feet of mammals than they were in ancestral tetrapods. The axis of the foot runs through the third digit in most species, and this one often is slightly larger than the others. The phalangeal formula of mammals is reduced to 2–3–3–3–3. Reduction begins in the therapsids, and stages are known in which the extra phalanges that were present in the last three digits of ancestral tetrapods are reduced to small nubbins of bone.

SUMMARY

1. The trunk skeleton consists of the vertebral column, ribs, median fins (in fishes), and the sternum (in terrestrial vertebrates). It forms the fundamental framework of the body and transfers body weight to the girdles and appendages in tetrapods. Rib movements help ventilate the lungs in tetrapods.

2. Vertebrae develop in an intersegmental position from concentrations of mesenchyme of sclerotomal origin. The mesenchyme first chondrifies in the neural and hemal arches, and in most species the cartilage is later replaced by bone. The centrum develops in different ways in different groups.

3. Ribs develop in the myosepta at their intersections with other skeletogenous septa. Fishes have many rib types. Tetrapod ribs are subperitoneal in position and usually attach both on the side of the centrum (often on a parapophysis) and on a conspicuous transverse process known as a diapophysis. In some body regions of tetrapods, the embryonic ribs fuse with the diapophysis, parapophysis, or both to form a compound transverse process known as a pleurapophysis.

4. The vertebral axis of a fish is an integral part of the locomotor apparatus. It resists compression of the body and converts the contraction of segmental, longitudinal muscles into lateral undulations. Since all parts of the vertebral column have nearly the same function, there is little regional differentiation. The first trunk vertebrae articulates with the skull, and caudal vertebrae bear hemal arches.

5. The notochord persists in the adults of agnathans, lungfishes, chondrosteans, and some other species that do not move rapidly through the water. Contemporary sharks and their allies have cartilaginous centra that often are calcified; advanced actinopterygians have well-ossified centra.

6. The vertebral axis of a fish ends in an expanded caudal fin that is supported internally by vertebral elements and distally by horny ceratotrichia (in chondrichthyans) or rows of bony, scalelike lepidotrichia (in osteichthyans). The types of caudal fins correlate with patterns of locomotion and buoyancy control.

7. Although the vertebral axis functions in lateral undulations of the trunk and tail in some amphibians, it is primarily a supporting beam. It must resist bending in the vertical plane so that these low-slung animals do not drag on the ground as they walk.

8. The vertebrae of early labyrinthodonts resembled those of rhipidistians. Each vertebral centrum consisted of an intercentrum and a pair of pleurocentra that flanked a notochord. Vertebrae became stronger in later species as the intercentrum, pleurocentrum, or both enlarged.

9. Ribs were well developed in labyrinthodonts and attached by two heads to most of the vertebrae. In contemporary amphibians they are very short and often contribute to pleurapophyses.

10. The vertebral axis of an amphibian is stronger and more complex than in a fish. A well-developed cervical vertebra (the atlas) permits movements between head and trunk, a single sacral vertebra and pair of ribs articulate the vertebral column with the pelvic girdle, and caudal vertebrae lie in the tail.

11. Evolutionary trends that began in the trunk skeletons of amphibians continued in reptiles. Intercentra are lost in most contemporary species, and the definitive centrum is an expanded pleurocentrum. The neck is longer and more mobile, and there are more cervical vertebrae. At least two sacral vertebrae and ribs articulate with the pelvic girdle. Tail length varies.

12. The anterior trunk ribs of reptiles connect by flexible costal cartilages with the sternum in most species, and rib and sternal movements are important in ventilating the lungs. Abdominal ribs may be present caudal to the sternum.

13. The structure of the trunk skeleton of a bird correlates with its specializations for flight and bipedal walking. A long and flexible neck permits the head and bill to be used in many manipulative activities that the specialized forelimbs can no longer do. Many trunk vertebrae are fused to form a firm fulcrum for wing action; others have united with the sacral vertebrae to form a solid synsacrum for the attachment of the pelvic girdle. The terminal caudal vertebrae have fused to form a pygostyle to which the large tail feathers attach.

14. The sternum of birds is ossified and usually bears a large keel that increases the area for the attachment of the large flight muscles.

15. Mammals are active, agile vertebrates with strong yet flexible trunk skeletons. Most have seven cervical vertebrae, of which the first two (the atlas and axis) are specialized to allow free movement of the head. The trunk region has differentiated into a thoracic region, which bears distinct ribs, and a lumbar region, in which the ribs have contributed to pleurapophyses. The thoracic region contains the lungs and transfers body weight to the pectoral girdle. Three or more sacral vertebrae usually unite, forming a strong sacrum. Tail length and functions vary among groups.

16. Paired fins may have evolved from the enlargement of parts of a lateral keel of bony plates. They helped provide stability as fishes adapted to a more active mode of life.

17. Early chondrichthyan fishes had large, broad-based, paired fins that provided stability and acted as hydroplanes. A narrowing of the base in contemporary species allows a shark to partially twist the fin and use it in maneuvering. No dermal bones are present in the pectoral girdle.

18. Actinopterygians have narrow-based, fan-shaped fins supported proximally by three basals and distally by bony fin rays that evolved from rows of bony scales. The pectoral fin attaches to a small, often unossified endoskeletal girdle, which in turn attaches to an arc of dermal bone connected to the back of the skull. The pelvic girdle is a small rod of cartilage or bone embedded in the ventral body wall.

19. Tetrapods evolved jointed legs that support the body and rotate forward and backward to develop propulsive thrusts.

20. The appendages of ancestral terrestial vertebrates were splayed. The humerus and femur moved back and forth close to the horizontal plane. The endoskeletal part of the pectoral girdle enlarged and ossified; the dermal part became smaller and lost its connection with the back of the skull. The pelvic girdle enlarged, ossified, and connected to a sacral vertebra.

21. The limbs of contemporary amphibians and reptiles are somewhat more slender than in ancestral terrestrial vertebrates but are otherwise similar. The ankle joint of reptiles is a mesotarsal joint. The endoskeletal pectoral girdle usually ossifies as a distinct scapula and anterior coracoid. The dermal part of the girdle is reduced further, and often only a clavicle and interclavicle remain.

22. Wings evolved from pectoral appendages three times during the evolution of

terrestrial vertebrates. The pterosaur wing was a membrane supported by an elongated fourth digit. The bird wing is composed of feathers attached to the arm and a modified hand containing three partially united digits. The bat wing is a membrane supported by the last four elongated digits.

23. Birds are bipeds on the ground and have a strong sacrum, pelvis, and hind legs. The loss of the pelvic symphysis permits a caudal displacement of the viscera and helps bring the body's center of gravity over the hind legs.

24. As synapsids and mammals became more active and agile animals than labyrinthodonts and stem reptiles, their limbs rotated at least partly under the body, and the humerus and femur moved forward and backward closer to the vertical plane.

25. As ventral muscles became less important and dorsal muscles more important in the swing of the legs and bracing them at the girdles, the dorsal parts of the girdles expanded and the ventral parts regressed to some extent. Changes are particularly pronounced in the pectoral girdle. The anterior coracoid was lost, and the posterior coracoid, which first appeared in synapsids, was reduced to a small, coracoid process.

26. The long bones of the limbs lengthened and became more slender. The radius is the main weight-supporting bone in the forearm of a mammal; the tibia is most important in the shin. Carpals and tarsals are reduced in number. The feet point forward, and the digits become more nearly equal in length.

REFERENCES

Ahlberg, P. E., 1992: Coelacanth fins and evolution. *Nature*, 358:459.

Balfour, F. M., 1876: The development of elasmobranch fishes. *Journal of Anatomy and Physiology*, 11:128–172.

Cave, A. J. E., 1975: The morphology of mammalian pleurapophyses. *Journal of Zoology* (London), 177:377–393.

Cracraft, J., 1971: The functional morphology of the hind limb of the domestic pigeon, *Columba livia. Bulletin of the American Museum of Natural History*, 144:173–267.

Gadow, H. F., 1933: *The Evolution of the Vertebral Column.* Cambridge, Cambridge University Press.

Gaudin, T. J., and Biewener, A. A., 1992: The functional morphology of xenarthrous vertebrae in the armadillo, *Dasypus novemcinctus* (Mammalia, Xenarthra). *Journal of Morphology*, 214:63–81.

Hebel, R., and Stromberg, M. W., 1976: *Anatomy of the Laboratory Rat.* Baltimore, The Williams & Wilkins Company.

Jenkins, F. A., Jr., 1969: The evolution and development of the dens of the mammalian axis. *Anatomical Record*, 164:173–184.

Jenkins, F. A., 1971: Limb posture and locomotion in the Virginia opossum (*Didelphis marsupialis*) and in other non-cursorial mammals. *Journal of Zoology* (London), 165: 303–315.

Kemp, T. S., 1969: The atlas-axis complex of the mammal-like reptiles. *Journal of Zoology* (London), 159:223–248.

Kummer, B., 1959: *Bauprinzipien des Saugerskelets.* Stuttgart, Georg Thieme Verlag.

Miller, E. M., Christensen, G. C., and Evans, H. E., 1964: *The Anatomy of the Dog.* Philadelphia, W. B. Saunders Company.

Norberg, U, M.: Flying, gliding, and soaring. *In* Hildebrand, M., Bramble, D. M., Liem,

K. F., and Wake, D. B, editors, 1985: *Functional Vertebrate Morphology*. Cambridge, The Belknap Press of Harvard University Press.

Nursall, J. R., 1962: Swimming and the origin of paired appendages. *American Zoologist,* 2:127–141.

Ostrom, J. H., 1979: Bird flight: How did it begin? *American Scientist,* 67:46–56.

Panchen, A. L., 1977: The origin and early evolution of tetrapod vertebrae. *In* Andrews, S. M., Miles, R. S., and Walker, A. D., editors: *Problems in Vertebrate Evolution.* Linnean Society of London, Symposium no. 4. London, Academic Press.

Patterson, C., 1977: Cartilage bones, dermal bones, and membrane bones, or the exoskeleton versus the endoskeleton. *In* Andrews, S. M., Miles, R. S., and Walker, A. D., editors: *Problems in Vertebrate Evolution.* Linnean Society of London, Symposium no. 4. London, Academic Press.

Pettingill, O. S., Jr., 1985: *Ornithology in Laboratory and Field,* 5th edition. Orlando, Academic Press.

Prange, H. D., Anderson, J. F., and Rahn, H., 1979: Scaling of skeletal mass to body mass in birds and mammals. *American Naturalist,* 113:103–122.

Rackoff, J. S., 1980: The origin of the tetrapod limb and the ancestry of tetrapods. *In* Panchen, A. L., editor: *The Terrestrial Environment and the Origin of Land Vertebrates.* Systematics Association, Special Volume, no. 15. London, Academic Press.

Romer, A. S., 1956: *The Osteology of Reptiles.* Chicago, University of Chicago Press.

Schaeffer, B., 1941: The morphological and functional evolution of the tarsus in amphibians and reptiles. *Bulletin of the American Museum of Natural History,* 78:395–472.

Schaeffer, B., 1967: Osteichthyan vertebrae. Linnean Society of London, *Zoology Journal,* 47:185–195.

Stephens, T. D., Sanders, D. D., and Yap, Y. F., 1992: Visual demonstration of the limb-forming zone in the chick embryo lateral plate. *Journal of Morphology,* 213:305–316.

Tabin, C., and Laufer, E., 1993: Hox genes and serial homology. *Nature,* 361:692–693.

Thacher, J. K., 1877: Median and paired fins, a contribution to the history of vertebrate limbs. *Transaction of the Connecticut Academy of Arts and Sciences,* 3:281–310.

Wake, D. B., and Lawson, R., 1973: Development and adult morphology of the vertebral column in the plethodontid salamander, *Eurycea bislineata,* with comments on vertebral evolution in the Amphibia. *Journal of Morphology,* 139:251–300.

Walker, W. F., Jr., 1981: *Dissection of the Frog.* San Francisco, W. H. Freeman and Company.

Westoll, T. S., 1958: The lateral fin-fold theory and the pectoral fins of ostracoderms and early fishes. *In* Westoll, T. S., editor: *Studies of Fossil Vertebrates.* London, University of London Press.

White, T. D., 1989: An analysis of epipubic bone function in mammals using scaling theory. *Journal of Theoretical Biology,* 139:343–357.

Williams, E. E., 1959: Gadow's arcualia and the development of tetrapod vertebrae. *Quarterly Review of Biology,* 34:1–32.

9 | The Muscular System

PRECIS

Muscles are the primary effector organs of vertebrates. Their actions cause most of an animal's responses as well as help support the animal and generate body heat. We will examine the structure and actions of muscles and how the various types are used. After considering the development of muscles and their grouping in fishes, we will trace their major evolutionary changes.

Outline

Most of the responses of vertebrates are brought about by the actions of muscles. Muscles not only propel the organism in its environment but also move food through the digestive tract, water or air across the respiratory surfaces, blood through the circulatory system, and excretory and reproductive products through their ducts. Muscles also help support the body by bracing the bones across joints, and muscle contraction is an important source of body heat. Muscles are the main **effectors** of vertebrates. Other effectors include the flagellated spermatozoa, chromatophores, the many glands, and ciliated cells. Ciliated cells move mucus in respiratory passages and help move excretory and reproductive products in certain passages. We will examine muscles at this time and consider the other effectors as we come to them in our discussion of the organ systems.

Types of Muscle Tissue

Muscle cells develop embryonically by the elongation of mesenchyme cells to form **myoblasts.** As the myoblasts elongate further, ultramicroscopic protein **actin** and **myosin myofilaments** develop within them. The myoblasts become the adult muscle cells, which are also known as **myocytes** or muscle fibers. Interactions between the myofilaments, discussed in more detail later in this chapter, lead to the development of tension along the longitudinal axis of the muscle fibers, and in this way muscles perform their functions. We recognize three broad categories of muscle tissue. **Skeletal muscles** are associated with the skeleton, **cardiac muscles** form the musculature of the heart wall, and **smooth muscles** contribute to the walls of blood vessels and many visceral organs.

Smooth Muscle

Smooth muscle fibers are elongated, spindle-shaped cells. Each has a single nucleus near the center of the cell (Fig. 9–1). The cells range in length from about 15 micrometers in the walls of small arteries to more than 500 micrometers in the wall of the uterus. Actin and myosin myofilaments are present in the cytoplasm, but they are not lined up in a regular manner, so they are difficult to see by light microscopy. Thus smooth muscle appears to have a homogeneous texture.

Smooth muscles are found in the walls of blood vessels and many visceral organs, and they also attach to the hairs in mammalian skin. Their actions are involuntary, and their contractions tend to be slow and sustained. They do not fatigue. Two types of smooth muscle fibers, which differ in their mode of action and innervation, have been recognized in amniotes.

Unitary smooth muscle fibers, which are found in the walls of the digestive tract, uterus, and urinary ducts, have spontaneous, rhythmic contractions that are usually initiated by the stretching of the muscle fibers. When a fiber becomes active, a redistribution of ions across its surface depolarizes the membrane and a wave of depolarization, called an **action potential,** spreads along its surface. The action potential of some fibers spreads slowly to others. Nerve cells or fibers, called **neurons,** terminate on some of the muscle fibers, and nerve impulses appear to

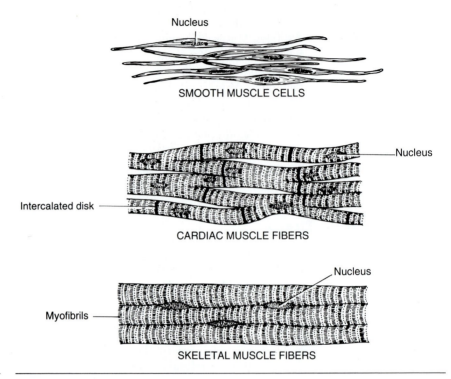

Nucleus

SMOOTH MUSCLE CELLS

Nucleus

Intercalated disk

CARDIAC MUSCLE FIBERS

Nucleus

Myofibrils

SKELETAL MUSCLE FIBERS

FIGURE 9–1   The three major types of muscle fibers. (Modified after Romer and Parsons.)

modulate their rate and force of contraction, but the contraction itself is primarily self-initiated, or **myogenic.** Unitary smooth muscle is well suited for the slow and sustained contractions needed to move food down the digestive tract, urine along the ureter, and so forth.

Multiunit smooth muscle is found in the walls of many blood vessels, in the iris of the eye, and in the walls of the sperm ducts. Nerve fibers terminate on most of the cells, and contraction is initiated by nerve impulses; that is, it is **neurogenic.** The degree of contraction of arteries and the pupil of the eye, and the ejaculation of sperm, are involuntary actions that must be more carefully regulated than the movements of the gut.

Cardiac Muscle

Cardiac muscle is composed of moderately elongated cells—about 80 micrometers long—that frequently branch (Fig. 9–1). Each cell contains a single, centrally placed nucleus. The myofilaments overlap and are arranged in such a way that the muscle cell appears striated. The individual cells are firmly united with each other, end to end, at specialized junctions known as **intercalated disks.** The contraction of cardiac muscle, like that of smooth muscle, is involuntary. Cardiac muscle also does not fatigue, but the arrangement of its myofilaments maximizes their overlap, so it contracts with greater force and speed than smooth muscle. Contraction is myogenic. A contraction originates in a region of specialized muscle known as the **pacemaker,** or **sinoatrial node** (p. 671), and the action potential spreads rapidly from cell to cell across the intercalated disks. Bundles of specialized cardiac muscles, the **Purkinje cells,** carry the action potential rapidly between more distant parts of the heart. Although we now know that cardiac muscle is not a single, multinucleated cell, as was formerly believed, functionally it acts like such a cell, or **syncytium** (Gr. *syn* = together + *kytos* = hollow vessel or cell). Nerve fibers terminate on the pacemaker and on many other cardiac muscle fibers, so the inherent myogenic rhythm of cardiac muscle is modulated by neurogenic control.

Skeletal Muscle

Skeletal muscle fibers develop embryonically by the end-to-end fusion of many myoblasts. Each adult skeletal muscle fiber is thus a syncytium that may contain several hundred nuclei. The nuclei are located peripherally around densely packed myofibrils in mammalian cells. As in cardiac muscle, the myofilaments of skeletal muscle overlap and are arrayed in such a way that the muscle appears striated (Fig. 9–1). Mammalian skeletal muscle fibers range in diameter from 10 to 100 micrometers, and some are many centimeters long. They may extend the length of a muscle or end in the connective tissue within a muscle. Contraction is initiated by nerve impulses. A single nerve cell branches and terminates at **motor end plates** on many muscle fibers. The activity of most skeletal muscles can be controlled voluntarily in the sense that an animal can decide whether to stand or move, but the complex series of muscle contractions needed for each activity is executed unconsciously by neuronal activity.

Muscle Structure

In all types of muscle tissue, the individual muscle fibers are enveloped by a thin layer of connective tissue, known as the **endomysium** (Gr. *endon* = within + *my-* = muscle), through which blood vessels and nerves that supply the fibers travel. Smooth and cardiac muscle fibers are organized as sheets or layers within the walls of the visceral organs. Their fibers do not have well-defined points of attachment, but pull on each other and the wall of the organs. They act on the contents of an organ by changing the organ's size and shape (as in the stomach or heart), or by changing the tension in its wall (as in some blood vessels).

Skeletal muscles are more complexly organized than smooth or cardiac (Fig. 9–2). Groups of muscle fibers from small bundles or **fasciculi** (L. *fasciculus* = small bundle) that are grouped together by a layer of connective tissue known as the **perimysium.** Many fasciculi, in turn, are surrounded by an **epimysium** and aggregated into units that we recognize as the individual **muscles.**

Skeletal muscles have distinct attachments to skeletal elements or to well-defined connective tissue septa by **tendons** (Fig. 9–3). Tendons are formed by the union of the connective tissue within the muscle with the connective tissue **periosteum** that surrounds a bone. Frequently, the connective tissue of the ten-

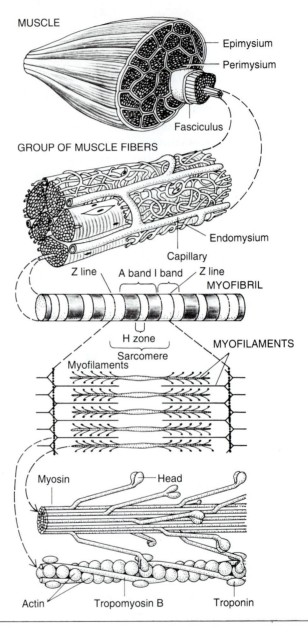

MUSCLE

— Epimysium

— Perimysium

Fasciculus

GROUP OF MUSCLE FIBERS

— Endomysium

Capillary

Z line   A band I band   Z line

MYOFIBRIL

H zone

Sarcomere

Myofilaments   **MYOFILAMENTS**

Myofilaments

Myosin   Head

Actin   Tropomyosin B   Troponin

FIGURE 9–2   The structure of a skeletal muscle. Successive stages of magnification have been used to show structure from an entire muscle to the ultramicroscopic myofilaments of actin and myosin that make up a myofibril within a muscle cell or fiber. (Partly after Williams et al.)

don penetrates the bone. Tendons may be conspicuous, cordlike structures, but often the muscle and bone are so close that the tendon is inconspicuous as in the humeral origin of the triceps. Sometimes tendons form broad, thin sheets, in which case they are called **aponeuroses** (Gr. *apo* = from + *neuron* = sinew).

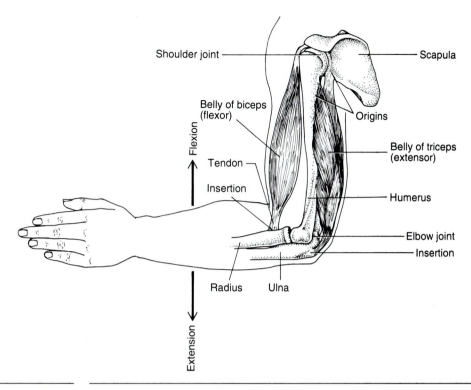

Flexion

Extension

Shoulder joint

Scapula

Belly of biceps
(flexor)

Origins

Belly of triceps
(extensor)

Tendon

Insertion

Humerus

Elbow joint

Insertion

Radius   Ulna

FIGURE 9–3   The attachments and actions of the major muscles of the upper arm.

Muscle Fine Structure and Contraction

Although the biochemistry of muscle contraction is beyond the scope of this book, it is important to recognize that muscle contraction results from the interaction of actin and myosin myofilaments. In skeletal muscle fibers, the myofilaments are arrayed in groups, the **myofibrils,** that can just be seen with the light microscope (Fig. 9–2). The myofibrils have a banded appearance with dark **A bands** alternating with light **I bands.** The striated appearance of a fiber is caused by the A and I bands of adjacent myofibrils being in register, that is, each type of band of one myofibril lies in the same transverse plane within a muscle fiber as the corresponding band in adjacent myofibrils. The banding of a myofibril, in turn, results from the arrangement of the myofilaments within it. Thicker myosin filaments overlap thin actin filaments in the darker A bands, and only actin filaments occur in the I bands. Another light area, the **H zone,** within an A band is a region where only myosin filaments are present. A thin, dark **Z line,** to which actin myofilaments attach, crosses the I band. The Z lines demarcate functional units of the myofibril, known as the **sarcomeres** (Gr. *sark-* = flesh + *meros* = part).

An action potential initiated on the membrane of a muscle fiber by a nerve impulse or other means initiates a series of biochemical changes that lead to the binding of the heads on the myosin filaments to sites that have become active on the actin filaments (Fig. 9–2). The heads swivel, and the actin filaments are pulled into the array of myosin filaments. The heads let go of the actin filaments, reattach at new sites, and the process continues. This mechanism is known as the **sliding filament hypothesis** of muscle contraction. The attachment and detachment of each myosin head requires one molecule of energy in the form of ATP (adenosine triphosphate). As the actin myofilaments slide between the myosin ones, the sarcomeres shorten (the Z lines come closer together) and the H zone becomes narrower or disappears.

The tension that results from the interaction of myosin and actin filaments is called muscle contraction. Depending on circumstances, the muscle fibers may or may not shorten. If the muscle shortens, it will cause bones or other structures to which it attaches to move, or the organ of which the muscle cells are a part may change size or shape, as when a bladder contracts. A muscle contraction that induces a shortening of a muscle against a constant load or force is called **isotonic contraction** (Gr. *isos* = equal + *tonos* = strain). Movements of the body and its parts are caused primarily by isotonic contractions. **Work,** as the term is used by physicists, is performed during isotonic contraction because work = the force developed × the distance through which the force works. In **isometric contraction** (Gr. *metron* = measure) tension develops, but little if any shortening of the muscle occurs because the ends of the muscle are held in place by other forces. Muscles that support an animal or hold a part in a fixed position contract isometrically, but we usually are not aware of their activity. We can sense the tension that develops during isometric contraction by trying to lift an object that is firmly attached to the floor. A great deal of force can be developed during isometric contraction, but no work in the physical sense is performed, for no movement occurs. Sometimes a muscle may increase in length as tension develops. This is called **negative work contraction.** An example is the action of the hamstring muscles (semimembranosus, semitendinosus, and biceps femoris) on the posterior surface of your thigh when you stand from a sitting position. Tension develops in these muscles as they help pull the thigh caudad, but they are stretched by other muscles as the leg straightens at the knee and the distance between their points of attachment increases.

Muscles perform their functions by one of these types of contraction. Skeletal muscles extend across one or more joints and either move skeletal elements relative to one another, as when you bend your knee, or stabilize the skeletal elements at the joint so they do not move, as when you stand still. We often describe muscles by the location of their main muscle mass, or **belly:** shoulder muscles, brachial muscles, antebrachial muscles. But the tendon of a skeletal muscle crosses at least one joint and acts on the skeletal element distal to its belly. It is convenient to describe the opposite attachments of a skeletal muscle as its origin and insertion (Fig. 9–3). The **origin** is often thought of as the attachment that remains fixed in position when the muscle shortens, and the **insertion** as the end attached to the element that moves. However, muscles develop the same tension at each end, and the movable end changes with circumstances. During the free movement of the forearm at the shoulder joint, the distal end of the triceps muscle moves as the forearm is extended by the shortening of the triceps (Fig. 9–3). But when a quadruped is walking, the foot remains on the ground during the propulsive phase of a step and the shoulder is moved forward relative to the foot. Under these circumstances the proximal end of the triceps may move more. By convention, the origin of a limb muscle is its proximal end, and the insertion its distal end.

Muscles perform their work by developing tension and often shortening. They are restored to their resting length upon relaxation by an **antagonistic force** that operates in a direction opposite to the direction of contraction. Frequently, muscles are arranged in antagonistic groups such that one muscle pulls a limb in one direction, and its antagonist pulls it in the opposite direction and restores the resting length. One example of an antagonistic set is the biceps and triceps muscles, which move the forearm in opposite directions (Fig. 9–3). The longitudinal and circular layers of muscle in the wall of the intestine are another example. The longitudinal muscles shorten the intestine and exert a force on the intestinal contents that causes the diameter of the intestine to increase; the contraction of circular muscles has the opposite effect, decreasing the diameter of the intestine and increasing its length. The antagonistic force need not be another muscle. The walls of blood vessels, for example, contain only circularly arranged muscle fibers. These work against and help control the hydrostatic pressure of the blood within the vessels.

Sets of terms define many antagonistic actions. The movement of a distal limb segment toward a more proximal one—the decrease in the angle between the two segments—is called **flexion. Extension** is the opposite movement. The term *flexion* may also be used in mammals to describe the forward movement of the brachium or thigh, but this action is usually called **protraction** in quadrupeds. **Retraction** is the opposite movement. Flexion also describes a ventral bending of the head or trunk. **Adduction** describes the movement of a part, such as a limb, toward some point of reference, in this case the midventral line of the body. **Abduction** is the opposite movement. **Rotation** is the movement of a bone around some longitudinal axis. Pronation and supination of the hand (p. 292) are special cases of rotation. A simple sliding of one bone on another, called **translation,** occurs at some joints, including many of those in the wrist and ankle. Some muscles are named according to the actions they cause. Most terms are self-evident: **levators, depressors, sphincters, dilators,** and so forth.

These are useful descriptive terms, but muscle action is more complex than they imply. Electromyographic studies have shown that muscles seldom act individually but rather in **synergistic** groups (Gr. *syn* = together + *ergon* = work). One muscle may initiate a movement and others become active at different times as skeletal elements move apart or toward each other. One muscle may restrict certain movements at a joint and allow other muscles to cause a more limited movement. For example, contracting flexor muscles in your forearm and hand will cause your hand to form a fist. However, the simultaneous contracting of certain extensor muscles on the back of the hand will prevent the proximal phalanges from flexing so you flex just the ends of your fingers. Movements at joints also require that the antagonistic muscles relax in synchrony with the contracting muscles.

Adaptations of Skeletal Muscles

Motor Units

The structure and function of skeletal muscles are carefully adapted to the conditions in which they act. The muscle fibers within a muscle are organized into **motor units.** Each motor unit consists of a motor neuron and the muscle fibers it supplies. A single motor unit is widely dispersed through a muscle, and many motor units, which are interspersed with each other, are present in a single muscle. Muscles differ with respect to the number of muscle fibers in a motor unit. Where a very fine regulation of contraction is needed, as in the small muscles that control the movements of the eyeball, each motor unit contains only a dozen or fewer muscle fibers. A delicate movement can be made by activating only one or a few motor units. Where a strong force must be generated, as in the large leg muscles that maintain posture, each motor unit may contain 2000 muscle fibers or more. During normal activity an ever-changing rotation of active, relaxing, and quiescent motor units occurs. As functional demands on a muscle change, the proportion of active motor units increases or decreases.

Tonic and Phasic Muscle

Skeletal muscles also vary in their biochemical and physiological properties. Two broad categories are recognized: tonic and phasic. Each muscle fiber in a **tonic** muscle receives multiple motor end plates from the neuron that supplies it. The action potential resulting from a single nerve impulse does not propagate far, so only sarcomeres near the motor end plates become active. A rapid succession of nerve impulses causes a more extensive propagation of the action potential and activates more sarcomeres. The extent and force of contraction are graded by the frequency of nerve stimulation. The rate of contraction of tonic muscles also is rather slow, and tonic muscles do not fatigue easily. These muscles are well adapted for slow, sustained, and carefully graded contraction. They are often small muscles. The muscles that move the eyeball are among the few tonic muscles in mammals, but some of the postural muscles of other vertebrates are tonic.

Most vertebrate skeletal muscles are **phasic.** Each muscle fiber in a phasic motor unit has a single motor end plate. One nerve impulse may not be adequate to depolarize the membrane and initiate an action potential, but several in rapid succession will do so. The action potential propagates easily and rapidly along the muscle fiber, so the contraction of individual fibers and motor units is not graded, as in tonic muscle, but **all-or-none.** If the stimulus is adequate to initiate contraction, the entire motor unit is activated, and a brief contraction, or muscle **twitch,** results (Fig. 9–4). Repetitive stimulation before the motor unit has relaxed from the first twitch can cause a second or third twitch to be superimposed, or **summated,** on the first. Very rapid stimulation can lead to a maximum **tetanic** contraction or spasm, of the motor unit. The force of contraction can be graded in part by the rate of nerve stimulation. But since phasic muscles are usually larger than tonic ones and contain many motor units, force is also graded by the number of motor units active at a given time.

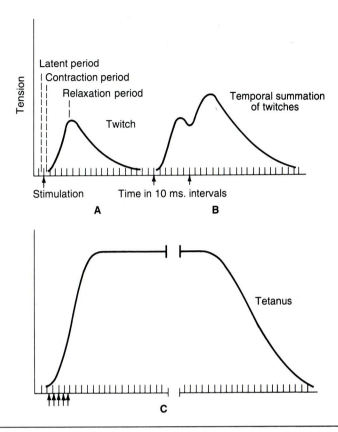

FIGURE 9–4   Tension development during the isometric contraction of a motor unit: **A**, A twitch. **B**, Temporal summation of two twitches. **C**, Tetanus. (After Dorit, Walker, and Barnes.)

Phasic muscle fibers differ in their metabolism and many aspects of their contraction. **Slow phasic fibers** derive their energy from oxidative metabolism. They are richly supplied with blood vessels, have a high mitochondrial content, and contain a large amount of **myoglobin,** which is a hemoglobin-like molecule that facilitates the transfer of oxygen from the blood. Their myoglobin content gives them a reddish color, and sometimes they are called red muscle. These muscle fibers contract relatively slowly and no faster than oxygen can be delivered to them. They do not go into oxygen debt or fatigue easily. They are particularly well adapted for isometric contraction and for slow, repetitive, isotonic contractions. **Fast phasic fibers** are adapted for rapid movements of brief duration. They derive their energy primarily from anaerobic glycolysis, so they can contract more rapidly than oxygen can be delivered to them. As would be expected, glycogen content is higher than in the slow fibers, and mitochondrial content is lower. Fast phasic fibers contain little myoglobin and appear white. Since little of the pyruvic acid resulting from glycolysis is oxidized during the bursts of activity, it is transformed into lactic acid. Fast phasic fibers fatigue quickly and go into oxygen debt until the lactic acid can be oxidized after their activity. Yet other phasic fibers have intermediate properties.

Many muscles contain populations of each fiber type. The slow phasic fibers are recruited first and are used in maintaining posture and in slow movements. When rapid, brief contractions are needed, the fast phasic fibers become active. In a sense, the muscle has at least two gears, low and high. Other muscles are composed entirely of one fiber type or the other. Many fishes have a band of dark, slow phasic fibers along each side of the flank that are used in slow swimming; white, fast phasic fibers that make up the rest of the trunk and tail musculature are recruited when bursts of speed are needed. The dark meat of a chicken's leg is composed of slow fibers well adapted for sustaining posture and walking. The white meat of the breast is composed of fast fibers used in rapid wing movements.

Muscle Architecture

Muscles also differ in the length and arrangement of the muscle fibers within them—that is, in their architecture (Fig. 9–5). **Strap-shaped muscles** contain long, parallel fibers and have relatively broad attachments. Some of the fibers in the muscle are nearly as long as the muscle. **Fusiform muscles** are similar, except that their fibers lead into narrow tendons at the ends of a muscle so the force of contraction is concentrated on a smaller area. **Pennate muscles** contain short, diagonally arranged fibers that insert on a tendon on one side of the muscle (uni-

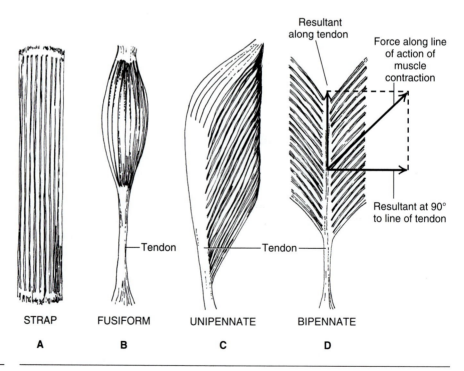

FIGURE 9–5   The architecture, or arrangement, of muscle fibers within a muscle. (After Williams et al.)

pennate muscles) or on one or more tendons more centrally located (bipennate or multipennate muscles). Examples of these types of muscles can be seen in the shoulder and arm of a cat (Fig. 9–18**C**): teres major (strap), biceps (fusiform), and subscapularis (multipennate).

These variables in muscle architecture affect the length a muscle can contract, the velocity of contraction, the force of contraction, and the power a muscle can develop. The degree to which a muscle can contract depends on the length of its muscle fibers. Most muscle fibers can contract about one third of their resting length before the actin myofilaments are completely pulled into the array of myosin ones. Strap and fusiform muscles contain longer fibers than pennate ones of the same size, contract a longer distance, and can induce a more extensive movement. They also contract faster because the rate of contraction or shortening is a function of the number of sarcomeres in series. Assuming that each sarcomere contracts at the same rate, the more sarcomeres in series the faster the muscle as a whole can contract. Long fibers have more sarcomeres than short ones.

Pennate muscles, on the other hand, can develop a greater force than strap or fusiform muscles of the same mass because the force a muscle develops is a function of the number of myosin-actin connections or cross bridges made at one time. Thus force depends on the number of myofilaments in a fiber and the number of muscle fibers in a muscle. Muscles vary somewhat with respect to the number of myofilaments in their fibers, and this number is increased with exercise. The number of fibers in a muscle can be estimated by a section that cuts most of the fibers at right angles. Such a section would be transverse to the long axis of a strap or fusiform muscle, but would have to curve around the periphery of a pennate muscle. Because pennate muscles contain many short fibers, they can generate more force per unit of muscle than muscles with other fiber arrangements. The useful force developed in a pennate muscle is not the force developed along the line of action of the muscle fibers, but the resultant of this force along the axis of the tendon (Fig. 9–5**D**). Force along the other resultant is lost, but the useful force far exceeds that in a parallel-fibered muscle of equal mass. Pennate muscles are well adapted for forceful isotonic contractions of short extent, and for isometric contractions. Since pennate muscles do not shorten greatly, they usually do not perform as much work as parallel-fibered muscles. (Recall that work = force × the distance through which the force acts.) Because they do not shorten much, pennate muscles also do not bulge much during contraction. They sometimes are found where space is limited, as inside a lobster's claw, or where body contours need to be maintained, as on the surface of a fish (p. 345). Most muscles develop maximum forces on the order of 3 to 6 kg per cm^2 of cross-sectional surface.

Force and velocity of contraction are inversely related. Muscles can move heavy loads (develop more force) when they contract slowly. This can be seen in a force-velocity curve (Fig. 9–6). The power output of a muscle is defined as the rate of doing work, and it is equal to force times velocity. The power output can be calculated from the muscle's force-velocity curve (Fig. 9–6). When force is low (but above zero) and velocity high, or force is high and velocity low, power is low. Maximum power is generated at about 30 percent of the maximum force. Although pennate muscles can develop more force than others of equal mass, they do not

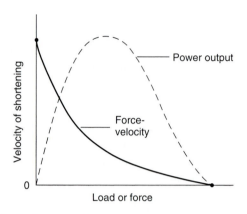

FIGURE 9–6 Force velocity and power output curves for the contraction of a skeletal muscle.

necessarily generate more power because their fibers are short and contraction velocity is low.

Muscle Development and Groups

A regional grouping of the muscles (e.g., those of the shoulder and arm), or a functional grouping (protractors, extensors, etc.), is useful in dissection or in functional studies, but such groups often include muscles of different phylogenetic origins. For example, some of the muscles acting on the shoulder of tetrapods are derived phylogenetically from appendicular, trunk, and gill arch muscles. To trace the evolution of the muscular system, we will group muscles according to their pattern of embryonic development and their nerve supply. Groups of muscles with a similar development and nerve supply are homologous and can be recognized in different evolutionary lineages.

The small muscles within the iris of the eye develop embryonically from the same ectodermal tissue (the optic cup) that gives rise to the iris and retina, but all other muscle tissue is derived from mesodermal mesenchyme. Following the useful, but sometimes artificial, division of the body into somatic and visceral parts (p. 204), we can divide these muscles into somatic and visceral groups (Table 9–1). Most **somatic muscles** lie in the "outer" tube of the body (that is, in the body wall and appendages), and most develop from the segmented myotomes of the embryonic somites (Fig. 9–7**A** and **B**). The ventral extensions of the myotomes peripheral to the lateral plate mesoderm form most of the flank musculature, but some flank muscles arise, at least in amniotes, from mesenchyme derived from the outer, somatic layer of the lateral plate. **Visceral muscles** develop in the "inner" tube of the body from the inner, or splanchnic, layer of the lateral plate. Visceral muscles accordingly are more deeply situated than somatic muscles; they contribute to the walls of most of the digestive tract and other visceral organs, and to the walls of the heart and blood vessels. Blood vessels begin their development in the splanchnic mesoderm but gradually extend to other parts of the body.

TABLE 9–1 **Groups of Somatic Muscles and Representative Muscles**

Group	Shark	Amphibian/Reptile	Mammal
I. Axial Muscles			
1. Extrinsic Ocular Muscles			
First two somitomeres (oculomotor nerve)	Dorsal rectus / Ventral rectus / Medial rectus / Ventral oblique	Dorsal rectus / Ventral rectus / Medial rectus / Ventral oblique	Dorsal rectus / Levator palpebrae superioris / Ventral rectus / Medial rectus / Ventral oblique
Third somitomere (trochlear nerve)	Dorsal oblique	Dorsal oblique	Dorsal oblique
Fifth somitomere (abducens nerve)	Lateral rectus	Lateral rectus / Retractor bulbi	Lateral rectus / Retractor bulbi
2. Branchiomeric Muscles			
Mandibular muscles (trigeminal nerve)	Adductor mandibulae / Levator palatoquadrati / Spiracularis / Preorbitalis	Levator mandibulae	Masseter / Temporalis / Pterygoideus / Tensor veli palati / Tensor tympani
	Intermandibularis	Intermandibularis	Mylohyoideus / Anterior digastric
Hyoid muscles (facial nerve)	Levator hyomandibulae / Dorsal constrictor	Depressor mandibulae / Branchiohyoideus	Stapedius / Platysma and facial muscles (part)
	Ventral constrictor	Interhyoideus	Platysma and facial muscles (part) / Posterior digastric / Stylohyoideus
	Interhyoideus	Sphincter colli	Trapezius complex
Branchiomeric muscles of remaining arches (glossopharyngeal, vagus, and, in amniotes, spinal accessory nerves)	Cucullaris	Cucullaris	
	Interarcuals / Branchial adductors	Levatores arcuum	Sternocleidomastoid complex
	Superficial constrictors and interbranchials	Dilator laryngis / Subarcuals / Transversi ventralis / Depressores arcuum	Intrinsic muscles of the larynx and certain muscles of the pharynx

TABLE 9–1 *Continued*

Group	Shark	Amphibian/Reptile	Mammal
3. Epibranchial Muscles (Dorsal rami of occipital and anterior spinal nerves)	Myomeres dorsal to gill region	Anterior part of dorsalis trunci	Anterior part of epaxial muscles
4. Hypobranchial Muscles (Ventral rami of spino-occipital nerves in anamniotes, hypoglossal nerve and cervical plexus in amniotes)			
Prehyoid muscles	Coracomandibular	Genioglossus Geniohyoid	Lingualis Genioglossus Hyoglossus Styloglossus Genioglossus
Posthyoid muscles	Rectus cervicis + coracobranchials	Rectus cervicis Omoarcaurals Pectoriscapularis	Sternohyoid Sternothyroid Thyrohyoideus
5. Trunk Muscles (Spinal nerves)			
Epaxial muscles (dorsal rami)	Epaxial portion of myomeres	Interspinalis Dorsalis trunci	Interspinalis Intertransversarii Occipitals — Transversospinalis Multifidi Spinalis Semispinalis Longissimus dorsi — Longissimus Splenius Iliocostalis — Iliocostalis

Hypaxial muscles (ventral rami)

- Hypaxial portion of myomeres
  - Subvertebralis
    - Longus colli — *Subvertebral*
    - Psoas minor
    - Quadratus lumborum
  - Levator scapulae
  - Thoraciscapularis
    - Omotransversarius — *Lateral*
    - Serratus ventralis (part)
    - Serratus ventralis (part)
    - Rhomboideus
    - Serratus dorsalis
    - Scalenus
  - External oblique
    - Rectus thoracis
    - External oblique
    - External intercostals
  - Internal oblique
    - Internal oblique
    - Internal intercostals
  - Transversus
    - Transversus abdominis
    - Transversus thoracis
    - Diaphragm muscles
  - Rectus abdominis
    - Rectus abdominis — *Ventral*

II. Appendicular Muscles
Pectoral Muscles

Dorsal group — Adductor

- Latissimus dorsi
  - Cutaneous trunci (part)
  - Latissimus dorsi
  - Teres major
- Subcoracoscapularis
  - Subscapularis
- Deltoid
  - Deltoid complex
- Scapulohumeralis anterior
  - Teres minor
- Triceps brachii
  - Triceps brachii
- Antebrachial extensors
  - Tensor fasciae antebrachii
  - Antebrachial extensors

Ventral group — Abductor

- Pectoralis
  - Cutaneous trunci (part)
  - Pectoralis complex
- Supracoracoideus
  - Supraspinatus
  - Infraspinatus
- Biceps brachii
  - Biceps brachii
- Brachialis
  - Brachialis
- Coracobrachialis
  - Coracobrachialis
- Antebrachial flexors
  - Antebrachial flexors

(Visceral muscles are considered with the organs of which they are a part.)

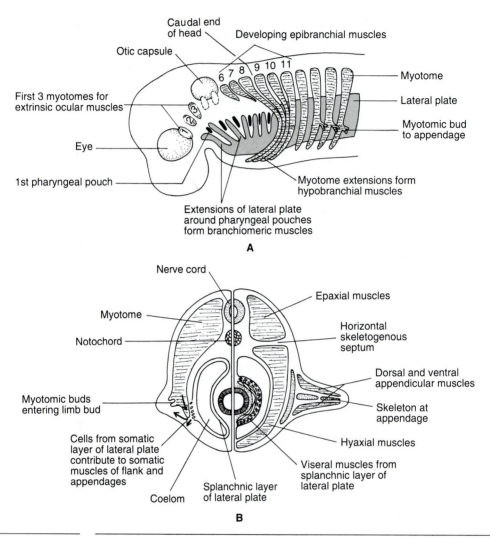

FIGURE 9–7   Vertebrate muscle development. **A,** Goodrich's proposal for the development of somatic and visceral muscles of the trunk and head, based on studies of a shark embryo. **B,** The development of somatic and visceral muscles as seen in a cross section of the trunk at the level of the pectoral appendage. An early stage is on the left, a later one is on the right.

The derivation of somatic and visceral muscles from the somites and lateral plate mesoderm, respectively, is quite clear in the trunk. For most of this century, zoologists believed that this was also the case in the head. Zoologists followed Goodrich's interpretation (1918, 1930) that the head was fundamentally segmented (Fig. 9–7A). Somites, often called head cavities, continued into the head, although one or more in the otic region were transitory. The somites gave rise to the somatic muscles of the head. A group of muscles, called **branchiomeric muscles** (Gr. *branchia* = gills + *meros* = part), developed in the pharynx wall in association with the visceral arches. Although techniques available at the time did not allow investigators to follow closely the migration of cells during development,

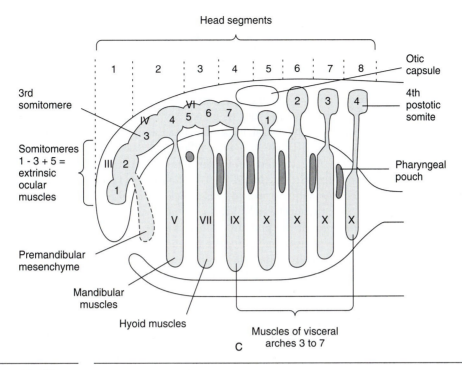

Head segments

FIGURE 9–7   *continued*
C, Northcutt's proposal for the segmentation of the vertebrate head and the development of cranial muscles, based on experimental studies of bird embryos by Noden, 1983. (**C**, After Northcutt.)

it was assumed that the branchiomeric muscles developed from a cranial extension of the lateral plate because the muscles could first be identified in the embryo in the wall of the pharynx. It was believed that the lateral plate, which contained no coelom in this region, continued forward beneath the pharynx and sent up tongues of premuscular tissue between the branchial pouches (Fig. 9–7**A**). Branchiomeric muscles, therefore, were assumed to be visceral.

Zoologists now realize that segmentation of the head is not as well defined as in the rest of the body, and the embryonic origin of muscles in the head is not as clear as in the trunk region. The **paraxial mesoderm,** which lies beside the neural tube and gives rise to the somites and myotomes in the trunk, does extend into the head, but only the part of it caudal to the otic capsule forms distinct somites (Fig. 9–7**C**). Four somites and body segments lie in this part of the head of gnathostomes. The fate of the paraxial mesoderm rostral to the otic capsule is only now being clarified. Recent studies by Meier (1981, 1982), Noden (1983, 1991), and Jacobson (1988) indicate that this part of the paraxial mesoderm does not become completely segmented. Rather it forms a series of seven **somitomeres,** or hillocks of mesenchymal cells, that are only partially separated from each other by slight indentations. Northcutt (1990) suggests that the first six somitomeres can be grouped by twos into units that appear comparable to the first three head segments of earlier investigators. The seventh somitomere appears to represent the remains of a fourth head segment (Fig. 9–7**C**). If couplets of somitomeres can be equated with well-defined segments, then the head of jawed vertebrates contains eight segments (four preotic and four postotic ones) as Goodrich believed. But Noden (1991) points out that it is risky to equate transient hillocks of mesenchyme with well-defined segments.

To determine the fate of cranial somites and somitomeres, Noden (1983) transplanted groups of Japanese quail head somites and somitomeres into embryos of the domestic chick. He could follow their subsequent development because tissue that differentiated from quail cells carried distinctive nuclear condensations of chromatin. He concluded that the first three somitomeres and the fifth gave rise to the somatic extrinsic ocular muscles that move the eyeball. Similarly, myoblasts from the most caudal head segment(s) and most cranial trunk ones migrate forward beneath the pharynx to form the hypobranchial muscles that lie in the floor of the pharynx and enter the tongue of tetrapods (Fig. 9–7**A**). The origin of ocular and hypobranchial muscles from somites and somitomeres confirms earlier studies that these are somatic muscles.

The surprising aspect of Noden's study is that the myoblasts that give rise to the branchiomeric muscles also migrate from somitomeres and somites, and not from a forward extension of the lateral plate. Branchiomeric muscles should, therefore, be considered to be somatic muscles and not visceral ones. In addition to their origin from somites and somitomeres, branchiomeric muscles resemble other somatic muscles in being striated and in being innervated by motor neurons that travel directly from the central nervous system. We restrict the term *visceral muscles* to those of the gut wall (except for the branchiomeric ones), other viscera, blood vessels, and heart. All are supplied by the autonomic nervous system, which involves a peripheral motor relay (p. 458), and all but those of the heart are smooth.

It may appear unusual that somatic muscles, normally restricted to the "outer" tube of the body, attach onto the visceral arches located deep in the pharynx wall. However, you will recall (p. 208) that these arches develop from ectodermal neural crest cells that migrate into the pharynx wall and not from cells in the wall. The connective tissue that permeates the branchiomeric muscles is also of neural crest origin. The somatic origin of branchiomeric muscles has been demonstrated so far only in birds. It might seem unwarranted to extend this generalization to other vertebrates, but this is the only hard evidence we have for the origin of branchiomeric muscles in any vertebrate. Their presumed origin from the lateral plate was an assumption based on their location and not on studies of migrating cells.

Visceral muscles will be discussed with the organs of which they form a part and will not be discussed further in this chapter. The somatic muscles can be subdivided into an axial group located along the longitudinal axis of the body and an appendicular group that develops from mesenchyme that migrates from the myotomes (and sometimes from the somatic layer of the lateral plate) into the limb buds, which are the primordia of the paired appendages (Fig. 9–7**B**). **Appendicular muscles** are defined as those that begin their differentiation within the limb bud. They remain associated with the paired appendages or girdles, but some migrate medially to take their origin from the trunk. Some trunk and branchiomeric muscles secondarily shift their insertions onto the girdles, but these are not considered part of the appendicular musculature as the term is used in this book. Appendicular muscles can be sorted into dorsal and ventral groups that primitively lie above and below the appendicular skeleton and girdles.

Axial muscles can be further subdivided according to the group of body segments from which they arise. The **extrinsic ocular muscles** develop from the first three somitomeres and the fifth. A specific cranial nerve is associated with each of these somitomeres and supplies the muscle(s) that develop from them (Fig. 9–7**C**): somitomeres 1 and 2 with the oculomotor nerve (III), somitomere 3 with the trochlear nerve (IV), and somitomere 5 with the abducens nerve (VI).

The **branchiomeric** group of muscles develops from the remaining somitomeres and the four postotic head segments. Again a specific cranial nerve supplies the muscles that develop from each (Fig. 9–7**C**). **Mandibular muscles** arise from somitomere 4 and are supplied by the trigeminal nerve (V). **Hyoid muscles** arise from somitomere 6 and are supplied by the facial nerve (VII). Muscles of the third visceral arch (first branchial arch) develop from somitomere 7 and are supplied by the glossopharyngeal nerve (IX). The remaining branchiomeric muscles, called **branchial muscles,** arise from the four postotic somites and are supplied by the vagus nerve (X).

The postotic somites also give rise, at least in fishes, to a small group of **epibranchial muscles** that lie above the gill region. **Hypobranchial muscles** below the gill region develop in all vertebrates from the migration of myoblasts from postotic and the more anterior trunk somites around the back of the gill region (Fig. 9–7**A**). Epibranchial and hypobranchial muscles are innervated in fishes by a group of occipital nerves, which leave the central nervous system from the occipital region of the skull, and by several anterior spinal nerves. Branches of the nerves that supply the hypobranchial muscles usually aggregate to form a conspicuous hypobranchial nerve. Epibranchial muscles retain their embryonic segmentation in anamniotes; hypobranchial muscles tend to fuse and form longitudinal bands, but traces of segmentation often remain.

The remaining myotomes form the muscles of the trunk and tail. These muscles remain segmented in fishes, but parts of many myotomes fuse to form longitudinal bundles and broad sheets in tetrapods. The myotomes become divided in all gnathostomes by the horizontal skeletogenous septum (p. 254) that extends from the vertebral column to the body surface at the lateral line (Fig. 9–7**B**). The portions of the myotomes that lie dorsal to the horizontal septum become the **epaxial muscles;** those ventral to it become the **hypaxial muscles.** Trunk and tail muscles are innervated by branches of the spinal nerves. Dorsal branches, or **rami** (L. *ramus* = branch), supply the epaxial group; ventral rami supply the hypaxial group. Appendicular muscles, which develop from mesenchyme that primitively comes from the ventral portions of certain myotomes, also are supplied by the ventral rami of spinal nerves.

The divisions of the muscular system, and important muscles within these groups, are summarized in Table 9–1. The groups can be recognized in the early embryos of all vertebrates by their embryonic origin, but in tetrapods muscles frequently migrate a considerable distance during later development. Fortunately, their motor nerve supply remains conservative, and the nerves usually follow the muscles during their developmental gymnastics. Nerve supply is a good criterion for muscle homology, so the pattern of innervation of the muscles should be studied with the muscles themselves. As one example, the muscles of the mammalian diaphragm, which lie caudal to the heart, develop from cervical myotomes because the developing heart and diaphragm are located far forward in an early embryo. As development continues, the heart and diaphragm shift caudad, and branches of the cervical nerves, known as the phrenic nerves, follow the diaphragm.

The Muscular System of Fishes

The basic pattern of the muscular system of vertebrates can be seen clearly in fishes because the muscles have not migrated far from their embryonic origins.

Extrinsic Ocular Muscles

The most rostral somatic muscles belong to the ocular group. In jawed fishes, as in most other vertebrates, they are represented by six strap-shaped muscles that attach to the surface of the eyeball. These muscles are responsible for eyeball movements as a moving animal continues to look at a fixed object, or as the animal shifts its field of vision. Four rectus muscles fan out from the posteromedial wall of the orbit to the eyeball (Fig. 9–8). They are named for their location and points of insertion: **dorsal** (or **superior**) **rectus, ventral** (or **inferior**) **rectus, medial rectus, lateral rectus.** Two oblique muscles originate on the anteromedial wall of the orbit and extend obliquely caudad to insert on the dorsal and ventral surfaces of the eyeball: **dorsal** (or **superior**) **oblique, ventral** (or **inferior**) **oblique.** The lateral rectus is innervated by the abducens nerve (VI), since it develops from the fifth somitomere; and the dorsal oblique is innervated by the trochlear nerve (IV), since it develops from the third somitomere. All other muscles in this group are innervated by the oculomotor nerve (III), since they develop from the first two somitomeres (Fig. 9–7**C**).

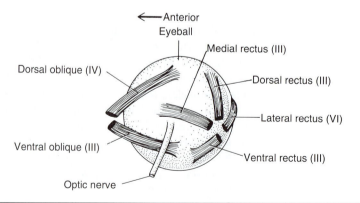

FIGURE 9–8   The extrinsic ocular muscles of a fish. The right eye has been removed from the orbit and is being viewed from the back, or medial side. The cranial nerves that supply the muscles are indicated by roman numerals.

Branchiomeric Muscles

The branchiomeric group of muscles develops caudad to the extrinsic ocular muscles. These muscles, and the visceral skeleton on which they act, are important and conspicuous in fishes because they are utilized both in feeding and in breathing movements. We will consider feeding and breathing in more detail in Chapters 16 and 18, but we will examine the basic pattern of these muscles at this time. We will use a shark as our example because the pattern of these muscles is evident. This pattern applies to other fishes as well, although details differ from group to group in accordance with differences in feeding and breathing. A shark's last five visceral arches, or branchial arches, lie adjacent to the pharyngeal lumen, and interbranchial septa extend from all but the last arch outward toward the side of the head. Most of each septum is composed of an **interbranchial muscle,** whose fibers have a circular arrangement (Fig. 9–9**A**). Skeletal branchial rays also extend from the arches into the septa, which they help stiffen. Gills lie on the anterior and posterior surface of a representative septum, and branchial chambers lie between successive septa. As the septa approach the body surface, the interbranchial muscle fibers change their direction and form superficial sheets that cover the dorsal and ventral parts of the branchial chambers, leaving only small openings, the external gill slits, to the body surface. These sheets of muscle are the **dorsal** and **ventral superficial constrictor muscles** (Fig. 9–9**B**). Primitively, each branchial arch probably had a **levator muscle** that extended from the fascia overlying the epibranchial muscles to the top of the arch, but in contemporary sharks, the levators of all the branchial arches have united to form a triangular muscle, the **cucullaris** (L. *cucullus* = hood), most of which inserts on the scapular region of the pectoral girdle. This is an example of a branchiomeric muscle that functionally has become associated with the appendicular skeleton, even though part of the muscle still attaches to the last branchial arch. Small **adductor** and **interarcual muscles** interconnect segments of the branchial arches where the arches form sharp angles (Fig. 9–9**A**). The muscles of the first branchial arch (or third visceral arch) are supplied by the glossopharyngeal nerve (IX); those of the remaining branchial arches are innervated by the vagus nerve (X).

Contraction of the branchial muscles (except for the levators) draws the parts of the branchial arches together and compresses the pharynx and branchial chambers. These muscles are responsible for discharging an expiratory current of water across the gills and out the external gill slits. Water is drawn into the pharynx through the mouth and spiracle by the elastic recoil of the branchial arches, assisted by the contraction of the hypobranchial muscles (p. 640).

Each visceral arch probably was a branchial arch in ancestral, jawless fishes, and it is likely that each one had a similar set of muscles. When the first visceral arch became incorporated into the jaws and the second arch became part of the jaw suspension mechanism, the muscles of these arches became modified. Most of the mandibular musculature of the first arch forms a powerful **adductor mandibulae** that closes the jaws (Fig. 9–9**B**). Another part, the **levator palatoquadrati,** helps lift the palatoquadrate cartilage during prey capture. This muscle is serially homologous to the branchial levators. The **spiracularis** of elasmobranchs controls the opening and closing of the spiracle, the **preorbitalis** protracts the

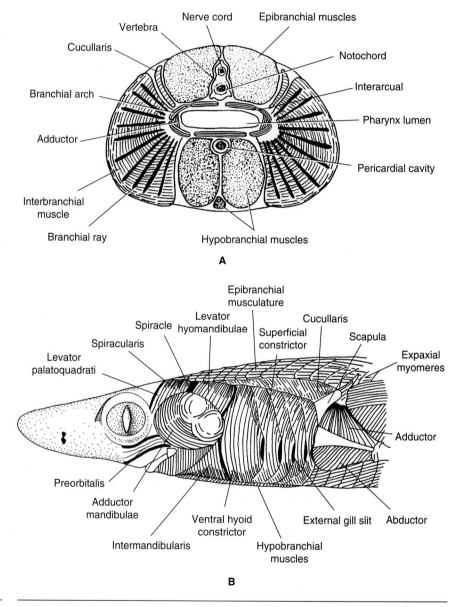

A

B

FIGURE 9–9 The branchiomeric muscles of a shark. **A**, A cross section through the pharynx to show the muscles associated with an interbranchial septum; gills have been removed. **B**, A lateral view of the head muscles. (After Walker and Homberger.)

jaws during feeding, and an **intermandibularis** helps compress the throat. The intermandibularis is in series with the ventral branchial constrictors. All the mandibular muscles are innervated by the trigeminal nerve (V). The hyoid muscles of the second arch include a **levator hyomandibulae, dorsal** and **ventral hyoid constrictors,** and an **interhyoideus.** All are supplied by the facial nerve (VII).

Epibranchial and Hypobranchial Muscles

The epibranchial muscles, which lie dorsal to the gill region, remain segmented (Fig. 9–9**B**). They represent a forward continuation of the epaxial trunk muscles and act with them. These muscles are innervated by the dorsal rami of the spino-occipital nerves.

The hypobranchial muscles, lying ventral to the gill region and representing a forward extension of the hypaxial trunk muscles, are more specialized, for they have assumed special functions associated with feeding and breathing. Hypobranchial muscles can be divided into prehyoid and posthyoid groups. The **prehyoid muscles** of tetrapods, as their name implies, lie between the jaws and the hyoid apparatus, but the group extends more caudad in fishes. The group is represented in sharks by a single pair of longitudinal muscle bands, called the **coracomandibulars,** that extend from the mandible toward the coracoid region of the pectoral girdle, although they do not reach the coracoid in many species (Fig. 9–10). **Posthyoid muscles** are represented by a **rectus cervicis** complex that extends between the hyoid arch and pectoral girdle. Deeper parts of the complex insert on the ventral surfaces of the branchial arches. Traces of the ancestral segmentation of myomeres can often be seen in the rectus cervicis complex, but muscle fibers in successive segments tend to fuse and form longitudinal bundles. Often, several distinctive muscles, which are named for their attachments, are recognized in the

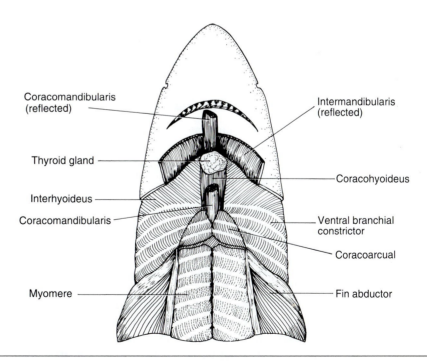

FIGURE 9–10   A ventral view of the branchiomeric and hypobranchial muscles of a dogfish.

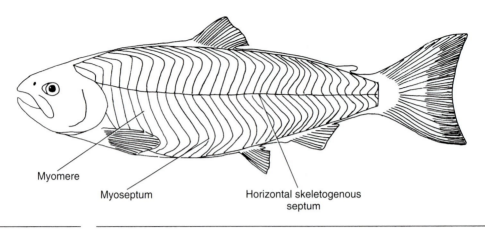

Myomere

Myoseptum

Horizontal skeletogenous septum

FIGURE 9–11   A lateral view of the trunk muscles of a salmon. (After Bond.)

group: **coracoarcuals** (L. *arculus* = small bow or arch), **coracohyoids, cora-cobranchials.** Hypobranchial muscles are innervated by the ventral rami of the spino-occipital nerves that unite to form the hypobranchial nerve.

Functionally, the hypobranchial muscles have become associated with the visceral skeleton because they attach to the mandible, hyoid arch, and branchial arches. Their contraction opens the mouth and expands the pharynx. These actions occur during feeding (p. 577) and the intake of a respiratory current of water (p. 640).

Trunk and Tail Muscles

The embryonic myotomes of the trunk and tail develop into a series of folded muscle segments, the **myomeres,** of adult fishes (Fig. 9–11). The sequential con-traction of the myomeres, acting with the vertebral column, causes a series of lateral undulations by which fishes swim (p. 347). Myomere structure is therefore correlated closely with the pattern of swimming. The individual myomeres are separated from each other by connective tissue **myosepta** that extend inward to the vertebral axis. A horizontal skeletogenous septum does not develop in cyclo-stomes, but one is present in all jawed fishes and divides the myomeres into epaxial and hypaxial parts.

Most of the embryonic myotomes transform into the myomeres, but in addi-tion, small buds that separate from the myotomes become associated with the median fins. The buds differentiate into small muscles that attach to the radial elements in the dorsal and anal fins and control the movements of the fins. Most of these movements help maintain stability, but some species of teleosts swim slowly by undulations that travel along the fins rather than down the trunk.

Appendicular Muscles

The paired appendages of most fishes do not generally deliver major propulsive thrusts. The fins may provide lift, and they are used in maintaining stability, braking, and maneuvering. Paired fin movements and appendicular muscles usually are quite simple (Fig. 9–9**B**). A single ventral **abductor** muscle is located on the anteroventral part of the girdle and ventral part of the fin. It pulls the fin ventrally and cranially. A dorsal **adductor** muscle, located on the posterodorsal part of the girdle and dorsal surface of the fin, pulls the fin dorsally and caudally. The simple abductor and adductor muscles become subdivided and more specialized in species whose fin movements are more complex. Skates and rays, for example, swim along the bottom by undulating their greatly enlarged pectoral fins rather than the trunk and tail.

The Evolution of the Axial Muscles

Hardly any activity of vertebrates was unaffected by the transition from water to land because the physical properties of water and land are so different. Methods of support, locomotion, feeding, and breathing all changed, and with them the muscles responsible for these functions.

Extrinsic Ocular Muscles

The extrinsic ocular muscles are among the few muscles not greatly affected by the shift from water to land. Tetrapods retain the six muscles seen in jawed fishes, unless their eyes have become degenerate. The eyeball is protected in the terrestrial environment by eyelids and often by a transparent, nictitating membrane that can be drawn across its surface. In most tetrapods, but not human beings, a **retractor bulbi** can pull the eyeball deeper into the orbit and so facilitate the movement of the nictitating membrane across its surface. Innervation of the retractor bulbi by the abducens nerve indicates its evolution from the lateral rectus. Part of the retractor bulbi of a bird forms a slip that attaches to and moves the nictitating membrane across the eyeball. Parts of the bird's superior rectus and inferior rectus have separated as a **levator palpebrae superioris** and a **depressor palpebrae inferioris** that act on the upper and lower eyelids, respectively. These muscles, as would be expected, are innervated by the oculomotor nerve. Mammals also have a levator palpebrae superioris, and some species have a depressor palpebrae inferioris (Table 9–1).

Branchiomeric Muscles

Changes in the branchiomeric musculature closely followed the extensive changes in the visceral skeleton that occurred during the transition from water to land and in the subsequent evolution of tetrapods. Many branchiomeric muscles were lost with gills, and others became associated with structures to which the visceral arches contribute: jaws, auditory ossicles, hyobranchial apparatus, and larynx. Some remained with the pectoral girdle.

Mandibular Muscles. Most of the mandibular muscles continue to act on the jaws. A sheet of muscle representing the fish intermandibularis remains as a compressor of the floor of the mouth (Fig. 9–12). It helps pump air into the lungs of amphibians (p. 653). This sheet is represented in mammals by a **mylohyoid**, which extends between the two sides of the lower jaw, and by the anterior part of the **digastric** muscle, whose posterior part develops from hyoid muscles (Fig. 9–13). Part of the mandibular levator group remains in some amphibians and reptiles with kinetic skulls, in which the palatoquadrate has a movable articulation with the braincase, but most levators have been lost in those species where the upper jaw is firmly united to the rest of the skull. The small **tensor veli palati** in the soft palate of a mammal is all that remains of this group.

The large mandibular adductor continues to be the most conspicuous of the jaw muscles. It is divided in different ways in the various lines of tetrapod evolution according to divergences in the methods of feeding. As jaws become stronger and their movements more complex in the line of evolution toward mammals, the adductor becomes divided into a **temporalis**, which arises from the temporal fossa, one or two **pterygoids**, which arise from the pterygoid processes on the underside of the skull, and a **masseter** (Gr. *maseter* = one that chews), which arises from the zygomatic arch (Fig. 9–18A). The evolution of these muscles is closely related to the evolution of temporal fenestration and to the jaw changes that occurred during the evolution of mammals (p. 239). In mammals all these muscles insert on the dentary bone, but one small part of the original adductor

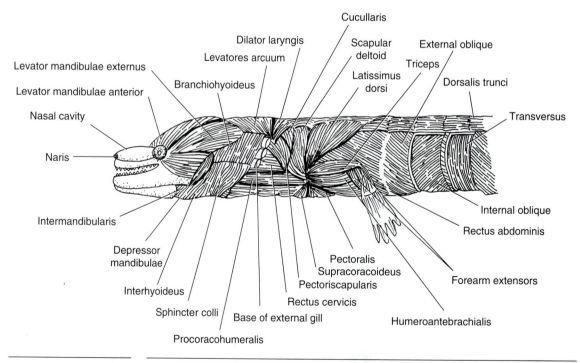

FIGURE 9–12   A lateral view of the head, pectoral, and trunk muscles of *Necturus*. (After Walker and Homberger.)

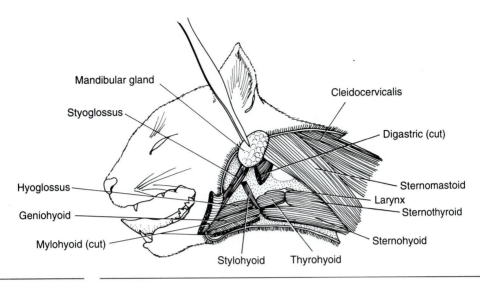

FIGURE 9–13   A lateral view of the hypobranchial and adjacent muscles of a cat. (After Walker and Homberger.)

remains attached to the mammalian derivative of the mandibular cartilage, that is, the malleus in the middle ear. This muscle is the **tensor tympani.** All mandibular muscles continue to be innervated by the trigeminal nerve (V).

Hyoid Musculature.   The hyoid arch is unusual among visceral arches in that it is acted on by the hypobranchial musculature as well as by its own branchiomeric musculature. Considerable hyoid musculature remains in amphibians. Superficial ventral sheets form an **interhyoideus** and **sphincter colli** (L. *collum* = neck) (Fig. 9–12). A **branchiohyoideus** and **depressor mandibulae** lie dorsally. The depressor mandibulae inserts on a retroarticular process of the lower jaw and, along with hypobranchial muscles, opens the jaws in non-mammalian tetrapods.

The remodeling of the jaw that occurred during the evolution of mammals has been accompanied by some changes in jaw-opening mechanisms. The geniohyoid and other hypobranchial muscles remain important, but the depressor mandibulae has been lost and is functionally replaced by the **digastric** muscle. The digastric extends between the caudal part of the skull and the ventral border of the mandible (Fig. 9–18**A**). In many species of mammals, the digastric is divided into two bellies, as its name implies, by a central tendon that attaches to the hyoid apparatus. The posterior belly develops from the hyoid musculature and is innervated by the facial nerve; the anterior belly develops from mandibular muscles and is supplied by the trigeminal nerve.

A small **stylohyoid** (Fig. 9–13), which extends from the styloid process of the skull to the hyoid, and a **stapedius,** which attaches to the stapes, are parts of the hyoid musculature that remain attached to the hyoid arch or its derivatives. (Recall that the stapes evolved from the hyomandibular cartilage.) The superficial sheets of the hyoid musculature lose their connection to the hyoid and spread out beneath the skin of the neck and face to form the **platysma** (Gr. *platysma* = flat object)

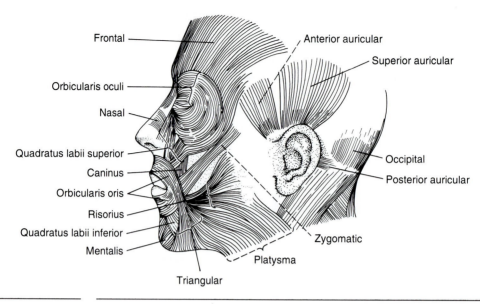

Frontal — Anterior auricular

Superior auricular

Orbicularis oculi —

Nasal —

Quadratus labii superior —

Caninus —

Orbicularis oris —

Risorius —

Quadratus labii inferior —

Mentalis —

Occipital

Posterior auricular

Zygomatic

Platysma

Triangular

FIGURE 9–14   The platysma and facial muscles of a human being. (After Neal and Rand.)

and **facial** muscles (Fig. 9–14). Facial muscles insert into the scalp and the base of the auricle, encircle the eye, and form the fleshy cheeks and lips of mammals. The evolution of fleshy cheeks and lips is coupled with the evolution of nipples on the mammary glands and the ability to suckle. Movements of the lips and face are also an important way of communicating in many mammals. The facial nerve (VII) is so called because it innervates these muscles, but it also supplies the rest of the hyoid musculature.

The Remaining Branchiomeric Muscles.   Deep muscles in the wall of the pharynx are of branchiomeric origin, as are the small **intrinsic muscles of the larynx** that extend between the laryngeal cartilages. Since these cartilages develop in mammals from the fourth, fifth, and sixth visceral arches, it should not be surprising that their muscles are innervated by a branch of the vagus nerve.

The cucullaris of fishes and amphibians expands considerably in mammals and divides to form a **trapezius** group of muscles on the back of the shoulder, and a **sternocleidomastoid** complex that extends from the base of the skull to the sternum and clavicle (Fig. 9–18**A**). Although branchiomeric in origin, these muscles no longer act on the visceral skeleton but help move the shoulder and head. Their motor innervation in amniotes is from the accessory nerve (XI), which evolved from the branch of the vagus that supplied the cucullaris. Branches of cervical spinal nerves also enter these muscles but are believed to carry a sensory feedback from muscle activity rather than motor neurons.

Hypobranchial Muscles

The relatively simple hypobranchial muscles of sharks are more complex in teleosts and other fishes that feed by rapidly expanding their buccopharyngeal cavity and sucking food in. They also are more complex in terrestrial vertebrates because food is not supported and carried by a current of water. The hyobranchial apparatus and a muscular tongue, both controlled by hypobranchial muscles, transport food in the mouth, swallow it, and sometimes even catch it.

The simple prehyoid coracomandibular of a shark differentiates in amphibians into a **geniohyoid** (Gr. *geneion* = chin), which extends from the jaw to the hyoid, and one or more slips that enter the newly evolved, muscular tongue. Often there is only a single **genioglossus** (Gr. *glossa* = tongue) that retracts the tongue and moves food back into the gullet. Posthyoid muscles of an amphibian are represented by a rectus cervicis that is little changed from that of a fish. An amphibian uses fishlike movements of the pharynx floor to move air into and out of the lungs rather than water across the gills (p. 652). In some species slips of the rectus cervicis, such as an **omoarcual** (Gr. *omos* = shoulder) and **pectoriscapularis,** attach to the pectoral girdle.

Hypobranchial muscles become more complex in amniotes, especially in mammals that use the tongue to manipulate food between their teeth during mastication (Fig. 9–13). In addition to the muscles present in amphibians and reptiles, mammalian prehyoid muscles include a **hyoglossus,** a **styloglossus,** and a **lingualis.** The lingualis is an intrinsic muscle of the tongue that is confined to the organ and constitutes much of its substance. Posthyoid muscles are a **sternohyoid,** a **sternothyroid,** a **thyrohyoid,** and sometimes an **omohyoid.** Studies by Crompton and his colleagues (1975) have shown that the actions of the hypobranchial muscles of mammals are very important in the complex movements of the hyoid apparatus and larynx that occur during jaw opening and food swallowing (p. 589). This group of muscles continues to be innervated by the equivalent of the ventral rami of spino-occipital nerves. The occipital component of amniotes emerges from the back of the skull as a distinct cranial nerve, the hypoglossal nerve (XII).

Epibranchial and Trunk Muscles

Major changes occurred in the trunk muscles of tetrapods because lateral undulatory movements of the trunk and tail became less important in locomotion than appendicular movements. The trunk muscles and the closely associated epibranchial muscles, however, became more important in mediating dorsoventral bending of the spine, an action that does not occur in fishes, and in supporting the body against gravity. They also control the movements of the head, which have become independent of trunk movements, and some are used to ventilate the lungs. Distinct myomeres remain in salamanders that still move by lateral undulations, but segmentation of the trunk muscles has been lost or reduced in most tetrapods.

Epaxial Muscles.   The epaxial muscles are particularly important in supporting the body and moving the vertebral column and head. They are represented in salamanders by a **dorsalis trunci** that remains segmented (Fig. 9–12). As trunk and head movements became more complex in reptiles and mammals, the epaxial muscles have become more specialized. Typically, three longitudinal bundles

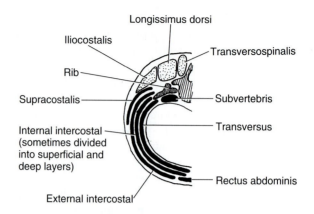

FIGURE 9–15 A transverse section through the trunk of a reptile to show the major groups of trunk muscles. Epaxial muscles are stippled; hypaxial ones are shown in black. (After Romer and Parsons.)

extend forward from the sacrum and pelvis to the head (Fig. 9–15). The most medial and deepest bundle in reptiles is a **transversospinalis** system, which consists of short fibers extending between adjacent vertebrae, or only across a few body segments. A more lateral **longissimus dorsi** consists of longer fibers, many of which extend from the sacrum to the neck. As the most lateral **iliocostalis** extends forward from the sacrum, it gives off slips that insert on the ribs or cervical vertebrae. All three of these bundles remain partly segmented in reptiles, which still use lateral trunk undulations in locomotion. Segmentation is lost in mammals, except in the deeper parts of the transversospinalis group. This group is represented in mammals in part by the **multifidi** and **spinalis dorsi** (Fig. 9–16A). The longissimus dorsi and iliocostalis of mammals are partially united in some species to form a large **erector spinae.** As these bundles approach the head, they become subdivided into many smaller muscles that support and move the head. All of the epaxial muscles are innervated by dorsal rami of spinal nerves.

Hypaxial Muscles. The hypaxial muscles of tetrapods are subdivided into three major groups (Fig. 9–15):

1. A **subvertebral** group lies ventral to the vertebral column and assists the epaxial muscles. It is large and powerful in mammals and differentiates into several muscles.
2. A **rectus abdominis** extends longitudinally on either side of the midventral line and helps support the abdomen.
3. A lateral group lies on the flank and typically forms three layers. These layers are particularly distinct in the abdomen, where we find (from the surface inward) an **external oblique,** an **internal oblique,** and a **transversus** (Figs. 9–12, 9–15, and 9–16**B**). Some reptiles have a partial fourth layer, the **supracostalis,** that lies superficial to the others. Fibers in each layer run at different angles, as do the layers of wood in plywood; this provides considerable strength

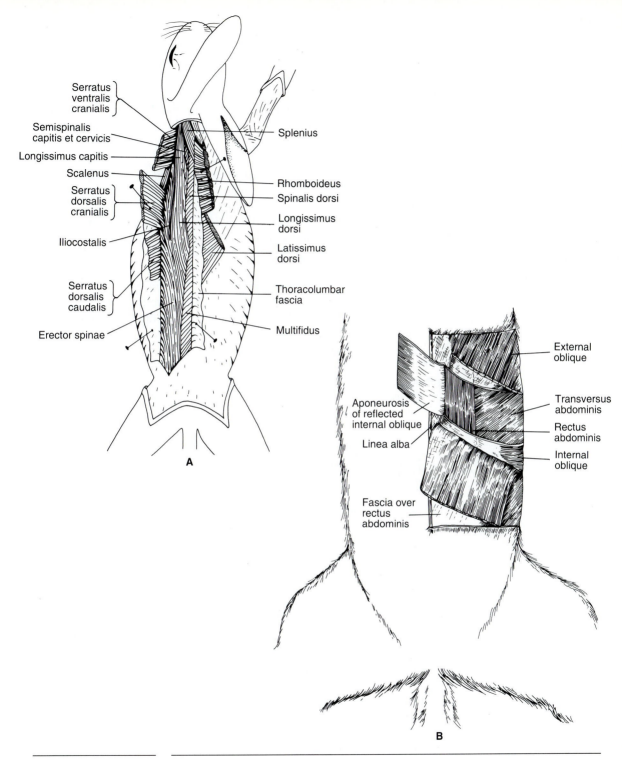

Serratus ventralis cranialis

Semispinalis capitis et cervicis

Longissimus capitis

Scalenus

Serratus dorsalis cranialis

Iliocostalis

Serratus dorsalis caudalis

Erector spinae

Splenius

Rhomboideus

Spinalis dorsi

Longissimus dorsi

Latissimus dorsi

Thoracolumbar fascia

Multifidus

A

External oblique

Aponeurosis of reflected internal oblique

Linea alba

Fascia over rectus abdominis

Transversus abdominis

Rectus abdominis

Internal oblique

B

FIGURE 9–16   Trunk muscles of amniotes: **A**, A dorsal view of the epaxial muscles of the rabbit. **B**, A ventral view of the hypaxial muscles of a cat. (After Walker and Homberger.)

to the abdominal wall. The layers are thick and continue to be segmented in salamanders, but they form broad, thin sheets in amniotes. These layers are represented by intercostal layers in the thoracic region of mammals and throughout most of the trunk in other tetrapods that have ribs. **External** and **internal intercostals** are well developed, but the transverse layer is found only in limited regions.

All the hypaxial muscles support the body wall, and many participate in ventilating the lungs (p. 662). Other respiratory muscles include a **serratus dorsalis** and a **scalenus** group, both of which separate from superficial hypaxial muscles. The primary respiratory muscle of mammals lies in the diaphragm, a partition that separates the thoracic and abdominal cavities (p. 560). It, too, derives from the hypaxial group, for it develops embryonically from the ventral part of cervical myotomes.

One or more hypaxial muscles extend from the trunk to the pectoral girdle (see Fig. 10–10). These muscles act on the girdle. There is no direct bony connection between the girdle and the vertebral column, although in amniotes the girdle usually has an indirect connection via the sternum and ribs. Muscles that extend between the trunk and the pectoral girdle also transfer body weight from the vertebral column to the girdle and appendages. Salamanders have a **levator scapulae** and a small **thoraciscapularis.** The thoraciscapularis is represented in mammals by a much larger **serratus ventralis** and **rhomboideus.** All hypaxial muscles continue to be innervated by the ventral rami of spinal nerves.

The Evolution of Appendicular Muscles

The structure and movements of the paired appendages of terrestrial vertebrates are far more complex than those of the paired fins of fishes because the limbs support the body and provide the major propulsive thrusts. Indeed, appendicular muscles constitute the bulk of the muscular system in many tetrapods. Despite their complexity, tetrapod appendicular muscles can be subdivided into dorsal and ventral groups on the basis of their embryonic development. In both fish and early tetrapod embryos, appendicular muscles begin as two premuscular masses that lie, respectively, above and below the developing appendicular skeleton. These masses differentiate into the simple abductor and adductor in the adult fish but cleave into many components during the later development of a tetrapod. Collectively, the muscles that develop from the dorsal mass are homologous to the fish adductor; those from the ventral mass, to the fish abductor.

We will discuss the functional anatomy of support and locomotion in Chapter 10. At this time, we can say that the dorsal muscles of both tetrapod pectoral and pelvic limbs abduct and extend the limbs as a whole and their various segments. These actions occur during the swing phase of a step when each limb in turn is removed from the ground and extended. Ventral muscles do the opposite—they adduct the limbs as a whole and flex their distal segments. These actions occur during the stance phase when each limb in turn is placed on the ground and develops a propulsive thrust. In addition to these actions, the pectoral and pelvic limbs as a whole are advanced or protracted during the swing phase. A force is developed that could retract or draw the limbs caudad during the stance phase, but since the feet remain on the ground, this force advances the trunk relative to the feet. Both dorsal and ventral muscles participate in these actions depending on whether their lines of action lie anterior (protraction) or posterior (retraction) to the shoulder and hip joints. Limb muscles also brace the bones across the joints, holding the body up on the legs.

The splayed limb position of amphibians and reptiles necessitates a powerful ventral musculature that adducts the humerus and femur and flexes the antebrachium and crus so as to raise the body from the ground. The mammalian body is supported more directly by bony columns because the limbs have rotated closer to the body axis, and movement of all limb segments is primarily back and forth in a parasagittal plane. Ventral muscles that act across the shoulder joint become much less important. Ventral parts of the girdle are reduced, as we have seen, and some ventral muscles are reduced or shifted dorsally. Dorsal muscles are better positioned to protract and retract the mammalian humerus. Some reduction of the ventral muscles that act across the hip joint has also occurred during the evolution of mammals, but reduction of the ventral parts of the pelvic girdle and its muscles is not so pronounced as in the shoulder. Limb movements became more complex as tetrapods evolved into more active and agile animals, and with this some subdivision and multiplication of both dorsal and ventral muscles have also occurred.

Appendicular muscles are exceedingly numerous and are best sorted out in the laboratory. We will consider the major muscles of the pectoral girdle and brachium as an example of the evolutionary changes that occur in appendicular muscles. Major changes can be appreciated by comparing these muscles in a reptile and mammal.

Muscles of the Pectoral Girdle and Appendage

Ventral Muscles.   The **pectoralis** is the largest and most superficial of the ventral muscles (Figs. 9–17**A** and 9–18**A**). It is a fan-shaped muscle that extends from the ventral surface of the trunk to the proximal end of the humerus. It is a major limb retractor in all tetrapods, and it also helps, particularly in amphibians and reptiles, adduct the limb and raise the body from the ground. A fan-shaped **supracoracoideus** lies deep and anterior to the pectoralis in amphibians and reptiles, and also acts across the shoulder joint. The pectoralis and supracoracoideus are enormous muscles in birds, for they are major flight muscles (p. 388). At first sight, mammals appear to lack the supracoracoideus, but studies have shown that such a muscle is present in an early embryo (Chen, 1955). As development continues, the muscle grows dorsally on either side of the newly evolved scapular spine to become the **supraspinatus** and **infraspinatus** (Fig. 9–18**B**).

Amphibians and reptiles have a large **coracobrachialis** that extends from the ventral surface of the large coracoid to the proximal and distal ends of the humerus (Fig. 9–17**B**). This muscle is greatly reduced in most mammals along with the coracoid, or lost entirely in some species (Fig. 9–18**C**). A notable exception is bats, whose large coracobrachiales assist in wing adduction.

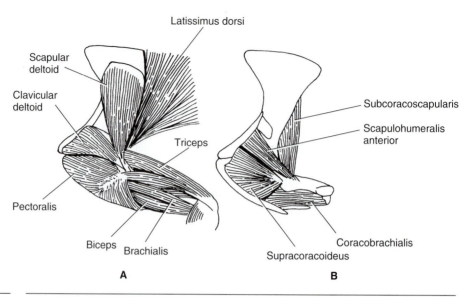

FIGURE 9–17   Lateral views of the shoulder and upper arm muscles of a lizard: **A**, Superficial muscles. **B**, Deeper muscles. (After Romer and Parsons.)

In both reptiles and mammals a **biceps brachii** and a **brachialis** lie on the underside of the humerus and act across the elbow joint to flex the antebrachium (Figs. 9–17**A** and 9–18**C**). Part of the biceps arises from the scapula, so this muscle also acts across the shoulder joint.

The ventral muscles on the forearm are a group of **antebrachial flexors** that arise primarily from the ventromedial surface of the distal end of the humerus and insert on the wrist and digits. Additional small **intrinsic flexors of the hand** are confined to the hand and control more delicate finger movements.

Dorsal Muscles.   A **deltoid** muscle, which is sometimes subdivided into two or more parts, arises in amphibians and reptiles from the anterior border of the scapula and from the clavicle and inserts on the proximal end of the humerus (Fig. 9–17**A**). It is an important limb protractor as well as an abductor. Part of its origin from the scapular spine in mammals supports the notion that the spine represents the original anterior border of the scapula.

A large, fan-shaped **latissimus dorsi** lies caudal to the deltoid. It arises from the trunk and also inserts on the proximal end of the humerus. It is a major limb retractor. A slip separates from the mammalian latissimus dorsi and shifts its origin to the caudal border of the scapula to become the **teres major** (Fig. 9–18**B**).

Amphibians and reptiles have a small **subcoracoscapularis,** which arises from the caudal border of the scapula deep to the latissimus dorsi, and a small **scapulohumeralis anterior,** which arises from the scapula deep to the deltoid (Fig. 9–17**B**). Both insert on the proximal end of the humerus. The scapulohumeralis anterior remains small in mammals, where it is known as the **teres minor.** The mammalian subcoracoscapularis, in contrast, expands greatly to take its origin

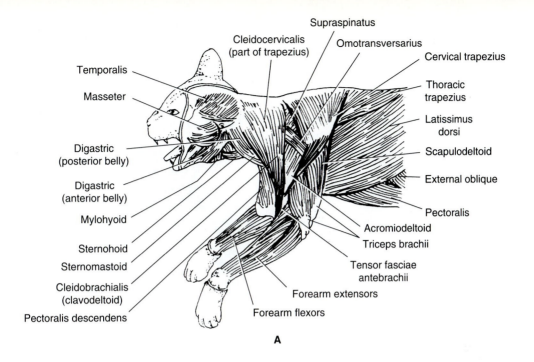

A

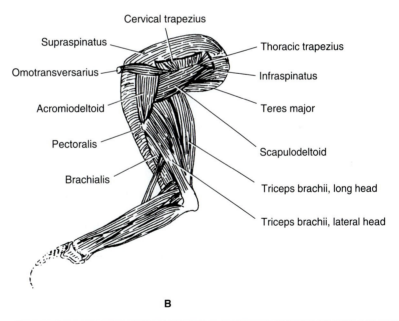

B

FIGURE 9–18 The shoulder and arm muscles of a cat: **A,** Lateral view of the superficial muscles. **B,** Lateral view of the deeper muscles after the removal of the trapezius and latissimus dorsi. *(continued on next page)*

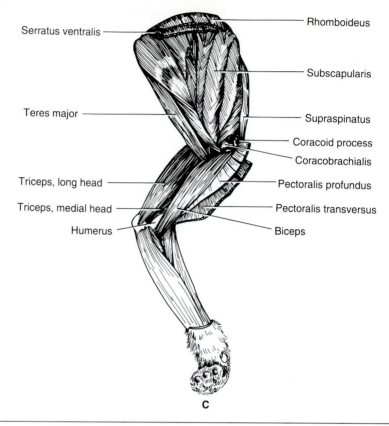

Serratus ventralis

Rhomboideus

Subscapularis

Teres major

Supraspinatus

Coracoid process

Coracobrachialis

Triceps, long head

Pectoralis profundus

Triceps, medial head

Pectoralis transversus

Humerus

Biceps

C

FIGURE 9–18   *continued*
C, Medial view. (After Walker and Homberger.)

from the entire median surface of the enlarged scapula; it is known as the **sub-scapularis** (Fig. 9–18C).

The large **triceps** is the only muscle on the dorsal surface of the brachium in all tetrapods (Figs. 9–17A and 9–18A). It arises from the humerus and scapula and inserts on the olecranon of the ulna. Its most obvious action is to extend the forearm in the swing phase of the step cycle, but when the hand is maintained in a fixed position on the ground during the stance phase, extension of the lower arm by the triceps helps push the body forward. The triceps also braces the elbow joint and helps support the body.

Antebrachial extensors in all tetrapods arise primarily from the dorsolateral surface of the distal end of the humerus and continue down the forearm to the wrist and digits. Small **intrinsic extensors of the hand** are confined to the hand.

Many mammals can wiggle the skin in order, for example, to shake off insects. An extensive but thin **cutaneous trunci** spreads out from the armpit to insert into the skin over the trunk. Its attachment to the bases of the latissimus dorsi and pectoralis is indicative of its evolutionary origin from appendicular muscles.

Electric Organs

Many tissues, such as muscles, glands, and nerves, generate weak electric currents when they become active. Such currents travel through water because water, even fresh water, contains ions. Many fishes and a few other aquatic vertebrates have evolved **electroreceptors** that detect these weak currents, and they utilize this information to home in on prey. About 250 other species of fishes have evolved specialized **electric organs** that amplify these weak currents. Since the tissue is usually modified muscle, we will consider electric organs at this time. Electroreceptors are discussed in Chapter 11.

Electric organs occur in many different groups, differ in many aspects of their construction, and must have evolved independently many times. Electric organs occur in some tropical teleosts: the gymnotids (relatives of minnows) of Central and South America, the mormyrids (osteoglossomorphs) of Africa, and the electric catfish, *Malapterurus,* of the Nile. The only electric saltwater fishes are several rays, of which *Torpedo* is the most famous, and the stargazer, *Uranoscopus.*

The electric organs of mormyrids and most of the gymnotids generate trains of relatively weak electric pulses. Some pulses are used in species and sex recognition; others form an electric field that the fish uses to navigate in the murky waters of its habitat. Objects whose electric conductive properties differ from that of the water distort the field, and the fish detects these distortions. The fish avoids distorting the electric field itself by keeping its trunk and tail straight and swimming slowly by undulating an elongated anal or dorsal fin.

The electric eel, electric catfish, and the marine species can generate electric currents sufficiently strong to stun, and sometimes kill, prey or predators. *Torpedo* and *Electrophorus* have the most powerful organs. *Torpedo* can generate a current flow as high as 50 amperes at 50 volts. (Amperage is a measure of the amount of current; voltage is the push or force driving the current.) This is a total power of 2500 watts (watts = amperes × volts). *Electrophorus* can also generate a power of 2000 watts or more, but it must do so at a higher voltage, up to 500 volts, since fresh water is a poorer conductor of electricity than salt water. The amperage in *Electrophorus* is much lower, about 4 amperes. It is probable that the powerful electric organs evolved from weaker ones used for communication and navigation. Once they attained enough power to stun other organisms, natural selection could favor this use. Darwin was hard put to explain the early stages in the evolution of these powerful organs, for fish with weaker organs were unknown in his time.

Different tissues or groups of muscles are utilized in the various fish groups (Fig. 9–19). The electric organ of the electric "eel," *Electrophorus* (a gymnotid), consists of modified hypaxial trunk muscles; that of *Malapterurus* forms a sheath directly beneath the skin of the trunk and tail. It may be specialized glandular tissue. Branchiomeric muscles form the electric organ of *Torpedo,* and extrinsic ocular muscles form that of *Uranoscopus.*

Most electric organs consist of a series of disk-shaped **electroplaques,** each of which represents a modified muscle cell or its motor end plate. In one gymnotid, *Sternarchus,* the electroplaque evolved from the motor neurons themselves rather than from the motor end plates. The structure and activity of an electric organ can be appreciated by examining that of *Electrophorus.* Each electroplaque is a

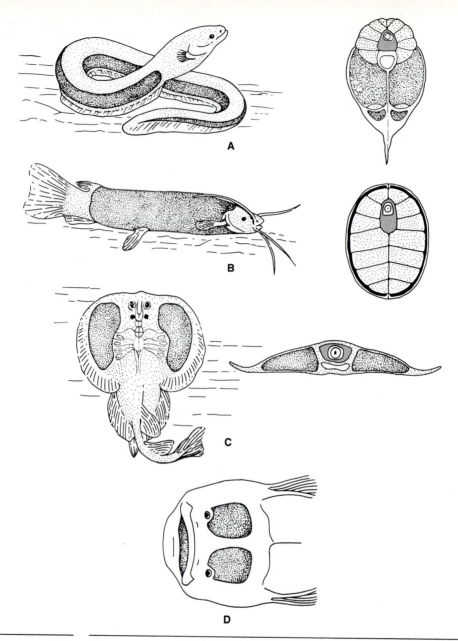

FIGURE 9–19 The location of the electric organ in representative fishes as seen in surface view and in transverse section: **A**, The electric "eel," *Electrophorus*. **B**, The electric catfish *Malapterurus*. **C**, The electric ray *Torpedo*. **D**, The stargazer *Uranoscopus*. (**A–C**, after Portmann. **D**, after Bond.)

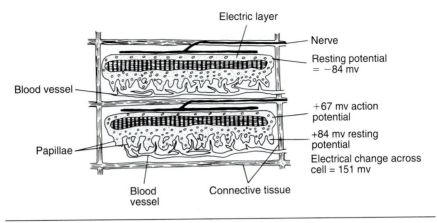

Electric layer

Nerve

Resting potential = −84 mv

Blood vessel

+67 mv action potential

+84 mv resting potential

Electrical change across cell = 151 mv

Papillae

Blood vessel   Connective tissue

FIGURE 9–20   Two electroplaques from the electric organ of *Electrophorus*. The resting potential is indicated in the top one; the action potential in the bottom one. Only the innervated surface becomes depolarized. (After Hoar.)

multinucleated, disk-shaped cell that is flat on the innervated side and highly folded on the other side (Fig. 9–20). Most myofibrils have been lost, although traces are sometimes found. At rest the outside of the plasma membrane is positive relative to the inside. The resting potential across the innervated membrane is −84 millivolts, about the same as in a normal muscle cell. When the organ discharges, only the innervated surface becomes depolarized, and its potential rises to +67 millivolts. This gives a total voltage range across the entire electroplaque of 151 millivolts: −67 on one surface to +84 on the other. The exceedingly high voltage of the entire organ in *Electrophorus* derives from the stacking of the electroplaques in long columns, each containing up to 10,000 electroplaques. About 70 such longitudinal columns are found on each side of the body. Since the electroplaques in each column are in series and discharge simultaneously, the voltage of each one is added to that of the others in the same series. The columns of electroplaques in *Torpedo* extend vertically, so they are much shorter, but as many as 2000 parallel columns are present on each side of the body. The parallel arrangement reduces the resistance to current flow and results in a higher amperage than in *Electrophorus,* but the shorter columns lower the voltage.

SUMMARY

1. Muscles are the major effectors of vertebrates, and they are responsible for body movements and the movement of most materials within the body.
2. Smooth muscles are composed of elongated, nonstriated, spindle-shaped cells. Their contraction tends to be slow and sustained. Unitary smooth muscles, which are found in most visceral organs, are myogenic, but their contraction is modulated by nerve impulses. Multiunit smooth muscle, found in the walls of blood vessels and the iris of the eye, is neurogenic.
3. Cardiac muscle is composed of moderately elongated, striated, and branching

cells that are tightly united with each other by intercalated disks. Its myogenic rhythm is modulated by nerve impulses.

4. Skeletal muscle is composed of extremely long, striated, and multinucleated cells.

5. Smooth and cardiac muscle fibers are arranged in sheets and layers within the walls of visceral organs and blood vessels. Skeletal muscle fibers usually attach to skeletal elements by the extension of the connective tissue within and around them. This tissue often forms tendons and aponeuroses.

6. Muscle cells or fibers contain protein myofilaments composed primarily of actin or myosin. The myofilaments are grouped into myofibrils that are barely visible with a light microscope. Muscles contract when cross bridges form between the actin and myosin myofilaments and actin filaments are pulled into the array of myosin filaments.

7. Contraction may be isotonic, isometric, or negative work.

8. Muscles contract against an antagonistic muscle or force, which restores the muscle to its resting length. Muscles usually act in synergistic groups.

9. A neuron and the skeletal muscle fibers it supplies constitute a motor unit. Motor units containing few muscle fibers are adapted for delicate movements; those with many, for generating force.

10. Tonic skeletal muscle fibers have multiple motor end plates. Their contraction is graded by the number of nerve impulses reaching them. Tonic muscles are usually small and found where slow, sustained, and carefully graded contractions are needed.

11. Most vertebrate skeletal muscles are phasic, or twitch, muscles. Phasic fibers have a single motor end plate. Their contraction is all-or-none, but the force of contraction of the entire muscle can be increased by the temporal summation of nerve impulses and by the recruitment of more motor units.

12. Slow phasic fibers derive their energy through oxidative metabolism. They contract slowly and no more rapidly than oxygen can be delivered to them, so they do not fatigue easily. Fast phasic fibers derive their energy by anaerobic glycolysis. They can contract rapidly but fatigue easily.

13. Many muscles are multigeared and contain populations of each type of phasic fiber. The slow ones are recruited first for slow and sustained contractions; the fast ones are used when a rapid but brief contraction is needed.

14. The arrangement of fibers within a muscle affects many aspects of its performance. Since muscle fibers can contract about one third of their resting length, strap- and fusiform-shaped muscles containing long, parallel fibers can induce a more extensive movement than pennate muscle with many short fibers. These muscles also contract faster because long fibers contain more sarcomeres in series.

15. Since the force that a muscle develops is a function of the number of myosin-actin cross bridges made at one time, a pennate muscle with many short fibers develops more force than other types of muscles of equal mass.

16. Force and velocity of contraction are inversely related, so muscles can develop more force when they contract slowly. Since power = force × velocity of contraction, maximum power is generated at intermediate forces and velocities of contraction.

17. Somatic muscles, most of which lie in the body wall and appendages, develop embryonically from the paraxial somites. The paraxial mesoderm continues rostral to the otic capsule as a series of less clearly defined somitomeres. Recent studies have shown that the branchiomeric muscles, which are associated with the visceral arches, also develop from cranial somites and somitomeres. Accordingly, branchiomeric muscles should also be considered to be somatic and not visceral muscles. True visceral muscles develop from the splanchnic layer of the lateral plate mesoderm. All contribute to the wall of visceral organs, are innervated by the autonomic nervous system, and, except for the cardiac muscles of the heart, are smooth.

18. Somatic muscles can be divided into axial and appendicular groups. The axial muscles include the extrinsic muscles of the eye, the branchiomeric muscles, the epibranchial and hypobranchial muscles, and the epaxial and hypaxial muscles of the trunk and tail.

19. The six extrinsic ocular muscles of fishes move the eyeball. They can be sorted into three groups innervated by the oculomotor (III), trochlear (IV), and abducens (VI) nerves.

20. Correlated with an elaborate visceral skeleton, the branchiomeric muscles of fishes are quite complex. The mandibular muscles, innervated by the trigeminal nerve (V), suspend, close, and protrude the jaws. The hyoid muscles, innervated by the facial nerve (VII), act on the hyoid arch. The branchial muscles, innervated by the glossopharyngeal (XI) and vagus (X) nerves, primarily compress the gill pouches during the expiration of a respiratory current of water. Their levators form the cucullaris, which inserts on the pectoral girdle.

21. The epibranchial muscles of fishes continue the epaxial trunk muscles to the head. The hypobranchial muscles act on both the jaws and visceral arches. They are used chiefly to open the mouth and to expand the pharynx during feeding and the intake of a respiratory current of water. They are innervated by the spino-occipital nerves.

22. The trunk muscles of fishes are segmented myomeres whose foldings and detailed structure correlate with the pattern of swimming. They are supplied by spinal nerves.

23. Since paired fin movements are not complex or strong, the appendicular muscles of many fishes are quite simple and consist only of a dorsal adductor and a ventral abductor. They are innervated by the ventral rami of spinal nerves.

24. Parts of the rectus complex of eye muscles attach to and move the upper and sometimes the lower eyelids of birds and mammals, but few other evolutionary changes have occurred in this group of muscles.

25. Most mandibular muscles close the jaws, and one, the digastric, helps open the jaws. One, the tensor tympani, follows a part of the mandibular arch (the articular) into the middle ear.

26. A part of the hyoid musculature contributes to the digastric, part forms the slender stylohyoid, and part forms the stapedius of the middle ear. The rest has moved to a superficial position to form the facial muscles and the platysma.

27. Some of the remaining branchiomeric musculature remains in the pharynx wall, and some gives rise to the intrinsic muscles of the larynx. The cucullaris

has become the sternocleidomastoid and trapezius complexes. The part of the vagus nerve that primitively innervated the cucullaris has become the accessory nerve (XI).

28. The hypobranchial muscles divide into many components as a muscular tongue evolves and feeding and swallowing movements become more complex. They are innervated by the hypoglossal (XII) and by cervical nerves in amniotes.

29. The epibranchial and epaxial trunk muscles tend to lose their segmentation, but they remain an important group that supports and moves the vertebral column and head.

30. Most hypaxial trunk muscles form three thin sheets that lie between the ribs and support the abdominal wall. A subvertebral group assists in the movements of the vertebral column. Several muscles of this group insert on the scapula and transfer body weight from the vertebral column to the pectoral girdle. The diaphragm belongs to this group.

31. Appendicular muscles become large and numerous as the appendages and girdles assume a major role in the support and locomotion of the body. Evolutionary changes in the appendicular groups parallel the changes in limb position and locomotion.

32. Muscle tissue of various origins forms the electric organs of fishes. These generate weak pulses used in navigation and species recognition and, in some species, stronger pulses that can stun predators or prey.

REFERENCES

Allen, E. R., 1978: Development of skeletal muscle. *American Zoologist,* 18:101–111.

Bennett, M. V. L., 1969: Electric organs. *In* Hoar, W. S., and Randall, D. J., editors: *Fish Physiology,* 5:493–574. New York, Academic Press.

Bourne, G. H., editor, 1972: *The Structure and Function of Muscle.* New York, Academic Press.

Chen, C. C., 1955: The development of the shoulder region of the opossum, *Didelphis virginiana,* with special reference to the musculature. *Journal of Morphology,* 97:415–472.

Edgeworth, F. H., 1935: *The Cranial Muscles of Vertebrates.* London, Macmillan Company.

Gans, C., and Bock, W., 1965: The functional significance of muscle architecture: A theoretical analysis. *Ergebnnisse der Anatomie und Entwicklungsgeschichte,* 38:115–142.

Goldspink, G., 1977: Design of muscles in relation to locomotion. *In* Alexander, R. McN., and Goldspink, G., editors: *Mechanics and Energetics of Animal Locomotion.* New York, John Wiley and Sons.

Goodrich, E. S., 1918: On the development of the segments of the head in *Scyllium. Quarterly Journal of the Microscopical Society,* 63:1–30.

Goodrich, E. S., 1930: *Studies on the Structure and Development of Vertebrates.* London, Macmillan.

Grundfest, H., 1960: Electric fishes. *Scientific American,* 203:115–124.

Huddart, H., 1975: *The Comparative Structure and Function of Muscles.* New York, Pergamon Press.

Jacobson, A. G., 1988: Somitomeres: Mesodermal segments of vertebrate embryos. *Development,* 104:209–220.

Lissmann, H. W., 1958: On the function and evolution of electric organs in fish. *Journal of Experimental Biology,* 35:156–191.

Meier, S. P., 1981: Development of the chick embryo mesoblast: Morphogenesis of the prechordal plate and cranial segments. *Developmental Biology*, 83:49–61.

Meier, S. P., 1982: The development of segmentation in the cranial region of vertebrate embryos. *Scanning Electron Microscopy*, 3:1269–1282.

Noden, D. M., 1983: The embryonic origins of avian cephalic and cervical muscles and associated connective tissues. *American Journal of Anatomy*, 168:257–276.

Noden, D. M., 1984: Craniofacial development: New views on old problems. *Anatomical Record*, 208:1–13.

Noden, D. M., 1991: Vertebrate craniofacial development: The relation between ontogenetic process and morphological outcome. *Brain, Behavior and Evolution*, 38:190–225.

Northcutt, R. G., 1990: Ontogeny and phylogeny: A reexamination of conceptual relationships and some applications. *Brain, Behavior and Evolution*, 36:116–140.

Romer, A. S., 1944: The development of tetrapod limb musculature—the shoulder region of *Lacerta*. *Journal of Morphology*, 74:1–41.

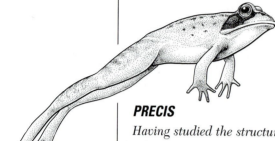

10 Functional Anatomy of Support and Locomotion

PRECIS

Having studied the structure and evolution of the skeletal and muscular systems, we now can address the functional interrelationships between the individual skeletal units and the muscles that act on them. We will examine the functional anatomy of support and locomotion in this chapter, and that of feeding mechanisms in Chapter 16. In a book of this scope, we cannot analyze all of the supporting mechanisms that have evolved among vertebrates nor all of their many patterns of locomotion. The ones we have selected are representative of most vertebrates, and they will illustrate the types of understanding of vertebrate structure that derive from functional analyses.

Now that you are acquainted with the structure and evolution of the skeletal and muscular systems, we can examine the functional interrelations between them. The individual bones, ligaments, and muscles form interrelated functional units that support the body and are responsible for the locomotion of the animal and the movements of most of its parts. By analyzing these units, we can learn much about the configuration of the musculoskeletal system and the way it works. We will examine the functional anatomy of support and locomotion in this chapter and that of feeding mechanisms in Chapter 16.

Swimming

Nearly all vertebrates can swim to some extent, and many are excellent swimmers. Swimming, of course, is the sole or primary pattern of locomotion in fishes and larval amphibians. They are **primary swimmers.** Terrestrial vertebrates that have returned to the water are **secondary swimmers.** They have readapted to an aquatic mode of life but still retain at least traces of their terrestrial ancestry. Even whales, who cannot survive on land, surface to breathe air. Most primary swimmers are also **undulatory swimmers,** for they propel themselves by lateral undulations that travel down the trunk and tail or, in a few cases, along fins that have long attachments to the body (Fig. 10–1). A few secondary swimmers also propel themselves with lateral undulations of a flattened tail (salamanders, crocodiles) or dorsoventral undulations of large caudal flukes (whales). But most secondary swimmers, and a few fishes, are **oscillatory swimmers** that propel themselves primarily with oscillations or paddle-like movements of their appendages, although undulations of the trunk help in some species. Frogs swim by oscillatory thrusts of their hind legs and large webbed feet, sea turtles use elongated pectoral flippers, penguins use paddle-like wings, and beavers paddle with large, webbed hind feet. We will analyze undulatory swimming in fishes as an example of an aquatic pattern of locomotion and examine the morphological features associated with this pattern.

Support in Water

According to Archimedes' principle, an object in a fluid (water or air) displaces a weight of fluid equal to its own weight, and the displaced fluid exerts an upward force on the object. Since water is a dense medium compared with air, the upward force exerted on an object in water is substantial, but this force by itself does not entirely compensate for the downward pull of gravity. Although water offers considerable support or buoyancy for an animal, it also offers more resistance to moving through it than air. A swimming fish must produce a lift force that overcomes the downward pull of gravity not compensated for by the buoyancy force, and a propulsive thrust that overcomes the drag of the water. It also must be able to maintain stability and to maneuver.

We have seen (p. 259) that the design of the vertebral column of fishes helps meet these problems. It is a compression strut that resists compression forces as

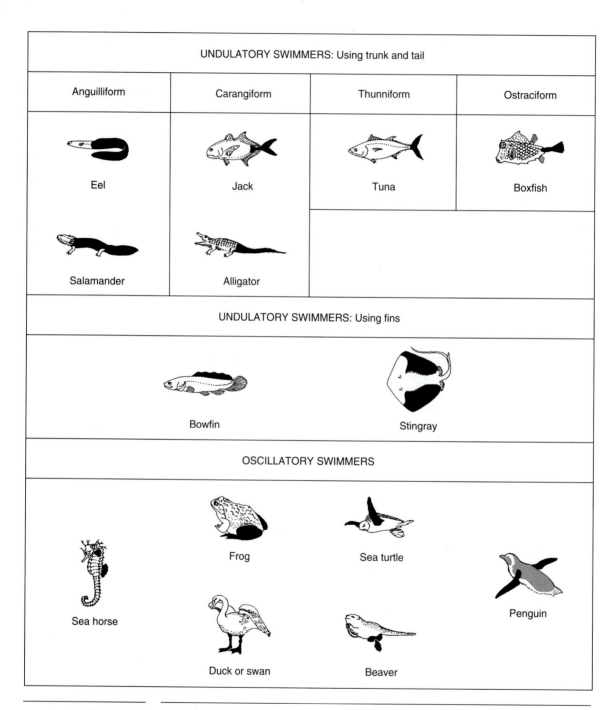

FIGURE 10–1 Major swimming patterns among vertebrates. The part of the body that generates a propulsive thrust has been shaded.

the animal pushes through the water and prevents the body from shortening when longitudinal muscle fibers contract. Usually the vertebral centra are strong disk- or spool-shaped elements of cartilage or bone. Neural and hemal arches articulate with the centra but are not always fused to them. The vertebral column, or parts of it, must be able to bend laterally during swimming, but forces that tend to bend it vertically are negligible. All or many of the joints between vertebrae allow lateral bending. Vertical bending need not be resisted, and only a few fishes with specialized patterns of swimming have zygapophyses.

Drag

Drag forces are substantial, and the design of many aspects of fish structure reduce them. Two major types of drag forces resist the forward motion of a fish: frictional drag and pressure drag. **Frictional drag** (Fig. 10–2**A**) is the force exerted on the surface of a fish due to the viscosity of the water. A thin **boundary layer** of water surrounds a moving fish. The innermost layer is carried along at nearly the same speed as the fish, but successive outer layers of water in the boundary zone move slower and slower, finally equaling the speed of the surrounding water. Shear forces between these layers are responsible for frictional drag. Frictional drag is lowest when the surface area is minimized relative to its mass, the fish is swimming slowly, and the water flows smoothly across its surface (that is, the flow is **laminar**). When a fish of the same mass and shape swims more rapidly, the boundary layer increases in thickness and the increased undulations of the trunk and tail disrupt the smooth flow of water (Fig. 10–2**B**). The boundary layer tends to separate from the surface, especially posteriorly, and this produces eddies. The thicker boundary layer and the turbulent flow increase drag. In general, a smooth surface also reduces frictional drag, but sometimes a small turbulence in the boundary layer prevents the separation of the layer and the formation of larger eddies. Spines or other small protuberances on the scales of some species create a small turbulence that prevents the separation of the boundary layer. The layer of mucus that covers most fishes further reduces frictional drag, for it reduces the viscosity of the water in the boundary layer.

Pressure drag results from differences in water pressure between the front and rear of the fish. Water is displaced as the fish moves forward and then falls in behind the fish, forming a wake (Fig. 10–2**B**). The reduced pressure and turbulent flow of water in the wake tend to hold the fish back. Pressure drag depends on body shape. A sphere moving through water would have a very high pressure drag but a low frictional drag because its surface area is small relative to its mass. A long, slender body of the same mass would have a low pressure drag but a high frictional drag because its surface is much larger. An optimal compromise between these shapes is a teardrop-shaped object having a maximum diameter about one third of the way back from the front that is equal to about one fifth the length of the object. Fishes that swim a great deal have such a streamlined or **fusiform** shape.

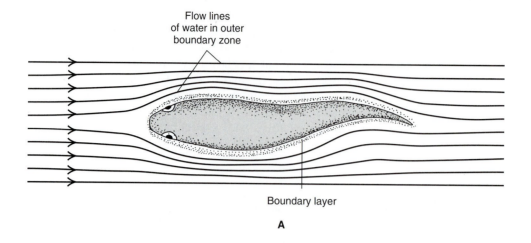

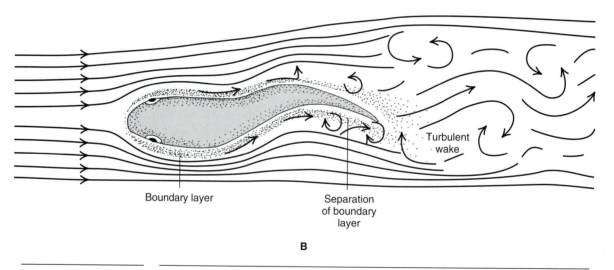

FIGURE 10–2 Drag forces on a fish. **A**, Frictional drag is low when the flow across the surface of the fish is smooth, or laminar. **B**, Frictional drag increases as a fish swims faster because the boundary layer thickens and begins to separate from the fish. Reduced pressure and turbulent flow in the wake cause pressure drag.

Propulsion

Most fishes generate propulsive forces by waves of undulations that thrust against the water and pass caudad along the trunk and tail. Undulatory swimming may be transient or periodic. Many reef fishes, freshwater sunfishes, and bass are examples of **transient swimmers.** They lie quietly in the water much of the time, but can accelerate rapidly as they dart forward and turn to escape predators or attack elusive prey. Body shape need not be highly streamlined, but must be adapted for acceleration and maneuvering. The body often is short, with a short turning radius, and is quite high from dorsal to ventral. The height presents a large surface area for thrust against the water. Many pelagic fishes, such as tunas and sharks, are examples of **periodic swimmers.** These fishes can develop sudden bursts of speed, but are adapted for cruising, often for long periods of time. Body shape is longer and more streamlined than in transient swimmers.

The amount of body that undulates varies considerably among species of periodic swimmers. Most of the trunk and tail move back and forth in eels and other long-bodied **anguilliform swimmers** (Fig. 10–1). The amount of the trunk and tail that undulates shifts caudad in other fishes in a continuum from undulations of the caudal half of the body, as seen in jacks (**carangiform swimming**), to undulations limited to little more than the tail, as occurs in tunas and some oceanic sharks (**thunniform swimming**). Boxfishes have a rigid, inflexible trunk and undulate only their tail (**ostaciform swimming**). Considerable lateral bending occurs throughout the vertebral column of anguilliform and carangiform swimmers, but most bending occurs near the tail base in thunniform swimmers. The vertebrae in this region are diplospondylous (p. 251). Centra are more numerous (two per body segment), and this increases flexibility. The joints between them are very flexible. Tunas have zygapophyses that restrict movement in more cranial vertebrae.

Thunniform swimmers are the most rapid fishes, and tunas can reach speeds of 75 km/hr. These fishes are highly streamlined, and drag is further reduced by limiting undulations to the caudal region of the body. The sickle-shaped tail can move back and forth rapidly, and its effectiveness is increased by being very high so it extends above and below the wake of the fish.

In a trout, which has a pattern of swimming intermediate between anguilliform and carangiform, waves of contraction begin near the anterior end of the trunk and sweep down the body, first on one side and then on the other (Fig. 10–3). As any region of the body moves from side to side and pushes obliquely backward against the water, it accelerates a certain mass of water. Because of its inertia, the water generates on the fish an opposite reaction force whose magnitude is the product of the mass of water moved and its acceleration. This reaction force can be resolved into forward and lateral components. The forward components from

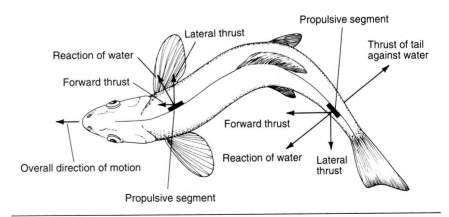

FIGURE 10–3   A dorsal view of a swimming trout showing forces developed by the undulating trunk and tail and their forward and lateral components. (After Webb.)

all regions of the body combine and propel the fish. The wave of contraction moves caudally along the body at a speed greater than the forward movement of the fish. As a result, the next posterior region of the body to exert a thrust will push on a mass of water already accelerated by the previous region of the body. The reaction of the water becomes greater as the waves of contraction travel caudally because of this summation effect and because the waves increase in amplitude. As they increase in amplitude, the waves push more directly in line with the overall direction of movement of the fish, so the forward components of the reaction of the water increase progressively. Since the caudal parts of the body and the tail move from side to side to a greater extent than the anterior regions, and the tail has a particularly broad, flat surface, the mass of water moved and its acceleration are accordingly greater.

The left and right lateral components of the reaction of the water cancel each other to some extent, but they also tend to rotate the fish about its center of gravity. That is, as the tail moves to the left, the front of the body would be expected to swing to the right. This happens only to a limited extent. The gradual tapering of the body toward the caudal end, especially at the narrow tail base, reduces the lateral force. Moreover, the shape of the fish offers great resistance to sideways movements because the heavy, front part of the body has more inertia than the tapering trunk and tail.

The segmented myomeres of fishes are well suited to generate undulatory waves of the trunk and tail. Since the vertebral column prevents the body from shortening, contraction of several myomeres on one side pulls the myosepta together and causes a curvature. The zigzag foldings of the myomeres and their overlapping, conelike extensions (Fig. 10–4**A** and **B**), which become more pronounced near the tail, allow one myomere to exert an influence over a greater body length than would otherwise be the case. The longer folds toward the tail, and tendon-like extensions from the apices of the cones into the tail, cause caudal undulations of increased amplitude and force. The overlap of myomeres ensures a smooth generation of force and flow of undulations. At a given transverse level of the body, several overlapping myomeres may be in different stages of contraction. Contraction of the surface myomere at one level may be reaching maximum force, while the conelike extension at this level of a more anterior myomere is beginning to relax, and the extension to this level of the myomere behind it is just starting to generate force.

In many fishes the superficial fibers of the myomeres insert into an **exotendon** that takes a helical course along the trunk and tail and is bound firmly to the skin. This tendon helps maintain body shape, transfers forces to the tail, and, like all tendons, can store energy. As a curvature passes down one side of the body, the exotendon on the opposite (convex) side is stretched to a slight extent and stores energy. This energy is released and assists bending when a wave of contraction passes down the previously convex side.

The direction of the muscle fibers is not alike in all parts of a myomere. Those near the center of the side of the body have a longitudinal orientation and are longer than the obliquely oriented fibers near the middorsal and midventral lines (Fig. 10–4**B**). Reasons for this are not well understood, but this difference may compensate for the varying distance the fibers lie from the plane of bending, which

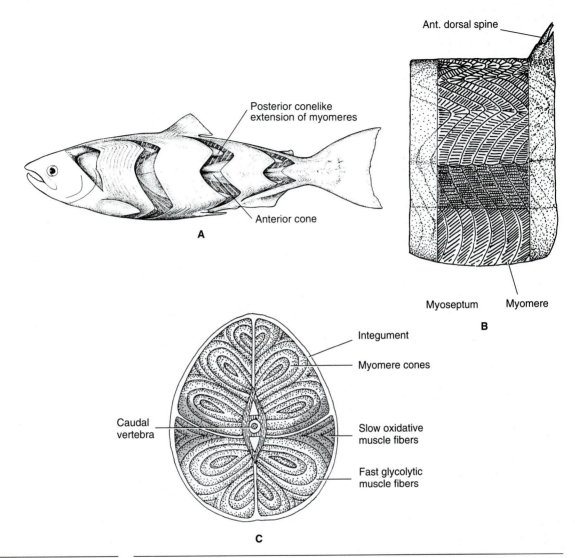

Posterior conelike
extension of myomeres

Anterior cone

A

Ant. dorsal spine

Myoseptum   Myomere

B

Integument

Myomere cones

Caudal
vertebra

Slow oxidative
muscle fibers

Fast glycolytic
muscle fibers

C

FIGURE 10–4   Myomeres of a fish. **A**, A salmon in which myomeres have been removed in several regions to show the deeper, conelike extensions of remaining myomeres. **B**, A surface view of the myomeres of a dogfish showing the directions of the muscle fibers within the myomeres. **C**, A transverse section through the tail of a dogfish showing overlapping myomere cones and the distribution of fiber types. (**A**, from Romer and Parsons, after Greene and Greene. **B**, from Walker and Homberger.)

is the sagittal plane of the body. Those fibers near the center of the trunk lie farther from the sagittal plane, and so must shorten more, than fibers near the middorsal and midventral lines. The obliquely oriented fibers near the middorsal and midventral lines shorten less and may contract nearly isometrically.

Muscle fibers in the myomeres are not all alike in their physiological properties (p. 307). The most lateral fibers are slow oxidative, or red, fibers (Fig. 10–4**C**) that are active during cruising or maintaining position in a current. When a burst of speed is needed, the more dorsal, ventral, and deeper fast glycolytic, or white, fibers are used. Fishes with these two types of muscle fibers have a two-geared system. Some species, including carp, have pink fibers with intermediate properties that are located between the white and red ones; these fishes have three gears.

Buoyancy

In addition to generating a propulsive thrust sufficient to overcome drag and move forward, a fish must float at an appropriate depth in the water and remain on an even keel as it swims. Because a fish displaces water equal to its volume, the water exerts a lift force or buoyancy that acts through its **center of buoyancy.** Buoyancy is opposed by the downward pull of gravity acting through the **center of gravity.** Since flesh is slightly denser than water (density = mass or weight divided by volume), the force of gravity is slightly greater than the buoyancy force, and the fish would sink unless buoyancy is increased in some way. A skeleton of cartilage reduces the density of cartilaginous fishes, and lipids stored in their livers increases buoyancy because lipids are lighter than flesh. Many sharks, especially pelagic ones, have sufficient lipid that they approach neutral buoyancy (the forces of gravity and lift are equal). Those sharks with less lipid tend to sink, and some species can be seen lying on the bottom of aquaria when not swimming. Actions of the fins and the shape of the head can counter the tendency to sink and keep such species afloat.

Sharks, in common with many early fishes, have heterocercal tails. The actions of a heterocercal tail are not entirely clear. Many investigators argue that as this type of tail moves back and forth, it generates a thrust that passes above the **center of balance** of the fish (Fig. 10–5**A**, line **A**). (The center of balance is essentially the resultant of the positions of the centers of buoyancy and gravity. Because the density of muscle, cartilage, and other tissues differ, and the tissues are not distributed evenly through the body, the centers of buoyancy and gravity do not lie at the same point.) Such a thrust produces a turning moment around the center of balance that raises the tail and drives the head down. The head, which is somewhat flattened ventrally, and the enlarged pectoral fins set low on the body have a planing effect and provide a lift at the front of the body. These two lift forces compensate for the tendency of the shark to sink and keep it horizontal in the water. As Thomson and Simanek (1977) and Thomson (1990) have pointed out, this argument assumes that all sharks are heavier than water and require lift forces to avoid sinking. Many species approximate neutral buoyancy yet have heterocercal tails. Thomson and Simanek also point out that the magnitude of the turning moment that lowers the head would increase with speed. As speed increases, more lift would have to be produced by the pectoral fins to keep the fish horizontal. This would be awkward, for it would increase drag at a time when drag should be reduced.

Thomson and Simanek have studied slow-motion pictures of many species of sharks swimming and analyzed tail action. Since the axis of the tail is inclined

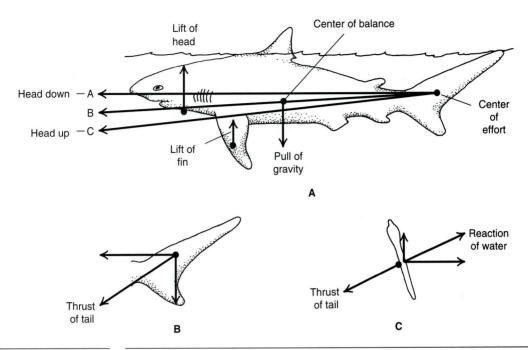

FIGURE 10–5   Action of the heterocercal tail of a shark. **A**, The heterocercal tail develops a forward thrust that passes through or near the center of balance as shown by lines **A, B**, and **C. B,** A lateral view of the tail showing forward and downward components of its thrust. **C,** A posterior view of the tail showing the reaction of the water and its upward and transverse components. (After Thomson.)

upward, the forward thrust that it delivers should be inclined downward (Fig. 10–5**B**). This thrust can be resolved into a horizontal forward component and a downward component. At the same time, the tail rotates around its longitudinal axis so that the reaction of the water to the transverse thrust resolves into a horizontal transverse component and an upward component (Fig. 10–5**C**). Relationships between downward and upward components determine where the net line of thrust of the tail will pass with respect to the center of balance (Fig. 10–5**A**). The optimal line of net thrust should pass through the center of balance because then there would be no turning moments in the sagittal plane (Fig. 10–5**A**, line **B**). But even a balanced thrust of this type requires some compensatory action of the pectoral fin and head. A balanced thrust is seldom delivered by the dorsal lobe of the caudal fin by itself. The ventral lobe is necessary to modify the line of action of the dorsal lobe, so the net thrust goes through the center of balance or close to it. The ventral lobe is, in effect, a "trimming device." Although the heterocercal tail gives a shark the potential of instability with respect to movements in the sagittal plane, this may be advantageous. Slight shifts in the axial rotation of the tail and action of the ventral lobe can raise or lower the net line of thrust of the tail. This would make it possible for the shark to change directions quickly and ascend or descend rapidly.

Early actinopterygian fishes have a lunglike sac of air in their body cavity. In

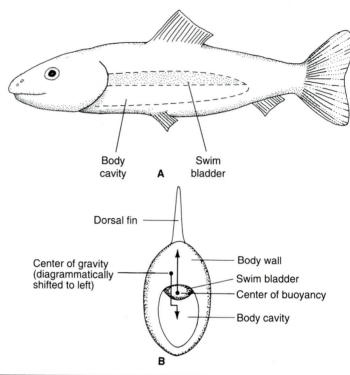

Body cavity **A** Swim bladder

Dorsal fin

Center of gravity (diagrammatically shifted to left)

Body wall

Swim bladder

Center of buoyancy

Body cavity

B

FIGURE 10–6   **A**, A lateral view of a teleost showing the location of the swim bladder in the dorsal part of the body cavity. **B**, A transverse section through the trunk showing the locations of its center of gravity and buoyancy. The center of gravity is located dorsal to the center of buoyancy but has been shifted slightly to the left in this figure for clarity. (**B**, After Alexander, 1977.)

later actinopterygians, including teleosts, the lung has evolved into a **swim bladder** (p. 648), which is located just ventral to the vertebral axis (Fig. 10–6). A sac of air, whether it be a lung or swim bladder, makes the fish less dense so it floats more easily. Although they have lungs, early actinopterygians retain the heterocercal tail characteristic of heavy-bodied fishes. The heterocercal tail may affect buoyancy, as it does in sharks, but it may have other functions as well. The tail is most obviously heterocercal in sturgeons that swim near the bottom. It is remarkably symmetrical externally (although the skeletal support is heterocercal) in the paddlefish, which filters food from higher in the water column. Teleosts can regulate the amount of gas in the swim bladder, and hence its size, very carefully. This gives them the ability to attain neutral buoyancy and float at any level in the water with little muscular effort. Forces acting on the body are different from other fishes. Since the lift of the water equals the pull of gravity, the head and fins need not generate lift. The head is not flattened ventrally, and the pectoral fins become smaller and shift more dorsally, a position better suited to their role in maneuvering. The primitive heterocercal tail becomes symmetrical externally and is braced internally by specialized hemal and neural arches. This condition is called homocercal (p. 261). The sac of gas places the center of buoyancy below the center of gravity. The fish is top heavy, and swimming motions, along with actions of the fins, must keep the center of gravity over the center of buoyancy or the fish will roll over. A dead teleost does this and floats belly up.

Stability

A swimming fish is subject to various displacement forces that must be countered if the fish is to remain on an even keel. The action of the tail, as we have seen, tends to cause the head to move from side to side, a motion called **yaw.** This is countered by the head and median fins. The large and relatively heavy head has considerable inertia and so does not move easily from side to side, and the surface area of the median fins resists lateral movement of the body. Any tendencies for the fish to rotate about its longitudinal axis (**roll**), or for its head to move up and down in the vertical plane (**pitch**), are also countered by the position and movements of the median and paired fins.

Support in Terrestrial Vertebrates

Vertebral Support

When vertebrates moved from water to land, support of body weight became a major problem because air is not a dense medium and affords little lift. Amphibians, reptiles, and most mammals rest lying on the ground in a sheltered location (Fig. 10–7**A**), yet even under these circumstances the vertebral column, girdles, and ribs must be capable of preventing body weight from collapsing the lungs and other internal organs. When terrestrial vertebrates walk, they raise their trunk off the ground. Amphibians and reptiles only raise themselves a short distance (Fig. 10–7**B**) because their humerus and femur project nearly laterally and move back and forth close to the horizontal plane. Mammals carry their trunk well off the ground, for their legs have rotated close to the trunk (Fig. 10–7**C**). Their humerus and femur move close to the vertical plane. The hind legs of birds, too, are carried under the body.

The vertebral column of terrestrial vertebrates is not a compression strut as it is in fishes, but a beam that supports all parts of the body against gravitational forces and transfers weight to the girdles and appendages. Vertical bending of the column must be resisted in amphibians and reptiles because they carry their trunk close to the ground. Yet lateral bending must occur because lateral undulations of the trunk and tail still play a role in the locomotion of most species. Vertical bending of the vertebral column occurs in many mammals during locomotion (see Fig. 10–21).

The vertebral column of terrestrial vertebrates is very strong, as we have seen (p. 263). Except for early labyrinthodonts, who spent a great deal of time in the water, and neotenic salamanders, the notochord has largely been replaced by solid, well-ossified, and unified centra. Intervertebral disks between successive centra often contain a remnant of the notochord known as the **nucleus pulposus.** This is surrounded by thick, circular layers of connective tissue fibers (Fig. 10–8). Inter-

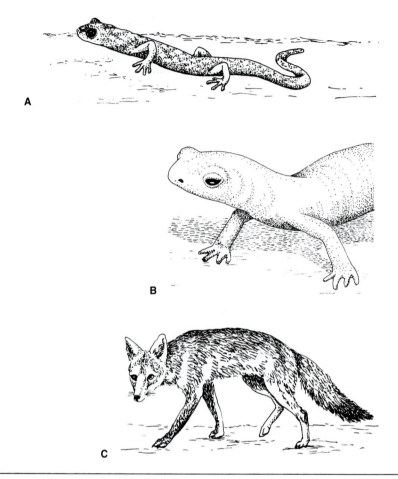

FIGURE 10–7   Limb positions of tetrapods. **A**, A salamander resting on the ground. **B**, A salamander walking. **C**, A fox walking. (**A** and **B**, after Halliday; **C**, after Macdonald.)

vertebral disks allow the vertebral column to bend, act as shock absorbers, and
distribute forces evenly over the surface of adjacent centra. If, for example, the
vertebral column bent dorsally, the dorsal edge of an intervertebral disk would be
compressed, but this force would be distributed through the semifluid disk to all
parts of the surface of adjacent centra. The neural arches are fused to the centra
and those of successive vertebrae are linked by zygapophyses. Depending on the
plane of their articular surfaces, zygapophyses restrict bending in some directions
(often in the vertical plane) but allow it in other directions.

The vertebrae are linked by strong ligaments (Fig. 10–8). Longitudinal ligaments extend across the dorsal and ventral surfaces of the centra, the tops of the
neural spines, and through the neural arch. Diagonal interspinal ligaments extend
between the neural spines, and the zygapophyses form synovial joints encapsulated
by articular capsules. Epaxial muscles link individual vertebrae (interspinalis) or

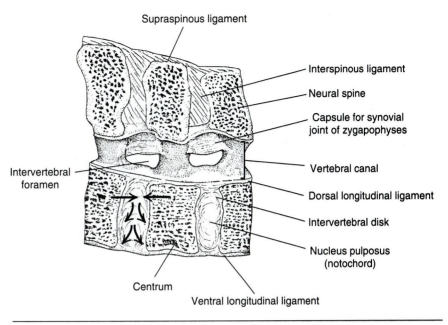

Supraspinous ligament

Interspinous ligament

Neural spine

Capsule for synovial joint of zygapophyses

Vertebral canal

Dorsal longitudinal ligament

Intervertebral disk

Nucleus pulposus (notochord)

Intervertebral foramen

Centrum

Ventral longitudinal ligament

FIGURE 10–8   A sagittal section through several lumbar vertebrae of a mammal showing intervertebral disks and some of the ligaments that unite the vertebrae. Anterior is to the left. The arrows in one disk show how the flexible disk would distribute forces resulting from a dorsal bending of the spine. (After Williams, et al.)

cross several vertebrae segments (multifidi, spinalis, erector spinae). The vertebral column and associated ligaments and muscles form a firm yet flexible complex.

Many early anatomists tried to analyze the structure of the vertebral axis of terrestrial vertebrates by applying the biomechanical principles of bridge construction. The vertebral axis resembles a bridge in some ways, but the spine is not a static support system. It is a dynamic structure that must support the head and body in many positions, receive the thrust of the legs during locomotion, and often participate during locomotion. A Dutch anatomist, Slijper (1946), has compared the trunk skeleton to an elastic bow. The arched trunk skeleton of many mammals resembles an archer's bow (Fig. 10–9**A**). The vertebral centra are comparable to the wood in the bow. They are supporting elements that are under compression. The distribution of the bone trabeculae within them follows the longitudinal stress lines (Fig. 10–8), and the joints between them permit the bow to bend. The bow is flexed by a "bowstring" composed of the sternum and ventral trunk muscles such as the rectus abdominis. The string is connected to the bow anteriorly by the short, stout anterior ribs, which are stabilized by the scalenus muscle, and posteriorly by the pelvic girdle. Subvertebral muscles (psoas and quadratus in mammals) form another partial bowstring. The curvature of the bow can change under the action of various trunk muscles as the animal changes positions and moves, but it cannot sag and permit the trunk to collapse as long as the bowstring is under tension. The trunk skeleton of amphibians and reptiles is flat, as it is in sway-back mammals such as a horse. This arrangement is more

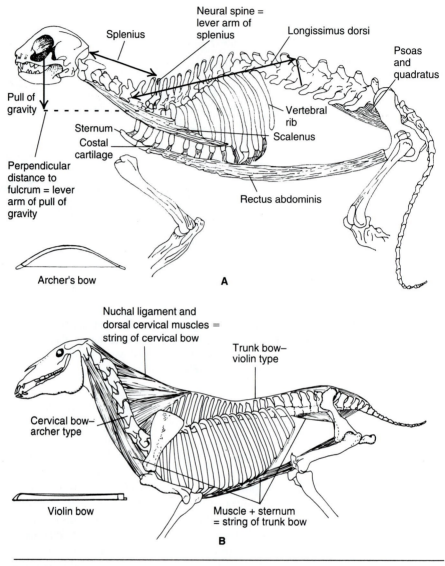

FIGURE 10–9 Diagrams of the biomechanics of the mammalian trunk skeleton. **A**, A cat, in which the bow is analogous to an archer's bow. **B**, A horse, in which the trunk is analogous to a violin bow. (**A**, after Slijper; **B**, After Wake.)

comparable to a violin bow, but the principles are the same (Fig. 10–9**B**). The head of mammals is raised, and the cervical vertebrae form another archer's bow that curves in the opposite direction from the trunk bow (Fig. 10–9**A**). The string of the cervical bow is formed by dorsal cervical muscles and nuchal ligaments.

The neural spines of the vertebrae are lever arms that transmit a force (the pull of muscles on them) to a center of rotation, or fulcrum, located between the centra (Fig. 10–9**A**). Cervical muscles, such as the splenius, acting through the neural spines of the anterior thoracic vertebrae, rotate the neck upward and support the head. Gravity acts to pull the head and neck downward. Here is another example of two sets of turning moments (p. 191). The neural spines of the cervical and anterior thoracic vertebrae are the lever arms for the in-force that raises the head, and they are oriented nearly perpendicularly to the lines of actions of muscles acting on them. When several muscles act on the same spines from different directions, the direction of the spines must be the resultant of all the forces. Since the splenius and similar oriented muscles are the major ones acting on the anterior thoracic neural spines, the spines tend to incline slightly caudad. Long neural spines increase the length of the lever arms, increase their mechanical advantage, and hence reduce the forces the muscles must develop. The magnitude of the in-moment is the product of the in-force (the pull of the splenius and similar muscles) and the in-lever arm (the length of the neural spines). The out-moment in this case is the product of the out-force (the downward pull of gravity from the center of mass of the head and neck) and the out-lever arm (the perpendicular distance from the line of action of gravity to the fulcrum at the base of the neck). No physical lever is present here. Since in- and out-moments must balance when the system is in equilibrium, it is not surprising that quadrupedal mammals with heavy heads have exceptionally long neural spines on their anterior thoracic vertebrae (Fig. 10–9**B**).

Ungulates with large heads that they must lower when they browse or graze also have particularly prominent nuchal ligaments (Fig. 10–9**B**). The ligaments are stretched by the weight of the head when the head is lowered, and the energy that stretched the ligaments is stored. When the animal raises its head, most of the stored energy is released and helps raise the head. Muscle action is still needed, but far less than would otherwise be the case.

Muscle pull on the neural spines is primarily in the sagittal plane or close to it. The situation is analogous to a carpenter's joist, which must resist bending in only one plane and whose strength is increased by increasing the depth of the joist (p. 196). Sagittal forces on the spines can be resisted by increasing their dimension in the sagittal plane. Since lateral forces are not strong, the spines are blade shaped and not thick from side to side.

In addition to the support and movements of the head, many other supporting functions and movements occur along the vertebral column. The complex interaction of many moments appears to explain the regional variation that we find in the length, direction of inclination, and anteroposterior dimensions of the neural spines. For example, the head can be supported by anterior thoracic vertebrae only if they, in turn, are supported by other back muscles, such as the longissimus dorsi, that arise from the lumbar vertebrae (Fig. 10–9**A**). The line of action of the longissimus dorsi in many mammals necessitates that the neural spines of the lumbar vertebrae incline anteriorly. Caudally directed neural spines near the front of the thoracic region and cranially directed spines in the lumbar region characterize many, but not all, quadrupedal mammals. This condition is referred to as **anticliny.**

Limb Support

When a terrestrial vertebrate is standing or moving, not only the vertebral column but also the legs must support body weight. Weight is transferred to the pelvic girdle and limb by way of one or more sacral vertebrae and ribs. The number of vertebrae and ribs involved, and the degree to which they are fused together, correlates with the forces they must transfer. Weight is transferred to the pectoral girdle and appendage not by bony connections but by a muscular sling extending between the trunk skeleton and the girdle and appendage (Fig. 10–10). Although the rhomboideus and pectoralis muscles participate to some extent, the major component of the sling in mammals is the serratus ventralis muscle, which extends between the distal ends of the bony ribs and the dorsal border of the scapula. When they are transferring weight, the ribs are compressed, but bone is well adapted to meet this stress. The costal cartilages, which extend from the ribs to

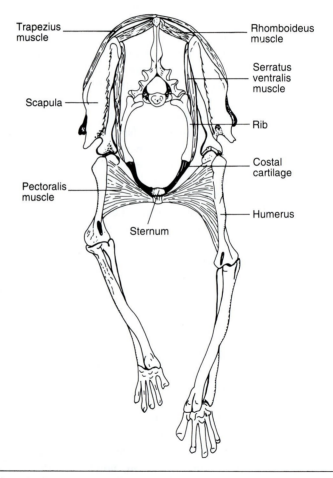

FIGURE 10–10   A cranial view of a transverse section through the thorax of a cat showing muscular connections between the trunk skeleton and the pectoral girdle and appendage. (From Walker and Homberger.)

the sternum, are not compressed but must be sufficiently flexible to allow the ribs to move during respiration.

The legs must be drawn under the body and extended by muscle action to support the body. Since the legs are not solid pillars but jointed struts, the continued contraction of muscles crossing the joints is needed to stabilize the moments around the joints and prevent the animal from collapsing, even when it is standing still. The amount of muscular energy needed to stabilize joints and hold up an animal is reduced in some mammal species by several factors. To the extent that the segments of the limbs can be aligned vertically, one directly above another, the moments at the joints are reduced. Weight is transferred directly through bones to the ground. Some dinosaurs, elephants, and other heavy terrestrial vertebrates tend to have pillar-like limbs of this type. The tibia of human beings also can be brought directly under the femur when standing. The configuration of joint surfaces and ligaments around joints restricts motion in certain directions. The olecranon of the ulna and calcaneus in the ankle form lever arms for muscle attachment, but they also prevent hyperextension at these joints. The cruciate and other ligaments behind the knee joint prevent hyperextension at this joint (Fig. 10–11). Muscle action is not needed to prevent joints of these types from bending backward.

Horses and some other ungulates have one or more passive **stay mechanisms** that hold limb joints in an extended position with minimal muscular effort, thereby enabling them to stand even when asleep. The way in which the knee (stifle) and ankle (hock) joints of a horse are stabilized is an example (Fig. 10–12). The superficial digital flexor muscle has become almost entirely tendinous and forms a strong cord that extends from near the distal end of the femur to the calcaneus, where it is firmly attached, and then onto the caudal surface of the phalanges. This cord forms one side of a parallelogram, the other side of which is formed by the tibia. The distal end of the femur and the taluscalcaneus form the ends of the parallelogram. As in any parallelogram, none of the angles at the corners can change without all others simultaneously changing. When the animal is resting, a muscle slip can pull the patella above a small crest on the medial side of the femoral trochanter. It is held in this position by strong patella ligaments extending to the tibial crest. The patella resting above the crest locks the angles of the parallelogram. No change can occur until other muscles pull the patella away from the crest and into the patella groove.

Terrestrial Locomotion

Walking and Running

The fundamental pattern of locomotion for terrestrial vertebrates is walking and running, although some slither (snakes), burrow, jump, or fly. A walking vertebrate advances by a succession of **steps,** in which one foot is first placed on the ground to develop a thrust that accelerates the body (the **propulsive phase** of a step cycle), and then is removed from the ground and advanced (the **swing phase**). The length of a step is the distance that the trunk is advanced during the propulsive phase of the cycle. Each leg goes through a step cycle, and one cycling of all the

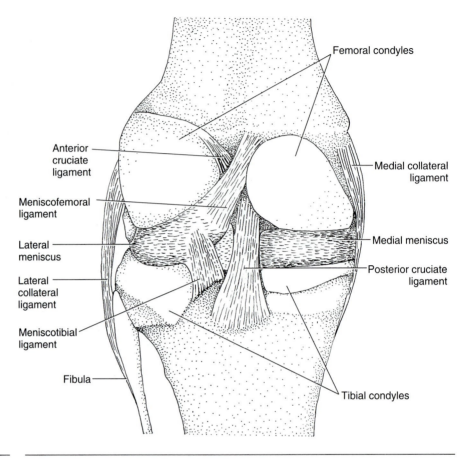

Femoral condyles

Anterior
cruciate
ligament

Medial collateral
ligament

Meniscofemoral
ligament

Lateral
meniscus

Medial meniscus

Lateral
collateral
ligament

Posterior cruciate
ligament

Meniscotibial
ligament

Fibula

Tibial condyles

FIGURE 10–11 A posterior view of the left knee joint of a horse showing the ligaments that prevent hyper-extension of the crus. The articular capsule has been removed. (After Getty.)

legs constitutes one **stride.** That is, a stride is measured from the first placement of one foot on the ground (often the left hind foot is taken as the reference foot) to the next placement of that same foot. A stride for a quadruped is composed of four step cycles, one for each leg. Throughout a stride, the oscillations of the limbs must be carefully coordinated to provide adequate thrust and speed, yet maintain stability to prevent the animal from falling even though one or more feet are off the ground during the swing phases of the steps. Moreover, this must be accomplished as efficiently as possible so the cost in energy expenditure can be sustained if the animal is traveling far, or can be made up soon if the animal must spurt for a short distance.

The Step Cycle of Amphibians and Reptiles. Of course, not all contemporary amphibians and reptiles walk in the same way. The pattern seen in many lizards is representative and probably is close to that used by ancestral tetrapods. As we have noted, most amphibians and reptiles carry their bodies close to the ground, for the limbs are splayed to the side. Brinkman (1981) has made a cineradiographic

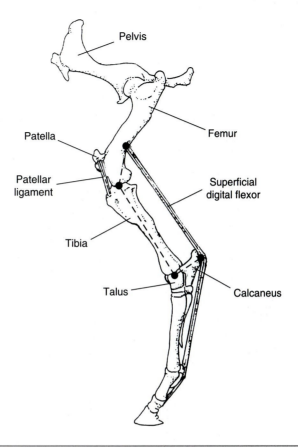

FIGURE 10–12   A lateral view of the pelvis and hind leg of a horse showing the stay mechanism that locks the femur and tibia in a fixed position when the animal is at rest. Anterior is to the left. Parts of the locking parallelogram are shown by dashed lines, and its four angles are shown by large dots. (Limb bones after Getty.)

study of hind limb movements in the iguana (*Iguana*), and Jenkins and Goslow (1983) have made a similar study of forelimb and girdle movements in the savannah monitor lizard (*Varanus exanthematicus*). The upper limb segment (humerus or femur) moves back and forth close to the horizontal plane (Figs. 10–13 and 10–14); slightly above the horizontal plane when the upper limb is protracted during the swing phase of the step; and slightly below this plane when the limb is retracted during the propulsive phase. The lower limb segment (antebrachium or crus) extends downward at nearly a right angle to the distal end of the humerus or femur and moves close to the vertical plane.

When a foot is placed on the ground at the beginning of the propulsive phase, the upper limb is protracted and the lower limb extended (Figs. 10–13 and 10–14). Extension of the antebrachium is more pronounced than of the crus. The hand or foot points forward and often slightly laterally. A propulsive thrust is developed by several actions. Appendicular muscles retract the humerus or femur, and, since the foot maintains its position on the ground, this draws the body

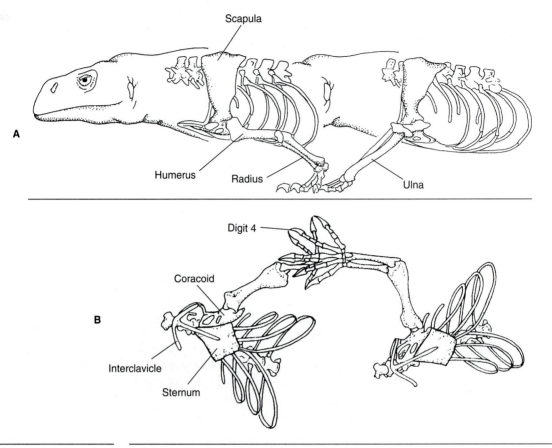

Scapula

A

Humerus   Radius   Ulna

Digit 4

Coracoid

B

Interclavicle

Sternum

FIGURE 10–13   The movements of the forelimb and pectoral girdle of *Varanus exanthematicus* at the beginning (left) and end (right) of the propulsive phase of a step. **A**, Lateral view. **B**, Dorsal view. (From Jenkins and Goslow.)

forward. The humerus or femur also rotates about its longitudinal axis, and this rotates the lower leg in the manner of a crank (Fig. 10–15**A**). While the humerus or femur are being retracted and rotated, the antebrachium or crus flexes as the distal end of the upper limb advances over the point of foot placement, and then the antebrachium or crus extends and helps push the body forward (Figs. 10–13 and 10–14).

The hand continues to point forward throughout the propulsive phase because the radius rotates on the humerus and ulna as the limb is drawn caudad. These are difficult motions to describe, but they can be visualized if you place your arm in the position assumed by an amphibian or reptile when the hand is first placed on the ground, and advance the body while maintaining the position of the hand on the ground. The situation is somewhat different in the hind leg because only a hinge action can occur at the knee joint. Rotation occurs in the ankle, primarily at the mesotarsal joint. The hind foot is directed forward at the beginning of the propulsive phase, but as the femur is drawn posteriorly and the body advances,

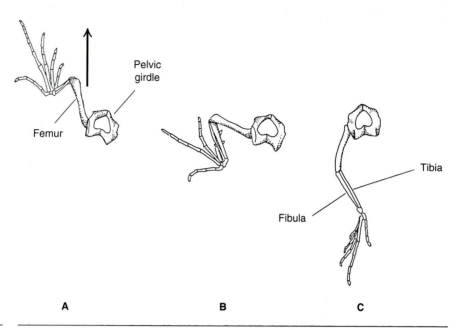

FIGURE 10–14 Dorsal view of the movements of the hindlimb and pelvic girdle of *Iguana* at the beginning (**A**), middle (**B**), and end (**C**) of the propulsive phase of a step. The arrow shows the direction of movement. (From Brinkman.)

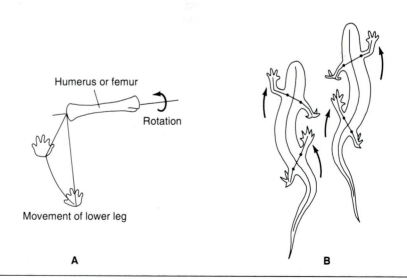

FIGURE 10–15 Locomotion in early tetrapods. **A**, The effect on the lower leg of the rotation of the humerus or femur. **B**, The contribution of lateral trunk oscillations to the rotation of the girdles and the advancement and retraction of the appendages. (After Rewcastle.)

the foot rotates laterally and points laterally or posteriorly at the end of the propulsive phase (Fig. 10–14). In both the hand and foot, duration of toe contact with the ground is maximized by digits that increase in length from the most medial or first one, to the fourth finger. The length of the fifth digit decreases abruptly. Plantar flexion of the hand or foot adds a final component to the thrust of the limb.

Swimming-like sinusoidal curvatures of the trunk and tail occur in most amphibians and reptiles (Fig. 10–15**B**). These motions rotate the girdles, which helps advance and retract the limbs and adds to step length. An extra increment is added to the length of the step of the front leg in lizards by a translational movement of the coracoid plate on the sternum (Fig. 10–13**B**). As the limb is retracted, the girdle also slides caudad on the sternum for a distance equal to about one fourth of the length of the coracoid plate.

The swing phase of a step is a simple recovery. The antebrachium or crus flexes as the hand or foot is removed from the ground. Humerus or femur are protracted and the forelimb extended.

All of these movements require a powerful ventral or adductor musculature to draw the humerus or femur ventrally to a slight extent and thus raise and keep the body off the ground. Strong forelimb muscles are also needed to bend the antebrachium or crus and maintain it in a vertical position. As we have seen (p. 331), the ventral parts of the girdles are greatly expanded and provide the needed surface for the origins of the adductor musculature. Large tuberosities, trochanters, and muscular ridges on the humerus and femur provide for the insertions. You can appreciate the muscular force needed by crawling with your arms splayed to the side rather than under your body.

The Step Cycle of Mammals. The limbs of mammals are no longer in the primitive splayed position, but have rotated far enough under the body so the humerus or femur move fore and aft closer to the vertical plane. The **stance,** or distance between the placement of left and right feet, is narrower than in amphibians and reptiles. This places the limbs closer to the projection of the body's center of gravity and provides better support with less muscular effort. Limbs situated more or less under the body swing through longer arcs, and this adds to step and stride length with no or little extra expenditure of energy.

The degree to which the limbs are beneath the body varies somewhat among mammal species. In running, or **cursorial,** species such as ungulates and carnivores, the limbs are well under the body and all parts of the limb move in the same parasagittal plane. But in many mammals they are not quite so far under. Jenkins (1971) and Jenkins and Weijs (1979) have studied limb movements in the opossum (*Didelphis virginiana*) by cineradiography. The elbow joint, which is directed posteriorly, lies slightly lateral to the shoulder joint at the beginning of the propulsive phase (Fig. 10–16**A**). The knee is directed forward and slightly laterally (Fig. 10–17**A**). In all species both the hand and the foot are directed forward, or nearly so, throughout the propulsive phase. As the propulsive phase continues, the humerus or femur are retracted, drawing the body forward, and the forelimb flexes and the feet dorsiflex as the shoulder or hip advances over the

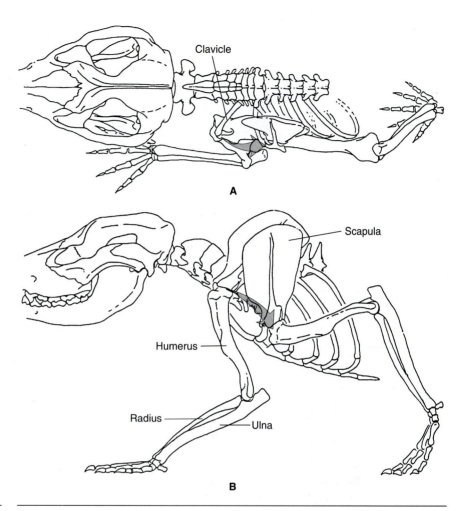

FIGURE 10–16 The movements of the forelimb and pectoral girdle of the Virginia opossum at the beginning (left) and end (right) of the propulsive phase of a step. **A**, Dorsal view. **B**, lateral view. (From Jenkins and Weijs.)

point of placement of the feet on the ground. The forelimbs extend and feet plantar flex near the end of the propulsive phase.

The scapula also rotates during the propulsive phase, so at the end of the phase the scapula is oriented more vertically than at the beginning and the glenoid fossa has moved caudally and ventrally (Fig. 10–16**B**). These movements add to step length. Although the hips are firmly articulated with the vertebral column, a small lateral movement of the vertebral column causes the hips to swivel slightly (Fig. 10–17**A**).

Muscle Recruitment in a Step Cycle. Many electromyographic studies have been made during locomotion in humans and quadrupeds. Dial, Jenkins, and Goslow (1991) have compared the sequence of actions of shoulder and brachial mus-

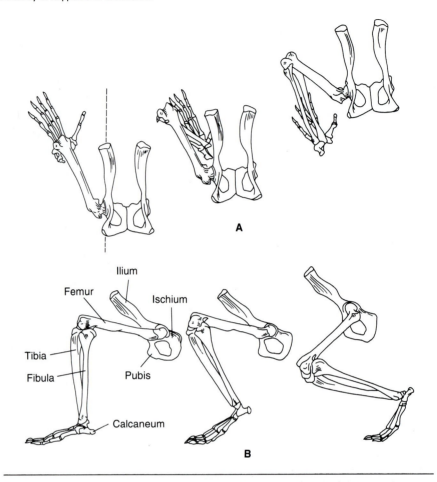

FIGURE 10–17 Three stages in the movements of the hindlimb and pelvic girdle of the Virginia opossum during the propulsive phase of a step. **A**, Dorsal view. **B**, Lateral view. (From Jenkins.)

cles during a step cycle in the savannah monitor lizard (*Varanus exanthematicus*), the Virginia opossum (*Didelphis virginiana*), and the European starling (*Sturnus vulgaris*). Dial et al. studied flight in starlings and drew on the work of Jenkins and Goslow (1983) for the lizard and Jenkins and Weijs (1979) for the opossum. Their results are summarized in Figure 10–18. Several points should be emphasized. Limb movements require the integrated action of many muscles, although more are used in the propulsive phase than in the swing phase. Many of the propulsive muscles become active near the end of the swing phase, presumably to check the forward movement of the limb and prepare for propulsion. Similarly, some muscles used in the swing phase become active near the end of the propulsion. The action of various muscles is staggered to some extent. Compare, for example, the period of activity of the biceps brachii and pectoralis in *Varanus*. Both are active in propulsion (the pectoralis adducts and helps retract the humerus and the biceps flexes the antebrachium), but the activity of the pectoralis begins sooner (near the end of the swing) and ends sooner than the activity of the biceps.

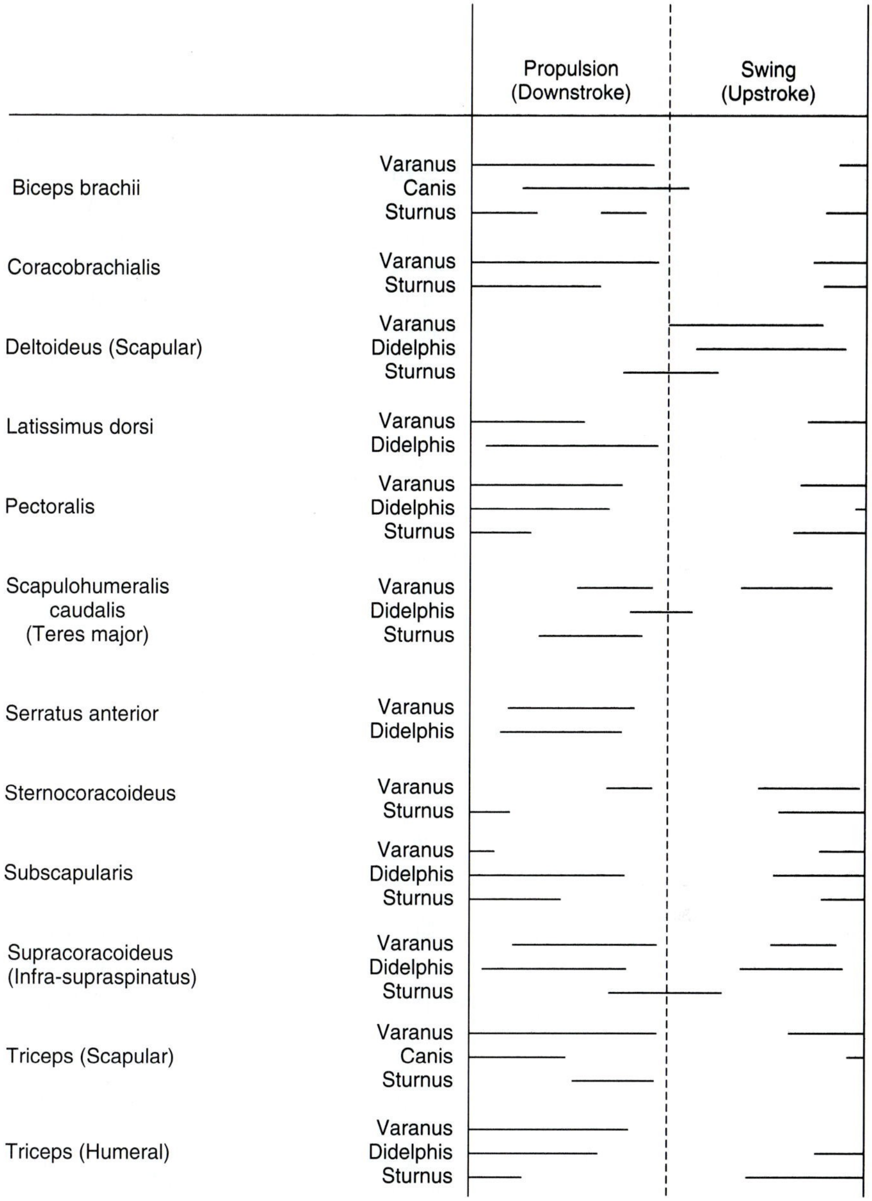

FIGURE 10–18   Periods of activity, as shown by electromyographic patterns, for selected shoulder and brachial muscles in a monitor lizard (*Varanus*), dog (*Canis*), opossum (*Didelphis*), and starling (*Sturnus*). (From Dial, Goslow, and Jenkins.)

Homologous muscles tend to be utilized in similar phases of the limb cycle in such diverse species as a reptile, mammal, and bird. The downstroke of the bird's wing (p. 386) is equated with the propulsive phase of walking. The evolution of the appendicular apparatus of tetrapods is conservative, with bones and muscles

---

FOCUS 10–1   Fiber-Type Recruitment in Tetrapods

Many of the appendicular muscles of tetrapods are composed of mixtures of two or three physiological types of striated muscle fibers. In mammals, slow oxidative fibers (red fibers) maintain posture and become active in slower movements that continue over long periods of time. Fast oxidative fibers (pink fibers) join the red fibers as speed increases. Fast glycolytic fibers (white fibers) become active only at high speeds. Since these are anaerobic fibers, high speed cannot be sustained.

These generalizations apply to ectothermic vertebrates as well, but the muscles of ectotherms must function over a wide range of body temperatures. This poses physiological problems because muscles contract more slowly at lower temperatures. What adaptations are made in a lizard that compensate for this and allow it to run at the same speed at different temperatures? Jayne, Bennett, and Lauder (1990) have studied this problem in *Varanus exanthematicus* using quantitative electromyography. They selected a thigh muscle to study, the iliofibularis, because red and white fibers are localized in different parts of the muscles. They found that red fibers are active at all speeds, but that at a given speed the intensity of the activity of the red fibers was greater at 25°C than at 35°C. White fibers become active only at high speeds, but the speed at which these fibers become active was less at the lower temperature. Thus lizards can run fast at lower temperatures by increasing the intensity of red fiber activity and recruiting white fibers at a lower temperature.

---

present in ancestral species being modified in their descendants and used in similar ways.

As in fishes (p. 350), many muscles of tetrapods contain mixtures of red and white fibers. The red fibers are active at slower speeds, and white fibers become active at higher speeds (Focus 10–1).

Gaits.   The particular combination of feet that are on and off of the ground during a stride is known as the **gait.** Gaits used by tetrapods depend on many variables, including limb posture and length, degree of stability or maneuverability needed, speed, and energetic costs. Since amphibians and reptiles move close to the ground, they must have gaits that provide good stability so their belly does not drag. Turtles frequently use a gait that is particularly stable (Fig. 10–19**A**). If the left hind foot is taken as the lead foot, the sequence of limb advancements and foot placements is left hind, left front, right hind, and right front foot. Each foot is on the ground for a relatively long time (about 0.75 of stride duration for a slowly walking turtle). This time, which is the same for each foot, is called the **duty factor.** Limb movement can be expressed in quantitative terms by stating the duty factor and the **relative phase** of the feet (Fig. 10–19**B**). The first foot to be placed (left hind) is given a relative phase of 0.00. The placement of the next (left front), which occurs slightly over one third of the stride interval, has a phase of 0.35. The next one (right hind) is placed at the middle of the stride and has a phase of 0.50. The placement of the last foot (right front) follows the right hind in a lag that is the same as that between the first two feet, so it has a phase of 0.85. This particular gait is called a **lateral sequence, diagonal couplet gait** because the placement of one foot (left hind) is followed by the placement of the

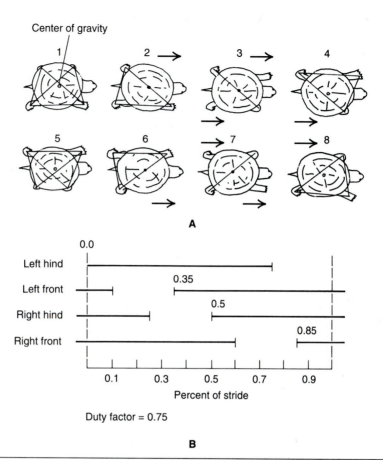

A

B

FIGURE 10–19 **A,** Drawings from cinephotographs showing limb movements and foot placements during one stride of a turtle. The stride begins with the placement of the left hind foot. Arrows indicate the feet that are off the ground and moving forward. Feet on the ground are connected by lines. Notice that the animal's center of gravity falls within triangles of supporting feet during most of the stride. **B,** A gait diagram of the stride shown in **A.** The horizontal lines indicate the period the feet are on the ground. Numbers give the lag in foot placement as a percent of stride interval. (After Walker, 1971.)

ipsilateral foot (left front), and the contralateral diagonal feet (left front and right hind or right front and left hind) are in closer phase in their movements and placements than any other pair of feet.

During most of the stride, the body is supported by either four or three feet (Fig. 10–19**A**), so the center of gravity falls within successive triangles of supporting feet. These triangles are quite wide from left to right because the splayed limb posture gives the animal a wide stance. The gait and wide stance provide the high degree of stability that these low-slung animals need. When the bases of the triangles of support shift between the sides of the body, there are brief periods when the body is supported by four feet (two overlapping triangles) or by two feet (a hind foot and its contralateral front foot). The latter is unstable, and some pitching and rolling occur as the center of gravity crosses the diagonal between supporting feet, but these periods are so brief that the animal does not drag on the ground.

Because mammals carry their limbs more nearly under the body than amphibians and reptiles, they have less stability. Mammals have a narrower stance, narrower triangles of support, and their center of gravity is higher off the ground. Stability, however, is not as critical as for amphibians and reptiles. They can rise and fall within each stride to a greater extent without dragging on the ground. Only an overall dynamic stability is necessary. This allows mammals to reduce duty factors more and have fewer feet on the ground at one time, and it gives them more gait options. Decreased stability also increases maneuverability; mammals can change directions more rapidly.

When moving slowly, many mammals use the lateral sequence, diagonal couplet walk seen in amphibians and reptiles. Duty factors are lower, however, so the body is supported during part of the stride by only two feet on the same side of the body (Fig. 10–20A). The same gait can be used at a moderate run when the duty factor for each foot falls below 0.50. Periods when the animal is supported by one or two feet increase in length. Most mammals shift to other gaits when they begin to run. A few long-legged mammals, such as giraffes and camels, change to a **pace** when speed increases (Fig. 10–20B). Legs on the same side of the body move together in nearly complete phase, and 0.50 out of phase with the legs on the other side of the body. In a running pace, two brief flight periods occur in each stride when no feet are on the ground. These occur when the supporting feet shift from one side of the body to the other. Horses can be trained to pace, but their natural gait at higher speeds is a **trot** (Fig. 10–20C). A contralateral pair of diagonal feet (left front and right hind feet, or right front and left hind feet) move nearly in unison and 0.50 out of phase with the other set. Again, if the trot is fast, two brief flight periods occur within each stride during the shift from one set of supporting feet to the other.

Walks, paces, and trots are described as **symmetrical gaits** because the left and right foot of each pair move 0.50 out of phase and are evenly spaced in time. At still higher speeds, most mammals shift to **asymmetrical gaits,** of which the **gallop** is one example (Fig. 10–20D). The movements of the left and right feet of each pair are no longer evenly spaced. Overlaps of the steps of different feet occur in most gaits, but duty factors are so low in a fast gallop that there is little overlap. The body is supported by a succession of one or two feet, so most of the distance that the body is carried forward by one foot is added to the steps of the other feet. Stride length is very long in a fast gallop, about 7 meters for a race horse.

A number of galloping mammals, such as a cheetah, add more to stride length by extending and flexing their back (Fig. 10–21). The plane of the articular surfaces

---

FIGURE 10–20 Some mammalian gaits as seen in a horse. Lines in the diagrams represent periods in ▶ one stride that the feet are on the ground. This can also be seen in the drawings above the graphs. The drawings for the walk go only through the first half of the stride; the last half would be mirror images of the four drawings shown. The drawings are complete for the other gaits. (After Gambaryan.)

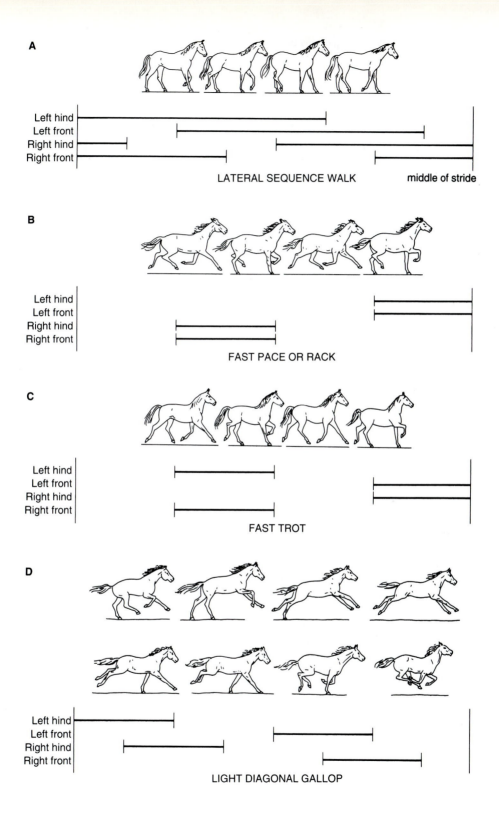

A

Left hind
Left front
Right hind
Right front

LATERAL SEQUENCE WALK   middle of stride

B

Left hind
Left front
Right hind
Right front

FAST PACE OR RACK

C

Left hind
Left front
Right hind
Right front

FAST TROT

D

Left hind
Left front
Right hind
Right front

LIGHT DIAGONAL GALLOP

371

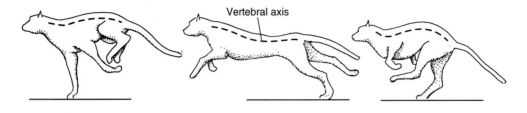

Vertebral axis

FIGURE 10–21   Dorsoventral flexion of the vertebral column adds to stride length in a running cheetah. (After Hildebrand, 1959.)

of some of the vertebral zygapophyses has shifted toward the vertical to make this possible.

The speed or **velocity** at which a tetrapod travels is equal to stride frequency times stride length. Increasing stride frequency involves oscillating the limbs more rapidly. Increasing stride length requires lowering the duty factors of the feet and changing gaits. When duty factors are lowered, the feet are off the ground for longer periods, during which they are carried forward by feet on the ground for a greater distance before being placed again. Indeed, as we have seen, all feet may be off the ground briefly during some fast gaits. Because stability is so important, low-slung amphibians and reptiles increase speed primarily by increasing stride frequency within a walk, although some can trot at moderate rates. Mammals, for whom stability is not so critical, increase speed first by increasing stride frequency but soon shift to gaits that increase stride length. Increasing stride length is less expensive in terms of energy consumed than increasing stride frequency (Focus 10–2).

Limbs as Lever Systems

An individual mammal can change speed by changing gaits and step length, but mammal limbs are also adapted to the ways in which the limbs are used. The limbs and the muscles that act on them are first- and second-order lever systems (p. 192). In the course of evolution, the length of limb segments, and the length and location of bony processes to which muscles attach, have been subject to selection pressures. In moles and armadillos, which use their front limbs for digging, the limb is adapted to maximize the force delivered at the end of a limb; whereas running ungulates have limbs adapted to maximize the speed or velocity with which the distal end of the limb moves (Fig. 10–22). For purposes of this comparison, we will treat the entire limb from the fulcrum at the shoulder joint to the ground as a single third-order lever that is retracted by muscles (such as the teres major) that extend from the scapula to the humerus. The out-lever arm is the distance from the shoulder joint to the ground, and the out-force is delivered downward and backward, perpendicular to the out-lever. The in-lever arm is the perpendicular distance from the shoulder joint to the line of action of the retractor muscle. Recall (p. 191) that at equilibrium the following relationship between lever arms and forces applies:

| FOCUS 10–2 | **Energy Costs of Locomotion** |

Much energy can be and is saved during locomotion by the tendons of muscles. For example, when the weight of a running digitigrade or unguligrade mammal bears down on a hind foot just placed on the ground, the toes and ankle flex. This stretches the long calcaneus tendon and others behind the foot and toes. The elastic recoil of the tendons then recovers most of the energy used in stretching the tendons, and the recovered energy helps to raise and advance the animal (Focus Figure).

Although recycling energy in this way is important, muscles must still contract. The energetic cost of muscle contraction is determined primarily by the energy expended to activate a muscle (pump calcium into the sarcoplasmic reticulum) and to generate a unit of force in a unit of time (cycling cross bridges between actin and myosin). Heglund and Taylor (1988) have shown that these costs increase much more rapidly with an increase in stride frequency than with an increase in stride length (change of gaits). They studied the energy costs of locomotion at different speeds in 16 species of mammals ranging in size from a mouse to a horse. To compare animals of such different sizes that travel at very different absolute speeds, they developed the concept of "equivalent speeds."

That is, they made comparisons at the preferred trotting speed (as selected by the animal), the speed at the trot-gallop transition, and at the preferred galloping speed. They found that the energy cost per gram of animal per stride was not dependent on size, and was therefore the same for a mouse and a horse. But the energy cost did increase as speed increased and the animal changed gaits: 5.0 joules° per kilogram per stride at the preferred trotting speed to 7.5 joules per kilogram per stride at the preferred galloping speed. This is only a 1.5-fold increase. Small animals, however, with their short legs, had much higher stride frequencies at equivalent speeds. At the trot-gallop transition, a mouse takes six times as many strides as a horse. Because it must cycle its limbs so rapidly, a mouse consumed energy on a per-gram basis six times the rate of a horse at equivalent speeds.

°A joule (J) is the unit of energy. It is equal to a force of 1 Newton (N) applied to an object and moving the object through a distance of 1 meter.

$$1\ N = 1\ kg\ m/s^2,$$
$$1\ J = 1\ kg\ m/s^2 \times m$$
$$1\ J = 1\ kg\ m^2/s^2$$

where k = kilograms, m = meters, s = seconds.

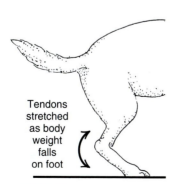
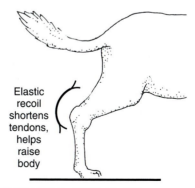

Tendons stretched as body weight falls on foot

Elastic recoil shortens tendons, helps raise body

Tendons behind the ankle are stretched when a dog lands on its hind foot and the foot flexes. Their elastic recoil extends the foot and helps raise and advance the animal.

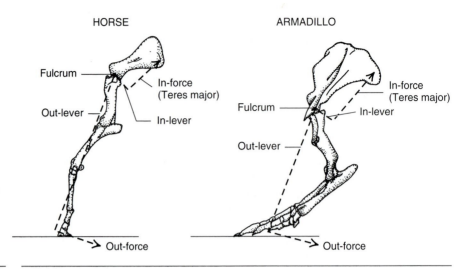

FIGURE 10–22 Limbs can be adapted to increase velocity or power by changes in the relative length of their in-lever and out-lever arms. An ungulate's pectoral girdle and limb (left) is adapted for high velocity; that of an armadillo (right) is designed for power. (After Smith and Savage.)

$$F_i \times L_i = F_o \times L_o$$

By solving this equation for the out-force, it is apparent that an increased value for the out-force can be obtained by increasing the length of the in-lever relative to the out-lever:

$$F_o = (F_i \times L_i)/L_o$$

This is what has happened in the evolution of the armadillo. The armadillo has a short out-lever arm relative to most mammals, and a distal shift in the point of attachment of the retractor muscle has increased the relative length of the in-lever.

Velocity also is related to the relative length of the lever arms. In a given unit of time, the part of a rotating lever that is located far from the fulcrum will move a greater distance than a point near the fulcrum. Since velocity = distance/time, the velocity of the distant point will be greater than for the point closer to the fulcrum. This relationship can be expressed as

$$V_o \times L_i = V_i \times L_o$$

where V_o and V_i represent the velocities at the ends of the out- and in-lever arms. By solving this equation for the out-velocity, it is apparent that an increased value for the out-velocity can be obtained by lengthening the out-lever arm and by shortening the in-lever arm through a proximal shift in the point of attachment of the retractor muscles:

$$V_o = (V_i \times L_o)/L_i$$

This is what has occurred in ungulates.

The relationship between force and velocity can also be expressed as a **gear**

ratio, which is the length of the out-lever arm divided by the length of the in-lever arm:

$$L_o/L_i$$

When the ratio is low, force is emphasized at the expense of velocity; when high, velocity at the expense of force. The ratio is low, about 4, for an armadillo, but much higher, about 10, for an ungulate.

Additional Adaptations for Speed

Limb lengths of mammals vary considerably among species because of the relationships between lever systems and the ways the limbs are used. Early mammals and many others that travel at moderate speeds walk flat footed. They place the soles of their feet flat on the ground, a foot posture termed **plantigrade** (L. *planta* = sole + *gradus* = step) (Fig. 10–23). Most primates retain a plantigrade foot, with the first digit articulated in such a way that the hands and feet can grasp branches. In this case the plantigrade foot is an arboreal adaptation. The human foot is also plantigrade, but has lost its grasping ability and is specialized for bipedal locomotion. Running species, as we have seen, evolved longer legs, which increases step and stride length and hence speed. Most carnivores walk on their digits, with

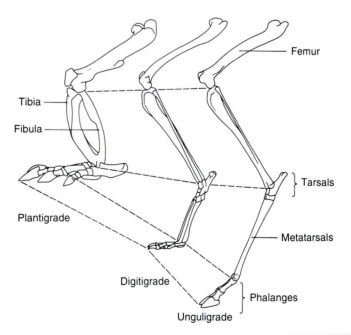

FIGURE 10–23   Limb types and foot postures of mammals. The femur of all has been drawn to the same length to emphasize the changes in proportions of the distal parts of the limbs. The plantigrade foot is of a powerful digger, the armadillo (*Dasypus*); the digitigrade foot is from a good runner, the coyote (*Canis latrans*); the unguligrade foot is from the extremely swift pronghorn (*Antilocapra*). (After Vaughan.)

the wrist and ankle carried off the ground, a foot posture termed **digitigrade.** Ungulates have very long legs and walk on the tips of those digits that reach the ground, a posture termed **unguligrade** (L. *ungula* = hoof). The terminal phalanx and claw are modified to form a hoof. Digits on each side of the supporting toes are reduced in size or lost. In perissodactyls the axis of the foot passes through the third digit, which is the largest (Fig. 10–24**A** and **B**). Rhinoceroses, adapted to run on relatively soft ground, retain three digits, the second to the fourth. This was also the case in ancestral, forest-dwelling horses, but only the third digit remains in contemporary, plains-dwelling species. The foot axis of artiodactyls passes between the third and fourth toes, which are equal in size (Fig. 10–24**C** and **D**). Pigs and other species with a primitive foot structure retain small second and fifth digits, but these are lost in more specialized species, including cattle and sheep. Metapodial and podial elements that supported lost toes are also reduced or lost. The two metacarpals and metatarsals that remain in cattle and sheep are fused to form a single **cannon bone.**

In the evolution of long legs as an adaptation for speed, the foot and distal segment of the limb elongate much more than the proximal segment (Fig. 10–23). This pattern of elongation minimizes the kinetic energy needed to oscillate the limbs. The kinetic energy needed to oscillate a limb segment can be expressed by the following formula:

$$\text{kinetic energy} = \tfrac{1}{2} \text{ mass of limb segment} \times \text{its velocity}^2$$

The proximal part of a limb moves through a shorter arc than the distal part and has less velocity. Since velocity is squared, it is to a cursor's advantage to have most of the limb mass concentrated proximally. There is less energetic cost in elongating the distal part of the limb because it contains primarily the long tendons of proximal muscles. Physicists refer to this distribution of mass as minimizing the **moment of inertia.**

A long foot also adds a segment of the limb that oscillates independently of the lower leg and upper limb segment (Fig. 10–25). This too increases speed because the total velocity at the distal end of the limb is the sum of the independent velocities of each segment. The situation is analogous to walking on a moving escalator. The total velocity of the person traveling on the escalator is the sum of the velocity of the escalator plus the velocity of the person's walk.

Other modifications of the appendicular skeleton, in addition to long legs, also occur in many cursorial mammals. The clavicle is essentially a strut that braces the shoulder joint and prevents a displacement of the scapula. This is the primitive condition in tetrapods and mammals. During the propulsive phase of the step cycle, when the front foot is on the ground and the trunk is pulled forward relative to the limb, the clavicle causes the shoulder joint to be deflected laterally because it acts as a "spoke" that fixes the distance between the sternum and acromion. Terrestrial mammals that retain a clavicle are species in which the legs have not

FIGURE 10–24   Front views of the right hind feet of representative ungulates. **A** and **B**, Perissodactyls: ▶
A, rhinoceros; **B**, modern horse. **C** and **D**, Artiodactyls: **C**, pig; **D**, cow.

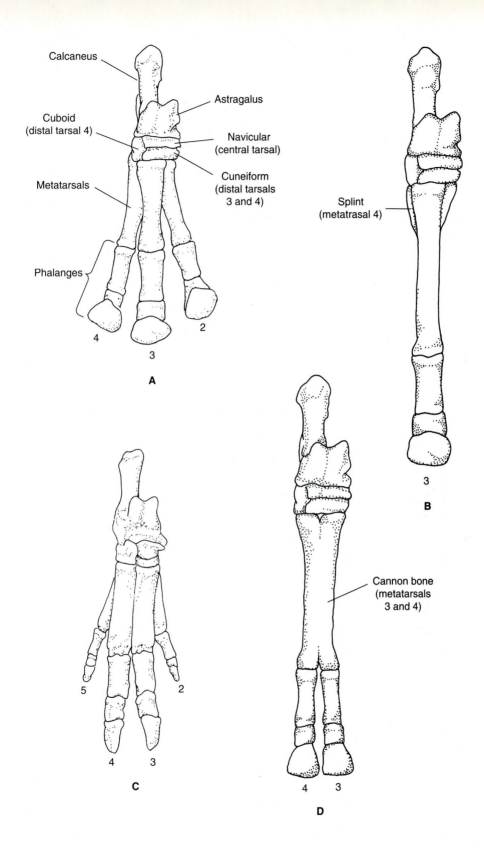

A

Calcaneus

Astragalus

Cuboid
(distal tarsal 4)

Navicular
(central tarsal)

Cuneiform
(distal tarsals
3 and 4)

Metatarsals

Phalanges

4

3

2

B

Splint
(metatrasal 4)

3

C

5

2

4

3

D

Cannon bone
(metatarsals
3 and 4)

4

3

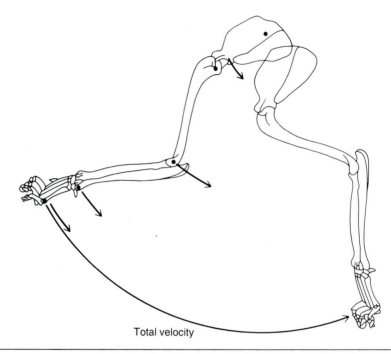

Total velocity

FIGURE 10–25 Rotation of the scapula and pectoral appendage segments during locomotion of the cat. The total velocity of the distal end of the limb is the summation of the independent velocities of the individual rotating limb segments.

rotated completely beneath the body so that the shoulder and elbow joints are not in the same parasagittal plane (Fig. 10–16**A**). The legs are further under the body in cursorial species, and the shoulder and elbow joints lie in the same parasagittal plane. The clavicle is reduced or lost in such species (see Fig. 8–27). The shoulder is not deflected laterally, with a consequent loss of some momentum of the trunk. The loss of the clavicle also permits the scapula to rotate more than in claviculate species. The scapula became, in effect, another limb segment that can add to step length (Fig. 10–25).

The long legs of ungulates are adapted for high speed and strength, and perform no other functions. The limbs move fore and aft in a single plane and, unlike the situation in cursorial carnivores, the primitive capacity of the forearm to rotate around its longitudinal axis is lost. The hand is permanently prone. The radius is the primary weight-supporting bone, and the distal end of the ulna frequently is reduced or even lost. The proximal end of the ulna and the olecranon, which are essential for the hinge joint at the elbow and for the insertion of forearm extensor muscles, are retained and may fuse to the radius (Fig. 10–26**A**). The tibia is the main weight-transferring bone in the shank, and the fibula is reduced or lost in many ungulates except for its proximal end, on which muscles attach, and its distal end (the lateral malleolus), which supports the ankle joint (Fig. 10–26**B**). Loss of unused toes, loss of the capacity to rotate the forearm, and loss of associated

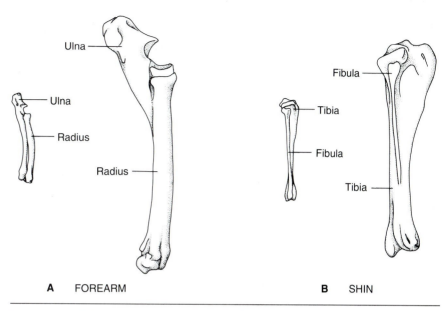

Ulna

Ulna

Radius

Radius

A   FOREARM

Fibula

Tibia

Fibula

Tibia

B   SHIN

FIGURE 10–26   Lateral views of the bones of the forearm and shin of an ancestral horse, *Hyracotherium* (left), and a modern horse, *Equus* (right). (After Simpson.)

muscles decrease the mass of the distal part of the limb and reduce the chance of dislocations.

Jumping

Most quadrupedal terrestrial vertebrates can jump or leap to some extent by rapidly extending their hind legs; a few species have become specialized for a jumping or **saltatorial** mode of locomotion. Among them are frogs and toads, kangaroos, the tarsier (an early, arboreal primate), some rodents, and rabbits. A saltatorial mode of life imposes certain requirements, and all saltators, regardless of the group of tetrapods to which they belong, share many features (Emerson, 1985). This is a good example of convergent evolution. The hind legs, and especially the distal parts, are greatly elongated. The legs are powerful and strongly constructed. The strong thrust of the hind legs during takeoff would twist the body of a jumper if its center of mass is not aligned with the line of action of the propulsive force. Saltators minimize this potential twist by having the center of mass shifted backward toward the sacrum, and by strengthening the vertebral column. This is accomplished by some combination of the following adaptations: shortened trunks, large zygapophyses that firmly unite the vertebrae, a fusion of some cervical vertebrae that reduces head bobbing, reduction in forelimb size, and sometimes by heavier tails. When not jumping, saltators walk, and modest-sized front legs remain. Frogs and toads do not have a tail, but the tail is long in most mammalian saltators. Kangaroo rats, which also have a tuft of long hair on the end of their tail, use the tail for balance and for guiding the leap. Tendons in the large tails and legs of kangaroos store energy when the animal lands on its feet and tail, and release the energy on the next leap. We will examine the specializations of frogs and toads in more detail as an example of saltatorial locomotion.

Frogs and toads not only use thrusts of their powerful hind legs for jumping and hopping, but also for swimming. When a frog is at rest on the ground, its limbs are flexed beneath it, and the elongated pelvic girdle bends downward at its joint with the vertebral column (sacroiliac joint). This gives the animal a somewhat humped-backed appearance (Fig. 10–27**A**). During a leap, both hind legs are extended simultaneously. At the same time, the pelvic girdle rotates upward at the sacroiliac joint, providing an extra increment of thrust and bringing the girdle and extending legs in line with the trunk. The animal lands on its front legs. When the animal lands, the pelvic girdle rotates downward, bringing the hind legs under the body. Many specializations accompany this mode of life. The hind legs are very long and strong. The tibia and fibula are fused (Fig. 10–27**B**), and the elongation of the two proximal tarsals and toes greatly lengthens the foot. The foot forms another limb segment that is extended during a leap. Several caudal vertebrae have united to form a long urostyle from which muscles extend to the elongated ilia of the pelvic girdle. This is the mechanism for rotating the pelvic girdle. The vertebral axis is short, the zygapophyses large, and little rotation occurs around the longitudinal axis of the vertebral column. The front legs act as shock absorbers when the animal lands, and the radius and ulna are fused. The adaptive advantages of saltation are discussed in Focus 10–3.

Parachuting, Gliding, and Flight

A number of lightweight, arboreal vertebrates, including tree frogs and some lizards, can spread their limbs, flatten their bodies, and **parachute** safely to the ground. Parachuting breaks an inadvertent fall and sometimes helps an animal escape a predator. Several lizards and mammals have evolved broad membranes that enable a falling individual to **glide** and travel a greater horizontal distance. In addition to the benefits of parachuting, glides enable a vertebrate to extend its foraging range with little extra expenditure of energy because the animal can move from tree to tree without the need to descend to the ground and climb up again. The gliding lizard, *Draco volans*, has a pair of lateral skin folds along its trunk that can be extended by six pairs of elongated ribs (Fig. 10–28**A**). Several marsupial mammals (phalangers), the colugo of the Indo-Pacific region, and several rodents (e.g., the "flying" squirrel) glide with lateral membranes that attach to the limbs and are extended when the animal spreads its limbs (Fig. 10–28**B**).

In true **flight** an animal uses wings to sustain itself in the air. As we have seen (p. 285), wings evolved from pectoral appendages independently in pterosaurs, birds, and bats. Flight requires light weight and the expenditure of a great deal of energy. Modifications in the digestive, respiratory, and other organ systems make

FIGURE 10–27   Adaptations of a frog for jumping. **A**, Movements of the pelvis at the sacroiliac joint during a leap. **B**, A dorsal view of the skeleton. (After Walker, 1981.)   ▶

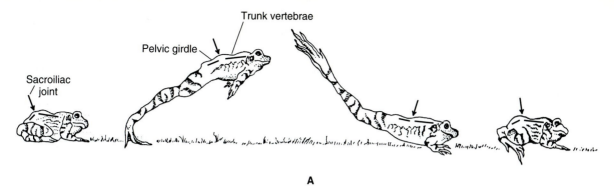

Trunk vertebrae

Pelvic girdle

Sacroiliac joint

A

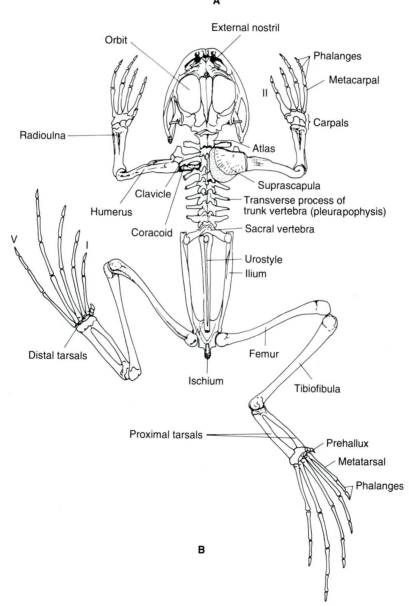

Orbit

External nostril

Phalanges

Metacarpal

II

Radioulna

Carpals

Atlas

Clavicle

Suprascapula

Humerus

Transverse process of trunk vertebra (pleurapophysis)

Coracoid

Sacral vertebra

Urostyle

Ilium

V

I

Femur

Distal tarsals

Ischium

Tibiofibula

Proximal tarsals

Prehallux

Metatarsal

Phalanges

B

possible the production of the needed energy. Birds and mammals are endothermic. We know little about the metabolism of pterosaurs, but indications of a furlike covering have been found in one fossil. The basic principles of flight are the same in pterosaurs, birds, and bats, but, since the wings are constructed slightly differently, it is not surprising that certain aspects of flight also differ. We will examine the principles of flight and the flight of birds as an example of flying vertebrates.

Lift and Drag

Birds glide, soar, and engage in flapping flight, but the principles that keep a bird aloft are the same. The wing as seen in cross section is streamlined and shaped like an airfoil (Fig. 10–29**A**). Since its anterior margin is thicker than its thin trailing margin, the air flows smoothly across it with a minimum of turbulence. Often the wing is cambered, for the underside, especially proximally, is slightly concave. Because of the wing's shape, its dorsal surface is slightly longer than the ventral surface. The airstream necessarily moves faster across the dorsal surface of the wing than the ventral surface, and this increased speed reduces the pressure on the dorsal surface relative to the ventral surface. This is in accordance with Bernoulli's principle. In the 17th century, Bernoulli demonstrated that pressure is least in a fluid stream (water or air) where flow is the fastest. The differential in air pressure generates a lift force that acts perpendicularly to the air stream from a center of pressure (X) on the wing. A drag force resulting from friction of the air across the bird and other factors acts parallel to the air stream and tends to hold the bird back. During level flight at a constant speed, the lift force must equal the weight of the bird and a propulsive force must overcome the drag.

The formula for lift is $L = \frac{1}{2}pV^2SC_1$ where p is the density of the air, V is the air speed, S is the area of the wing, and C_1 is the coefficient of lift. We can ignore p because it is essentially a constant, but the other variables can change and they have a large effect on lift. Air speed is particularly important because this factor is squared. Birds that soar slowly, such as vultures, need wings with a large surface area to compensate for the low air speed; whereas rapidly flying birds, such as swifts, can and do have shorter and narrower wings. The coefficient of lift depends on several variables, of which the most important for us is the angle of attack of the wing (that is, the extent to which the leading edge of the wing is elevated above the horizontal). Lift increases as the angle of attack increases, but a trade-off is an increased air turbulence above the wing that increases drag (Fig. 10–29**B**). The turbulence becomes so great as the angle of attack approaches 15° that lift is suddenly lost and the bird stalls. Turbulence can be reduced and lift maintained by creating slots in the wing through which air must flow rapidly (Fig. 10–29**C**). A familiar analogy of the effect of slots is the rapid flow of water between boulders that make "slots" in a river. The river may be flowing slowly, but speed increases if it has to flow between boulders because the volume of water is the same (volume = cross-sectional area of the river × velocity). If cross-sectional area decreases, velocity must increase. The alula (p. 287) at the front of a bird's wing can be elevated to make one slot. This is particularly helpful during takeoff and landing, when air speed is low. Additional slots can be formed by separating

---

FOCUS 10–3   Metabolic Costs of Amphibian Saltation

Walton and Anderson (1988) and Anderson, Feder, and Full (1991) have studied the energetic cost of locomotion in Fowler's toad (*Bufo woodhousii fowleri*) on a treadmill. The gait of these toads combines walking and hopping, with hops becoming more frequent as speed increases. In most terrestrial vertebrates, the energetic cost of locomotion (as measured by oxygen consumption) increases linearly with speed up to a maximum speed that can be sustained by aerobic metabolism. Speed can increase further, but this increase cannot be sustained. Toads are an exception. Oxygen consumption reaches a maximum at a speed of 0.27 km per hour, but toads can sustain speeds up to 0.45 km per hour without additional oxygen consumption. Since the anatomy of toads makes it unlikely that they can store much elastic energy in tendons, they must either have the ability to sustain anaerobic metabolism for a prolonged time, or their switching from walking to hopping at higher speeds conserves energy. There is no evidence that they switch to anaerobic metabolism at higher speeds, because the amount of phosphocreatine (an energy reserve) in their muscles does not fall and lactic acid (a metabolic byproduct that accumulates in the absence of sufficient oxygen) does not accumulate. As would be expected, the metabolic cost of a single hop is more than for a single walking stride. However, a toad covers a far greater distance in a single hop than in a single walking stride, and this more than compensates for the greater cost of a hop. The metabolic cost to cover a given distance is 1.9 times as much for walking as hopping. By shifting to hopping at higher speeds, toads can go faster with no more energy consumption. This situation is analogous to many mammals changing gaits from a walk to a trot to a gallop as speed increases. Hopping for toads, and saltatorial locomotion in general, also has other advantages: the rapid acceleration from rest, and the ability to cover a great distance in one thrust of the hind legs. Both of these attributes of a leap are important in escaping predators, as anyone who has tried to catch a toad or frog can attest.

---

feathers along the trailing edge of the wing and wing tips, where turbulence is particularly pronounced.

Drag has two components. **Profile drag,** which resembles the frictional drag of fishes, results primarily from the friction of the airflow across the surface of the bird and wings, and from turbulence above the wing. Friction depends on body shape and streamlining, factors that an individual bird cannot control, but turbulence can be reduced by factors we have discussed. **Induced drag** results from the flow of air from the underside of the wing into the reduced pressure area above the wings. This occurs particularly along the posterior border of the wing and at the wing tips, producing a **tip vortex.** Tip vortices are especially strong in rapidly flying birds because air speed and the pressure difference between the two surfaces of the wing are so great. Tip vortices are less in birds with narrow and pointed wings than in ones with broad and blunt ones because there is less surface area near the wing tip for the air to flow around. Induced drag resembles the pressure drag in fishes in some ways, but the reduced pressure in birds is above the wing, not behind the animal as in fishes.

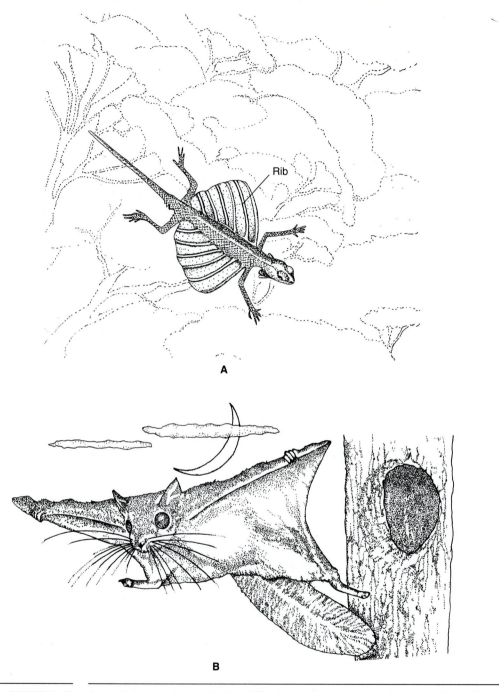

Rib

A

B

FIGURE 10–28 Some gliding vertebrates. **A**, The gliding lizard, *Draco volans*. **B**, A flying squirrel, *Glauco-mys volans*.

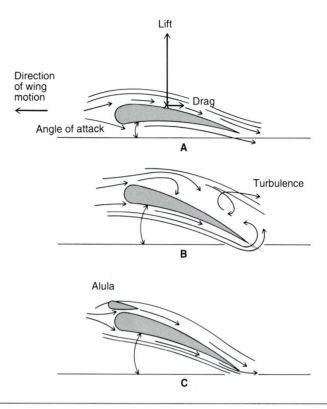

FIGURE 10–29   **A**, Forces generated by the flow of air across a wing. **B**, An increased angle of attack causes turbulence. **C**, Turbulence is reduced by a slot formed by elevating the alula.

Flapping Flight

In forward flapping flight, the propulsive force that overcomes drag derives to a large extent from the downward and forward movement of the wings that occurs during the downstroke (Fig. 10–30**A**). The distal part of the wing, which is formed by the feathers attached to the hand, rotates so its leading edge is directed slightly downward. As this part of the wing moves through the air, it generates a local lift force that is perpendicular to the local air flow. Since the wing is inclined, the local lift force is also inclined forward. The local lift force can be resolved into a vertical component that helps keep the bird aloft, and a forward propulsive component. This part of the wing acts much like a propeller in generating a forward thrust. Since the distal part of the wing is at a maximum distance from the wing's fulcrum at the shoulder, it moves downward more than the proximal part of the wing. Its velocity accordingly is greater. This increases the air speed and the forward thrust. In some species the flight feathers on the distal part of the wing, which are shaped like airfoils (p. 158), separate and act as individual propellers. The upstroke is a simple recovery stroke and generates no useful forces.

As the flow of air leaves the trailing edge of the flapping wings, it is shed as a series of vortex rings (Fig. 10–30**B**). These move downward and backward in the wake of the bird. Some investigators believe that the upward and forward

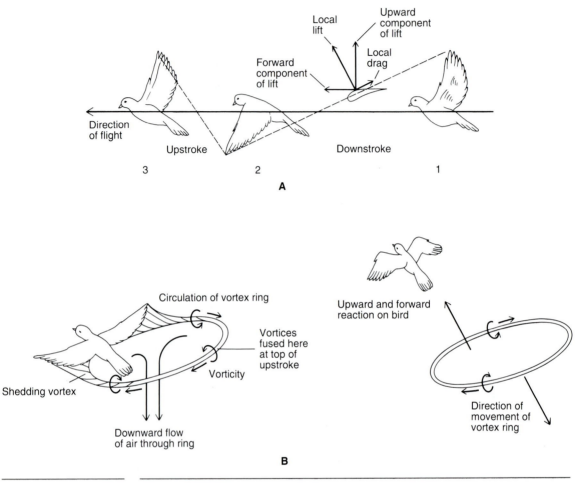

FIGURE 10–30   **A**, Flapping flight in a bird. **B**, The contribution to flight of vortex rings of air shed from the wings. (After Rayner.)

reaction of the air to these rings, in accordance with Newton's third law of motion (for every action there is an opposite and equal reaction), helps keep the bird aloft and push it forward.

Jenkins, Dial, and Goslow (1988) have studied the movements of the skeletal elements during flapping flight by making cineradiographs at 200 frames per second of starlings (*Sturnus vulgaris*) flying in a wind tunnel. During the upstroke (Fig. 10–31**D**), the wing is folded. The humerus lies close to the horizontal plane and is retracted close to the trunk. Forearm and hand bones are flexed at the elbow and wrist. As the upstroke ends and the downstroke begins (Fig. 10–31**A** and **B**), the humerus is protracted and elevated 80° to 90° above the horizontal. Forearm and hand are maximally extended. The humerus is lowered to about 20° below the horizontal near the end of the downstroke (Fig. 10–31**C**); however, the distal part of the wing continues downward because of a slight rotation of the

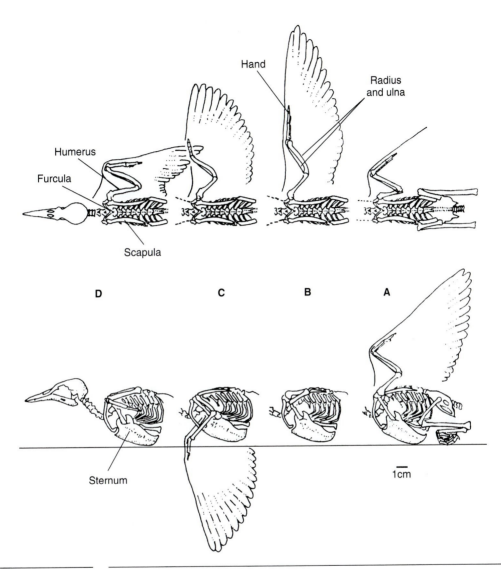

Hand

Radius
and ulna

Humerus

Furcula

Scapula

D   C   B   A

Sternum   1cm

FIGURE 10–31   Limb and girdle movements in a wingbeat cycle of a starling as seen in dorsal (top row) and lateral views (bottom row). **A**, Upstroke-downstroke transition. **B**, Mid-downstroke. **C**, End of downstroke. **D**, Mid-upstroke. (From Jenkins, Dial, and Goslow.)

humerus. The dorsal ends of the fused clavicles (the wishbone or furcula) bend laterally during the downstroke and recoil on the upstroke (dotted lines in Fig. 10–31, top row). The sternum also moves slightly, ascending and moving posteriorly during the downstroke, and descending and advancing during the upstroke (Fig. 10–31, bottom row). Movements of the furcula and sternum may affect the bellow-like air sacs that extend from the lungs and help ventilate the lungs (p. 658).

The radius and ulna lie parallel to each other and are articulated to the

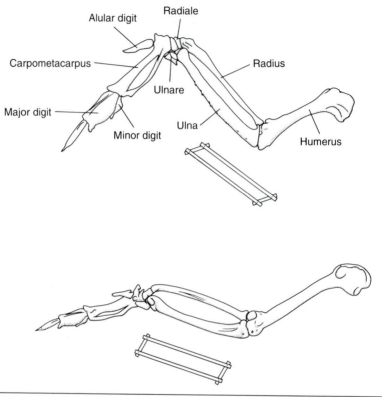

Radiale

Alular digit

Carpometacarpus

Radius

Ulnare

Major digit

Minor digit

Ulna

Humerus

FIGURE 10–32   The radius and ulna of a bird's wing are articulated to adjacent elements in a manner that resembles and acts like a set of parallel rulers.

humerus and carpometacarpus in such a way that they act as a parallel ruler mechanism (Fig. 10–32). Flexion or extension at the elbow automatically causes a comparable flexion or extension at the wrist. These actions are assisted by proximal hand extensor and flexor muscles whose long tendons extend to the hand; however, the parallel ruler mechanism allows some reduction of distal muscles and the lightening of the distal part of the wing.

Many muscles, of course, are involved in the downstroke and upstroke of the wings (Focus 10–4). The two major ones, the pectoralis and supracoracoideus, arise chiefly from the broad, keeled sternum (Fig. 10–33). The pectoralis, a downstroke muscle, inserts on the ventral surface of the humerus, as one would expect. Perhaps unexpectedly, the supracoracoideus, an upstroke muscle, also arises ventrally on the sternum, but its insertion tendon passes through a **foramen triosseum** near the shoulder joint to attach on the dorsal surface of the humerus. The foramen triosseum, as its name implies, is located where the scapula, coracoid, and furcula come together. The supracoracoideus is a ventral appendicular muscle, but the shift of its tendon of insertion allows it to function as a dorsal muscle. The use of a ventral muscle for the downstroke helps maintain the low center of gravity needed for stability.

FOCUS 10–4   ## Muscle Activity During a Wing Cycle

As an illustration of functional anatomy in Chapter 1, we discussed briefly an analysis of electromyographic activity of bird flight muscles made by Dial, Goslow, and Jenkins (1991). We return to this study now in more detail. The major muscles involved are shown in Figure 1–1 on p. 3. Two parts are recognized to the pectoralis: sternobrachialis (SB) and thoracobrachialis (TB). By referring to Figure 1–2**B** on p. 5, you will see a diagram of muscle activity during a complete wing cycle, which lasts on average 72 milliseconds. The average duration of electromyographic activity of the muscles is shown by the portion of the muscle arcs that is filled in. Upstroke muscles are stippled, downstroke ones are hatched, and transitional ones are unmarked. Notice that downstroke muscle activity begins in the late upstroke in preparation for the ensuing downstroke. The shoulder joint moves outward, spreading the furcula (sternocoracoideus); the distal wing is extended (humerotriceps); the humerus is protracted (pectoralis and biceps);

and the upward movement of the wing is slowed (pectoralis). Many muscles contribute to the downstroke. The scapulohumeralis caudalis and scapulotriceps become active during the downstroke-upstroke transition. The scapulohumeralis caudalis rotates and retracts the humerus, and the scapulotriceps appears to stabilize the elbow at the end of the downstroke. Upstroke muscles become active late in the downstroke, decelerating the wing and preparing for the upstroke. The pectoralis (on the upstroke) and the supracoracoideus (on the downstroke) are actively stretched before they begin to shorten. These stretch-shorten cycles may first store elastic energy that is later released.

Dial (1992) has extended this research to different modes of flight in pigeons (takeoff, ascending, and level flight). Gatesy and Dial (1993) also have studied tail muscle activity, and Dial and Biewener (1993) have calculated the power output of the pectoralis muscle during different modes of flight.

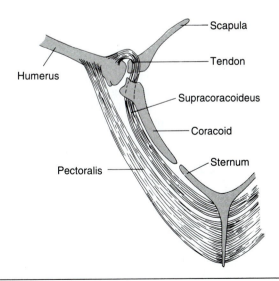

FIGURE 10–33   A cross section through the shoulder region and sternum of a bird showing the arrangement of two major muscles, the pectoralis and supracoracoideus, that move the wing down and up, respectively. (After Storer.)

Soaring

Many birds can soar (that is, maintain their position in the air or even rise) by holding their wings extended and not flapping them. Land birds, such as hawks and vultures, are **static soarers** that use rising air currents. Static soarers glide downward, but the air around them is rising faster than they are falling so they stay aloft and may increase their elevation. Rising air currents are formed when the wind is deflected upward by a cliff, hill, or other obstacle. Solar heating of the ground also forms thermals—large hills of air in which convection currents carry the air upward in the center. As the air rises and cools, it flows downward on the periphery of the thermal. Thermals often separate from the ground and expand as they continue to rise upward, sometimes for thousands of feet, into an area of lower pressure. Static soarers fly slowly and must constantly maneuver and turn to stay in the center of the thermal. Since air speed is low, they must have large, broad wings to generate adequate lift (Fig. 10–34**A**). Wing loading, which is the amount of weight supported by each unit of wing surface, is low. Separating the feathers at the tips and trailing margin of the wings lowers induced drag and lift-reducing turbulence.

Although sea birds engage in static soaring when air is deflected upward by a ship, coastline, or even by a line of waves, they also use **dynamic soaring.** As the wind blows steadily across the surface of a large body of water, friction between the air and water causes a gradient in air speed. Speed is lowest at the surface and increases up to about 15 m, where the speed becomes constant. Dynamic soarers use this differential. They glide steeply downward with the wind, bank and turn into the wind, and use the momentum gained from their fall to start to climb. As they ascend, air speed and lift increase and this carries them upward. (Remember that lift is proportional to the square of the air speed.) Potential energy is converted to kinetic energy as the birds glide downward, and kinetic energy is converted back to potential energy as they soar upward. Dynamic soarers have long, narrow, and pointed wings (Fig. 10–34**B**). Wing loading is higher than for static soarers, but increased air speed compensates for the reduced lift generated by the relatively smaller wing area. The wing shape reduces tip vortices and induced drag. Gliding and soaring conserves energy relative to flapping flight, but some muscle effort is required to hold the wings in a protracted and extended position against the force of air currents, the wind, and the downward pull of gravity on the bird (Meyers, 1993).

SUMMARY

1. The vertebral column of fishes need not resist large, vertical bending forces because fishes receive a great deal of support from the water. However, their vertebral column must allow lateral bending and also act as a compression strut that resists shortening of the body when longitudinal muscle fibers in their body wall contact.

2. The teardrop shape of fishes reduces pressure drag; and the surface area of the fish, which causes frictional drag, is not excessive relative to its mass.

3. Most fishes propel themselves by waves of lateral undulations that thrust against the water. Myomere structure is well suited to generate the undulatory

A

B

FIGURE 10–34 Soaring in birds. **A**, A vulture, a static soarer, has broad wings highly slotted at the wing tips. **B**, A sea gull, a dynamic soarer, has long, narrow wings pointed at the tips.

waves. Slowly contracting, oxidative (red) fibers in the myomeres are used during normal cruising; fast-contracting, glycolytic fibers (white) are used for short bursts of speed.

4. Although fishes are supported to a large extent by the water they displace, their flesh is denser than water. Other factors are also needed to keep them afloat. The skeleton of cartilage that characterizes cartilaginous fishes, and lipids stored in their livers, reduce their density somewhat. The ventrally flattened head of sharks, the wide pectoral fins set low on the body, and the action of the heterocercal tail can generate additional lift forces. The swim bladder of teleosts enables them to attain neutral buoyancy. Their caudal fin has become symmetrical externally, or homocercal.

5. The median and paired fins of a fish provide stability against roll, pitch, and yaw.

6. The trunk skeleton of terrestrial vertebrates must prevent body weight from collapsing internal organs when the animals rest on the ground. When they walk, the vertebral column resists vertical bending and transfers body weight to the girdles and appendages. Individual vertebrae are strong and firmly linked together.

7. Biomechanically the vertebral column resembles an archer's or violin bow. The vertebral column cannot collapse as long as the bowstring (chiefly the sternum and ventral abdominal muscles) is intact.

8. Weight is transferred from the trunk skeleton to the pelvic girdle by way of one or more sacral vertebrae and their ribs, and to the pectoral girdle by a muscular sling that attaches to the ribs and scapula.

9. Limbs transfer weight from the girdles to the ground. Since the limbs are jointed struts, muscle action is needed to stabilize the joints and prevent the limbs from collapsing, even when the animal is standing still. The amount of muscle action needed can be reduced by a vertical orientation of the limbs, or by passive stay mechanisms that lock the joints in a fixed position.

10. Terrestrial vertebrates walk by a succession of steps. One cycling of all the limbs constitutes one stride.

11. The limbs of amphibians and reptiles are splayed so their humerus and femur move back and forth close to the horizontal plane, and their lower limbs move close to the vertical plane. Since a foot remains on the ground during the propulsive phase, the action of retractor and flexor muscles, and the cranklike rotation of the humerus and femur, advances the trunk. Lateral, sinusoidal curvatures of the trunk assist the limbs in locomotion.

12. The limbs of mammals are more nearly under the body, so all segments of the limbs move close to the vertical plane. The limbs support the body more effectively in this position and swing through longer arcs.

13. Homologous muscles are utilized in similar phases of a forelimb cycle in such diverse species as a lizard, opossum, and starling. Propulsive muscles become active near the end of the swing phase, presumably to check the forward movement of the limb. Similarly, swing phase muscles become active near the end of propulsion.

14. Many appendicular muscles contain at least two types of muscle fibers. Slow

oxidative (red) fibers maintain posture and are used in normal movement; fast glycolytic (white) fibers become active only at high speeds.

15. The combination of feet on and off the ground during a stride is known as a gait. Amphibians and reptiles, which carry their bodies close to the ground, utilize very stable gaits in which successive triangles of feet support the body during most of the stride (e.g., a lateral sequence, diagonal couplet walk). Mammals also can walk slowly, but they carry their bodies well off the ground and can utilize less stable gaits with fewer feet on the ground when they move faster (pace, trot, gallop).

16. The velocity at which a tetrapod travels is equal to stride frequency times stride length. Tetrapods increase speed first by oscillating their limbs more rapidly, which increases stride frequency. As speed increases, mammals shift to gaits that have fewer feet on the ground at one time, and this increases stride length. Increasing stride length is less expensive in energy terms.

17. In the course of evolution, the limbs of mammals have become adapted for power or speed by changing the relative length of the lever arms for the in- and out-forces.

18. As limbs become adapted for higher speed, limb length increases and foot posture changes from plantigrade through digitigrade to unguligrade. The distal parts of the limbs elongate more than the proximal parts. This keeps most of the muscle mass of the limb proximally and reduces the amount of energy needed to oscillate the limb.

19. Limbs with long feet have, in effect, another limb segment. This increases velocity because the total velocity of the limb equals the sum of the independent velocities of each segment.

20. The ability to rotate the limb around its longitudinal axis is lost in many cursors. Forelimb bones fuse, and the distal parts of the ulna and tibia (except for the part supporting the ankle joint laterally) are often lost.

21. Jumping or saltatorial tetrapods are characterized by strong, powerful, and long hind legs, by shortened trunks, by firmly articulated vertebrae, and by short front legs. The adaptations of frogs and toads illustrate this pattern of locomotion.

22. Parachuting vertebrates break a fall from a tree by flattening their bodies. Gliding vertebrates can spread broad membranes and travel a greater horizontal distance. Flying vertebrates have evolved wings and can sustain themselves in the air. Birds are an example of flying vertebrates.

23. The differential in air speed and pressure across a wing generates a lift force that acts vertically perpendicular to the air stream. The lift is proportional to the area of the wing, the square of the air speed, and the coefficient of lift. Increasing the angle of attack of the wing increases the coefficient of lift, but also increases lift-reducing turbulence above the wing. Slots, which are formed by raising the alula and separating feathers at the wing tip and along the trailing margin, increase lift and reduce the turbulence. Lift must equal the bird's weight in level flight.

24. Drag results from air friction across the bird's surface, and from air flowing from under the wing into the reduced pressure area above the wing.

25. In flapping flight, a propulsive force that overcomes drag results from the downward and forward movement of the wings, especially the distal parts that act as propellers. Additional lift and forward thrust may come from the reaction on the bird of vortex rings of air that are shed from the wings and travel downward and backward.

26. Many birds also soar. Static soarers soar in rising air currents. Adequate lift at low air speeds derives from broad wings that are highly slotted at the tip. Oceanic birds are dynamic soarers. They glide downward with the wind, turn into the wind, and use the momentum of their glide to start to ascend. The increasing air speeds they encounter as they rise over the ocean provide additional lift. They have long, narrow, and pointed wings.

REFERENCES

Alexander, R. McN., 1977: Swimming. *In* Alexander, R. McN., and Goldspink, G., editors: *Mechanics and Energetics of Animal Locomotion.* New York, John Wiley and Sons.

Alexander, R. McN., 1977: Terrestrial locomotion. *In* Alexander, R. McN., and Goldspink, G., editors: *Mechanics and Energetics of Animal Locomotion.* New York, John Wiley and Sons.

Alexander, R. McN., 1984: Elastic energy stores in running vertebrates. *American Zoologist,* 24:85–94.

Anderson, B. D., Feder, M. E., and Full, R. J., 1991: Consequence of gait change during locomotion in toads *(Bufo woodhousii fowleri). Journal of Experimental Biology,* 158: 133–148.

Brinkman, D., 1981: The hind limb step cycle of *Iguana* and primitive reptiles. *Journal of Zoology (London),* 181:91–103.

Dial, K. P., 1992: Activity patterns of the wing muscles of the pigeon *(Columba livia)* during different modes of flight. *Journal of Experimental Zoology,* 262:357–373.

Dial, K. P., and Biewener, A. A., 1993: Pectoralis muscle force and power output during different modes of flight in pigeons *(Columbia livia). Journal of Experimental Biology,* 176:31–54.

Dial, K. P., Goslow, G. E., Jr., and Jenkins, F. A., Jr., 1991: The functional anatomy of the shoulder in the European starling *(Sturnus vulgaris). Journal of Morphology,* 207:227–244.

Elder, H. Y., and Trueman, E. R., editors, 1980: *Aspects of Animal Movement.* Society for Experimental Biology, Seminar Series, no. 5. Cambridge, Cambridge University Press.

Emerson, S. E., 1985: Jumping and Leaping. *In* Hildebrand, M., Bramble, D. M., Liem, K. F., and Wake, D. B., editors: *Functional Vertebrate Morphology.* Cambridge, Harvard University Press.

Gambaryan, P. P., 1974: *How Mammals Run.* Translated by H. Hardin. New York, John Wiley and Sons.

Gatesy, S, M., and Dial, K. P., 1993: Tail muscle activity patterns of walking and flying pigeons *(Columba livia). Journal of Experimental Biology,* 176:55–76.

Getty, R., 1975: *Sisson and Grossman's The Anatomy of Domestic Animals,* 5th edition. Philadelphia, W. B. Saunders Company.

Goldspink, G., 1977: Design of muscles in relation to locomotion. *In* Alexander, R. McN., and Goldspink, G., editors: *Mechanics and Energetics of Animal Locomotion.* New York, John Wiley and Sons.

Gray, J., 1968: *Animal Locomotion.* London, Weidenfeld and Nicolson.

Heglund, N. C., and Taylor, C. R., 1988: Speed, stride frequency and energy cost per stride: How do they change with body size and gait? *Journal of Experimental Biology,* 138:301–318.

Hildebrand, M., 1959: Motions of the running cheetah and horse. *Journal of Mammalogy,* 40:481–495.

Hildebrand, M., 1980: The adaptive significance of tetrapod gait selection. *American Zoologist,* 20:255–267.

Hildebrand, M., 1985: Walking and running. *In* Hildebrand, M., Bramble, D. M., Liem, K. F., and Wake, D. B., editors: *Functional Vertebrate Morphology.* Cambridge, Harvard University Press.

Jayne, B. C., Bennett, A. F., and Lauder, G. V., 1990: Muscle recruitment during terrestrial locomotion: How speed and temperature affect fiber type use in the lizard. *Journal of Experimental Biology,* 152:101–128.

Jenkins, F. A., Jr., 1971: Limb posture and locomotion in the Virginia opossum (*Didelphis virginiana*) and in other non-cursorial mammals. *Journal of Zoology* (London), 165:303–315.

Jenkins, F. A., Jr., Dial, K. P., and Goslow, G. E., Jr., 1988: A cineradiographic analysis of bird flight: The wishbone in starlings is a spring. *Science,* 241:1495–1498.

Jenkins, F. A., Jr., and Goslow, G. E., Jr., 1983: The functional anatomy of the shoulder of the savannah monitor lizard (*Varanus exanthematicus*). *Journal of Morphology,* 175:195–216.

Jenkins, F. A., Jr., and Weijs, W. A., 1979: The functional anatomy of the shoulder of the Virginia opossum (*Didelphis marsupialis*). *Journal of Zoology* (London), 188:379–410.

Macdonald, D., editor, 1984: *The Encyclopedia of Mammals.* New York, Facts on File Publications.

McMahon, T. A., 1984: *Muscles, Reflexes, and Locomotion.* Princeton, Princeton University Press.

Meyers, R. A., 1993: Gliding flight in the American kestrel (*Falco sparverius*): An electromyographic study. *Journal of Morphology,* 215:213–224.

Norberg, U. M., 1985: Flying, gliding, and soaring. *In* Hildebrand, M., Bramble, D. M., Liem, K. F., and Wake, D. B., editors: *Functional Vertebrate Morphology.* Cambridge, Harvard University Press.

Nursall, J. R., 1956: The lateral musculature and the swimming of fish. *Proceedings of the Zoological Society of London,* 126:127–143.

Rayner, J. M. V., 1980: Vorticity and animal flight. *In* Elder and Trueman, *q.v.*

Simpson, G. G., 1961: *Horses.* New York, Doubleday & Company and the American Museum of Natural History.

Slijper, E. J., 1946: *Comparative Biologic-Anatomical Investigations on the Vertebral Column and Spinal Musculature of Mammals.* Koninklijke Nederlandsche Akademie van Wetenschappen, afd. Natuurkunde, Tweede Sectie, Deel XLII, no. 5. Amsterdam, N. V. Noord-Hollandsche Uitgevers Maatschappij.

Smith, J. M., and Savage, R. J. G., 1954: Some locomotory adaptations of mammals. *Zoological Journal,* Linnean Society of London, 42:603–622.

Snyder, R. C., 1954: The anatomy and function of the pelvic girdle and hindlimb in lizard locomotion. *American Journal of Anatomy,* 95:1–46.

Storer, T. I., 1943: *General Zoology.* New York, McGraw–Hill Book Company.

Sukhanov, V. B., 1974: *General System of Symmetrical Locomotion of Terrestrial Vertebrates and Some Features of Movement of Lower Tetrapods.* Translated from Russian. Published for the Smithsonian Institution and National Science Foundation. New Delhi, Amerind Publishing Company Pvt. Ltd.

Taylor, C. R., 1978: Why change gaits? *American Zoologist,* 18:153–161.

Thomson, K. S., 1990: The shape of a shark's tail. *American Scientist,* 78:499–501.

Thomson, K. S., and Simanek, D. E., 1977: Body form and locomotion in sharks. *American Zoologist,* 17:343–354.

Walker, W. F., Jr., 1971: A structural and functional analysis of walking in the turtle *Chrysemys picta marginata. Journal of Morphology,* 134:195–214.

Walker, W. F., Jr., 1981: *Dissection of the Frog,* 2nd edition. San Francisco, W. H. Freeman and Company.

Walton, M., and Anderson, B. D., 1988: The aerobic cost of saltatory locomotion in Fowler's toad (*Bufo woodhousii fowleri*). *Journal of Experimental Biology,* 136:273–288.

Webb, P. W., 1984: Form and function in fish swimming. *Scientific American,* 251:no.1: 72–82.

Webb, P. W., and Blake, R. W., 1985: Swimming. *In* Hildebrand, M., Bramble, D. M., Liem, K. F., and Wake, D. B., editors: *Functional Vertebrate Morphology.* Cambridge, Harvard University Press.

11 | The Sense Organs

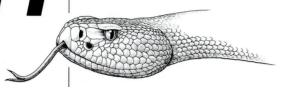

PRECIS

Vertebrates receive information about changes within their bodies and in their outside world through free nerve endings and a wide variety of receptor cells. The major receptor mechanisms of vertebrates and the ways they changed during evolution will be examined in this chapter.

Outline

To avoid being eaten, to find shelter, food, and mates—in short, to survive in a changing world—vertebrates must be able to detect changes in their external and internal environments and make the appropriate responses. The ability to detect and respond to changes is a basic property of life, and to a degree all living cells have it, but as animals became more active and complex, certain cells, known as **receptors,** become specialized to monitor the environment for the benefit of the entire organism. Cells in the nervous and endocrine systems are stimulated, sort out this information, and send nerve impulses, or secrete chemical messengers known as hormones, that activate the appropriate muscles, glands, and other effectors. Nervous and hormonal integration overlap in many ways, but nervous integration deals with more rapid and discrete types of activity than hormonal integration. In this and the next two chapters, we will examine the receptors and the nervous system; endocrine glands will be introduced later.

Receptors

The receptive region of a nerve cell or neuron may be stimulated directly by environmental changes within or outside the body and generate a nerve impulse. This is a common method of reception among invertebrates. Vertebrates retain free nerve endings of this type, but usually there are distinct receptor cells and neurons. **Receptor cells** act as transducers, instruments that convert one form of energy into another. They are specialized to detect a minute energy change in a specific environmental signal, such as light, pressure, sound, or taste, and then initiate a nerve impulse in a sensory neuron. Receptor cells have an electric **resting potential** derived from an unequal distribution of ions across their plasma membrane. The resting potential is maintained in **phasic receptors** until they are stimulated. Upon stimulation, ion channels in the plasma membrane open, and ionic depolarization of the membrane occurs, and the cell develops a **receptor potential.** This response is graded; that is, the receptor potential is proportional to the magnitude of the stimulus. A few receptors are **tonic** and are active all the time. Stimulation increases or decreases the receptor potential. If the receptor potential attains a certain magnitude, it initiates a nerve impulse in the sensory neuron. Nerve impulses do not vary in magnitude with the magnitude of the receptor potential. A nerve impulse is said to be all-or-none; it occurs or it does not occur. The magnitude of sensory stimulation above the threshold level needed to initiate the nerve impulse is encoded by the frequency of nerve impulses and sometimes by their pattern. The decoding and perception of particular sensations is a function not of the nerve impulse but of the specificity of the receptors and the connections of the neurons within the nervous system.

Animals have evolved receptor mechanisms to detect environmental changes that are important for their survival; many physiochemical changes that are not critical go undetected. Human beings, for example, cannot sense small changes in electrical fields, but many fishes can. It is convenient to classify the numerous receptors according to the type of sensory signal to which they are attuned (Table 11–1). It sometimes also is useful to group them, like bones and muscles, according to their locations within the body. Somatic receptors, sometimes called **extero-**

TABLE 11–1 **Major Vertebrate Receptors**

Chemical receptors
    Olfactory cells (smell)
    Taste buds (taste)
Mechanical receptors
  Cutaneous receptors
    Free nerve endings (pain and other modalities)
    Meissner's corpuscles (touch and pressure)
    Merkel's disks (touch and pressure)
    Pacinian corpuscles (touch and pressure)
    Ruffini endings
  Proprioceptors
    Tendon and joint receptors (tension)
    Muscle spindles (degree and rate of muscle contraction)
  Lateral line and ear receptors
    Hair cells (vibrations, gravity, electric fields for certain modified hair cells)
Electromagnetic spectrum and thermal receptors
    Rods and cones (visible light)
    Pit organs of some snakes (infrared)

---

receptors, are located in the outer tube of the body; visceral receptors (**enteroreceptors**) lie deeper within the body.

Many free nerve endings and individual receptor cells are scattered through the skin and tissues of the body. Others are aggregated and combined with cells that support and protect them, amplify the environmental stimulus, and help localize the source of the stimulus. We call such aggregations **sense organs.** We can examine only some of the numerous receptors and sense organs of vertebrates.

Chemoreceptors

All animals detect many chemical changes in their external and internal environments. Vertebrates receive chemosensory signals in four ways. (1) Neuron endings in serous and mucous membranes of the eyes, mouth, and nose detect noxious chemical stimuli. (2) Internal chemoreceptors continuously monitor certain aspects of the internal environment. The **carotid** and **aortic bodies,** located on major arteries near the heart, detect changes in the oxygen and carbon dioxide content of the blood. The pH of cerebrospinal fluid, which decreases (becomes acid) when carbon dioxide levels in the blood rise, is monitored by hydrogen ion receptors in the fourth ventricle of the brain. The rates of breathing and blood circulation are reflexly adjusted by information provided by all these receptors. Finally, vertebrates have special (3) smell and (4) taste receptors that detect odors and food in the external environment. We have **olfactory receptors** (L. *olfactus* = smell, from *olfacere* = to smell) in our noses and **gustatory receptors** (L. *gustatus* = tasted, from *gustare* = to taste) on our tongues. Although taste and smell have much in common, we think of smell as chemical information carried in the air and taste as chemical information primarily from food in contact with parts of the mouth. This distinction becomes blurred in a fish, whose body surface, nose, and mouth are all bathed in water. In general, olfactory receptors can detect a lower concentration of a substance than gustatory ones, so they can detect faint traces of distant substances.

Olfactory Receptors

Olfactory receptors are specialized neurons that detect chemical substances called **odorants.** The olfactory neurons are essentially similar in all vertebrates, and they develop embryonically from paired ectodermal thickenings, or **placodes,** that invaginate to form a pair of **nasal sacs.** The cell body of the neuron, which contains the nucleus, lies in the olfactory epithelium and sends a receptive process, the dendrite, to the surface (Fig. 11–1). Nonmotile cilia on the end of the dendrite, which range in number from about 5 to 20, increase the receptive surface several hundredfold. A long process, or axon, extends from the cell body to the main olfactory bulb of the brain. Small groups of these neurons constitute the olfactory nerve (cranial nerve I). Olfactory impulses have a particularly short route to the brain.

Odorants in solution in a mucous sheet flowing across the olfactory epithelium bind with receptor molecules in the plasma membrane of the olfactory cilia and initiate receptor potentials. Olfactory neurons are extraordinarily sensitive. Dogs can detect a substance known as diacetal in concentrations as low as 1.7×10^{-18} molar (Meisami, 1991). Vertebrates can detect a wide range of odors, possibly as

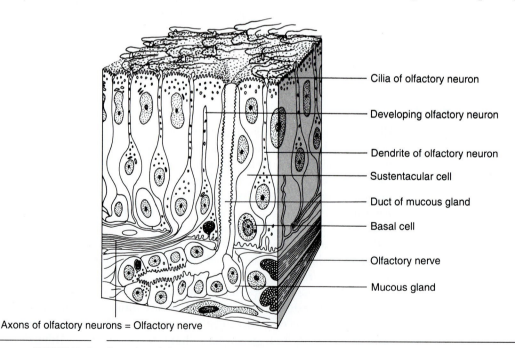

Cilia of olfactory neuron

Developing olfactory neuron

Dendrite of olfactory neuron

Sustentacular cell

Duct of mucous gland

Basal cell

Olfactory nerve

Mucous gland

Axons of olfactory neurons = Olfactory nerve

FIGURE 11–1   A portion of the human olfactory epithelium. (After Williams et al.)

many as 10,000 in the case of humans, but it is not entirely clear how the distinction is made. Odorant molecules have distinctive molecular configurations and bind with and activate receptor molecules in the plasma membrane of the cell. Buck and Axel (1991) have demonstrated that the complex genetic control of the synthesis of receptor proteins makes possible the production of a great variety of receptor proteins. These proteins can be grouped into families that probably recognize major differences among odorants. One family, for example, may recognize aromatic compounds such as benzene and its derivatives, another aliphatic compounds, and so on. Subtle variations within the benzene family of receptor proteins enable distinction of different benzene derivatives: toluene, xylene, or phenol. But it is still unlikely that there is a different receptor protein for each possible odorant. Each olfactory cell can bind with one or a small group of structurally related odorants. Perception of different odors probably depends on processing in the brain of the combination of information from stimulated olfactory neurons, the intensity of stimulation, and the synaptic pattern that their projections make. An analogy is color vision, in which the processing of information from only three types of photoreceptors (red, green, and blue) enables us to distinguish among several hundred hues.

Olfactory cells show considerable sensory adaptation. They are very sensitive to a new odor, but their activity slows down or stops in the continued presence of the same odor. Thus, we notice an odor on first entering the kitchen but fail to detect it after a few minutes. The olfactory cells are interspersed and supported by **sustentacular cells.** Mucus is secreted by the sustentacular and by goblet cells in fishes, and by multicellular glands in tetrapods. Ciliated cells are also present in the olfactory epithelium of tetrapods. Basal cells in the epithelium can differentiate into new cells of any type to replace cells that are slowly lost.

Vertebrates move water or air across the olfactory epithelium in different ways. A lamprey has a single nostril on the top of the head that leads to an olfactory sac (embryonically paired) and to the hypophyseal sac (Fig. 11–2**A**). Respiratory movements of the pharynx vary the pressure on the hypophyseal sac, and water is alternately sucked into and forced out of the adjacent olfactory sac. The paired olfactory sacs of most jawed fishes have distinct incurrent and excurrent external nostrils, or **nares,** on the surface of the head (Fig. 11–2**B**). Pleats inside the sacs increase the olfactory surface area.

The excurrent openings from olfactory sacs or nasal cavities of rhipidistian fishes and tetrapods are internal nostrils, or **choanae** (Gr. *choane* = funnel), that open in the roof of the mouth (Fig. 11–3**A**). The nasal cavities of tetrapods are also part of the airways to and from the lungs (p. 651). The olfactory epithelium is restricted to the dorsal part of the nasal epithelium and is ventilated by the respiratory movements. In amphibians and reptiles, which do not ventilate their lungs as frequently as birds and mammals, these respiratory movements are supplemented by additional pumping movements of the floor of the buccopharyngeal cavity. The surface area of the olfactory epithelium is increased in amniotes by scrolls of bone, collectively called the **turbinates** (Fig. 11–3**B**). Turbinates are particularly well developed in mammals. They bear the olfactory epithelium and greatly increase the olfactory surface area. Their presence in ancestral mammals suggests that these animals had an acute olfactory sense that probably reflected a

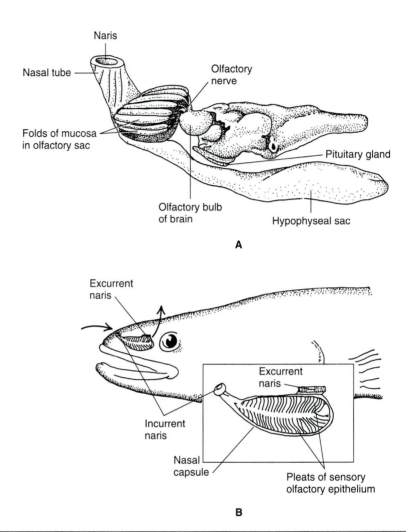

Naris

Nasal tube

Olfactory nerve

Folds of mucosa in olfactory sac

Pituitary gland

Olfactory bulb of brain

Hypophyseal sac

A

Excurrent naris

Excurrent naris

Incurrent naris

Nasal capsule

Pleats of sensory olfactory epithelium

B

FIGURE 11–2   Olfactory organs of fishes. **A,** The olfactory organ of a lamprey. **B,** The olfactory organ of an eel. (After Kleerekoper.)

nocturnal mode of life. One of the turbinates lies at the front of the nasal cavity directly in line with the respiratory current. It helps cleanse, moisten, and warm inspired air, and it cools expired air. As expired air cools, water condenses and is reused to moisten inspired air. It is likely that such a mechanism evolved in relation to the elevated ventilation rates and endothermy that characterized the evolution of mammals (Hillenius, 1992).

The sense of smell provides most vertebrates with considerable information about their surroundings. It helps them find food, recognize members of their own species, and avoid enemies. It aids in homing in some species. Young salmon become imprinted to the odors of the stream in which they hatched, and years later olfactory clues help them return to the same stream to spawn.

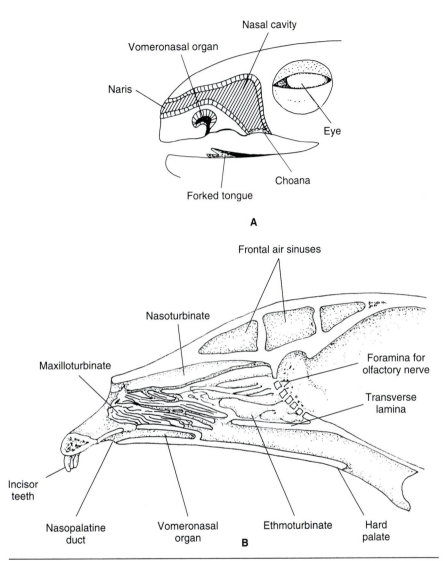

FIGURE 11-3   The nasal cavity of tetrapods: **A,** A lizard. **B,** A mammal. (**A,** after Bellairs. **B,** after Hillenius.)

The sense of smell in many vertebrates also is a part of a communication system based on the secretion into the external environment of species-specific chemical messengers called **pheromones** (Gr. *pherein* = to bear + *hormaein* = to excite). Pheromones have many advantages. Most are small molecules that are easy to synthesize. They may last for hours, they are carried around obstacles, and they are effective in the dark. The amount of information one pheromone can convey is limited, but some vertebrates secrete several pheromones with different meanings. Pheromones frequently are used to warn conspecifics of danger. Injured minnows produce an alarm substance that induces others to flee. Other pheromones indicate social status in a hierarchy, are used to establish territories, and signal sexual readiness. As one example of the last point, female goldfish produce a hormone that promotes oocyte maturation. Some of this hormone is released as a pheromone into the environment, sexually arouses conspecific mature males, and causes them to liberate sperm (Soresen et al., 1987).

The sense of smell is less important for arboreal or flying vertebrates than for aquatic and terrestrial ones because olfactory trails do not always cross the gap from tree to tree. Only a few birds, including the nocturnal and ground dwelling kiwi of New Zealand and the vulture and other scavengers, have well-developed senses of smell. This sense is poorly developed in primates, for they became arboreal animals early in their evolutionary history. Only vestiges of olfactory organs remain in cetaceans, which close their external nostrils under water.

The Vomeronasal Organ

In most tetrapods the medioventral part of the olfactory epithelium is distinct from the rest of this epithelium and forms a pair of **vomeronasal organs,** or **organs of Jacobson.** These organs are absent from aquatic species, including larval amphibians, most turtles, crocodiles, and aquatic mammals. Birds, most bats, and many primates, including human beings, also lack them, although vestiges sometimes appear during embryonic development. The receptive neurons resemble the olfactory ones except that the cilia are replaced by microvilli. The axons of the vomeronasal neurons terminate in an **accessory olfactory bulb** of the brain, which has different projections within the brain than the main olfactory bulb.

The paired vomeronasal organs are not completely separated from the main olfactory chambers in amphibians and primitive reptiles, and odorants enter via the external or internal nostrils. In a lizard or snake, the vomeronasal organs usually form a pair of distinct, saclike structures that have their own entrances into the mouth (Fig. 11–3**A**). A snake uses them in combination with its forked tongue. Odorants adhere to the tongue as it is darted in and out of the mouth, and the tips of the tongue are brought close to the palatal entrances of the vomeronasal organs. Snakes use this mechanism to follow prey trails and in sexual recognition. In mammals the organs are cul-de-sacs that open into the front of the mouth through the **nasopalatine duct,** the nasal cavity, or through both the nasal cavity and the mouth (Fig. 11–3**B**). They appear to detect pheromones important to the animal's social and sexual interactions. Males of some species can determine whether a female is in heat by the pheromones she produces. When next you walk the family dog, you may notice it "mouthing" but not swallowing certain objects. This activity probably helps odorants enter the vomeronasal organs.

The Terminal Nerve

All vertebrates except for cyclostomes have a minute **terminal nerve** that is usually closely applied to the surface of the olfactory nerve. The distal ends of its neurons terminate in the rostral part of the nasal mucosa, and their cell bodies are scattered along the nerve. The functions of the terminal nerve are not entirely certain, but in jawed fishes, amphibians, and mammals, its fibers contain luteinizing hormone releasing hormone (LHRH). This hormone, which is also found in many brain cells (notably those of the hypothalamus), helps regulate reproduction (as we discuss in Chapter 21). The presence of LHRH in the terminal nerve suggests that it may be part of a chemosensory system regulating some aspects of reproduction via the nasal detection of pheromones.

Gustatory Receptors

Taste is detected by barrel-shaped clusters of 20 to 30 receptor and sustentacular cells of endodermal origin that are called **taste buds** (Fig. 11–4**A**). The surface of the taste cells bears microvilli that contain the receptive molecules. Taste buds open to the surface by pores. Because taste buds are exposed and subject to wear, mature cells have a life span of only a week or two; undifferentiated cells within the buds continue to divide and transform into replacement cells. The receptive cells are supplied by distinct sensory neurons that return to the brain in cranial nerves from the mouth and pharynx. The facial nerve (VII) supplies taste buds in the oral cavity; the glossopharyngeal (IX) and vagus (X) nerves supply those in the pharynx. Taste buds are considered visceral sensory organs because of their derivation from endoderm.

Although taste buds are used primarily to find and recognize food, they are also important in sexual and other behavioral interactions in many species. Taste buds do not respond to as low a concentration of substances as olfactory cells; substances must be in contact with the buds. Taste buds also respond to a narrower spectrum of chemical substances. Areas of the human tongue are particularly sensitive to salt, sour, sweet, and bitter substances, but you do not need to be a gourmet to detect more than this. As with olfactory cells, it is likely that different categories of taste buds are sensitive to a spectrum of substances. The distinction among tastes probably results from the particular combination of taste buds that are activated and the pattern of their projection in the brain. Our final perception of taste depends on a complex mixture of signals not only from the taste buds but also from olfactory cells and tactile receptors. It is not easy to sort out the contributions of each. The same is probably the case in other vertebrates.

Taste buds are distributed throughout the oral cavity and pharynx in fishes and amphibians. They also spread onto the skin in many fishes and aquatic amphibians. They occur over the entire body surface in catfishes and minnows but are particularly abundant on the barbels around the mouth (Fig. 11–4**B**). Buds on the body surface arise from endodermal cells that migrate during development. They are supplied by the facial nerve. In amniotes taste buds are limited to the oral cavity and pharynx. Many reptiles and birds have taste buds on the back of the tongue and palate. Taste buds are more abundant in mammals. Most are associated with papillae on the tongue, but some are found on the palate, pharynx, and epiglottis.

Mechanoreceptors of the Skin

Mechanoreception is an important sense in vertebrates. It is used for sensing the surfaces of objects or other individuals and for registering the activity of muscles. This is essential in providing information for coordinating muscle activity. Mech-

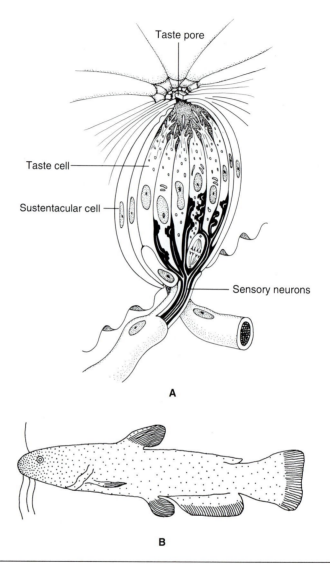

Taste pore

Taste cell

Sustentacular cell

Sensory neurons

A

B

FIGURE 11–4   Gustatory organs: **A,** A human taste bud. **B,** The distribution of taste buds on the surface of a catfish. Each dot represents 100 taste buds. Buds are too numerous on the barbels to show as individual dots. (**A,** after Williams et al. **B,** after Atema.)

anoreception is also used in sensing positional and acceleration changes in the body and for detecting displacements, vibrations, and pressure changes in the medium (water or air) in which the animal lives. In all cases, a very slight mechanical deformation of some part of the plasma membrane of a neuron or receptor cell initiates the receptor potential. We will begin by examining the mechanoreceptors of the skin.

Fishes receive considerable information about their aqueous environment from a special group of mechanoreceptors known as the lateral line system, which lie in or just beneath the skin (p. 409). In addition, fishes have branching **free nerve endings** in their dermis, and these may penetrate the epidermis. These endings are activated by vibrations, touch, injuries, abrupt temperature changes, and other external stimuli. How a fish perceives these stimuli is still being investigated. When vertebrates moved onto the land, they entered a more diverse and variable environment. Since a lateral line system is specialized to detect water movements, it is of no use on land and it has been lost by tetrapods. More specialized cutaneous receptors began to evolve in amphibians and reached a complex development in birds and mammals. Most cutaneous receptors are confined to the skin and are supplied by spinal nerves. A few spread into the mouth. These are supplied by branches of the fifth cranial nerve, which also supplies the cutaneous receptors over the surface of the head. All cutaneous receptors are regarded as somatic receptors.

Mammals also have many free nerve endings in the skin and in the cornea of the eyes. Their primary function is to alert the animals to cuts, burns, and other injuries that we perceive as pain. Beyond this, mammal skin contains a bewildering array of touch and pressure receptors. Some are nerve endings layered or laminated with connective tissue fibers. Other nerve endings are not laminated, but are typically associated with receptive cells (Mathews, 1972; Ray and Kruger, 1983). Probably because they are protected from their surroundings, laminated receptors adapt rapidly to changes in a stimulus and give a response when the stimulus begins or ends. They do not respond to a continuing stimulus. Endings that are not laminated adapt more slowly and remain active for the duration of the stimulus. Those endings, laminated or not, that lie just beneath the epidermis detect stimuli in their immediate vicinity (i.e., their receptive fields are small and well defined). Examples are **Meissner's corpuscles** (laminated and rapidly adapting) and **Merkel's disks** (not laminated and slowly adapting) (Fig. 11–5). Receptors lying deeper in the dermis detect stimuli over a wider and less clearly defined area. Examples are **Pacinian corpuscles** (laminated and rapidly adapting) and **Ruffini endings** (not laminated and slowly adapting).

In addition to the array of receptors we have discussed, mammals have neuron endings entwined around hair follicles so the hairs act as lever arms. A slight movement of the hair tip is magnified at the hair follicle and initiates a nerve impulse. The long **vibrissae** on the snout on many mammals are so sensitive that they can respond to air currents as well as to light touch. Cats, rats, and other nocturnal species use their vibrissae to provide information for moving about in the dark.

Less is known about the cutaneous mechanoreceptors of nonmammalian tetrapods. Frogs, snakes, and some other species are extraordinarily sensitive to ground (seismic) vibrations and use this information to help detect the presence of predators or prey (Hartline, 1971). Seismic receptors have been found in the skin, although the ear also plays a role. The lateral margins of a duck's bill are supplied with Pacinian-like corpuscles that are useful in detecting food items in the muddy water in which ducks feed. Mechanoreceptors on or near the feather follicles may enable birds to detect an imminent stall and measure airspeed across wings (Brown and Fedde, 1993).

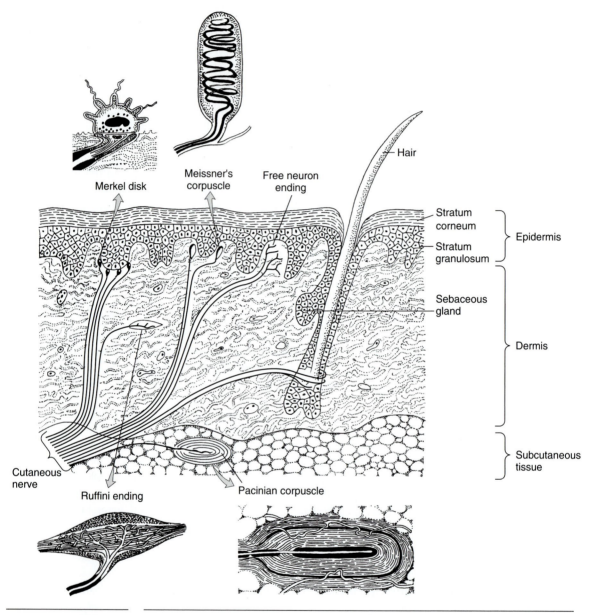

FIGURE 11–5 Cutaneous receptors as seen in a vertical section of mammalian skin. (Receptors after Williams et al.)

Proprioceptors

The contraction and relaxation of locomotor and other muscles at the appropriate time, in the correct sequence, and with the needed force and velocity require some sensory feedback from the muscles to centers in the spinal cord and brain

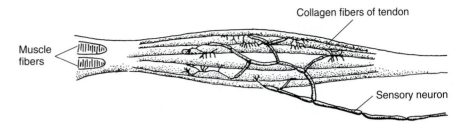

Muscle fibers

Collagen fibers of tendon

Sensory neuron

FIGURE 11–6   Proprioceptors: A tendon organ. (After Williams et al.)

that control their activities. A category of mechanoreceptors known as **proprio-ceptors** (L. *proprius* = one's own + *ceptus* = taken, from *capere* = to take), which are located in muscles, tendons, and joints, continuously provide this information, although we are seldom aware of their activity. Most proprioceptors are associated with somatic muscles, so they are considered somatic receptors.

Tendon organs consist of groups of encapsulated tendon fibers that are entwined by sensory nerve endings (Fig. 11–6). They detect tensions developed by the muscles and provide information on which muscles are active and the magnitude of the forces developed. If forces become dangerously great, sensory information reaching the central nervous system will initiate reflexes that reduce the number of motor impulses going to the muscles.

The regulation of muscle action is different for tetrapods than for fishes because they are subject to greater gravitational forces and their locomotor movements are more complex. Adjustments must be made continuously in the degree and rate of muscle contractions as limb angles and the loads on the muscles change. Tendon organs are more numerous than in fishes, and **muscle spindles** evolve within the skeletal muscles. Muscle spindles provide information by which the degree and rate of muscle contraction can be adjusted to meet the changing forces to which the muscles are subjected (Focus 11–1).

Lateral Line Receptors

The Lateral Line System

Fishes and larval amphibians, including neotenic species, have a unique somatic sensory system in their skin that enables them to detect water disturbances. This is the **lateral line system.** It is lost at metamorphosis in amphibians and never appears in amniotes, even those, such as the cetaceans, that have readapted to an aquatic mode of life. The sense organs within the system are small clusters of mechanoreceptor and sustentacular cells called **neuromasts** (Fig. 11–7A). The individual receptor cells are termed **hair cells** because each bears a single, long **kinocilium** that is followed by a cluster of 15 to 30 **stereocilia** of decreasing length. The kinocilium is a modified cilium containing the characteristic 9 + 2 pattern of microtubules. The stereocilia lack microtubules and are modified microvilli. The cilia project into an overlying gelatinous secretion, the **cupula.**

FOCUS 11–1 Muscle Spindles

In essence, a muscle spindle is a small, fusiform group of specialized muscle fibers inside a sheath. These fibers are called **intrafusal fibers** (L. *fusus* = spindle) because they lie within the sheath. The surrounding, normal muscle fibers are referred to as **extrafusal.** Most of the intrafusal fibers are known as **nuclear bag fibers** because their nuclei are concentrated in the swollen equatorial region of the fibers. The equatorial region lacks contractile myofibrils and is encircled by **annulospiral endings** of sensory neurons. Other, more slender intrafusal fibers, known as **nuclear chain fibers,** receive branching sensory neuron ends nearer their poles. Motor neurons also terminate near the polar ends of each type of intrafusal fiber.

As muscles stretch because of increasing loads on them, the spindles are passively stretched. This is detected by the sensory endings on the nuclear chain fibers. Sensory information returning to the spinal cord initiates nerve reflexes that increase the motor output to the extrafusal fibers in the stretched muscles. The increase causes the extrafusal fibers to con-

tract enough to compensate for their increased load. If, for example, you stand flat-footed and bend your knees, the gastrocnemius and soleus muscles on the back of the shin and the hamstring muscles on the back of the thigh are stretched. If this were not compensated for by an increased contraction of these muscles, you would fall.

During muscle contraction the polar regions of the nuclear bag fibers also are stimulated and contract. If their rate of contraction matches the rate of contraction of the surrounding extrafusal fibers, tension will not develop in the nucleated, noncontractile region of the nuclear bag fibers. If the rate of contraction of the extrafusal fibers lags behind that of the nuclear bag fibers, the nuclear region of the nuclear bag fibers becomes stretched. This is detected by the annulospiral neuron endings. Sensory impulses returning to the spinal cord will initiate an increase in the rate of nerve impulses going to the extrafusal fibers until the rates of contraction of intrafusal and extrafusal fibers are again the same.

During embryonic development the cells that will form the hair cells migrate from ectodermal placodes that are adjacent to the one that will give rise to the receptive cells of the inner ear. Some fishes have as many as three placodes rostral to the otic placode and three caudal to it (Northcutt, 1989). In cyclostomes and amphibians the neuromasts lie on the skin surface or in linearly arranged pits, known as **pit organs.** Most neuromasts are on the head. Jawed fishes retain some surface neuromasts and pit organs, but most of their neuromasts are in water-filled skin canals that open to the surface by pores (Fig. 11–7**B**). In chimaeras the neuromasts lie in open grooves on the skin. The canals and grooves have a distinct pattern. One canal extends along the flank from head to tail while others ramify on the head (Fig. 11–8).

The hair cells are tonic mechanoreceptors that generate a constant base rate of nerve impulses; bending the cupula alters the rate. The neuromasts enable the animal to detect water movements in different directions because movements that bend the cupula toward the kinocilium increase the rate of nerve impulses that are generated, whereas an opposite movement decreases the rate. Efferent neurons that extend from the brain to many hair cells modulate their sensitivity and

FOCUS 11–1 *continued*

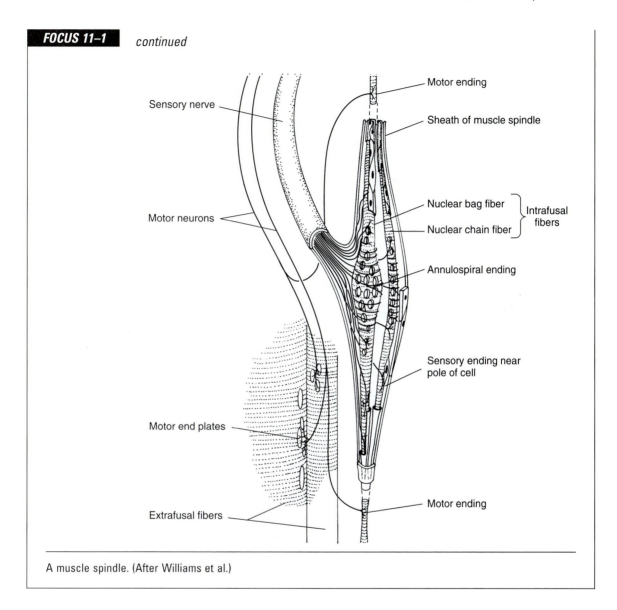

Sensory nerve

Motor neurons

Motor end plates

Extrafusal fibers

Motor ending

Sheath of muscle spindle

Nuclear bag fiber

Nuclear chain fiber

Intrafusal fibers

Annulospiral ending

Sensory ending near pole of cell

Motor ending

A muscle spindle. (After Williams et al.)

may suppress background "noise." The sensory neurons on which impulses travel from the neuromasts to the medulla of the brain are known as lateralis neurons. Traditionally, they have been regarded as components of the facial, vagus, and sometimes the glossopharyngeal cranial nerves, but many neuroanatomists now believe that the lateralis neurons constitute a distinct **lateralis nerve** (p. 477).

The polarity of the neuromasts within a canal is not the same. The kinocilia in one set of neuromasts may all be on the caudal edge of the cells; those in another set may be rotated 180 degrees and be on the rostral edge. This arrangement,

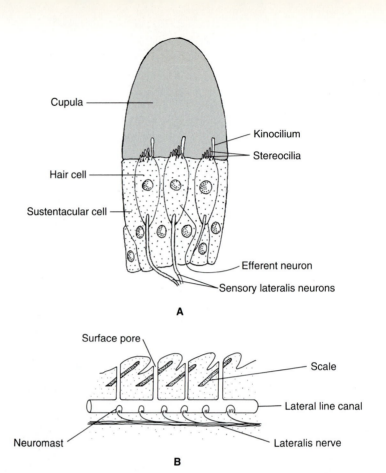

A

B

FIGURE 11–7   The lateral line sensory system of a fish: **A,** A neuromast. **B,** A vertical section through the skin and lateral line canal.

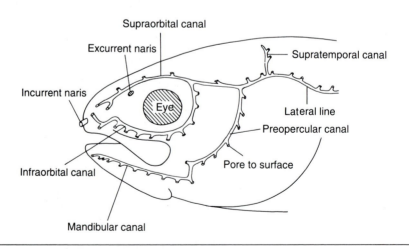

FIGURE 11–8   The distribution of lateral line canals on the head of a teleost. (After Jarvik.)

together with the pattern of distribution of the canals, enables a fish to detect both water disturbances and to determine their source. "Water flowing from head to tail" may be the message of one set; "water flowing from tail to head" may be the message of the other set. Many sorts of water disturbances can be detected: currents, the movements of nearby animals or of the fish itself, disturbances caused when a fish approaches a rock or other stationary object, and low-frequency vibrations generated by a nearby sound source. The lateral line system has aptly been defined as "distant touch."

Electroreceptors

Water conducts electricity well, and electroreceptors have evolved in many fish groups as well as in larval urodeles and gymnophiones. They appear to have evolved from lateral line organs and are supplied by lateralis neurons. Electroreceptors are an ancient sensory system of vertebrates. The presence and shape of parts of a pore-canal system in the cephalic plates of ostracoderms suggest their presence in these early vertebrates. The electroreceptors of lampreys, cartilaginous fishes, early actinopterygians, and sarcopterygians are **ampullary organs.** In a shark, groups of ampullary organs (ampullae of Lorenzini), oriented in different directions, are clustered on the head (Fig. 11–9**A**). Each organ consists of a subcutaneous tube that lies tangential to the surface. One end opens by a pore on the body surface, and the other terminates in a slight enlargement that contains modified hair cells (Fig. 10–9**B**). The entire tube is filled with a gelatinous, mucopolysaccharide secretion, so a cupula is absent. The jelly has the properties of a capacitor. It has a low electrical resistance and readily holds and conducts a voltage gradient. The rate of discharge of the tonic receptive cells is altered by an electric current oriented parallel to the jelly-filled tubes. The electrical resistance of the skin and tube walls prevents electric currents from reaching the receptive cells except through the jelly. Impulses are sent to the brain on lateralis neurons. Behavioral studies have shown that cartilaginous fishes use the ampullary organs in the **passive electrolocation** of prey organisms. They can detect the weak electric currents generated inadvertently by the contraction of the cardiac and respiratory muscles of prey organisms buried in the sand, even when other senses have been blocked. The lower limit of detection is a gradient of 0.01 microvolts per centimeter. Raschii (1986) has shown that the size and distribution of the ampullae of Lorenzini in skates correlates with differences in their feeding strategies.

Many organisms, including sharks, can orient themselves in magnetic fields, and some use this ability in navigation and migration. The ampullary organs of cartilaginous fishes are sensitive enough to detect the minute voltage gradient induced by swimming in the earth's magnetic field, and in some species the use of this sense in navigation has been demonstrated experimentally (Kalmijn, 1978, 1988).

In addition to ampullary organs, bony fishes that generate electric pulses and fields (p. 335) have phasic **tuberous organs** adapted to detect these currents. Among contemporary fishes, tuberous organs have been found in lungfishes, sturgeons, electric catfish, the gymnotids of South America, and the mormyrids of Africa. They may have been secondarily lost by other teleost groups. Studies on

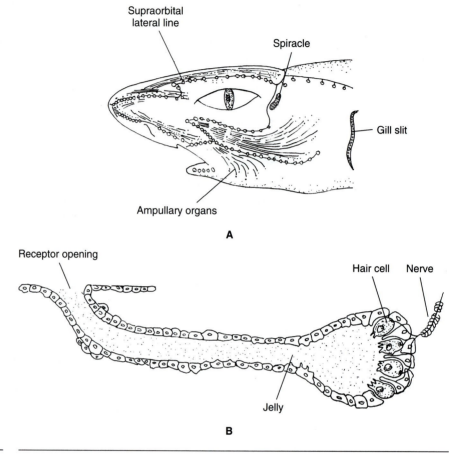

FIGURE 11–9 Ampullary organs of chondrichthyan fishes: **A,** The distribution of the lateral line and ampullary organs on the head of a shark. **B,** A diagram of a single ampullary organ. (**A,** after Dijkgraaf and Kalmijn. **B,** after Szabo.)

the gymnotids and mormyrids have shown that these fishes engage in **active electrolocation,** detecting nearby objects in the turbid waters in which they live by the distortions they produce in the fishes' own electric fields. They also use electric pulses in **electrocommunication.** This is important in species and sex recognition, territoriality, and probably in other social interactions as well.

Air does not conduct electricity well, and electroreceptors were lost in terrestrial vertebrates. However, ampullary-like organs have evolved in the skin on the ducklike bill of the platypus, a monotreme mammal that searches under water for food with its bill. The organs appear to be modified cutaneous glands and are supplied by the trigeminal nerve. This is an interesting example of convergent evolution.

The Ear

The ear has much in common with the lateral line system (Platt, Popper, and Fay, 1989). The receptor cells of both are hair cells that are stimulated when the movement of liquid or other material across their surface bends their cilia. The two systems develop from adjacent ectodermal placodes. The neurons from the eighth, or statoacoustic, nerve from the ear and the lateralis neurons from lateral line organs terminate in adjacent areas of the medulla of the brain. Some investigators regard the lateral line and ear as different parts of a single **octavolateralis system.** The lateral line part of this system detects water disturbances around primitively aquatic vertebrates. The part of the ear containing the hair cells has invaginated and become isolated from external aquatic disturbances. It has, instead, become specialized to detect internal liquid disturbances caused by changes in the orientation and movement of the body (balance and acceleration) and by external sound waves that are able to reach it.

All vertebrates have an inner ear embedded in the otic capsule of the skull, and this is where the receptive cells are located. Only an inner ear is present in fishes, but tetrapods also have external and middle ears that are specialized to receive airborne sound waves and transmit them to the inner ear.

Inner Ear Structure

The inner ear develops embryonically in all vertebrates as an invagination of the ectodermal **otic placode** to form an **otic vesicle** (Fig. 11–10). The vesicle gradually differentiates into a series of membranous sacs and ducts that are filled with a lymphlike fluid called **endolymph.** The entire complex is known as the **membranous labyrinth** (Fig. 11–11). It lies within a system of parallel canals and chambers in the cartilage or bone of the otic capsule of the skull. These spaces are the **osseous labyrinth.** The space between the membranous labyrinth and the osseous labyrinth is crisscrossed by strands of connective tissue and filled with a liquid called **perilymph.**

The stalk by which the otic placode invaginated to form the otic vesicle persists

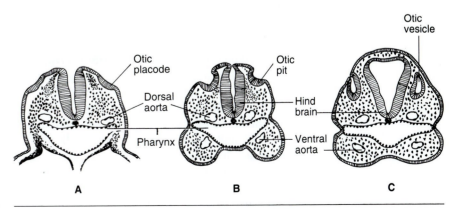

FIGURE 11–10   Three stages in the development of the inner ear. (After Arey.)

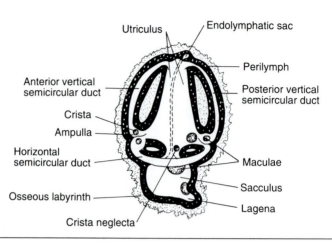

Utriculus

Endolymphatic sac

Perilymph

Anterior vertical
semicircular duct

Posterior vertical
semicircular duct

Crista

Ampulla

Horizontal
semicircular duct

Maculae

Sacculus

Osseous labyrinth

Lagena

Crista neglecta

FIGURE 11–11   The inner ear of a fish. (After Kingsley.)

in an adult cartilaginous fish as an **endolymphatic duct** that opens on the surface of the head. In other vertebrates it either is lost or forms a small, deeply seated **endolymphatic sac.** Each membranous labyrinth usually has three **semicircular ducts** that connect with a chamber known as the **utriculus** (L. *utriculus* = small bag). These ducts are sometimes called canals, but technically the term **semicircular canal** applies to the spaces in the osseous labyrinth in which the semicircular ducts lie. Two of the ducts lie in the vertical plane perpendicular to each other; the third lies in the horizontal plane at right angles to the other two. A swelling (the **ampulla**), which contains receptive cells, is located at one end of each duct. The semicircular ducts and utriculus are remarkably similar in most vertebrates (Fig. 11–12). Lampreys lack the horizontal duct, hagfishes have only a single duct, and cartilaginous fishes have a utriculus that is divided into two parts.

The utriculus connects ventrally with a larger sac, called the **sacculus** (L. *sacculus* = small sac), from which a caudoventral evagination of some type arises (Fig. 11–11). The saccular region of the ear varies greatly among vertebrates (Fig. 11–12). In fishes, amphibians, and many reptiles, the caudoventral evagination forms a small **lagena** (L. *lagena* = flask), but in some reptiles, birds, and mammals it develops into a longer, often coiled **cochlear duct** (Gr. *kochlias* = snail).

Several types of hair cell groups are located within the membranous labyrinth. An enlarged neuromast, termed a **crista** (L. *crista* = crest), occurs in the ampulla of each semicircular duct. Another, called the **crista neglecta** because it was long overlooked, is found in the utriculus of fishes (Fig. 11–11). Larger patches of hair cells, termed **maculae** (L. *macula* = spot), occur in the utriculus and sacculus of all vertebrates and in the lagena of many animals. The maculae often appear as small white spots because each is overlain by small calcareous crystals, termed **statoconia,** which are secreted into a gelatinous membrane. The statoconia are loosely organized in cartilaginous fishes and mixed with sand grains that enter through the endolymphatic duct. Some of the mineral particles are magnetic. It was once proposed that these magnetic particles enabled these fishes to detect the earth's magnetic field, but it has been shown that this is not the case (Hanson,

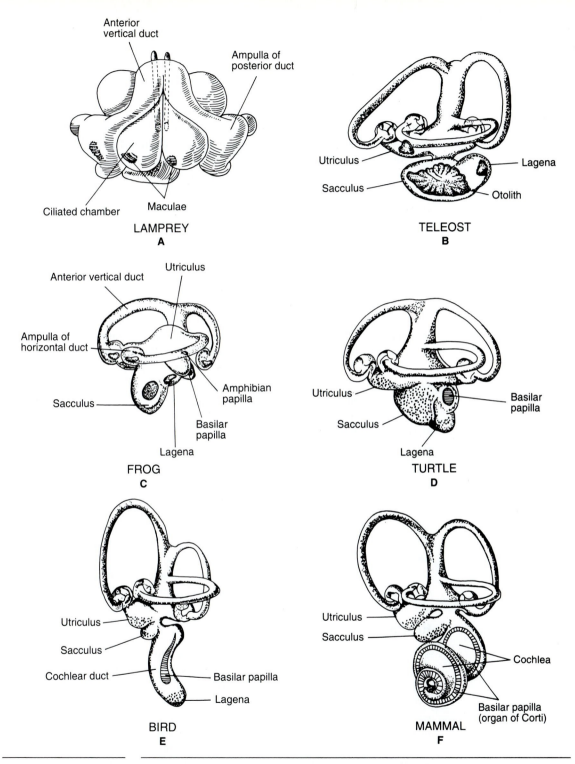

FIGURE 11–12 Lateral views of the membranous labyrinth of representative vertebrates: **A,** A lamprey. **B,** A teleost. **C,** A frog. **D,** A turtle. **E,** A bird. **F,** A mammal. (**A** and **C,** after Retzius. Others, after von Frisch.)

Westerberg, and Öblad, 1990). Often the statoconia are consolidated into larger **otoliths** (Gr. *ot-* = ear + *lithos* = stone). The saccular otolith of the teleost ear is exceptionally large, nearly fills the sacculus, and has a distinctive shape in many species. It appears to function in sound detection (p. 419). The saccular otolith of some species of teleosts grows by the accretion of layers of calcareous crystals, and the number of layers can be used as an indication of the fish's age.

Sensory papillae are patches of hair and sustentacular cells that are found only in the sacculus, lagena, or cochlear duct of tetrapods. The cilia of their hair cells impinge on an overlying **tectorial membrane** (L. *tectum* = roof). Movement between the cilia and the membrane activates the receptor cells.

Equilibrium

Hair cells within the ampullae, utriculus, and sacculus of all vertebrates detect changes in position and movement and so provide information that helps an animal maintain equilibrium. Additional information about position and movement comes from sight, proprioceptors, and, in tetrapods, tactile and pressure receptors in the feet. The pull of gravity on the statoconia or otoliths, particularly the large one in the sacculus, registers **static equilibrium,** that is, the present orientation of the body in space. If the animal rolls to one side or pitches, the changed pull of gravity on the otoliths in the sacculus and utriculus creates shear forces on the cilia of the hair cells and alters their rate of discharge. Since the hair cells in the maculae have different polarities, they can detect displacements in different directions. Stops, starts, and other changes in **linear acceleration** also affect these otoliths because they have considerable inertia and their movements will lag behind those of the body as a whole.

Angular acceleration, or turning movements of the head, is detected by the cristae in the semicircular ducts. If, for example, a vertebrate turns to the left or to the right, the crista in the ampulla of a horizontal duct moves at the same rate as the body, but movement of the endolymph in the narrow duct lags slightly and so bends the cupula, which nearly blocks one entrance into the duct. Turning movements in the vertical plane will affect some combination of the vertical ducts. Presumably, the large diameter of the ducts in lampreys enables these animals to detect horizontal movements (McVean, 1991).

Hearing in Fishes

To us, hearing is the detection of airborne pressure waves, but to a fish, it is the detection of certain water disturbances generated by an underwater sound source. Sound waves generated by a vibrating object spread much more rapidly through a dense medium, like water, than through air. Sound travels at 1500 meters per second in water, compared with 330 meters per second in air. The waves have two components: (1) low-frequency **particle motion** or **displacement waves,** which are somewhat analogous to the ripples produced when a pebble is dropped into the water; and (2) higher frequency **pressure waves,** which result from the alternate compression and rarefaction of molecules in the water. The amplitude of the displacement waves decays very rapidly (proportional to the square of the distance from the sound source) regardless of the sound frequency. The pressure waves decay more slowly (linearly with distance) in a frequency-dependent manner; low-frequency sounds travel farther than high-frequency ones. Because of differences in their decay rates, displacement waves predominate close to the sound source (that is, in the **near-field**), whereas the pressure waves are more important in the **far-field.**

Since hair cells are stimulated by the displacement of their stereocilia, displacement waves directly affect the hair cells that they can reach, but pressure waves must be converted to a small displacement to activate the hair cells. The lateral line of a fish is well suited to detect and localize the source of low-frequency water particle displacements in the near-field. Maximum sensitivity is 50 to 150 Hz. Some of these displacement waves may also affect the inner ears of a few species. There is evidence that such waves can pass through the skin and jelly-filled parietal fossa in the skulls of some cartilaginous fishes and affect the crista neglecta, which lies in part of the utriculus (Corwin, 1989).

Mechanisms whereby fishes detect the higher frequency, far-field pressure waves are complex. Since the density of a fish is close to that of water, pressure waves easily pass through and move the fish at nearly the same amplitude and frequency as they move the water. In a sense, a fish is transparent to sound. For a sound to be detected, the waves must induce a movement over certain hair cells that differs from that of the rest of the body. There is considerable neurophysiological evidence that the macula in the sacculus is more sensitive to far-field pressure waves than other parts of the inner ear (Tavolga, Popper, and Fay, 1981). When a pressure wave passes through the body of a fish, the movements of the large, dense saccular otolith lag the movements of the rest of the body. This would cause a relative movement between the otolith and the hair cells on which it rests, causing them to bend. Considerable evidence indicates that fishes, like tetrapods, can detect the sound source. Tetrapods detect direction by differences in the time and intensity of the arrival of the pressure waves at the two ears, but it is unlikely that fishes can do so because pressure waves travel so fast in water, and the distance between the two inner ears of most fishes is so small. Directional discrimination in fishes must depend on differences in the polarity of groups of hair cells in the macula.

Carps, minnows, and other members of the teleost superorder Ostariophysi can use the swim bladder, a sac of air in the body cavity, as a hydrophone (p. 650). The swim bladder has a different density from the rest of the fish and amplifies some of the sound waves passing through the fish. This vibrates the walls and internal gas of the bladder and changes the pressure within it. The amplified waves are transferred to the inner ear by paired sets of small bones, the **weberian ossicles,** that are derived from ribs (Fig. 11–13). The ossicles extend from the anterior end of the swim bladder to a perilymphatic sac that lies beside the sacculus. The system is analogous to the tympanic membrane and auditory ossicles of tetrapods. Ostariophysians have the lowest auditory thresholds and can detect the highest frequencies (up to 5000 Hz) of any group of fishes. They are most sensitive to frequencies between 200 and 600 Hz.

The world beneath water is a silent place to a diver, whose ears are adapted to detect airborne sound waves. But what do fishes hear? Many sorts of noises have been detected with underwater listening instruments. Waves, surf, and currents all generate sounds, as do rapid changes in the direction or velocity of swim-

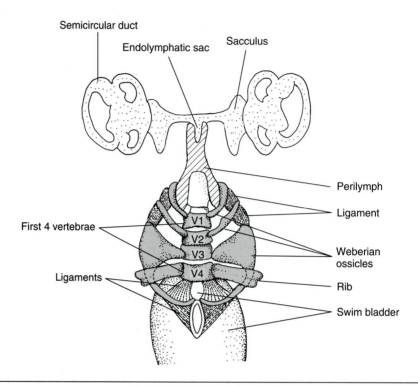

FIGURE 11–13 A dorsal view of the weberian apparatus of an ostariophysan teleost. (After Popper.)

ming animals. Animals feeding on coral, shellfish, and other coarse material make noises. Many fishes also produce sounds at will by gnashing their teeth or rubbing spines or certain pectoral bones together. Others have muscles that pluck on the swim bladder as one might pluck a bass fiddle, or they drum on the body wall overlying the swim bladder with their pectoral fins. Some of the sounds are very loud. A chorus of marine catfish can generate a noise of 20 decibels—equivalent to a subway train 10 meters away. A fish could learn a great deal about its surroundings and the activities of its neighbors from sounds. There is evidence that some species use sounds to call mates, startle predators, and warn conspecifics of danger.

Hearing in Terrestrial Vertebrates

Although fishes have no difficulty in receiving pressure waves, this is not the case for tetrapods because the sound pressure waves are in air, a medium that is far less dense than water. The force of the pressure waves must be amplified considerably to enter the denser liquids of the inner ear and produce a displacement wave that can move certain hair cells relative to overlying structures. Low-frequency sound waves (below 1000 Hz) of high enough intensity, which may travel through the ground as well as air, can cause a slight displacement of superficial skull bones. Caecilians, terrestrial salamanders, some lizards, and all snakes detect low-frequency waves in this way. Terrestrial vertebrates that detect higher frequency sound waves have a delicate **tympanic membrane** on or near the surface of the head. An air-filled **middle ear,** or **tympanic, cavity** lies on the inside of the tympanic membrane and connects to the pharynx by way of the **auditory,** or **eustachian, tube** (Figure 11–14**B**). The tympanic cavity and auditory tube equalize the air pressure on the medial and lateral sides of the tympanic membrane, and this enables the membrane to respond quickly to high-frequency sounds. The tympanic cavity and auditory tube of an amniote develop from the first embryonic pharyngeal pouch, so they are homologous to the reduced first gill pouch, or spiracle, of a fish. We are uncertain whether this homology applies to the amphibian middle ear cavity and auditory tube, which show certain peculiarities in their development.

The conversion of a pressure wave in the air to a displacement wave is made by the response of the tympanic membrane, or certain skull bones when a membrane is absent. Three other structures, found in all terrestrial vertebrates, are essential (Fig. 11–14):

1. There must be at least one auditory ossicle, the **stapes,** the lateral end of which connects either to a tympanic membrane or to superficial skull bones on which airborne sound waves impinge. Its medial end bears a foot plate that fits into an **oval window** on the lateral surface of the otic capsule. The stapes is homologous to the hyomandibular of fishes (p. 217). The large size of the tympanic membrane relative to the size of the foot plate of the stapes forms a pressure amplification mechanism. All the energy impinging on the tympanum is concentrated on the small foot plate. The resulting increase in energy per unit of area on the foot plate enables the high-frequency vibrations in the air to overcome the inertia, or impedance, of the liquid in the perilymphatic duct and set up displacement waves in it. The external and middle ears are often described as an impedance-matching device.

2. There must be at least one specialized **perilymphatic duct** that receives displacement waves from the stapes and carries them past a receptive part of the membranous labyrinth to some point where they can be dissipated. This is either the cranial cavity or a **round window** that opens into a middle ear cavity, if such a cavity is present.

3. Finally, there must be at least one sensory area within the membranous labyrinth specialized to receive the displacement waves from the perilymphatic duct. Tetrapods have one or more unique groups of hair cells known as **papillae** that are specialized to receive these waves. They consist of hair cells overlain by a tectorial membrane rather than by otoliths. There are many differences in the location of the papilla or papillae among tetrapods.

It should be clear from what we have said that tetrapods detect airborne sound waves in different ways. They utilize some structures that are homologous and some that are not. This suggests considerable independent evolution of auditory mechanisms. Many investigators now believe that a tympanic membrane was not present in all ancestral amphibians, and that its absence from caecilians and salamanders is a retention of a primitive condition. An ear utilizing a tympanic membrane evolved independently three times: (1) in some labyrinthodonts and their

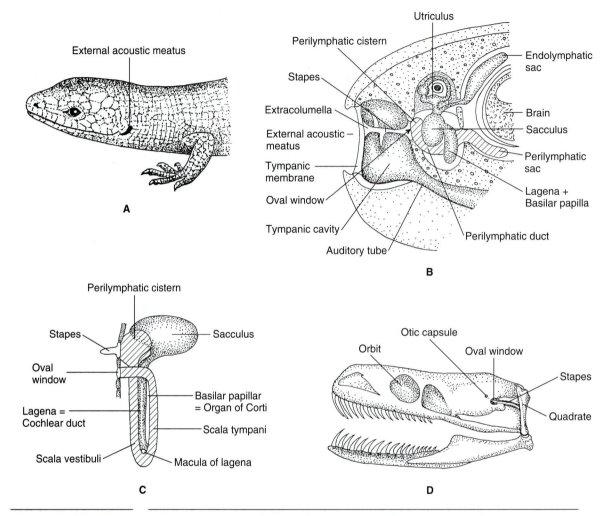

FIGURE 11–14   The ears of reptiles and birds: **A,** A surface view of the ear of a lizard. **B,** A transverse section through the ear of a lizard. **C,** The auditory part of the ear of a bird. **D,** The middle ear of a snake. (**B,** after Portmann. **C,** after Romer and Parsons.)

frog descendants, (2) in the line of evolution to contemporary reptiles and birds, and (3) in the late synapsids and the mammals.

The Ear of Reptiles and Birds

We begin with the ears of reptiles and birds because the amphibian ear is very unusual. In a reptile the tympanic membrane is usually located at the bottom of a recess, the **external acoustic meatus** (Fig. 11–14A). The membrane is described as postquadrate in position because it lies just caudal to the quadrate bone, which usually partly encases the membrane and middle ear cavity. A **stapes** crosses the tympanic cavity and transfers displacement waves from the membrane through the oval window to a **perilymphatic cistern** (Fig. 11–14**B**). A **perilymphatic duct** extends from the cistern across the lagena, which contains the receptive **basilar papilla,** to a **perilymphatic sac,** where the waves are dissipated. This sac may lie beside a round window opening back into the tympanic cavity, but it often connects with the cranial cavity. The ear is very sensitive and responds to high-frequency sounds. Some lizards can detect sound waves with a frequency as high as 10 KHz.

The bird ear is similar. The main differences are that the lagena forms an elongated **cochlear duct,** and the basilar papilla is an equally long **organ of Corti** (Fig. 11–14**C**). Songbirds can detect frequencies as high as 21 KHz.

A few lizards and all amphisbaenians and snakes lack tympanic membranes; the stapes connects instead with the quadrate bone (Fig. 11–14**D**). These animals detect chiefly ground vibrations, but their ears are also sensitive to low-frequency airborne vibrations. It is probable that the tympanic membrane has been secondarily lost in these species as an adaptation to a burrowing mode of life: lizards that lack it are, for the most part, burrowers; all amphisbaenians live beneath ground; and we believe that snakes passed through a burrowing phase early in their evolution (most snakes now live above ground, but the most primitive species still burrow).

The Ear of Amphibians

Frogs have a large tympanic membrane located high up on the head surface, dorsal to the quadrate bone (Fig. 11–15**A**). Many labyrinthodonts had an otic notch in a similar position that may have held a tympanic membrane. A stapes continues in a frog across the tympanic cavity to an oval window in the otic capsule. Frogs also have an additional auditory ossicle, known as the **operculum,** that lies in the oval window and fits into a notch on the caudal surface of the footplate of the stapes. The operculum develops embryonically from a part of the wall of the otic capsule and is not homologous to the fish operculum, which is a gill covering. An **opercularis muscle** extends from the operculum to the suprascapular element of the pectoral girdle, thereby coupling the inner ear to the girdle, leg, and ground. The opercularis muscle is homologous to the levator scapulae muscle. A perilymphatic cistern lies just inside the oval window and continues past the sacculus to a perilymphatic sac that extends into the cranial cavity. Two sensory papillae are present in frogs: a **basilar papilla** near the entrance to the lagena, and a more dorsal **amphibian papilla** (Fig. 11–12**C**). Larval salamanders have a stapes that connects to superficial skull bones, but this is lost or becomes reduced later in development so adult salamanders have only the opercularis system (Fig. 11–15**B**).

Frogs and salamanders are secretive creatures that are extraordinarily sensitive to low-frequency seismic vibrations that can alert them to potential danger. Early in this century, Kingsbury and Reed (1909) proposed that the opercularis system detected ground vibrations of this type. Hetherington, Jaslow, and Lombard (1986) corroborated this hypothesis and found that the opercularis is a tonic muscle that is tense and rigid, exactly what would be expected if it was part of a vibration detecting system. Low-frequency vibrations of this type are detected by the amphibian papilla. The tympanic system of frogs is adapted to receive much

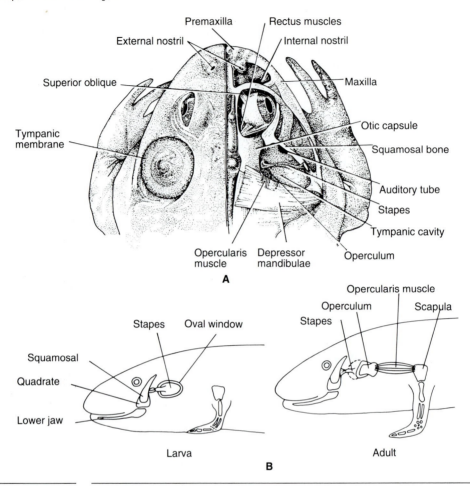

FIGURE 11–15   The ears of amphibians: **A,** The ear of a bullfrog in surface view (left) and dissected (right). **B,** The auditory mechanisms of a larval (left) and adult (right) salamander. (**A,** after Walker. **B,** after Kingsbury and Reed.)

higher sound frequencies, such as their mating cells. These are detected by the basilar papilla. Salamanders do not have mating calls and, not surprisingly, they lack the tympanic system, including the basilar papilla. Hetherington et al. (1986) also found that movements of the stapes and operculum in frogs are independent and are not linked in such a way that the opercular system could modulate movements of the operculum, as indicated by earlier researchers (Lombard and Straughan, 1974).

The Ear of Mammals and Its Evolution

Mammals have a third type of tympanic ear. An external flap, the **auricle** or pinna, helps funnel sound waves down an external acoustic meatus to the tympanic membrane (Fig. 11–16**A**). The human auricle is not very large or important, but the auricle of many mammals is large and can be moved and directed toward a sound

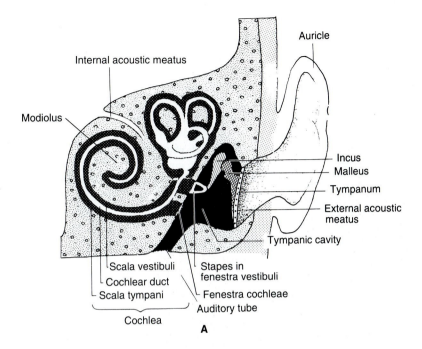

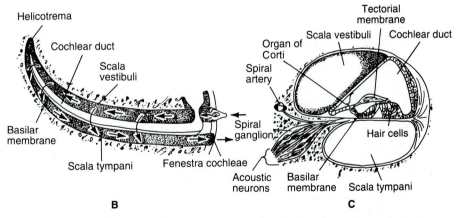

FIGURE 11–16   The mammalian ear, based on the human ear: **A,** A transverse section through the ear. **B,** A diagram of an uncoiled cochlea. **C,** A transverse section through the cochlea. (**A,** from Walker and Homberger; **B** and **C,** after Dorit, Walker, and Barnes.)

source. The auricle amplifies sound waves slightly because the waves it gathers are concentrated on the relatively small tympanic membrane.

A set of three auditory ossicles, the **malleus, incus,** and **stapes,** extends from the tympanic membrane across the middle ear cavity to the oval window, technically known as the **fenestra vestibuli.** The ossicles derive their names from their shapes (L. *malleus* = hammer, L. *incus* = anvil, L. *stapes* = stirrup). As we have seen (p. 242), the malleus evolved from the quadrate bone; the incus, from the articular bone. The quadrate and articular bones bear the jaw joint in non-mammalian vertebrates. As in other vertebrates with tympanic ears, the major amplification of sound wave force derives from the large size of the tympanic membrane relative to the foot plate of the stapes. The three auditory ossicles form a lever system that further increases this amplification about one and a half times, but as in any lever system, the increase in force at one end of the system is accompanied by a reduction in the extent of movement at that end. In human beings total force on the foot plate of the stapes is 22 times that on the tympanic membrane, but the extent of movement of the foot plate is only about one third that of the membrane.

Two small muscles, the **tensor tympani** and **stapedius,** attach to the malleus and stapes, respectively. Since the malleus derives from the mandibular arch, the tensor tympani is a mandibular arch muscle and is innervated by the trigeminal nerve. The stapes is a hyoid arch derivative, and its stapedius is innervated by the facial nerve. Tension in these muscles dampens oscillations of the ossicles and so protects delicate inner ear structures from large movements caused by loud noises, including the sounds that the mammal itself makes, like the ultrasonic sounds of bats and cetaceans (p. 427), which are emitted at very high energy levels.

The hypothesis that the mammalian ear, with its three ossicles, evolved from the reptile tympanic ear, with its single ossicle, presents some serious problems. How could the articular and quadrate, when they became redundant as jaw joint bones in the reptile-to-mammal transition, move into the middle ear without disrupting its function? What would be the selective advantage of adding them to an effective ear? These problems evaporate if, as we now believe, mammals evolved from ancestral amniotes that did not have tympanic ears (captorhinids and early synapsids). In these animals a rather large stapes was connected to the quadrate. There is no modification of the quadrate, as there is in modern reptiles, that would accommodate a tympanic membrane and middle ear behind it. Presumably, captorhinids and early synapsids detected ground vibrations and low-frequency air-borne vibrations through the skull, including the lower jaw. These were transmitted through the articular, quadrate, and stapes to the inner ear. As the postdentary jaw joint bones and stapes became smaller and lighter (p. 242), they became more efficient in transmitting sound. Indeed, selection for improved sound transmission may have been another factor in addition to feeding mechanics, in the changes in jaw mechanisms. Advanced synapsids had an unusual flange, the **reflected lamina,** that extended ventrally and caudally from the angular bone (see Fig. 7–22). Many investigators believe that the reflected lamina of the angular held a "mandibular" tympanic membrane that lay adjacent to an air-filled space. Such a space would further increase the efficiency of the articular, quadrate, and stapes in sound transmission. With the final shift of the jaw joint to the dentary and squamosal bone in early mammals, the articular, quadrate, and stapes could be acted on by natural selection to function only as auditory ossicles. If early mammals occupied an insectivorous, nocturnal niche, the ability to detect high-frequency sounds would have a great selective advantage. Rather than trying awkwardly to fit the articular and quadrate into a functional ear, we now believe that these bones and the stapes were sound-transmitting bones throughout the early history of tetrapods in the line of evolution to mammals. Selective forces favoring changes in jaw mechanics made it possible for other selective forces to convert these bones into efficient auditory ossicles.

The lagena and perilymphatic duct found in ancestral terrestrial vertebrates are elongated in mammals to form a snaillike **cochlea** (Fig. 11–16A). The cochlea spirals around a core of bone, the **modiolus,** that contains the cochlear branch of the eighth cranial nerve, which is called the vestibulocochlear nerve in mammals. The cochlea becomes progressively narrow as it approaches its apex. Its spiral ranges from a one-quarter turn in the platypus to four turns in the guinea pig. The human cochlea makes three and a half turns. The presence in ancestral mammals of spaces in the otic capsule for an elongated and partly coiled cochlea is additional evidence that these mammals had a keen sense of hearing.

The lagena itself forms the **cochlear duct** that contains an elongated **basilar papilla** or **membrane** known as the **organ of Corti** (Fig. 11–16B and **C**). The hair cells of the organ of Corti rest on a specialized basilar membrane, and their cilia impinge on an overlying tectorial membrane. Displacement waves are received from the stapes at the **fenestra vestibuli** by a perilymphatic duct called the **scala vestibuli.** This duct passes along one side of the cochlear duct and returns on the other as the **scala tympani.** The scala vestibuli and scala tympani connect at the apex of the modiolus by a small opening known as **helicotrema.** Displacement waves are released through the scala tympani back into the tympanic cavity at the round window, or **fenestra cochleae.** When displacement waves cross the cochlear duct from the scala vestibuli to the scala tympani, they cause a slight movement of the basilar membrane. Shear forces develop between the hair cells and the tectorial membrane, activating the hair cells, which in turn initiate nerve impulses. Since the dimensions and other properties of the cochlea and basilar membrane change from base to apex, traveling waves of different frequencies cause a maximum displacement of the membrane at different levels. High frequencies are detected near the base of the cochlea; low ones register near its apex. Most mammals can detect frequencies as high as 20 KHz.

Our common bats have evolved a very sensitive auditory system that enables them to avoid obstacles in the dark and to find their insect prey by sending out high-frequency noises and listening to the echoes. A large, ossified larynx allows a high tension to be developed on the vocal cords and sounds with frequencies as high as 150 KHz to be generated. The sounds are emitted through the mouth in some species, through the nose in others. Because these high frequencies have very short wavelengths, and hence a high resolution, bats can discriminate between objects that are close together. The sounds are emitted as pulses lasting 1 to 4 milliseconds. Pulses can be repeated up to 100 times per second as a bat homes in on an insect. Frequencies drop about an octave during the course of each pulse. Evidence indicates that different species of bats listen to different aspects of the echoes: (1) time lag in echo return, (2) decrease in loudness between the emitted sound and its echo, and (3) differences in frequency between the emitted sound and its echo, which are sensed as a "beat note." There has been considerable independent evolution of these systems among bats.

Cetaceans have evolved an underwater sonar system that also is based on emitting high-frequency sounds and listening to the echoes. Considerable modification of the mammalian ear, originally sensitive to airborne sounds, has occurred so that underwater sounds can be detected. Many cetaceans have also evolved communication systems based on low-frequency "songs" (about 20 Hz) that can travel hundreds and perhaps thousands of kilometers.

Photoreceptors

The ability to detect changes in light is important for nearly all animals because their periods of feeding and reproduction and many other aspects of their physiology and behavior are closely attuned to the diurnal cycle and to seasonal changes in day length. The additional ability to detect the source of light can provide information on the approach of predators, location of shelter, and so forth. For those animals with groups of photoreceptive cells numerous enough to discriminate between different light intensities (or wave lengths) in the environment, and a neuronal circuitry capable of processing these data, light provides much information about the outside world. The informational content of a visual image—its location, size, shape, movement—is greater than that provided by other senses.

To receive light, an animal must have **photoreceptive cells** containing a pigment that absorbs quanta of light energy, which initiates chemical changes that generate nerve impulses. The light energy of most value to vertebrates falls between wavelengths of 380 nanometers (violet) and 760 nanometers (red). Shorter wavelengths extending into the ultraviolet are quickly absorbed by water and so are of no value to aquatic animals. They also contain so much energy per quantum that they are potentially destructive to tissue in terrestrial vertebrates. The integument protects deeper tissues from this radiation. Infrared and longer wavelengths contain too little energy to be biologically useful in most cases, although some vertebrates have special receptors that detect infrared radiation as heat (p. 441).

The body surface of most metazoans, including many vertebrates, can detect changes in light, although specific light-sensitive cells have not always been identified (Zimmerman and Heatwolfe, 1990). Image-forming eyes are limited to animals that move about a great deal in well-illuminated environments: many arthropods, cephalopods, and most vertebrates.

Median Eyes

In many vertebrates the detection of light and the physiological adjustments to changes in light levels are apparently mediated by eyes different from those that form images. In addition to image-forming lateral eyes, anamniotes and many reptiles have one or two light-receptive median eyes on top of the head. These develop, as does the retina of image-forming eyes, as outgrowths from the diencephalic region of the brain.

An adult lamprey has a nonpigmented spot of skin on the top of its head beneath which lies a **pineal eye**. A **parietal eye** lies deep to the pineal (Fig. 11–17A). A slight right-left asymmetry in the position of these eyes and in their connections to the brain suggests that they may originally have been paired organs. Each is a hollow sphere of cells. The deep cells are photoreceptive and partly

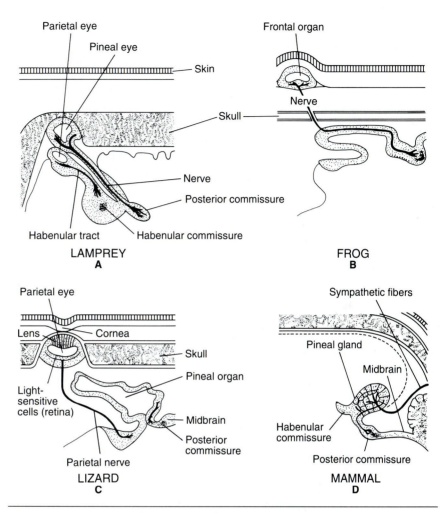

FIGURE 11–17   The median eye complex of representative vertebrates. (After Studnicka.)

shielded by pigment. Sensory neurons extend from them to the brain. Experiments on the ammocoetes larva of the lamprey have shown that the median eye complex generates a low level of neuronal activity in dim light (Eddy, 1972). This appears to be correlated with the larva's nocturnal activity. Strong light inhibits neuronal activity, and the larva rests during the day. Beyond this, the median eye complex is a neuroendocrine transducer that translates light signals into chemical messages. In the absence of light, the complex produces an enzyme that converts serotonin into melatonin. Melatonin causes the pigment in the melanophores to concentrate, and thus the animal blanches in the dark. Light inhibits the formation or activity of this enzyme, pigment is dispersed, and the animal is dark during the day. Experiments have shown that the median eye complex is essential for metamorphosis and influences the development of sexual maturity. All this indicates that the complex has a role in the daily and seasonal rhythms of the animal, but the nature of the involvement is not well understood.

Although not universally present, the median eye complex is an ancient and widespread feature of vertebrates. A skull foramen for it has been found in ostracoderms, placoderms, early cartilaginous and bony fishes, and in ancestral amphibians and reptiles. Contemporary species do not always have a foramen for it, but a reduced pineal organ, often called the **epiphysis,** is present in chondrichthyan fishes and most actinopterygians. Tadpoles and frogs have a small **frontal organ,** which represents the parietal eye (Fig. 11–17**B**). *Sphenodon* and many lizards have a well-developed parietal eye complete with cornea-like and lenslike structures and a reduced pineal organ (Fig. 11–17**C**). The parietal eye monitors the level of solar radiation and affects the animal's orientation to the sun and its movements into the sun or shade. The complex is represented in birds and mammals by an endocrine **pineal gland** whose activities also are affected by light (Fig. 11–17**D** and p. 546).

The Structure and Function of Image-Forming Eyes

Except for those species living in dark habitats—some deep-sea fishes and many cave-dwelling or burrowing vertebrates, whose eyes have been secondarily reduced or lost—all vertebrates have a pair of image-forming eyes. The eyes are essentially little biological cameras. They are usually located on the side of the head in such a position that there is little overlap in the two visual fields. The eyes are directed more rostrally in some birds and many mammals, and the visual fields overlap to a greater extent. A high degree of overlap of the visual fields provides depth perception, or stereoscopic vision, in many mammals.

Various structures are found in the orbit around each eyeball. Extrinsic ocular muscles (p. 318) attach to the eyeball and control its movements. Eyes must be directed at objects in the visual field despite movements of the body and head, and the movements of the two eyeballs usually are synchronized very precisely. Tear glands and movable eyelids moisten and protect the eye surface in terrestrial vertebrates (p. 439).

Although the eyeball varies greatly in adaptive details among vertebrates, its basic structure is the same in all. The human eyeball is representative (Fig. 11–18). Its wall consists of three layers of tissue: an outer supportive **fibrous tunic,** a middle nutritive **vascular tunic,** and an inner **retina** containing the light photoreceptive cells. Part of the fibrous tunic is modified as a cornea that allows light to enter. Parts of the vascular tunic and retina are modified to form the iris, which controls the amount of light entering, and a ciliary body, which supports and focuses the **lens.** A watery **aqueous humor** fills the spaces in the eyeball in front of the lens, and a more gelatinous **vitreous body** (L. *vitreus* = glassy) lies behind the lens. The speed of light slows as it passes through the different parts of the eyeball, so the light rays bend, or refract, toward the optic axis and cast an inverted image on the retina.

The Fibrous Tunic

The fibrous tunic is a dense connective tissue that forms the essential supportive framework for the eyeball and maintains eyeball shape. Most of it is an opaque,

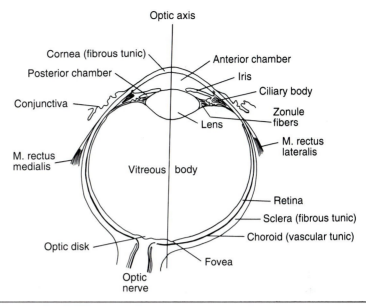

FIGURE 11–18   A mammalian eye.

white **sclera** (Gr. *skleros* = hard)—the "white" of our eyes. The fibrous tunic forms the **cornea** (L. *corneus* = horny) in front of the lens. The cornea is avascular and otherwise modified to facilitate the passage of light. A delicate, epithelial **conjunctiva** covers the surface of the cornea, turns onto the inner surface of the eyelids, if they are present, and becomes continuous with the surface layers of the skin. A ring of cartilage or **sclerotic bones** develop in the sclera of many vertebrates. Usually, they are located close to the level of the lens, where they support and strengthen the point of origin of lens muscles. They also help maintain eyeball shape, especially in some birds and other species with nonspherical eyeballs.

The Vascular Tunic, Iris, and Ciliary Body

The next layer, the vascular tunic, is richly supplied with blood vessels and also contains some pigment. This layer, called the **choroid** (Gr. *chorion* = skinlike membrane enclosing the fetus), in the posterior part of the eyeball helps nourish the underlying retina. Choroid folds sometimes extend into the vitreous body. They form the **falciform process** in some fish, the **papillary cone** in lizards, and the **pecten** in birds (see Fig. 11–22**C** and **D**). The pecten is sometimes elaborate and may contain as many as 30 folds. Its destruction lowers greatly the oxygen content within the eyeball. Some investigators have postulated that by casting a shadow on the retina, these processes also help the animal detect the movement of an image across the retina. Pigment in the choroid and in the adjacent pigmented layer of the retina prevents a scattering of light and a blurring of the visual image.

It is advantageous that as much light as possible reaches the photoreceptive cells of the retina in vertebrates that are active in dim light: many fishes, a few reptiles (crocodiles), and many mammals. These animals have a **tapetum lucidum** (L. *tapete* = carpet + *lucidus* = shining) behind part of the retina; it reflects light that has passed through the receptive cells back onto them. The receptive cells are thus adequately stimulated, but the trade-off is some blurring of the image. The tapetum may be located in the choroid or in the adjacent part of the retina. The reflective layer is composed of specially arranged collagen fibers, extracellular plates of guanine, or purine rods within certain cells. The eyeshine of nocturnal animals caught in automobile headlights is light reflected by the tapetum.

The vascular tunic and accompanying nonnervous retinal tissue continue beside the lens and turn in front of it to form the **iris** (Gr. *iris* = rainbow). Muscle fibers within the iris regulate the size of its opening, called the **pupil**, and hence the amount of light passing through the lens (Fig. 11–18). These muscle fibers, of retinal origin, come from ectoderm and usually are smooth. Both circularly arranged sphincter and radially arranged dilator muscle fibers are present in most vertebrates. The pupil usually is a round opening, but its shape is slitlike in many vertebrates, such as cats and some lizards, which are active under a wide range of ambient light levels. The halves of the iris on either side of the slit can be drawn far apart, like curtains, in dim light and almost completely closed in bright light to protect the very sensitive retina. The "bunching up" of contracting tissues around a circular pupil prevents it from being closed as completely.

The vascular tunic and associated nonnervous retinal tissue adjacent to the lens form the **ciliary body.** Delicate **zonule fibers** extend from it to the lens, which they help hold in place (Fig. 11–19). Smooth muscle fibers of retinal origin in the ciliary body focus the lens in different ways in different groups of vertebrates. **Ciliary processes** of the ciliary body also secrete the lymphlike aqueous humor into the eyeball's **posterior chamber,** the space between the lens and the

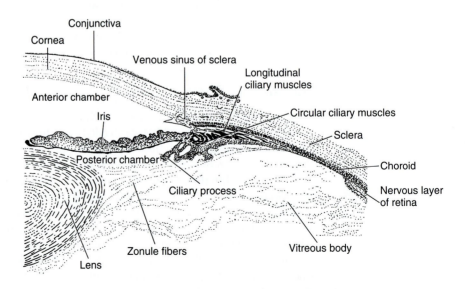

FIGURE 11–19   Detailed structure of the front of a mammalian eye. (After Fawcett.)

iris. From here the aqueous humor flows into the **anterior chamber,** located between the iris and the cornea. It finally drains into the bloodstream through the **venous sinus of the sclera** that encircles the eye at the junction of the iris and cornea. The aqueous humor nourishes the avascular cornea and lens and develops an intraocular pressure that helps maintain the shape of the eyeball.

The Retina

The retina is the third tissue layer of the eyeball. Its most peripheral portion and its **pigmented layer** become associated with the choroid, ciliary body, and iris, as we have seen. Its light-receptive and nervous portion is limited to the posterior part of the eyeball and consists of five layers of cells (Fig. 11–20**A**). The photo-receptive **rods** and **cones** lie deep in the retina next to the pigment layer. Processes of the pigment cells extend around the receptive region of the rods and cones in many nonmammalian vertebrates. Pigment migrates into the processes during bright illumination, thereby partly shielding the cells, and withdraws when light levels fall. Four neuron layers lie between the rods and cones and the vitreous body: an **outer plexiform layer** composed of horizontal cells, a **layer of bipolar neurons,** an **inner plexiform layer** formed by amacrine cells, and a **ganglion cell layer** next to the vitreous body. The axons of the ganglion cells course along the surface of the retina and turn inward at the **optic disk** to form the **optic nerve** (Fig. 11–18). Since there are no receptors in the optic disk, this is a **blind spot.** Because of the way the eye evolved and develops embryonically (p. 435), light must pass through all of the neuron layers of the retina and through the bases of the rods and cones to activate the outer segments of the rods and cones. The outer segment of a photoreceptive cell is considered a modified cilium because it is connected to the rest of the cell by a narrow, cilia-like stalk (Fig. 11–20**B**). The stalk contains the characteristic nine peripheral microtubules but lacks the central two.

Membranes within the outer segments of the rods and cones contain the pigments that absorb light energy and convert it into chemical energy that activates the cell and initiates a nerve impulse. All the pigments in vertebrate eyes consist of a protein, known as an **opsin,** that is combined with an aldehyde of vitamin A_1 or A_2, called **retinene.** The proteins vary in different pigments and determine the spectrum of light energy that is absorbed. The pigment in the rods, known as **rhodopsin,** or **porphyropsin** in some species, absorbs light over a wide spectral range, but the range varies with the species (p. 438). Cones contain pigments, known as **iodopsins,** with more restricted bands of absorption. Multiple pigments, sometimes in one cell but more often in different cells, are a prerequisite for color vision. Mammals with color vision have three types of cones, each having an absorption maximum in a different part of the spectrum: 450 nonometers (blue), 525 nanometers (green), and 550 nanometers (red). Retinal photopigments that are sensitive to ultraviolet light have been found in many vertebrates, especially in fishes and birds. Studies indicate that many of these animals can detect ultraviolet light, but the adaptive significance of this is not clear (Jacobs, 1992).

Rods are insensitive to colors, but they are very sensitive to light and have a very high degree of convergence within the retina. That is, a great many rods

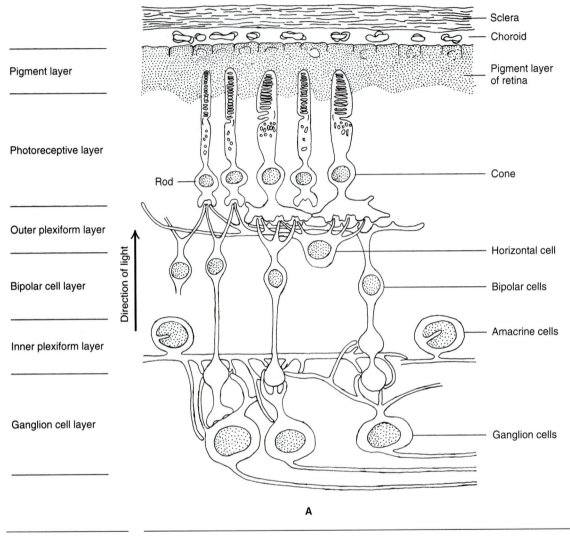

Pigment layer

Photoreceptive layer

Outer plexiform layer

Bipolar cell layer

Inner plexiform layer

Ganglion cell layer

Direction of light

Sclera

Choroid

Pigment layer of retina

Rod

Cone

Horizontal cell

Bipolar cells

Amacrine cells

Ganglion cells

A

FIGURE 11–20 Retinal structure: **A,** The layers of the eyeball and of the connections among cells within the retina.

synapse with a few bipolar cells, and these converge on still fewer ganglion cells. The human retina contains approximately 120 million rods and 5 million cones, but only the axons of 1 million ganglion cells form the optic nerve. A small amount of light falling on many rods in the same convergent pathway can summate and generate nerve impulses in the bipolar and ganglion cells. The trade-off, however, is a decrease in visual acuity because a distinction cannot be made between spots of light falling on the same convergent pathway. Rods can function in dim light, but they cannot distinguish colors, and rod vision is slightly blurred. Cones distinguish colors but require brighter illumination. They have far less convergence and so form much sharper images.

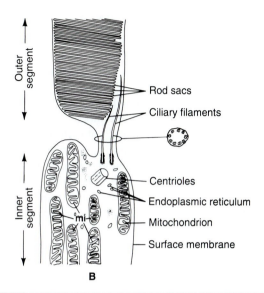

B

FIGURE 11–20   *continued*
B, An enlargement of the junction between the outer and the inner segment of a rod. (**A**, after Dowling and Boycott. **B**, after Giese.)

The eye of humans and many other vertebrates contains both rods and cones, but these receptors are not uniformly distributed. A small central area of the retina has a high concentration of cells with little convergence, and the **fovea** within it contains only cones. This is the region of highest visual acuity, and we adjust our eyes so that images fall here when we wish to discriminate fine points of light. The retina is thinned out in the fovea, so the receptors lie in a little pit. The pit may act as a diverging lens and further enhance visual acuity. The number of cones decreases rapidly from the fovea toward the periphery of the retina, whereas the number of rods increases to a maximum about 20° from the fovea.

Many additional types of interconnections occur within the retina. The retinal regions from which ganglion cells receive signals partly overlap. Horizontal connections among rods, cones, and bipolar neurons are made by the horizontal cells in the outer plexiform layer. Horizontal interconnections among bipolar and ganglion cells are made by the amacrine cells of the inner plexiform layer. Some of these interconnections inhibit and others facilitate the transmission of impulses. The significance of these arrangements is not entirely clear, but some sharpen image boundaries and facilitate the detection of motion. Obviously, a great deal of signal processing occurs in the retina—not unexpectedly, since the retina is a part of the brain. (Dowling, 1987.)

The Origin and Development of the Eye

How did an organ as complex as the eyeball evolve? Clues can be found among some of the protochordates and in the embryonic development of the eyeball. Amphioxus has ciliated, photoreceptive cells in the lumen of its central nervous system. Most lie on the floor or lateral walls of the lumen, and they are particularly numerous anteriorly. Although shielded by pigment from stimuli from beneath the animal, they can be activated by light from above because light easily passes through the thin integument and body wall. These photoreceptive cells probably are homologous to the vertebrate rods and cones because the rods and cones develop embryonically from ciliated **ependymal epithelial** cells that lie in the lining of the neural tube.

As the ancestors of vertebrates became larger and more active animals, selection would favor a concentration of photoreceptive cells near the anterior end of the body, in the lumen of part of the evolving brain. An evagination of this region would bring these cells closer to the body surface. The selective advantage of having the photoreceptive part of the brain near the body surface would be enhanced if adjacent tissues were modified to help light reach it. The cornea and lens may have evolved in this way.

This scenario is hypothetical, but a similar sequence of events occurs during the embryonic development of the vertebrate eye. The first indication of eye development is the evagination of an **optic vesicle** from the lateral wall of the diencephalon (Fig. 11–21). As this vesicle grows toward the body surface, its proximal part narrows as an **optic stalk,** and its distal part invaginates to form a two-layered **optic cup.** A deep **choroid fissure** on the ventral surface of the stalk and cup lodge blood vessels supplying the developing eye. The outer layer of the optic cup becomes the pigment layer of the retina, and the inner layer differentiates into the photoreceptive cells and neuronal layers of the retina. The outer, receptive segments of the rods and cones develop from cilia of the ependymal epithelium lining the optic cup. Because of the way the optic cup develops, these cilia are still directed toward the lumen. After the lumen has narrowed later in develop-

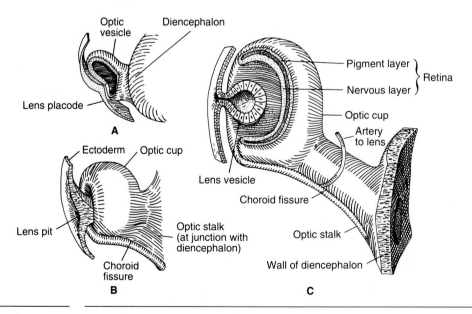

FIGURE 11–21   The development of the eyeball. (After Arey.)

ment, the receptive segments of the rods and cones lie next to the pigment layer of the retina, and light must pass through the nervous layers of the retina to reach them.

As the optic cup approaches the body surface, it induces the surface ectoderm first to thicken as a **lens placode** and then to invaginate and form a **lens vesicle** that will differentiate into the lens. Adjacent mesodermal tissues encapsulate the lens and optic cup and form the fibrous and vascular tunics. Tissue from the ectodermal optic cup contributes to the iris and ciliary body and forms their muscles. Embryonic blood vessels that cross the vitreous body to supply the developing lens atrophy after birth; otherwise, they would cast shadows on the retina.

Adaptations of the Eye

The Eye of Fishes

Fishes live in an aquatic environment, and water continuously bathes and cleanses the cornea surface. Tear glands are unnecessary and never evolved. Most fishes also lack movable eyelids, but a few have stationary skin folds above and below the eye.

Light levels change more slowly in most aquatic environments than on land, and the pupillary response of many fishes is slow. Dilator fibers are absent in the iris of many species, and the pupil slowly expands when the sphincter muscles relax.

Since the refractive index of water is close to that of the cornea, light waves pass through the cornea of a fish without being bent or refracted. The denser lens, therefore, becomes the primary refractive structure in a fish's eye, and it must be thick to provide adequate refraction (Fig. 11–22**A**). The spherical lens is composed of modified layers of epithelial cells. Successive layers have different curvatures and slightly different optical properties, so light is gradually refracted toward the optic axis as it passes through them. This avoids the image distortion due to spherical aberration that would occur if the lens were a homogeneous sphere.

Water rapidly absorbs most light waves, restricting visibility. Most fishes are somewhat nearsighted because the lens is close to the cornea and far from the retina. To focus sharply on a distant object, certain muscles move the fish lens closer to the retina. Lens-to-retina distance is analogous to lens-to-film distance in a camera. (To focus on a near object, the lens-to-film distance is increased by moving the lens away from the film; to focus on a distant object, this distance is decreased.) A lamprey accommodates for a more distant object by contracting a **corneal muscle** that pulls the cornea inward and pushes the lens back. An actinopterygian has a small **retractor lentis** muscle of ectodermal origin that attaches to the lens and pulls it toward the retina. Appropriately for a large predator, the eyes of a shark at rest are focused on somewhat more distant objects. To see a nearer object, an ectodermal **protractor lentis** muscle pulls the lens away from the retina. In all cases, intraocular pressures restore the lens to its resting position when the focusing muscle relaxes.

Light intensity is reduced beneath many bodies of water, and long wavelengths of light, especially reds and oranges, are quickly absorbed. Correlated with

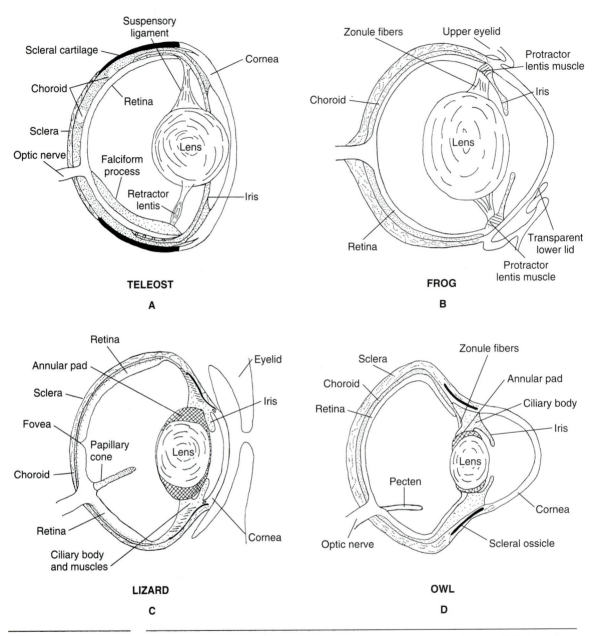

FIGURE 11–22 The eyeballs of representative vertebrates: **A,** A teleost. **B,** A frog. **C,** A lizard. **D,** An owl. (After Walls.)

this, the retina of many fishes consists primarily of rods. The maximum sensitivity for pigments in the photoreceptive cells of fishes living in marine coastal and fresh water is in the green-yellow range, corresponding to the maximum available light (Crescitelli, 1972). Only blue light penetrates far into the water, and species living in deeper oceanic waters have a pigment that absorbs maximally in the blue part of the spectrum. Only fishes living in brightly illuminated habitats, such as coral reefs and many clear freshwater lakes, have many cones. As would be expected, many fishes have a tapetum lucidum that reflects light back onto the retina.

The Eye of Terrestrial Vertebrates

Since they are not constantly bathed in water, the eyes of terrestrial vertebrates must be protected and kept moist in other ways. The tetrapod eye usually has one or more **eyelids** that can move across its surface and protect and cleanse it. The amphibian's eye has a stationary upper eyelid but a movable and transparent lower one (Fig. 11–22**B**). Amphibians also can retract the eyeball deeper into the orbit so that the lids completely close over it. In amniotes, both upper and lower lids are movable, and a third, transparent eyelid, the **nictitating membrane** (L. *nictare* = to wink), is usually present. The nictitating membrane normally is retracted into the median corner where upper and lower lids come together, but it can be flicked across the surface of the cornea. In human beings it is reduced to a vestigial **semilunar fold. Eyelashes** are associated with the eyelids of mammals. Tetrapods also have evolved **lacrimal, harderian,** or other **tear glands** whose secretions flow across the cornea or are spread across it by the movement of the eyelids (Fig. 11–23). The tears of mammals drain into a **lacrimal duct** that opens into the nasal cavity. Modified sebaceous glands, known as **tarsal glands,** open on the edges of the eyelids; their secretions tend to prevent the tears from flowing onto the face.

The cornea of the terrestrial vertebrate eye is important in light refraction because its index of refraction is much greater than that of air. Its curvature is critical for light refraction. The lens is less important in refraction than in fishes, is thinner along the optic axis, and has less of a curvature. In an amphibian the lens changes shape during metamorphosis from a spherical shape in the aquatic larva to a more oval shape in the adult (Fig. 11–22**B**). It is even thinner in amniotes (Fig. 11–22**C** and **D**). Distant vision is important for most terrestrial vertebrates. The lens is positioned closer to the retina than in fishes, so distant objects are in focus when the eye is at rest.

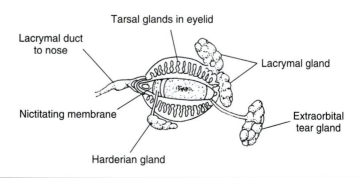

Tarsal glands in eyelid

Lacrymal duct to nose

Lacrymal gland

Nictitating membrane

Extraorbital tear gland

Harderian gland

FIGURE 11–23   The lacrimal apparatus of a mammal. (After Portmann.)

Considerable independent evolution of the eye has occurred among terrestrial vertebrates, and their methods of accommodation differ considerably. The amphibian eye accommodates for a close object by contracting a **protractor lentis** muscle that pulls the lens away from the retina and toward the cornea. Amniotes have elastic lenses and accommodate for near objects by changing lens shape. This is analogous to changing lenses in a camera. Except in snakes, the lens of a reptilian eye is encircled by an **annular pad** that lies adjacent to the ciliary body (Fig. 11–22**C**). During accommodation for close objects, a sphincter-like muscle within the ciliary body exerts a force on the annular pad, the lens bulges, its refractive powers are increased, and the image is brought into focus. The eyes of snakes degenerated to some extent during a period when their ancestors were burrowing, and then re-evolved as snakes readapted to live on the ground surface. Ciliary muscles were lost, and accommodation for distant objects was accomplished by the contraction of iris muscles that push the lens deeper into the eyeball.

The flying mode of life requires a keen sense of sight and quick adjustments to changes in light and distance. A bird's eye is extraordinarily large and is strengthened by a ring of sclerotic bones. The eyes of some hawks and owls are larger in absolute size than ours. A bird's iris muscles are striated rather than smooth and respond rapidly to changes in levels of light. (This is also the case in a few lizards.) Birds retain the reptilian annular pad mechanism for accommodation, but the ciliary muscles also act in other ways. One set of muscles pulls the lens forward into a narrowing ring of sclerotic bones, and the lens bulges; another set pulls on the cornea and changes its curvature slightly. The sclerotic bones provide a firm point of origin for these muscles.

In a mammalian eye at rest, intraocular pressure tends to push the wall of the eyeball peripherally, and this force is transferred to the elastic lens by the ciliary body and zonule fibers. The lens is under tension and somewhat flattened, and distant objects are in focus on the retina. Contraction of circular and longitudinal muscles within the ciliary body bring it closer to the lens (Fig. 11–19). Tension is released, the lens bulges slightly, and nearer objects come into focus. The lens is restored to its resting shape in amniotes by the relaxation of ciliary muscles and the action of the intraocular pressure.

Since the level of illumination can be much higher on land than in water, terrestrial vertebrates frequently have more cones and a higher visual acuity than many fishes, and they are more likely to have color vision. Most amphibians, which are secretive and shade-dwelling creatures, do not have many cones in their retinas, but some frogs are an exception and have been reported to have color vision. Cones are abundant in the retinas of reptiles and birds, and most of these animals have well-developed color vision and high visual acuity. Diurnal and predacious hawks have exceptionally keen sight. Their eyes are directed forward, so the two visual fields overlap to a large extent, but not completely. Two foveae—a central one on the optic axis, and one placed more laterally—contain as many as a million cones per square millimeter. Images of objects that are straight ahead of the hawk fall on the central fovea of each eye, and concurrent images of objects to either side fall on the lateral foveae. Unlike reptiles and birds, early mammals are nocturnal creatures, have few cones but many rods, and are colorblind. Color vision has evolved in primates and a few other primarily diurnal groups.

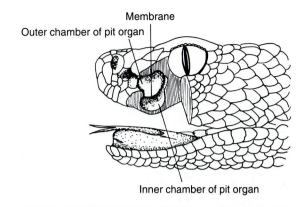

Membrane

Outer chamber of pit organ

Inner chamber of pit organ

FIGURE 11–24   A section through the pit organ of a rattlesnake. (After Gamow and Harris.)

Thermoreceptors

Many vertebrates are able to detect thermal changes in their environment. Such an ability is critical for thermal regulation in endothermic birds and mammals, but the mechanism for detecting temperature changes is less certain than for detecting changes in the visual part of the electromagnetic spectrum. Neurophysiologists have demonstrated small areas of mammalian skin that are sensitive to heat or cold (warm and hot spots), and certain cells in the mammalian hypothalamus respond to changes in blood temperature. Some investigators have proposed that specific encapsulated neuron endings in the skin are hot or cold receptors. Others believe that temperature changes are detected by the combination and pattern of stimulation of receptors and free neuron endings.

Well-formed heat sensors are present in some snakes. In boa constrictors and pythons these are shallow pits on the scales bordering their mouths, and a rattlesnake has a pair of deep **pit organs** (hence the animal's other name, pit viper) on the face between the eyes and nostrils (Fig. 11–24). Highly branched free neuron endings cover a delicate membrane within the pit. These pits are sensitive to slight differences in infrared radiation between the warm-blooded prey of the snake and the prey's surroundings. If the pits have not been obstructed, a rattlesnake can strike at a bird or mouse with uncanny accuracy, even in the dark.

SUMMARY

1. Free nerve endings, receptor cells, and aggregations of receptor cells called sense organs provide vertebrates with essential information about changes in their internal and external environments.
2. The sense of smell, or olfaction, is detected by olfactory neurons. These cells are located in the nasal cavities, and their processes to the brain constitute the olfactory nerve. Vertebrates differ in the ways that they move water or air across these cells.
3. Most terrestrial vertebrates have a distinct group of olfactory cells that form

the vomeronasal organs. Often, pheromones and other odors of importance in the animal's social interactions are detected by this mechanism. The terminal nerve may also be part of a system for detecting pheromones dealing with reproduction.

4. Gustatory receptors are taste buds of endodermal origin that are usually located within the oral cavity and pharynx. They are supplied by the facial, glossopharyngeal, and vagus nerves. Many migrate over the body surface in some fish species.

5. The skin of vertebrates contains many free neuron endings that are stimulated by injuries. In addition, mammals have specialized touch and pressure receptors.

6. Proprioceptors include tendon organs and muscle spindles. They monitor the force, the degree, and the rate of contraction of skeletal muscles. They provide information by which these factors are adjusted during muscle contraction.

7. The lateral line system of fishes and larval amphibians consists of linearly arranged neuromasts usually located within canals in the skin. They provide a fish with information on currents, its own movements, and low-frequency vibrations. Lateralis neurons return to the brain in the facial, glossopharyngeal, and vagus nerves.

8. The electroreceptors found in many fishes are modified parts of the lateral line system. Many fishes can sense the electric activity of the muscles of other animals as they search for prey; some others also generate and receive electric pulses in electrolocation and electrocommunication systems.

9. The inner ear is closely related to the lateral line system. It consists of a membranous labyrinth that is surrounded by perilymph and lodged in an osseous labyrinth in the otic capsule. The receptive cells lie within the membranous labyrinth.

10. Certain groups of hair cells within the membranous labyrinth detect the movements and orientation of an animal in space (equilibrium) and others detect sound waves.

11. A sound source in water generates low-frequency particle displacement waves and higher frequency pressure waves. Fishes detect low frequencies primarily by the lateral line; they detect higher frequencies by the sacculus. Some fishes use the swim bladder and a set of weberian ossicles as a hydrophone.

12. Low-frequency sound waves can be detected by tetrapods through superficial skull bones, but a tympanic membrane beside an air-filled middle ear cavity is needed to convert pressure waves above 1000 Hz into displacement waves. Movements of the tympanic membrane are transmitted by one or more auditory ossicles across the middle ear to the inner ear. Here they set up waves in a special perilymphatic channel that carries them to the receptive cells.

13. The tetrapod ear shows a great deal of variation. A tympanic membrane has probably evolved independently in at least three lines: frogs, most reptiles and birds, synapsids and mammals.

14. In most reptiles and birds the ear has a tympanic membrane and a single auditory ossicle, the stapes. Amphisbaenians and snakes have secondarily lost the tympanic membrane, and the stapes abuts on the quadrate bone.

15. A frog's ear has a tympanic membrane, which connects to the stapes, and an

operculum, which connects by a muscle to the pectoral girdle and front leg. The tympanic membrane detects high-frequency mating calls; the opercular system detects low-frequency environmental noises and seismic vibrations. Adult salamanders have only the opercular system.

16. The mammalian ear has a tympanic membrane and three auditory ossicles, two of which bore the jaw joint in synapsids.

17. The lagena has become the cochlear duct of the inner ear, and perilymphatic channels form the rest of the cochlea.

18. Most early fishes, amphibians, and reptiles had a median eye complex consisting of a pineal eye, parietal eye, or both. This complex is retained in many contemporary fishes, amphibians, and reptiles. The median eye monitors ambient light and appears to initiate physiological adjustments to light levels. The complex is represented in birds and mammals by the pineal gland.

19. The image-forming eyes of vertebrates are essentially alike in all species and consist of three layers of tissue. The fibrous tunic forms the cornea and sclera; the vascular tunic forms the choroid, ciliary body, and iris; and the retina is the deepest layer.

20. Light is received by the rods and cones. Rods function in dim light, but they cannot distinguish color, and rod vision is slightly blurred. Cones distinguish colors but require brighter illumination. They form much sharper images.

21. Since light does not travel far in water, most fishes have nearsighted eyes with a spherical lens located close to the cornea. Most fishes focus on more distant objects by contracting a muscle that pulls the lens toward the retina.

22. The surface of the tetrapod eyeball is protected by movable eyelids and bathed by tears secreted by tear glands.

23. The eye of most tetrapods at rest is focused on more distant objects than a fish's eye. An amphibian accommodates for near objects by contracting a muscle that pulls the lens toward the cornea; except for snakes, amniotes do it by increasing the curvature of the lens.

24. Vertebrates can detect temperature changes, but we are uncertain as to the receptors in many cases. Boa constrictors, pythons, and pit vipers have one or more pairs of thermoreceptive pits on their heads that help them detect their warm-blooded prey.

REFERENCES

Allin, E. F., 1975: Evolution of the mammalian middle ear. *Journal of Morphology*, 147: 403–438.

Arey, L. B., 1957: *Developmental Anatomy*. 6th edition. Philadelphia, W. B. Saunders Company.

Atema, J., 1980: Smelling and tasting under water. *Oceanus*, 23:4–18.

Bertmar, G., 1981: Evolution of vomeronasal organs in vertebrates. *Evolution*, 35:359–366.

Blaxter, J. H. S., 1987: Structure and development of the lateral line. *Biological Reviews of the Cambridge Philosophical Society*, 62:471–514.

Boord, R. L., and McCormick, C. A., 1984: Central lateral line and auditory pathways: A phylogenetic perspective. *American Zoologist*, 24:765–774.

Brown, R. E., and Fedde, M. R., 1993: Airflow sensors in the avian wing. *Journal of Experimental Biology,* 179:13–30.

Buck, L., and Axel, R., 1991: A novel multigene family may encode odorant receptors: A molecular basis for odor recognition. *Cell*, 65:75–187.

Chouchkov, Ch., 1978: *Cutaneous Receptors*. Berlin, Springer-Verlag.

Cock Buning, T. de, 1983: Thermal sensitivity as a specialization for prey capture and feeding in snakes. *American Zoologist*, 23:363–375.

Coombs, S., Gorner, P., and Munz, H., editors, 1989: *The Mechanosensory Lateral Line, Neurobiology and Evolution*. New York, Springer-Verlag.

Corwin, J. T., 1981: Audition in elasmobranchs. *In* Tavolga, W. N., Popper, A. N., and Fay, R. R., editors: *Hearing and Sound Communication in Fishes*. New York, Springer-Verlag.

Crescitelli, F., 1972: The visual cells and visual pigments of the vertebrate eye. *In* Dartnall, H. J. A., editor: *Handbook of Sensory Physiology*, vol. 7. Berlin, Springer-Verlag.

Crescitelli, F., 1977: *The Visual System in Vertebrates*. New York, Springer-Verlag.

Dowling, J. E., 1987: *The Retina, an Approachable Part of the Brain*. Cambridge, Harvard University Press.

Dowling, J. E., and Boycott, B. B., 1966: Organization of the primate retina: electron microscopy. *Proceedings of the Royal Society of London, Series B*, 166:80–111.

Dijkgraaf, S., and Kalmijn, A, J., 1966: Function of ampullae of Lorenzini. *Z. Vergl. Physiol.*, 53:187–194.

Eakin, R. M., 1973: *The Third Eye*. Berkeley, University of California Press.

Eddy, J. M. P., 1972: The pineal complex. *In* Hardistry, M. W., and Potter, I. C., editors: *The Biology of Lampreys*. London, Academic Press.

Fay, R. C., and Popper, A. N., 1985: The octavolateralis system. *In* Hildebrand, M., Bramble, D. M., Liem, K. F., and Wake, D. B., editors: *Functional Vertebrate Morphology*. Cambridge, Harvard University Press.

Feng, A. S., and Hall, J. C., 1991: Mechanoreception and phonoreception. *In* Prosser, C. L., editor: *Neural and Integrative Animal Physiology*. New York, Wiley-Liss.

Fritzsch, B., and Wahnschaffe, U., 1983: The electroreceptive ampullary organs of urodeles. *Cell Tissue Research*, 229:483–503.

Gamow, I., and Harris, J. F., 1973: The infrared receptors of snakes. *Scientific American*, 228:94–100.

Giese, A. C., 1979: *Cell Physiology*, 5th edition. Philadelphia, W. B. Saunders Company.

Goldsmith, T. H., 1991: Photoreception and vision. *In* Prosser, C. L., editor: *Neural and Integrative Animal Physiology*. New York, Wiley-Liss.

Gould, S. J., 1990: An earful of jaw. *Natural History*, March, 12–23.

Hanson, M., Westerberg, H., and Öblad, M., 1990: The role of magnetic statoconia in dogfish (*Squalus acanthias*). *Journal of Experimental Biology*, 151:205–218.

Hartline, P. H., 1971: Physiological basis for detecting sound and vibrations in snakes. *Journal of Experimental Biology*, 54:349–371.

Hetherington, T. E., Jaslow, A. P., and Lombard, R. E., 1986: Comparative morphology of the amphibian opercularis system: I. General design features and functional interpretation. *Journal of Morphology*, 190:42–61.

Hillenius, W. J., 1992: The evolution of nasal turbinates and mammalian endothermy. *Paleobiology* 18:17–29.

Hodgson, E. S., and Mathewson, R. F., editors, 1978: *Sensory Biology of Sharks, Skates, and Rays*. Arlington, Va., Office of Naval Research, Department of the Navy.

Hueter, R. E., symposium editor, 1991: Vision in elasmobranchs. *Journal of Experimental Zoology*, Supplement 5:1–141.

Jacobs, G. H., 1992: Ultraviolet vision in vertebrates. *American Zoologist*, 32:544–554.

Johansson, R. S., and Vallbo, A. B., 1983: Tactile sensory coding in the globrous skin of the human hand. *Trends in Neurosciences*, 6:27–32.

Kalmijn, A. J., 1978: Experimental evidence of geomagnetic orientation in elasmobranch fishes. *In* Schmidt-Koenig, K. and Keeton, W. T., editors: *Animal Migration, Navigation and Hearing*. New York, Springer-Verlag.

Kalmijn, A. J., 1988: Electromagnetic orientation, a relativistic approach. *In* O'Conner, M. E., and Lovely, R. H., editors: *Electromagnetic Fields and Neurobehavioral Functions*. New York, John Wiley and Sons.

Kingsbury, B. G., and Reed, H. D., 1909: The columella auris in Amphibia. *Journal of Morphology*, 20:549–628.

Kleerekoper, H., 1969: *Olfaction in Fishes*. Bloomington, Indiana University Press.

Levine, J. S., 1985: The vertebrate eye. *In* Hildebrand, M., Bramble, D. M., Liem, K. F., and Wake, D. B., editors: *Functional Vertebrate Morphology*. Cambridge, Harvard University Press.

Lombard, R. E., and Bolt, J. R., 1979: Evolution of the tetrapod ear: An analysis and reinterpretation. *Biological Journal of the Linnean Society*, 11:17–76.

Lombard, R. E., and Straughan, I. R., 1974: Functional aspects of anuran middle ear structures. *Journal of Experimental Biology*, 61:71–93.

Meisami, E., 1991: Chemoreception. *In* Prosser, editor: *Neural and Integrative Animal Physiology*. New York, Wiley-Liss.

Northcutt, R. G., 1989: The phylogenetic distribution and innervation of cranial mechanosensory lateral line. *In* Coombs, S., Gorner, P., and Munz, H., editors: *The Mechanosensory Lateral Line, Neurobiology and Evolution*. New York, Springer-Verlag.

Northcutt, R. G., and Davis, R. E., editors, 1983: *Fish Neurobiology*. Ann Arbor, University of Michigan Press.

Parrington, F. R., 1979: The evolution of the mammalian middle and outer ears: A personal review. *Biological Reviews*, 54:369–387.

Parsons, T. S., 1967: Evolution of nasal structure in the lower tetrapods. *American Zoologist*, 7:397–413.

Parsons, T. S., editor, 1966: The vertebrate ear. *American Zoologist*, 6:368–466.

Platt, A. C., Popper, A. N., and Fay, R. R., 1989: The ear as part of the octavolateralis system. *In* Coombs, S., Gorner, P., and Munz, H., editors: *The Mechanosensory Lateral Line, Neurobiology and Evolution*. New York, Springer-Verlag.

Popper, A. N., 1980: *Comparative Studies of Hearing in Vertebrates*. New York, Springer-Verlag.

Popper, A. N., and Fay, R. R., 1977: Structure and function of the elasmobranch auditory system. *American Zoologist*, 17:443–452.

Ralph, C. L., Firth, B. T., Gern, W. A., and Owens, D. W., 1979: The pineal complex and thermoregulation. *Biological Reviews*, 54:41–72.

Raschii, W., 1986: A morphological analysis of the ampullae of Lorenzini in selected skates. *Journal of Morphology*, 189:225–247.

Ray, P. H., and Kruger, L., 1983: Spacial properties of receptive fields of mechanosensitive primary afferent fibers. *Federation Proceedings, Federation of American Societies of Experimental Biology*, 42:2536–2541.

Retzius, G., 1881–1884: *Das Gehororgan der Wirbelthiere. Morphologischhistologische Studien*, 2 volumes. Stockholm, Samson and Wallin.

Sorensen, P. W., Hara, T. J., and Stacey, N. E., 1987: Extreme olfactory sensitivity of goldfish to a steroidal pheromone. *Journal of Comparative Physiology*, 160:305–314.

Studnicka, F. K., 1905: Die Parietalorgane. *In* Oppel, A., editor, *Lehrbuch der vergleichenden mikroskopischen Anatomie der Wirbeltiere*, part 5. Jena, Gustav Fishcher.

Szabo, T., 1974: Peripheral and central components in electroreception. *In* Fessard, A., editor: *Handbook of Sensory Physiology*, volume 3, part 3: 13–58. Amsterdam, Elsevier.

Tavolga, W. N., Popper, A. N., and Fay, R. C., 1981: *Hearing and Sound Communication in Fishes*. New York, Springer-Verlag.

van Bergeijk, W. A., 1967: The evolution of vertebrate hearing. *In* Neff, W., editor: *Contributions to Sensory Physiology*. New York, Academic Press.

Waldvogel, J. A., 1990: The bird's eye view. *American Scientist*, 78:342–353.

Walker, W. F., Jr., 1981: *Dissection of the Frog*, 2nd edition. San Francisco, W. H. Freeman and Company.

Walls, G. L., 1942: *The Vertebrate Eye and Its Adaptive Radiation*. Bloomfield Hill, Mich., Cranbrook Institute of Science.

Webster, D. B., Fay, R. R., and Popper, A. N., editors, 1992: *The Evolutionary Biology of Hearing*. New York, Springer-Verlag.

Wever, E. G., 1978: *The Reptilian Ear: Its Structure and Function*. Princeton, Princeton University Press.

Wever, E. G., 1985: *The Amphibian Ear*. Princeton, Princeton University Press.

Wurtman, R. J., Axelrod, J., and Kelley, D. E., 1968: *The Pineal*. New York, Academic Press.

Zimmerman, K., and Heatwolfe, H., 1990: Cutaneous photoreception, a new sensory mechanism for reptiles. *Copeia*, 1990:860–862.

12

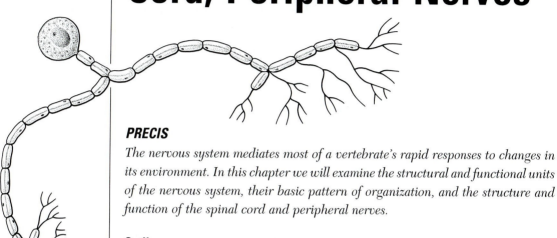

The Nervous System: Organization, Spinal Cord, Peripheral Nerves

PRECIS

The nervous system mediates most of a vertebrate's rapid responses to changes in its environment. In this chapter we will examine the structural and functional units of the nervous system, their basic pattern of organization, and the structure and function of the spinal cord and peripheral nerves.

Outline

Most of the rapid responses to changes in a vertebrate's external or internal environment are mediated by the nervous system. Activating the correct combination of effectors so that appropriate responses are made requires a processing of signals within the nervous system that is called **coordination** or **integration.** The nervous system is constantly bombarded with sensory signals. Many are filtered as they pass to higher brain centers so that some are suppressed and others enhanced. Often, sensory inputs from different types of receptors are combined to provide a more complete "picture" of changes taking place. The sensory input may be combined with stored information (memories) about what happened under similar circumstances in the past experience of the individual or species before a response is made. A sheep, for example, senses another animal nearby. If this is another grazing ewe, the sensory signals may be ignored, but if the approaching animal is a wolf or a ram, the ewe should run away or perhaps assume a mating stance. Some responses are affected by the activity of endocrine glands. A ewe under the influence of sex hormones in the mating season will behave very differently toward a ram than at other times. The motor responses themselves require a complex pattern of signals. Escaping a predator requires not only the sequential activation of certain muscles and the inhibition of antagonistic ones, but also many physiological adjustments. Rates of breathing and circulation must increase, energy reserves must be made available to the limb muscles, and so on. Often, information about the outcome of the response is fed back into the nervous system and reinforces or alters the memories stored there. Conditioned reflex experiments have shown that all groups of vertebrates can be trained to perform simple discrimination tasks. Although much vertebrate behavior is innate and based on inherited neuronal connections, vertebrates do learn from experience.

Components of the Nervous System

Neurons and Synapses

The primary structural and functional units of the nervous system are the nerve cells, or **neurons** (Gr. *neuron* = nerve). All are ectodermal, for they develop embryonically from the neural tube, from the neural crest, or from ectodermal placodes. They are arrayed in such a way that we can recognize a **central nervous system,** consisting of the brain and spinal cord, and a **peripheral nervous system,** consisting of nerves extending between the central structures and the receptors and effectors. Many morphological types of neurons are known, but all contain four essential parts: (1) a cell body or trophic segment, (2) dendrites or a receptive segment, (3) an axon or conductive segment, and (4) terminal telodendria or a transmissive segment (Fig. 12–1**A**).

The **cell body** contains the nucleus and metabolic machinery of the cell. Proteins and neurotransmitter substances that carry signals between cells are

FIGURE 12–1   Representative neurons of vertebrates: **A,** Sensory and motor neurons. The segments of a ▶ neuron are indicated for the motor neuron. **B,** A Purkinje cell of the cerebellum. (After Ramon y Cajal.)

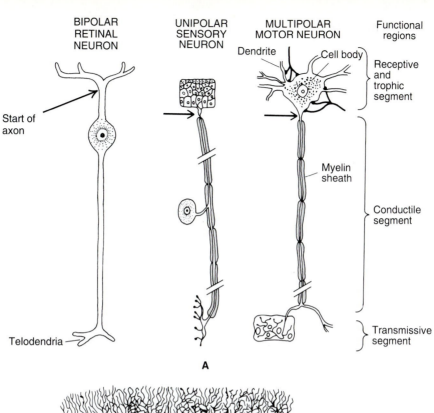

BIPOLAR
RETINAL
NEURON

UNIPOLAR
SENSORY
NEURON

MULTIPOLAR
MOTOR NEURON

Dendrite

Cell body

Functional
regions

Receptive
and
trophic
segment

Start of
axon

Myelin
sheath

Conductile
segment

Transmissive
segment

Telodendria

A

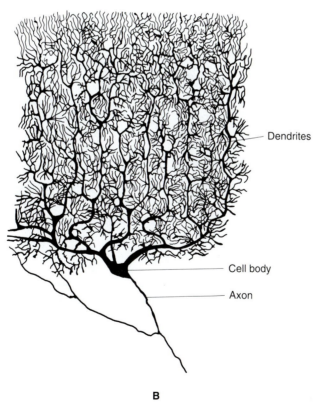

Dendrites

Cell body

Axon

B

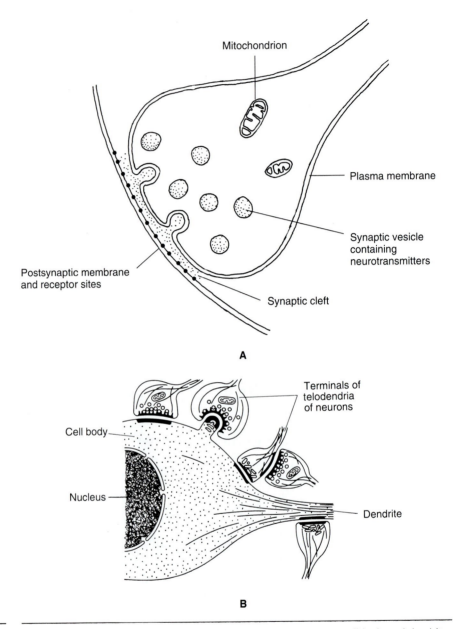

Mitochondrion

Plasma membrane

Synaptic vesicle containing neurotransmitters

Postsynaptic membrane and receptor sites

Synaptic cleft

A

Terminals of telodendria of neurons

Cell body

Nucleus

Dendrite

B

FIGURE 12–2 Synapses: **A,** The structure of a synapse. **B,** Multiple synapses on a cell body and dendrite. (**A,** after Noback. **B,** after Williams et al.)

synthesized here. In a multipolar motor neuron, the cell body lies very close to the dendrites, and a long axon extends outward from the cell body. But the cell body is not always next to the dendrites. It may lie near the middle of an axon, as in a bipolar cell from the retina, or it may be set off on one side of the axon, as in a unipolar sensory neuron. Biologists previously defined dendrites and axons with respect to their location relative to the cell body (as in the multipolar neuron), but they are now defined by their functions. Dendrites receive stimuli and initiate nerve impulses; axons conduct nerve impulses.

The **axon** (Gr. *axon* = axle or axis) is a long, slender, cytoplasmic process that is often called a **nerve fiber.** Axons conduct sensory impulses from the dendrites to the central nervous system, impulses within the central nervous system, and motor impulses from the central nervous system to the effectors. Axons frequently branch. As we have seen (p. 306), one axon branches and innervates many motor fibers to form a motor unit. The axon is specialized to conduct a nerve impulse rapidly, often over a long distance, without a diminution of signal amplitude. Its plasma membrane has an electric **resting potential** of about 60 millivolts that derives from an unequal distribution of ions across it. As with muscle cells (p. 303), a high concentration of extracellular, positively charged sodium ions causes the outside of the membrane to be positive relative to the inside. The biochemistry of a nerve impulse will not be discussed in detail, but you need to understand the main events. Briefly, when an axon receives a threshold stimulus, ion channels in the membrane open, membrane permeability changes, and the axon develops an **action potential.** Sodium ions rush into the axon, and membrane polarity is momentarily reversed. Biologists say the membrane is depolarized. This affects adjacent parts of the membrane, and a **nerve impulse** spreads as a wave of depolarization. Depolarization of any segment is followed quickly by the pumping out of sodium ions and the restoration of the resting potential. The action potential of an axon is an all-or-none response because it does not vary in amplitude with the strength of the stimulus. Stimulus strength is coded only by whether the axon is active or inactive (the binary on-off system) and by the number and pattern of nerve impulses in a unit of time.

The branching **telodendria** (Gr. *telos* = end + *dendria* = trees) at the end of an axon make contact with the receptive segments of other neurons at special junctions known as **synapses** (Gr. *synapsis* = union), or with effectors at synapse-like **myoneural** or **axoglandular junctions** (Fig. 12–2**A**). Telodendria contain many minute **synaptic vesicles,** in which transmitter substances are stored. Many types of **neurotransmitters** are known, among them acetylcholine, noradrenaline, serotonin, and dopamine. When a nerve impulse reaches the telodendria, some of the neurotransmitter substance is released, crosses the narrow **synaptic cleft,** binds with receptor sites on the postsynaptic cell, and excites or inhibits this cell. An axon can be stimulated experimentally at its center, and a nerve impulse will travel away from the point of stimulation in both directions. The impulse cannot reverse itself, however, because it cannot go back over previously stimulated membrane until the resting potential of the membrane is restored. The polarity of the nervous system depends on the synapses because nerve impulses normally cross them in only one direction, from the presynaptic neuron containing the synaptic vesicles to the postsynaptic cell containing receptive binding sites for the neurotransmitters.

The receptive segment of the neuron includes the branching **dendrites** (Gr. *dendria* = trees) and often the cell body. Dendrites of some neurons form extensive, branching trees (Fig. 12–1**B**). This segment can be stimulated directly, as it is at free nerve endings, by receptor cells, or by the telodendria of other neurons.

Thousands of synapses of many different neurons may occur on the dendrites and cell body of a single neuron (Fig. 12–2**B**). Synapses between telodendria may also be present. The magnitude of the electric potential of the plasma membrane of the dendrites and cell body may be in constant flux as different neurotransmitter substances impinge on it. Some neurotransmitters are excitatory because they partially depolarize the postsynaptic membrane; others, released by different neurons, hyperpolarize the membrane and have an inhibitory effect. A threshold level of depolarization must be reached for a nerve impulse to be initiated in the axon. Signals may be aborted, delayed, accelerated, or processed in other ways at the receptive region of a neuron. Whether a nerve impulse is generated in the postsynaptic axon depends on the number of excitatory and inhibitory impulses impinging on the receptive region, on their frequency, and on the nature of the neurotransmitter receptive sites.

Schwann Cells

All axons in the peripheral nerves are surrounded during embryonic development by **neurilemma** or **schwann cells** of neural crest origin. Usually, an inner, tonguelike process of these cells grows around a segment of the axon many times so that many layers of the Schwann cells' lipid plasma membrane are laid down and form a fatty **myelin sheath** (Gr. *myelos* = marrow) that is visible by light microscopy (Fig. 12–3**A**). Some axons have as many as a hundred layers of plasma membrane. The myelin sheath is interrupted between successive Schwann cells by the **nodes of ranvier** (Fig. 12–3**B**). The Schwann cells on some of the peripheral axons, especially those of many of the autonomic nerves, envelop but do not continue to grow around the axons. These axons are described as nonmyelinated because the layers of Schwann cell plasma membrane are too thin to be detected by light microscopy. Since the nodes of Ranvier are the only regions where the plasma membrane of a myelinated axon is close to the extracellular fluid, the wave of depolarization that constitutes the nerve impulse jumps from node to node. Impulse transmission is much more rapid in myelinated than in nonmyelinated axons. (Transmission velocity is also affected by temperature and by axon diameter, being faster at higher temperatures and in neurons of large diameter.) Mesodermal connective tissue envelops individual and groups of peripheral neurons to form a peripheral **nerve** (Fig. 12–3**C**). Nutritive blood vessels follow the connective tissue into the larger nerves.

Neuroglia

Many nonnervous cells, collectively called the **neuroglia** or **glia** (Gr. *glia* = glue), occur in the central nervous system. They are 30 to 50 times more numerous than the neurons. **Oligodendrocytes** send out processes that wrap around and myelinate central axons (Fig. 12–4). Each oligodendrocyte on average myelinates 15 axons. This is a slow process, and myelination is not completed in the human brain until many months after birth. Regions of the central nervous system containing myelinated axons are whitish in cord and brain sections that have not been stained and are called **white matter.** Most of the white matter consists of bundles of

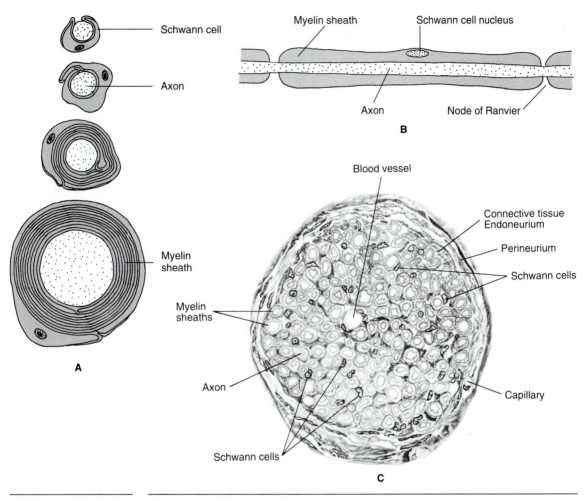

FIGURE 12–3   Axons and nerves: **A,** The myelination of an axon. **B,** A myelinated axon. **C,** A transverse section of a peripheral nerve. (**A,** after Williams et al. **C,** after Fawcett.)

axons extending between parts of the central nervous system. Such bundles are called **tracts** and not nerves. Regions containing nonmyelinated axons, dendrites, and cell bodies are **gray matter.**

Astrocytes are star-shaped glial cells that are interposed among the neuron cell bodies (Fig. 12–4). Astrocytes perform many functions. They are supportive cells. They regulate the ionic composition and electrical balance of extracellular fluids. In particular, they take up excess potassium ions that flow out of very active neurons and extrude potassium ions through their end feet to surfaces that they contact: capillaries, the pia mater on the brain surface, and the specialized **ependymal epithelium** lining the nervous system. Astrocytes also degrade metabolites, including transmitter substances released into synapses. Since they contact both capillaries and neuron cell bodies, they may have a nutritive role. Oligodendrocytes

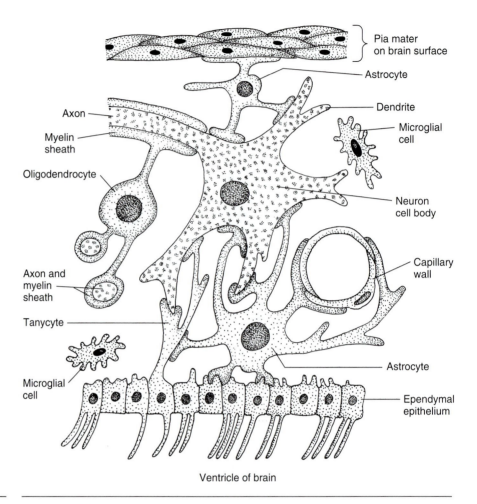

Pia mater
on brain surface

Astrocyte

Axon

Dendrite

Myelin
sheath

Microglial
cell

Oligodendrocyte

Neuron
cell body

Capillary
wall

Axon and
myelin
sheath

Tanycyte

Astrocyte

Microglial
cell

Ependymal
epithelium

Ventricle of brain

FIGURE 12–4   A section of the brain to show neuroglia, other nonnervous cells, and their relationships to central neurons and blood vessels. (After Williams et al.)

and astrocytes develop from the neural tube ectoderm. Collectively they are sometimes called **macroglia.**

Smaller **microglia** develop from macrophages of mesodermal origin (Fig. 12–4). They have a phagocytic role, removing foreign and degenerative products. **Tanycytes** in the ependymal epithelium also extend processes inward to many neuron cell bodies and presumably participate in the metabolism of nerve cells.

Organization of the Nervous System

The basic structure and plan of organization of the nervous system is the same in all vertebrates. Because of the way they develop (p. 125), the spinal cord and brain of a vertebrate are hollow and lined by a non-nervous and partly ciliated ependymal epithelium (Fig. 12–4). The adult spinal cord has a small **central canal** (Fig. 12–5) that enlarges within parts of the brain to form a series of interconnected **ven-**

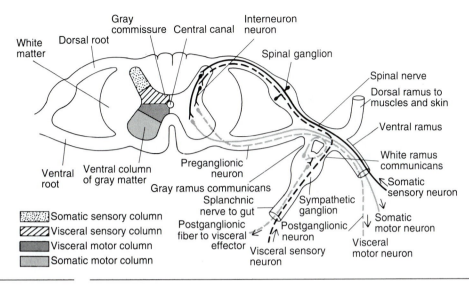

FIGURE 12–5   A transverse section through the spinal cord and nerve of a mammal to show the types of neurons that constitute the nervous system and their interrelationships.

tricles. **Cerebrospinal fluid** slowly circulates within these spaces. **Spinal nerves** attach to the cord at segmental intervals by **dorsal** and **ventral roots.** In most vertebrates the roots of each body segment unite slightly peripheral to the cord to form a spinal nerve from which **dorsal, ventral,** and **communicating rami,** or branches, extend to different parts of the body. Most of the cranial nerves appear to be serially homologous to either a ventral or a dorsal root of a spinal nerve.

Neuron Types and Their Distribution

The entire nervous system consists of three categories of neurons (Fig. 12–5). **Primary sensory,** or **afferent, neurons** (L. *ad* = toward + *ferens* = carrying, from *ferre* = to carry) carry impulses from free nerve endings or receptor cells into the central nervous system. **Motor,** or **efferent, neurons** (L. *ex* = from + *ferens* = carrying, from *ferre* = to carry) carry signals from the central nervous system to the effectors. **Interneurons** (L. *inter* = between) are confined to the central nervous system. They receive signals from the primary sensory neurons, integrate this information through their complex connections with other interneurons, and ultimately send signals to the motor neurons. Sensory and motor neurons are mixed in the spinal nerves, but they segregate beside the spinal cord. Sensory neurons always enter the cord through the dorsal roots of the spinal nerves, and in amniotes the motor neurons leave through the ventral roots.

Sensory neurons of spinal nerves develop from neural crest cells. The cell bodies of sensory neurons usually remain peripheral to the central nervous system in small swellings called **spinal ganglia** that are located on the dorsal roots of spinal nerves. Sensory neurons form synapses with the dendrites and cell bodies of interneurons or directly with the motor neurons. Interneurons and most motor neurons develop from ectodermal neural tube cells, and their cell bodies remain

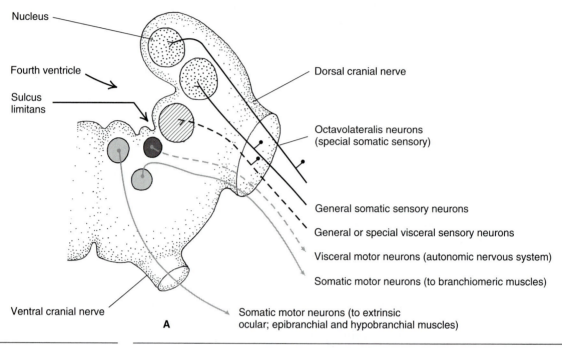

Nucleus

Fourth ventricle

Sulcus limitans

Dorsal cranial nerve

Octavolateralis neurons (special somatic sensory)

General somatic sensory neurons

General or special visceral sensory neurons

Visceral motor neurons (autonomic nervous system)

Somatic motor neurons (to branchiomeric muscles)

Ventral cranial nerve

A

Somatic motor neurons (to extrinsic ocular; epibranchial and hypobranchial muscles)

FIGURE 12–6 Diagrams showing the location of the nuclei of the cranial nerves in the brain stem: **A**, A composite transverse section through the posterior part of the brain.

in deep masses of gray matter that primitively are located beside the central canal and ventricles.

In a cross section of the spinal cord the gray matter has the appearance of a butterfly (Fig. 12–5). A pair of longitudinal **dorsal columns** and a pair of **ventral columns** are interconnected across the midline by a **gray commissure.** The dorsal columns contain the dendrites and cell bodies of interneurons that receive many of the terminations of the primary sensory neurons. These interneurons relay information to other parts of the central nervous system. The ventral columns contain the dendrites and cell bodies of the motor neurons. Motor neurons receive impulses from some primary sensory neurons and from many interneurons that originate in the dorsal columns or in higher brain centers. The dorsal columns are sensory in function, whereas the ventral ones are motor. Dorsal sensory and ventral motor regions are particularly distinct in an embryo because a longitudinal groove, the **sulcus limitans,** is present between them on the lateral wall of the central canal and in many of the cavities of the brain. The sulcus limitans can be seen in the caudal parts of the adult brain of many vertebrates, but most of it is obscure (Fig. 12–6**A**).

Functional Groups

The sensory and motor neurons and the gray columns where they terminate or originate can be further subdivided into functional groups associated with the somatic and visceral parts of the body. **Somatic sensory neurons** from somatic

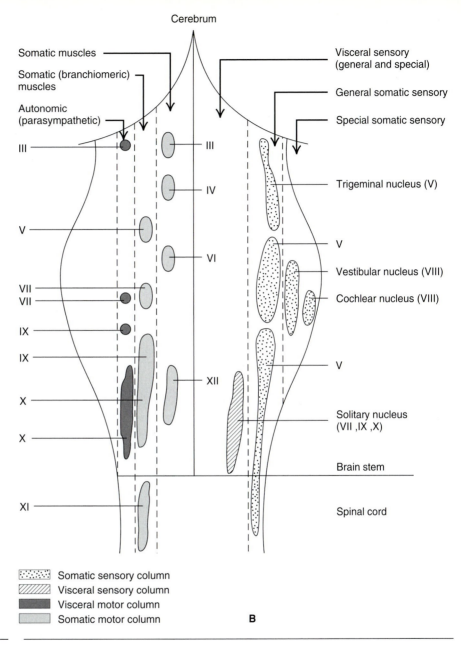

Cerebrum

Somatic muscles

Somatic (branchiomeric) muscles

Autonomic (parasympathetic)

Visceral sensory (general and special)

General somatic sensory

Special somatic sensory

III

III

IV

V

VI

VII
VII

IX

IX

XII

X

X

XI

Trigeminal nucleus (V)

V

Vestibular nucleus (VIII)

Cochlear nucleus (VIII)

V

Solitary nucleus (VII ,IX ,X)

Brain stem

Spinal cord

- Somatic sensory column
- Visceral sensory column
- Visceral motor column
- Somatic motor column

B

FIGURE 12–6 *continued*
B, A projection in dorsal view of the brain stem and nuclei. Motor nuclei are shown on the left; sensory ones on the right. Cranial nerves are identified by Roman numerals.

receptors enter and many terminate in a **somatic sensory column** located in the most dorsal part of the dorsal column (Fig. 12–5). **Visceral sensory neurons** from visceral receptors terminate in a **visceral sensory column** located more ventrally in the dorsal column. **Visceral motor neurons** to visceral muscles in the gut, heart, and other internal organs and to exocrine glands arise in a **visceral motor column** located in the dorsal part of the ventral column, and **somatic motor neurons** to somatic muscles arise in a **somatic motor column** located in the most ventral part of the ventral column.

Anatomists used to divide the visceral motor neurons into autonomic ones supplying glands and visceral organs, and branchiomeric ones going to the branchiomeric muscles. As we have discussed (p. 314), recent evidence indicates that the branchiomeric muscles are somatic. Neurons supplying them are also somatic (p. 474). Thus visceral motor neurons supply only glands and the muscles in the wall of visceral organs. These neurons form the **autonomic nervous system.** We will return to the autonomic nervous system later in this chapter, but at this time you should notice that visceral motor neurons differ from somatic motor ones in not extending directly from the central nervous system to the organs they supply. They always relay in some **peripheral ganglion** (Fig. 12–5). **Preganglionic visceral motor neurons** extend from the spinal cord (or brain) to the peripheral ganglion, where they synapse with a **postganglionic visceral motor neuron** that continues to the effectors. The postganglionic neurons develop from neural crest cells. In mammals, the preganglionic neurons leave the spinal nerve to a sympathetic ganglion in a **white ramus communicans,** so called because these fibers are myelinated and white. Some postganglionic neurons return to the spinal cord, and these unmyelinated fibers enter through a **gray ramus communicans.**

The generalizations we have made about spinal columns and neurons apply with some modifications to the cranial nerves and to the **brain stem** to which they attach. The brain stem is the "stalk" of the brain—that is, the brain less the cerebellum and forebrain (cerebrum and diencephalon). As the gray columns in the spinal cord continue into the brain, they break up into islands of gray matter called **nuclei** (Fig. 12–6). The nuclei have the same relationship to each other as the columns did in the spinal cord (that is, sensory nuclei are dorsal and motor ones are ventral).

As we discuss later (p. 473), most of the cranial nerves are serially homologous to the type of spinal nerve we find in lampreys, in which the roots form separate dorsal and ventral spinal nerves that do not unite. Like spinal nerves, some cranial nerves contain somatic sensory neurons from the skin and visceral sensory neurons from receptors in the gut wall and other visceral organs. These are called, respectively, **general somatic sensory** and **general visceral sensory neurons** to differentiate them from special sensory neurons that occur in certain cranial nerves but not in spinal nerves. The sensory fibers unique to certain cranial nerves are **octavolateralis neurons** from the lateral line and ear (**special somatic sensory neurons**) and **gustatory fibers** from taste buds (**special visceral sensory neurons**). Octavolateralis and gustatory neurons develop embryonically from ectodermal placodes, not from neural crest cells as do other sensory neurons. As would be expected, all of the sensory neurons enter the brain through dorsal nerves.

They terminate in dorsal nuclei that have a characteristic location (Fig. 12–6**A**). Special somatic sensory neurons terminate in the most dorsal nuclei, and general somatic sensory ones in a nucleus ventral to it (but still in a dorsal position). Both general and special visceral sensory neurons terminate in a nucleus located ventral to the somatic sensory one, but the special visceral sensory (gustatory) neurons terminate in the rostral part of this nucleus and the general visceral sensory neurons in the caudal part. As in lampreys and some other early vertebrates, motor neurons originate in ventral nuclei and leave through both the dorsal and ventral nerves.

Basic Neuronal Circuitry

The activities of the nervous system are controlled by many types of circuits among sensory neurons, interneurons, and motor neurons. Neurologists are only beginning to understand the complex pathways that mediate the integrative functions of the brain, including learning and memory. Many other circuits are well known, and we will consider some of them in the next chapter on the brain. At this time we will examine three basic types of neuronal pathways common to all vertebrates: reflexes, pathways from lower centers to higher brain centers, and pathways from the brain to lower centers.

Much activity is controlled by **reflexes** in the spinal cord (Fig. 12–7). If you

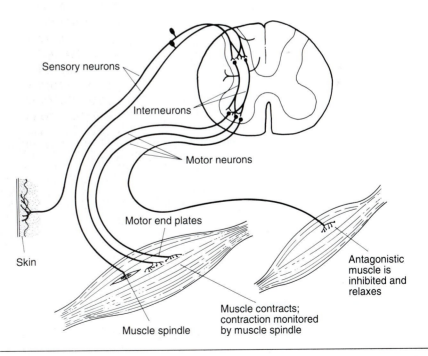

FIGURE 12–7   A reflex arc.

touch a hot stove, for example, impulses travel on primary sensory neurons and synapse with interneurons in the dorsal column of the spinal cord. These interneurons stimulate and inhibit the right combination of motor neurons and you jerk your hand away. An integrated movement such as this requires that certain muscles contract with the appropriate force and at the correct time, while the tension in antagonistic muscles is reduced in a comparable way. This reflex is a three-neuron reflex arc, for it includes three types of neurons: a sensory neuron, an interneuron, and a motor neuron. The familiar knee-jerk reflex requires only two types of neurons because the sensory neurons synapse directly with the motor ones. Reflexes such as this involve neurons on only one side of the body and one body segment. Other reflexes include neurons that cross, or **decussate,** and affect the other side of the body. Neuronal tracts that decussate are called **commissures,** and the gray commissure is the decussation within the spinal cord. Reflexes may also involve neurons that ascend or descend and affect many body segments. These are described as **intersegmental reflexes.** The coordinated movements of swimming and walking are integrated reflexly by pools of intersegmental and decussating neurons.

Swallowing, the rate of breathing, the rate of heartbeat, and many other vital processes are controlled reflexly by centers in the brain stem. Many reflexes in the cord and brain stem are the product of a long evolutionary history. They are innate and are the same in all individuals of a species. Others, called **conditioned reflexes,** develop during the lifetime of an individual as a result of an animal's repetitive experiences. When learning to drive a car, you must think about each action, but many of these actions become conditioned with experience. Conditioned reflexes are not inherited directly, although the capacity for them to develop is.

Many sensory impulses ascend from the spinal cord or brain stem to higher centers in the brain (Fig. 12–8**A**). Sensory impulses usually ascend on interneurons, but impulses from many proprioceptors and some touch receptors that enter the cord ascend on the primary sensory neurons as far as the caudal end of the brain stem. Groups of axons with the same function terminate in the same part of the brain and ascend together in common tracts that are described by their points of origin and termination—for example, the **spinocerebellar tract** from the spinal cord to the cerebellum, the **spinothalamic tract** to the thalamus in the forebrain, and so on (Fig. 12–8**B**). These tracts are not distinguishable morphologically in the spinal cord other than by their general location in major bundles, or **funiculi,** of white matter. Most sensory impulses decussate on their way to higher centers so that impulses originating on the left side of the body terminate in the right side of the brain, but the decussation is not always complete. The reasons for decussation are not well understood, other than the obvious need for the two

FIGURE 12–8   Pathways to and from the brain: **A,** A surface projection of the pathway taken by an ▶ impulse from a temperature receptor, and the descending pathway to an effector. **B,** The location in the spinal cord of the major ascending and descending pathways. Descending tracts are shown on the left; ascending ones on the right. (**A,** after Dorit, Walker, and Barnes. **B,** after Williams et al.)

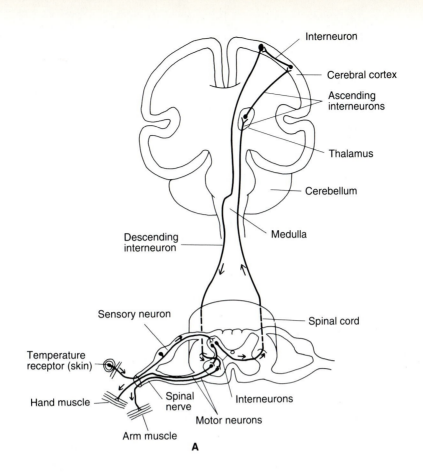

Interneuron

Cerebral cortex

Ascending
interneurons

Thalamus

Cerebellum

Medulla

Descending
interneuron

Sensory neuron

Spinal cord

Temperature
receptor (skin)

Hand muscle

Spinal
nerve

Interneurons

Motor neurons

Arm muscle

A

DESCENDING TRACTS

ASCENDING TRACTS

Fasciculus
gracilis

Fasciculus
cuneatus

Dorsal funiculus,
spinomedullary tracts

Lateral corticospinal
tract

Lateral funiculus

Spinocerebellar
tract

Lateral
reticulospinal
tract

Tectospinal
tract

Vestibulospinal
tract

Spinothalamic
tract

Ventral reticulospinal
tract

Ventral corticospinal
tract

Ventral funiculus

B

halves of the body to function as an integrated unit and not as isolated left and right sides.

Many ascending interneurons of mammals terminate in nuclei in the **thalamus** (Fig. 12–8A). Sensory signals may be filtered here, and certain ones continue on other internuncial neurons to specific parts of the gray matter of the cerebrum, which is the highest brain center. When they reach specific parts of the **cerebral cortex** in human beings, we become aware of the sensation. You have already jerked your hand away from that hot stove you touched, but now you sense that you burned your fingers. You may decide to do something, perhaps turn down the stove and fetch the first aid kit. Impulses descend through the central nervous system on other interneurons, decussate on commissures at some level of the brain, and continue down the cord in specific tracts, such as the **corticospinal tract** (Fig. 12–8B), to terminate in neuron pools that will activate and inhibit the appropriate muscles.

The Spinal Cord and Spinal Nerves
Spinal Cord

The spinal cord lies between the brain and the spinal nerves, but the cord is far more than a passageway for nerve impulses between receptors and effectors and the brain. Considerable integration occurs in the spinal cord. The rhythmic and coordinated movements of swimming and walking, for example, are controlled and integrated reflexly by pools of intersegmental and decussating spinal neurons. Grillner and Wallen (1985) have shown that these spinal circuits act as **central pattern generators.** A cat whose spinal cord has been transsected behind the skull can walk on a moving treadmill with a near normal pattern of limb extension and flexion, but the animal does need to be supported. The passive movements of the limbs caused by the treadmill activate receptors, and this is sufficient to start the central pattern generators. A continued sensory input is not necessary to maintain rhythmic patterns, but the synergy among muscles is less well coordinated without it. The details of locomotor integration are controlled in the spinal cord, but centers in the brain normally are responsible for initiating activity of the central pattern generators, for maintaining balance, for modulating speed and force of muscle contraction, and, of course, for integrating goal-directed movements.

The complexity of the spinal cord, and the degree to which the brain exerts control over spinal activity, increases during the course of vertebrate evolution. Lampreys and hagfishes have a very primitive spinal cord. Since none of the axons are myelinated, a distinction cannot be made between white and gray matter in their spinal cord or brain (Fig. 12–9A). The cell bodies of the motor neurons are located, as one would expect, in the ventral part of the spinal cord, but their dendrites ramify into the peripheral part of the cord, and it is here that synapses occur with sensory and interneurons. Although the cell bodies of a few large sensory neurons lie dorsally in the peripheral part of the cord, the cell bodies of most sensory neurons are located in spinal ganglia. A few interneurons decussate and ascend and descend within the cord, but tracts are not well developed. The best-

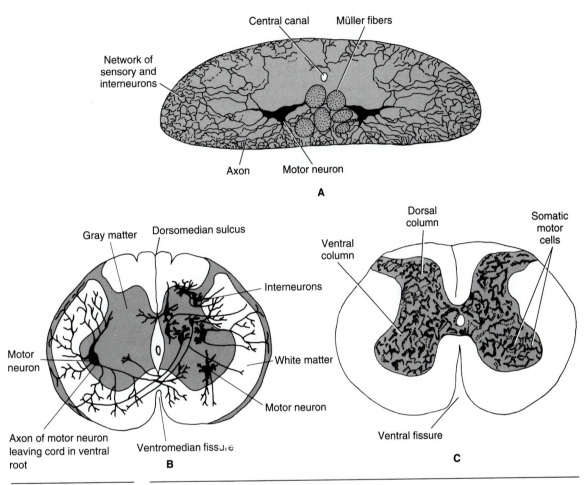

Central canal   Müller fibers

Network of
sensory and
interneurons

Axon   Motor neuron

A

Gray matter   Dorsomedian sulcus

Interneurons

Motor
neuron

White matter

Motor neuron

Axon of motor neuron
leaving cord in ventral
root

Ventromedian fissure

B

Dorsal
column

Somatic
motor
cells

Ventral
column

Ventral fissure

C

FIGURE 12–9   Transverse sections of the spinal cord of representative vertebrates: **A,** A lamprey. **B,** A frog. **C,** An alligator. (**A,** after Tretjakoff. **B,** after Gaupp.)

defined one is composed of the axons of giant **Müller cells.** These axons descend from cell bodies near the caudal end of the brain and synapse with the dendrites of motor neurons. This mechanism initiates rapid escape reactions. Although Müller fibers have a diameter of 50 micrometers, they are not myelinated, and conduction velocity is only 5 meters per second. This is faster than in other lamprey neurons, but far slower than in the smaller but myelinated fibers of other vertebrates. The spinal cord is relatively thin and avascular. Its flattened shape facilitates the diffusion of gases, nutrients, and other products.

The spinal cord is larger, well vascularized, and rounded in other fishes and amphibians (Fig. 12–9**B**). The early evolution of a highly vascular nervous system has been an important factor in vertebrate evolution, for it allows a large and complex nervous system to develop. The cell bodies of interneurons and motor neurons lie in the gray matter, but many of their dendrites, and the terminations of sensory neurons, continue to ramify in the peripheral part of the cord, where many of the synapses among them are made. As more ascending and descending fiber tracts evolve, the white matter increases and bulges outward except near the midline, where a **dorsomedian sulcus** and a **ventromedian fissure** remain.

The gray matter in the spinal cord of an amniote has a characteristic butterfly shape when seen in cross section (Fig. 12–9**C**). Synapses among neurons are now confined to the gray matter; white matter contains only ascending and descending fiber tracts. The brains of birds and mammals exert more control over body activity than in other vertebrates, and tracts to and from the brain are more numerous and larger than in other vertebrates. The degree of vascularization of the cord increases, and the number of neuroglia, which service the neurons, also increases (Sawro, 1990).

The spinal cord is nearly as long as the vertebral column in most vertebrates, although in mammals, frogs, and a few teleosts the cord is much shorter because during development it grows more slowly than the rest of the body (Fig. 12–10). As a consequence, the more caudal spinal nerves form a bundle, known as the **cauda equina,** as they extend from their attachment to the cord caudally through the vertebral canal to the intervertebral foramina through which they leave. A thin **terminal filament,** composed of the connective tissues surrounding the spinal cord, extends from the end of the cord to the caudal end of the vertebral canal. Vertebrates with well-developed limbs have a large number of neurons supplying them, and the diameter of the spinal cord is greater in the cervical and lumbar regions, where most of the nerves to the limbs originate.

Spinal Nerves

As we have seen, the spinal nerves of vertebrates usually attach by roots to the spinal cord, but in lampreys the homologues of the dorsal and ventral roots form distinct **dorsal** and **ventral spinal nerves.** The ventral nerves are segmentally spaced and lead directly into the myomeres. They carry only somatic motor neurons (Fig. 12–11**A** and **B**). The dorsal nerves are intersegmental and pass between the myomeres to the body surface and gut regions. They contain both somatic and visceral sensory neurons and the small number of visceral motor neurons that leave the central nervous system through the spinal nerves. Most visceral motor neurons reach the viscera of lampreys through a cranial nerve, the vagus. The presence of distinct dorsal and ventral spinal nerves, each with its own types of neurons, is believed to be a primitive vertebrate character.

In all other vertebrates, including hagfishes, the distinct dorsal and ventral spinal nerves of lampreys, and probably ancestral vertebrates, have united to form a definitive spinal nerve. The originally separate nerves are now roots of spinal nerves. All sensory fibers continue to enter through the dorsal root, and somatic motor fibers continue to leave through the ventral root. In most fishes and amphibians, many visceral motor fibers continue to leave through the dorsal roots, but some also leave through the ventral root. All visceral motor fibers leave through the ventral roots in amniotes along with the somatic motor neurons (Fig. 12–11**B**).

The number of spinal nerves varies with the number of segments in the trunk and tail. A frog, specialized to thrust powerfully with its hind legs as it swims or jumps, has a short trunk, a reduced tail, and only ten pairs of spinal nerves. A

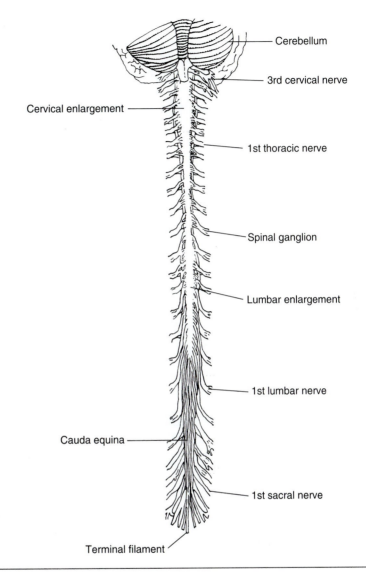

Cerebellum

3rd cervical nerve

Cervical enlargement

1st thoracic nerve

Spinal ganglion

Lumbar enlargement

1st lumbar nerve

Cauda equina

1st sacral nerve

Terminal filament

FIGURE 12–10   A dorsal view of the spinal cord and nerves of a human being.

snake, which moves by the lateral undulations of its long trunk and tail, may have several hundred pairs of spinal nerves. Spinal nerves are named only when body regions are well defined, as they are in mammals: **cervical nerves, thoracic nerves, lumbar nerves, sacral nerves,** and **caudal nerves.**

Epaxial muscles remain segmented in most vertebrates, and the dorsal rami of spinal nerves supplying these muscles retain their primitive segmentation. Even in mammals, where longitudinal muscle bundles evolve, the deeper epaxial muscles remain segmented. Correlated with the complexity of embryonic development of the hypaxial muscles, and especially of the appendicular muscles, many ventral

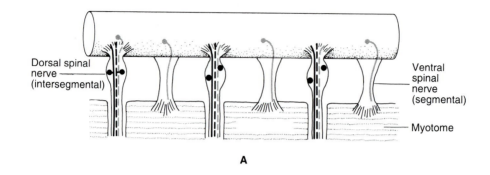

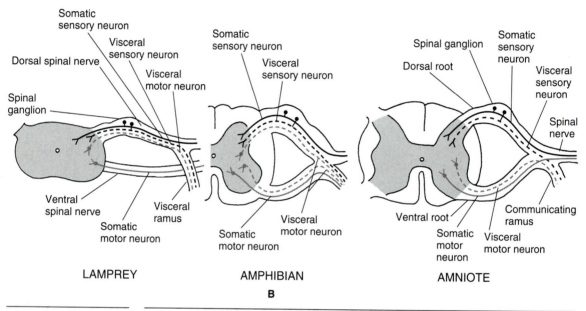

LAMPREY **AMPHIBIAN** **AMNIOTE**

B

FIGURE 12–11   The evolution of the spinal nerves: **A,** A dorsal view of the spinal cord and nerves of a lamprey. **B,** Transverse sections through the spinal cord and nerves of a lamprey, an amphibian, and an amniote. (Modified after Romer and Parsons.)

rami unite to form complex networks or **plexuses** before nerves are distributed to the muscles (Fig. 12–12). The innervation of muscles is conservative. Motor neurons establish a connection very early with developing muscle fibers in the myotomes, and they tend to follow this tissue if it migrates and unites with tissue from other myotomes. One muscle in a limb, for example, may have developed from muscle tissue that came from myotomes 8 to 10; another muscle, from tissue derived from myotomes 9 to 12. To reach these muscles, some of the neurons in spinal nerves 8 to 12 must come together and then diverge. Plexuses occur in all gnathostomes, and they reach their highest complexity among mammals. A **cervical plexus** supplies many ventral neck muscles; a **brachial plexus,** the pectoral appendage; a **lumbosacral plexus,** the pelvic appendage; and a **coccygeal plexus,** some of the pelvic muscles.

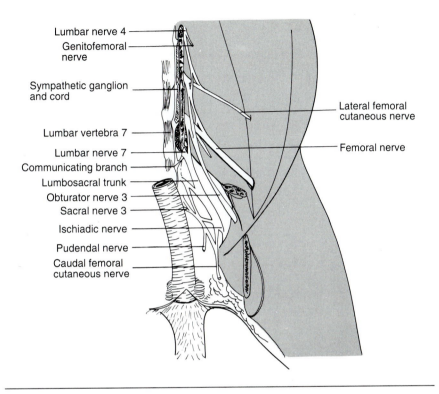

Lumbar nerve 4

Genitofemoral nerve

Sympathetic ganglion and cord

Lumbar vertebra 7

Lumbar nerve 7

Communicating branch

Lumbosacral trunk

Obturator nerve 3

Sacral nerve 3

Ischiadic nerve

Pudendal nerve

Caudal femoral cutaneous nerve

Lateral femoral cutaneous nerve

Femoral nerve

FIGURE 12–12   The lumbosacral plexus of a cat. (After Walker and Homberger.)

Cranial Nerves

The Cranial Nerves of Fishes

The distribution of the cranial nerves of fishes is shown in Figure 12–13, and their functional components are listed in Table 12–1. The nerves have names descriptive of some aspect of their structure or function, and they are given a Roman numeral that corresponds to their sequence in human beings. This confusing array of nerves can be simplified somewhat if we interpret some as unique sensory nerves to the head, others as serially homologous to the ventral spinal nerves of lampreys and probably ancestral vertebrates, and others as serially homologous to dorsal spinal nerves.

Sensory Nerves Unique to the Head.   This group includes four sensory nerves found only in the head and composed only of special somatic sensory fibers.

Terminal Nerve (0)

As we have seen (p. 404), this nerve appears to be part of a chemosensory system regulating certain aspects of reproduction in response to olfactory pheromones. It is given the designation 0 because it was discovered after the conventional numbering system had been established.

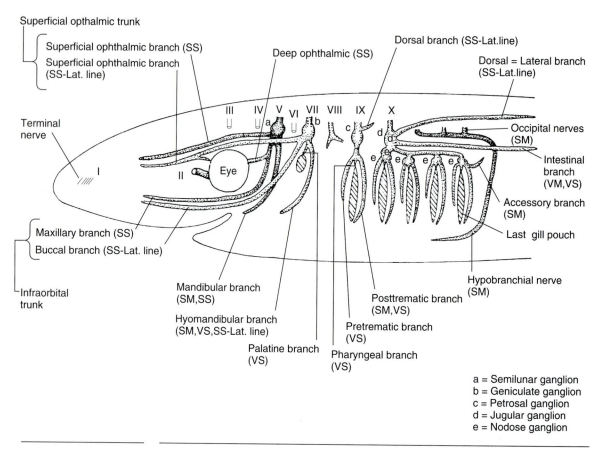

Superficial opthalmic trunk

Superficial ophthalmic branch (SS)

Superficial ophthalmic branch (SS-Lat. line)

Dorsal branch (SS-Lat.line)

Deep ophthalmic (SS)

Dorsal = Lateral branch (SS-Lat.line)

Terminal nerve

III IV V VI VII VIII IX X

Occipital nerves (SM)

I

II Eye

Intestinal branch (VM,VS)

Accessory branch (SM)

Last gill pouch

Maxillary branch (SS)

Buccal branch (SS-Lat. line)

Infraorbital trunk

Mandibular branch (SM,SS)

Hyomandibular branch (SM,VS,SS-Lat. line)

Palatine branch (VS)

Pharyngeal branch (VS)

Pretrematic branch (VS)

Posttrematic branch (SM,VS)

Hypobranchial nerve (SM)

a = Semilunar ganglion
b = Geniculate ganglion
c = Petrosal ganglion
d = Jugular ganglion
e = Nodose ganglion

FIGURE 12–13   A diagram of the cranial nerves and their major branches in a dogfish. Nerves are identified by Roman numerals and their functional components by abbreviations: SM = somatic motor, SS = somatic sensory, VM = visceral motor (autonomic), VS = visceral sensory. Ganglia are identified by small Arabic letters: a = semilunar ganglion, b = geniculate ganglion, c = petrosal ganglion, d = jugular ganglion, e = nodose ganglia.

Olfactory Nerve (I)

The olfactory nerve returns chemosensory fibers from the nasal sac. It is not a long nerve in most vertebrates because the nasal sac and the olfactory bulb of the brain, to which its axons go, usually lie next to each other.

Optic Nerve (II)

Since the retina develops as part of the brain (p. 435), the optic nerve is technically a brain tract and not a nerve. Its axons are those of the ganglion cells of the retina. These are interneurons that receive impulses from the bipolar neurons and carry them to the optic lobes of the brain.

Statoacoustic Nerve (VII)

Since the fish ear lacks a cochlea, a fish's eighth nerve is often called the statoa-

TABLE 12–1 Cranial Nerves and Their Major Functional Components

Chondrichthyan Fishes

Nerves	Somatic Sensory General and Proprioceptive	Somatic Sensory Special	Visceral Sensory General	Visceral Sensory Special (Taste)	Visceral Motor (Autonomic)	Somatic Motor	Mammals
0 Terminal		X[3]					No change
I Olfactory		X[3]					Vomeronasal component usually added
II Optic		X[4]					Less decussation of fibers
III Oculomotor					X	X	No change
IV Trochlear						X	No change
V Trigeminal	X[1]					X[6]	No change
VI Abducens						X	No change
VII Facial		X[5]	X	X		X[6]	Lateral line fibers lost; autonomic fibers to lacrimal and rostral salivary glands added
VIII Statoacoustic Vestibulocochlear		X[5]					No change

TABLE 12-1 *continued*

Chondrichthyan Fishes

Nerves	Somatic Sensory		Visceral Sensory		Visceral Motor (Autonomic)	Somatic Motor	Mammals
	General and Proprioceptive	Special	General	Special (Taste)			
IX Glossopharyngeal		X[5]	X	X		X[6]	Lateral lines fibers lost; autonomic fibers to parotid glands added
X Vagus		X[5]	X	X	X	X[6]	Lateral line fibers lost; reduction of somatic motor fibers; accessory branch becomes cranial nerve XI
Occipitals	2					X	Becomes cranial nerve XII

[1]Probably includes proprioceptive fibers from extrinsic ocular muscles.
[2]Proprioceptive fibers from hypobranchial muscles probably travel in spinal nerves.
[3]Chemosensory fibers.
[4]The optic nerve is a brain tract.
[5]Acousticolateralis fibers.
[6]Supplies branchiomeric muscles, which are now considered to be somatic.

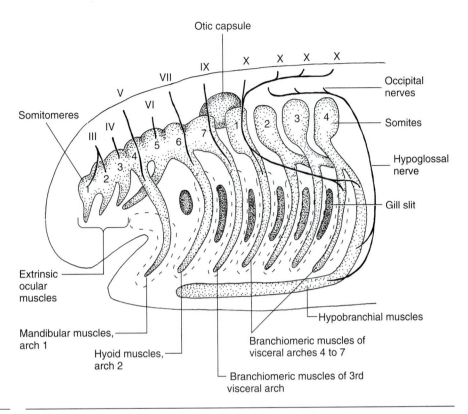

FIGURE 12–14   Relationship between cranial nerves, somitomeres, somites, and visceral arches during the development of the head of an early vertebrate. Nerves are identified by Roman numerals, somitomeres and somites by Arabic numerals, and visceral arches are outlined by dashes. (After Northcutt, 1990.)

coustic instead of the vestibulocochlear nerve, the mammalian term. It returns octaval fibers from all parts of the inner ear.

Ventral Cranial Nerves.   Those cranial nerves interpreted as ventral cranial nerves resemble the ventral spinal nerves because they contain primarily somatic motor neurons going to many of the somatic muscles in the head (see also Fig. 12–6A). This group consists of those nerves going to the extrinsic ocular muscles and to the epibranchial and hypobranchial muscles. Some also contain somatic sensory proprioceptive neurons, which return information on the degree of contraction of the muscles. The derivation of these muscles from somitomeres and somites is shown in Figure 12–14.

Oculomotor Nerve (III)

The oculomotor nerve supplies those extrinsic ocular muscles that develop from the first two somitomeres: dorsal rectus, medial rectus, ventral rectus, and ventral oblique. Anatomists are not sure of the location of proprioceptive neurons returning from these muscles in fishes. In human beings these neurons travel in the trigeminal nerve. The oculomotor nerve has a small **ciliary branch** that carries autonomic neurons (visceral motor) into the eyeball to supply the ciliary and iris muscles.

Trochlear Nerve (IV)

The trochlear nerve (L. *trochlea* = pulley), so named because the muscle it supplies passes through a connective tissue pulley in mammals, supplies the dorsal oblique muscle, the one extrinsic ocular muscle that develops from the third somitomere.

Abducens Nerve (VI)

The abducens nerve (L. *abducens* = leading away) arises from the ventral surface of the brain and supplies the lateral rectus muscle, the one muscle that develops from the fifth somitomere.

Occipital Nerves

The epibranchial and hypobranchial muscles of fishes, which are located in the caudal region of the head and beneath the pharynx, are supplied by three or four occipital nerves and by a variable number of anterior spinal nerves. The occipital nerves are called occipital rather than cranial or spinal because the location of the caudal border of the skull varies. In some fish species these nerves emerge within the skull from the caudal end of the brain; in other species they emerge just behind the skull from the spinal cord. Their dorsal rami supply the epibranchial muscles, and their ventral rami unite to form the **hypobranchial nerve** that follows the hypobranchial musculature into the floor of the pharynx.

Dorsal Cranial Nerves.   The remaining four cranial nerves appear to be comparable to dorsal spinal nerves. They are mixed nerves that contain some combination of sensory and motor neurons (see also Fig. 12–6**A**). All supply somatic motor neurons to the branchiomeric muscles of the visceral arches, and for this reason they are sometimes called branchiomeric nerves. Beyond this, some have visceral motor autonomic neurons, and all have somatic sensory, visceral sensory, or both types of sensory neurons.

Glossopharyngeal Nerve (IX)

The glossopharyngeal nerve may be taken as a prototype for this group of nerves. Its primary branch is a **posttrematic branch** (Gr. *tremat-* = hole or cleft), which passes caudad to the first complete gill pouch, carries somatic motor neurons to the muscles of the third visceral arch, and returns visceral sensory neurons from the adjacent part of the gill pouch. Other branches are a **pretrematic branch** that goes in front of the first gill pouch, and a small **pharyngeal branch** to the roof of the pharynx, both of which carry only visceral sensory neurons. In mammals these receptors lie on the back of the tongue and adjacent parts of the pharynx; hence the name for this nerve (Gr. *glossa* = tongue). Finally, a small **dorsal branch** returns general somatic sensory fibers from a small segment of the skin, and special somatic sensory lateral line neurons from a segment of the lateral line. The cell bodies of its sensory neurons are contained in the **petrosal ganglion**

(Gr. *petros* = stone), so called because it lies within the stony, petrous part of the mammalian temporal bone.

Vagus Nerve (X)

The vagus nerve (L. *vagus* = wandering) has four branches that supply the branchiomeric muscles of the last four visceral arches (4 to 7) and return visceral sensory neurons from the last four gill pouches. Since each of these branches resembles the glossopharyngeal nerve, it is likely that the vagus nerve evolved by the union of four dorsal cranial nerves. The dorsal branch of the vagus (called the **lateral branch**) returns special somatic sensory lateralis neurons from most of the lateral line along the side of the trunk. In addition, the vagus has a short **accessory branch,** which carries somatic motor neurons to the cucullaris muscle, and a long **intestinal branch,** which carries autonomic (visceral motor) neurons to the abdominal viscera and returns visceral sensory neurons. The cell bodies of the sensory neurons in this nerve lie in the **jugular ganglion** and in a series of **nodose ganglia** (L. *nodus* = knob).

Facial Nerve (VII) and Trigeminal Nerve (V)

Gill pouches are reduced rostral to the glossopharyngeal nerve, so it is not surprising that more rostral dorsal cranial nerves have lost the pretrematic branch. The facial nerve retains a small pharyngeal branch (called the **palatine branch**), but the trigeminal has lost this, too. The posttrematic branch of the facial nerve is the **hyomandibular nerve,** and that of the trigeminal nerve is the **mandibular nerve.** These nerves carry somatic motor neurons to the branchiomeric muscles of the hyoid and mandibular arches, respectively. The hyomandibular nerve also carries visceral sensory neurons, and the mandibular nerve carries general somatic sensory neurons from the skin over the lower jaw. The dorsal branches of the facial and trigeminal nerve bifurcate and pass dorsal and ventral to the orbit. The dorsal branches of the facial nerve (**superficial ophthalmic** and **buccal branches**) return special somatic sensory lateral line neurons from most of the head. The dorsal branches of the trigeminal nerve (**superficial ophthalmic** and **maxillary branches**) return general somatic sensory neurons from the skin over the head. The cell bodies of the sensory neurons in the facial nerve lie in the **geniculate ganglion** (L. *geniculum* = little knee); those of the trigeminal nerve lie in the **semilunar ganglion.**

Evolution of the Cranial Nerves

The Origin of the Ventral and Dorsal Cranial Nerves: Head Segmentation.   We have interpreted certain cranial nerves as the serial homologues of ventral spinal nerves, and others as the serial homologues of dorsal cranial nerves. This assumption is valid if the vertebrate head, or a substantial part of it, evolved from a series of segments comparable to the trunk segments. Segmentation of the vertebrate trunk is fundamentally a segmentation of the locomotor myomeres, but this imposes a segmentation on the axial skeleton to which the myomeres attach, and on the spinal nerves and blood vessels that supply them. The head may also have been segmented in ancestral vertebrates, as the front of the body is in amphioxus.

As activity increased and a high degree of cephalization evolved in early vertebrates, the head acquired new sensory, integrative, feeding, and respiratory functions. Because of these new functions, and because the head develops embryonically under the influence of neural crest cells and ectodermal placodes, we should
not be surprised to see differences among segments in the head and trunk. However, if the head is segmented, we should be able to account for a ventral and a
dorsal nerve for each of the head segments. We have discussed (p. 315, and Fig.
9–7) the segmental nature of the cranial paraxial mesoderm. Four distinct somites
lie caudal to the otic capsule, and seven less clearly demarcated somitomeres lie
rostral to the otic capsule (Fig. 12–14). Northcutt (1990) has proposed that two
somitomeres are the equivalent of a head segment. If one somitomere in the otic
region has been lost, and there is evidence in cartilaginous fish embryos for the
loss of premuscular tissue in this region, then vertebrates have four preotic head
segments and four postotic segments (Table 12–2). Do some, at least, of the cranial
nerves relate to these segments in the way that separate dorsal and ventral spinal
nerves of lampreys (and probably ancestral vertebrates) relate to trunk segments?
We cannot be certain at this time, but Goodrich argued for serial homology in his
classic synthesis (1930), and Northcutt (1990) has tentatively supported this view.

The cranial nerves we have interpreted as ventral cranial nerves do resemble
ventral spinal nerves. They contain only (or primarily) somatic motor neurons, and
they lack sensory neurons to the skin or viscera. They attach ventrally on the brain
stem, with the exception of the trochlear nerve, which leaves dorsally; but the
nucleus of the trochlear is in a ventral position. As you can see in Figure 12–14
and Table 12–2, the oculomotor (III), trochlear (IV), and abducens (VI) nerves
appear to be the ventral nerves of the first three head segments. Segments 4 and
5 lack ventral nerves, probably because the expansion of the otic capsule in this
region has eliminated one somitomere and possibly part of a somite. The segmented ventral series of cranial nerves resumes with the occipital nerves. These
appear to be the ventral nerves of segments 6, 7, and 8.

The trigeminal nerve (V), the facial nerve (VII), the glossopharyngeal nerve
(IX), and the vagus nerve (X) appear to be serially homologous to the dorsal spinal
nerves because they have most of the characteristics seen in dorsal spinal nerves.
They attach more dorsally on the brain stem than the ventral cranial nerves, they
contain general somatic sensory and general visceral sensory neurons, and they
carry motor neurons to the striated branchiomeric muscles associated with the
visceral arches (Fig. 12–14). As we have discussed (p. 316), the branchiomeric
muscles have the characteristics of somatic muscles in the one group of vertebrates
(birds) in which the origin of these muscles is clear (they arise from paraxial mesoderm and are striated). We are provisionally calling them somatic muscles. Motor
neurons supplying the branchiomeric muscles have traditionally been called special visceral motor neurons, but, since they supply somatic muscles, they should
be called somatic motor neurons (Northcutt, 1990). They resemble other somatic
motor neurons (and differ from the visceral motor neurons of the autonomic nervous system) in traveling all the way to the muscles they supply without a peripheral relay. The nuclei of these neurons are also located far ventrally in the brain
stem (Fig. 12–6**A**), a location consistent with the view that they are somatic motor
neurons. The hypothesis that the dorsal cranial nerves are serially homologous to

TABLE 12–2 **Segmentation of Cranial Muscles and Nerves**
The probable scheme of head segmentation and related structures in an ancestral vertebrate. An ancestral vertebrate probably had distinct octavolateralis and chemosensory nerves, in addition to ventral and dorsal cranial nerves. An X indicates that sensory fibers of these types are now in the adjacent dorsal nerve.

Head Segment	Origin	Visceral Arch	Muscles	Ventral Nerves (Somatic Motor)	Dorsal Nerves (Somatic Motor General Somatic and Visceral Sensory)	Octavo-lateralis Nerves	Chemosensory Nerves
							0 Terminal
							I Olfactory (olfactory)
1	Somitomeres 1 2		Most extrinsic ocular muscles	III Oculomotor	Deep ophthalmic branch of V? (profundus)		
2	3 4	1	Mandibular Dorsal oblique	IV Trochlear	V Trigeminal		
3	5 6	2	Hyoid Lateral rectus	VI Abducens	VII Facial	X	X (Gustatory) X
						VII Stato-acoustic	
4	7 Lost?	3	1st branchial arch	—	IX Glosso-pharyngeal	X	X (Gustatory)

TABLE 12-2   continued

Head Segment	Origin	Visceral Arch	Muscles	Ventral Nerves (Somatic Motor)	Dorsal Nerves (Somatic Motor General Somatic and Visceral Sensory)	Octavo-lateralis Nerves	Chemosensory Nerves
5	Somites 1	4	2nd branchial arch	—	X Vagus	X	X (Gustatory)
6	2	5	3rd branchial arch Hypobranchial	Occipital	X Vagus	X	X (Gustatory)
7	3	6	4th branchial arch Hypobranchial	Occipital	X Vagus	X	X (Gustatory)
8	4	7	5th branchial arch Hypobranchial	Occipital	X Vagus	X	X (Gustatory)

dorsal spinal nerves requires that somatic motor neurons in ancestral vertebrates left the central nervous system through both dorsal and ventral nerves. If this were also the case in spinal nerves, they were subsequently lost from dorsal spinal nerves. Some dorsal nerves of amphioxus contain somatic motor neurons, which supports the view that originally somatic motor fibers were present in all dorsal and ventral nerves.

As you can see in Figure 12–14 and Table 12–2, the trigeminal nerve (V) is the dorsal nerve of the second head segment; the facial nerve (VII) that of the third segment; the glossopharyngeal nerve (IX) that of the fourth segment; and the four branchiomeric branches of the vagus are the dorsal nerves of the last four head segments. Some investigators interpret the deep ophthalmic branch of the trigeminal nerve, which is a distinct nerve in lampreys and the embryos of some fishes, as the missing dorsal nerve of the first segment, but this is not certain.

The Origin of Special Sensory Neurons. Most of the cranial nerves appear to be serially homologous to primitive ventral and dorsal spinal nerves, but cranial nerves are complicated by the presence in the head of sensory systems and their special sensory neurons that probably never occurred in the trunk and therefore are not represented in spinal nerves. We can ignore the retina and optic nerve because these are really parts of the brain. But the head has chemosensory (gustatory and olfactory) systems and a group of special somatic sensory neurons that come from the ear and lateral line organs. As we have seen (p. 415), neurons from the ear and lateral line have much in common: they receive stimuli from hair cells and have similar connections within the brain. Biologists often call them the **octavolateralis system.** Many of the special sensory neurons now travel with other sensory neurons in dorsal cranial nerves, but were they always a part of these nerves? Cole (1896) and Landacre (1910) pointed out many years ago that the octavolateralis and chemosensory neurons differ from other sensory neurons in important ways, and their work is supported by Northcutt (1990). Octavolateralis and chemosensory neurons develop from ectodermal placodes and not from neural crest cells, which give rise to all other sensory neurons. Moreover, the cell bodies of octavolateralis and chemosensory neurons tend to lie more distally in ganglia than those of other sensory neurons; in some cases they lie in distinct ganglia. The octavolateralis and chemosensory systems also project to different areas within the brain than do other sensory neurons (see Fig. 12–6). Northcutt (1989 and 1990) proposes that the octavolateralis and chemosensory neurons were not originally components of the dorsal nerves but formed distinct nerves. The inner ear and octaval neurons develop embryonically from one ectodermal placode, and in some fishes the lateralis neurons develop from as many as three preotic and three postotic placodes. It is therefore likely that ancestral vertebrates had as many as six distinct lateral line nerves as well as the eighth nerve (octaval fibers). The olfactory nerve and probably the terminal nerve develop from another placode, and the gustatory fibers develop from at least three others. Ancestral vertebrates probably had many placodal nerves, which were distinct from the dorsal and ventral cranial nerves. Of these, the terminal, olfactory, and eighth nerve remain distinct, but the other placodal nerves have become so closely associated morphologically with dorsal cranial nerves that it is convenient to consider them as parts of these nerves

(Table 12–2). Lateral line and gustatory fibers now travel with the facial, glosso-pharyngeal, and vagus nerves.

Changes in the Cranial Nerves of Tetrapods.

The pattern of the cranial nerves we have seen in fishes persists with predictable modifications in tetrapods (Table 12–1 and Fig. 12–15). Few changes have occurred in the nerves returning from the special sensory organs. Among the changes are the following:

1. Most terrestrial vertebrates have a vomeronasal organ and they acquire with it a vomeronasal component to the olfactory nerve.
2. In fishes and most nonmammalian vertebrates, the fibers of the optic nerve completely decussate in the **optic chiasma,** located at the point of attachment of the optic nerves to the brain. A few fibers do not decussate in amphibians and some reptiles, and in many mammals only half decussate and the others remain on the same side, or ipsilateral. This appears to correlate with overlapping visual fields and the development of stereoscopic vision (p. 515).
3. A cochlea is well developed in some reptiles and in birds and mammals, so the eighth nerve is known as the vestibulocochlear nerve.

Few changes have occurred in the ventral cranial nerves. The oculomotor, trochlear, and abducens nerves have not changed except for a slight increase in the number of extrinsic ocular muscles that they supply (p. 323). The level of the head-trunk transition is stabilized in amniotes, and the occipital nerves attach to the back of the brain and emerge through a skull foramen as the **hypoglossal**

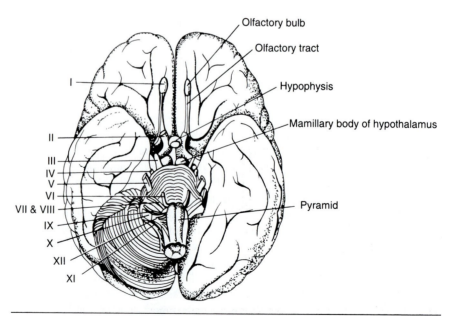

FIGURE 12–15   A ventral view of a human brain, showing the locations of the cranial nerves. (After Noback et al.)

nerve (XII). The hypoglossal extends beneath the tongue (Gr. *hypo* = under + *glossa* = tongue) and supplies the muscle of the tongue and other hypobranchial muscles.

Since more changes have occurred in those structures that are supplied by the dorsal cranial nerves, these nerves have changed more than the ventral cranial nerves.

The trigeminal nerve of amniotes consists of ophthalmic, maxillary, and mandibular branches. Its ophthalmic branch incorporates both the superficial ophthalmic and the deep ophthalmic branches of fishes. The other branches are essentially the same as in a fish. All the branches carry general somatic sensory fibers from receptors in the skin of the head, in the teeth, in the mouth, and on the tongue. Those on the tongue are from general receptors and not from taste buds. Some investigators believe that most, if not all, of the proprioceptive fibers from the extrinsic ocular muscle also return in the trigeminal nerve because the cell bodies of some of these neurons have been found in the human semilunar ganglion. The motor fibers of the trigeminal nerve are confined to its mandibular branch. They are somatic motor fibers to the mandibular muscles and tensor tympani. (Recall that the malleus, to which the tensor tympani attaches, is a mandibular arch derivative.)

Most adult amphibians and all amniotes have lost the lateral line organs and, with them, the dorsal branches of dorsal cranial nerves in which the lateralis neurons travel.

Although lateralis fibers no longer travel with it, the facial nerve has kept its other components. It continues to carry somatic motor fibers to the hyoid musculature. Most of this musculature forms the platysma and facial muscles in mammals, but the stapedius, stylohyoid, and posterior belly of the digastric remain attached to hyoid arch derivatives. Important cranial glands evolve in tetrapods, and the facial nerve acquires visceral motor (autonomic) neurons that supply the lacrimal glands, the mandibular and sublingual salivary glands, and glands in the palatal and nasal mucosa. The major sensory fibers in the facial nerve are special visceral sensory ones from taste buds on the palate and anterior two thirds of the tongue. Those on the tongue leave with somatic sensory fibers of the trigeminal in its lingual branch but then separate from the lingual, cross the tympanic membrane in the **chorda tympani,** and enter the brain with the facial nerve.

The glossopharyngeal nerve has lost the lateralis fibers that traveled with it in fishes, but the rest of the nerve continues to supply visceral receptors, primarily taste buds on the caudal parts of the pharynx and tongue, and a few muscles in the wall of the pharynx. It has acquired visceral motor (autonomic) neurons to the parotid salivary gland.

The vagus nerve has also lost its lateralis fibers. Its major branch is homologous to the intestinal branch of fishes. This branch follows the trachea and esophagus carrying visceral motor (autonomic) and general visceral sensory neurons to the heart, lungs, and abdominal viscera. It has lost most of its taste neurons and somatic motor ones, apart from those to the larynx.

The accessory branch of the vagus of a fish separates from the vagus in an amniote and becomes a distinct cranial nerve, the **accessory nerve (XI).** The accessory carries somatic motor fibers to derivatives of the cucullaris: the trapezius and sternocleidomastoid complexes of muscles. Proprioceptive fibers from these muscles are believed to return in the spinal nerves that also innervate them.

The Autonomic Nervous System

The skeletal and branchiomeric muscles of the body are supplied by somatic motor neurons that travel directly from the central nervous system to the effectors. As we have seen (p. 458), exocrine glands and the muscles of the gut wall, heart, and other internal organs are innervated by visceral motor neurons (previously called general visceral motor neurons) that undergo a peripheral relay. This group of neurons constitutes the autonomic nervous system.° Lightly myelinated preganglionic neurons leave the central nervous system and synapse in peripheral ganglia with the dendrites and cell bodies of postganglionic neurons. The axons of the postganglionic neurons, which are not myelinated, continue to the effectors. There is considerable divergence and convergence within the autonomic ganglia. A single preganglionic fiber usually synapses with many postganglionic neurons. Conversely, a single postganglionic neuron may receive many preganglionic fibers from different sources. The principal neurotransmitter secreted by the preganglionic neurons is acetylcholine, but a variety of peptide neurotransmitters may also be released. These appear to modulate the activity of acetylcholine. The autonomic ganglia are not simply relay stations but points where divergence and integration of impulses can occur. General visceral sensory fibers, which return from the viscera to the central nervous system in the same nerves that carry the motor fibers, provide essential feedback for the central regulation of autonomic activity. Nevertheless, these fibers are not considered a part of the autonomic system, which is defined in functional terms as a visceral motor system that regulates visceral activity and helps the body maintain homeostasis.

The Autonomic Nervous System of Mammals

The mammalian autonomic nervous system, which has been more thoroughly studied than that of other vertebrates, is subdivided into three divisions: **sympathetic** (Gr. *syn* = with + *pathos* = feeling), **parasympathetic** (Gr. *para* = beside or accompanying), and **enteric** (Gr. *enteros* = intestine) **systems.** These systems differ in the location of their preganglionic neurons and in the organization and activity of their postganglionic neurons.

The enteric system is restricted to the gut wall, but most visceral organs, including the digestive tract, receive both a sympathetic and a parasympathetic innervation. Because their postganglionic fibers release different transmitters at

---

°When it was assumed that branchiomeric muscles were visceral, it was necessary to distinguish the neurons supplying them (special visceral motor neurons) from the neurons of the autonomic nervous system (general visceral motor neurons). Since we are treating the branchiomeric muscles as somatic, and their neurons as somatic motor, this distinction no longer needs to be made. The only visceral motor neurons are those of the autonomic nervous system.

TABLE 12–3   **Examples of the Effects of Sympathetic and Parasympathetic Stimulation in Mammals**

Organs	*Sympathetic Stimulation*	*Parasympathetic Stimulation*
Endocrine glands		
Adrenal medulla	Secretes	—*
Cardiovascular system		
Heart rate	Increases	Decreases
Force of ventricular contraction	Increases	—
Coronary arteries	Dilate	Constrict
Arteries in skeletal muscles	Active ones dilate	—
Lungs		
Muscles in bronchioles	Relax	Contract
Digestive organs		
Salivary glands	Secrete some mucus	Secrete enzymes
Gastric glands	—	Secretion
Pancreas	—	Secretion
Liver	Releases sugar into blood	Bile flows
Visceral blood vessels	Constrict	—
Intestinal muscles	Relax	Contract
Anal sphincter	Contracts	Relaxes
Urogenital organs		
Bladder sphincter	Contracts	Relaxes
Arteries of external genitals	Constrict	Relax
Skin		
Hair muscles	Contract	—
Sweat glands	Secrete	—
Cutaneous arteries	Constrict	—
Eye		
Iris sphincter	—	Contracts
Iris dilator	Contracts	—
Ciliary muscles	Relax	Contract

*Innervation not present.

the neuron-effector junctions, the sympathetic and parasympathetic systems have different effects: one system stimulates and the other inhibits the activity of different organs (Table 12–3).

Postganglionic parasympathetic fibers secrete acetylcholine and so are described as **cholinergic fibers.** Postganglionic sympathetic fibers secrete noradrenaline (also called norepinephrine) and so are **adrenergic fibers.** Noradrenaline is chemically similar to the adrenaline secreted by the medullary cells of the adrenal gland (p. 548). Indeed, sympathetic stimulation and adrenal medulla secretion supplement each other in enabling a vertebrate to adjust to stresses, such as escaping a predator or chasing prey. This is often called the **fight or flight** reaction. In essence, this reaction provides more energy to the skeletal muscles of the body. The rate and force of cardiac muscle contraction increases, blood pressure rises, bronchioles in the lungs dilate, more blood is sent to the active skeletal muscles and less to the skin and viscera, and blood sugar levels increase. Hairs are also elevated, sweating increases, and the pupils dilate. Gut motility, digestive enzyme secretion, and sexual activity are inhibited. Parasympathetic stimulation has the opposite effects and may be called **rest and digest.** Energy is conserved and restored. Heart rate and force of contraction decrease, blood pressure falls, bronchioles constrict, more blood is diverted to the digestive tract, gut motility and digestive enzyme secretion increase, bile flows, sugars are stored, and defecation and urination are promoted. The pupils contract, and sexual activity may be stimulated.

The morphological distinction between the sympathetic and parasympathetic systems is not always as clear as their pharmacological and physiological differences. Morphologists describe the mammalian sympathetic system as being a **thoracolumbar outflow** from the central nervous system and the parasympathetic system as a **craniosacral outflow.** Preganglionic sympathetic fibers in human beings leave through the thoracic and first three or four lumbar spinal nerves, while preganglionic parasympathetic fibers leave through the oculomotor, facial, glossopharyngeal, vagus, and sacral nerves 2 through 4 (Fig. 12–16). But there are exceptions, and a few sympathetic fibers occur in the vagus and sacral nerves of some mammalian species.

Preganglionic sympathetic fibers that send impulses to cervical, cranial, and cutaneous organs (hair muscles, sweat glands, cutaneous blood vessels) enter the thoracolumbar portion of the paired **sympathetic cords** and relay in **sympathetic ganglia** along the cord. The cords extend through the neck to the base of the head and bear cervical ganglia in which the relays to cervical and cranial organs occur, but the cervical extensions of the sympathetic cords do not receive fibers from cervical spinal nerves. Most of the preganglionic sympathetic fibers to abdominal viscera pass through the sympathetic cords and continue in **splanchnic nerves** to relay in ganglionic masses that are located at the bases of major abdominal arteries. These ganglia have the same names as the arteries: **coeliac, superior mesenteric,** and **inferior mesenteric ganglia.** Postganglionic sympathetic fibers to all cutaneous structures and to the blood vessels in skeletal muscles re-enter the spinal nerves; those going to the skin of the head travel with the trigeminal nerve. Postganglionic fibers to other organs in the head, thorax, and abdomen follow along the arteries.

Preganglionic parasympathetic fibers leaving through the oculomotor nerve carry impulses to the ciliary body and iris of the eye; those in the facial and glossopharyngeal nerves, to the lacrimal and salivary glands and to glands in the mucous membranes of the nose, mouth, and pharynx; those in the vagus, to the thoracic and abdominal viscera; and those in the sacral nerves, which unite to form a small **pelvic nerve,** to pelvic viscera. Preganglionic parasympathetic fibers are much longer than those of the sympathetic system because they extend nearly all the way to the effectors; the postganglionic fibers are very short. The peripheral relays occur in minute ganglia in the walls of the organs they supply. No parasympathetic fibers go to the body wall or extend into the limbs, so cutaneous organs and smooth muscles in the walls of most blood vessels do not have a parasympathetic supply.

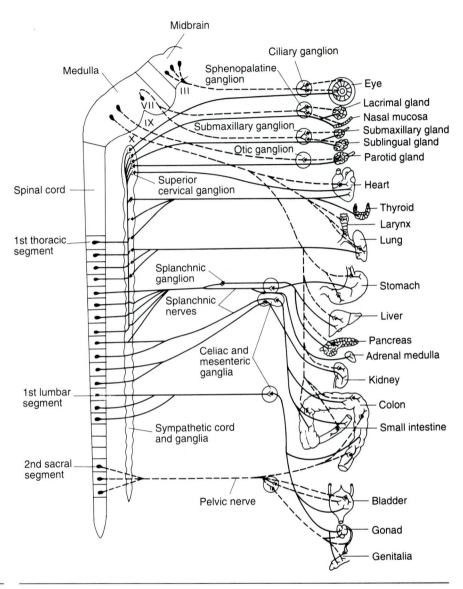

FIGURE 12–16   The mammalian autonomic nervous system. Sympathetic neurons are shown in solid lines; parasympathetic ones in broken lines. Sympathetic fibers to the skin are not shown. (After Dorit, Walker and Barnes.)

The degree of divergence also differs between these two systems. Each preganglionic sympathetic fiber synapses with about ten postganglionic neurons, so divergence is extensive and sympathetic effects diffuse. Parasympathetic stimulation is more narrowly targeted because each of its preganglionic fibers synapses with only about three postganglionic neurons.

The enteric system is composed of two interconnected neuronal networks: a **myenteric plexus** lies between the longitudinal and circular muscle layers of the gut wall, and a **submucosal plexus** lies just beneath the mucosa. Sensory neurons that are restricted to the gut wall detect changes in gut muscle tension and in the chemical environment in the gut lumen. Interneurons and motor neurons that are confined to the gut wall control gut motility, blood vessel tone, and the secretion of gastric and intestinal glands. The enteric system can function autonomously, but a second level of control is exercised by sympathetic and parasympathetic neurons that terminate in the plexuses.

The Evolution of the Autonomic Nervous System

The autonomic system of other amniotes appears similar to that of mammals. Anamniotes have ways of regulating visceral activity, but only parts of their autonomic system have been studied. In general, the distinction between sympathetic and parasympathetic systems is not so clear as in amniotes, and few organs have a dual innervation.

The autonomic nervous system of lampreys and hagfishes is rudimentary. Fibers believed to be sympathetic supply some blood vessels, but sympathetic ganglia are absent. The vagus contains parasympathetic fibers, but no parasympathetic fibers go to the eyeball in the oculomotor nerve.

Cartilaginous fishes do not have a well-organized sympathetic cord, but have segmental, sympathetic ganglia in the trunk, and these are interconnected by a few plexus-like, longitudinal fibers (Fig. 12–17). Nearly all the peripheral relays occur in these ganglia, and postganglionic fibers follow arteries to the viscera. A

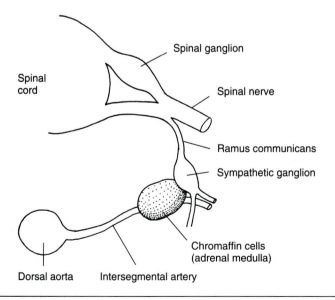

FIGURE 12–17   A sympathetic ganglion of a dogfish and its connections. The adrenal medula is part of an endocrine gland (p. 548).

number of sympathetic fibers also travel in the vagus, and these relay close to the organs they supply. There is no extension of sympathetic fibers into the head or to the cutaneous chromatophores and blood vessels of a cartilaginous fish (whose skin, of course, does not have the thermoregulatory role that it has in birds and mammals). Parasympathetic fibers to the eye leave in the oculomotor nerve; those going to the heart and stomach travel in the vagus nerve. Parasympathetic neurons are not present in other cranial nerves because large glands are not present in the head. Parasympathetic neurons have an inhibitory effect on the rate of heartbeat, as in amniotes. Only the stomach receives both a parasympathetic and a sympathetic supply. The primary excitatory fibers that help cause gastric secretion are the sympathetic ones—just the opposite of the effect of sympathetic stimulation on the stomach of amniotes. The function of the parasympathetic fibers to the stomach is less clear.

The autonomic system is more clearly organized in teleosts and amphibians. The sympathetic ganglia lie along a pair of sympathetic cords that extend to the base of the head. Sympathetic fibers re-enter the spinal nerves of teleosts, innervate the chromatophores, and cause the pigment in these cells to concentrate. A pituitary hormone has the opposite effect. The heart receives a sympathetic innervation in amphibians, and its effect is opposite that of the parasympathetic fibers in the vagus. Indeed, the antagonistic roles of sympathetic and parasympathetic innervation were first discovered from studies on frog hearts. Amphibians begin to evolve tear and salivary glands, and a few parasympathetic fibers travel to them in the facial and glossopharyngeal nerves. A few frogs have a parasympathetic outflow through their sacral nerves, but there is no outflow in this region in teleosts or salamanders.

SUMMARY

1. The nervous system integrates sensory information and initiates most of a vertebrate's rapid responses to changes in its environment.
2. Neurons are the structural and functional units of the nervous system. There are many types, but each consists of a trophic segment (the cell body), a receptive segment (dendrites), a conducting segment (axon), and a transmissive segment (telodendria).
3. Nerve impulses are usually carried in one direction across synapses between neurons, and across myoneural and axoglandular junctions, by transmitter substances. These substances may have excitatory or inhibitory effects.
4. Most peripheral neurons become myelinated during development by Schwann cells.
5. Several types of nonnervous neuroglia are abundant in the central nervous system. Oligodendrocyctes myelinate central axons. Astrocytes regulate ionic balances in extracellular fluid, degrade metabolites, and may have a nutritive role. Microglia are phagocytes.
6. Myelinated axons form the white matter of the central nervous system; nonmyelinated axons and neuron cell bodies lie in gray matter.
7. The central cavity of the nervous system expands into ventricles within the brain. These contain cerebrospinal fluid.

8. The nervous system consists of only three broad categories of neurons: sensory or afferent, motor or efferent, and interneurons, which are confined to the central nervous system. The cell bodies of most sensory neurons are located in peripheral ganglia; most of those of motor neurons are in central gray matter.

9. Sensory and motor neurons can be sorted into functional groups: general somatic sensory neurons from cutaneous receptors and proprioceptors, special somatic sensory neurons from special sense organs, general visceral sensory neurons from most visceral receptors, special visceral sensory neurons from taste buds, visceral motor (autonomic) neurons to gut and heart muscles and to exocrine glands, and somatic motor neurons to somatic muscles. All have characteristic terminations or origins within the central nervous system.

10. Neurons form reflex arcs within the spinal cord and brain stem that are responsible for much locomotor and visceral activity. Neurons also form pathways that ascend to the brain and descend from it. Most pathways decussate on the way to and from the brain. Ascending pathways relay in the thalamus before projecting to the cerebrum.

11. During the course of vertebrate evolution, the spinal cord became more vascular and enlarged as more neuronal tracts developed between the spinal cord and brain.

12. The dorsal and ventral roots of spinal nerves form separate nerves in lampreys, but these unite in all gnathostomes. Sensory neurons always enter through dorsal roots of spinal nerves, and somatic motor ones always leave through the ventral roots. Visceral motor neurons leave through the dorsal roots of cyclostomes but have shifted during evolution, so all lie in the ventral roots of amniotes.

13. The cranial nerves of fishes can be sorted into three groups, each with distinctive characteristics. (1) The terminal, olfactory, optic, and statoacoustic nerves are unique to the head and contain only special somatic sensory neurons. (2) The oculomotor, trochlear, abducens, and occipital nerves supply somatic motor neurons to the extrinsic ocular, epibranchial, and hypobranchial muscles. They rarely contain other types of neurons. They appear to be serially homologous to the ventral spinal nerves of lampreys. (3) The trigeminal, facial, glossopharyngeal, and vagus nerves supply somatic motor neurons to the branchiomeric muscles, and they also contain a variety of sensory neurons. They appear to be serially homologous to dorsal spinal nerves of lampreys.

14. The head of ancestral vertebrates probably contained eight segments. The nerves we have interpreted as serially homologous to ventral and dorsal cranial nerves can be related to these segments.

15. The special sensory fibers that form the terminal, olfactory, and statoacoustic nerves, and contribute to the dorsal nerves, probably originated in a distinct series of octavolateralis nerves and chemosensory (olfactory and gustatory) nerves. The optic nerve does not belong to any of the groups of cranial nerves because it is a brain tract.

16. During the evolution from fishes to mammals, few changes occurred in the ventral cranial nerves other than the emergence of the occipital nerves from

within the skull as the hypoglossal nerve (XII). Major changes occurred in the dorsal cranial nerves. Lateral line fibers were lost from the facial, glossopharyngeal, and vagus nerves. The facial and glossopharyngeal nerves acquired visceral motor neurons (autonomic) to tear, salivary, and other glands of the head. The vagus nerve lost most of its somatic motor neurons apart from those to the larynx. The accessory branch of the vagus nerve separated as a distinct nerve, the accessory nerve (XI).

17. The autonomic nervous system is the motor system for gut, heart, and other visceral muscles, and for exocrine glands. It is characterized by a motor relay in a peripheral ganglion where some divergence and integration of impulses occurs.

18. The autonomic nervous system of amniotes consists of separate sympathetic and parasympathetic systems, and an enteric system, which is confined to the gut wall. Most visceral organs, including the enteric system, receive both a sympathetic and parasympathetic innervation. In general, sympathetic innervation prepares an animal for fight or flight, and parasympathetic innervation prepares an animal to rest and digest. Sympathetic and parasympathetic systems leave from different parts of the central nervous system and relay at different sites. The autonomic nervous system is poorly developed in cyclostomes and cartilaginous fishes, but is more evident in bony fishes and amphibians.

REFERENCES

All references on the nervous system are collected at the end of Chapter 13, p. 531.

13 The Nervous System: The Brain

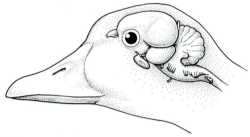

PRECIS

After examining the development of the brain and its meninges, we will consider the structure and organization of the fish brain as an example of an early vertebrate brain. We will then summarize the major evolutionary changes seen in amphibians, reptiles, and birds before considering the structure and functional organization of the mammalian brain in some detail.

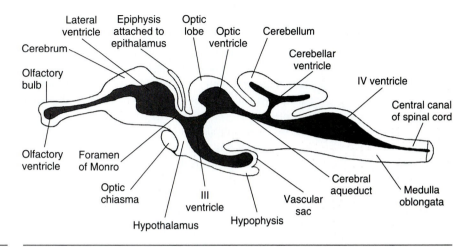

Lateral ventricle · Epiphysis attached to epithalamus · Optic lobe · Optic ventricle · Cerebellum · Cerebrum · Cerebellar ventricle · Olfactory bulb · IV ventricle · Central canal of spinal cord · Olfactory ventricle · Foramen of Monro · Optic chiasma · III ventricle · Vascular sac · Cerebral aqueduct · Medulla oblongata · Hypothalamus · Hypophysis

FIGURE 13–1   A sagittal section of the brain of a dogfish, showing the major parts of the brain and its ventricular system (After Walker and Homberger.)

Vertebrates differ from protochordates in having evolved a brain, a process called **encephalization.** We may never be certain as to the circumstances that led to the evolution of a brain, but the front of the body of an active animal—and we believe that ancestral vertebrates were far more active than the protochordates—is the first part to encounter environmental changes. Natural selection favors the concentration of sense organs here. Ancestral vertebrates probably had a concentration of olfactory receptors, photoreceptors, and neuromasts of the lateral line and ear on their heads. Processing the input from these receptors and directing the activity of segmental and branchial structures would require an aggregation of neurons and the enlargement of the rostral end of the neural tube. The brain probably began to evolve in this way. The brain stem integrates the activity of the cranial muscles. Together with a higher center of the brain, known as the hypothalamus (Fig. 13–1), it also integrates visceral activity. The evolution of other parts of the brain may have been related initially to the input from the special sense organs on the head. The olfactory bulbs and cerebral hemispheres receive olfactory impulses; the optic lobes, visual impulses; and the cerebellum, the input from the lateral line and vestibular part of the ear. Although these structures have other connections in living vertebrates, their size in fishes is correlated closely with the amount of information they receive from the special sense organs.

Determining Brain Pathways and Functions

The brain is a vital and complex organ. Investigators can see gross changes that have occurred during evolution by dissecting the brain of representative vertebrates, but these changes tell us little about brain functions. To begin to understand function, it is first necessary to make a "road map," or wiring diagram, of neuronal organization within the brain. This will show us the pathways neurons take and give us some indication of which brain regions may influence other brain regions. Many techniques have been used. Favored current techniques involve the injection into parts of the brain of isotopically labeled amino acids such as horse-radish peroxidase or various dyes. These can be introduced at a synapse, or injected directly into the cell body or axon of a neuron. The injected material will travel forward along an axon to its termination or backward into the cell body. The stained axons or cell bodies can be identified in serial sections of the brain. Some substances will cross one or more synapses and label an entire pathway.

Researchers can refine the wiring diagrams and gain some indications of function by electrophysiological experiments. They implant electrodes precisely into specific nuclei and tracts, stimulate the center, and search for electrical signals, called **evoked potentials,** in other parts of the brain.

Neurologists also have learned a great deal about human brain function by observing the changes in behavior caused by trauma or pathological lesions, and then locating the damaged area in postmortem examination. They have determined other functions during surgical procedures, which often are done under local anesthesia because the brain contains no pain receptors. As a surgeon probes for a tumor, he or she can observe a twitch of a muscle or other changes in behavior, and the patient can comment on sensations.

Many additional brain functions can be determined in experimental animals by observing the changes in behavior that result from the electrical stimulation of centers, the microsurgical destruction of areas by cautery, and the local administration with micropipets of neurotransmitters and other drugs. Immunocytological procedures are also being used to localize neurons containing particular neurotransmitters.

Recently advancements in imaging techniques have allowed neurologists to explore the structure and function of a living brain without using disruptive invasive procedures. A particularly interesting example is **positron emission tomography** (PET). PET produces computer-generated images of a series of planes or sections through a brain. Unlike X-ray computerized tomography (CT scans), which identifies different densities in the brain, PET shows the distribution of isotopes that emit radiation in the form of positrons. A positron-emitting isotope can be bound to an important biological compound, given to a subject, and various biological processes can be explored. For example, an analogue of glucose, which can be taken up by cells but not leave them, can be labeled with a radioactive isotope and given to a subject. Since the analogue of glucose is taken up by particularly active cells, investigators can compare the sites and degree of glucose utilization of a person at rest and then performing various tasks. In this way, investigators have determined which parts of the brain become active when the subject is viewing an object, listening to a sound, memorizing something, or doing other things. Neurologists are learning a great deal about the internal organization of the brain and its functions, but our knowledge is far from complete. Neurobiology is one of the most rapidly developing fields of biology.

The Development of the Brain

The vertebrate brain develops from the enlargement and differentiation of the rostral end of the neural tube. Three primary swellings begin to appear as the neural folds close (Fig. 13–2 and Table 13–1): a forebrain or **prosencephalon** (Gr. *pros* = before + *enkephalos* = brain), a midbrain or **mesencephalon** (Gr. *mesos* = middle), and a hindbrain or **rhombencephalon** (Gr. *rhombos* = lozenge). The prosencephalon first gives rise to the **optic vesicles** that will form the retina (p. 436), and to a midventral **infundibulum**, which forms the part of the

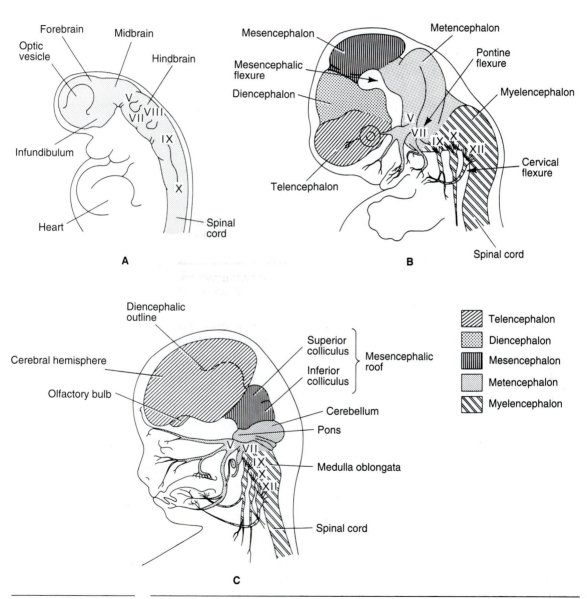

FIGURE 13–2 Diagrams of three stages in the development of the human brain to show the differentiation of the principal brain regions. Certain cranial nerves are identified by Roman numerals. (After Corliss.)

TABLE 13–1   **Brain Divisions and Their Major Derivatives in Mammals**

Primary Divisions	Secondary Divisions	Major Adult Derivatives	Ventricles
Prosencephalon	Telencephalon	Olfactory bulbs Cerebral hemispheres Basal ganglia Hippocampus	Lateral ventricles
	Diencephalon	Optic vesicles Infundibulum Neurohypophysis Pineal gland Epithalamus Thalamus Hypothalamus	Ventricle III
Mesencephalon	Mesencephalon	Superior colliculus Inferior colliculus Tegmentum	Cerebral aqueduct
Rhombencephalon	Metencephalon	Cerebellum Pons	Ventricle IV
	Myelencephalon	Medulla oblongata	Ventricle IV

pituitary gland known as the **neurohypophysis.** Another brain region, the **telencephalon** (Gr. *telos* = end) develops soon after as a pair of rostrolateral extensions from the prosencephalon (Fig. 13–2). These differentiate into the paired **cerebral hemispheres,** which together constitute the **cerebrum,** into the paired **olfactory bulbs,** and into deep masses of gray matter. In all vertebrates, except for actinopterygian fishes (Focus 13–1), the gray matter of the cerebrum lies deeply next to the paired **lateral ventricles.** These ventricles are lined by the ependymal epithelium, as are all cavities within the central nervous system.

The rest of the prosencephalon remains in the midline and differentiates as the **diencephalon** (Gr. *dia* = through). Its cavity, the **third ventricle,** remains connected to the lateral ventricles by the paired **interventricular foramen of Monro** (Figs. 13–1 and 13–3). The lateral walls of the diencephalon thicken to form the **thalamus** (Gr. *thalamos* = inner chamber), the floor becomes the **hypothalamus,** and the roof, most of which remains very thin, is the **epithalamus.** The optic stalks and pituitary gland attach to the hypothalamus, and the median eye complex, or pineal gland, develops from the epithalamus.

The mesencephalon remains undivided. A narrow **cerebral aqueduct of Sylvius** passes through it and connects the third ventricle with the **fourth ventricle** in the rhombencephalon. In anamniotes, the dorsal part of the mesencephalon, known as its **tectum** (L. *tectum* = roof), enlarges as a pair of **optic lobes** (Fig.

FOCUS 13–1   **The Inside-Out Cerebrum of Actinopterygians**

As we pointed out, the cerebral hemispheres of most vertebrates develop as a pair of rostroventral extensions from the rostral part of the prosencephalon. These extensions are simple *evaginations*. As the rostrolateral extensions develop in actinopterygian fishes, the lateral ventricles open and the gray matter and ependymal epithelium are exposed to the surface (the Focus Figure). In short, the cerebral hemispheres have turned inside out and curved ventrally on themselves; they have *everted*. There are no true lateral ventricles.

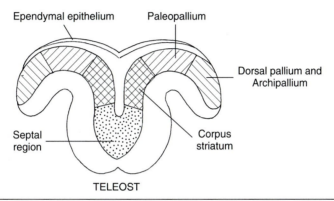

A transverse section through the everted cerebrum of an actinopterygian. Notice that the ependymal epithelium and gray matter lie on the surface. See Figure 13–3 for code to gray regions.

13–1). They usually contain optic ventricles. The optic lobes differentiate into the paired **superior** and **inferior colliculi** in mammals (Fig. 13–2**C**). The ventral part of the mesencephalon is known as its **tegmentum** (L. *tegmentum* = covering).

The rhombencephalon divides into a **metencephalon** (Gr. *meta* = after) and a **myelencephalon,** which leads to the spinal cord (Gr. *myelos* = marrow, or spinal cord). The **cerebellum,** which may contain an extension of the fourth ventricle in fishes, develops from the dorsal part of the metencephalon. In birds and mammals the floor of the metencephalon includes nuclei that relay impulses between the cerebrum and cerebellum and a bridge of transverse fibers. This region is known as the **pons** (L. *pons* = bridge). In other vertebrates both the floor of the metencephalon and the myelencephalon form the **medulla oblongata.**

The major divisions of the brain lie along the same horizontal plane in most adult vertebrates (Fig. 13–1). Brain regions become folded on one another in birds and mammals because the head extends forward from an upward-curving neck. **Mesencephalic, cervical,** and **pontine flexures** develop sequentially in the positions shown in Figure 13–2**B**.

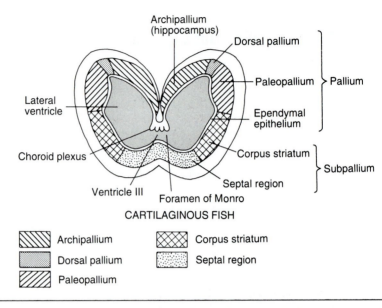

CARTILAGINOUS FISH

⧄ Archipallium	⧉ Corpus striatum
▨ Dorsal pallium	⬚ Septal region
⧄ Paleopallium	

FIGURE 13–3   A transverse section through the cerebrum of a cartilaginous fish showing the deep location of the gray matter and its subdivisions in an evaginated cerebrum. The code given for the subdivisions of the gray matter will also be used in Focus Figures of sections through the cerebrum.

The Meninges and Cerebrospinal Fluid

The brain and spinal cord become covered by one or more layers of connective tissue that develop partly from mesoderm and partly from neural crest cells. These are the **meninges** (plural of Gr. *meninx* = membrane). Fishes have a single layer, the **primitive meninx,** that closely invests the central nervous system. Strands of connective tissue extend from it to a layer of connective tissue that lines the cranial cavity and vertebral canal. A gelatinous material fills the space between the primitive meninx and the surrounding cartilage and bone. Cerebrospinal fluid circulates in the cavities within the central nervous system, and some is secreted into spaces around it. In amphibians and reptiles the primitive meninx is divided into a dense **dura mater** (L. *dura mater* = hard mother), which unites with the connective tissue lining the cranial cavity, and a more delicate and vascular **secondary meninx,** which covers the brain and spinal cord.

Mammals and birds have three meninges because the secondary meninx has differentiated into two layers (Fig. 13–4A). A **pia mater** (L. *pia mater* = tender mother) closely invests the surface of the central nervous system and follows all the convolutions of the brain. An **arachnoid** (Gr. *arachne* = spider + *eidos* =

FIGURE 13–4   The meninges and cerebrospinal fluid of a mammal: **A,** The meninges as seen in a trans- ▶ verse section through the center of the cerebrum. **B,** A sagittal section of the central nervous system to show choroid plexuses, the meninges, and the circulation of the cerebrospinal fluid. (**A,** from Walker and Homberger. **B,** modified from various sources.)

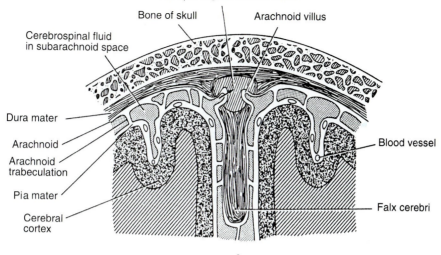

Superior sagittal venous sinus

Bone of skull

Arachnoid villus

Cerebrospinal fluid
in subarachnoid space

Dura mater

Arachnoid

Arachnoid
traculation

Pia mater

Cerebral
cortex

Blood vessel

Falx cerebri

A

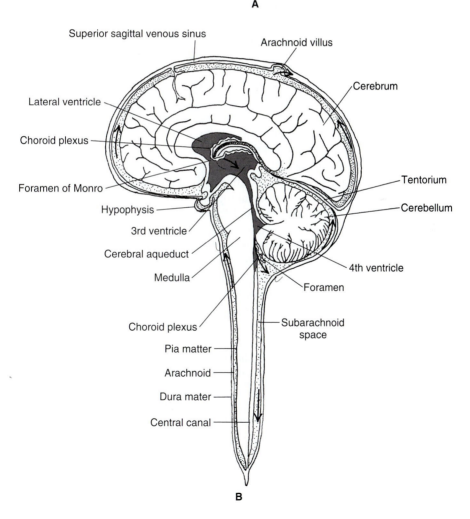

Superior sagittal venous sinus

Arachnoid villus

Lateral ventricle

Choroid plexus

Foramen of Monro

Hypophysis

3rd ventricle

Cerebral aqueduct

Medulla

Choroid plexus

Pia matter

Arachnoid

Dura mater

Central canal

Cerebrum

Tentorium

Cerebellum

4th ventricle

Foramen

Subarachnoid
space

B

form) lies peripheral to the pia mater and crosses many of the crevasses on the brain surface. Strands of connective tissue (arachnoid trabeculations) extend like spider webs from the arachnoid across a **subarachnoid space** and attach to the pia. The cerebrospinal fluid that circulates around the central nervous system of birds and mammals is confined to the subarachnoid space and cavities of the brain. Many small blood vessels lie in the strands of the arachnoid and in the pia and follow the pia into the convolutions of the brain. The dura mater lies peripheral to the arachnoid. It is fused with the endosteum lining the cranial cavity. In mammals a longitudinal, sickle-shaped septum of dura, known as the **falx cerebri,** extends between the cerebral hemispheres, and a transverse dural septum, the **tentorium,** extends between the cerebrum and the cerebellum (Fig. 13–4**B**). The latter is ossified in many species. These septa help stabilize the brain and hold it in place, especially during rapid rotations of the head.

Part of the wall of many ventricles is very thin, consisting only of the ependymal epithelium and the adjacent vascular meninx. An area of this type is known as a **tela choroidea.** Vascular tufts that extend from the tela choroidea form **choroid plexuses** that secrete much of the cerebrospinal fluid. Additional cerebrospinal fluid appears to be produced by the ependymal epithelium itself. The choroid plexuses extend into the ventricles in mammals, but in anamniotes and reptiles some are everted into the space around the brain. Mammals have choroid plexuses in the floor of the lateral ventricles and in the roofs of the third and fourth ventricles (Fig. 13–4**B**). Cerebrospinal fluid flows toward the fourth ventricle, partly by secretion pressure and partly by the action of cilia on the ependymal cells. Cerebrospinal fluid escapes through pores in the thin roof of the fourth ventricle and circulates slowly in the subarachnoid space. It eventually returns to the blood, primarily by diffusion through **arachnoid villi** that project into large venous sinuses in the dural septa.

The cerebrospinal fluid forms a liquid cushion around the brain and spinal cord, helping support these delicate structures and buffering them from blows to the head or body. Since the subarachnoid spaces follow blood vessels deep into the brain, the cerebrospinal fluid also enters the brain and becomes continuous with the extracellular fluid. The extracellular fluid bathing brain neurons and glial cells thus has two origins: capillaries within the brain and the choroid plexuses.

Blood-brain barriers exist between the extracellular fluids and the blood in the choroid plexuses and in the capillaries of the brain. The endothelial cells of capillaries in the choroid plexuses and brain have tight junctions between them and are less permeable than other capillaries, and the capillaries in the brain are often surrounded by feet of astrocyte cells (see Fig. 12–4). The choroid plexuses and cerebral capillaries act as selective barriers and active transport sites for many substances, thereby carefully regulating the composition of cerebral extracellular fluid. Lipid-soluble molecules, such as oxygen, carbon dioxide, and some drugs (heroin), easily cross these barriers; but proteins, many hormones, many enzymes, and waste products in the blood and other substances are held back. Glucose, amino acids, and other nutrients are actively transported across. Active transport is bidirectional, and many waste products and metabolites are removed in this way.

Since the brain has no lymphatic system, the cerebrospinal fluid also acts as a subsidiary drainage system for the brain, carrying off excess water, carbon dioxide, and metabolites. These are discharged into the blood at the arachnoid villi. The cerebrospinal fluid makes an important contribution to brain nutrition and drainage because the volume of the cerebrospinal fluid, about 150 ml in an adult human, is turned over every 3 or 4 hours.

The Brain of Fishes

The number of neurons in the brain and the complexity of their interrelations correlate in a general way with the range of behavior and other abilities of vertebrates. Accordingly, the size of the brain and its parts varies considerably among vertebrates, depending on their level of activity and mode of life. As might be expected, lampreys and hagfishes have the smallest brain relative to body weight of any group of vertebrates. Other fishes have higher ratios of brain weight to body weight, comparable to those of amphibians and reptiles. Surprisingly, most chondrichthyan fishes have substantially higher ratios than teleosts, ones that approach those of birds and mammals.

Parts of the Fish Brain

Fish have large olfactory bulbs and only a moderate-sized cerebrum (Fig. 13–5). Much of the cerebrum receives and integrates olfactory information, although other senses are projected here as well. Smell is important in the fish mode of life.

The epithalamus includes a choroid plexus, a pineal/parietal complex, and a small habenula, which is an olfactory center. The choroid plexus is everted in some species, and some of the cerebrospinal fluid is discharged around the brain. The degree of development of the pineal/parietal complex varies among species. A long and thin epiphysis remains in cartilaginous fishes (Fig. 13–5**B**).

The thalamus is a relay center to and from the cerebrum and many other centers of the brain. It is not large in most fishes because the cerebrum receives less information from other centers and has less influence on body activities than in amniotes.

The hypothalamus is an important integration center for visceral activity in all vertebrates, and it is relatively as large in fishes as in other groups. It includes a pair of **inferior lobes.** The pituitary gland, or hypophysis, attaches to its ventral surface. Fishes have an unusual **vascular sac** at the caudal border of the hypothalamus. Some investigators have proposed that it detects changes in water pressure, and hence indicates depth, but there is little experimental evidence for this. The sac has been shown to absorb dyes injected into the cerebrospinal fluid around the central nervous system, so it may be an important site for the reabsorption of this fluid.

Lampreys have poor sight, their optic lobes are relatively small (Fig. 13–5**A**). The optic lobes are well developed in other fishes because they are centers for the integration of visual, tactile, chemosensory, and other sensory modalities.

The cerebellum receives information from the lateral line, electroreceptors, vestibular parts of the ear, and many proprioceptors. Impulses leaving the cere-

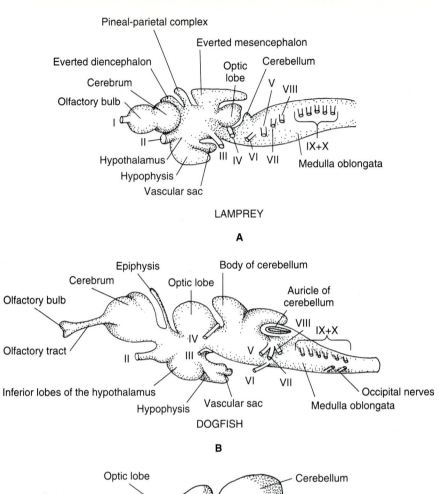

LAMPREY

A

DOGFISH

B

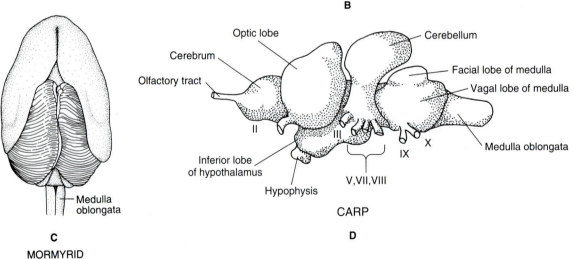

MORMYRID

C

CARP

D

FIGURE 13–5   The brains of fishes, most in lateral view: **A**, A lamprey. **B**, A dogfish. **C**, A mormyrid (in dorsal view); the enlarged cerebellum nearly covers the entire brain. **D**, A carp. (**C**, after Hopkins. **D**, after Bond.)

bellum modify the locomotor activity that is controlled reflexly in the spinal cord. The cerebellum is a **suprasegmental center** for motor coordination. The term *suprasegmental* means that levels of control are superimposed on the basic pattern of segmental reflexes occurring within the brain stem and spinal cord. Hagfishes, alone among vertebrates, do not have a cerebellum, and it is very small in lampreys. However, the cerebellum is well developed in most fishes, and it is enormous in those species in which part of the lateral line is modified for electrolocation (Fig. 13–5**C**).

The medulla oblongata is well developed and contains centers for the control of feeding, respiratory, and cardiac movements. The special visceral sensory nuclei of the facial and vagus nerves form conspicuous lateral bulges, the **facial** and **vagal lobes,** in carp, catfishes, and other species that rely heavily on their gustatory sense (Fig. 13–5**D**).

Functional Organization of the Fish Brain

Although less is known about the organization and functional pathways in the brains of nonmammalian vertebrates than in those of mammals, neuroanatomists have learned a great deal in recent years. Many studies have been made on cartilaginous fishes, and especially on teleosts. Those on actinopterygians have been reviewed by Northcutt and Davis (1983) and Davis and Northcutt (1983). We will briefly discuss the general activities of major centers. This will serve as a point of departure for tracing major trends in encephalization among vertebrates.

Medullary and Tegmental Centers.   A primitive and fundamental level of control is a network of short, branching interneurons in the medulla and tegmentum, called the **reticular formation** (Fig. 13–6). The reticular formation lies between the sensory and the motor nuclei of the cranial nerves attached to the brain stem. It interconnects them, forms ascending and descending pathways, and has extensive connections with other brain centers. Clusters of functionally related neurons tend to come together, and a few nuclei begin to organize in the reticular formation of fishes. The major sensory input of the reticular formation comes from the visceral and somatic sensory nuclei of the brain stem, but it also receives some other sensory input, including a limited amount that ascends from the spinal cord on **spinoreticular tracts.** Sensory information is relayed from the reticular formation to other parts of the brain, including the optic tectum, hypothalamus, and thalamus. The reticular formation is part of a primitive pathway for sending somatic sensory information forward to the thalamus and the cerebrum. Chondrichthyan fishes have a few direct **spinothalamic fibers** that bypass the reticular formation. The reticular formation, in turn, receives input from other parts of the brain, including the cerebrum, thalamus, hypothalamus, optic tectum, and cerebellum. The major efferents of the reticular formation lead to the motor nuclei of the cranial nerves, and these mediate feeding and swallowing movements, respiratory movements, cardiac rate, and other vital activities. A few efferents leave by **reticulospinal tracts** to the motor columns of the spinal cord. Motor activity that is controlled in the spinal cord can be modulated by the extensive input into the

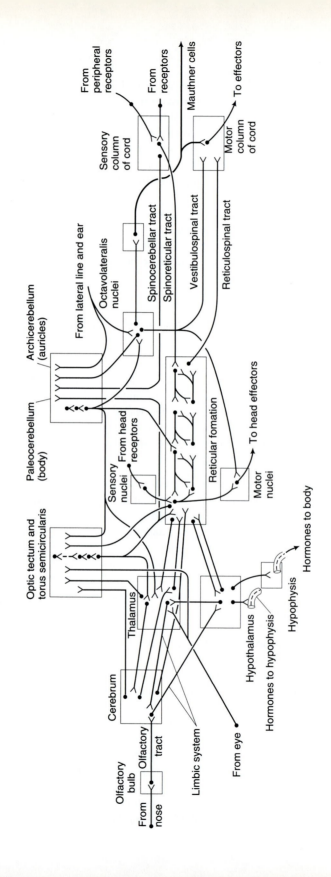

From peripheral receptors

From receptors

Mauthner cells

To effectors

Sensory column of cord

Motor column of cord

From lateral line and ear

Octavolateralis nuclei

Spinocerebellar tract

Spinoreticular tract

Vestibulospinal tract

Reticulospinal tract

Archicerebellum (auricles)

Paleocerebellum (body)

Sensory nuclei

From head receptors

Reticular formation

To head effectors

Motor nuclei

Optic tectum and torus semicircularis

Thalamus

Hypothalamus

Hormones to hypophysis

Hypophysis

Hormones to body

Cerebrum

Limbic system

From eye

Olfactory bulb

Olfactory tract

From nose

reticular formation. Swimming and eye movements, for example, can be directed toward food.

Another important brain center that affects locomotion is a pair of **acousticolateral** or **octavolateralis nuclei** in the medulla (Fig. 13–6). Each receives information from lateral line organs and the ear indicating body movement and position. This nucleus can be divided into several parts. The most dorsal group of neurons receives impulses from electroreceptors, when present; an intermediate group, from other lateral line organs; and the most ventral group, from the ear. Both vestibular and acoustic fibers from the ear terminate in this last area, but the subsequent projection of acoustic fibers is not certain in fishes. Neurons that begin in the octavolateralis nuclei project to different parts of the cerebellum, directly to the motor columns of the spinal cord by a **vestibulospinal tract,** and to the motor nuclei of nerves innervating the extrinsic ocular muscles. Eye and body movements are adjusted to changes in body position and to information provided by the lateral line.

Actinopterygians have a pair of giant **Mauthner cells** whose cell bodies lie close to the octavolateralis nuclei and doubtless receive impulses from them as well as from other brain centers. Their axons decussate and descend the full length of the spinal cord, synapsing with motor neurons on the way. Impulse transmission is rapid on these large-diameter, myelinated axons, and the system mediates an extremely fast reflex resulting in escape swimming movements.

The Hypothalamus.   Much visceral activity is coordinated in the reticular formation, but the hypothalamus also is a major center for visceral integration. It is particularly important for activity mediated through the autonomic nervous system and through many of the endocrine glands. Although the hypothalamus has connections that enable it to receive most types of sensory information, its primary sensory inputs are gustatory and other visceral sensory inputs relayed from the reticular formation, and somatic sensory olfactory signals relayed from parts of the cerebrum (Fig. 13–6). Some of its cells also respond directly to the levels of glucose, water, salts, hormones, and other factors in the blood. In all vertebrates, the hypothalamus influences the feeding, swallowing, respiratory, and cardiac centers of the brain stem. It also regulates periods of rest and activity, the amount of food and water taken in, gut movements and digestion, blood sugar levels, water and salt balance, and reproductive activity. Some parts of the hypothalamus have a stimulating and others an inhibitory effect. It controls the various functions directly, via efferent neurons to the reticular formation and thalamus, and thence to motor centers, and indirectly, via its influence over the secretion of the hypophysis or pituitary gland. Secretory neurons, whose cell bodies are located in the

◀ *FIGURE 13–6*   A composite wiring diagram showing some of the major neuronal interconnections in the brain of jawed fishes. Although the outline of the brain has not been added, the diagram is essentially a lateral projection, with anterior toward the left. All connections that are known cannot be shown in a diagram of this sort. The diagram is simplified and highly schematic, but introduces the complexity of the fish brain and illustrates points made in the text.

hypothalamus, synthesize hormones that affect water balance and smooth muscle contraction. The hormones travel along axons of these cells to the posterior part of the hypophysis, where they are stored and released (p. 545). Other hypothalamic neurons synthesize hormones known as releasing and inhibiting hormones. These travel in blood vessels to the anterior part of the hypophysis and regulate the blood level of other hormones synthesized there (p. 542).

The Cerebellum.   The cerebellum, optic tectum, and cerebrum are suprasegmental centers that affect the other nervous centers we have considered. Although these suprasegmental centers have some influence on visceral activity, they are primarily somatic integration centers. The cerebellum does not initiate swimming, but it is important for overall motor coordination. A pair of laterally placed **cerebellar auricles,** which constitute the **archicerebellum,** are closely associated with the octavolateralis nuclei and probably evolved as outgrowths from them (Fig. 13–5**B**). The cerebellar auricles receive vestibular and lateral line impulses from the octavolateralis nuclei and also some vestibular input directly from the ear (Fig. 13–6). A medial **cerebellar body,** or **paleocerebellum,** receives proprioceptive and tactile information from the trunk and tail and electroreceptive information in certain species. Most of the impulses from the spinal cord ascend directly on **spinocerebellar tracts.** Visual information that is important in body balance and movements is sent to the cerebellum from the tectum by way of the thalamus. Afferent neurons terminate among cerebellar neurons that form a three-layered cerebellar cortex in all vertebrates except lampreys. Interactions among the cortical neurons ensure that the degree and timing of muscle contractions occur with reference to body position and movement and the current state of muscle contraction. Efferent neurons from the cerebellar cortex go to deep cerebellar neurons, which in fishes are not well organized into nuclei. Efferents from the deep cerebellar neurons continue to the octavolateralis nuclei, the reticular formation, and the tectum; a few project to the thalamus.

The Mesencephalic Roof.   Most of the mesencephalic roof forms a conspicuous pair of bulges called the **optic tectum** or **optic lobes.** Caudally the roof contains a group of deeper nuclei called the **torus semicircularis.** The torus semicircularis receives inputs from medullary centers associated with the octavolateralis system. The optic tectum is a very important integrating center in fishes, and the optic lobes are usually the largest part of the brain. Most of the fibers from the retina terminate here, although a few go to the thalamus (Fig. 13–6). Perception of form appears to occur here. If parts of the tectum are destroyed, a fish cannot recognize food (assuming other sensory clues have been destroyed), and it fails to pursue or catch food pellets. Beyond this, the tectum and torus semicircularis receive projections directly or indirectly from every major brain region, so they receive virtually all types of sensory information: visual, olfactory, gustatory, tactile, and octavolateralis. Cell bodies of tectal neurons migrate during embryonic development from the area around the cerebral aqueduct toward the surface, where they form a complex, multilayered configuration that is somewhat analogous to the mammalian cerebral cortex. Six layers of neurons are recognized in some species. The connections and complexity of the optic tectum suggest that motor impulses responding to a combination of sensory inputs can be initiated here. Direct electrical stimulation of the tectum can initiate a variety of coordinated eye and body movements: the eyes may turn in a certain direction, the trunk flexes so that the body would move in this direction, and one pectoral fin is protracted and one retracted in ways to assist in this movement (Vanegas, 1983).

Olfactory Bulbs, Cerebrum, and Thalamus.   Olfactory signals are sent on the processes of olfactory cells directly to the olfactory bulbs. From there, signals continue on **olfactory tracts** to parts of the cerebrum (Fig. 13–6). Various regions of the cerebral wall of fishes can be recognized in a cross section (Fig. 13–3). The roof of a cerebral hemisphere is known as the **pallium** (L. *pallium* = mantle); the floor is the **subpallium.** The most ventral and lateral part of the pallium, known as the **paleopallium,** receives olfactory impulses from the olfactory bulb. The most mediodorsal part, known as the **archipallium** or **hippocampus,** so called because of its shape in mammals (Gr. *hippokampos* = seahorse), receives olfactory input from the paleopallium and probably inputs from the hypothalamus and dorsal thalamic nuclei. In tetrapods, and probably to some extent in fishes, this region is part of the **limbic system,** which plays an important role in feeding, reproductive, and emotional behavior. A small **dorsal pallium,** which lies between the paleopallium and archipallium, receives some sensory information that is relayed forward through dorsal thalamic nuclei.

A **septal region** and **corpus striatum** are recognized in the subpallium. The septal region is part of the limbic system. In many fishes, the corpus striatum appears to be the major integration center of the cerebrum. It is the main part of the cerebrum to which dorsal thalamic nuclei project sensory information, and from which ventral thalamic nuclei relay motor output. Fishes have no direct pathways from the cerebrum to motor nuclei in the brain stem and spinal cord.

The cerebrum of fishes integrates olfactory information, it is essential for olfactory learning, and it initiates motor responses to olfactory stimuli. At one time investigators believed that the cerebrum was primarily an olfactory center, but it is now clear that it has many other functions. Different parts of the cerebrum receive many other sensory stimuli, and experiments have shown that, at least in teleosts, the cerebrum modulates defensive reactions and some aspects of reproductive behavior. For example, the response of males to sexual stimuli, and their role in nest building, are disrupted following bilateral removal of the hemispheres (Davis and Kassel, 1983).

Major Trends in Tetrapod Brain Evolution

The hypothalamus and reticular formation are conservative parts of the brain that changed less in the evolution of amphibians and other tetrapods than other regions. The hypothalamus continues as a major center regulating visceral activity. The reticular formation continues as an important center and passageway between the spinal cord, cranial nerves, and higher brain centers. Somatic centers, including the cerebrum, optic lobes, and cerebellum, are the parts of the brain that were

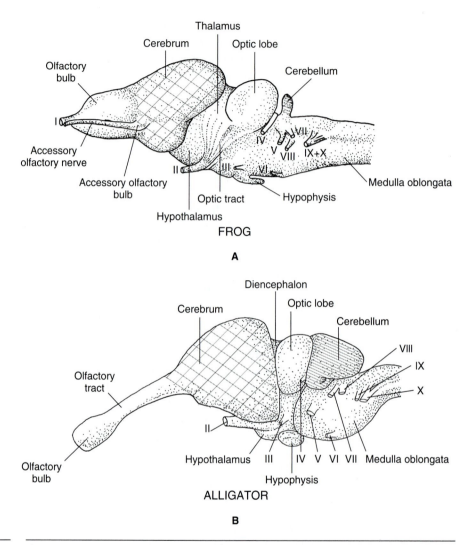

FIGURE 13–7 Lateral views of the brains of representative tetrapods. **A**, A frog. **B**, An alligator.

most affected as vertebrates adapted to terrestrial life and eventually became agile and inquisitive creatures with a wide range of behavioral responses.

The Amphibian Brain

The amphibian brain remains similar in many ways to the brain of many fishes (Fig. 13–7**A**). It has not assumed as dominant a role in behavior as it eventually does in amniotes. Locomotion, and the clasping reflex by which a male frog grasps a female during the breeding season, are centered in spinal reflexes. A decapitated frog can even jump if appropriately stimulated. Mauthner cells used in rapid

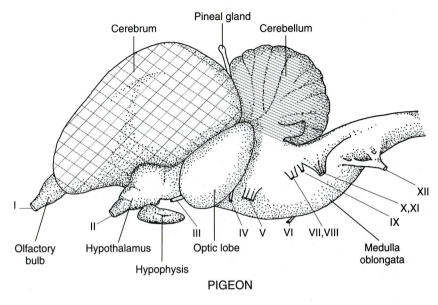

PIGEON

C

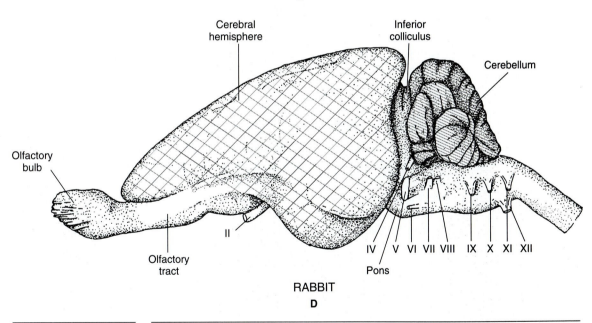

RABBIT

D

FIGURE 13–7   *continued*
C, A pigeon. **D**, A rabbit. (**A**, after Gaupp. **B**, after Weidersheim. **C**, after Pettingill. **D**, after Barone et al.)

escape reactions are still present in salamanders and frog tadpoles, but they are lost in adult frogs.

The major sensory input into the cerebrum continues to be olfactory, although some visceral, visual, and probably other sensory stimuli are projected to it from dorsal thalamic nuclei. As in fishes, the corpus striatum and other deep masses of gray matter are the primary integrating region of the cerebrum (Focus 13–2). A few sensory impulses from caudal parts of the body reach the thalamus directly on spinothamalic tracts, but most continue to be relayed forward through the reticular formation. Cerebral efferents extend to the hypothalamus, ventral thalamic nuclei, and reticular formation. None go directly to the spinal cord. The cerebrum certainly mediates responses to olfactory stimuli, but we know little about its other functions. A decerebrated frog does not seek shelter or feed spontaneously, but it will snap at an insect dangled in front of it.

The optic tectum remains the primary integration center of the body. It continues to project to the reticular formation, but tectospinal tracts have evolved that extend directly to motor columns in the spinal cord. Electrical stimulation to parts of the tectum elicits many coordinated locomotor and feeding movements. Stimulation of certain parts of the tectum also inhibits certain activities, including the clasping reflex in a male frog.

Locomotor movements are not as complex as in many fishes, and the cerebellum is smaller than in most fishes. Larval and neotenic species, such as the mudpuppy (*Necturus*), retain a lateral line system, but this system, together with lateral line centers in the brain, is lost in adult amphibians.

The Reptile Brain

The cerebrum of reptiles is substantially larger than that of amphibians, and it has expanded caudally, covering much of the diencephalon and reaching the optic lobes (Fig. 13–7**B**). Although much of the cerebrum continues to mediate olfactory responses, certain masses of deep gray matter receive auditory, visual, and somatosensory information from the general body surface (Focus 13–2). This information is relayed forward to the cerebrum through dorsal thalamic nuclei, some of which receive direct and larger spinothalamic tracts than in amphibians. Clearly more functions are being integrated in the cerebrum, but the optic tectum is still dominant. As in fishes and amphibians, efferent fibers leave the cerebrum through ventral thalamic nuclei and the reticular formation. None go directly to the spinal cord.

The optic tectum interacts with the cerebrum in ways not fully understood. Destruction of the cerebrum does not interrupt responses to visual stimuli, for these are initiated in the tectum, but cerebral destruction does increase the training time needed for visual learning. The tectum has more direct connections with motor centers than in amphibians.

The cerebellum is much larger in reptiles than in amphibians. Efferents from the cerebellum continue to go to the vestibular nuclei and reticular formation, from where they are sent to motor centers. Some efferents from the cerebellum also project to the thalamus, but it is not known whether the thalamic nuclei receiving cerebellar impulses project to the cerebrum. There are no major

The Evolution of Tetrapod Cerebral Centers

The cell bodies of the cerebral neurons of amphibians lie in gray matter that surrounds the lateral ventricle (see the figure, part **A**). This distribution and the divisions of this gray matter into regions are very similar to its divisions in a fish (p. 494 and Fig. 13–3). The **paleopallium** and **archipallium** continue to be the primary olfactory regions. A few impulses relayed from the thalamus reach the small **dorsal pallium.** But, as in fishes, the **corpus striatum** receives most of the impulses relayed from the thalamus

and appears to be the primary nonolfactory integrating region of the cerebrum. Efferent impulses leave through the **septum** and corpus striatum.

Ulinski (1990) has made a cladistic analysis of forebrain structure, searching for features that were probably present in ancestral amniotes. His reconstruction of the ancestral amniote cerebrum is very similar to the pattern we have described for contemporary amphibians (see the figure, part **A**). Contemporary reptiles represent

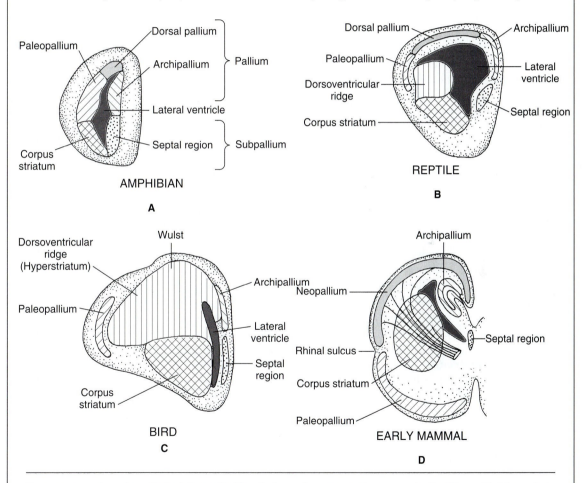

Transverse sections through the left cerebral hemisphere of representative tetrapods. See Figure 13–3 for code to gray regions. **A,** A frog. **B,** A lizard. **C,** A bird. **D,** An early mammal.

continued

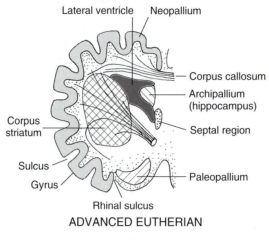

Lateral ventricle   Neopallium

Corpus callosum

Archipallium (hippocampus)

Corpus striatum

Septal region

Sulcus

Gyrus

Paleopallium

Rhinal sulcus

ADVANCED EUTHERIAN

E

E, An advanced eutherian. (Adapted from various sources.)

one evolutionary lineage from ancestral amniotes. In a cross section through a reptile cerebral hemisphere, one can recognize a thickening of the ventral wall above the corpus striatum to form a new region known as the **dorsoventricular ridge** (see the figure, part **B**). This is the region that receives most of the auditory, somatosensory, and visual information relayed from dorsal thalamic nuclei. The dorsal pallium has expanded and also receives visual information. Although much of the cerebrum remains olfactory, more functions are being integrated.

Cerebral enlargement in birds has resulted largely from the expansion and change in orientation of the reptilian dorsoventricular ridge (see the figure, part **C**). The dorsoventricular ridge (or **hyperstriatum**) forms a thickening of the lateral wall of the pallium that reduces the lateral ventricle to a narrow slit (Ulinski, 1990). A part of the dorsoventricular ridge forms a dorsal swelling, which is known by its German name, the **Wulst** (bulge). The Wulst largely replaces the dorsal pallium. The Wulst and much of the dorsoventricular ridge are the visual centers. Other sensory information, including somatosensory and auditory, is projected via dorsal thalamic nuclei to deeper parts of the dorsoventricular ridge and to an expanded corpus striatum. Many behavioral responses and much learning occur in the corpus striatum.

The mammalian cerebrum is characterized by the large surface **neopallium** (See the figure, parts **D** and **E**). For a long time neuroanatomists believed that the neopallium had no precursor among other amniotes. More recently neuroanatomists have sought for precursors. Karten (1991) argues that part of the mammalian neopallium evolved from cells that migrated from the reptilian dorsoventricular ridge and part from an expansion of the dorsal pallium. Ulinski (1990) points out that the dorsoventricular ridge is a contemporary reptilian structure that probably was not present in ancestral amniotes. Since the line of evolution to mammals diverged from the line to contemporary reptiles at the time of the earliest amniotes, the dorsoventricular ridge could not have contributed to the mammalian neopallium. Their neopallium must have evolved solely by an expansion of the dorsal pallium, which Ulinski believes was present in ancestral amniotes.

pathways, such as we find in birds and mammals, between the cerebrum and cerebellum.

The Bird Brain

Birds evolved from reptiles, and their brains retain many reptilian features. However, their brains are considerably larger than those of reptiles. Relative to body weight, they are as large as the brains of many mammals. Both birds and mammals independently evolved complex neuronal circuits, which were made possible in part by the evolution of endothermy and the concomitant rapid rate of nerve impulse transmission. High processing speeds are prerequisite for both complex brains and advanced computers.

The large cerebrum of birds has grown caudad and contacts the expanded cerebellum (Fig. 13–7**C**). The cerebrum covers the diencephalon, but the optic lobes have shifted laterally and protrude from beneath the cerebral hemispheres. Cerebral enlargement has resulted primarily from the expansion of deep masses of gray matter located dorsal to the corpus striatum (Focus 13–2). Much of this gray matter is related to vision. Visual input is important and extensive in birds, and birds have two visual pathways. Most fibers in the optic nerve terminate in the tectum, from which many visual impulses are relayed through dorsal thalamic nuclei to visual centers in the cerebrum. Other optic fibers go directly to other dorsal thalamic nuclei, from which they are relayed to different areas in the cerebrum. Decussation of the optic fibers is nearly complete in the optic chiasma of birds, so each eye projects to the optic tectum and dorsal thalamus of the contralateral side. However, the subsequent projection of the thalamus to the cerebrum is bilateral, so visual centers in the cerebrum receive information from both eyes. This is essential for stereoscopic vision.

Other sensory information, including auditory and somatosensory impulses from the body surface, is projected via dorsal thalamic nuclei to other deep masses of gray matter in the cerebrum. Olfactory centers vary among birds. Aquatic species and those that are meat eaters or scavengers have relatively large olfactory centers, but in most species the olfactory bulbs and olfactory centers in the cerebrum are small.

Some efferent fibers from the cerebrum go directly to the tectum, and a few go directly to the spinal cord, but many efferent impulses continue to be relayed to motor centers on ancestral pathways through ventral thalamic nuclei and the reticular formation. Visual learning and much behavior, including feeding, courtship, nest building, mating, and care of the young, are integrated in the deep masses of cerebral gray matter (Focus 13–2). These centers also affect locomotion, but this function is subordinate to the locomotor activity of the tectum. A decerebrated pigeon cannot stand, but it can walk if supported and will fly if tossed into the air.

The tectum continues to be an important integrating center. It receives most of the optic fibers and has the usual connections with the brain stem. Fiber tracts descend directly from the tectum to certain nuclei in the brain stem and spinal cord.

The cerebellum of a bird is very large. The archicerebellum and paleocerebellum are well developed, and a new area has evolved that has interconnections with the cerebrum. Fibers from the cerebrum project to the **pons,** a center that differentiates in the floor of the metencephalon. Fibers project from the pons to the new cerebellar region. Some cerebellar efferents return to the cerebrum via deep cerebellar nuclei and the thalamus. The interconnections with the cerebrum, as well as the usual inputs from the vestibular part of the ear, proprioceptors, and tectum, enable the cerebellum to coordinate a bird's complex motor activities, such as flight, feeding, and nest building. Efferents from the cerebellum lead through deep cerebellar nuclei to the vestibular nuclei and reticular formation. Fibers descend from these centers to the motor columns of the cord.

The Mammal Brain

Both the cerebrum and cerebellum of mammals are very large (Fig. 13–7**D**). Cerebral enlargement has resulted from the evolution and great expansion of a surface area known as the **neopallium** (also called **isocortex**), which evolved from the expansion of the small dorsal pallium present in the brains of ancestral amniotes (Focus 13–2). The large cerebral cortex is a distinctive mammalian feature. As the neopallium enlarged between the paleopallium and archipallium, these primitive olfactory areas were pushed apart. The archipallium, or hippocampus, moved further medially and, in eutherians, bulges into the floor of the lateral ventricle. The paleopallium, which forms the **piriform lobe,** shifted laterally and ventrally (Fig. 13–8), and is separated from the neopallium by a conspicuous **rhinal sulcus** (Gr. *rhin-* = nose + L. *sulcus* = furrow).

The older, olfactory cortical areas are characterized by three neuronal layers, but the neopallium has a complex, six-layered organization. Although the neopallium has connections with olfactory and visceral centers, it is primarily a nonolfactory, somatic integration region (Fig. 13–9). It has become the dominant center of the brain for these functions.

The surface of the neopallium probably was smooth in ancestral mammals, as it is now in such divergent species as the duckbilled platypus, opossum, rabbits, and many rodents. As mammals increased in size in different evolutionary lineages, the neopallium expanded and became highly convoluted, forming surface folds, called **gyri** (Gr. *gyros* = circle), that are separated by grooves, the **sulci** (Fig. 13–9). A convoluted surface is found in the spiny anteater, kangaroos, most primates, the larger carnivores and herbivores, and in cetaceans. A convoluted surface provides more surface area for the increased number of neurons needed to process the increased sensory input and motor output. This increase in the number of neurons probably could not be accommodated by increasing the thickness of the cortex because this would disrupt its intricate layered structure.

The dominance of the neopallium has affected many other parts of the brain. The thalamus receives and processes more sensory information, much of which is sent to the neopallium. It is much larger than in other vertebrates. Large motor pathways have evolved that extend directly from the neopallium to motor centers, but older pathways from the cerebrum remain. These pass through deep masses of gray matter in the cerebrum (called **basal nuclei**), the ventral thalamus, and

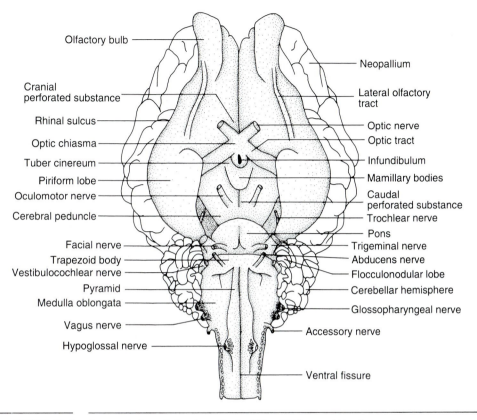

FIGURE 13–8   A ventral view of the sheep brain. (After Walker and Homberger.)

the reticular formation. New pathways interconnect the cerebrum and cerebellum and play a major role in motor coordination.

Many functions of the optic tectum are shifted to the cerebrum, but a small tectum continues to be an important center for certain optic and auditory functions. The optic lobes are reduced to a pair of **superior colliculi** (L. *colliculus* = little hill); the torus semicircularis of nonmammalian vertebrates, which are auditory centers, form a small pair of additional bulges, the **inferior colliculi.**

Major Sensory and Motor Pathways in the Mammalian Brain

It is convenient to consider the further structure and function of the mammalian brain by examining functionally related groups of neurons.

Ascending Sensory Pathways

Many sensory impulses continue to ascend to higher brain centers by relays through the reticular formation, but mammals, to a greater extent than other vertebrates, have evolved ascending **lemniscal systems** that carry sensory

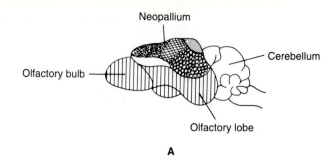

A

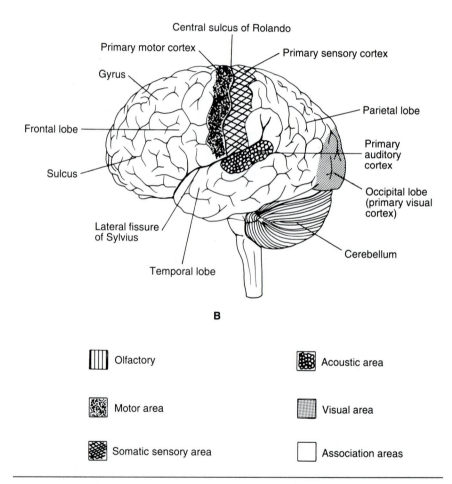

B

	Olfactory		Acoustic area
	Motor area		Visual area
	Somatic sensory area		Association areas

FIGURE 13–9 Areas of the cerebral cortex of mammalian brains: **A,** A shrew. **B,** A human being. (**A,** after Romer and Parsons.)

impulses more directly and rapidly to the neopallium. Some of the lemniscal systems form conspicuous bundles on parts of the brain (Gr. *lemniskos* = ribbon). Although different sensory stimuli have different pathways, those coming from pain, temperature, pressure, touch, and taste receptors have several common features. Impulses ascend on three neuron chains. The **primary sensory neurons** from the receptors usually terminate in one of the sensory columns of the cord, or sensory nuclei of the brain stem, at the level at which they enter the central nervous system (Fig. 13–10). The sensory neurons from receptors for tactile discrimination on the trunk and limbs, and from proprioceptors, are an exception to this generalization because they ascend in the dorsal funiculi of the spinal cord all the way to the **cuneate** and **gracile nuclei** in the medulla. Tactile discrimination pathways are particularly well developed in primates and other groups that use their hands to handle objects. **Second-order neurons** continue from the sensory columns or nuclei, usually decussate, and ascend on lemniscal tracts to specific

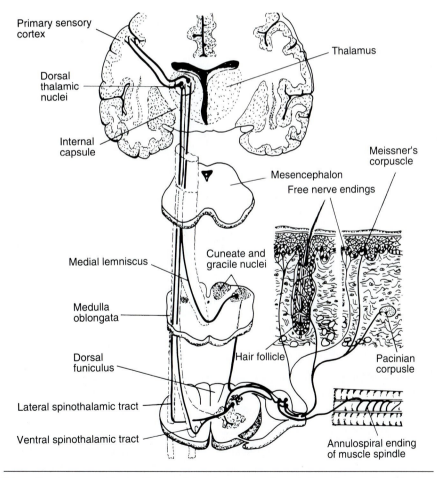

FIGURE 13–10 Ascending sensory pathways from peripheral receptors through the spinal cord and brain stem to the cerebral cortex. (Adapted from various sources.)

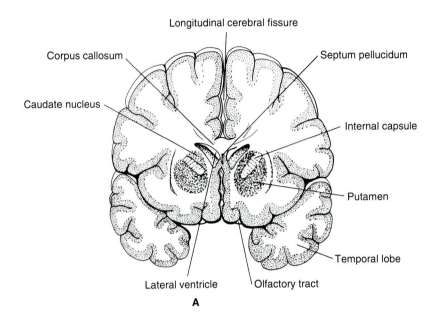

Longitudinal cerebral fissure

Corpus callosum

Septum pellucidum

Caudate nucleus

Internal capsule

Putamen

Temporal lobe

Lateral ventricle   Olfactory tract

A

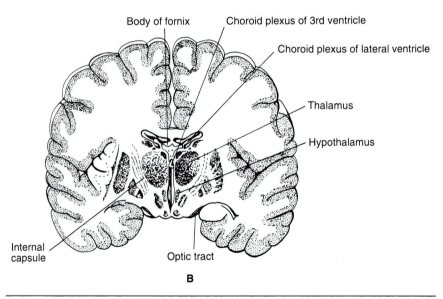

Body of fornix   Choroid plexus of 3rd ventricle

Choroid plexus of lateral ventricle

Thalamus

Hypothalamus

Internal
capsule   Optic tract

B

FIGURE 13–11   Two transverse sections through the human brain: **A,** Through the cerebrum and corpus striatum. **B,** Slightly posterior to **A** through the cerebrum and thalamus. (After Noback et al.)

dorsal thalamic nuclei on the opposite, or contralateral, side of the body. A few fibers do not cross and remain ipsilateral. **Third-order neurons** extend from the thalamic nuclei through the **internal capsule** (Fig. 13–11), a bundle of fibers in the corpus striatum, to terminate in very specific parts of the **primary sensory**

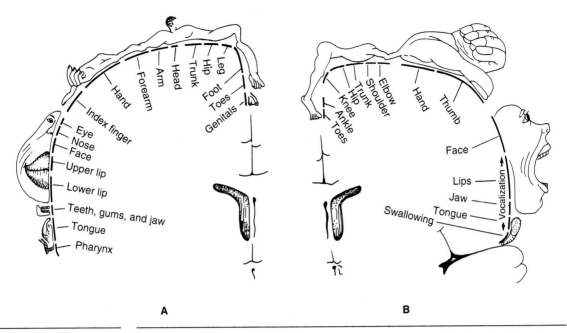

A   B

FIGURE 13–12   Vertical sections through **(A)** the primary sensory cortex that receives impulses from the skin and **(B)** the primary motor cortex of a human being. The extent of the areas devoted to various sensory and motor functions is indicated. (After Penfield and Rasmussen.)

cortex. This cortex is located just caudal to the **central sulcus of Rolando** (Fig. 13–9).

Impulses are processed in each of the nuclei through which they pass. Among other things, extraneous information ("noise") tends to be suppressed, and the important information ("signal") is enhanced. Except for pain and certain combinations of signals that are interpreted as pleasurable, which are centered in the thalamus, consciousness of stimuli is attained when the impulses reach the cerebral cortex. The amount of the primary sensory cortex that receives signals from different parts of the body is related not to the absolute size of this part of the body but to the density of receptors. Impulses from the fingers, lips, and tongue occupy a disproportionately large share of the primary sensory cortex of primates (Fig. 13–12**A**).

The Optic System

Optic, auditory, and olfactory pathways are different from those of other senses. The fibers in the optic nerve are those of retinal **ganglion cells.** These are not primary sensory neurons but interneurons of a higher order. Many of the optic fibers of a mammal remain ipsilateral, unlike the optic fibers in other vertebrates, all or most of which decussate in the optic chiasma (Fig. 13–13). Cetaceans are one exception; all of their optic fibers decussate. The degree of decussation correlates with the degree to which the visual fields of the two eyes overlap. Herbivores and other species subject to predation tend to have laterally placed eyes,

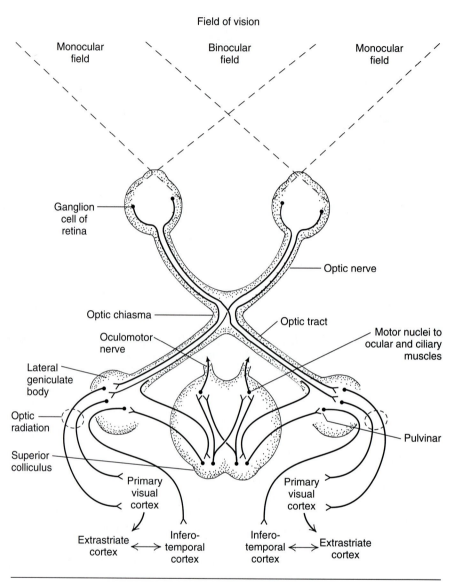

Field of vision

Monocular field   Binocular field   Monocular field

Ganglion cell of retina

Optic nerve

Optic chiasma

Oculomotor nerve

Optic tract

Motor nuclei to ocular and ciliary muscles

Lateral geniculate body

Optic radiation

Pulvinar

Superior colliculus

Primary visual cortex

Primary visual cortex

Extrastriate cortex   Infero-temporal cortex   Infero-temporal cortex   Extrastriate cortex

FIGURE 13–13   A diagram of the mammalian optic system.

which give them wide fields of vision that allow them to see approaching predators. The fields of vision of the two eyes overlap only slightly, and most of their optic fibers decussate. In carnivores, and especially in advanced primates, the eyes have rotated forward. The visual fields of the two eyes overlap extensively, and many optic fibers are ipsilateral. As many as 50 percent of the optic fibers project ipsilaterally in some primates; the others continue to decussate. Objects in the center of the visual field are seen by each eye but from slightly different angles. The image in each eye projects to both sides of the brain, and the processing of these images permits stereoscopic vision and depth perception. This is important for a carnivore, which must judge prey distance, and for a primate that scampers through the trees.

After passing through the optic chiasma, the optic fibers form an **optic tract** that continues to the thalamus and tectum (Fig. 13–13). Unlike the pattern in anamniotes and early amniotes, most of the optic fibers of mammals go to a dorsolateral thalamic nucleus known as the **lateral geniculate body,** rather than to the tectum. Fibers beginning in the lateral geniculate body pass through the **optic radiation** of the internal capsule (Fig. 13–13) and terminate in the multilayered **primary visual,** or **striate cortex,** which is located in the occipital pole of the neopallium. This is the primary visual pathway. Processing of the visual signals occurs at all levels: retina, lateral geniculate body, and primary visual cortex. This processing analyzes pattern, color, and depth of the visual image and leads to perception. The processing of the visual maps is hierarchically organized, with the neurons at each level "seeing" more and more. The primary visual cortex projects to the **extrastriate cortex** (visual association area) lying rostral to the striate cortex. This area is essential for conceptualizing visual relationships (position) of objects and for learning to identify objects by their appearance.

Other fibers in the optic tract go directly to the superior colliculi of the tectum (Fig. 13–13). The superior colliculi also receive an input from the extrastriate part of the cortex. Following processing of the retinal and cortical inputs, the superior colliculi send fibers to another dorsal thalamic nucleus, the **pulvinar,** from which impulses are sent to the **inferotemporal cortex** in the temporal lobe of the cerebrum. The inferotemporal area also interacts directly with the extrastriate cortex. This complex system is believed to be involved with maintaining visual attention and analyzing motion.

The superior colliculi are also centers that participate in many visually related reflexes. Among these are the congruent movements of the eyes to follow a moving object, eye movements that keep the eyes fixed on an object when the head moves, and movements of the head and eye in the direction of an unexpected stimulus. Eye movements involve inputs to the superior colliculi from the retina, the visual cortex, and the vestibular system. Output is to the motor nuclei of the extrinsic ocular muscles (Fig. 13–13). The superior colliculi, and a small pretectal area just rostral to them, also mediate pupillary and accommodation reflexes through their effect on autonomic nuclei. A few tectal fibers go to a small nucleus that sends efferent fibers back to the retina through the optic nerve. These fibers, which also are present in some other vertebrates, may modulate retinal sensitivity and may affect image processing in the retina.

The Auditory System

The primary sensory neurons of the auditory system begin in the organ of Corti of the cochlea, travel in the vestibulocochlear nerve, and terminate in **cochlear nuclei** in the medulla (Fig. 13–14). Most second-order neurons from the cochlear nuclei decussate in the **trapezoid body,** located on the ventral surface of the medulla (Fig. 13–8), and terminate in the **superior olivary complex of nuclei** in the medulla and pons. Some second-order neurons remain ipsilateral. Third-order neurons that originate in the superior olivary complex, with second-order

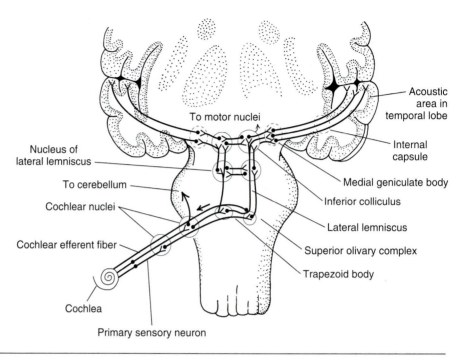

Acoustic
area in
temporal lobe

To motor nuclei

Internal
capsule

Nucleus of
lateral lemniscus

Medial geniculate body

To cerebellum

Inferior colliculus

Cochlear nuclei

Lateral lemniscus

Cochlear efferent fiber

Superior olivary complex

Trapezoid body

Cochlea

Primary sensory neuron

FIGURE 13–14   The mammalian auditory pathways projected on a dorsal view of the brain stem. (After Noback et al.)

fibers from the cochlear nuclei that bypass the olivary complex, continue on a tract called the **lateral lemniscus** to the **inferior colliculi** of the tectum. Some tectal fibers project from here to the reticular formation and motor nuclei. The tectum mediates reflexes that move the head and sometimes the auricles toward the sound sources. Other tectal fibers project to the **medial geniculate body** of the thalamus. Thalamic fibers continue through the internal capsule to the **primary auditory cortex,** which is located in the temporal lobe of the neopallium (Fig. 13–9). As is the case with the optic system, processing of signals occurs at all levels. Two points should be emphasized. First, the cochlear nuclei and other ascending centers are **tonotopically** organized, that is, fibers carrying different sound frequencies from the cochlea separate and end in different parts of the centers. Sound is "dissected" into its separate components, and these are finally synthesized in the primary auditory cortex and adjacent areas where the awareness and interpretation of sounds occur. Second, auditory signals from each ear ascend on each side of the brain. This enables the brain to localize the source of sounds by comparing differences in the timing and intensity of sounds arising in each ear.

Efferent fibers return from the superior olivary nuclei through the vestibulocochlear nerve to the organ of Corti. These fibers modulate the activity of the organ of Corti by enhancing the meaningful signal and suppressing extraneous "noise."

The Olfactory and Limbic Systems

The olfactory and limbic systems include the olfactory bulbs, paleopallium, archipallium, and septum. These centers probably evolved in early vertebrates as olfactory centers. They continue to process olfactory signals, but many of them have also acquired other functions.

Olfactory neurons from the nasal cavities extend the short distance to the **main olfactory bulb,** where they terminate in hundreds of complex tangles of neurons called **glomeruli** (Fig. 13–15A). Among other connections in the glomeruli, the olfactory neurons synapse with large **mitral cells,** whose axons are the primary ones leaving the olfactory bulb. Each glomerulus may receive as many as 25,000 (in rabbits) olfactory neurons and the terminations of nearly 100 other neurons. Many of these are from short neurons within the glomerulus, collaterals of mitral cells that feed back to the glomerulus, and neurons coming from the contralateral olfactory bulb. Obviously a great deal of signal processing, some known to be inhibitory, occurs here. Considerable convergence also occurs because far fewer mitral axons leave the glomeruli than olfactory neurons enter.

The axons of mitral cells form the **olfactory tract,** which leads to the cerebrum. Some mitral cells, or their collaterals, enter a small **anterior olfactory nucleus,** where they synapse with neurons that pass through the **anterior commissure** to the contralateral olfactory bulb (Fig. 13–15A). The main target of the mitral cells is the **piriform lobe** (L. *pirum* = pear + *forma* = shape), which is located on the lateroventral surface of the cerebrum (Fig. 13–8). The piriform lobe is homologous to the paleopallium of nonmammalian vertebrates, and it is the primary olfactory area. Olfactory discrimination, awareness, and learning probably occur here. Conditioned olfactory reflexes are lost when this area is destroyed. The piriform lobe projects to the dorsal thalamic nuclei and to the habenula in the epithalamus. Efferents from the habenula lead to the reticular formation and motor nuclei.

Other mitral fibers from the main olfactory bulb project to the septal region, to the hippocampus, and to an almond-shaped group of nuclei known as the **amygdala** (Gr. *amygdale* = almond) (Fig. 13–15A). A different part of the amygdala receives fibers from the vomeronasal organ via the accessory olfactory bulb. The amygdala is located deeply in the ventral part of the cerebrum lateral to the optic chiasma. It evolved from part of the corpus striatum but is now a part of the limbic system. The amygdala receives other fibers from the thalamus, cerebral cortex, and reticular formation. It influences many aspects of behavior, primarily through efferent fibers that go to the hypothalamus. Among its functions are mediating sniffing, licking, and other olfactory-based reflexes. It also sends fibers to parts of the thalamus and neopallium.

The **hippocampus** is an important part of the limbic system. As we have seen, it represents the archipallium, which has been shifted medially by the expansion of the neopallium (Focus 13–2 Figure). It receives olfactory fibers from the **entorhinal cortex** on the caudal part of the piriform lobe (Fig. 13–15A and **B**). The hippocampus also receives gustatory and other visceral sensory fibers from the reticular formation, fibers from the **cingulate cortex** on the medial side of a cerebral hemisphere, and auditory, visual, and general somatic sensory projections.

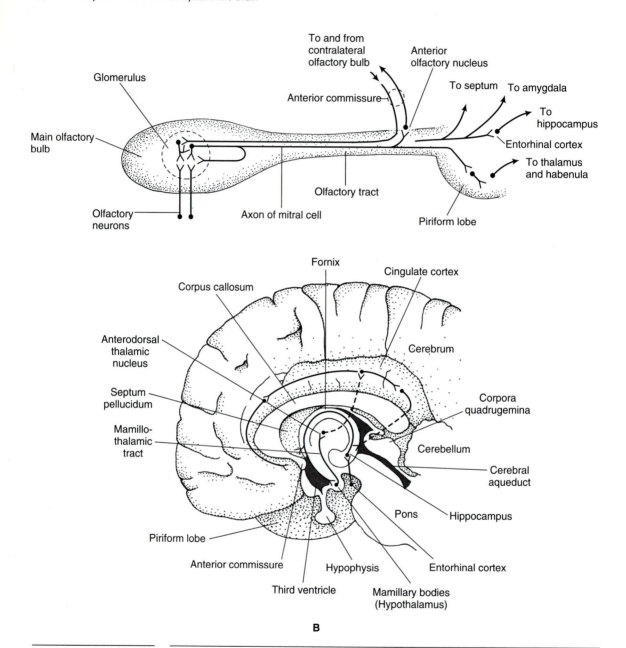

FIGURE 13–15 Diagrams of major components of the olfactory and limbic systems. **A,** The major olfactory pathways in ventrolateral view. **B.** The limbic system as projected on a dissected sagittal-section of the brain. (Modified after Noback.)

It retains all of the visceral integrating functions that it has in fishes. Beyond this, the hypothalamus is an important center in both birds and mammals for controlling body temperature and the high level of metabolism that characterize these endothermic vertebrates. Efferents from the hippocampus aggregate to form a prominent tract known as the **fornix** (L. *fornix* = arch or vault), which makes an arc in the base of the cerebrum and then turns ventrally to the **mamillary bodies** of the hypothalamus. A small section of the fornix goes to septal nuclei, which in turn project to the hypothalamus. A prominent **mamillothalamic tract** from the mamillary bodies returns to the thalamus. Relays in anterodorsal thalamic nuclei and the cingulate cortex return impulses to the hippocampus. The complex pathway from hippocampus to mamillary bodies, to anterodorsal thalamic nuclei, to the cingulate cortex, and back to the hippocampus is called the **Papez circuit** (Fig. 13–15**B**). This circuit integrates the hypothalamus into the limbic system.

Because of their extensive connections, the hippocampus, amygdala, hypothalamus, and other parts of the limbic system influence many aspects of behavior, especially motivational and emotional behaviors related to self- and species preservation. These include feeding, drinking, fighting, fleeing, reproduction, and care of the young. The limbic system exerts much of its influence by inhibiting the hypothalamus and tegmental part of the reticular formation. For example, electrical stimulation of parts of the limbic system cause a mammal to stop an activity in which it is engaged. Conversely, destruction of parts of the limbic system releases the mammal from normal inhibitory stimuli and leads to an overreaction to stimuli. Destruction also causes some behaviors to become repetitive, and complex sequences are not completed in an orderly fashion. A female rat may continuously pick up and drop a newborn infant, apparently not sure what to do next.

The limbic system also has been implicated in the formation of short-term memories. A human being with lesions in the limbic system can remember events of times long past, but because he or she is unable to bring together and reinforce the signals needed to establish new memories, he or she cannot recall events that occurred a few minutes ago. The formation of some memories may require an emotional input, especially of the type that is essential to survival.

Motor Control and Pathways

The motor systems allow vertebrates to maintain posture and balance and to move the body and its parts in response to sensory clues or (in many vertebrates) the desires of the animal. Many reflexes, the rhythmic, stereotyped movement of body parts during locomotion, and other movements are controlled by pools of neurons in the spinal cord (p. 462).

Two levels of motor control from the brain are superimposed on the spinal level in mammals: the brain stem level and the cortical level. The **brain stem level** of control is composed primarily of three important pathways (Fig. 13–16). Ipsilateral **reticulospinal** and **vestibulospinal tracts** descend from the reticular formation and vestibular nuclei to the motor pools of the spinal cord, and modulate their activity in such a way as to maintain posture and balance. A contralateral **tectospinal** tract descends from the superior colliculi into the cervical region of the spinal cord. This system integrates head and eye movements with other body movements. In anamniotes and reptiles, most higher levels of motor control are in the brain stem because relatively few tracts descend from the cerebrum through the corpus striatum and ventral thalamic nuclei. These phylogenetically older brain stem pathways have been described as the extrapyramidal system to distinguish them from the cortical or pyramidal system of mammals, which descends directly

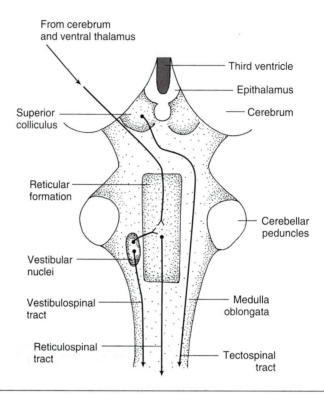

FIGURE 13–16   A diagram of important brain stem motor control pathways as seen in a dorsal view; the cerebellum has been removed.

from the neopallium to motor centers. But the extrapyramidal and pyramidal systems become intertwined functionally, so this dichotomy is not completely valid.

The cortical level of motor control, commonly called the **pyramidal system,** evolved along with the neopallium (Fig. 13–17). It is the highest of the hierarchically arranged levels of motor control. The pyramidal system allows mammals to execute complex, voluntary motor activities with precision. For human beings, this includes speech, manipulating tools with hands, playing a piano, or batting a ball. These activities are purposeful and may be initiated at will. They also are learned to a large extent, so performance improves with practice.

Most of the fibers of the pyramidal system begin with large, pyramidal-shaped cell bodies located in the precentral gyrus, or **primary motor cortex** of the neopallium (Fig. 13–9). The amount of cortex occupied by cells supplying different parts of the body is not proportional to the size of the part, but to the number of motor units it contains (Fig. 13–12**B**). Neurons supplying the lips, tongue, and hands occupy a disproportionately large share of the motor cortex in primates. Other pyramidal cells begin in the premotor cortex lying rostral to the primary motor cortex, and, surprisingly, a few begin in the somatic sensory cortex. The latter help regulate the transmission of sensory signals to control centers in the brain. Axons of pyramidal cells form **corticospinal** and **corticobulbar tracts** that descend through the internal capsule and contribute to conspicuous bulges, the

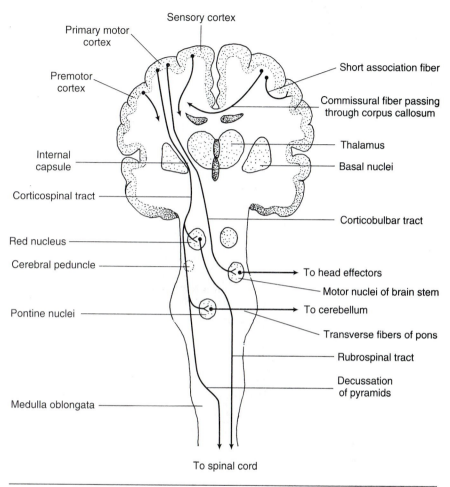

Primary motor cortex

Premotor cortex

Sensory cortex

Short association fiber

Commissural fiber passing through corpus callosum

Thalamus

Basal nuclei

Internal capsule

Corticospinal tract

Red nucleus

Cerebral peduncle

Pontine nuclei

Medulla oblongata

Corticobulbar tract

To head effectors

Motor nuclei of brain stem

To cerebellum

Transverse fibers of pons

Rubrospinal tract

Decussation of pyramids

To spinal cord

FIGURE 13–17   The pyramidal motor pathway and important cerebral connections.

cerebral peduncles, that can be seen on the lateroventral part of the mesencephalon (Fig. 13–8). The corticobulbar fibers decussate and terminate in the motor nuclei of cranial nerves, but the corticospinal fibers form the **pyramids** on the underside of the medulla, decussate, and continue to the motor columns of the spinal cord. Pyramidal cells send collateral axons into the **pontine nuclei,** whence impulses are sent to the cerebellum. Additional collateral axons enter the red nucleus and other parts of the brain stem and integrate cortical and brain stem levels of control.

Basal Nuclei and Cerebellum

Two other parts of the brain, the basal nuclei and cerebellum, do not directly initiate motor activity, but have a significant influence on the motor centers that do. The corpus striatum of mammals consists of several nuclei situated deeply in the base of the cerebrum. These nuclei are interspersed with fiber tracts (such as the internal capsule) that pass between them (Fig. 13–11). Collectively the nuclei are called **basal nuclei.** The larger ones are the **lentiform nucleus,** which is divided into the **globus pallidus** and **putamen,** and the **caudate nucleus.**

The basal nuclei receive inputs from virtually all parts of the cerebral hemispheres, including sensory, motor, and limbic areas. Since their output through thalamic nuclei is primarily back to the primary motor cortex, premotor cortex, and to other parts of the frontal lobes, their actions are mediated largely through the pyramidal system. Lesions in the basal ganglia, or disruption in the synthesis of transmitter substances by certain of their neurons (chiefly dopamine in Parkinson's disease), cause tremors, an increase in muscle tone, and slowness of movement. Beyond affecting these aspects of muscle activity, the basal nuclei are involved in cognitive aspects of motor control, such as the planning and execution of complex movements.

The cerebellum fine tunes motor control primarily by comparing motor directives originating elsewhere with performance, and adjusting the output of the motor systems. Accuracy of movement is very important in mammals because they have an extraordinarily wide range of intricate motor behavior. Their cerebellum is large and complex. A folding of the cerebellar cortex forms many leaf-shaped **folia** (L. *folium* = leaf), providing the surface area necessary for the large number of cerebellar cell bodies (Figs. 13–8 and 13–9). Myelinated axons within the cerebellum have a branching, treelike configuration known as the **arbor vitae** (L. *arbor vitae* = tree of life). The primitive archicerebellum is represented by the **flocculonodular lobes** (Fig. 13–18**A**). The paleocerebellum is represented by all but the central part of the **vermis** (L. *vermis* = worm) and by small anterior and posterior parts of the **cerebellar hemispheres.** In addition to these, a **neocerebellum** has evolved that includes the central part of the vermis and most of the cerebellar hemispheres. The cerebellum is attached to the rest of the brain by three pairs of stalks, or **cerebellar peduncles,** through which fibers enter and leave (Fig. 13–18**C**).

Afferent fibers coming primarily from vestibular and proprioceptive centers, and from the motor cortex, terminate in different regions of the cerebellar cortex, as discussed later. The cortex is composed of three layers, as in most other vertebrates: a basal layer of small **granular cells,** a middle layer that contains the cell bodies of the large **Purkinje cells,** and a superficial **molecular layer** composed primarily of synapses among all these neurons (Fig. 13–18**B**). Afferent neurons terminate either on the extensive dendritic trees of the Purkinje cells, which ramify in the molecular layer (see Fig. 11–1**B**), or on the granular cells. The granular cells, in turn, project to the molecular layer. Other cells establish additional interconnections, and the Purkinje cells themselves have recurrent axons

FIGURE 13–18   The mammalian cerebellum. **A,** Cerebellar areas as seen on a flattened, dorsal view of the cerebellum. **B,** Important neuronal connections within the cerebellar cortex. **C,** Major cerebellar connections as seen on a lateral projection of the brain.

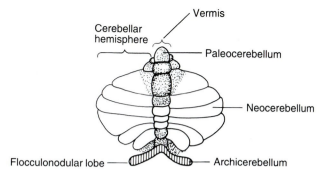

A

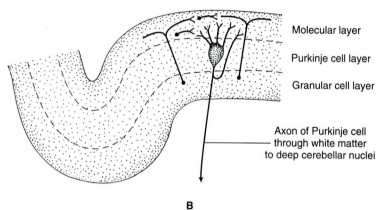

B

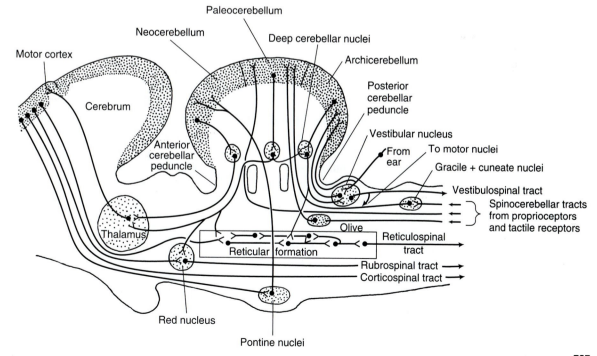

C

that feed back into the molecular layer. Efferent impulses leave on the primary axons of the Purkinje cells to one of three pairs of **deep cerebellar nuclei** (Fig. 13–18**B**). Efferents from these nuclei establish connections with motor pathways. The Purkinje cells have an inhibitory effect on their targets, but the degree of inhibition is modulated by the numerous impulses impinging on the Purkinje cells. Some impulses reinforce the inhibition; others reduce it, a phenomenon called **disinhibition.**

Archicerebellum, paleocerebellum, and neocerebellum have different inputs and affect different aspects of motor coordination. As in other vertebrates, vestibular impulses from the inner ear are projected to the archicerebellum, either directly or after relay in the vestibular nuclei (Fig. 13–18**C**). They enter through the posterior peduncle. The archicerebellum also receives an input from visual centers (thalamus and superior colliculi). Output from the archicerebellum relays in one of several pairs of deep cerebellar nuclei and goes back to the vestibular nuclei. Acting through the vestibular nuclei, whose fibers descend into the spinal cord or ascend to the nuclei of the extrinsic ocular muscles, the archicerebellum adjusts the tone of muscles that maintain balance and coordinates head and eye movements.

Proprioceptive and tactile impulses reach the paleocerebellum on spinocerebellar tracts. The primary tract is composed of second-order sensory neurons that ascend directly from the spinal cord. Another spinocerebellar tract is composed of second-order neurons that have relayed with primary sensory neurons in the cuneate and gracile nuclei at the caudal end of the medulla, and yet another tract is relayed in the olivary complex. All these tracts decussate at some level on their ascent to the cerebellum. The paleocerebellum projects via a different pair of deep cerebellar nuclei to the pair of **red nuclei** in the tegmentum from which rubrospinal tracts descend to the motor columns. The paleocerebellum also projects to the thalamus, from which impulses are sent back to the cerebrum (Fig. 13–18**C**). Beyond this, the paleocerebellum has connections by way of the cerebellar nuclei of the archicerebellum with the vestibular nuclei and reticular formation. This part of the cerebellum monitors evolving movements and affects the execution of the movements and muscle tone through its connections with the motor centers from which the vestibulospinal, reticulospinal, rubrospinal, and corticospinal tracts emerge.

The neocerebellum receives impulses primarily from the motor cortex of the neopallium, which are relayed in pontine nuclei and decussate in the **transverse fibers of the pons** (Fig. 13–18**C**). This decussation forms a conspicuous bulge on the ventral surface of the metencephalon (Fig. 13–8). The cerebellum also has a few connections with the tectum, but these are less important than in other vertebrates because so many functions have been transferred from the tectum to the neopallium. The neocerebellum projects via another pair of deep cerebellar nuclei only to the thalamus. From here, impulses go back to the motor cortex. The precise timing and duration of muscle contractions for the skilled, voluntary movements initiated in the motor cortex appear to require an interaction between the motor cortex and the neocerebellum. Lesions to parts of the neocerebellum, or its connections, upset the timing and duration of such contractions.

Cortical Integration

The neopallium of mammals is divided into **frontal, parietal, temporal,** and **occipital lobes** that underlie skull bones bearing the same names (Fig. 13–9**B**). The boundaries between the lobes are not well defined in many mammals, but a prominent **lateral fissure of Sylvius** separates the human temporal lobe from the others, and a **central sulcus of Rolando** separates the frontal and parietal lobes. The **primary sensory cortex** lies in the parietal lobe just posterior to the central sulcus; the **primary motor cortex** lies in the frontal lobe just rostral to the central sulcus; the **primary auditory cortex** lies in the temporal lobe; and the **primary visual cortex** is in the posterior part of the occipital lobe. Cortical regions between these sensory and motor cortices are called **association areas.** Apart from differences in their afferent and efferent connections, the neopallial cortical areas differ in cell density, thickness of their six layers, types of interconnections, and in other cytological details. On the basis of these differences, Brodmann, an early 20th-century German anatomist, subdivided the cortex into nearly 50 areas. His numbering system is still used in detailed analyses.

The different areas of the cortex have many interconnections. **Short association fibers** pass from one gyrus to adjacent ones, **long association fibers** interconnect more distant lobes within one hemisphere, and **commissural fibers** pass between the two hemispheres (Fig. 13–17). The phylogenetically older olfactory and limbic parts of the two hemispheres are interconnected by the **anterior commissure** (Fig. 13–15**A**) and by the **commissure of the fornix.** The neopallium of the two hemispheres is interconnected in eutherian mammals by a very large commissure called the **corpus callosum** (L. *corpus callosum* = hard body) (Figs. 13–15**B** and 13–17). Monotremes and marsupials lack the corpus callosum, but marsupials have an exceptionally large anterior commissure that includes neopallial commissural fibers. In this, as in many other ways, marsupials are not inferior to eutherians but have evolved similar capabilities in different ways.

The size of the neopallium varies greatly among mammals. In insectivores and other early eutherians it constitutes about 20 percent of the telencephalon. Association areas are small, and most of the cortex is occupied by the primary sensory and motor areas (Fig. 13–9**A**). In later eutherians the neopallium is much larger, and in human beings it forms about 80 percent of the telencephalon (Fig. 13–9**B**).

Some of the association areas supplement the activity of the adjacent primary sensory or motor area. The region preceding the primary visual cortex, for example, is necessary for conceptualizing visual relationships and learning to recognize objects. The association area rostral to the primary motor cortex interacts with the primary motor cortex in such a way that a complex motor activity, such as playing a piano, is not a random series of movements but forms an integrated pattern. The most rostral part of the frontal lobe, sometimes called the **prefrontal lobe,** has extensive connections with other parts of the neopallium and with the limbic system. If these connections are severed, as neurosurgeons formerly did in treating certain neuroses by an operation known as a prefrontal lobotomy, a person's motivation, ability to formulate goals and plan for the future, and ability to concentrate decrease.

Association areas also integrate separate visual, tactile, olfactory, and other sensory signals to provide an overall perception of the environment and of changes taking place. Association areas are essential for numerous mental processes, including learning, memory, and, in human beings, communication and such processes as reasoning and conceptual and symbolic thought. The two cerebral hemispheres of human beings perform somewhat different functions. In right-handed people the left hemisphere has a dominant role in communication skills, such as speech, reading, and writing; the right hemisphere dominates in spacial recognition and other artistic skills. In left-handed people these functions are reversed.

The human brain probably contains 10 billion or more neurons, and it processes millions of signals simultaneously. Neurologists have learned a great deal about brain structure and function by a process of reducing the brain to its elementary pathways and localizing many functions. But the brain is far more than sets of pathways and functions. It is now necessary to put the units together and learn how the brain functions as an integrated whole. This will require more complex methods of analysis. Studies have begun, but neurology will continue to be a challenging and fruitful field for research long into the future.

SUMMARY

1. Wiring diagrams of centers and pathways within the brain are determined by histochemical and neurophysiological experiments. Brain functions are determined by observations and experiments involving the selective destruction or stimulation of centers and tracts. New imaging techniques allow investigators to explore the activities of neurons in subjects engaged in various activities without disruptive invasion.

2. The brain develops from five embryonic enlargements of the neural tube: telencephalon, diencephalon, mesencephalon, metencephalon, and myelencephalon. Cavities within the brain enlarge to form a series of interconnected ventricles.

3. The brain is covered by connective tissue meninges, the innermost of which contributes to the choroid plexuses that secrete most of the cerebrospinal fluid. The cerebrospinal fluid helps support, protect, and nourish the brain.

4. Feeding, respiratory movements, and cardiac rate are among the activities mediated in the reticular formation of the fish brain stem.

5. The hypothalamus is also an important visceral center, controlling many activities mediated by the autonomic nervous systems and pituitary gland, or hypophysis.

6. The cerebellum of fishes consists of paired auricles (the archicerebellum), which are associated with the octavolateralis area, and a central body (the paleocerebellum), which receives tactile, proprioceptive, and electroreceptive information. The cerebellum interacts with other centers in motor coordination.

7. The tectum, or optic lobes, is often the largest part of the fish brain, for it is the major integration center.

8. The telencephalon of fishes is characterized by relatively large olfactory bulbs and small cerebral hemispheres. The cerebrum is an olfactory center,

although it receives and integrates some other sensory information as well.

9. The gray matter of a fish cerebral hemisphere lies around the lateral ventricle. It can be divided into a dorsolateral paleopallium and a dorsomedial archipallium, both of which are olfactory centers. A ventromedial septal area (also olfactory) and a ventrolateral corpus striatum with thalamic connections are present as well.

10. The reticular formation and hypothalamus continue to be important visceral centers in all vertebrates and have changed little in the course of evolution. The parts of the brain related to somatic sensory and somatic motor integration, however, changed greatly as vertebrates became more agile and inquisitive animals. Suprasegmental control of body activity increased. Many functions shifted from the tectum to the expanding cerebral hemispheres. The increasing importance of the cerebrum affected many other brain regions, especially the thalamus and cerebellum. Direct ascending and descending pathways became more numerous and important.

11. The brains of amphibians have changed little compared with those of fishes. A few more types of sensory impulses are projected to the cerebrum from the thalamus. More direct connections evolved between the spinal cord and the thalamus and from the tectum to the spinal cord.

12. The cerebral hemispheres of contemporary reptiles are larger than those of amphibians. The increased size results primarily from a thickening of the cerebral wall dorsal to the corpus striatum. More integration occurs in the cerebrum than in anamniotes, but the optic tectum remains an important center.

13. The brains of birds are as large relative to body size as the brains of many mammals. The great enlargement of the cerebrum has resulted primarily from an increase in size of deep masses of gray matter that lie above the corpus striatum. Much of this gray matter forms visual centers, but other sensory information is relayed to the corpus striatum. The cerebrum of birds is important in integrating feeding, nest building, and many other activities, but the optic tectum remains important. The cerebellum has also expanded, and new pathways between it and the cerebrum have evolved.

14. Mammals also have very large cerebral hemispheres, but their great enlargement in mammals resulted from the expansion of the small dorsal pallium to form a neopallium. As the neopallium expanded, it formed a surface cortex and pushed the older paleopallium and archipallium apart. The neopallium has become the dominant somatic integration center.

15. The evolution and expansion of the neopallium affected many other parts of the brain. The thalamus enlarged. Most tectal functions shifted to the neopallium. A neocerebellum with neopallial connections has evolved.

16. Many ascending sensory pathways have evolved. All those from lower centers go to the thalamus, where some processing of signals occurs before they are sent to the primary sensory cortex of the neopallium.

17. Most of the optic fibers of mammals project via the geniculate bodies of the thalamus to the primary visual cortex of the neopallium. Processing of the visual image occurs in the retina, thalamus, and primary visual cortex. The primary visual cortex, in turn, projects to the extrastriate cortex, an area essen-

tial for learning to recognize objects. Some optic fibers continue to go to the optic lobes (superior colliculi), which are linked by pathways to the primary visual, extrastriate, and inferotemporal cortices. This system appears to be involved in maintaining visual attention. The superior colliculi also participate in congruent eye movements, pupillary reflexes, and accommodation.

18. Auditory fibers go first to the cochlear nuclei in the medulla. From there they are projected on several pathways to the inferior colliculi of the tectum. Most continue from the tectum to the thalamus and to the primary auditory cortex of the neopallium and adjacent areas, where the awareness and interpretation of sounds occur.

19. Olfactory neurons terminate in clusters of neurons in the olfactory bulbs, where considerable processing of the impulses appears to occur. The primary target of neurons leaving the bulbs is the paleopallium, or piriform lobe. This is the primary olfactory center, and olfactory discrimination awareness and learning probably occur here.

20. The archipallium still receives some olfactory signals. It has turned inward on the medial surface of the cerebrum to form the hippocampus. The hippocampus is a central part of the limbic system that integrates the olfactory centers, hypothalamus, thalamus, and part of the cerebrum into a system that is important in motivational and emotional behaviors related to survival. The limbic system has also been implicated in short-term memory formation.

21. Motor activity in mammals is initiated and regulated by several hierarchically arranged levels of control. Basic locomotor activity is controlled by pools of neurons in the spinal cord. A brain stem level of control modulates the activity of the spinal cord in such a way as to maintain posture and balance. This level, sometimes called the extrapyramidal system, is also present in nonmammalian vertebrates. Mammals have a third level of control, the pyramidal system, that descends directly from motor centers in the cerebral cortex to the brain stem and spinal cord. This system allows mammals to execute complex, voluntary activities.

22. The basal nuclei do not directly initiate motor activity, but they have extensive connections with the cerebral cortex and affect the output of the pyramidal system. Disturbances in these nuclei may cause tremors and other abnormalities in muscle activity.

23. The cerebellum fine tunes motor control by comparing directives initiated elsewhere with performance, and adjusting the output of the motor systems as needed. The archicerebellum adjusts the tone of muscles that maintain balance and regulates head and eye movements. The paleocerebellum adjusts evolving movements. The neocerebellum, new to mammals, affects the timing and duration of muscle contractions for the skilled, voluntary movements executed by the pyramidal system.

24. Areas within the neopallium can be localized for specific sensory and motor functions. The rest of the neopallium consists of association areas, where considerable processing and integration of sensory and motor signals occur and where complex neuronal processes for learning, memory, and conceptual thought are believed to occur. In early mammals these association areas are small, but in advanced primates they occupy most of the cerebrum. Numerous

interconnections occur between areas in a hemisphere and between hemispheres on large commissural tracts.

REFERENCES FOR CHAPTERS 12 AND 13

Adelman, G., editor, 1987: *Encyclopedia of Neurosciences*. Boston, Birkhauser.

Barone, R., Pavaux, C., Blin, P. C., and Cuq, P., 1973: *Atlas d'Anatomie du Lapin*. Paris, Masson & Cie.

Benzo, C. A., 1986: The brain. *In* Sturkie, P. D., editor: *Avian Physiology*, 4th edition. New York, Springer-Verlag.

Cole, F. J., 1896: On the cranial nerves of *Chimaera monstrosa* (Linn.); with a discussion of the lateral line system and the morphology of the chorda tympani. *Transactions of the Royal Society of Edinburgh*, 38:630–680.

Crosby, E. C., and Schnitzlein, H. N., 1982: *Comparative, Correlative Neuroanatomy of the Vertebrate Telencephalon*. New York, Macmillan.

Davis, R. E., and Kassel, J., 1983: Behavioral functions of the teleostean telencephalon. *In* Davis, R. E., and Northcutt R. G., editors: *Fish Neurobiology, Volume 2*:237–263. Ann Arbor, The University of Michigan Press.

Davis, R. E., and Northcutt, R. G., editors, 1983: *Fish Neurobiology, Volume 2, Higher Brain Areas and Functions*. Ann Arbor, The University of Michigan Press.

Demski, L. S., 1984: Evolution of the neural systems in the vertebrates: Functional-anatomical approaches. *American Zoologist*, 24:689–833.

Dorit, R. L., Walker, W. F., Jr., and Barnes, R. D., 1991: *Zoology*. Philadelphia, Saunders College Publishing.

Ebbesson, S. O. E., and Northcutt, R. G., 1976: Neurology of anamniotic vertebrates. *In* Masterton, R. B., Campbell, C. B. G., Bitterman, M. E., and Hotton, N., editors: *Evolution of Brain and Behavior*, vol. 1. Hillsdale, N.J., Lawrence Erlbaum and Associates.

Gaupp, E., 1891–1904: *Anatomie des Frosches*. Braunschweig. Friedrich Vieweg und Sohn.

Goodrich, E. S., 1918: On the development of the segments of the head in *Scyllium*. *Quarterly Journal of Microscopical Sciences*, 63:1–30.

Goodrich, E. S., 1930: *Studies on the Structure and Development of Vertebrates*. London, Macmillan. (Reprinted Chicago, University of Chicago Press, 1986.)

Goslow, G. E., Jr., 1985: The neural control of locomotion. *In* Hildebrand, M., Bramble, D. M., Liem, K. F., and Wake, D. B., editors: *Functional Vertebrate Morphology*. Cambridge, Harvard University Press.

Grillner, S., and Wallen, P., 1985: Central pattern generators for locomotion, with special reference to vertebrates. *Annual Review of Neuroscience*, 8:233–261.

Hopkins, C. D., 1983: Functions and Mechanisms in Electroreception. *In* Northcutt, R. G., and Davis, R. E., *q.v.*

Kandel, E. C., Schwartz, J. H., and Jessell, T. M., editors, 1991: *Principles of Neural Science*, 3rd edition. New York, Elsevier.

Kappers, C. U. A., Huber, G. C., and Crosby, E., 1936: *The Comparative Anatomy of the Nervous System of Vertebrates, Including Man*. New York, Macmillan.

Karten, H. J., 1991: Homology and the evolution of the "neocortex." *Brain, Behavior and Evolution*, 38:264–272.

Landacre, F. L., 1910: The origin of the cranial ganglia in *Ameiurus*. *Journal of Comparative Neurology*, 20:309–411.

Larsell, O., 1967–1972: *The Comparative Anatomy and Histology of the Cerebellum*. Minneapolis, University of Minnesota Press.

Nieuwenhuys, R., 1982: An overview of the organization of the brain of actinopterygian fishes. *American Zoologist*, 22:287–310.

Noback, C. R., Strominger, N. L., and Demarset, R. J., 1991: *The Human Nervous System, Introduction and Review*, 4th edition. Philadelphia, Lea & Febiger.

Noden, D. M., 1991: Vertebrate craniofacial development: The relation between ontogenetic process and morphological outcome. *Brain, Behavior and Evolution*, 38:190–225.

Norris, H. W., and Hughes, S. P., 1920: The cranial, occipital and anterior spinal nerves of the dogfish. *Journal of Comparative Neurology*, 31:293–395.

Northcutt, R. G., 1985: Evolution of the vertebrate central nervous system. *American Zoologist*, 24:701–716.

Northcutt, R. G., 1985: Brain phylogeny, speculations on pattern and cause. *In* Cohen, M. J., and Strumwasser, J., editors: *Comparative Neurobiology, Modes of Communication in the Nervous System*. New York, John Wiley & Sons.

Northcutt, R. G., 1987: Evolution of the vertebrate brain. *In* Adelman, G., editor: *Encyclopedia of Neurosciences*. Boston, Birkhauser.

Northcutt, R. G., 1989: The phylogenetic distribution and innervation of the mechanosensory lateral line. *In* Coombs, S., Gorner, P., and Munz, H., editors: *The Mechanosensory Lateral Line, Neurobiology and Evolution*. New York, Springer-Verlag.

Northcutt, R. G., 1990: Ontogeny and Phylogeny: A re-evaluation of conceptual relationships and some applications. *Brain, Behavior and Evolution*, 36:116–140.

Northcutt, R. G., and Davis, R. E., editors, 1983: *Fish Neurobiology, Volume 1, Brain Stem and Sense Organs*. Ann Arbor, The University of Michigan Press.

Penfield, W., and Rasmussen, T., 1950: *The Cerebral Cortex of Man*. New York, Macmillan Company.

Pettingill, O. S., Jr., 1985: *Ornithology in Laboratory and Field*, 5th edition. Orlando, Academic Press.

Ramón y Cajal, S., 1909: *Histologie du Système Nerveux de l'Homme & des Vertébrés*. Translated by L. Azoulay. Paris, Maloine. (Reprinted Madrid, Instituto Ramon y Cajal, 1952).

Sawro, Von W. A., 1990: Vergleichende Mikrostruktur der ventralen Ruckenmarkshorner einiger Wirbeltiere. *Zoologische Jahrbuch Abteil Anatomie Ontogonie Tiere*, 120:143–162.

Stensio, E. A., 1963. The brain and cranial nerves in fossil, lower craniate vertebrates. *Skrifter Norske Videnskaps-Akademi 1 Oslo I. Mat.-Naturv. Klasse. Ny Serie*, 13:1–120.

Tretjakoff, D., 1927: *Das periphere Nervensystem des Flussneunauges*. Zeitschrift Wissenschaftliche Zoologie., 129:359–952.

Ulinski, P. S., 1990: Nodal events in forebrain evolution. *Netherlands Journal of Zoology*, 40:215–240.

Vanegas, H., 1983: Organization and physiology of the teleostean optic tectum. *In* Davis, R. E., and Northcutt, R. G., editors: *Fish Neurobiology, Volume 2*:43–90. Ann Arbor, The University of Michigan Press.

Wiedersheim, R., 1906: *Vergleichende Anatomie der Wirbeltiere*. Jena, Gustav Fischer.

Young, J. Z., 1933: The autonomic nervous system of selachians. *Quarterly Journal of Microscopical Science*, 75:571–624.

14 | Introduction to the Endocrine Glands

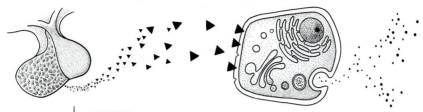

PRECIS

The endocrine glands are part of the control system for the body. We will examine the similarities and differences between nervous and endocrine control and their interrelationship, and discuss those glands that arise in part from nervous tissue. Other endocrine glands will be considered with the organ systems from which they arise.

Outline

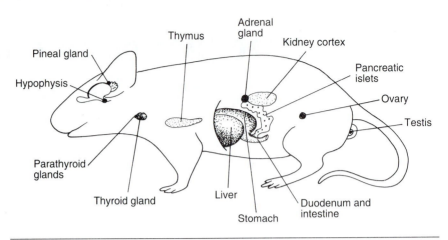

FIGURE 14–1   The location of the major endocrine glands in a rat.

Much activity of the body is regulated by the nervous system, but many other functions are controlled by approximately two dozen ductless, endocrine glands that are widely scattered throughout the body. **Endocrine glands** (Gr. *endon* = within + *krinein* = to separate) differ from **exocrine glands** by discharging their secretions, known as **hormones** (Gr. *hormaein* = to excite), into body fluids rather than onto epithelial surfaces. This fluid is usually the blood, but some hormones enter the lymph and cerebrospinal fluid. Most endocrine glands develop from epithelial surfaces, as do all exocrine glands, but some are formed from modified neurons.

An overview of the location of the endocrine glands of mammals is given in Figure 14–1, and a list of the major vertebrate hormones and their primary effects is given in Table 14–1. We know far more about mammalian hormones than those of other vertebrates, but comparisons will be made when possible. We will first consider similarities and differences between nervous and endocrine control. Then we will examine those endocrine glands that are closely related to the nervous system and place much of the endocrine system under nervous control. These glands are the hypophysis, the pineal gland, the urophysis, and the adrenal gland. Part of the adrenal gland develops from the same type of neural crest cells that give rise to the postganglionic neurons of the sympathetic nervous system. Other endocrine glands that affect metabolism and osmoregulation are considered along with the respiratory, digestive, and excretory systems, and those that regulate reproduction with the reproductive system.

Endocrine and Neuronal Integration

Most multicellular organisms have a nervous system and a set of endocrine glands. Both systems have somewhat similar functions. Both neurons and endocrine glands exert their effects by stimulating or inhibiting target cells, usually by means of chemical messengers. Hormones are one kind of chemical messenger; transmitter substances, synthesized by neurons and released when nerve impulses reach

TABLE 14-1   **Vertebrate Hormones and Their Primary Effects***

Adenohypophysis

Melanophore-stimulating hormone (MSH)	Promotes dispersion of melanin in melanophores of fishes, amphibians, and reptiles
Thyrotropic hormone (TSH)	Stimulates thyroid gland
Adrenocorticotropic hormone (ACTH)	Stimulates adrenal cortex
Growth hormone (GH)	Promotes synthesis and release of liver somatomedin
Follicle-stimulating hormone (FSH)	Promotes development of ovarian follicles and seminiferous tubules of testis, estradiol synthesis
Luteinizing hormone (LH)	Promotes development of corpus luteum from a follicle (which produces progesterone), production of estradiol and testosterone, ovulation
Prolactin (P)	Affects osmoregulation in fishes, promotes maternal behavior, milk synthesis

Hypothalamus via hypophyseal portal system

Thyrotropin releasing hormone (T-RH)	Releases TSH
Adrenocorticotropin releasing hormone (C-RH)	Releases ACTH
Follicle-stimulating hormone releasing hormone (TSH-RH)	Releases FSH
Luteinizing hormone releasing hormone (LH-RH)	Releases LH
Melanophore-stimulating hormone releasing hormone (MSH-RH)	Releases MSH
Melanophore-stimulating hormone release-inhibiting hormone (MSH-RIH)	Inhibits release of MSH
Growth hormone releasing hormone (GH-RH)	Releases GH
Growth hormone release-inhibiting hormone (GH-RIH)	Inhibits release of GH
Prolactin releasing hormone (P-RH)	Releases P
Prolactin release-inhibiting hormone (P-RIH)	Inhibits release of P

Hypothalamus via axons to neurohypophysis

Antidiuretic hormone (ADH) (vasopressin)	Promotes water reabsorption
Oxytocin	Releases milk, stimulates contraction of uterine muscles

(Continued on p. 536)

TABLE 14–1 *continued*

Pineal gland	
Melatonin	Inhibits release of MSH and gonadotropins (FSH, LH)
Urophysis	
Urotensins	Appear to affect osmoregulation in teleosts
Adrenal cortex	
Glucocorticoids (cortisol)	Control protein and carbohydrate metabolism
Mineralocorticoids (aldosterone)	Control electrolyte and water balances, blood pressure
Cortical androgens	Promote muscular development, pubic hair growth, libido
Adrenal medulla (chromaffin cells)	
Noradrenaline and adrenaline	Augment action of sympathetic nervous system
Thyroid gland	
Thyroxine and tri-iodothyronine	Promote oxidative metabolism, amphibian metamorphosis; inhibit release of TSH
Ultimobranchial bodies (or "C" cells of mammalian thyroid gland)	
Calcitonin	Prevents excessive withdrawal of calcium from bone
Parathyroid glands	
Parathormone (PTH)	Promotes calcium reabsorption from kidneys and release from bones
Thymus	
Thymosin	Promotes development of immunological capacity
Islets of Langerhans in pancreas	
Insulin	Reduces blood glucose level by promoting entry of glucose into certain cells and its conversion to glycogen and fat
Glucagon	Increases blood sugar level by converting glycogen stored in liver to glucose
Liver	
Somatomedin	Promotes synthesis of DNA and protein, cell growth
Mucosa of stomach	
Gastrin	Stimulates gastric secretion
Mucosa of duodenum & jejunum	
Secretin	Stimulates secretion of water, salts, and bicarbonate from pancreas
Cholecystokinin-pancreozymin	Stimulates secretion of pancreatic enzymes and contraction of gall bladder

Enterogastrone?	May inhibit gastric secretion
Kidney cortex	
Renin	Increases blood pressure, promotes secretion of adrenal cortex
Dihydroxycholecalciferol	Affects absorption of calcium and bone calcification
Ovary	
Estrodiol	Promotes development of female secondary sex characters, prepares uterine lining for embryo, affects growth of follicles, inhibits release of FSH
Progesterone	Supplements estrogens in development of female secondary sex characters, maintains uterine lining, inhibits release of LH and FSH
Relaxin	Relaxes pelvic ligaments before birth, promotes growth of uterine muscles, stimulates development of glandular tissue in mammary glands
Testis Leydig cells	
Testosterone	Promotes growth of male reproductive organs, development of male secondary sex characters, male behavior
Placenta	
Chorionic gonadotropin (CG)	Maintains corpus luteum until placenta produces gonadal steroids
Estradiol	Same effect as ovarian hormone
Placental lactogen (somatomammotropin)	Milk synthesis
Relaxin	Same effect as ovarian relaxin
Skin	
Vitamin D	Promotes absorption of mineral ions from gut, release of calcium from bones

*These are the hormones produced by the major vertebrate endocrine glands; not all are present in all vertebrates.

synapses or neuron-effector junctions, are another. In a few cases hormones and transmitters are the same products. A major difference between nervous and hormonal integration is the way the chemical messengers are transmitted, whether along neurons or through the circulatory system (Fig. 14–2**A** and **B**). Intermediate conditions of **neurosecretion** are known, in which neurons synthesize and transmit a hormone to blood vessels, through which it is carried to its target organs (Fig. 14–2**C**). Often, the neuron endings and blood vessels form a cluster called a **neurohemal organ.**

The nervous and endocrine control systems differ in other important ways. Neurons are activated by receptor cells or by other neurons. A few endocrine glands also produce and release their hormones in response to a nerve impulse,

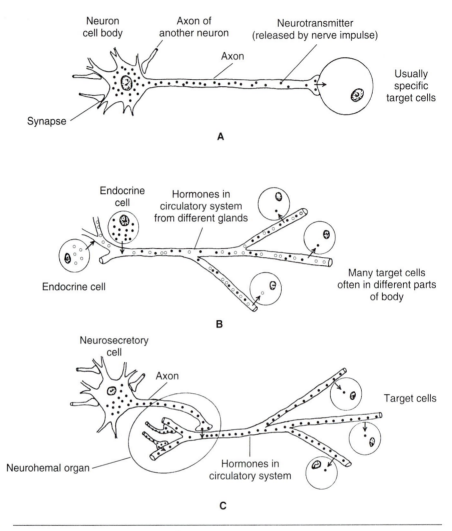

FIGURE 14–2 Neuronal and endocrine control systems: **A,** Action of a neuron on its target cell. **B,** Transportation of two hormones through the circulatory system: each activates a different endocrine gland. **C,** Release of hormones by a neurosecretory cell into the circulatory system.

but most are activated differently. Some endocrine glands respond to changes in the level of a substrate. For example, the entrance of food into the intestine from the stomach promotes the release of secretin from duodenal cells, and this hormone stimulates a copious secretion of pancreatic juice (p. 628). Most endocrine glands respond to changes in the blood level of their own hormone or to that of some other endocrine gland. Often, complex feedback mechanisms between endocrine glands control the hormone levels in the blood.

Once activated, neurons and endocrine glands send their signals to the target cells, but hormones do not go directly to their target cells, as nerve impulses do. Hormones are carried passively in the blood, either in solution or loosely bound to carrier plasma proteins. They must pass into and out of capillaries, so transit time is relatively long. The blood carries the entire spectrum of hormones throughout the body, but only target cells with receptor molecules specific to a particular hormone will respond. Nearly all the cells of the body respond to some hormones, such as thyroxin, which is produced by the thyroid gland; other cells respond to a narrower range of hormones. Only the cells responsible for the aqueous and alkaline portion of the pancreatic juice respond to secretin.

The effect of a nerve impulse on the target cell is momentary because enzymes that enter the synaptic or neuron-effector junctions rapidly metabolize and degrade the transmitter substances. Additional nerve impulses are needed for a continued response. Hormones, in contrast, have much longer effects because they are relatively stable compounds. They must, however, be continuously synthesized because they are lost in various ways. They are taken out of circulation when they bind with receptors, and the hormone-receptor complex is eventually metabolized. Many hormones are also lost through the excretory system.

Because of these differences, endocrine integration is slower and has longer lasting effects than neuronal integration. Hormones are particularly well suited to regulate growth, metamorphosis, slow skin-color changes, metabolism, water and mineral balances, sexual development, and reproduction. They play a major role in the maintenance of homeostasis. Hormones are less effective than nerve impulses in rapidly and briefly stimulating specific organs. Although endocrine and neuronal integration are distinct control systems, they interact with each other in many ways.

The Nature of Hormones and Hormonal Action

Hormones have specific actions because of their differences in chemical structure, but they all belong to only a few categories of organic compounds. Hormones produced by cells of ectodermal or endodermal origin are proteins, peptides, or other amino acid derivatives. The hypothalamic hormones, which control the release of hormones from the adenohypophysis (a part of the pituitary gland), are small neuropeptides secreted by neurons and consisting of 10 to 44 amino acids. These and other neuropeptides are closely related biochemically, and many have common ancestral molecules (Sherwood and Parker, 1990). Insulin, the first hormone whose structure was determined, is a small protein consisting of 51 amino acids. Most of the proteins produced by the adenohypophysis are large proteins with molecular weights of 25,000 to 30,000. Noradrenaline and adrenaline, which are produced by the adrenal medulla, are catecholamine derivatives of the amino acid tyrosine. Thyroxine and tri-iodothyronine, produced by the thyroid gland, combine tyrosine with iodine. Hormones produced by cells of mesodermal origin, including those of the adrenal cortex, testis, ovary, and placenta, are steroids.

The mechanism of hormone action on target cells is not completely understood. Investigators do know that hormones act by affecting enzyme systems, and two ways they do so have been identified. There may be others. Protein and peptide hormones and the catecholamines are water soluble proteins carried in solution in the blood plasma. They do not pass through the lipid plasma membrane of cells; rather they bind with specific receptor molecules on the cell surface (Fig. 14–3**A**). This activates an enzyme on the inner surface of the membrane, which in turn activates a second messenger. The second messenger activates a succession of intracellular enzymes that increase or decrease a critical cell process. This mechanism has been elucidated by research on how adrenaline promotes the conversion of glycogen to glucose in liver cells. The adrenaline-receptor complex activates **adenylate cyclase** on the inner side of the plasma membrane, and this activates **cyclic adenosine monophosphate** (cAMP), which is the second messenger.

Steroid and thyroid hormones are not soluble in water, so they are carried loosely bound to carrier proteins in the plasma. Since they are lipid soluble, they easily enter cells where they unite with receptor molecules (Fig. 14–3**B**). The hormone-receptor complex enters the nucleus and binds to an acceptor site on a chromosome. This activates a gene and leads to the synthesis of proteins (enzymes) that produce the biological effect of the hormone.

Hormones can be effective in very small amounts. Many are present in the blood at concentrations of only 10^{-12} molar. Considerable amplification occurs during their action. A peptide hormone-receptor complex may activate only a few molecules of an enzyme, but these in turn activate more and more at each step in an enzyme cascade. A steroid hormone-receptor complex activates a gene, but the gene can synthesize many protein molecules.

The Hypophysis and Hypothalamus

Development and Structure of the Hypophysis

The pituitary gland, technically known as the **hypophysis cerebri** (Gr. *hypo* = under + *physis* = growth), attaches to the ventral surface of the hypothalamus of the brain. It develops embryonically from two ectodermal evaginations that meet and unite (Fig. 14–4**A**). An **infundibulum** grows ventrally from the diencephalon, and **Rathke's pouch** extends dorsally from the roof of the developing mouth, or stomodaeum. The infundibulum remains connected to the brain and gives rise to the part of the gland known as the **neurohypophysis.** This consists in mammals of a **medial eminence,** next to the brain, and a narrow **infundibular stalk,** which leads to an expanded **pars nervosa** (Fig. 14–4**B**). Rathke's pouch loses its connection with the stomodaeum and gives rise to the rest of the gland, the **adenohypophysis.** The lumen of Rathke's pouch may persist as a narrow cleft that divides the adenohypophysis into a thin **pars intermedia,** located beside the pars nervosa, and a thicker **pars distalis. A pars tuberalis** may extend from the pars distalis along the infundibular stalk. Some mammals, including cetaceans, lack the pars intermedia. The neurohypophysis is called the posterior lobe of the pituitary in the older literature; the adenohypophysis, the anterior lobe. These terms accurately describe their positions in mammals but not in many other vertebrates.

The neurohypophysis contains no secretory cells, but axons of neurosecretory neurons in the hypothalamus carry hormones to the medial eminence and pars nervosa, from which they are released into networks of blood vessels. Much of the neurohypophysis is a neurohemal organ (Fig. 14–4**C**). The adenohypophysis

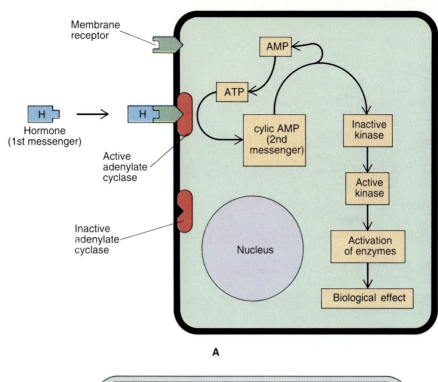

A

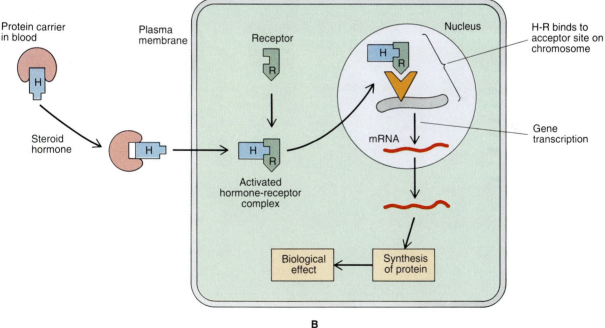

B

FIGURE 14–3   Methods of hormone action. **A,** A protein or peptide hormone binds with a membrane receptor, which in turn activates an enzyme (the second messenger) on the inside of the membrane. **B,** A steroid hormone enters the cell and binds with a receptor. The receptor-hormone complex enters the nucleus and activates a gene. (After Dorit, Walker, and Barnes).

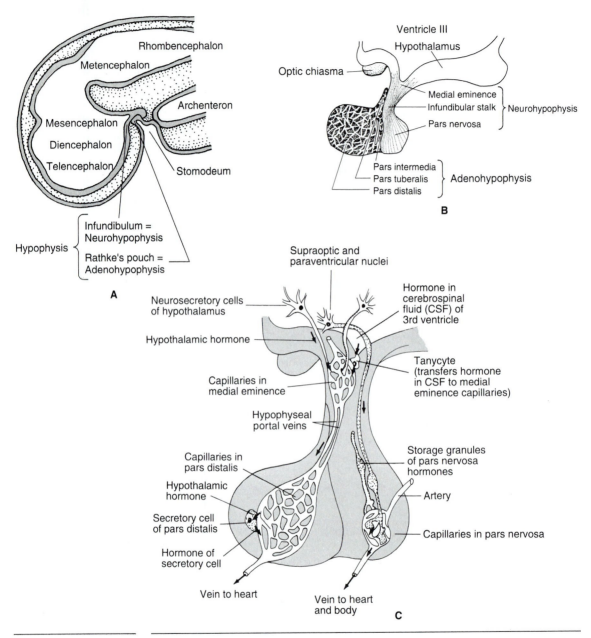

A

Rhombencephalon

Metencephalon

Mesencephalon

Diencephalon

Telencephalon

Archenteron

Stomodeum

Hypophysis {
  Infundibulum = Neurohypophysis

  Rathke's pouch = Adenohypophysis
}

B

Ventricle III

Hypothalamus

Optic chiasma

Medial eminence
Infundibular stalk } Neurohypophysis
Pars nervosa

Pars intermedia
Pars tuberalis } Adenohypophysis
Pars distalis

C

Supraoptic and paraventricular nuclei

Neurosecretory cells of hypothalamus

Hypothalamic hormone

Hormone in cerebrospinal fluid (CSF) of 3rd ventricle

Capillaries in medial eminence

Tanycyte (transfers hormone in CSF to medial eminence capillaries)

Hypophyseal portal veins

Capillaries in pars distalis

Hypothalamic hormone

Storage granules of pars nervosa hormones

Artery

Secretory cell of pars distalis

Capillaries in pars nervosa

Hormone of secretory cell

Vein to heart

Vein to heart and body

FIGURE 14–4 The mammalian hypophysis: **A,** Embryonic development. **B,** Parts of the adult gland. **C,** Connections between the hypothalamus and the hypophysis.

contains many types of secretory cells, which account for its name (Gr. *aden* = gland). Few neurons enter the adenohypophysis, but in most vertebrates it is functionally connected to the hypothalamus by a unique vascular supply known as the **hypophyseal portal system.** Hypothalamic hormones are discharged by axons into capillaries of the medial eminence, and these are carried by small portal

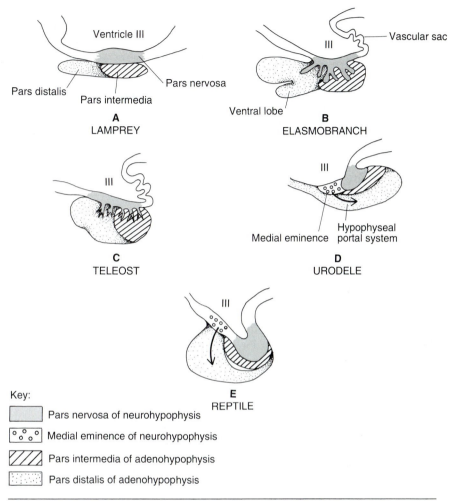

FIGURE 14–5   Hypophyses of representative vertebrates: **A,** A lamprey. **B,** An elasmobranch. **C,** A teleost. **D,** An urodele. **E,** A reptile.

veins to capillaries in the adenohypophysis, where they control the synthesis and release of its hormones. (Portal veins lie between two capillary beds rather than returning directly to the heart.) The adenohypophysis also can be affected by hormones in the cerebrospinal fluid because specialized ependymal epithelial cells, called **tanycytes,** can transfer hormones from this fluid into the capillaries of the medial eminence (see Fig. 12–4).

All vertebrates have pituitary glands, and those of most tetrapods are similar to the mammalian gland (Fig. 14–5). Birds, however, lack the pars intermedia. In fishes the neurohypophysis does not form as distinct a lobe, and in a lamprey it is little more than a part of the floor of the diencephalon. A distinct median eminence and pars nervosa are present in lungfishes but not in other fish species. Fishes lack the hypophyseal portal system, but in most species finger-like extensions of the neurohypophysis project into the adenohypophysis and they may perform some of the functions of a portal system. The infundibulum of all fishes, except cyclostomes, also gives rise to a thin **vascular sac** of uncertain function (p. 497).

Hormones of the Adenohypophysis

At least seven hormones are produced by the mammalian adenohypophysis (Table 14–1). They have also been demonstrated in many other vertebrates. **Melano-phore-stimulating hormone** (MSH) is synthesized by cells in the pars intermedia, or by pars intermedia cells that become incorporated in the pars distalis. Its primary effect is to cause the skin of many fishes, amphibians, and reptiles to darken by promoting the dispersal of melanosomes in the melanophores. In the absence of this hormone, melanosomes become concentrated near the center of the cells and the skin becomes lighter in color. These physiological color changes help adapt the animal to changes in the tone of its background. Temperature and emotional state in some reptiles also affect the release of MSH. Skin pigment in birds and mammals is deposited outside the pigment-producing cells, so these animals do not have physiological color changes, but MSH is one of several hormones that control the synthesis of melanin.

The other six adenohypophyseal hormones are synthesized by distinctive cell types in the pars distalis and pars tuberalis. In many vertebrates **thyrotropic hormone** (TSH) stimulates the thyroid gland to synthesize hormones that increase the rate of oxidative metabolism (p. 605). The thyroid gland is located ventral to the pharynx or near the larynx. **Adrenocorticotropic hormone** (ACTH) stimulates the synthesis of hormones produced in the cortex of the adrenal gland, an organ located near the kidney. Adrenal cortical hormones, in turn, affect a wide range of metabolic processes (p. 549).

Growth hormone (GH) channels amino acids into protein synthesis and has other effects that promote body growth. Some of the effects of GH appear to result from its stimulating the release of other growth factors (**somatomedins**) from liver cells. A deficiency of GH during human childhood leads to certain types of dwarfism, and an excess, to giantism. GH and somatomedins have been found in all tetrapod groups and some fishes.

Follicle-stimulating hormone (FSH) and **luteinizing hormone** (LH) were first discovered in females, where they promote the development of the ovarian follicles and their transformation after ovulation into corpora lutea. The ovarian follicles and corpora lutea, in turn, produce estradiol and progesterone, hormones essential for the development of female secondary sex characters and the female reproductive cycles (Chapter 21). Later discoveries have shown that these adenohypophyseal hormones also are present in males. FSH promotes the development of the seminiferous tubules in the testis, which produce sperm. LH (sometimes called interstitial cell stimulating hormone in males) promotes the synthesis of the male sex hormone, testosterone, by the interstitial cells of the testis.

Prolactin (P) was named from the discovery that it stimulates the synthesis (but not the release) of milk by the mammary glands. It is closely related genetically and biochemically to growth hormone, and both probably evolved from a common ancestral hormone by gene duplication (prolactin is a larger molecule) followed by divergence. Prolactin occurs in all groups of vertebrates, with the possible exception of lampreys and hagfishes, and it appears to have the widest range of actions of the adenohypophyseal hormones. Along with growth hormone, prolactin stimulates the secretion of somatomedins by liver cells. It is essential in many ways for reproduction. Prolactin helps sustain the production of testosterone by the testis and progesterone by the corpus luteum. It stimulates reproductive migrations in many animals, including the movement of certain salamanders to water. It promotes nest building and maternal care in some teleosts and many birds. Indeed, roosters that have been given sufficient amounts of prolactin will brood the chicks. In some birds prolactin interacts with other hormones to produce brood patches—ventral skin regions that lose feathers and transfer maternal body heat more efficiently to the eggs. In pigeons it stimulates the desquamation of cells in the crop and the formation of a crop milk that is fed to newly hatched young. The effects of prolactin are not limited to reproduction. In some fishes it helps control water and salt balances and is thus essential for certain saltwater species to enter fresh water during their spawning runs. Similar effects have been found in nonmammalian tetrapods, but prolactin's role in mammalian osmoregulation is not yet clear. Finally, prolactin is involved in modulating the immune response.

Neurosecretions of the Hypothalamus

The synthesis and release of adenohypophyseal hormones are regulated by at least ten small, neuropeptide hormones that are synthesized in the hypothalamus and carried in the hypophyseal portal system to their target cells in the adenohypophysis (Table 14–1). The release of thyrotropic hormone, adrenocorticotropic hormone, follicle-stimulating hormone, and luteinizing hormone is affected by the blood level of the hormones of their target organs. For example, a deficiency of the thyroid hormones (thyroxine and tri-iodothyronine) in the blood stimulates the release of a specific hypothalamic hormone (thyrotropin releasing hormone) that causes specific adenohypophyseal cells to release more thyrotropic hormone. An excess of thyroid hormones in the blood causes a reduction in the release of thyrotropic hormone. Both effects occur via the hypothalamus and the portal system. Control mechanisms of this type are called **negative feedback systems.** The production and release of melanophore-stimulating hormone, growth hormone, and prolactin are more complex because both releasing and inhibiting hormones are involved. Neurons from the hypothalamus extend into the pars intermedia and terminate on cells that produce melanophore-stimulating hormone, at least in amphibians, so it is likely that these cells are also affected by direct nerve stimuli.

Two chemically related peptide hormones are synthesized in the **paraventricular** and **supraoptic nuclei** in the hypothalamus of mammals (Fig. 14–4C). They are carried by axon tracts to the pars nervosa of the hypophysis, where they are stored and released. **Antidiuretic hormone** (also called ADH and vasopressin) prevents diuresis, the excretion of a copious and dilute urine, by promoting the reabsorption of water from parts of the kidney tubules (p. 738). **Oxytocin** stimulates the contraction of certain smooth muscles. It is important in initiating uterine contractions at birth and releasing milk from the mammary glands in response to the suckling stimulus of the infant (p. 782). Homologous hormones, ones with a few amino acid substitutions, occur in all vertebrates. In other tetrapods they also promote water reabsorption from kidney tubules, the urinary bladder, or skin (in amphibians), and in birds, the contraction of oviduct muscles at egg laying. Their biological functions in fishes are not well known.

We have seen that the hypothalamus is critical for the neuronal integration of visceral functions. Because of its neurosecretory relationships with the pituitary gland, the hypothalamus also is the major control center for the pituitary, which in turn is a central part of the endocrine system because its hormones affect so many other glands and physiological processes. The hypothalamus responds to the blood level of many hormones, but it is also a major **neuroendocrine transducer,** for it converts nerve impulses to hormonal signals and thereby places much of the endocrine system under some degree of nervous control. Endocrine secretion can be affected by emotional states, by changes in many physiological processes, and by light and other environmental changes detected by sense organs, which then transmit impulses through the nervous system. Much nervous activity, in turn, is influenced by hormonal levels. A female dog, for example, comes into heat under the influence of reproductive hormones and behaves differently toward male dogs than when she is not in heat. The two control systems of the body, nervous and endocrine, are thus closely integrated.

The Pineal Gland

As we have seen (p. 428), many fishes, amphibians, and reptiles have a median pineal eye, a parietal eye, or both. The photoreceptive cells in these eyes, as well as those of the lateral image forming eyes, produce some **melatonin** when light is not impinging on them. Melatonin is synthesized from the amino acid tryptophan via various intermediate products, one of which is serotonin, and light inhibits certain enzymes in this pathway. Gern, Duvall, and Nervina (1986) have proposed that melatonin in ancestral vertebrates was not a hormone and acted only locally in the eyes, facilitating the photoreceptive process (i.e., melatonin did not act on distant tissues). However, some melatonin spilled over into the blood, and, since there was a regular nightly surge of melatonin, it could be co-opted to time physiological processes that would be most advantageous if they occurred at regular intervals. Natural selection apparently favored such a linkage. Melatonin became a hormone that acted on more distant organs involved in anamniotes and reptiles in diurnal color changes, degree of exposure to sunlight, and other rhythms (p. 548).

Of the original pineal-parietal complex, only the pineal remains in birds and mammals, and it has become the **pineal gland** in most species. Cetaceans lack the organ. Like the retina and neurohypophysis, the pineal gland develops from the diencephalon of the brain and remains attached to the posterior part of the diencephalic roof (Fig. 14–6). The ancestral photoreceptive cells became converted into secretory **pinealocytes.** That pinealocytes develop from photorecep-

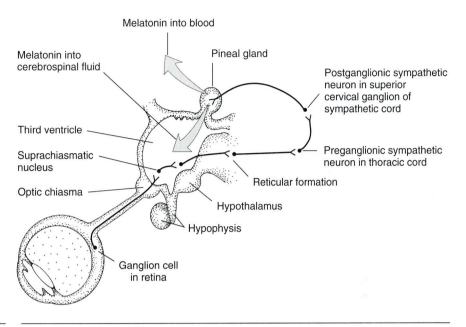

Melatonin into blood

Melatonin into
cerebrospinal fluid

Pineal gland

Postganglionic sympathetic
neuron in superior
cervical ganglion of
sympathetic cord

Third ventricle

Preganglionic sympathetic
neuron in thoracic cord

Suprachiasmatic
nucleus

Reticular formation

Optic chiasma

Hypothalamus

Hypophysis

Ganglion cell
in retina

FIGURE 14–6   A diagram of a sagittal section through the mammalian diencephalon showing the neuronal pathway by which light inhibits the synthesis of melatonin by the pineal gland, and the route through which melatonin reaches other organs when it is synthesized during dark periods.

tive cells is evident in some neonatal mammals, where photoreceptive ciliary processes persist for a while (Zimmerman and Tso, 1975). Activity of the pinealocytes is still affected by light. Light may be received directly in birds because some photoreceptive cells remain in their pineal gland. In mammals, and to a lesser extent in birds, light affects the pineal gland via neuronal connections with the lateral image forming eyes. The mammalian pineal gland is a neuroendocrine transducer because nervous signals control the synthesis of an enzyme.

The axons of some ganglion cells in the retina leave the optic pathway at the optic chiasma, enter the hypothalamus, and terminate in a suprachiasmic nucleus, from which impulses go to another hypothalamic nucleus (Fig. 14–6) (Turek, 1985). Impulses are relayed through the reticular formation and sympathetic nervous system to the pineal gland. In rats, and probably human beings and many other mammals, the suprachiasmic nucleus appears to act as a "biological clock." It has an inherent circadian rhythm (L. *circa* = approximately + *dies* = day) that affects the gland by the pathway described. Melatonin levels in the blood rise at night and fall during the day. Light impulses reaching the suprachiasmic nucleus modulate this endogenous rhythm and adjust the mechanism to changes in day length that occur seasonally or experimentally by manipulating light and dark periods. It has been proposed that "jet lag" may result from the melatonin rhythm getting out of phase with the light-dark cycle (Fevre-Montage et al., 1978).

Melatonin is released into the cerebrospinal fluid of the third ventricle, from which it is transferred to the medial eminence by tanycytes. Thus, it can influence the activity of the adenohypophysis. It also is released into the blood, which carries it throughout the body. Because of its cyclic production and wide distribution, melatonin has the potential for affecting many physiological processes and adjusting them to diurnal and seasonal cycles, but it is not clear how widespread its effects are. Melatonin acts in opposition to melanophore-stimulating hormone and causes the pigment in the melanophores of some fishes and amphibians to concentrate (hence the name for this hormone). Its higher level at night causes these animals to lighten. Studies on rodents have shown that melatonin has an inhibiting effect on gonad development, but an increase in light levels reduces this effect. Through its effects on the medial eminence and adenohypophysis, the pineal gland may play a critical role in the control of seasonal reproductive cycles.

The Urophysis

Teleost fishes have a neurohemal organ known as the **urophysis** (Gr. *oura* = tail + *physis* = growth) located beneath the caudal end of their spinal cord. Neurosecretory neurons in the caudal part of the cord send axons into the urophysis, where they come close to blood vessels. Peptide hormones synthesized by these cells have been called **urotensins** because they can increase blood pressure in teleosts. The urophysis resembles a neuroendocrine transducer. It appears to play a role in osmoregulation, but more research is needed to clarify its biological function. Chondrichthyan fishes have a comparable group of neurosecretory cells, but their axons do not aggregate to form such a well-defined organ.

The Adrenal Gland

The paired **adrenal glands** are also called **suprarenal glands** from their location in mammals, beside or above the kidneys (Fig. 14–7**A**). Each mammalian gland consists of two distinct parts, the cortex and the medulla, which have different origins, structures, and functions. The **adrenal medulla** of mammals is composed of **chromaffin cells,** which are arrayed around the periphery of venous sinuses, into which they discharge their secretions (Fig. 14–8). The chromaffin cells synthesize **noradrenaline,** but in mammals most of this is methylated to form **adrenaline,** a similar but somewhat more potent hormone. These catecholamine hormones are released when the chromaffin cells are stimulated by *preganglionic* sympathetic neurons. Noradrenaline is the same product as the neurotransmitter released by the *postganglionic* sympathetic neurons at their junctions with effectors, so the medulla reinforces the action of the sympathetic nervous system in adjusting the body to stress (p. 481).

The similarity of their products points to an affinity between chromaffin cells and postganglionic sympathetic neurons. This notion is supported by the innervation of both by preganglionic sympathetic neurons, and by embryonic studies showing that both are derived from neural crest cells. At one stage during their migration to the adrenal medulla, the precursors of the chromaffin cells are associated with the sympathetic ganglia that contain the cell bodies of developing postganglionic sympathetic neurons. A few chromaffin cells remain in the sym-

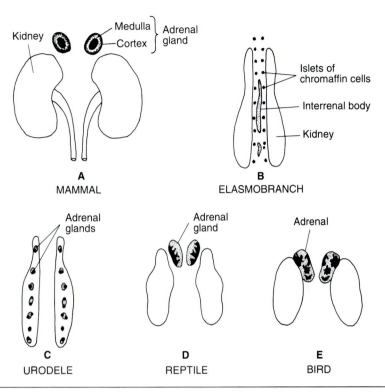

FIGURE 14–7 The adrenal glands of representative vertebrates. The location of the chromaffin cells is shown in black; cortical material is shaded. **A**, A mammal. **B**, An elasmobranch. **C**, An urodele. **D**, A reptile. **E**, A bird.

pathetic ganglia, or in small, adjacent clusters called **paraganglia.** It appears that chromaffin cells are modified postganglionic sympathetic neurons.

Other chromaffin-like cells occur in the lining of the gastrointestinal and respiratory tracts, and in the connective tissue of the gut, liver, and pancreas. All perform certain common biochemical steps (decarboxilation of tyrosine) in the synthesis of noradrenaline, adrenaline, and related hormones. Some investigators believe that these dispersed chromaffin and chromaffin-like cells are a **diffuse neuroendocrine system** that supplements or amplifies the actions of the sympathetic nervous system and adrenal medulla.

The **adrenal cortex** develops by the proliferation of coelomic epithelial cells into the mesenchyme adjacent to the developing kidneys and gonads. The adult mammalian cortex consists of irregular cords of epithelial cells interspersed with venous sinusoids (Fig. 14–8). The cortex produces many steroid hormones that chemically are similar to those produced by the ovary and testis, which also develop partly from a proliferation of coelomic epithelium adjacent to the embryonic kidneys (p. 749). The functions of the cortical hormones overlap to some extent, but they can be sorted into three groups: glucocorticoids, mineralocorticoids, and cortical androgens. **Glucocorticoids,** of which cortisol is the most important, primarily affect carbohydrate metabolism. Cortisol stimulates the breakdown of

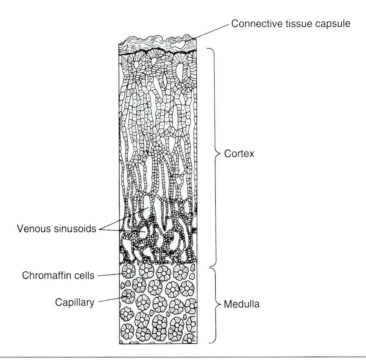

FIGURE 14–8   The histological structure of the mammalian adrenal gland.

protein and the conversion of its products into carbohydrate. This conversion occurs within liver cells. Blood sugar levels increase, and excess sugar is stored as glycogen in muscle and liver cells. The increased availability of sugar and glycogen help a vertebrate adjust to stress. High levels of glucocorticoids also suppress many of the body's normal inflammatory and immunological responses. They inhibit capillary dilation, decrease capillary permeability, and inhibit the mobilization of lymphocytes. These reactions, too, help an animal handle certain aspects of stress. Mammals that are experimentally maintained in stressful situations develop enlarged adrenal cortices.

Aldosterone, the most important **mineralocorticoid,** affects electrolyte balances and blood pressure. It promotes the reabsorption of sodium ions by the kidney tubules and the excretion of potassium and hydrogen ions. Because of the osmotic effect of sodium reabsorption, the amount of water in the interstitial fluid and blood increases, and this can lead to an increase in blood pressure.

Cortical androgens resemble the male sex hormone, testosterone, and at high blood levels they have a masculinizing effect on females. At normal levels they promote protein synthesis and muscle growth in both sexes. In the absence of disease, cortical hormones are maintained in the blood at appropriate levels by complex negative feedback mechanisms that affect, via the hypothalamus and hypophyseal portal system, the release of adrenocorticotropic hormone by the adenohypophysis. Cortical hormones also interact with hormones from many other endocrine glands.

All vertebrates have groups of cells homologous to the mammalian adrenal cortical and chromaffin tissue, but they are not so intimately associated. Many fishes have an elongated **interrenal body** between the kidneys, which is comparable to the cortex, and a series of **islets of chromaffin cells** beside the sympathetic ganglia in a position similar to the mammalian paraganglia (Figs. 14–7**B** and 12–17). Cortical and chromaffin tissues come together in most amphibians to form a series of islets along the ventral surface of the kidneys (Fig. 14–7**C**). The association of cortical and chromaffin cells is more intimate in amniotes (Fig. 14–7**D** and **E**), and in mammals the chromaffin cells are completely encased by the cortex. Noradrenaline is the major secretion of the chromaffin cells in nonmammalian vertebrates and in embryonic mammals. Noradrenaline is methylated in adult mammals to produce large amounts of adrenaline. Methylation requires cortisol, produced in the cortex, to synthesize a medullary enzyme. The close association of the cortex and medulla in mammals allows cortisol to reach the medulla directly since some blood flows through the cortex to the medulla.

SUMMARY

1. Neurons transmit their chemical messengers along axons; endocrine glands discharge theirs into the circulatory system. Neurosecretory cells transmit messengers along axons and then into the circulatory system.
2. Neurons are activated by receptors or by other neurons. A few endocrine glands are also activated by neurons, but most respond to changes in substrate levels or to changes in blood level of their own hormones or those of other glands.
3. Hormones do not go directly to their target cells but are carried passively in the circulatory system throughout the body. Only cells with appropriate receptor molecules respond to the different hormones.
4. The effects of hormonal stimulation last longer than those of neuronal stimulation.
5. Hormones are particularly well suited to regulate metabolic processes, sexual development, and reproduction. They are less effective than neurons in rapidly and briefly stimulating a specific organ.
6. Most hormones are proteins, peptides, or other amino acid derivatives. Those produced by the adrenal cortex, gonads, and placenta are steroids.
7. Peptide and protein hormones affect cells by activating a second messenger within the cell, which in turn activates a succession of intracellular enzymes. Steroid hormones enter the nucleus and activate specific genes that synthesize proteins.
8. The neurohypophysis of the pituitary gland develops from the infundibulum of the diencephalon; the adenohypophysis, from an ectodermal evagination from the stomodaeum (Rathke's pouch).
9. The neurohypophysis contains no secretory cells but receives neurosecretory axons from the hypothalamus. The adenohypophysis contains several types of secretory cells. The release of their hormones is controlled by hypothalamic hormones that reach the adenohypophysis via a hypophyseal portal system of vessels.

10. The neurohypophysis does not form as distinct a lobe in fishes as it does in tetrapods. A hypophyseal portal system is not present in fishes.

11. The adenohypophysis produces seven hormones, most of whose major actions are indicated by their names: melanophore-stimulating hormone, thyrotropic hormone, adrenocorticotropic hormone, growth hormone, follicle-stimulating hormone, luteinizing hormone, and prolactin. Prolactin is widespread among vertebrates and affects many processes: milk synthesis, growth, the production of sex hormones, reproductive behavior, water and salt balances, and the immune system.

12. The hypothalamus synthesizes antidiuretic hormone (vasopressin) and oxytocin. Oxytocin promotes the contraction of certain smooth muscles. These hormones are stored in and released from the neurohypophysis.

13. The hypothalamus is a neuroendocrine transducer that is affected by nerve impulses and in turn affects the endocrine system. It brings much of the endocrine system under nervous influences. Hormones, in their turn, affect the activity of much of the nervous system.

14. The pineal gland of a bird or mammal is another neuroendocrine transducer that has been implicated in regulating certain activities in relation to diurnal cycles and seasonal changes in day length.

15. Teleosts and some other fishes have a urophysis at the caudal end of the spinal cord that also resembles a neuroendocrine transducer. It may have a role in osmoregulation.

16. The adrenal medulla is composed of chromaffin cells of neural crest origin that have an action similar to that of postganglionic sympathetic neurons. They release similar substances and help adjust the body to stress.

17. The adrenal cortex arises from coelomic epithelial cells and produces three groups of steroid hormones: glucocorticoids, which affect carbohydrate metabolism and the immune system; mineralocorticoids, essential for salt and water balance; and androgens, which promote protein synthesis and growth.

18. The cortical and medullary cells are separate in fishes but gradually come together in tetrapods. Glucocorticoids may assist in the synthesis of the hormones of the medullary cells.

REFERENCES

Barrington, E. J. W., 1975: *An Introduction to Comparative and General Endocrinology*, 2nd edition. New York, Oxford University Press.

Bentley, P. J., 1976: *Comparative Vertebrate Endocrinology*. Cambridge, Cambridge University Press.

Bern, H. A., 1985: The elusive urophysis: Twenty-five years in pursuit of caudal neurohormones. *American Zoologist*, 25:763–769.

Carmichael, S. W., and Winkler, H., 1985: The adrenal chromaffin cell. *Scientific American*, 235:2:40–49, August.

Clark, N. B., Norris, D. O., and Peter, R. E., symposium organizers, 1983: Evolution of endocrine systems in lower vertebrates: A symposium honoring Professor Aubrey Gorbman. *American Zoologist*, 23:593–748.

DeGroot, L. J., editor, 1989: *Endocrinology*, 2nd edition. Philadelphia, W. B. Saunders Company.

Ellis, LeG. C., symposium organizer, 1976: Endocrine role of the pineal gland. *American Zoologist*, 16:3–101.

Fevre-Montage, M., Von Cauter, E., Refetoff, S., et al., 1978: Effects of "jet lag" on hormonal patterns. II. Adaptation of melatonin circadian periodicity. *Journal of Clinical Endocrinology and Metabolism*, 52:642–649.

Gern, W. A., Duvall, D., and Nervina, J. M., 1986: Melatonin: A discussion of its evolution and actions in vertebrates. *American Zoologist*, 26:985–996.

Gorbman, A., Dickhoff, W. W., Vigna, S. R., Clark, N. B., and Ralph, C. L., 1983: *Comparative Endocrinology*, New York, John Wiley and Sons.

Pang, P. K. T., and Epple, A., editors, 1980: *Evolution of Vertebrate Endocrine Systems*. Lubbock, Tex., Texas Tech Press.

Reiter, R. J, editor, 1981: *The Pineal Gland*. Boca Raton, CRC Press.

Schreibman, M. P., and Pang, P. K. T., symposium organizers, 1973: The current status of fish endocrine systems. *American Zoologist*, 13:710–936.

Sherwood, N. M., and Parker, D., 1990: Neuropeptide families: An evolutionary perspective. *Journal of Experimental Biology*, supplement, 4:63–71.

Tamarkin, L., Curtis, J. B., and Almeida, O. F. X., 1985: Melatonin: A coordinating signal for mammalian reproduction? *Science*, 277:714–719.

Turek, F. W., 1985: Circadian neural rhythms in mammals. *Annual Reviews of Physiology*, 47:49–64.

Turner, C. D., and Bagnara, J. T., 1976: *General Endocrinology*, 6th edition. Philadelphia, W. B. Saunders Company.

Zimmerman, B. L., and Tso, M. O. M., 1975: Morphological evidence of photoreceptor differentiation of pinealocytes in the neonatal rat. *Journal of Cell Biology*, 66:60–75.

15 | The Body Cavity and Mesenteries

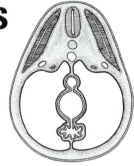

PRECIS

We turn now from the organ systems that support, move, and integrate the body to those that provide the body with nutrients and other needed materials and remove the waste products of metabolism. Large parts of the digestive, respiratory, and related systems lie in the body cavity, or coelom, and are supported by mesenteries. We will examine the development and structure of the coelom and mesenteries in this chapter.

Outline

The Coelom and Mesenteries of Fishes and Their Development
The Coelom and Mesenteries of Tetrapods

The cells of the body need raw materials for their growth, maintenance, myriad activities, and replacement. High-energy organic molecules as well as other nutrients and water must be obtained, oxygen and carbon dioxide must be exchanged, nitrogenous wastes must be eliminated, salt and water balances must be maintained, and materials must be transported between the cells and the body's sites of entry and exit. These activities are performed by the digestive, respiratory, excretory, and circulatory systems, major parts of which are located within the body cavity.

The body cavity, or **coelom,** is a space within the mesoderm that surrounds the visceral organs. It is lined by the **serosa,** a thin membrane consisting of a simple squamous epithelium, which is called the coelomic epithelium or **mesothelium,** and a thin underlying layer of loose connective tissue. The serosa secretes a small amount of watery, serous fluid into the cavity. A coelom allows room for the beating of the heart, changes in lung volume, the filling and emptying of the digestive tract, and other functional changes in the size and shape of the organs. Thin membranes, the **mesenteries,** extend between the organs and from the organs to the body wall. Most mesenteries are identified by the prefix **meso-** followed by the name of the organ to which they go (e.g., the mesogaster extends between the dorsal body wall and the stomach). Some mesenteries are called ligaments (e.g., the falciform ligament extends between the ventral body wall and the liver). Although such ligaments are supporting structures, they are quite different from ligaments extending between bones. Mesenteries consist of two layers of mesothelium with a thin layer of loose connective tissue between them. Mesenteries keep the organs in proper relationship to each other and provide passageways for ducts, blood vessels, and nerves going from one organ to another and between the organs and body wall. Fat is stored in some mammalian mesenteries.

The Coelom and Mesenteries of Fishes and Their Development

The coelom develops as a space within the part of the lateral plate mesoderm that extends from the level of the heart to the caudal end of the intestine (Fig. 15–1**A**). Coelomic spaces may enter the somites early in development, but these extensions are ephemeral. As development continues, the lateral plates enfold the gut and its derivatives (Fig. 15–1**B**). These organs thus come to be surrounded by the coelom, and they become suspended by **dorsal** and **ventral mesenteries** formed by the coming together of the lateral plates above and below the gut. The liver grows from the gut into the ventral mesentery, where it expands and divides the ventral mesentery into a **lesser omentum** that extends from the stomach and intestine to the liver, and a **falciform ligament** that extends from the liver to the ventral body wall (Fig. 15–1**C**). The pancreas grows into the dorsal mesentery. The heart develops by the fusion of a pair of blood vessels in the ventral mesentery just anterior to the liver (Fig. 15–1**D**). Continuous dorsal and ventral mesenteries are present early in development, but those supporting the heart, called **mesocardia,** as well as most of the ventral mesentery caudal to the liver are soon lost (Fig. 15–1**E**). The originally separate left and right halves of the coelom then become continuous anterior and caudal to the liver.

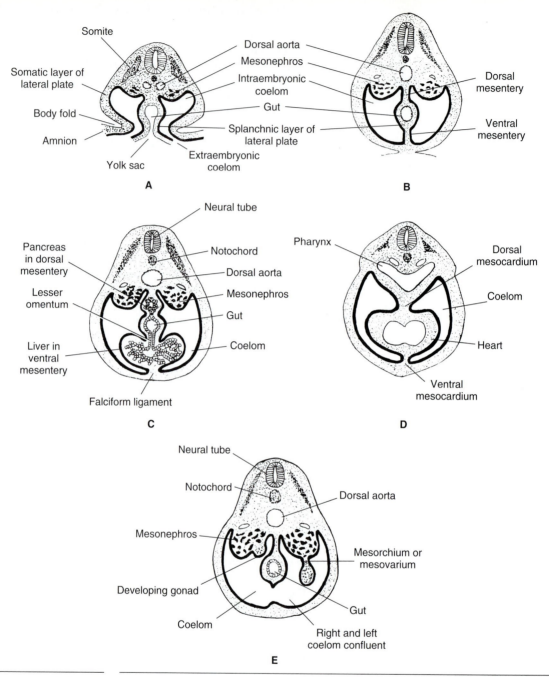

FIGURE 15–1 The development of the coelom and mesenteries as seen in a series of transverse sections of vertebrate embryos: **A,** Lateral body folds separate the embryo and intraembryonic coelom from the yolk sac and extraembryonic coelom. **B,** A portion of the coelom lies on each side of the gut and a continuous dorsal and ventral mesentery. **C,** The liver grows into the ventral mesentery, and the pancreas grows into the dorsal mesentery. **D,** The heart develops by the fusion of a pair of veins in the ventral mesentery (mesocardia) cranial to the liver. **E,** Much of the ventral mesentery disappears caudal to the liver, and left and right coeloms become confluent. (After Corliss.)

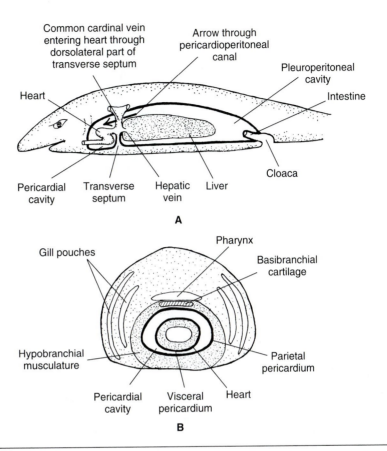

A

B

FIGURE 15–2   The coelom and its divisions in a primitive fish, such as a shark: **A,** Lateral view. **B,** A transverse section through the pharynx and pericardial cavity.

Most of the embryonic dorsal mesentery persists in the adult. The parts suspending different organs have different names: a **mesogaster** supports the stomach; a **mesentery,** the intestine. (Note that the term *mesentery* may be used in a generic sense for any of these membranes and in a strict sense for the one supporting the intestine.) The stomach becomes J shaped in adult fishes. Part of the mesentery (*sensu strictu*) is pulled beside the mesogaster; these membranes may fuse. A **mesocolon** suspends the caudal end of the intestine. Additional mesenteries support the reproductive organs, which push into the coelom from the dorsal body wall: a **mesorchium** goes to each testis; a **mesovarium,** to each ovary; and a **mesotubarium,** to each oviduct (Fig. 15–1E, right side).

In all vertebrates a transverse partition known as the **transverse septum** develops between the liver and the heart. It divides the coelom into an anterior **pericardial cavity** around the heart and a posterior **pleuroperitoneal cavity** around the abdominal viscera and lungs, when lungs are present (Fig. 15–2A). Coelomic epithelium within the pericardial cavity is known as pericardium; **visceral pericardium** covers the surface of the heart, **parietal pericardium** lies

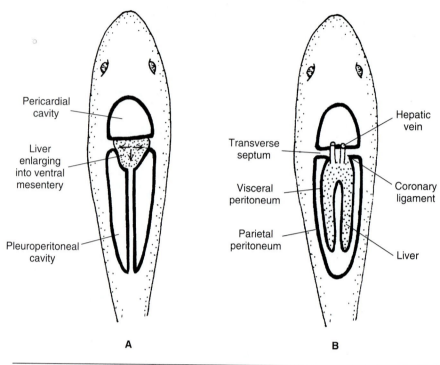

FIGURE 15–3 **A,** A frontal section of a fish to show the expansion of the liver in the ventral mesentery to form the ventral part of the transverse septum. **B,** The subsequent caudal growth of the liver out of the transverse septum. Arrows indicate the directions of liver expansion.

adjacent to the body wall (Fig. 15–2**B**). The coelomic epithelium within the pleuroperitoneal cavity is called **peritoneum;** that in the pleural cavities containing the lungs, when these cavities become distinct in higher vertebrates, is known as the **pleura.**

Several complex processes are involved in the formation of the transverse septum. To simplify slightly, visualize the liver expanding laterally in the ventral mesentery (Fig. 15–1**C**) and carrying the coelomic epithelium covering it to the body wall where the visceral and parietal layers unite (Fig. 15–3**A**). This forms a partition between the ventral parts of the pericardial and the pleuroperitoneal cavities. The liver extends caudally in the ventral mesentery during its subsequent enlargement, leaving the partition it formed as the ventral part of the transverse septum (Fig. 15–3**B**). The ventral mesentery caudal to the liver disappears, but the liver remains connected to the transverse septum by a mesentery known as the **coronary ligament,** through which hepatic veins from the liver enter the heart. While these processes are going on, a pair of common cardinal veins that carry blood from the dorsal body wall to the heart push in medially from the dorsolateral parts of the body wall to the caudal end of the heart. They carry with them sheets of coelomic epithelium, which form the dorsal part of the transverse septum (Fig. 15–2**A**). In cartilaginous fishes a small passage, the **pericardioper-**

itoneal canal, remains between the folds carrying the common cardinals and interconnects the two parts of the coelom (Fig. 15–2A).

The separation of the pericardial from the pleuroperitoneal cavity is partly a byproduct of the developmental processes we have described. However, the complete, or nearly complete, separation of the two cavities also allows the organs in one cavity to undergo their functional movements independently of those of the organs in the other cavity. As we will discuss later (p. 679), the pericardial cavity also allows the development of a reduced pressure around the fish heart, and this plays a role in the hemodynamics of blood circulation.

The heart and pericardial cavity of an adult fish are located far forward, ventral to the caudal end of the pharynx, and they are surrounded by the firm hypobranchial musculature and pharynx floor (Fig. 15–2A and **B**). This position brings the heart close to the gills, through which the heart must pump blood before it is distributed to the body. The transverse septum is located about the level of the pectoral girdle and lies in the transverse plane.

In some species of fishes a pair of small **abdominal pores** lead from the caudal end of the pleuroperitoneal cavity to the cloaca. Their significance is uncertain, but they may represent an ancestral passage for sperm and eggs from the coelom to the outside. Gametes are discharged in a somewhat similar way in contemporary lampreys.

The Coelom and Mesenteries of Tetrapods

The heart and pericardial cavity of early embryonic tetrapods are also located far forward in the body. During later development they migrate caudally, closer to the lungs, which replace the gills as the site for gas exchange. The pericardial cavity thus comes to lie ventral to the anterior part of the pleuroperitoneal cavity, and the transverse septum assumes an oblique orientation. This is the condition characteristic of adult amphibians and early reptiles (Fig. 15–4A). The lungs are located ventrolateral to the digestive tract and its supporting mesenteries in the part of the pleuroperitoneal cavity overlying the pericardial cavity (Fig. 15–4B). The areas containing the lungs are often called the **pleural recesses.** They frequently extend ventrally on each side of the pericardial cavity, thereby partly separating the pericardial cavity from the body wall. All of the membrane between pleuroperitoneal and pericardial cavities is homologous to the transverse septum of fishes, but the part of it separating the pleural recesses from the pericardial cavity is often called the **pleuropericardial membrane.**

Additional folds of coelomic epithelium separate the pleural recesses from the rest of the pleuroperitoneal cavity in many reptiles (some lizards, snakes, and crocodiles) and in birds and mammals. The coelom of these animals thus consists of four compartments: a pericardial cavity, two pleural cavities, and a peritoneal cavity (Fig. 15–5A). The location of the lungs in distinct pleural cavities allows them to expand and contract independently of other organs. In reptiles and birds the folds separating the pleural cavities from the peritoneal cavity form the **oblique septum** (Fig. 15–6).

In mammals the separation between the two pleural cavities and the perito-

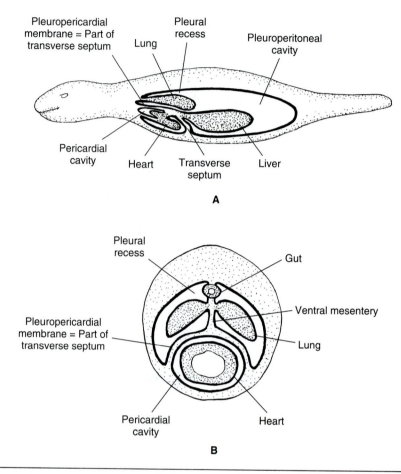

FIGURE 15–4   The coelom and its divisions in a primitive tetrapod, such as a salamander: **A,** Lateral view. **B,** A transverse section at the level of the pericardial cavity.

neal cavity is formed by the **pleuroperitoneal membranes,** which push in from the dorsolateral body wall, and by other folds that extend laterally from the mesenteries and medially from the body wall in this region and meet the pleuroperitoneal membranes (Fig. 15–5**A**). These membranes, with the part of the transverse septum separating the pericardial and peritoneal cavities, become invaded by somatic muscles and form the **diaphragm** (Fig. 15–7). This is the primary respiratory muscle. Since these developmental processes occur early in the embryo when the pericardial cavity and heart are located far forward, the somatic muscles entering the diaphragm are of cervical origin. When these structures shift caudally later in development, branches of cervical spinal nerves follow the diaphragmatic musculature as the **phrenic nerves** (Fig. 15–5**B**).

The enlarging lungs and pleural cavities of mammals grow laterally and ventrally around the pericardial cavity and heart, often meeting ventral to the pericardial cavity (Fig. 15–5**B**). The pericardial cavity becomes completely separated from the body wall. In some species it is also separated from the diaphragm by a

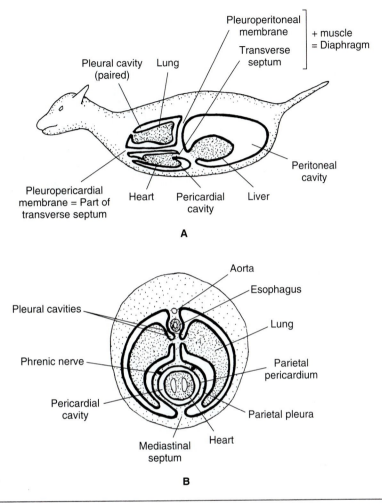

FIGURE 15–5   The coelom and its divisions in a mammal: **A,** Lateral view. **B,** A transverse section at the level of the pericardial cavity.

caudal and ventral growth of the lungs and pleural cavities. The wall of the pericardial cavity, now separated from the body wall (and sometimes from the diaphragm), consists of the parietal pericardium and parietal pleura with a thin layer of connective tissue sandwiched between them. This combined wall often is called the **pericardium** or **pericardial sac.**

Many organs lie between the pleural cavities of mammals: the pericardial cavity and heart, the esophagus, major arteries and veins, the phrenic and other nerves, and, in young mammals, the thymus. The area between the two pleural cavities that contains these structures is called the **mediastinum.** The mesentery formed where the medial walls of the two pleural cavities meet above and below these structures is the **mediastinal septum.**

Mammalian mesenteries within the peritoneal cavity are similar to those of

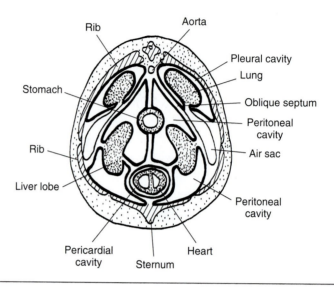

FIGURE 15–6   A transverse section through the pleural, peritoneal, and pericardial cavities of a bird. (Partly after Feduccia and McCrady.)

---

other vertebrates except for rotations of the stomach and intestine that affect the dorsal mesentery. During embryonic development the stomach rotates about its longitudinal axis so that its original dorsal border, now called the greater curvature, comes to lie on the left side of the body (Fig. 15–8A). The spleen, which lies in the dorsal mesentery, is also pulled to the left side. The mesogaster becomes disproportionately elongated to form a double-walled sac, the **greater omentum,** or **omental bursa,** which grows caudad. In some species of mammals, the greater omentum forms a large apron that covers much of the intestinal region. Its wall contains a great deal of fat, which is a fuel reserve for the body and probably

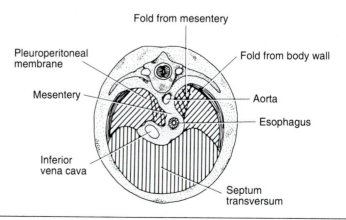

FIGURE 15–7   A diagram of the mammalian diaphragm to show the membranes that contribute to its formation. (After Corliss.)

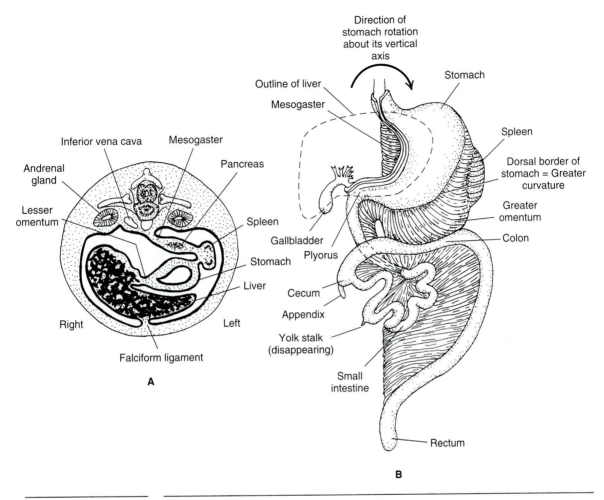

FIGURE 15–8 Rotations of the stomach and intestine during mammalian development, and the formation of the omental bursa. **A,** A transverse section through the peritoneal cavity. **B,** A ventral view of the abdominal viscera. (After Corliss.)

important insulation. Recall that most of the body heat of endotherms is generated by the visceral organs (p. 90).

Simultaneously with rotating about its longitudinal axis, the stomach rotates clockwise (as seen from its ventral surface) about its vertical axis so that its greater curvature and omental bursa shift caudad (Fig. 15–8**B**). This pulls the pylorus craniad to the right side of the body. Fusion of the expanding liver to the dorsal part of the diaphragm and adjacent body wall reduces the originally wide communication between the cavity in the omental bursa and the rest of the peritoneal cavity to a small **epiploic foramen.**

The intestine elongates faster than the rest of the body, so it forms a series of coils (Fig. 15–8**B**). It also rotates about its vertical axis so that part of the colon is pulled cranially toward the stomach, and the colon loops around the small intestine. The coiling and rotation allow the long intestine to be accommodated in the relatively short peritoneal cavity.

SUMMARY

1. The definitive coelom is a space lined with epithelium in the lateral plate mesoderm. It surrounds the visceral organs and allows their functional movements.

2. Mesenteries consist of a double layer of coelomic epithelium. These membranes hold the visceral organs in place and provide routes for the passage of ducts, blood vessels, and nerves from the body wall to the organs and between the organs.

3. Early in development a left and a right coelom are separated from each other by continuous dorsal and ventral mesenteries. Much of the ventral mesentery later disappears caudal and cranial to the liver.

4. The coelom of a fish becomes divided into an anterior pericardial cavity and a posterior pleuroperitoneal cavity by a transverse septum, through which blood vessels enter the heart.

5. The heart and pericardial cavity of a primitive tetrapod shift caudally, from a position beneath the gills closer to the lungs. As a consequence, the pericardial cavity underlies the cranial part of the pleuroperitoneal cavity, and the transverse septum assumes an oblique orientation.

6. In some reptiles and in birds and mammals, folds separate the portions of the pleuroperitoneal cavity containing the lungs from the part containing the other visceral organs. The coelom of these tetrapods is thus divided into two pleural cavities—a peritoneal cavity and a pericardial cavity.

7. In a mammal the pleuroperitoneal folds and the ventral part of the transverse septum are invaded by somatic musculature of cervical origin and form the diaphragm.

8. The pleural cavities of mammals extend ventrally during development, separating the pericardial cavity from the body wall. The area between the pleural cavities, which contains the pericardial cavity and other organs, is known as the mediastinum.

9. The part of the mesogaster supporting the stomach of a mammal forms a large coelomic sac known as the greater omentum. Considerable fat may be stored in the omental wall.

10. Rotations of the stomach and intestine during development allow the relatively short peritoneal cavity to accommodate a large stomach and long intestine.

REFERENCES

References for Chapter 15 are included with those for Chapters 16 and 17 at the end of Chapter 17, p. 630.

16 | The Digestive System: Oral Cavity and Feeding Mechanisms

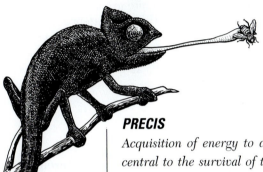

PRECIS

Acquisition of energy to drive metabolism and other physiological processes is central to the survival of the individual. Competition for food resources is often intense in the animal's natural surroundings. Increased efficiency of the acquisition and processing of food is at a premium in natural selection. Many adaptations in the feeding mechanisms of vertebrates are correlated with their relative efficiency in competitive interactions. Many animals are predacious. Other factors such as elusiveness and numerous defense strategies of the prey may also have played a role in the evolution of various adaptive features in the vertebrate feeding apparatus.

Outline

For most vertebrates, metabolic rate and all life processes depend on regular ingestion of food items, which in many cases must be mechanically broken down in the oral cavity prior to chemical digestion. Thus, food is taken into the mouth **(ingestion),** processed **(mastication),** and then swallowed **(deglutition).** Chewing, or mastication, serves two functions: (1) food or prey items are mechanically broken down to a condition suitable for swallowing; and (2) the resulting increase in surface area of the masticated food facilitates the penetration of digestive enzymes and so increases the rate of chemical breakdown. Foods with resistant cell walls, such as leaves, stems, and grasses, require extensive mechanical maceration if the digestive enzymes are going to be effective. Vertebrates have evolved many different designs with which a vast array of food resources are exploited. In this chapter we provide a brief review of the functional design of the feeding apparatus of aquatic and terrestrial feeding within an evolutionary perspective.

The Development of the Digestive Tract

Most of the digestive tract develops from the embryonic **archenteron** (Gr. *arche* = beginning + *enteron* = gut), which is continuous in many vertebrates with a yolk sac (Fig. 16–1**A**). The endoderm of the archenteron forms only the lining of the digestive tract and its various derivatives. The derivatives are the lining of the respiratory passages, which evaginate near the front of the archenteron, and the secretory cells of the liver, pancreas, and other glandular outgrowths of the archenteron. Connective tissues and muscles in the walls of those organs located in the body cavity, and the coelomic epithelium covering them, develop from the adjacent splanchnic layer of the lateral plate mesoderm. In the head and branchial region, the muscles are of somitic and somitomeric origin and the connective tissue comes from neural crest cells, as do the visceral arches (p. 127). At first the archenteron has a broad connection with the yolk sac, but the formation of head and tail folds and lateral body folds gradually separates the embryo from the yolk sac. The archenteron becomes differentiated into a **foregut** extending toward the head, a **hindgut** extending toward the tail, and a **midgut** that remains connected to the yolk sac.

As the embryo continues to take shape and separate from the underlying yolk mass, the digestive tract becomes tubular and lengthens (Fig. 16–1**B**). The pharynx differentiates from the anterior part of the archenteron. A series of lateral diverticula from the pharynx become the pharyngeal pouches. The first of these forms the auditory tube and middle ear cavity of a tetrapod that has these structures (p. 421). Epithelial buds from certain pouches form the parathyroid gland and thymus. The thyroid gland is an evagination from the rostral part of the pharynx floor; the lungs and respiratory passages are evaginations from the caudal part of the floor. We will return to these pharyngeal derivatives later. An esophagus connects the pharynx with the developing stomach, and the intestine follows the stomach. The liver develops as a ventral evagination of the archenteron just caudal to the stomach and heart, and the pancreas develops from one or more outgrowths with or near the liver evagination. A urinary bladder grows out from the floor of the

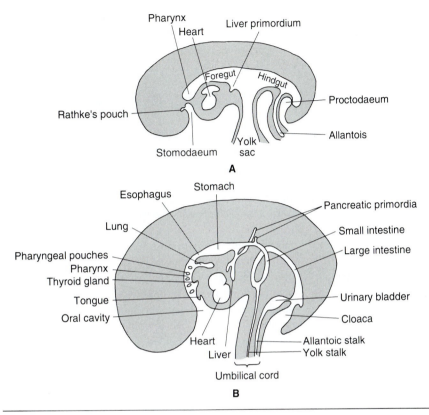

FIGURE 16–1   Lateral views of the development of the digestive tract of a terrestrial vertebrate: **A,** An early stage, in which the archenteron has differentiated into a foregut and hindgut but a yolk sac is still present. **B,** A later stage, in which gut regions are differentiating.

hindgut in most tetrapods. In amniote embryos this diverticulum enlarges as the allantois, one of the extraembryonic membranes (p. 132).

An ectodermal pocket, the **stomodaeum** (Gr. *stoma* = mouth + *hodaion* = way), invaginates at the front of the embryo and extends toward the archenteron. It forms the oral or buccal cavity. When the oral plate between the stomodaeum and the archenteron breaks down, the oral cavity and pharynx become continuous. A similar ectodermal **proctodaeum** (Gr. *proktos* = anus) invaginates at the caudal end of the embryo. It is separated from the archenteron for a short period by the cloacal membrane. When this breaks down, the proctodaeum and caudal end of the archenteron form a chamber called the **cloaca** (L. *cloaca* = sewer), which receives the terminations of the digestive, urinary, and reproductive tracts.

The Mouth and Oral Cavity

The mouth opening and the **oral** or **buccal cavity** are variable parts of the digestive tract because vertebrates gather and ingest many kinds of food in numerous ways. Even the position of the mouth opening is not the same (Fig. 16–2). The

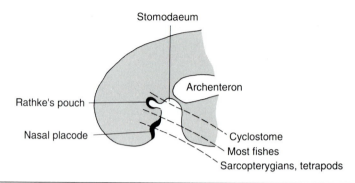

FIGURE 16–2   The stomodaeum and associated structures in a vertebrate embryo, indicating the level at which the mouth opening develops in different vertebrate groups.

mouth opening of a cyclostome develops near the caudal end of the stomodaeum so that Rathke's pouch, a dorsal stomodaeal evagination that contributes to the pituitary gland (p. 540), lies rostral to the mouth opening. In most fishes the mouth opening develops at a level between the hypophyseal sac and the pair of embryonic nasal placodes. Both the entrance and the exit from each nasal sac thus come to lie rostral to the mouth opening. In rhipidistian fishes and tetrapods the mouth opening is at the level of the nasal placodes. An external nostril leading into each nasal sac lies rostral to the mouth, and an internal nostril from each sac lies within the oral cavity.

Basic Modes of Feeding

A brief discussion of feeding methods will provide perspective, since oral structures are adapted to the way food is gathered. (Additional aspects of feeding will be examined with the teeth, jaws, and tongue.) The protochordates and the ammocoetes larva of the lamprey are jawless **suspension** or **filter feeders.** Some combination of ciliary currents, the movement of paired velar flaps at the entrance to the pharynx, and an alternate expansion and contraction of the buccopharyngeal cavity moves a current of water through the mouth and pharynx. Suspended food particles in the water are entrapped by mucus and carried back into the esophagus. Methods of entrapment vary. In the ammocoetes larva mucus is secreted by goblet cells in the laterally placed pharyngeal pouches, which contain the gills. Tributary strands of mucus from each pouch move into the pharynx lumen and unite to form a longitudinal mucous cord to which food particles adhere. This cord is carried caudally by ciliary action. The **endostyle** in the floor of the pharynx is not the source of the mucus, as formerly believed, but may contribute digestive enzymes. Ancestral vertebrates lacked jaws, but their increased activity in comparison with protochordates, and their better developed sensory apparatus, may have allowed them to be predators of small, soft-bodied animals. Some ostracoderms had bony plates bearing denticles on the borders of their mouths. These may represent the evolutionary beginning of teeth.

The evolution of teeth and jaws enabled vertebrates to feed in other ways, grasping and macerating larger prey. Many jawed fishes, however, have reverted to filter feeding.

Most fishes are **suction feeders.** They capture prey by drawing the water with the prey into the mouth cavity. However, some fishes are **ram feeders.** They simply overtake the prey with their wide-open mouth. Many primitive terrestrial vertebrates transport food within the mouth by rapidly advancing the head relative to the food, the inertia of which keeps it stationary. Gans (1969) has called this **inertial feeding,** for the inertia of the food carries it back during a succession of rapid forward darts of the head. Other terrestrial vertebrates use their tongue in food transport. Finally, most mammals and some reptiles chew or **masticate** their food.

Teeth

Living jawless vertebrates lack bony teeth, but horny cones called teeth are associated with the mouth and tongue of adult cyclostomes. Enamel-like proteins have been demonstrated in the horny teeth of hagfishes. Horny teeth are part of the lamprey's highly specialized method of feeding in which the animal clings to its prey, rasps its flesh, and sucks in the blood and other body fluids.

Bony teeth evolved with jaws and are present in some part of the oral cavity of gnathostomes, unless they have been secondarily lost. A representative adult tooth is composed of bonelike dentine covered on the exposed surface by a layer of hard enamel (Fig. 16–3**A**). The structure of these materials is similar to that in bony scales (p. 151). The blood vessels and nerves that maintain the tooth enter its base and lie in the **pulp cavity.** The entrance into the pulp cavity is wide during tooth development and growth but becomes relatively narrow in a mature tooth. The part of the tooth above the gum and subject to wear is its **crown;** that embedded in the gum and sometimes in the jaw is its **root.**

The development of the dentine and enamel of teeth is the same as their development in bony scales. Indeed, teeth evolved from the denticular parts of scales (Fig. 5–6**A**) as jaws began to form. The first indication of tooth formation in the embryo is the ingrowth into the gum of a longitudinal ridge of epithelial cells called the **dental lamina** (Fig. 16–3**B**). Underlying **dental papillae** push into the dental lamina and interact with the epidermal cells to form a series of **tooth germs.** The mesenchyme within these papillae is of neural crest origin. The dental lamina soon atrophies, but tooth germs remain in the jaws throughout the life of most vertebrates and during the embryonic period of mammals. The individual teeth develop from tooth buds that are derived secondarily from the tooth germs (Fig. 16–3**C**). The portion of the tooth bud derived from the epidermal dental lamina forms a cap-shaped **enamel organ** that overlies the dental papilla. The central cells of the enamel organ secrete glycoproteins into the intercellular spaces and separate to form a diffuse reticulum of star-shaped cells, but the surface of the enamel organ remains epithelial in nature. The inner layer of the epithelium differentiates as column-shaped **ameloblasts,** which lay down enamel. The adjacent cells of the dental papilla form columnar **odontoblasts,** which produce the dentine.

In fishes teeth may be distributed throughout the oral cavity and pharynx——

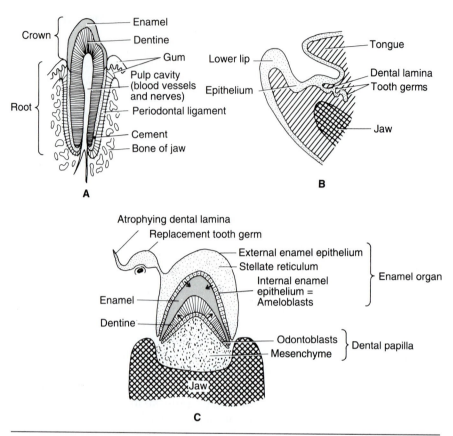

FIGURE 16–3 Tooth structure and development: **A,** A mature mammalian tooth. **B,** An early stage in tooth development. **C,** A later stage, in which the tooth bud is producing enamel and dentine.

on the jaws, palate, tongue, and some of the branchial arches. Tetrapod teeth are usually limited to the jaw margins and sometimes the palate. They are attached to underlying structures by connective tissue fibers which form a **periodontal ligament** and sometimes also by cement. **Cement** is an acellular and avascular type of bone. As can be seen in Figure 16-4, the root is frequently attached loosely to the top or side of the jaw margin. A superficial attachment to the jaw is called **acrodont** or **pleurodont.** Acrodont teeth differ from pleurodont teeth-

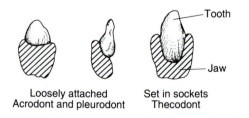

FIGURE 16–4 Vertical sections through the teeth and lower jaws to show methods of tooth attachment.

primarily in being attached to the top or inside edge of the jaw rather than to the outside edge, and having a more limited replacement. Teeth may also be set within deep sockets in the jaw, an attachment termed **thecodont** (Gr. *theke* = case), which enables the teeth to withstand strong forces. The thecodont teeth of late synapsids and mammals, together with many other changes in oral structures, are associated with the evolution of increased activity. More food must be ingested, and it must be processed more rapidly. Food is therefore masticated to increase the surface area available for the action of the digestive enzymes.

In most vertebrates new teeth can develop from the retained tooth germs throughout the animal's life as older teeth fall out or wear out. Such vertebrates are described as **polyphyodont** (Gr. *polyphyes* = manifold + *odont-* = tooth). A new tooth begins to form before an old one is lost. As the new tooth matures, the root of the old one is reabsorbed. The tooth loosens and eventually falls out. Loss and replacement are not random processes but follow a complex cycle. Waves of activation inducing new tooth formation travel slowly along the jaws from anterior to posterior. Successive waves of replacement, and sometimes overlapping waves, follow each other closely. Thus, at any one time there are areas where old teeth have been lost, where newly formed ones are just coming in, and where fully formed teeth are present. Empty sockets tend to be flanked by fully formed or newly erupted teeth. The mechanism ensures that half or more of the teeth are always functional. Not all vertebrates are polyphyodont. Most mammals are **diphyodont** and have only two sets: milk (or deciduous) and permanent. The toothed whales are **monophyodont** and have only a single set.

Although tooth size may vary, all the teeth of most fishes and most amphibians and reptiles have a similar shape, a condition termed **homodont.** The shape depends on how the teeth are used (Fig. 16–5). In most fishes and early tetrapods they are simple cones, for they function primarily to prevent caught prey from escaping. Shark teeth are often triangular with sharp and sometimes serrated edges, for sharks slice their prey into large chunks. Some shark species have broad and strong teeth, while in others the teeth are hooklike. The fishes and the few early tetrapods that feed on shellfish and plants have flattened teeth that may fuse to form crushing tooth plates. Snakes that prey on small mammals that are capable of biting back have evolved methods of prey immobilization. One method utilizes large grooved or hollow teeth—fangs—to inject the victim with a poison secreted by modified salivary glands.

Teeth have been reduced or lost where they have no value to a species' method of feeding. Frog tadpoles feed primarily on plant material. Instead of teeth, a tadpole has horny papillae, called labial teeth, around the mouth opening that help it cling to plants. Small bits of plant food are scraped up by horny beaks at the front of the jaws and filtered from the respiratory current within the buccopharyngeal cavity. Carnivorous larval salamanders, in contrast, are suction feeders and have small teeth to hold their prey. Many adult terrestrial frogs and salamanders lack teeth on the jaw margins. Insect prey is caught by a flick of the tongue and held in the mouth by palatal teeth. A horny sheath on the jaw margins replaces teeth in turtles and contemporary birds.

The ingestion and mastication of food by mammals has been accompanied by the evolution of teeth specialized for different functions. Such a dentition with

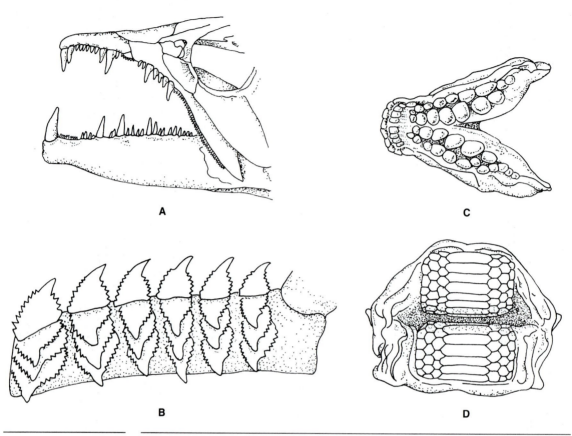

A

C

B

D

FIGURE 16–5 Examples of different types of teeth in various homodont vertebrates: **A,** Conical teeth of a pikelike characin (an order of South American and African fishes). **B,** Serrated functional and replacement teeth of a shark. **C,** Peglike teeth of a porgy (a perciform fish). **D,** Crushing tooth plates of a ray. (**A** after Gregory. **C** and **D,** after Lagler, Bardach, and Miller.)

teeth that differ in shape is said to be **heterodont** (Fig. 16–6). Primitive placental mammals have three small **incisor** teeth (abbreviated I) at the rostral end of each side of the upper and lower jaws. These are followed by a single **canine** (C), four **premolars** (Pm), and three **molars** (M). The number of each type of tooth can be expressed as a dental formula in which the numerator indicates the number on each side of the upper jaw and the denominator, the number in a lower jaw: I 3/3, C 1/1, Pm 4/4, M 3/3. The number tends to be reduced in more advanced placentals. Our dental formula is I 2/2, C 1/1, Pm 2/2, M 3/3. Other dental formulas characterize marsupials and extinct groups of early mammals.

In a contemporary mammal, teeth begin to emerge as the suckling period ends. As the jaw grows larger, deciduous incisors, canines, and premolars emerge. As the jaws continue to grow, these teeth are replaced by larger, permanent ones. The good occlusion that is needed for mastication is maintained. Later in life the molar teeth erupt in sequence, in our case at approximately 6, 12, and 18 years of age. The molars are not replaced, for final jaw length is attained by the time the

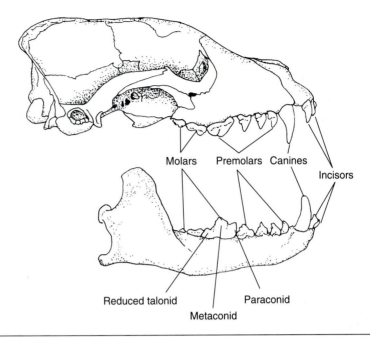

Molars   Premolars Canines

Incisors

Reduced talonid   Paraconid

Metaconid

FIGURE 16–6   A lateral view of the skull of the African hunting dog, *Lycaon*, to show the types of teeth in eutherian mammals. This dog has lost the last upper molar. The last upper premolar and the first lower molar are specialized in most carnivores as shearing carnassial teeth. (After Vaughan.)

last one erupts. Precise occlusion and limited tooth replacement are among the diagnostic features of mammals.

The configurations of particular teeth are adapted to their use. Incisors are typically small, spade-shaped teeth used for cutting, cropping and picking up food. Rodents have only two pairs of greatly enlarged incisors specialized for gnawing plant food. The pulp cavity of each one remains open and the tooth grows throughout life at the same rate that it wears away at the tip. Enamel is limited to the front surface and forms a sharp cutting edge as the softer dentine behind it wears away more rapidly. The tusks of elephants and spiraled tusks of narwhals are examples of highly modified incisors.

Canines usually are large, conical teeth used in seizing, piercing, and killing prey. They are particularly large in carnivores (Fig. 16–6). The canines of some male pigs are modified as defensive tusks. The crowns of human canine teeth resemble and function as incisors, but the roots are canine-like and much larger than those of incisors.

Primitively, the premolars are puncturing teeth, and the molars have a combined cutting and crushing action. These teeth, collectively called the **cheek teeth,** are the most complex and variable. Except for the first premolar, each has two or more roots, and the crown bears a complex pattern of conical cusps connected by sharp crests. The molar of an ancestral triconodont had one large, central cusp flanked by a pair of smaller ones (Fig 16–7**A**). In triconodonts the cusps were

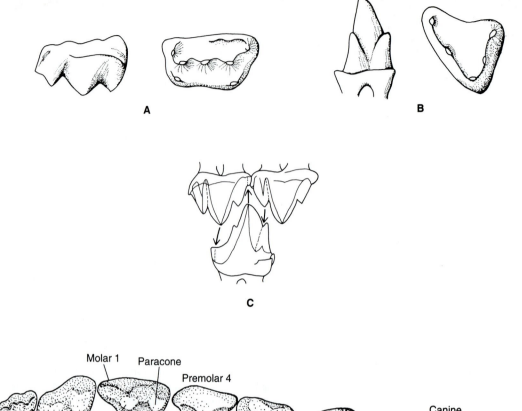

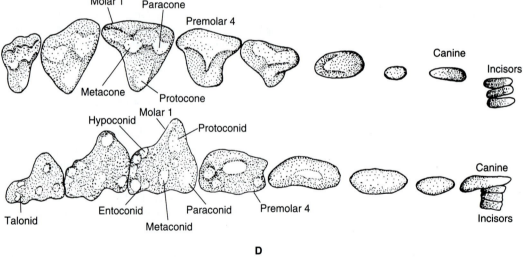

A

B

C

Molar 1 Paracone

Premolar 4

Metacone Protocone

Canine

Incisors

Hypoconid Molar 1

Protoconid

Canine

Entoconid Paraconid Premolar 4

Talonid Metaconid

Incisors

D

FIGURE 16–7   The evolution of mammalian molar teeth: **A,** The upper molar of a triconodont; lateral view (left), occlusal view (right). **B,** The molars of an early therian (a symmetrodont); lateral view of lower left molar (left), occlusal view of upper right molar (right). **C,** The occlusion of upper and lower molars of an ancestral therian. Shearing surfaces are outlined, and the way the cusps pass each other is indicated by arrows and dotted lines. **D,** The tribosphenic molars of an insectivore. The right upper teeth are shown in the upper row, and the left lower teeth in the bottom row. (**A** and **B** after Romer and Parsons, **C** after Crompton.)

in a linear sequence, but in early therian mammals, the symmetrodonts, the cusps had shifted to form a triangle with the primary cusp at the apex (Fig. 16–7**B**). The apex of the upper molar was directed lingually; that of the lower molar turned labially. The teeth were spaced such that a lower molar fit between two upper ones, and their crests formed good shearing surfaces as the teeth slid past each other (Fig. 16–7**C**). The triangular configuration, called a **trigon** in the upper molar and a **trigonid** in the lower one, increased the number of shearing facets compared with the more primitive, linear arrangement.

An additional cone, the **protocone,** evolved on the lingual apex of the upper molar in the ancestors of marsupials and placental mammals. Two of the original three cusps remained near the base of the trigon as the **paracone** (rostrally) and **metacone** (caudally) (Fig. 16–7**D**). Small accessory cones may be present. The original three cusps of the lower molar remained as the **protoconid, paraconid, and metaconid;** a low heel, known as the **talonid,** evolved from a ridge on the caudal border of the trigonid. The central part of the talonid formed a shallow basin that was flanked by small cusps: an **entoconid** on the lingual side, an **ecto-conid** on the labial side. During tooth occlusion, the protocone of the upper molar fell on the talonid basin, thereby providing some crushing action that supplemented the shearing action as the crests of the trigon and trigonid slid past each other. The molars of these mammals are called **tribosphenic.** Tribosphenic molars still occur in primitive insectivorous eutherians but became modified in many derived orders as diets changed.

The shearing action of the cheek teeth is accentuated in carnivores (Fig. 16–6). In contemporary species the fourth upper premolar and the first lower molar are specialized as cutting **carnassials** (L. *carnis* = flesh). The cusps of the premolar are linearly arranged and form a good shearing facet on the lingual side of the tooth. This crosses the shearing surface on the labial side of the molar, which is formed by the paraconid and metaconid. The protoconid and talonid are reduced in most carnivores and have been lost in cats. Postcarnassial teeth are reduced or lost.

The crushing action of the molars is greater in omnivores and herbivores. A **hypocone** evolves on the caudolingual corner of the upper molar beside the protocone. This converts the trigonid into a square-shaped tooth (Fig. 16–8). The lower molar also becomes squared by the loss of the paraconid and the elevation of the talonid to the height of the rest of the trigonid. The cusps of both upper and lower molars become round hillocks rather than sharp cones. Crushing and grinding surfaces are thus formed. Human beings, pigs, and primitive herbivores have **bunodont molars** of this type. The overall height of the tooth does not increase, and the tooth is described as **low crowned** (Fig. 16–9**C**).

The four cusps of the bunodont molar, originally separate, and certain accessory cusps fuse to form ridges in more specialized herbivores. **Lophodont teeth** of this type occur among perissodactyls, elephants, and rodents (Fig. 16–9**A**). In artiodactyls the original four cusps remain independent but become crescent shaped. This is a **selenodont tooth** (Fig. 16–9**B**). Frequently, many of the premolars in these groups of mammals become molarized and assume the structure of molars. Sometimes premolars and molars can only be distinguished by their pattern of embryonic development (recall that deciduous premolars are replaced

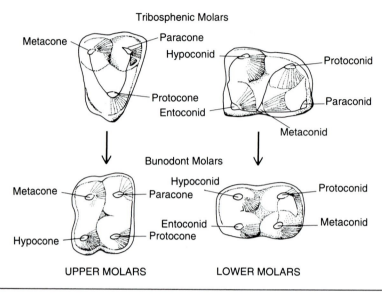

FIGURE 16–8   The evolution of bunodont molars: Top row, primitive tribosphenic molars; bottom row, bunodont molars. Right upper molars are shown on the left; left lower molars are on the right. (After Romer and Parsons.)

by permanent ones; molars are not replaced). The cheek teeth of species that feed on gritty grasses or other abrasive plant food are subject to considerable wear. Adaptations for this are an increase in the height of the ridges or crescent-shaped cusps, and a migration of cement from the root over the surface of the tooth and into the valleys between the ridges and cusps. This is a high-crowned or **hypsodont tooth** (Fig. 16–9**C**). The tooth becomes more resistant to wear, and more tooth is available to wear away. As the ridges wear away, alternating layers of cement, enamel, and dentine are exposed. Since these materials have different degrees of hardness, enamel being the hardest and dentine softer, there is differential wear. Sharp ridges of enamel become flanked on one side by cement and on the other by dentine.

The Feeding Mechanisms of Vertebrates

The study of feeding mechanisms in vertebrates has undergone great progress due to the introduction of high-speed video to determine bone movements, simultaneous electromyographic recording of muscle activity, and the use of strain gauges to permit accurate determination of deformation. These new experimental approaches have yielded a more comprehensive understanding of not only the functions but also the functional transformations of the skull and associated muscles during vertebrate evolution. We will discuss the vertebrate head from a functional perspective.

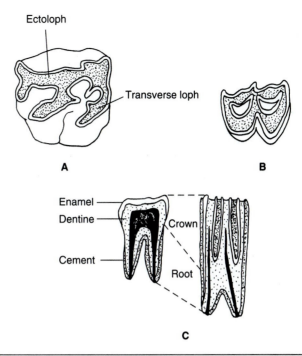

FIGURE 16–9   Some adaptations of molar teeth for grinding: **A,** Occlusal view of the lophodont molar of a rhinoceros. **B,** Occlusal view of the selenodont molar of a deer. **C,** Vertical sections through a low-crowned molar (left) and a high-crowned, or hypsodont, molar (right). (After Vaughan.)

Feeding in the Aquatic Medium: Fishes and Amphibians

The starting point of the evolution of the feeding mechanism in ray-finned fishes (Actinopterygii) is the pattern found in the primitive paleoniscoid fishes (Fig. 16–10) in which the mouth is opened by epaxial muscles that lift the head (16–10**D**). At the same time ventral body muscles (hypaxial) and the sternohyoideus muscle, which runs from the pectoral girdle to the hyoid, pull down the lower jaw. These actions produce a large gape but no increase in volume of the mouth cavity. It is hypothesized that these fish captured their prey by overtaking it with their widely opened mouth and a sudden forward movement. Once the prey was overtaken, the mouth was closed by a substantial set of jaw-closing muscles, the adductor mandibulae (Fig. 16–10**C**). Because the predator overtakes the prey by a sudden forward movement, this mode of feeding is called **ram feeding.**

Sharks also open their jaws by lifting the head with epaxial muscles while the mandible is pulled down by the ventral muscles, which run between the hyoid and pectoral girdle (rectus cervicis or sternohyoideus) and between the mandible and pectoral girdle (the coracomandibularis) (Fig. 16–11**A**). Some sharks can produce an enormous gape while overtaking their prey. Most will also protrude their upper jaws while simultaneously enlarging the volume of the mouth and pharynx. Such movements produce suction, during which the prey is drawn into the mouth

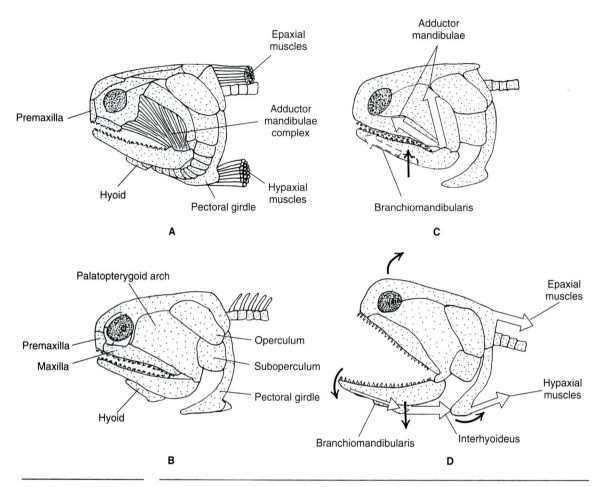

FIGURE 16–10   Lateral views of the head of a fossil primitive actinopterygian fish. **A,** With reconstructed muscles. **B,** Major components of the skull, **C,** Closing of the mouth. **D,** Opening of the mouth. Open arrows depict muscle actions. Solid arrows depict movements of skull components. (Modified after Lauder, 1982.)

cavity. Most sharks capture their prey by a combination of ram and suction feeding. Adduction of the jaw results in a formidable bite because of the very large adductor mandibulae and preorbitalis muscles (Fig. 16–11**B**), while during the recovery phase the volume of the mouth and pharynx is restored by the levator palatoquadrati, interhyoideus, and intermandibularis muscles, which pull the hyoid and palatoquadrate forward and upward. The insertion site of the adductor mandibulae in relation to the jaw joint is such that it produces a great torque around the jaw joint and therefore an optimized out-force at the teeth. The dogfish (*Squalus acanthias*) and some other sharks (especially the predacious great white shark, which is the largest carnivorous marine living fish in the world) can produce very large gapes and biting forces (Fig. 16–12) with sharp, serrated teeth.

In the actinopterygian fishes, a new skull design evolved, characterized by a

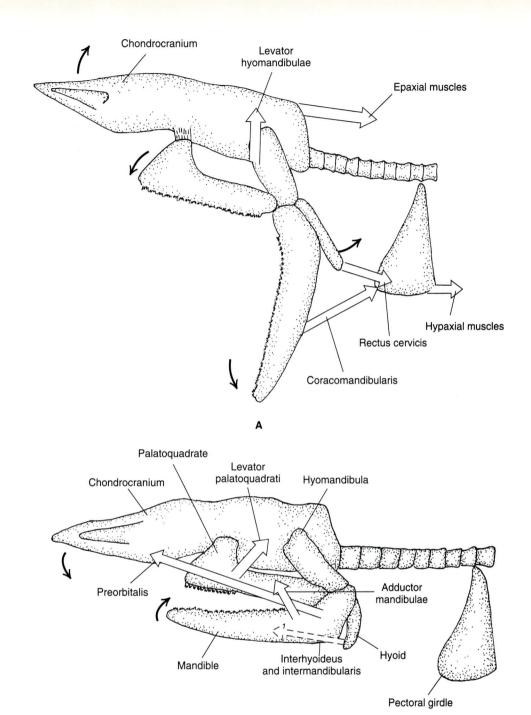

FIGURE 16–11   Lateral views of the head of a shark. Open arrows indicate muscle action. Width of arrow correspond with level of activity. Solid arrows indicate movements of skull components. **A,** Opening of mouth. **B,** Closing of mouth.

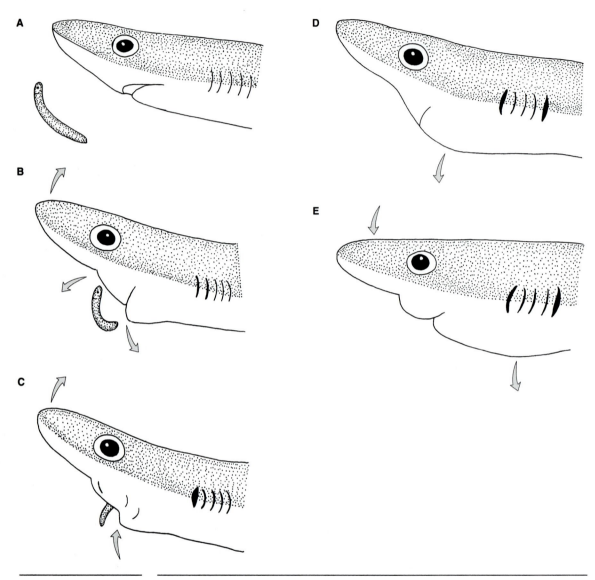

FIGURE 16–12   Tracings of a high-speed video of the dogfish (*Squalus acanthias*) feeding on a sand lance. **A,** Approaches the prey. **B,** Head lifting and jaw depression produce maximum gape. **C,** Maximum snout lift. **D,** Maximum hyoid depression. **E,** Maximum hypobranchial depression. (Unpublished, courtesy of Cheryl Wilga, 1993.)

progressively greater number and mobility of bony elements (Fig. 16–13). During the evolution of the ray-finned fishes, the following structural changes occurred (Fig. 16–13): (1) In the upper jaw the premaxilla becomes greatly enlarged, highly mobile, and the only toothed element, while the highly mobile maxilla loses its teeth and can make forward and backward swinging movements (Fig. 16–13). (2) The palatopterygoid arch, which forms the jaw suspension, changes from an

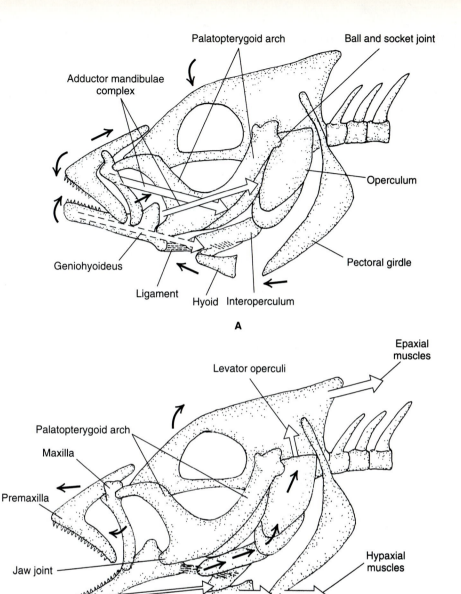

Palatopterygoid arch

Ball and socket joint

Adductor mandibulae complex

Operculum

Geniohyoideus

Pectoral girdle

Ligament

Hyoid **Interoperculum**

A

Levator operculi

Epaxial muscles

Palatopterygoid arch

Maxilla

Premaxilla

Hypaxial muscles

Jaw joint

Geniohyoideus

Sternohyoideus

B

FIGURE 16–13 Lateral views of the head of an advanced teleost fish. Open arrows indicate muscle actions. Width of arrow represents level of activity as determined by electromyography. Solid arrows indicate movements of skull components. **A,** Closing of jaws. **B,** Opening of mouth.

oblique (Fig. 16–10**A, B**) to a vertical position so that the volume of the mouth cavity is substantially increased. (3) Concomitant with the vertical orientation of the palatopterygoid arch, the jaw joint was shifted from a position posteroventral to the orbit (Fig. 16–10**B**) to an anteroventral location (Fig. 16–13). (4) The gill cover (operculum), which was originally rigidly attached to the skull (Fig. 16–10**A**), is now separate from the skull and articulates with a ball-and-socket joint to the palatopterygoid arch (Fig. 16–13). It can rotate around its joint and flare outward. Its movements are transferred to the other two bones of the gill cover (suboperculum and interoperculum) and via a ligament to the posterior corner of the mandible. (5) The hyoid becomes connected to the gill cover by a ligament, so that movement and forces of the hyoid are also indirectly transferred to the lower jaw. In this advanced design the mouth is opened by lifting of the head by action of the epaxial muscles, while simultaneous action of the hypaxial and sternohyoideus muscles pulls the hyoid backward and downward (Fig. 16–13**B**). This downward and backward movement of the hyoid is transmitted via a ligament to the interoperculum, which pulls via a ligament on the lower, posterior corner of the mandible, causing it to drop. A third mechanism to open the mouth involves the rotation of the gill cover by the action of a muscle that lifts the operculum (levator operculi). This rotation of the gill cover is again transmitted by the ligament to the posterior corner of the mandible, causing it to drop (Fig. 16–13**B**). While the lower jaw is lowered, the premaxilla slides forward. Such a movement is called jaw protrusion. At the same time the volume of the buccal cavity increases significantly because the palatopterygoid arch, which forms the sidewalls of the mouth cavity, flares out while the floor of the buccopharynx, formed by the hyoid, drops. Such a sudden increase in volume of the buccopharyngeal cavity when the mouth is opened draws water into the mouth. In this way prey is sucked into the mouth. Once the prey has entered the mouth cavity, the jaws are closed by actions of the adductor mandibulae muscles (Fig. 16–13**A**). This is followed by a recovery phase in which the hyoid is drawn forward and up to its original position by the geniohyoideus muscle. The original volume of the mouth is restored and water escapes from under the gill cover while the prey is retained.

This basic functional design with highly mobile components in the skull operated by a multitude of muscles is retained throughout the adaptive radiation of the over 21,000 species of ray-finned fishes even though numerous variations have evolved. The unsurpassed efficiency and versatility of the combination of suction and ram feeding in water have enabled these fishes to exploit any conceivable food resource from the 9000-meter depths of the abyss to the 3000-meter heights of the mountain streams. Bottom feeders have their gape directed ventrally, while surface feeders have their mouths directed upward. Fish eaters have long jaws, while algae scrapers have short jaws. The vast majority of fishes have evolved a nearly circular gape, which is the most efficient design for suction, analogous to a pipette.

Aquatic salamanders also feed by suction. Suction feeding patterns in salamanders resemble those of actinopterygian fishes. The skull is lifted by the epaxial muscles coincident with the lowering of the lower jaw by the depressor mandibulae muscle, which runs from the quadrate to insert on the articular bone of the mandible caudal to the jaw joint (Fig. 16–14). A depressor mandibulae is a new muscle,

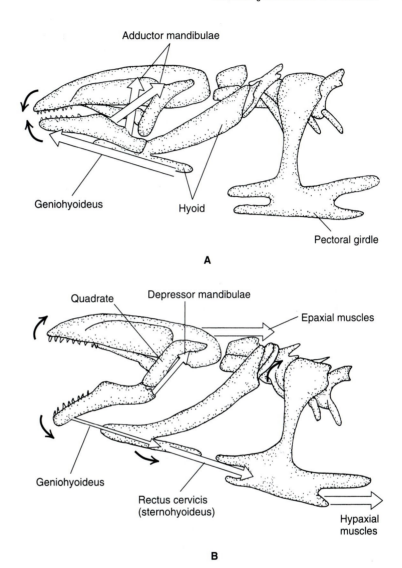

Adductor mandibulae

Geniohyoideus

Hyoid

Pectoral girdle

A

Quadrate

Depressor mandibulae

Epaxial muscles

Geniohyoideus

Rectus cervicis
(sternohyoideus)

Hypaxial
muscles

B

FIGURE 16–14   Lateral views of the skull of a salamander. Open arrows indicate muscle action. Width of
arrow corresponds with level of activity. Solid arrows indicate movements of skull components. **A,** Closing of the mouth. **B,** Opening of the mouth. (Modified from Reilly and Lauder.)

which is not present in fishes. Mouth opening is further assisted by contractions
of the hypaxial and rectus cervicis (sternohyoideus) muscles that greatly enlarge
the volume of the mouth and pharyngeal cavities (Fig. 16–14**B**). The sudden
increase in mouth and pharyngeal volume draws the water and prey into the
mouth. The mouth is then closed by the adductor mandibulae muscle complex,
which pulls up the lower jaw. As in fishes, mouth closure is followed by a recovery
phase during which the position of the hyoid is restored by the geniohyoideus
muscle (Fig. 16–14**A**) and the water is forced out of the mouth.

Swallowing in Aquatic Anamniotes

Once prey or food has been captured, it must be transported from the jaws to the pharynx and esophagus. Aquatic anamniotes use a hydraulic transport. In both fishes and aquatic amphibians, prey is transported by water currents within the mouth and pharynx. These currents are created by repeated, cyclical movements of the jaws and hyoid that closely resemble the mouth opening, closing, and recovery phases during prey capture.

In teleosts the dorsal and most posterior ventral elements of the gill arches are specialized to form toothed structures, the pharyngeal jaws, with retractor and protractor muscles. These pharyngeal jaws can rake prey and food into the gullet (esophagus) working in conjunction with hydraulic transport. In many fishes the pharyngeal jaws and muscles become specialized so they can function not only for transport of prey and food, but also for mastication prior to swallowing.

Terrestrial Feeding: Amniotes

One of the major features to change during the aquatic-to-terrestrial transition in vertebrate evolution is the feeding mechanism. Significantly different designs of the feeding apparatus are required because of the different dynamics of prey or food particles when in water or air. As we have seen, suction is almost universal in aquatic vertebrates, but it has no role in the prey capture mechanism of terrestrial tetrapods. The essential difference between aquatic and terrestrial feeding is that in land-dwelling vertebrates coordinated movements of the mandible and tongue replace water flow.

The kinematics in a lizard serve as a model of a generalized terrestrial vertebrate feeding cycle (Fig. 16–15). In all terrestrial vertebrates, including amphibians, reptiles, and birds with kinetic skulls, and mammals, which lack cranial kinesis, the feeding cycle consists of four stages: slow opening, fast opening, fast closing, and slow closing/power stroke. The cycle begins with *slow opening* of the mandible. The snout of a lizard lifts up in relation to the braincase. This movement of one part of the skull in relation to the braincase is called **cranial kinesis,** which is made possible by the presence of a transverse hinge across the skull roof. In lizards this transverse joint is located just above the posterior margin of the orbit (Fig. 16–15A). Kinesis is effected by actions of the protractor pterygoidei and levator pterygoidei muscles, which draw the pterygoid bones forward and upward relative to the braincase. Because the pterygoids are firmly articulated to the anterior components of the skull, the snout is lifted. The mandible is slightly lowered by moderate contraction of the depressor mandibulae, while the hyoid is pulled forward by the mandibulohyoideus (geniohyoideus) muscle. This stage is followed by the *fast opening* stage, during which there is a sudden and rapid opening of the mouth to maximum gape (Fig. 16–15**B**). Fast opening is caused by strong contractions of the depressor mandibulae, sternohyoideus, and mandibulohyoideus muscles. After fast opening, the *fast closing* mechanism is activated by strong contractions of the adductor mandibulae and pterygoideus muscles (Fig. 16–15**C**). The mandible is lifted and the gape decreases rapidly. The final stage is *slow closing/power stroke*, during which the snout is depressed by the strong action of

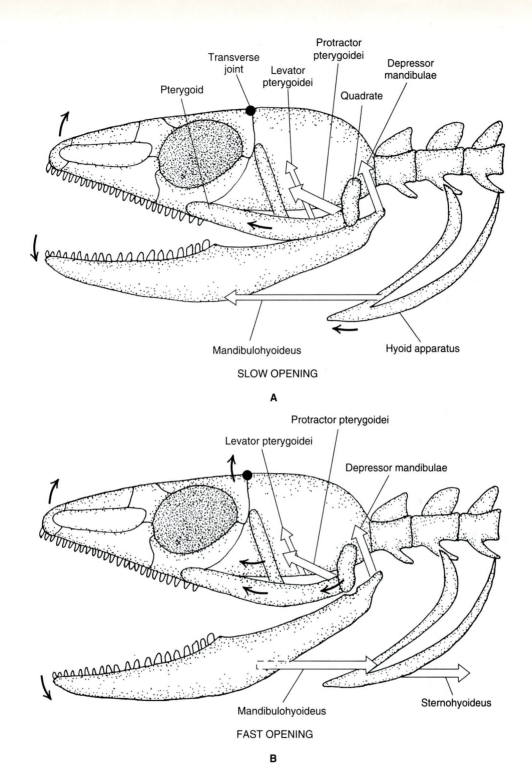

Transverse joint

Pterygoid

Levator pterygoidei

Protractor pterygoidei

Quadrate

Depressor mandibulae

Mandibulohyoideus

Hyoid apparatus

SLOW OPENING

A

Protractor pterygoidei

Levator pterygoidei

Depressor mandibulae

Mandibulohyoideus

Sternohyoideus

FAST OPENING

B

FIGURE 16–15   Lateral views of the skull of a lizard. Open arrows indicate muscle action. Width of arrow corresponds with level of activity. Solid arrows indicate movements of skull components. **A,** Slow opening. **B,** Fast opening. (*continued*)

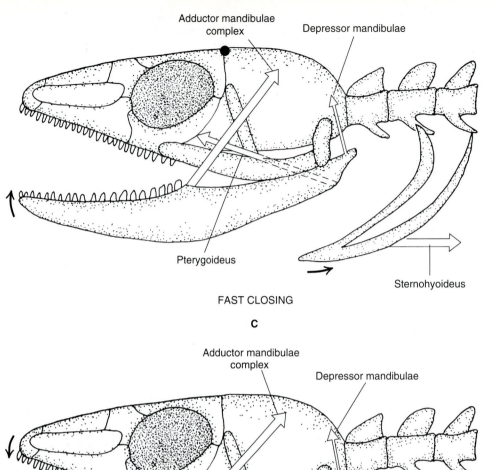

FAST CLOSING

C

SLOW CLOSING/POWER STROKE

D

FIGURE 16–15 continued
C, Fast closing. D, Slow closing/power stroke.

FOCUS 16–1 **Cranial Kinesis or Intracranial Mobility**

As we have seen, cranial kinesis is a term for the mechanism of relative motion between adjacent parts of the skull (e.g., between braincase and muzzle; Fig. 16–15**B**). Cranial kinesis is widespread among a great many vertebrates but is lost in mammals. It is a primitive feature of tetrapods and it has undergone many elaborations in a great number of evolutionary lineages. In vipers and pit vipers (Focus Figures **A** and **B**) cranial kinesis is most extremely expressed. The braincase itself is solid, but the loss of the temporal roof frees the squamosal bone to swing laterally and rostrally. The quadrate articulates movably with the squamosal. The rostral ends of the two mandibles are connected by a ligament that permits them to separate. Contraction of the protractor pterygoidei and levator pterygoidei muscles will pull the palatopterygoid forward and up. The highly specialized fang-bearing maxilla is pushed forward and rotates, thereby erecting the fang. At the same time the squamosal moves upward and the quadrate swings anteriorly and laterally while the mandible is lowered by the depressor mandibulae, producing an enormous gape. Closing of the mouth results in depression and a retraction of the palatopterygoid arch. During swallowing the palatopterygoid arch is alternately protracted and retracted, with the effect of a ratchet and advancing the large prey into the esophagus. Each side can be protracted and retracted independently. Cranial kinesis enables snakes to swallow whole prey of enormous size after it has been immobilized by constriction or by injecting poison.

Kinetic skulls also reach a high degree of development during the spectacular adaptive radiation of all modern birds. Functionally the avian skull consists of four units: (1) the braincase, (2) the bony palate and the quadrate, (3) the upper jaw, and (4) the mandible (Focus Figures **C** and **D**). The quadrate allows anteroposterior movements. The bony palate also slides back and forth. A hinge joint is present at the frontonasal junction. The protractor pterygoidei activates kinesis by imparting a forward push to the base of the upper jaw. This causes the bill to rotate upward about the transverse hinge joint at the frontonasal junction. The depressor mandibulae lowers the mandible. Closing of the mandible and lowering of the upper jaw is effected by the pterygoideus and adductor mandibulae muscles.

It is hypothesized that cranial kinesis in birds enhances the manipulatory repertoire of the jaws. A versatile and precise grip is especially valuable in birds for feeding by opening seeds and shucking out the edible parts, nest building, preening, etc. It also offers shock absorption in birds such as woodpeckers. Kinesis also allows the bird to maintain a steady line of sight while feeding.

continued

the pterygoideus muscle, which pulls the pterygoid bones caudally and downward (Fig. 16–15**D**). At the same time continuous strong contraction of the adductor mandibulae complex causes a strong bite against the prey or food. In some reptiles the power stroke includes cutting, slicing, and crushing of the prey. Cranial kinesis in snakes and birds is discussed in Focus 16–1.

The four stages in the feeding cycle are also found in all mammals studied, including humans. *Slow opening* begins with actions of both the anterior and posterior bellies of the digastric muscle (Fig. 16–16**A**). The role of the digastric in jaw opening is a new feature characteristic of mammals. It replaced the older system of jaw opening in reptiles and amphibians in which the depressor mandibulae muscle pulls on a process of the articular bone of the mandible. In mammals

FOCUS 16-1   *continued*

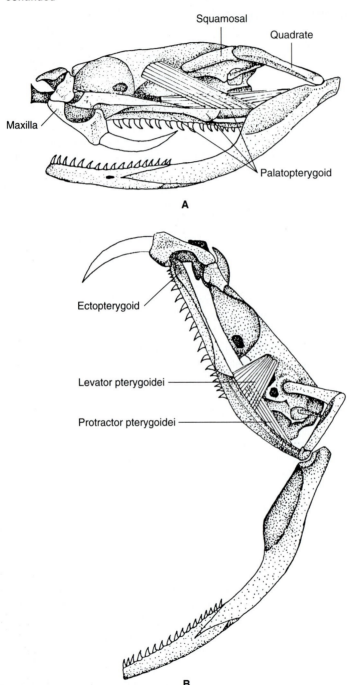

Squamosal

Quadrate

Maxilla

Palatopterygoid

A

Ectopterygoid

Levator pterygoidei

Protractor pterygoidei

B

Cranial kinesis in a rattlesnake (**A** and **B**) and a bird (**C** and **D**).

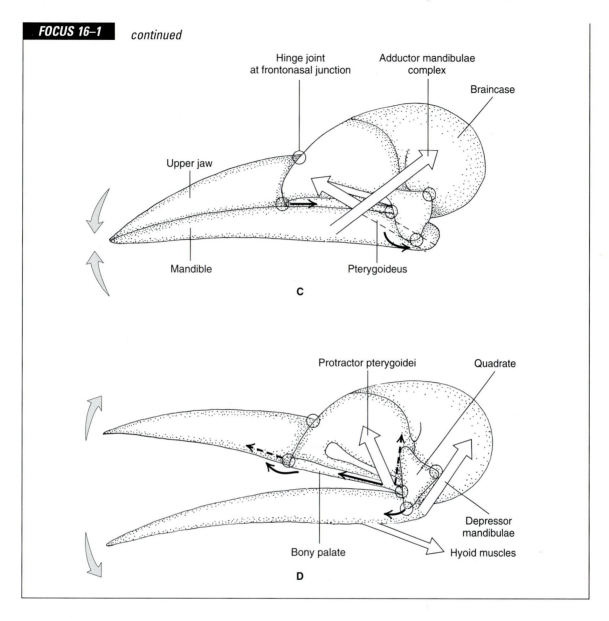

the articular bone is transformed into an auditory ossicle. Slow opening is primarily effected by the digastric. The hyoid moves forward because of contraction of the geniohyoideus. As in reptiles, slow opening is followed by *fast opening*, in which the gape increases rapidly and the sternohyoideus and geniohyoideus muscles become active, pulling down the mandible over a large distance (Fig. 16–16**B**). This action is made possible because the hyoid is stabilized by simultaneous contractions of the anterior and posterior bellies of the digastric. In the next stage (*fast closing*), the mandible is lifted rapidly over a great distance by the contractions

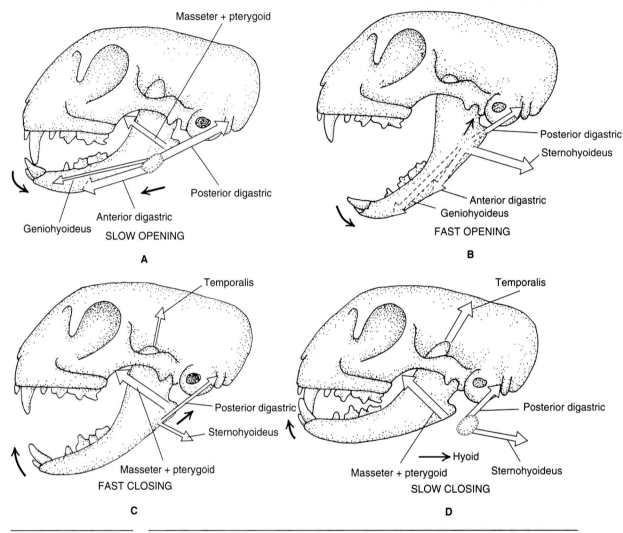

FIGURE 16–16   Lateral views of the skull of a cat. Open arrows indicate muscle action. Width of arrow corresponds with level of activity. Solid arrows indicate movements of skull components. **A,** Slow opening. **B,** Fast opening. **C,** Fast closing. **D,** Slow closing/power stroke.

of the masseter and pterygoideus muscles (Fig. 16–16**C**). The final stage, *slow closing/power stroke*, is characterized by strong contractions of the masseter, the pterygoideus, and temporalis muscles, while the hyoid moves backward (Fig. 16–16**D**). It is during this stage that food is cut into smaller pieces.

Mastication requires a precise positioning of the lower teeth relative to the upper ones, and the positions may change as food is gathered and processed. A forward movement of their lower jaw enables rabbits and rodents to engage their incisor teeth when they gnaw food; then a backward movement of the lower jaw enables them to engage their cheek teeth and grind the food. Incisors and cheek teeth are not engaged at the same time. Some species of herbivores grind their food by fore and aft movements of the lower jaw, others by side-to-side move-ments, and yet others by a combination. A mammal usually chews on only one side of its mouth at a time. A carnivore moves its lower jaw laterally to engage the carnassial teeth first on one side and then on the other.

The shape and size of the jaws, the shape and position of the jaw joint, and the size and arrangement of the jaw muscles are adapted to the type of food eaten. A comparison of the jaw mechanics of a representative carnivore and herbivore affords a good example (Fig. 16–17). The jaw joint of the carnivore is on the same horizontal plane as that of the teeth. This is effective for cutting food because it brings the cheek teeth together like scissor blades. The carnassial teeth, which are nearest the joint, engage first. The mandibular joint is a hinge with a cylindrical mandibular condyle that fits within a transverse, groove-shaped mandibular fossa. The configuration of the joint restricts fore and aft movements of the lower jaw but allows it to move from side to side as well as up and down. The jaw joint of a herbivore, in contrast, lies well above the plane of the teeth. As a consequence, all the cheek teeth on one side engage concurrently, like the two surfaces of a nutcracker. The mandibular condyle and fossa form relatively flat surfaces, and this allows the freedom of movement of the lower jaw needed in grinding food.

The temporal muscle is large in carnivores, and constitutes more than half the adductor mass. It arises in a large temporal fossa and inserts on a long coronoid process of the mandible. This configuration gives the temporalis a long lever arm and hence a good mechanical advantage. The lever arm is the perpendicular dis-tance between the muscle's line of action and the fulcrum (jaw joint) about which it acts (Fig. 16–17**A, B,** and **C**). The vertical component of the muscle's line of action provides a powerful bite force, and the horizontal component pulls the mandible backward into the mandibular fossa and maintains the integrity of the joint. This backward pull is particularly important for a carnivore, for it compen-sates for the large forward force on the front of the jaw as the animal pulls on its prey with its canine teeth. The more modest masseter muscle on the lateral surface of the jaw and the pterygoid muscle on the medial surface contribute to the bite force and form a muscular sling that moves the lower jaw from side to side to engage the carnassial teeth.

In sharp contrast, the herbivore's masseter and pterygoid muscles form most of the adductor mass. The enlarged angular process of the mandible increases the length of the lever arm and mechanical advantage of these muscles (Fig. 16–17**D, E,** and **F**). The masseter usually is composed of two layers of fibers with lines of action nearly at right angles to each other. These lines of action provide strong fore and aft as well as vertical forces. The masseter and pterygoid position the teeth and provide most of the complex forces needed to grind the food. There is no strong forward force directed at the front of the jaws. The temporal muscle, temporal fossa, and coronoid process are small.

The Palate

The roof of the oral cavity is called the palate. In aquatic vertebrates the palate is flat and smooth, while in terrestrial vertebrates it is vaulted with relief. Many fishes and early tetrapods have **palatal teeth.** A pair of internal nostrils, the **choanae,**

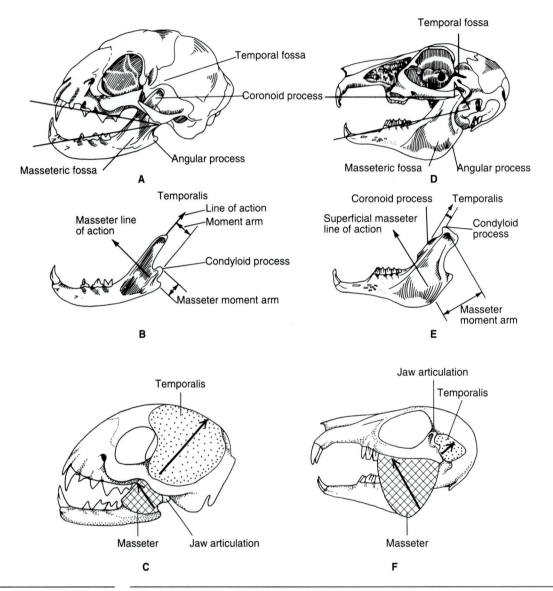

FIGURE 16–17 The jaw mechanics of a carnivore, based on a cat (**A, B,** and **C**), and a herbivore, based on a rabbit (**D, E,** and **F**). Lines through the teeth indicate the way the teeth come together as the jaws close: as in a pair of scissors (**A, B**) or as in a nutcracker (**D, E**). Location of jaw articulations and relative sizes of masseter and temporalis muscles differ significantly (**C** and **F**). (**A, B, D,** and **F** after Walker and Homberger.)

open near the rostral end of the palate in rhipidistian fishes, amphibians, reptiles, and birds (Fig. 16–18**A** and **B**). The choanae of rhipidistians are thought to be the outlets for the olfactory currents of water circulating through their nasal cavities. Those of tetrapods are part of the air passages to and from the lungs. Most tetrapods have vomeronasal organs, and the entrances to them are associated with

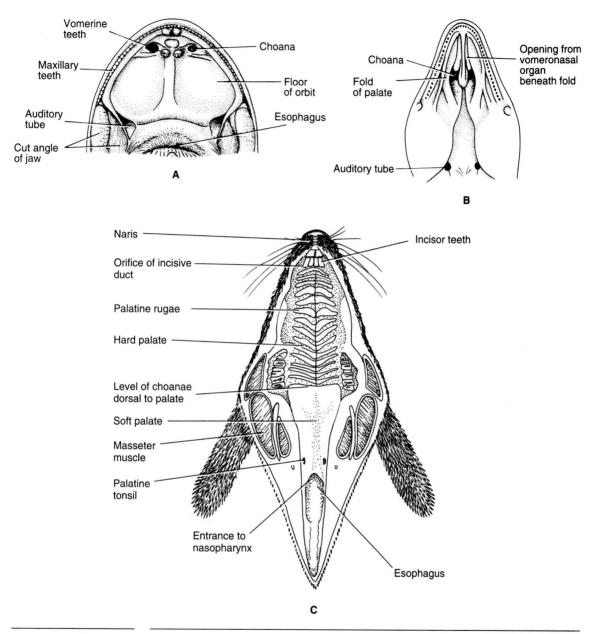

Vomerine teeth

Maxillary teeth

Auditory tube

Cut angle of jaw

Choana

Floor of orbit

Esophagus

A

Choana

Fold of palate

Opening from vomeronasal organ beneath fold

Auditory tube

B

Naris

Orifice of incisive duct

Palatine rugae

Hard palate

Level of choanae dorsal to palate

Soft palate

Masseter muscle

Palatine tonsil

Incisor teeth

Entrance to nasopharynx

Esophagus

C

FIGURE 16–18   Ventral views of the palates of representative tetrapods: **A,** A frog. **B,** A lizard. **C,** A rabbit. (**A** and **C,** after Walker. **B,** after Romer and Parsons.)

the choanae or perforate the palate independently (p. 404). In many reptiles and birds small **palatal folds** lie lateral and caudal to the choanae and form grooves that continue the air passages further caudad (Fig. 16–18**B**). A bony **hard palate** evolved from similar folds in therapsids and mammals and further separated the oral cavity and nasal passages (p. 238). A fleshy **soft palate** continues caudad from the mammalian hard palate and divides the rostral part of the pharynx into nasal and oral parts. Together, the hard and soft palate constitute the **secondary palate** (Fig. 16–18**C**). The hard palates of many mammals bear cornified, transverse palatal rugae that may help hold the food as it is chewed. The cornified baleen plates used in filter feeding by the toothless whales develop from similar ridges.

Tongue, Cheeks, and Lips

The main difference between aquatic and terrestrial feeding is the mechanism of food transport. In aquatic feeding, water currents created by movements of the bones of the skull and hyoid transport prey within the mouth and pharyngeal cavity. In contrast, terrestrial vertebrates rely on movements of the tongue. A tongue develops in the floor of the buccopharyngeal cavity in many vertebrates. A lamprey has a protrusible tongue that bears horny teeth and is used to rasp its prey's flesh. In most fishes the basihyal, a median ventral element of the hyoid, forms a tonguelike structure called the **primary tongue,** which is not muscular. In some species it bears teeth. The fish's primary tongue does not play a role in transporting or manipulating the food.

Tetrapods have evolved mobile muscular tongues that are supported by the hyoid and anterior branchial arches and invaded by hypobranchial musculature (p. 327). Only the most caudal part of the tetrapod tongue is homologous to the fish's primary tongue. The rest of the tongue of an amphibian is called the gland field, and it develops as a ventromedial elevation between the hyoid and the mandibular arches. In amniotes a pair of more rostral **lateral lingual swellings** are added to the gland field, now called the **tuberculum impar,** and primary tongue (Fig. 16–19**A**). The muscular tongue of tetrapods plays an important role in positioning the food between the teeth of upper and lower jaws and in transporting and swallowing the food. Tongue movements have been studied by X-ray motion pictures of tongues implanted with radio-opaque markers (Fig. 16–20). In both reptiles and mammals, tongue movements are correlated with movements of the hyoid and jaws in the four stages we have discussed. During slow opening the lower jaw opens slightly and the tongue and hyoid move forward and upward. There are also contractions and expansions within different parts of the muscular tongue (Fig. 16–20). First the anterior region expands and moves forward under the food. It then contracts back under the food and the tongue as a whole moves forward, holding the food against the palate. At the end of slow opening the tongue and hyoid are in their most anterior position and the tongue is under the food (Fig. 16–20**B**). During fast opening the gape increases rapidly while the tongue and hyoid are drawn posteriorly, carrying the food back in the mouth cavity (Fig. 16–20**C**). Fast closing is characterized by rapid adduction of the jaw while the tongue and hyoid continue to move backward. The cycle concludes with slow closing without a strong force being applied by the adductor muscles. The tongue and hyoid reverse their directions and move forward (Fig. 16–20**D**). A varying number of transport cycles will carry food through the oral cavity. Transport is regularly interrupted by biting cycles, during which the tongue moves forward or sideways during fast opening to place the food between the teeth for tooth contact in fast closing.

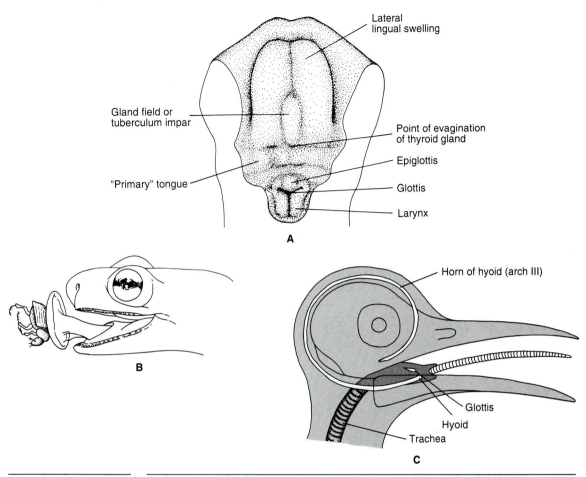

Lateral
lingual swelling

Gland field or
tuberculum impar

Point of evagination
of thyroid gland

Epiglottis

Glottis

"Primary" tongue

Larynx

A

B

Horn of hyoid (arch III)

Glottis

Hyoid

Trachea

C

FIGURE 16–19   Tongues: **A,** A dorsal view of the development of the mammalian tongue. **B,** The tongue of a plethodont salamander catching an insect. **C,** The tongue and supporting hyoid apparatus of a woodpecker. (**A,** after Corliss. **B,** modified after Noble.)

In the cat the movements of the tongue are also the result of movements of the mandible and hyoid, and movement produced within the body of the tongue itself. Tongue movements in the cat are associated with characteristic profiles of jaw and hyoid movements, as described in the four stages of the transport cycle in reptiles (Fig. 16–20), with the exception of the long duration (> 45 percent of cycle duration) of the slow opening stage. The tongue moves forward during the slow opening stage and backward during most of the rest of the transport cycle. During this extended slow opening stage, the cat's tongue expands and protrudes under the food. The extension of the tongue itself is about 60 percent longer than the retracted tongue. The forward-moving tongue with the food on it is elevated to the palatal rugae. Such contact with the rugae would hinder the forward movement of the food on the advancing tongue and would leave it more posteriorly placed. Retraction and shortening of the tongue in the other steps of the cycle

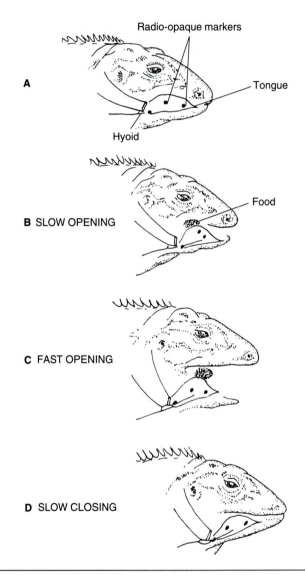

Radio-opaque markers

A

Tongue

Hyoid

B SLOW OPENING

Food

C FAST OPENING

D SLOW CLOSING

FIGURE 16–20   Food transport in the mouth with the tongue in the lizard *Ctenosaura*. Dots represent radio opaque markers in the tongue and hyoid. From a cineradiographic film. (From Smith.)

take place with the tongue away from the palate. Protraction and extension of the tongue against the palatal rugae followed by tongue retraction and shortening away from the palate form the basis of the food transport mechanism in the oral cavity of mammals (Thexton and McGarrick, 1989).

The tongue is often used in food gathering. Most frogs and salamanders and some lizards, such as the African chameleon, can rapidly project part of their tongue from the mouth and capture an insect (Fig. 16–19**B**). The tip of the tongue is covered by a sticky mucus. A woodpecker has a long and spiny tongue that is supported by an elaborate hyobranchial apparatus, part of which is housed in a bony tube that curves around the back of the skull (Fig. 16–19**C**). After a woodpecker has chiseled a hole in a tree with its bill, the long tongue is protruded to shuck out an insect or grub. Ant- and termite-eating mammals also gather their food with long, sticky tongues. Certain of the tongue muscles of the giant anteater extend as far caudad as the posterior end of the sternum. Spiny papillae on the tongues of cats and other carnivores help the animals rasp flesh from a bone.

The tongue also is used in many other ways. It bears taste buds and other receptors, so is an important sense organ. Snakes and some lizards have protrusible, forked tongues that are used in association with the vomeronasal organ to detect certain olfactory clues (p. 404). Evaporation of water from the surface of the large, moist tongues of dogs and other panting mammals helps regulate body temperature. We, of course, use our tongue in speech and in the manipulation and swallowing of food.

The fleshy lips and cheeks that characterize marsupial and eutherian mammals assist the muscular tongue in manipulating food within the mouth, and they are used by the newborn in sucking milk from the mother's nipples or teats. Monotremes have horny beaks and cannot suckle; their young lap up milk that is secreted onto hairs rather than discharged through nipples. In many species of rodents, part of each cheek evaginates beneath the facial muscles to form a pair of large cheek pouches in which food can be temporarily stored.

Oral Glands

Most fishes lack oral glands, apart from scattered mucus-secreting cells. Lampreys are a notable exception. They have a pair of large glands that secrete the anticoagulant needed to keep their prey's blood flowing freely as they feed on it. Oral glands, including multicellular salivary glands, are well developed in terrestrial vertebrates. Their mucous and serous secretions lubricate the food and facilitate food manipulation and swallowing in the terrestrial environment. Modified salivary glands of some snakes produce the hemolytic or neurotoxic poisons that are injected by fangs to immobilize their prey. The secretion of digestive enzymes is not an important function of the salivary glands of amphibians, reptiles, or birds, although a few amphibians synthesize a small amount of **salivary amylase,** a starch-splitting enzyme.

The oral cavities of mammals are particularly well equipped with scattered secretory cells and large salivary glands (Fig. 16–21). Most mammals have well-developed **parotid, mandibular,** and **sublingual glands.** Some species also have a **zygomatic gland** in the floor of the orbit, and **buccal** and **molar glands** beneath the mucous membrane of the lips. Saliva is composed primarily of serous and mucous secretions that are important for lubricating the food during mastication and swallowing. Only a few mammals, including human beings, secrete salivary amylase. Starch digestion is initiated during mastication and continues in the stomach until the high acidity of the gastric secretions stops the action of amylase. The swallowed saliva of a cow provides a medium in a chamber of the stomach, the rumen, in which bacteria multiply and cellulose is broken down (p. 616). A cow may produce more than 100 liters of saliva a day. Toxins are produced

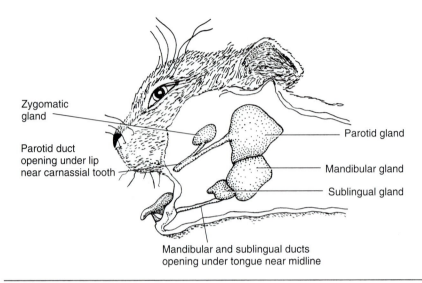

Zygomatic gland

Parotid duct opening under lip near carnassial tooth

Parotid gland

Mandibular gland

Sublingual gland

Mandibular and sublingual ducts opening under tongue near midline

FIGURE 16–21   The salivary glands of a mammal, based on those of a cat. (After Walker and Hornberger.)

by the salivary glands of some insectivores, and an anticoagulant is secreted by the salivary glands of vampire bats.

Evolution of the Vertebrate Feeding Mechanism

Major changes in the feeding apparatus and behavior have emerged during vertebrate evolution (Fig. 16–22). Reilly and Lauder (1990) have identified the following five characteristics as primitive:

1. Prey transport is effected by water currents within the oral cavity.
2. The gape remains constant prior to fast opening.
3. Mouth opening is produced by both pronounced head elevation (epaxial muscles) and lower jaw depression (sternohyoideus muscle).
4. A recovery phase follows prey ingestion.
5. Hyoid retraction by the sternohyoideus is coincident with the fast open stage.

These characteristics are present in all bony fishes and sharks. The only feature that remains constant during the evolution of all vertebrates is characteristic number 5: hyoid retraction by the sternohyoideus is coincident with fast opening. It is in the amphibians that a manipulative tongue first evolved for prey transport within the oral cavity. This feature becomes further elaborated and perfected in the amniotes. Three more features become modified in the amniotes: the chewing and prey transport cycle begins with a slow opening stage prior to fast opening, the gape is produced by lowering the mandible rather than by head lifting, and the recovery phase is no longer distinguishable. Two evolutionary innovations emerged in amniotes: inertial feeding and a four-stage masticatory cycle for food processing in the mouth. In many reptiles and birds, especially those with reduced or specialized tongues, inertial feeding takes place. Once the prey is captured, the

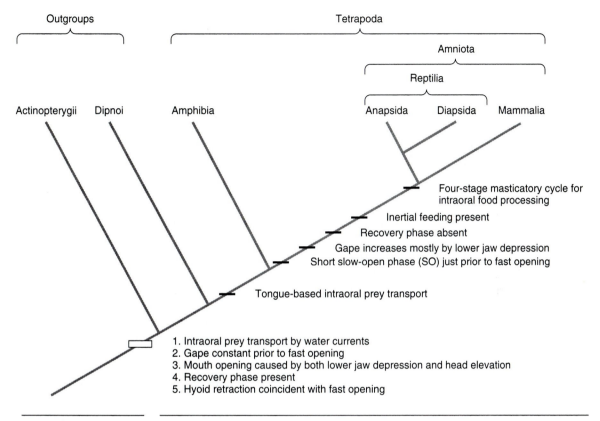

Outgroups   Tetrapoda

Amniota

Reptilia

Actinopterygii   Dipnoi   Amphibia   Anapsida   Diapsida   Mammalia

Four-stage masticatory cycle for intraoral food processing

Inertial feeding present
Recovery phase absent
Gape increases mostly by lower jaw depression
Short slow-open phase (SO) just prior to fast opening

Tongue-based intraoral prey transport

1. Intraoral prey transport by water currents
2. Gape constant prior to fast opening
3. Mouth opening caused by both lower jaw depression and head elevation
4. Recovery phase present
5. Hyoid retraction coincident with fast opening

FIGURE 16–22   Evolution of feeding patterns of vertebrates. (Modified from Reilly and Lauder.)

head is pulled rapidly backward, imparting a posteriorly directed motion to the prey (Smith, 1986). The jaws are then opened very fast and the head thrusted forward to surround the prey. These inertial thrusts are repeated until the prey reaches the esophagus. Most birds, snakes, and carnivorous lizards employ inertial feeding in which within a few tenths of a millisecond there is complete acceleration-deceleration of the skull and cervical complex.

With the emergence of the complex four-stage masticatory cycle, amniotes do not use their jaws and tongue simply to transport whole prey or large items of food to the esophagus, but they bite, align, masticate, pack, and transport the food to the esophagus. This highly efficient processing of food enhances the rate of digestion of a wide variety of foods.

SUMMARY

1. Vertebrates obtain the food and other nutrients needed to sustain their metabolism through the digestive system. Food must be ingested, sometimes broken down mechanically, transported through the mouth and swallowed, chemically digested, and residues must be egested.

2. The lining and glandular derivatives of the digestive tract develop from the endodermal archenteron; the rest of the wall develops from mesenchyme, most of which comes from splanchnic mesoderm. An ectodermal stomodaeum forms the oral cavity, and an ectodermal proctodaeum contributes to the cloaca.

3. Protochordates and larval lampreys are filter feeders. Many types of feeding mechanisms have evolved. Most fishes and aquatic tetrapods are suction feeders. Many fishes are ram feeders. In general, many primitive terrestrial vertebrates are inertial feeders, and mammals masticate their food.

4. The enamel and dentine of teeth are similar to these products on bony scales. Enamel is produced by ameloblasts of the enamel organ; dentine by odontoblasts.

5. In most vertebrates the teeth are loosely bound to the jaws or to underlying skeletal elements (acrodont and pleurodont), but where forces acting on the teeth are large, they are set in sockets (thecodont).

6. Successive waves of new tooth formation replace teeth throughout the life of most vertebrates, but in most mammals replacement is limited to a deciduous and a permanent set.

7. All the teeth have a similar shape in most vertebrates (homodont), and the shape is related to the type of food eaten and to the method of obtaining it. In late therapsids and in most mammals teeth differentiate and serve different functions (heterodont).

8. Mammalian teeth begin to emerge as the suckling period ends, and their pattern of eruption and replacement ensures the good occlusion needed for mastication.

9. Incisors typically are small teeth used for cutting, cropping, and picking up food. The canines usually are large, conical teeth used in seizing, piercing, and killing prey. Primitively, the premolars are puncturing teeth, and the molars have a combined cutting and crushing action.

10. The molars of ancestral mammals were triconodont, with three cusps in a linear sequence. A shift of the cusps to a triangular configuration in symmetrodonts increased the number of shearing facets. Ancestral marsupials and eutherians had tribosphenic molars, in which a low heel was added to the lower molars.

11. Teeth, especially the premolars and molars, became modified as mammals adapted to different diets. Contemporary carnivores have a set of shearing carnassials that evolved from the last upper premolar and first lower molar. Omnivores and primitive herbivores have low crowned, bunodont molars that typically bear four rounded cusps. The cusps form ridges (lophodont molars) or crescents (selenodont molars) in more specialized herbivores. Herbivores that feed on rough food have evolved high-crowned molars.

12. Primitive ray-finned fishes (paleoniscoids) are ram feeders. A large gape is produced by raising the head and lowering the lower jaw, and the fish overtakes its prey. The jaws are then closed.

13. Most sharks capture prey by a combination of ram and suction feeding. They can form a large gape by raising the skull and lowering the lower jaw.

14. Other ray-finned fishes have become suction feeders. Many parts of the skull are highly mobile, the size of the buccopharyngeal cavity is rapidly increased, and food is sucked in along with a current of water. Evolutionary modifications in the head for suction feeding include the following: (1) the enlargement and increased mobility of the premaxilla, accompanied by changes in the maxilla that enable it to swing backward and forward; (2) an increasing vertical orientation of the palatopterygoid arch; (3) a forward shift of the jaw joint; (4) separation of the operculum from the skull roof, and the evolution of a mobile articulation between the operculum and palatopterygoid arch; and (5) the evolution of mechanisms that transmit the movement and forces developed on the hyoid to the lower jaw.

15. Aquatic salamanders are also suction feeders. Patterns of feeding resemble those of actinopterygian fishes, except that the lower jaw is lowered by a depressor mandibulae muscle.

16. Food is hydraulically transported through the buccopharyngeal cavity of aquatic anamniotes by a current of water. The current is produced by repeated movements of the jaws and hyoid that resemble the movements used in prey capture.

17. Coordinated movements of the mandible and a muscular tongue replace water flow through the mouth of terrestrial vertebrates. The feeding cycle always involves (1) slow mouth opening, (2) rapid mouth opening, (3) slow mouth closing, and (4) rapid closing with the development of sufficient force to firmly grasp, cut, crunch, or masticate the food. These movements are accompanied by cranial kinesis in most reptiles and birds but not in mammals.

18. The fleshy cheeks and lips of marsupials and eutherians assist in food manipulation and are used by the newborn in suckling.

19. Apart from scattered mucus-producing cells, most fishes lack oral glands, but these glands are well developed in tetrapods. The salivary glands of mammals include the parotid, mandibular, sublingual, zygomatic, molar, and buccal glands. The glands of some mammals produce salivary amylase.

20. Choanae open onto the primary palate in rhipidistian fishes, amphibians, reptiles, and birds. In mammals a hard palate carries the choanae to the pharynx, and a soft palate divides the rostral part of the pharynx into nasal and oral parts.

REFERENCES

References for this chapter are combined with others on the coelom and digestive system at the end of Chapter 17, page 630.

17 | The Digestive System: Pharynx, Stomach, and Intestine

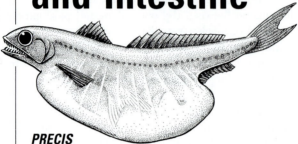

PRECIS

Vertebrates must chemically break down their food to absorb the contained nutrients that are needed to sustain their metabolism. The structure and evolution of the digestive system are affected by the type of food eaten, by the animal's level of metabolism, and by body size. Many endocrine glands develop from parts of the digestive tract.

Outline

The Pharynx and Its Derivatives
    Development
    The Thyroid Gland
    The Parathyroid Glands and Ultimobranchial Bodies
    The Thymus and Tonsils
Gut Tube Structure
The Esophagus
The Stomach
The Intestine and Cloaca
The Liver and Pancreas
    The Liver
Focus 17–1 *The Functional Anatomy of Large Mammalian Herbivores*
    The Pancreas
Gastrointestinal Hormones

Food that has been transported through the oral cavity, and is masticated there in some species, next enters the pharynx. From there it passes through the esophagus, stomach, and intestine. Digestion occurs primarily in the stomach and intestine. Many digestive enzymes acting in sequence are needed to break down the food into small molecules. These molecules must then be absorbed along with minerals and other nutrients. The large quantities of water that are released into the digestive tract as a component of the digestive secretions must be reabsorbed. Finally, the undigested residues, cells sloughed off the digestive tract lining during the passage of the food, and the bacteria that have multiplied in parts of the digestive tract must be eliminated, or **egested.** During these processes the lining of the digestive tract is subjected to wide changes in pH, to the action of powerful proteolytic enzymes, and to considerable abrasion. It must be protected and repaired.

The structure of the digestive tract and its evolution are affected by many factors, including the type of food eaten, the level of activity and metabolism of the animal, and the size of the animal. Since no vertebrate can synthesize the enzyme **cellulase** needed to digest the woody cellulose in plant food, herbivorous vertebrates must have colonies of micro-organisms in some part of their digestive tract that can synthesize cellulase. These micro-organisms may be lodged in special compartments of the stomach or intestine, or in an exceptionally long intestine. Levels of activity and metabolism are coupled to the rate at which food must be digested and absorbed. An active, endothermic animal needs a large digestive and absorptive surface area to process an increased amount of food. Regardless of other factors, small animals also need more surface area per unit of body mass than large ones because their metabolism is relatively higher.

The Pharynx and its Derivatives

Development

The **pharynx** arises from the anterior part of the archenteron and is the part of the digestive tract from which the paired endodermal **pharyngeal pouches** arise (Fig. 17–1). Jawed fishes have six pairs of pouches that extend laterally, meet ectodermal furrows, and open to the surface. The first pouch is reduced to a narrow passage, called the **spiracle,** or has been lost. The remaining ones develop into the branchial chambers that contain the gills. Because of its importance in respiration (Chapter 18), the pharynx of fishes is a large chamber. As food passes through the pharynx of a fish, it is prevented from entering the branchial chambers by a filter of **gill rakers** attached to the bases of the branchial arches.

The lungs of tetrapods develop as ventral outgrowths from the floor of the caudal part of the pharynx. In a tetrapod, the pharynx is quite short, and in the adult it is little more than a connecting segment between the oral cavity and the esophagus. Although pharyngeal pouches develop in embryonic tetrapods, the caudal ones are crowded together and reduced in number. Mammals have only four pouches, but the last one incorporates the fifth pouch. The pharyngeal pouches of tetrapods do not open to the body surface, except in larval amphibians.

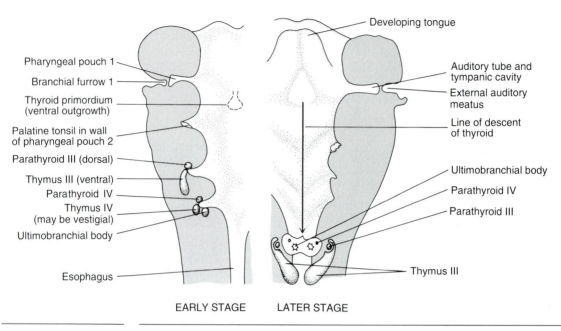

Pharyngeal pouch 1
Branchial furrow 1
Thyroid primordium (ventral outgrowth)
Palatine tonsil in wall of pharyngeal pouch 2
Parathyroid III (dorsal)
Thymus III (ventral)
Parathyroid IV
Thymus IV (may be vestigial)
Ultimobranchial body
Esophagus

Developing tongue
Auditory tube and tympanic cavity
External auditory meatus
Line of descent of thyroid
Ultimobranchial body
Parathyroid IV
Parathyroid III
Thymus III

EARLY STAGE   LATER STAGE

FIGURE 17–1   Frontal sections of the development of the mammalian pharynx and its derivatives. An early stage is shown on the left side; a later one is on the right. (After Corliss.)

Food passing through the pharynx is prevented from entering the air passages to and from the lungs by the closure of a slitlike **glottis** in the floor of the pharynx. In addition, mammals have a flaplike **epiglottis,** which lies rostral to the glottis (Fig. 18–21), and this deflects food around the glottis. Several endocrine glands and glandlike structures also develop from the pharyngeal epithelium.

The Thyroid Gland

The **thyroid** gland forms as a midventral outgrowth from the pharynx floor between the first and second pairs of pouches (Fig. 17–1). The thyroid primordium of a gnathostome becomes bilobed, loses its connection with the pharynx, and migrates a variable distance caudad. In a fish it remains near its site of origin as a median gland in the floor of the pharynx. In an amphibian it forms a pair of glands in the pharynx floor. The thyroid gland of reptiles and birds migrates far caudally, near the posterior end of the trachea, and may remain as a median organ or become paired. In mammals it is a bilobed gland that lies near the anterior end of the trachea or, in human beings, ventral to the thyroid cartilage of the larynx (whence its name).

The thyroid gland is composed of many small, nearly spherical follicles embedded in a vascularized connective tissue (Fig. 17–2). As is common in endocrine glands, many of the capillaries are **fenestrated.** That is, exceptionally thin, window-like areas in their walls facilitate the passage of larger molecules. The single

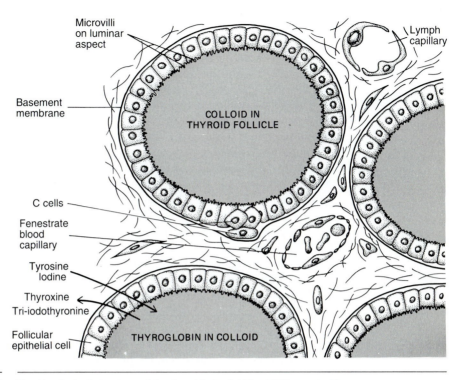

FIGURE 17–2   The histological structure of the thyroid gland. (After Williams, et al.)

layer of epithelial cells that forms the follicle wall synthesizes a **colloid** that is secreted into the lumen of the follicle and stored there. The colloid contains **thyroglobulin,** a tyrosine-rich protein bound to iodine. As thyroid hormones are needed by the body, microvilli on the lumen side of the epithelial cells elongate and envelop bits of colloid. The colloid is digested within the cells and discharged from the opposite surface as the thyroid hormones, **thyroxine** and **tri-iodothyronine.** These are composed of small groups of iodinated tyrosine. The thyroid gland is unique among endocrine glands in being able to store its hormones. Both the formation of colloid and its digestion to form the hormones can occur simultaneously under the influence of the thyrotropic hormone (TH) of the adenohypophysis (Table 14–1). Release of the thyrotropic hormone is promoted by thyrotropic releasing hormone of the hypothalamus.

It is difficult to generalize on the functions of the thyroid hormones because they affect so many processes and interact in complex ways with other hormones, but three major categories of effects have been identified in birds and mammals (not all are applicable to ectothermic vertebrates). First, the thyroid hormones accelerate oxidative metabolism throughout the body, and they are needed to maintain the high level of metabolism that characterizes endotherms. The production of these hormones decreases greatly in hibernating mammals. Production peaks when these mammals begin to emerge from hibernation and generate considerable body heat. Attempts to demonstrate a similar function for ectotherms emerging from hibernation have not been consistently successful. Second, thyroid hormones are essential for normal growth and development, presumably because of their influence on protein synthesis. A low level of these hormones in human infants causes a severe retardation of mental development known as cretinism. Thyroid hormones have a dramatic effect on amphibian development, for metamorphosis occurs only when the adenohypophysis begins to secrete thyrotropic hormone and the thyroid gland becomes active. Early metamorphosis can be induced by providing extra levels of these hormones or, in some cases, extra iodine. Many cases of neoteny in salamanders result from low levels of either thyrotropic or thyroid hormones. *Necturus,* however, does not respond to extra thyrotropic or thyroid hormones. Possibly, the cell receptors needed for a normal response are somehow blocked in this genus. Third, the normal development of the skin and its various derivatives (pigment, feathers, hair) requires thyroid hormones in a wide range of vertebrates. Low levels interfere with normal skin molting in amphibians and reptiles.

Every adult vertebrate has a thyroid gland, but the capacity to iodinate tyrosine and synthesize iodinated proteins is not confined to vertebrates. Many invertebrate groups can do so, and the iodinated proteins usually are a component of their skeletons, pharyngeal teeth, and other hard parts. The significance of this ability is not clear, for there is no clear evidence that invertebrates make use of these proteins. Protochordates synthesize these proteins in certain cells of their endostyle. They are released into the pharynx and digested and absorbed as iodinated tyrosines in the intestine. This pattern of production, release, and absorption implies a function, but its nature is not clear. The endostyle-like **subpharyngeal gland** of the ammocoetes larva of the lamprey also produces and releases iodinated proteins into the pharynx. Part of this gland transforms into the thyroid gland in adults. We see here a transformation of an exocrine gland into an endocrine gland. It appears that the vertebrate thyroid gland evolved as a means of storing iodinated proteins and digesting them in such a way that iodinated tyrosines are released directly into the blood.

The Parathyroid Glands and Ultimobranchial Bodies

Mineral metabolism is regulated in vertebrates by parathyroid glands and parafollicular or C cells. Most tetrapods have two pairs of **parathyroid glands** that develop as epithelial buds from the third and fourth pharyngeal pouches (Fig. 17–1), but only a single pair is present in some lizards, rats, and other groups. The epithelial buds usually form on the ventral surface of the pouches, but those of mammals are more dorsal. They separate from the pharyngeal pouches and come to lie near the thyroid gland or, as in many mammals, become embedded in its dorsal surface. Fishes lack parathyroid glands, but, in common with other vertebrates, they do have clusters of **parafollicular** or **C cells.** The C cells arise from the neural crest, but usually are lodged in **ultimobranchial bodies** that develop from the posterior surface of the last pair of branchial pouches. The ultimobranchial bodies are vestigial in most mammals, but the C cells lodge within the thyroid gland (Fig. 17–2).

The parathyroid glands consist of anastomosing cords of epithelial cells interspersed with vascular connective tissue. These cells synthesize the gland's hormone, **parathormone** (PTH), which contributes to the maintenance of calcium and phosphate balances. Maintenance of mineral balances requires controlling the rate of mineral absorption from the gut, its interchange between body fluids and bones, and its excretion through the kidneys. Mineral homeostasis is in a dynamic equilibrium that is affected by several hormones. The parathyroid cells of mammals and, presumably, other tetrapods monitor blood calcium level. If the level falls, a release of parathormone restores it to normal, primarily by activating osteoclasts that transfer calcium from bone to the blood, and also by reducing the excretion of calcium. Since calcium is essential for nerve and muscle metabolism, prolonged low levels cause muscle tetany and death from suffocation. An abnormal increase in blood calcium level inhibits parathormone release, and more calcium is stored in bones and excreted. Parathormone also affects phosphate levels but appears to do so indirectly, and its effects are the opposite of those on calcium.

Another hormone, **1,25-dihydroxycholecalciferol,** acts closely with parathormone. It maintains normal blood calcium levels by facilitating the synthesis of calcium-binding proteins that are essential for the movement of calcium across the intestinal mucosa during its absorption. 1,25-dihydroxycholecalciferol is a complex steroid hormone whose synthesis begins in the skin by the action of ultraviolet light on vitamin D (cholecalciferol); further metabolism in the liver and kidneys transforms it to its active form.

Calcitonin, which is secreted by the C cells in the ultimobranchial bodies or mammalian thyroid gland, also affects mineral metabolism. Its effects oppose those of parathormone. Elevated blood calcium levels caused by parathormone secretion are reduced by calcitonin because it promotes an increased deposition of calcium in bone. How fishes maintain mineral homeostasis is not entirely clear. They have no parathyroid glands, but they do possess ultimobranchial bodies, and some of the enzymes needed for the synthesis of 1,25-dihydroxycholecalciferol have been found.

The Thymus and Tonsils

A **thymus** develops in all vertebrates from the endodermal epithelium of certain pharyngeal pouches and from the adjacent ectodermal epithelium. In fishes all the pouches, or the first four, contribute to thymus formation, but in tetrapods the number is more restricted. In mammals, only the third and fourth are involved and the contribution of the third is by far the greater (Fig. 17–1). Thymic epithelium is derived from the dorsal surface of the pouches in question except in mammals, where it comes from the ventral surface. After separating from the pouches, the epithelial thymic buds are invaded by stem cells from the spleen and bone marrow. These cells differentiate into **T lymphocytes** (often simply called **T cells**), which are released, circulate in the blood, and populate lymph nodes and other lymphoid organs, where they continue to multiply and differentiate under the influence of at least one thymic hormone **(thymopoietin).** T cells, of which there are several types, play a key role in certain of the body's immune responses (p. 706). Many T cells participate in a **cell-mediated response** in which the cells directly attack virus-infected cells, fungi, and other foreign cells. Other T cells assist in the proliferation and activation of other types of lymphocytes (B lymphocytes, p. 706), and others have yet other functions. The thymus is relatively large in newborn individuals, but it continues to grow at a slower rate until puberty. Thereafter it is invaded by fat cells and begins to regress, but some thymic tissue persists and remains active into old age.

Tonsils are also lymphoid organs that participate in the body's immune responses. They develop around the wall of the anterior part of the pharynx, a position that exposes their cells to invading antigens. The paired **palatine tonsils** of mammals begin their development as epithelial derivatives of the second pair of pharyngeal pouches (Fig. 17–1), and they, too, are invaded by lymphocytes.

Gut Tube Structure

The postpharyngeal digestive tract is a tubular structure, parts of which are often enlarged to form saccular chambers. The lining of the gut tube is a stratified squamous epithelium in those parts subject to considerable abrasion: oral cavity, pharynx, esophagus, and cloaca. It is a simple columnar epithelium in all or most of the stomach and in the intestine. A loose, fibrous connective tissue and muscle form the rest of the wall. The structure and arrangement of these layers in the mammalian intestine are representative (Fig. 17–3).

The innermost layer, called the **mucous membrane** or **mucosa,** is the most complex. It consists of the simple columnar epithelium lining the intestine, and a fibrous connective tissue containing many blood and lymphatic vessels that receive absorbed materials. Aggregations of white blood cells in the connective tissue may form **lymphocyte nodules.** They are part of the body's immune system (p. 706). Electron microscopy has shown that the surface of the epithelial cells bears numerous **microvilli** that greatly increase the digestive and absorptive surface area. The mucosal surface is further increased by larger folds, the nature of which varies greatly among vertebrates and between gut regions. The mammalian small intestine has many circular folds, the **valves of Kerckring,** and the lining is studded with numerous minute, finger-like projections called **villi.** Many of the epithelial cells, known as **goblet cells,** secrete mucus that lubricates the lining and protects it against autodigestion. Other lining cells produce the intestinal enzymes. These enzymes are not free within the lumen but appear to act among the microvilli at the cell surface. Digestive products thus are concentrated at the cell surface; this facilitates absorption. Many simple tubular glands, the **crypts of Lieberkühn,** invaginate between the bases of the villi into deeper parts of the mucosa. Undifferentiated stem cells multiply in the crypts and spread over the intestinal lining, replacing cells that are sloughed off. Zymogenic cells deep in the crypts may secrete lysozyme, an antibacterial substance. Scattered **argentaffin cells** in the crypts secrete peptide hormones that help regulate digestive activities (p. 627). A thin layer of smooth muscle, the **muscularis mucosae,** forms the outer border of the mammalian mucosa. Some of the muscle cells enter the villi. The muscularis mucosae mediates those movements of the villi and mucosa that are independent of the movement of the gut as a whole.

A second layer, the **submucosa,** which is composed of a very vascular, fibrous

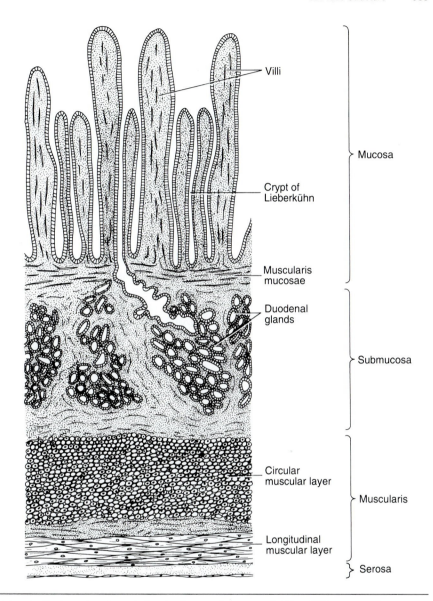

Villi

Crypt of
Lieberkühn

Muscularis
mucosae

Duodenal
glands

Circular
muscular layer

Longitudinal
muscular layer

Mucosa

Submucosa

Muscularis

Serosa

FIGURE 17–3 A longitudinal section through the duodenum of a mammal. (After Williams, et al.)

connective tissue, lies deep to the mucosa. Compound glands invaginate into the submucosa of the mammalian duodenum from the mucosa (Fig. 17–3). A layer known as the **muscularis** lies peripheral to the submucosa. It is composed of two layers of muscle that spiral around the intestine and are antagonistic to each other. The outer, longitudinal layer forms such an open spiral that its fibers are seen primarily in lateral view in a longitudinal section; the inner, circular layer forms such a tight spiral that its fibers are seen in cross section. These muscles are smooth. Their activity is integrated by the enteric part of the autonomic nervous system, which consists of a myenteric plexus located between the two layers of the muscularis and a submucosal plexus located between muscularis and submucosa (p. 484). Not all of the smooth muscle cells receive a direct innervation from the enteric system, but action potentials spread slowly from cell to cell. Peristaltic contraction of the muscle layers propels the food mass through the digestive tract and churns it, thereby helping break it up mechanically and mix it with digestive secretions. Churning is particularly pronounced in the stomach. The circular muscles also form sphincters, such as the **pyloric sphincter** at the caudal end of the stomach, that retain the food for a certain time. Connective tissue covered by coelomic epithelium forms a thin layer known as the **serosa** on the surface of the wall of those parts of the digestive tract lying within the body cavity.

The Esophagus

The "esophagus" of the lamprey represents the dorsal part of the larval pharynx, the ventral part of which forms the respiratory tube. In other vertebrates the **esophagus** (Gr. *oisophagos* = gullet) is a connecting segment between the pharynx and the stomach or, in those vertebrates that lack stomachs, between the pharynx and the intestine. The esophagus primarily transports food, but performs additional functions in some species. It is specialized to crush eggs in egg-eating snakes, and in some fishes, reptiles, and birds it serves for the temporary storage of partly swallowed prey. Grain- and seed-eating birds and a few other vertebrates have a **crop,** a sac that develops from the caudal portion of the esophagus. Food can be gathered quickly and stored here, and seeds may be softened by water and some bacterial fermentation. Under the influence of prolactin secreted by the adenohypophysis (p. 544), the crops of pigeons and doves of both sexes produce a milky secretion that is fed to the young.

The crop and lower esophagus of the hoatzin, a South American bird, is specialized as a fermentation chamber, analogous to the chambered stomach of cows and other ruminants (p. 616) (Grajal et al., 1989). This bird feeds on soft, young leaves. These are ground up by cornified epithelial ridges in the highly muscular and very large crop. A bacterial colony in the crop produces cellulase. Cellulose in the leaves is digested into short-chain organic acids that can be absorbed from the crop or more distally along the digestive tract. As the bacteria multiply, they produce microbial protein that the bird digests in its stomach and intestine. A large fermentation chamber is an unusual specialization for a bird, since most birds conserve weight as an adaptation for flight, but the hoatzin is a poor flier.

The esophagus is lined by a stratified squamous epithelium that is slightly cornified in some turtles, most birds, herbivorous mammals, and other species that swallow coarse food. Mucous- and serous-secreting cells usually are present, but larger glands seldom are found. The lining of the esophagus frequently bears papillae or longitudinal folds. Unlike more caudal parts of the gut tube, the musculature in its wall often is striated. Normally its lumen is collapsed, but it can be distended as food is swallowed. The esophagus is short in fishes and amphibians, but much longer in amniotes. This is because the heart has moved caudad, other viscera have also shifted posteriorly, and a well-defined neck is present (Figs. 17–4, 17–5, and 17–6).

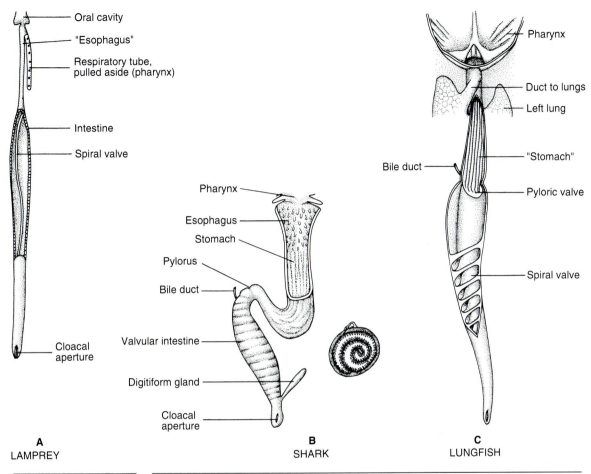

FIGURE 17–4   Ventral views of the digestive tracts of representative fishes. The drawings extend from the caudal end of the pharynx to the cloaca or anus. The liver and pancreas have not been included. **A,** A lamprey. **B,** A shark (insert is a cross section of the spiral valve in the intestine). **C,** A lungfish, *Protopterus*. (*continued*)

The Stomach

Protochordates, the ammocoetes larva, and adult lampreys lack stomachs (Fig. 17–4**A**). This may have been the ancestral vertebrate condition. The small food particles on which these animals fed—probably more or less continuously—could have been processed directly by the intestine. Presumably, a stomach evolved as vertebrates began to feed on larger organisms that were caught at less frequent intervals. The stomach is first of all a chamber for the storage of food. Some investigators have postulated that hydrochloric acid production by the gastric glands evolved in the context of killing bacteria and helping preserve the food. The synthesis of pepsin, an **endopeptidase** that cleaves long protein chains into shorter ones, may have evolved later.

A stomach is not essential for digestion because endopeptidases are also pro-

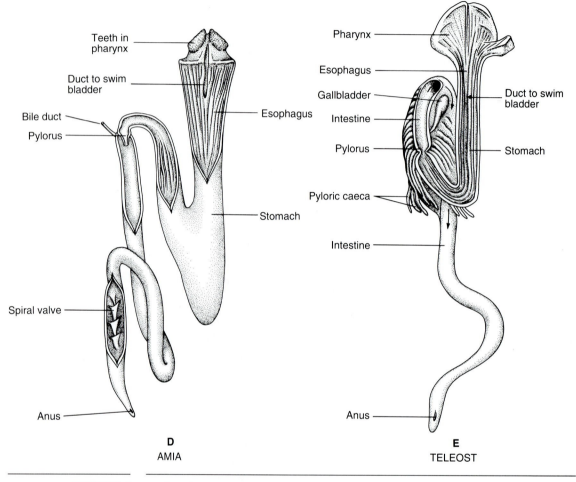

FIGURE 17–4 *continued*
D, A neopterygian, *Amia*. **E**, A teleost, *Salmo*. (Mostly after Bolk et al.)

duced by the pancreas. If an animal feeds on very small particles or the food is finely ground in its mouth or pharynx, a stomach may be secondarily lost. This presumably occurred in chimaeras, lungfishes (Fig. 17–4**C**), and carp-like fishes and minnows whose stomachs are poorly developed or absent.

The stomach usually is a J-shaped sac (Figs. 17–4**B** and **D**, 17–5, 17–6), although it is straight in some slender, long-bodied vertebrates: snakes and many salamanders and lizards. A pyloric sphincter, or **pylorus** (Gr. *pyloros* = gate-keeper), at its posterior end normally is contracted, so food is retained in the stomach until it has been broken down—mechanically by churning movements and chemically by pepsin—into a consistency that can be processed by the intestine. A **cardiac sphincter** sometimes is present at the esophageal end. Topographically, the stomach can be divided into a pyloric region, adjacent to the pylo-

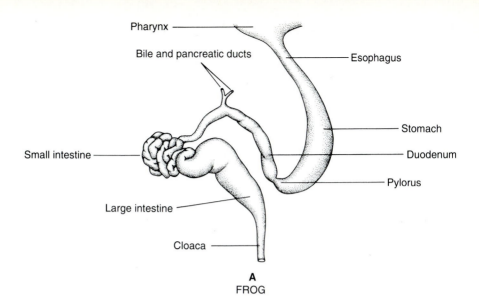

Pharynx

Bile and pancreatic ducts

Esophagus

Small intestine

Stomach

Duodenum

Pylorus

Large intestine

Cloaca

A
FROG

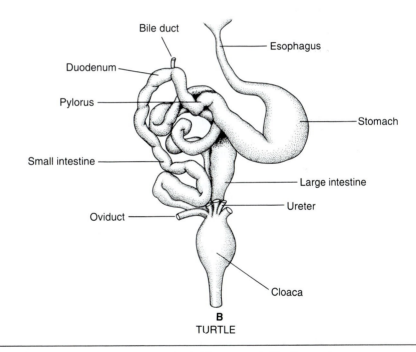

Bile duct

Esophagus

Duodenum

Pylorus

Small intestine

Stomach

Oviduct

Large intestine

Ureter

Cloaca

B
TURTLE

FIGURE 17–5 Ventral views of the digestive tracts of primitive tetrapods (the liver and pancreas are not shown): **A,** A frog, *Rana.* **B,** A turtle, *Emys.* (**A,** after Walker. **B,** after Bolk et al.)

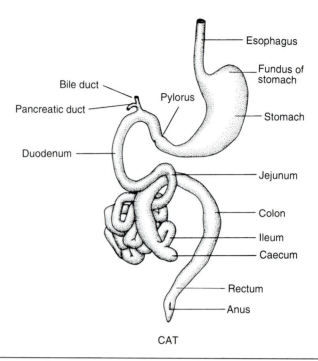

FIGURE 17–6 Ventral view of the digestive tract of a cat.

rus, and a body, or corpus. In mammals it also has a cardiac region, nearest the heart, and a dome-shaped bulge, the fundus, adjacent to the esophageal entrance (Fig. 17–6).

A stratified, squamous, esophageal type of epithelium may extend far into the stomach in some turtles, rodents, ruminants, and other vertebrates that feed on coarse food, but usually the stomach is lined by a simple columnar epithelium. Goblet cells are abundant throughout the lining and, together with branched tubular glands in the pyloric and cardiac regions, secrete a copious amount of mucus that helps protect the stomach from autodigestion. Simple tubular **gastric glands,** which are found primarily in the body and fundus, secrete **hydrochloric acid** and **pepsinogen.** Part of the pepsinogen molecule is cleaved from the rest of the molecule by hydrochloric acid in the stomach lumen. This forms the smaller but active enzyme, **pepsin.** Both acid and pepsinogen are synthesized by the same cells in most vertebrates, but a division of labor has evolved in mammals. Parietal cells produce the acid; chief cells produce the pepsinogen.

Other enzymes may also be synthesized in the stomachs of some vertebrates. Many amphibians, reptiles, birds, and some mammals produce **chitinase,** an enzyme that hydrolyzes the insect cuticle. Mammals, especially young ones, secrete **renin,** an enzyme that curdles milk and causes milk protein to remain in the stomach long enough to be acted on by pepsin.

Some vertebrates have unusual specializations of the stomach. A **gizzard** that can grind up food has evolved from all or much of the stomach in gizzard shad and certain other fishes, some reptiles, including crocodiles, and all birds

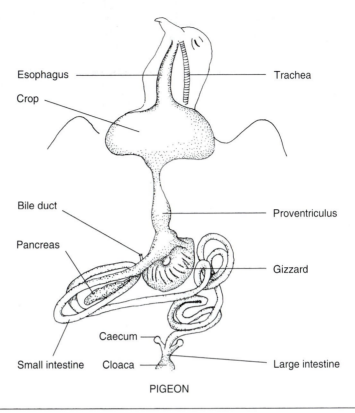

Esophagus — Trachea

Crop

Bile duct — Proventriculus

Pancreas

— Gizzard

Caecum

Small intestine   Cloaca — Large intestine

PIGEON

FIGURE 17–7   Ventral view of the digestive tract of a pigeon.

(Fig. 17–7). The gizzard has a thick muscular wall and a tough proteinaceous lining, and it usually contains small stones that have been swallowed. It is important for birds because they have lost their teeth and cannot otherwise mechanically break down their food. The bird's gizzard develops from the posterior part of the stomach; the anterior part, known as the **proventriculus,** produces gastric secretions that mix with the food as it is ground up.

In puffer and porcupine fishes the stomach, or a diverticulum from it, can be inflated with water, or sometimes with air. The fish assumes a globular shape with erect spines and is less appealing to a predator.

A number of mammals have evolved complex, chambered stomachs that enable them to process plant food. Plant food is abundant and easier to gather than catching prey, but its energy and protein content are low. A large volume of plant food must be processed to overcome these disadvantages. This requires that some part of the gut, such as the stomach, have a large capacity and that passage time of the digesting food mass be slow. This provides space and allows time for colonies of micro-organisms to ferment and break down cellulose. Hippopotamuses, deer, camels, and ruminants (sheep and cattle) have large, multichambered stomachs. However, these types of stomachs are not confined to artiodactyls but are also found in some other plant-eating mammals including kangaroos, sloths, and some monkeys. A chambered stomach is also found in many whales, but its functions are not clear. Fermentation of chitin in the skeletons of invertebrate food may occur in the whale's stomach.

Details of the morphology of chambered stomachs vary, but the stomach of a cow is representative. A cow's stomach consists of four chambers in linear sequence: **rumen, reticulum, omasum,** and **abomasum** (Fig. 17–8). Only the abomasum is lined by a simple columnar epithelium and contains gastric glands. The other chambers are lined with a stratified, esophageal type of epithelium. The rumen and reticulum are the largest chambers, have a wide connection between them, and function more or less as a unit. The **reticulorumen** of a large cow may contain 300 liters of material. Plant food and saliva accumulate in the reticulorumen, and the food is acted on by colonies of anaerobic bacteria and ciliated protozoa that synthesize the enzyme **cellulase.**

Cellulose is fermented and broken down into acetic, butyric, and other short-chain organic acids, and carbon dioxide and methane. The saliva, which is chiefly a weak sodium bicarbonate solution, buffers the acids and serves as a medium in which the micro-organisms grow and multiply. Simple nitrogenous compounds in the food are converted to ammonia, from which the micro-organisms synthesize new protein. Additional nitrogen for protein synthesis is derived from swallowed air and by diffusion of nitrogen and urea from the blood. Urea is a particularly important nitrogen source for bacterial protein in camels, goats, and other species that feed on a low-nitrogen diet. The bacteria also synthesize most of the B complex of vitamins. Periodically, a ruminant belches, releases some of the gas, and regurgitates some of the food. The animal remasticates and reswallows the food, that is, it ruminates or chews its cud (L. *ruminare* = to chew the cud). This process is repeated several times. Most of the organic acids are absorbed directly from the rumen, whose internal surface is increased greatly by numerous, small, leaf-shaped folds, and through the reticulated surface of the reticulum (chefs know the wall of this chamber as tripe). Up to 70 percent of a cow's caloric requirements are met by the direct absorption of organic acids from the reticulorumen. Only food that is broken down into very fine particles, liquid, and micro-organisms can pass between the deep, leaflike folds of the omasum and go into the abomasum, where normal protein digestion begins. Many of a ruminant's amino acids are derived from microbial proteins harvested from the reticulorumen.

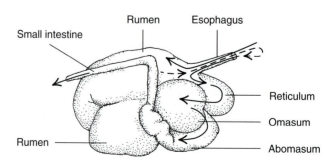

FIGURE 17–8 The stomach of a ruminant. Arrows show the course of the food. (After Schmidt-Nielsen.)

The Intestine and Cloaca

Apart from initial starch digestion in the mouth of some vertebrates and initial protein digestion in the stomach, the intestine is the primary site for both digestion and absorption in all but a few vertebrates, notably the ruminants. Secretions of both the liver and pancreas are added to the intestinal secretions and play an important role in digestion. Although bile released by the liver contains no enzymes, it has a twofold digestive role. Its alkalinity neutralizes the acidity of the gastric contents as they are discharged into the intestine and produces a pH favorable for the action of pancreatic and intestinal enzymes. Its bile salts emulsify fats, breaking them down into small globules that collectively have a large surface area on which pancreatic and intestinal lipase can act.

Both bile and pancreatic secretions are released at a continuous, slow rate in many herbivores that feed much of the time. Bile and pancreatic secretions are released only when the gastric contents enter the intestine in intermittent feeders. Control of the release is by a combination of nerve reflexes and hormonal action (p. 627). Bile may accumulate in a **gallbladder** in some vertebrates. A gallbladder is more likely to be found in intermittent feeders than in continuous feeders, but some herbivores (some ruminants, rabbits) do have one. Carnivores, whose diet includes considerable fat, have a conspicuous gallbladder where bile is concentrated between meals by the reabsorption of water. Little bile concentration occurs in those herbivores that have a gallbladder.

Both the pancreas and intestine produce **lipase,** which hydrolyzes the small fat globules formed by the action of bile into fatty acids and glycerol. **Pancreatic amylase** continues the breakdown of starch (initiated in mammals by salivary amylase) into double sugars or disaccharides. Several intestinal **disaccharidases** degrade different disaccharides into single sugars or monosaccharides. The intestine produces **enterokinase,** an enzyme that activates trypsin and chymotrypsin secreted by the pancreas. **Trypsin** and **chymotrypsin** are endopeptidases that resemble gastric pepsin in cleaving long protein chains into shorter fragments, but they act at different biochemical sites. Several **exopeptidases** produced by both the pancreas and intestine degrade polypeptides into the individual amino acids. **Nucleases** from the pancreas hydrolyze nucleic acids.

Digested organic materials, mineral ions, water (much of which was released in the digestive secretions), and other small molecules are absorbed from the intestinal lumen. Fatty acids and fat-soluble vitamins simply **diffuse** through the lipid plasma membrane of intestinal cells following their concentration gradients. Water, and water-soluble materials such as glycerol, cannot pass directly through the plasma membrane but diffuse through small, protein-lined pores in the membrane. Some minerals, including calcium and iron, bind to a carrier molecule that moves across the membrane in a process called **facilitated diffusion.** Sodium ions bind to a different carrier molecule that moves across the cell as sodium is actively pumped out of the opposite surface of the cell. Most of the amino acids and monosaccharides also bind to the same carrier and are actively transported across the plasma membrane with the sodium in a process called **cotransport.** Undigested residues, together with bacteria that multiply in the intestine, are egested as feces.

The surface area of the intestine available for these activities, like the sites at which they occur, varies considerably among vertebrates according to the type of food eaten and the animal's level of activity and metabolism. Cyclostomes, chondrichthean fishes, and some bony fishes have short, nearly straight intestines that extend from the stomach to the end of the trunk (Fig. 17–4**A** to **C**). The intestine usually opens into a **cloaca,** a chamber that also receives the excretory and reproductive ducts. The cloaca opens to the body surface by a **cloacal aperture.** In cyclostomes the surface area of the intestine is increased slightly by a longitudinal fold that makes a low-pitched spiral along the intestine. In cartilaginous and early bony fishes it is increased much more by a tightly pitched fold, the **spiral valve.** The spiral valve considerably increases both transit time and the absorptive surface compared with a straight intestine. This type of intestine shows little regional differentiation, and the entire intestine is called the **valvular intestine.** The terminal part, which in cartilaginous fishes receives the contents of a salt-excreting digitiform or rectal gland (p. 731), is sometimes called the rectum, even though it is different from the mammalian rectum.

In more advanced bony fishes, the spiral valve is reduced or lost, and in some species the intestine increases in length and coils (Fig. 17–4**D**). Herbivorous fishes have particularly long intestines, increasing transit time of the food, and in some species the intestine lodges a bacterial colony that digests cellulose. Internal longitudinal and transverse folds and, in a few species (mullets), small finger-like villi further increase surface area.

In addition to a moderately long and partly coiled intestine, most teleosts have **pyloric caeca** that evaginate from the intestine just distal to the pylorus (Fig. 17–4**E**). Caeca range in number from three or four to several hundred, and they further increase digestive and absorptive surface area. Sardines and anchovies, which feed on waxy copepods, have caeca that secrete a special **wax lipase.** A cloaca is absent from most teleosts, and the intestine terminates in an **anus** on the body surface.

Lampreys, salmon, and some other fishes stop feeding as they move toward their breeding grounds. Their intestines undergo autodigestion and become greatly reduced.

The intestines are moderately long in most amphibians and reptiles (Fig. 17–5) and very long in herbivorous species or herbivorous developmental stages, in which bacterial fermentation of cellulose occurs. The intestine of the herbivorous tadpole is so long that it must be tightly coiled like a watch spring to fit in the body cavity; that of the carnivorous adult frog is much shorter. Surface area also is increased in amphibians and reptiles by internal folds and occasionally by a few villi. The intestine can be divided into a **small intestine** and a slightly wider **large intestine.** The large intestine is short in amphibians and primarily reabsorbs water and temporarily stores feces. Herbivorous lizards and turtles have a large and often sacculated large intestine. A short **intestinal caecum** may be present at the proximal end of the large intestine. The large intestine lodges a bacterial colony that ferments plant food in the same way as occurs in the chambered stomach of some mammals. The resulting organic acids are absorbed directly. A cloaca is present.

As would be expected, the intestines of endothermic birds and mammals are longer and have much more internal surface area than those of amphibians and reptiles (Figs. 17–6 and 17–7). The length of the intestine as a whole is several body lengths in insectivorous and carnivorous species of birds and mammals, and much longer in herbivorous and granivorous species. The length of the intestine of some herbivorous mammals that feed on bulky, low-caloric foods may be 25 body lengths. Most birds do not feed on such low-caloric foods and do not have such long intestines, whose weight would impair flight. Grouse, who fly only short distances, and rheas, who do not fly at all, do feed on plant food. They have a pair of exceptionally long intestinal caeca that lodge bacterial colonies, and much digestion and absorption of plant food occurs here (Fig. 17–9**A**).

The internal surface area of the small intestine of birds and mammals is increased greatly by numerous finger- or leaf-shaped **intestinal villi** (Fig. 17–3). Villi extend into the large intestines of birds but not of mammals. The small intestine can be differentiated, primarily by type and number of glands and degree of vascularity, into a **duodenum, jejunum,** and **ileum.** These regions are more clearly defined in mammals than in birds. Although functions are not completely segregated, somewhat more digestive activity occurs in the anterior parts of the small intestine; somewhat more absorption occurs in the posterior parts. The large intestine, or **colon,** continues to discharge into a cloaca in birds and monotreme mammals (Fig. 17–10). A copulatory organ is located in the ventral cloacal wall of the cloaca, or tail base, in male reptiles and a few birds. Birds have a dorsal cloacal diverticulum, the **bursa of fabricius,** which is the site for the maturation of B lymphocytes. These cells are involved in certain immune reactions. In marsupials and eutherians the embryonic cloaca becomes divided during development. The dorsal part contributes to the **rectum,** which opens at the anus on the body surface; the ventral part, to the urogenital passages (p. 775).

The caecum and colon are of moderate length and size in carnivorous and omnivorous mammals such as the cat or human being (Fig. 17–6). A **vermiform appendix** may lie at the end of the caecum of some mammalian species, including humans. Its walls contain many nodules of lymphocytes, and it probably is a part of the body's immune system. The mammalian colon contains bacteria that produce cellulase, and the resulting organic acids are absorbed. Mammals with relatively short colons may utilize a moderate amount of cellulose in this fashion. Some vitamins (including vitamin K, needed for blood clotting) also are synthesized by colic bacteria and absorbed from the colon. But the colon of most mammals is primarily a site for water reabsorption and for the consolidation of the feces.

In perissodactyls (rhinoceros, horse), rodents, and lagomorphs (rabbits), the caecum, colon, or both are very long, large, and sacculated (Fig. 17–9**B** and **C**). These mammals do not have a chambered stomach. They are often called **hindgut fermenters,** in contrast to the **foregut fermenters** with a chambered stomach. The caecum and colon lodge the bacterial flora that ferment plant food, and absorption of the breakdown products occurs here. Bacteria multiply in the caecum and colon, utilizing as nitrogen sources residual nitrogen in the food and nitrogen and urea that diffuse in from the blood. However, most hindgut fermenters harvest less microbial protein than foregut fermenters because the hindgut lies posterior to the primary digestive and absorptive surfaces of the gut. Rodents and lagomorphs do harvest considerable bacterial proteins by eating part

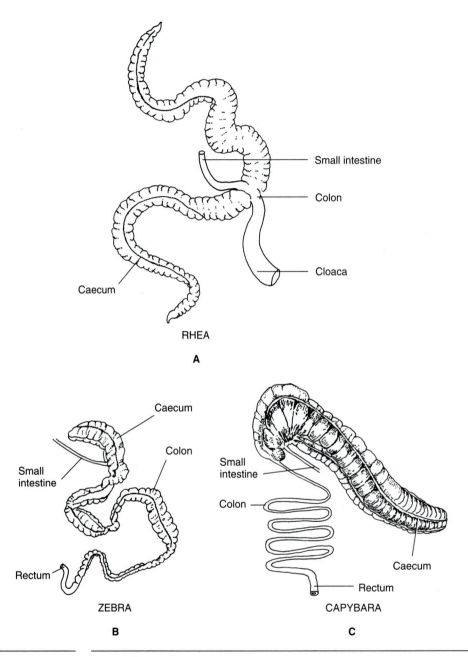

Small intestine

Colon

Cloaca

Caecum

RHEA

A

Caecum

Colon

Small intestine

Rectum

ZEBRA

B

Small intestine

Colon

Caecum

Rectum

CAPYBARA

C

FIGURE 17–9 Modifications of the caecum and colon in hindgut fermenters. **A,** *Rhea*, an Australian flight-less bird. **B,** The zebra. **C,** The capybara, a large South American rodent. (After Stevens.)

of their feces and recycling it through their digestive tract, a practice called **coprophagy.** Rabbits eat primarily the soft feces that are discharged from the caecum (where most of their plant digestion occurs) encased in a thick mucous coat. Foregut and hindgut fermenters represent different solutions among mam-

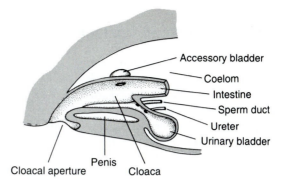

FIGURE 17–10   A longitudinal section through the cloaca of a male turtle.

mals to the effective utilization of plant food. Each method has certain advantages and disadvantages (Focus 17–1).

The specializations of the digestive tract of birds and mammals enable them quickly to digest and absorb the large amount of food that they need to sustain their high level of metabolism. They grind their food in the mouth or gizzard, digestive glands are numerous, a high body temperature accelerates enzyme action, their intestinal surface area is large, the walls of the intestine are very vascular, gut muscles are well developed, and gut mobility is high. Transit time through the gut is far faster in birds and mammals than in other vertebrates. A snake may take a week or more to digest a mouse, but an owl or fox can digest a mouse in a day.

The Liver and Pancreas

We have seen that the liver and pancreas have important digestive roles. These organs also have additional vital functions. The liver develops as a ventral evagination from the archenteron at the level of the transverse septum and just caudal to the stomach and heart. It remains connected to the intestine by the **common bile duct.** The pancreas develops from one or more primordia. Usually, a **ventral pancreatic bud** arises from the liver primordium, and one or sometimes two **dorsal pancreatic buds** arise as separate archenteron evaginations (Fig. 17–11). All the pancreatic primordia usually extend into the dorsal mesentery and unite. Either the stalk of the ventral pancreas bud, that of the dorsal pancreatic bud, or both may persist in the adult as pancreatic ducts.

The Liver

The adult liver is the largest organ of the body, and it occupies most of the space in the peritoneal cavity. Microscopically, the liver consists of many groups, or **lobules,** of **hepatic plates** (Gr. *hepat-* = liver) composed of **hepatic cells** (Fig. 17–12). The plates of each lobule usually are organized around a **central vein,** which is a tributary of the **hepatic vein** (p. 675) and so leads directly or indirectly

> ### FOCUS 17–1   The Functional Anatomy of Large Mammalian Herbivores
>
> Zoologists recognize two groups of hoofed mammals or ungulates: the odd-toed perissodactyls (tapir, rhinoceros, horse) and the even-toed artiodactyls (pigs, deer, camels, sheep, antelope, and cattle). The perissodactyls evolved in the Paleocene, when widespread tropical forest provided a wide choice of plant food. Perissodactyls were at first predominant, but the increasing spread of grasslands in the Miocene favored the artiodactyls.
>
> The foregut fermenting artiodactyls need only partly chew their food as they gather it. They then regurgitate and thoroughly masticate the food, often at a time and in a place where they can keep alert to predators. A relatively long time is needed for food to be broken down mechanically and chemically in the reticulorumen to the very fine particles that can pass through the omasum to the abomasum. Transit time through the digestive tract of a cow averages about 80 hours. Plant alkaloids and other toxins can be detoxified as the food ferments in the reticulorumen. Bacterial protein synthesized in the reticulorumen can be nearly fully utilized as the food passes through the abomasum and intestine.
>
> Hindgut fermenting perissodactyls masticate their food as they eat. Plant toxins are absorbed as the food mass passes through the stomach and intestine and must be detoxified in the liver. The protoplasm of broken plant cells contains considerable protein and other nutrients, and these are processed in the stomach and intestine. Digestion of cellulose occurs in the caecum and/or colon, but is only 70 percent as
>
> efficient as in foregut fermenters. (You may have noticed the relatively high fiber content of horse feces compared with cow dung.) Little of the microbial protein synthesized in the caecum and colon can be utilized. Hindgut fermenters have the advantage of a relatively fast transit time, averaging 48 hours in a horse. A hindgut fermenter can obtain a lot of energy relatively quickly by processing a large volume of plant food, but the food must contain the needed proteins.
>
> The differences between foregut and hindgut fermenters have ecological and behavioral implications. Both groups do well on young grasses that contain a fair amount of protein (low fiber/protein ratio) (Focus Figure). When feeding on dried grasses or herbage with a higher fiber content, a horse does better than a cow. If enough water is available, a horse simply increases its intake of this material and processes a sufficient volume to get the needed energy and nutrients. Because of the slow passage time of material through its stomach, a cow cannot consume low-quality food fast enough to obtain enough energy to survive successfully.
>
> It has been argued that perissodactyls were inferior and were displaced by the artiodactyls, which evolved in the late Eocene. Functional anatomy has put a different perspective on this evolutionary question. The two groups possess different anatomical and functional designs that have led to very different strategies of energy resource utilization. Perissodactyls can exploit resources with high fiber and low protein-glucose contents that are not accessible to artiodac-

to the heart. Branches of the **hepatic artery,** containing oxygenated blood, and branches of the **hepatic portal vein,** which has drained the stomach, intestine, and spleen, lie peripheral to each hepatic lobule. **Hepatic sinusoids** pass between and through the hepatic plates as they carry blood from the hepatic artery and hepatic portal vein to the central vein. Although the hepatic sinusoids are the size of capillaries, they lack the complete endothelial lining of capillaries, so blood in them comes into direct contact with many hepatic cells and with phagocytic cells

FOCUS 17–1   *continued*

tyls. The two groups have coexisted successfully since the late Eocene by engaging in resource partitioning, which is firmly rooted in their contrasting functional designs. A comparison of foregut and hindgut fermentators is an excellent

example of the profound influence that divergent anatomical and functional designs have on the basic biology, ecology, and distributional and evolutionary patterns of the ungulates.

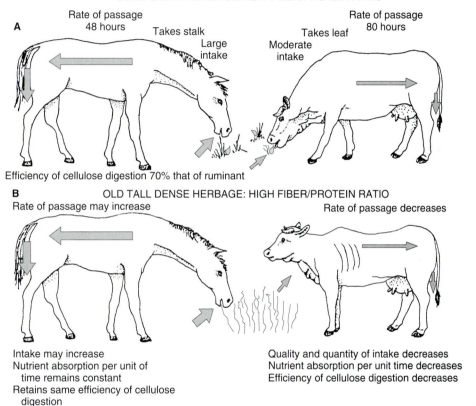

A comparison of the feeding strategies of a horse (a hindgut fermenter) and a cow (a foregut fermenter). **A,** The horse and cow do equally well on young, fresh herbage, but they select different parts of the plants. **B,** The horse also does well on older and woodier herbage, but the cow does not. (After Janis.)

within the sinusoids. Minute **bile canaliculi** lie between hepatic cells and drain **bile,** the secretion of the hepatic cells, into **hepatic ducts** that leave the liver and unite to form the common bile duct (Fig. 17–11**A**).

Often, a **gallbladder** evaginates from the common bile duct and remains connected to it by the **cystic duct** (Fig. 17–11**B**). A sphincter at the intestinal entrance of the common bile duct normally is contracted so that bile backs up

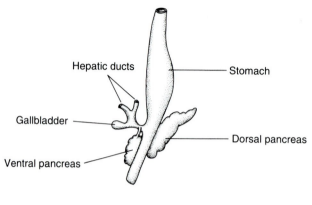

EARLY STAGE

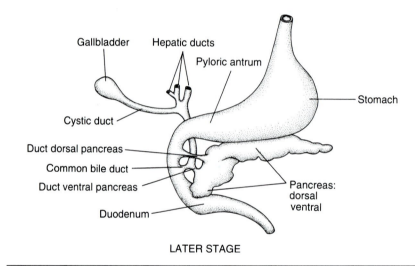

LATER STAGE

FIGURE 17–11   Two stages in the embryonic development of the pancreas and liver. (After Corliss.)

into the gallbladder, where it is stored and sometimes concentrated by water reabsorption. When the gastric contents are discharged into the intestine, this sphincter relaxes, the gallbladder contracts, and bile flows into the intestine.

Liver structure and functions are related closely to both the digestive and the circulatory systems. Bile is a very alkaline solution containing **bile pigments,** which are excretory products derived from the breakdown of hemoglobin in the liver, and **bile salts,** which, as we have discussed, emulsify fats.

Other liver functions derive from its intimate circulatory connections. Worn-

FIGURE 17–12   The liver: **A,** A microscopic section of a mammalian liver to show liver lobules and asso- ▶ ciated blood vessels and ducts. **B,** A stereodiagram of a portion of a liver lobule. (**A,** after Hammersen. **B,** after Elias.)

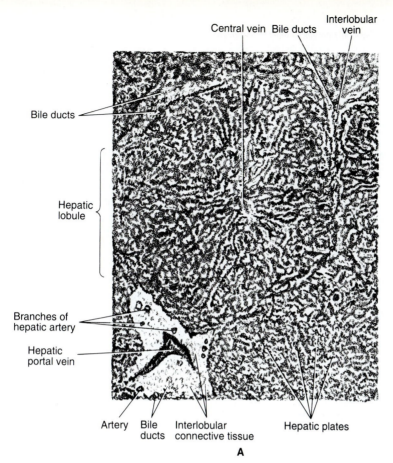

Central vein · Bile ducts · Interlobular vein

Bile ducts

Hepatic lobule

Branches of hepatic artery

Hepatic portal vein

Artery · Bile ducts · Interlobular connective tissue · Hepatic plates

A

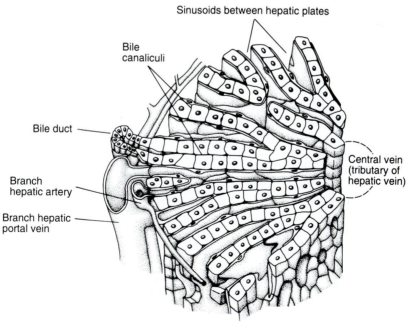

Sinusoids between hepatic plates

Bile canaliculi

Bile duct

Branch hepatic artery

Branch hepatic portal vein

Central vein (tributary of hepatic vein)

B

out red blood cells are destroyed in the spleen, but the hemoglobin that is released is carried to the liver and taken up by hepatic cells. The iron part of the molecule is salvaged, but the rest of the molecule is converted to bile pigments and excreted in the bile. New blood cells develop within the connective tissues of the liver in the embryos of all vertebrates and in the adults of anamniotes. All products that are absorbed from the intestine (except for much of the fat, which enters the lymphatic system) are carried to the liver by the hepatic portal vein and pass through the hepatic sinusoids before being distributed to the body. Many harmful materials are phagocytosed and detoxified. Numerous interconversions of glucose, amino acids, and fats occur. Many food molecules in excess of the animal's immediate needs are converted to glycogen or, in many fishes, lipids and are stored in the hepatic cells. When amino acids undergo conversion into other materials, the amino group is removed first and converted within the hepatic cells into ammonia, urea, or other waste products. These are carried by the blood to the kidneys or other excretory surfaces and excreted. Between meals, when blood sugar levels fall, food stored in the hepatic cells is reconverted to glucose and released into the circulatory system. Most proteins in the blood plasma are synthesized in the liver, as is the yolk that is transferred by female vertebrates to their eggs. Under the influence of pituitary growth-stimulating hormone, liver cells also synthesize the hormone **somatomedin,** which stimulates growth.

The Pancreas

Every vertebrate has a pancreas. In lampreys, lungfishes, and teleost fishes, however, the pancreatic tissue is embedded in the wall of the intestine or scattered in the mesentery and is not a grossly visible organ. The pancreas is a complex gland, containing both exocrine and endocrine portions. The exocrine portion, which constitutes most of the organ, is a compound acinar gland (Fig. 17–13). Numerous minute, roundish acinar units, each of which is made up of several types of cells, synthesize the many digestive enzymes that are released into the intestine in an alkaline solution when the gastric contents are discharged.

The endocrine portion of the pancreas consists of many scattered clusters of cells known as the **pancreatic islets of Langerhans** (Fig. 17–13). The islets develop embryonically by budding from the pancreatic acini and smaller ducts. Most mammals have many thousands of them, and larger species, including human beings, have more than a million. Several distinct types of secretory cells have been identified. **Alpha cells** secrete the hormone **glucagon; beta cells** secrete **insulin.** Insulin lowers blood sugar levels by increasing the permeability of the plasma membrane of some cells (primarily muscle and adipose tissue) to glucose, and by activating enzymes that convert glucose to storage products: glycogen in liver and muscle cells, fat in adipose tissue. An abnormally low level of insulin leads to the disease **diabetes mellitus,** which is characterized by a very high level of blood sugar and other metabolic disturbances. Insulin also promotes the conversion of amino acids to protein. Glucagon acts primarily on liver cells. It activates enzymes that reconvert glycogen to glucose and so increases blood sugar levels. Both hormones respond directly to changes in blood sugar levels. An increase in blood sugar triggers a release of insulin; a decrease triggers a release of glucagon.

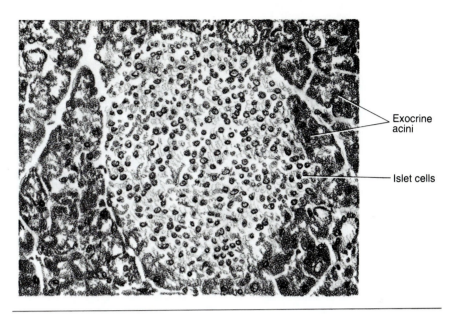

Exocrine acini

Islet cells

FIGURE 17–13   An endocrine pancreatic islet of a mammal surrounded by exocrine acinar tissue. (After Fawcett.)

Sympathetic and parasympathetic neurons also terminate on many islet cells and appear to modulate their responses to blood sugar.

Pancreatic endocrine tissue occurs in all vertebrates, but the cell composition varies, and the tissue does not always form discrete islets. Lampreys and some urodeles have only insulin-producing beta cells. Insulin appears to have evolved very early, for it also is found in the intestinal mucosa of protochordates and some invertebrates. Both alpha and beta cells occur in other vertebrates, but the product of the alpha cells, glucagon, has not been identified in many fishes and some amphibians. Since a lamprey does not have a distinct pancreas, the islet cells are lodged in the wall of the anterior part of the intestine. The endocrine tissue of a chondrichthean fish forms one or more cell layers around the periphery of the acini and some ducts. This may be primitive. It suggests, as does the embryonic origin of the islet cells, that the islets evolved by a gradual separation and modification of the exocrine tissue. Other vertebrates have islets, although those of a few teleosts (goosefish and toadfish) and lungfishes are aggregated to form one or two large **principal islets** that are separated from the exocrine tissue in distinct lobes of the pancreas.

Gastrointestinal Hormones

The sequential processing of foods passing through the digestive tract necessitates a precise integration between food movements and the release of the digestive secretions. This control is effected partly by the autonomic nervous system and partly by hormones secreted by scattered argentaffin cells in the epithelial lining of the stomach, duodenum, and jejunum (Fig. 17–14). In general, parasympathetic

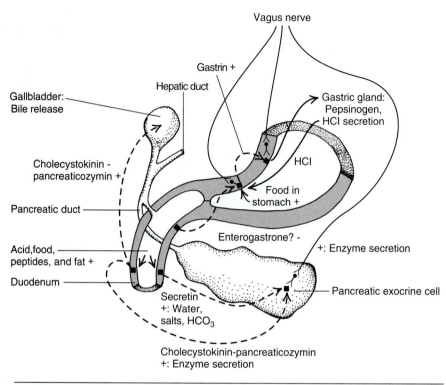

FIGURE 17–14 The nervous and endocrine control of gastric and pancreatic secretion. The effects of hormone secretion are shown by dashed arrows. A plus (+) indicates stimulation of secretion (or gallbladder contraction); a minus sign (−) indicates inhibition.

stimulation via the vagus nerve promotes gut movements and secretion, and sympathetic stimulation inhibits these actions. The hormones involved have been studied primarily in mammals, but at least one (secretin) has been found in the lamprey intestine. By using electron microscopy and immunocytochemical techniques, investigators have identified nearly a dozen cell types and products in human beings, but the actions of only three or four are understood. All these cells produce amines and peptides as their hormones, and there is some evidence that the secretory cells are of neural crest origin. They show an affinity to certain nerve cells and chromaffin cells of similar origin. All lie next to the gut lumen and can respond directly to the presence of digestive products in the tract.

The secretion of hydrochloric acid and pepsinogen by the gastric glands is caused partly by vagal stimulation and partly by the secretion of **gastrin** by cells in the pyloric region of the stomach. Gastrin secretion is promoted by vagal stimulation and the presence of food in the stomach. Its release is inhibited as acid levels increase in the stomach.

The entry of acid food into the intestine stimulates intestinal cells to secrete **secretin.** This hormone, discovered by Bayliss and Starling in 1902, was the first hormone whose effects could be clearly demonstrated. Subsequent investigations have shown that secretin promotes only the release of water, salts, and bicarbonate.

The release of the pancreatic enzymes requires another hormone, **cholecysto-kinin-pancreozymin,** whose secretion is stimulated by the presence of fat and peptides in the intestine. This hormone also promotes the contraction of the gall-bladder. Parasympathetic stimulation supplements hormonal enzyme release.

Enterogastrone was formerly identified as another intestinal hormone whose release was triggered by fat entering the intestine. It was believed to inhibit gastric secretion, but the existence of this hormone is now uncertain.

SUMMARY

1. The pharynx of a fish is a large and important part of the respiratory system. In a terrestrial vertebrate it is a relatively short food passage.

2. The thyroid gland develops from the floor of the rostral part of the pharynx. It synthesizes, stores, and releases the hormones thyroxine and tri-iodothyronine. These hormones in mammals accelerate oxidative metabolism and are essential for normal growth and maintenance of the skin and its derivatives. They are also essential for amphibian metamorphosis and for some aspects of skin maintenance in non-mammalian vertebrates.

3. Tetrapods have parathyroid glands that usually develop from the third and fourth pharyngeal pouches and produce parathormone. Fishes lack this gland but, like terrestrial vertebrates, have C cells that produce calcitonin. Parathormone, calcitonin, and other hormones interact in mineral homeostasis.

4. The thymus and palatine tonsils begin their development as epithelial buds from certain pharyngeal pouches. They soon are invaded by lymphocytes and become part of the body's immune system.

5. The gut tube is composed of a mucosa that contains mucus-, enzyme-, and hormone-producing cells, a connective tissue submucosa, two layers of smooth muscles, and a connective tissue and epithelial serosa.

6. The esophagus is particularly long in terrestrial vertebrates, for the pharynx is short and a neck has developed. It is modified as a crop in grain-eating birds, but normally it is a simple food-conducting passage.

7. Stomachs were probably absent from ancestral vertebrates. They first evolved as storage chambers when vertebrates began to feed intermittently on larger food. Stomach glands produce hydrochloric acid and the endopeptidase pepsinogen. They also produce chitinase in some tetrapods and renin in young mammals.

8. The stomach is modified as a gizzard that grinds food in some fishes, some reptiles, and all birds. The stomach of ruminants and some other mammals is a multichambered organ in which cellulose is digested by cellulase, which is produced by a colony of bacteria and other micro-organisms. Considerable absorption occurs in the reticulorumen, and much of a ruminant's proteins are derived from harvesting bacteria.

9. Secretions of the liver and pancreas are added to intestinal secretions and play an important role in digestion. Bile from the liver neutralizes the acid gastric contents when they enter the intestine, and the liver's bile salts emulsify fats. Many pancreatic and intestinal enzymes break down all categories of food particles into small molecules that can then be absorbed along with

water, minerals, and other needed materials. Undigested residue and a considerable volume of bacteria are eliminated as feces.

10. The configuration and divisions of the intestine vary greatly among vertebrates. They are closely related to the animal's type of food, body size, and level of activity. In general, ectotherms have shorter and simpler intestines with less surface area than endotherms. The intestine and urogenital ducts of most vertebrate groups terminate in a common chamber, the cloaca. In therian mammals the cloaca becomes divided.

11. Intestinal bacteria, like those in the chambered stomach of some mammals, can ferment cellulose, and the resulting organic acids can be absorbed. This process is particularly pronounced in herbivorous tetrapods, and their intestinal caecum and/or parts of the colon are greatly enlarged to accommodate the bulky plant food and bacterial colony. Fermentation in the foregut (stomach) or hindgut has certain advantages and disadvantages.

12. The liver produces bile that emulsifies fat and facilitates its absorption. Blood returning from the digestive tract passes through liver sinusoids. Hepatic cells detoxify many noxious substances, interconvert many absorbed food products, store excess food as glycogen or lipid, release food into the blood as needed, synthesize urea, synthesize many plasma proteins, assist in iron metabolism, and produce the hormone somatomedin.

13. The pancreas develops as outgrowths of the liver diverticulum and adjacent parts of the intestine. Pancreatic tissue remains embedded in the intestinal wall in lampreys, lungfishes, and teleosts but forms a distinct organ in other vertebrates. Many of the digestive enzymes that act in the intestine are synthesized in the pancreas.

14. Endocrine cells of the pancreas form the pancreatic islets. They produce the hormones glucagon and insulin, which are essential for carbohydrate metabolism.

15. The sequential processing of food passing through the digestive system is integrated by the autonomic nervous system and by gastrointestinal hormones produced by certain mucosal cells.

REFERENCES FOR CHAPTERS 15, 16, AND 17

Alexander, R. McN., 1970: Mechanics of feeding action of various teleost fishes. *Journal of Zoology* (London), 162:145–156.

Barrington, E. J. W., 1945: The supposed pancreatic organs of *Petromyzon fluviatilis* and *Myxine glutinosa. Quarterly Journal of Microscopical Science*, 85:391–417.

Bels, V. L., and Baltus, I., 1989. First analysis of feeding in *Anolis* lizards. *Progress in Zoology*, 35:141–145.

Bramble, D. M., and Wake, D. B., 1985: Feeding mechanisms of lower tetrapods. *In* Hildebrand, M., Bramble, D. M., Liem, K. F., and Wake, D. B., editors: *Functional Vertebrate Morphology.* Cambridge, Harvard University Press.

Butler, P. M., and Joysey, K. A., editors, 1978: *Development, Function and Evolution of Teeth.* New York, Academic Press.

Chivers, D. J., and Hladik, C. M., 1980: Morphology of the gastrointestinal tract in primates; Comparisons with other mammals in relation to diet. *Journal of Morphology*, 166:337–386.

Clark, R. B., 1964: *Dynamics of Metazoan Evolution; The Origin of the Coelom and Its Segments*. Oxford, Oxford University Press.

Crompton, W. A., 1971: The origin of the tribosphenic molar. *In* Kermack, D. M., and Kermack, K. A., editors: *Early Mammals*. Supplement 1 to the *Zoological Journal of the Linnean Society*, vol. 50. London, Academic Press.

Crompton, W. A., Thexton, A. J., Hiiemae, K. M., and Cook, P., 1975: The movement of the hyoid apparatus during chewing. *Nature*, 258:69–70.

Dahlberg, A. A., editor, 1971: *Dental Morphology and Evolution*. Chicago, University of Chicago Press.

Denison, R. H., 1961: Feeding mechanisms of Agnatha and early gnathostomes. *American Zoologist*, 1:177–181.

Doran, G. A., 1975: Review of the evolution and phylogeny of the mammalian tongue. *Acta Anatomica*, 91:118–129.

Elias, H., 1949: A re-examination of the structure of the mammalian liver. II. The hepatic lobule and its relation to the vascular and biliary systems. *American Journal of Anatomy*, 85:379.

Elias, H., and Sherrick, J. C., 1969: *Morphology of the Liver*. New York, Academic Press.

Frazzetta, T. H., 1962: A functional consideration of cranial kinesis in lizards. *Journal of Morphology*, 111:287–320.

Gans, C., 1969: Comments on inertial feeding. *Copeia*, 1969:855–857.

Grajal, A., Strahl, S. D., Parra, R., Dominguez, M. G., and Neher, A., 1989: Foregut fermentation in the hoatzin, a Neotropical leaf-eating bird. *Science*, 245:1236–1238.

Hammersen, F., 1980: *Sobotta/Hammersen Histology, A Color Atlas of Cytology, Histology, and Microscopic Anatomy*. Baltimore, Urban & Schwartzenberg.

Hiiemae, K. M., and Crompton, A. W., 1985: Mastication, food transport, and swallowing. *In* Hildebrand, M., Bramble, D. M., Liem, K. F., and Wake, D. B., editors: *Functional Vertebrate Morphology*. Cambridge, Harvard University Press.

Janis, C., 1976: Evolutionary strategy of the Equidae and origins of the rumen and cecal digestion. *Evolution*, 30:757–774.

Jennings, J. B., 1972: *Feeding, Digestion and Assimilation in Animals*, 2nd edition. London, Macmillan, St. Martin's Press.

Lagler, K. F., Bardach, J. E., Miller, R. E., 1962: *Ichthyology*. New York, John Wiley and Sons.

Langer, P., 1988. *The mammalian herbivore stomach. Comparative anatomy, function and evolution*. G. Fischer, New York.

Lauder, G. V., 1982. Patterns of evolution in the feeding mechanism of actinopterygian fishes. *American Zoologist*. 22:275–285.

Lauder, G. V., 1979: Feeding mechanisms in primitive teleosts and the halecomorph fish *Amia calva*. *Journal of Zoology* (London), 187:543–578.

Lauder, G. V., 1985: Aquatic feeding in lower vertebrates. *In* Hildebrand, M., Bramble, D. M., Liem, K. F., and Wake, D. B., editors: *Functional Vertebrate Morphology*. Cambridge, Harvard University Press.

Liem, K. F., 1978. Modulatory multiplicity in the functional repertoire of the feeding mechanism in cichlid fishes. I. Piscivores. *Journal of Morphology*, 158:323–360.

Liem, K. F., 1980: Adaptive significance of intra- and interspecific differences in the feeding repertoires of cichlid fishes. *American Zoologist*, 20:295–314.

Lombard, E., and Wake, D. B., 1976: Tongue evolution in the lungless salamanders, Family Plethodontidae. I. Introduction, theory, and a general model of dynamics. *Journal of Morphology*, 148:265–286.

Mitchell, P. C., 1901: On the intestinal tract of birds. *Transactions of the Linnean Society of London*, series 2, 8:173–275.

Mitchell, P. C., 1916: Further observations of the intestinal tracts of mammals. *Proceedings of the Zoological Society of London*, 1916:183–251.

Moss, M., 1970: Enamel and bone in shark teeth; with a note on fibrous enamel in fishes. *Acta Anatomica*, 77:161–187.

Nickel, R., Schummer, A., and Seiferle, E., 1979: *The Viscera of the Domestic Mammals*, 2nd edition. New York, Springer-Verlag.

Noble, G. K., 1931: *The Biology of the Amphibia*. New York, McGraw–Hill Book Company.

Oguri, M., 1964: Rectal glands of marine and fresh water sharks: Comparative histology. *Science*, 144:1151–1152.

Osborn, J. W., and Crompton, A. W., 1973: The evolution of mammalian from reptilian dentitions. *Brevoria*, 399:1–18.

Peyer, B., 1968: *Comparative Odontology*. Translated by R. Zangerl. Chicago, University of Chicago Press.

Pough, F. H., symposium organizer, 1983: Adaptive radiation within a highly specialized system: The diversity of feeding mechanisms in snakes. *American Zoologist*, 23:337–460.

Preuss, F., and Fricke, W., 1979: Comparative schemata on the histology of the liver with consequences in terminology. *Journal of Morphology*, 162:211–220.

Reilly, S. M., and Lauder G. V., 1990. The evolution of tetrapod feeding behavior: Kinematic homologies in prey transport. *Evolution*, 44:1542–1557.

Rogers, T. A., 1958: The metabolism of ruminants. *Scientific American*, 198:2:34–38.

Ruckebusch, Y., and Thivend, P., editors, 1980: *Digestive Physiology and Metabolism of Ruminants*. Westport, Conn., AVI.

Schwenk, K., and Throckmorton G. S., 1989. Functional and evolutionary morphology of lingual feeding in squamate reptiles: Phylogenetics and kinematics. *Journal of Zoology*, (London), 219:153–175.

Skoczylas, R., 1978: Physiology of the digestive tract. *In* Gans, C., et al., editors: *Biology of the Reptilia*, vol. 8. New York, Academic Press.

Smith, K. K., 1984. The use of the tongue and hyoid apparatus during feeding in lizards (*Ctenosaura similis* and *Tupinambis nigropunctatus*). *Journal of Zoology*, (London), 202:115–143.

Stevens, C. E., 1988: *Comparative Physiology of the Vertebrate Digestive System*. Cambridge, Cambridge University Press.

Thexton, A. J., and McGarrick J. D., 1989. Tongue movement in the cat during the intake of solid food. *Archives of Oral Biology*, 34:239–248.

Walker, W. F., Jr., 1981: *Dissection of the Frog*. San Francisco, W. H. Freeman and Company.

Wells, L. J., 1954: Development of the human diaphragm and pleural sacs. *Carnegie Contributions to Embryology*, 35:107–134.

18 | The Respiratory System

PRECIS

The metabolic energy used by animals is derived from the cellular oxidation of food molecules. The oxygen that is needed, like the carbon dioxide produced as a byproduct, enters and leaves the body by diffusion at a respiratory membrane, but water or air must be moved across the respiratory membrane by some ventilation process. Many evolutionary changes occurred as vertebrates moved from water to land and eventually became endothermic animals.

Most of the body's energy needs are supplied by the cellular oxidation of absorbed food products. In very small animals the oxygen needed for this, and the carbon dioxide produced as a byproduct, can be carried between the external environment and the cells by diffusion, but diffusion is a slow process, and it is efficient only for short distances. Diffusion plays a role in vertebrates in the movement of gases over short distances, but systems also are needed for the **bulk flow,** or movement by a muscular pump, of the medium containing the gases over longer distances. First, some sort of a pump must move the medium containing the gases—water or air—across a thin, moist, and vascular membrane through which the gases can diffuse and be exchanged with those in the blood. A membrane of this type is called a **respiratory membrane.** Second, another pump, the heart, must move the blood through vessels that extend between the respiratory membrane and the vicinity of the cells, where diffusion of gases again takes place. We will consider the first set of problems in this chapter, that is, the nature of the respiratory membrane and the bulk flow of the external medium across it, a process called **ventilation.**

An efficient respiratory system optimizes the diffusion and exchange of gases between the body and the external environment, but many factors affect the nature of the respiratory membrane and the way it is ventilated. The rate of diffusion, R, follows the general equation for diffusion: $R = D \times A \times (\Delta p/d)$ in which D is a diffusion constant whose value depends on the properties of the medium (density, temperature) through which diffusion takes place, A is the surface area across which diffusion is occurring (i.e., the respiratory membrane), Δp is the difference in partial pressures of the gas on either side of the respiratory membrane, and d is the distance (i.e., the thickness of the respiratory membrane) along which diffusion occurs. High rates require a large surface area (A) in the respiratory membrane and a short diffusion distance (d) between the external medium and the blood. Flow rates of the medium across the respiratory membrane, and of blood through it, must be slow enough to juxtapose the medium and blood long enough for diffusion to occur, and fast enough to maintain a difference in the concentration of gases in the medium and blood, that is, to maintain a diffusion gradient (Δp). Ventilation rates or the amount of respiratory membrane in use at any particular time, or both, should be adjustable so that gas exchange matches the needs of the animal as they vary with the animal's level of activity. It would be uneconomical to provide more respiratory surface or ventilation than an animal needs during its greatest activity. The quantity of gas in the water or air also affects the size of the membrane and its rate of ventilation. The density and viscosity of the medium affects the amount of energy required to move it across the respiratory membrane. The movement of other diffusible molecules, such as water, salts, and nitrogenous wastes, across the membrane may need to be reduced or enhanced. Body size is an important factor because of its effect on surface-volume relationships. It follows from these considerations that respiratory membranes and the methods of ventilating them differ greatly among vertebrates.

The Respiratory System of Fishes

A fish's respiratory system must be adapted to two major constraints of life in water. First, the amount of oxygen dissolved in water is much less than the concentration of oxygen in air. The quantity dissolved in water depends on the partial pressure of the oxygen in the air, which is about 160 mm Hg at sea level, because this is the force that drives oxygen into the water. The amount also depends on solubility, and oxygen is not very soluble in water. Although 20°C air at sea level contains 210 ml of oxygen per liter, fresh water under the same conditions contains only 6.6 ml/l; salt water contains somewhat less, 5.3 ml/l. Solubility of oxygen increases slightly in cold water, so fresh water at 0°C holds 10.3 ml/l. The second problem for a fish is that the water in which the oxygen is dissolved is much denser and more viscous than air. Because of these two constraints, a fish must have a rather large respiratory surface and move a large volume of water across it at considerable energetic expense. Any design features of the respiratory system that reduce these energetic costs would be to a fish's advantage.

Unloading carbon dioxide is less of a problem because it is highly soluble in water. It does, however, combine with water to form carbonic acid (H_2CO_3). The pH of the medium may fall, but most fishes live in sufficiently large bodies of near neutral water (pH 7) that this is of no consequence.

Several other problems derive from the close proximity of blood and water across the respiratory membrane. Heat is quickly exchanged, so most fishes cannot maintain an overall body temperature different from the water in which they live (p. 52). Water, valuable salts, and nitrogenous wastes will also diffuse through the membrane.

Gills

Protochordates are sufficiently small and inactive animals for the body or the pharynx wall to serve as a respiratory membrane. A feeding and respiratory current of water flows through the mouth, into the pharynx, and out numerous pharyngeal slits; gills are not present. Ancestral vertebrates became larger and more active animals. Gills, which provide a large surface area for the respiratory membrane, evolved where they could be ventilated by a current of water.

The larvae of many fish species have **external gills,** which are highly vascularized filamentous processes with large surface areas attached to the lateral surface of the head between certain gill slits. Adult fishes have **internal gills.** These consist of a great many vascularized plates, the **primary gill lamellae,** attached to the walls of the gill or branchial pouches (Gr. *branchia* = gills) or to the gill arches. Minute and tightly packed **secondary gill lamellae,** where gas exchange occurs, attach perpendicularly and transversely to the primary lamellae (see Fig. 18–3).

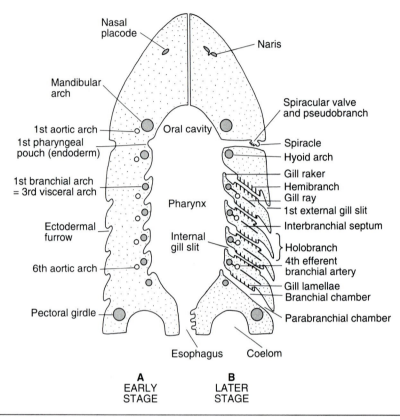

Nasal placode

Naris

Mandibular arch

Spiracular valve and pseudobranch

1st aortic arch
1st pharyngeal pouch (endoderm)

Oral cavity

Spiracle
Hyoid arch

1st branchial arch = 3rd visceral arch

Gill raker
Hemibranch
Gill ray
1st external gill slit

Pharynx

Ectodermal furrow

Interbranchial septum

Internal gill slit

Holobranch
4th efferent branchial artery

6th aortic arch

Gill lamellae
Branchial chamber

Pectoral girdle

Parabranchial chamber

Esophagus   Coelom

A
EARLY
STAGE

B
LATER
STAGE

FIGURE 18–1 A frontal section through the pharynx of an elasmobranch. An early developmental stage is shown on the left (**A**); the adult condition is on the right (**B**).

The Structure and Development of Internal Gills

The pharyngeal pouches develop embryonically as lateral, endodermal evaginations from the portion of the archenteron that will become the pharynx (Fig. 18–1**A**). Ectodermal furrows push inward from the body surface to meet the endodermal pouches, and the intervening tissue soon breaks down. In most fishes the epithelium covering the gills appears to be of ectodermal origin. The plates of tissue between successive pouches, to which the primary gill lamellae attach, are called **interbranchial septa** (Fig. 18–1**B**). A skeletal visceral arch lies in each interbranchial septum. The first visceral arch (the mandibular arch of gnathostomes) lies rostral to the first embryonic pouch, and the last one (the seventh visceral arch) lies caudal to the last pouch. Skeletal, supporting **gill rays** usually extend peripherally from the visceral arches into the interbranchial septa or primary gill lamellae. Muscles and nerves are associated with the skeletal elements, and each of the first six interbranchial septa of a jawed fish contains an embryonic artery, known as the **aortic arch,** that will supply the gills.

Gill pouches were numerous in ancestral vertebrates. Some ostracoderms had 15 pairs. One contemporary species of hagfish has 14 pairs of pouches, but the

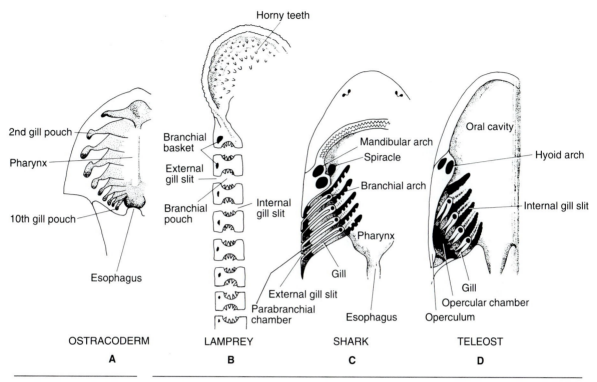

FIGURE 18–2   Frontal sections through the pharynx of four fishes to show the configuration and numbers of branchial pouches: **A,** An ostracoderm. **B,** A lamprey. **C,** A shark. **D,** A teleost. (After Lagler, Bardack, and Miller.)

lamprey has only seven (Fig. 18–2). Most chondrichthyan fishes and early bony fishes have six pairs of pouches, the first of which is reduced to a pair of **spiracles** (Figs. 18–1**B** and 18–2). Teleosts have lost the spiracles and so have only five pairs of pouches (Fig. 18–2). All the gill pouches are lined by gill lamellae in contemporary cyclostomes, and we believe that this resembled the ancestral vertebrate condition. The distribution of lamellae is more limited in jawed fishes. Interbranchial septa that bear gill lamellae on both surfaces constitute a complete gill, or **holobranch.** Jawed fishes usually have four of these. In addition, some sharks, some lungfishes, and some chondrosteans have a half gill, or **hemibranch,** on the posterior surface of the hyoid septum (Fig. 18–1**B**). Gill lamella are seldom present on the posterior surface of the last gill pouch, for there is no aortic arch here to supply them. The African lungfish, *Protopterus,* is an exception, for it has a hemibranch in this position that is supplied by a branch of the sixth aortic arch.

Only a small, gill-like structure lies in the spiracle. Since it receives oxygenated blood from other gills (p. 680), it is called a **pseudobranch** (Fig. 18–1**B**). Its function in elasmobranchs is uncertain. In teleosts, most of whom have retained a pseudobranch even though they have lost the spiracle, it may be a sense organ, or it may function in salt balance.

Gills of Agnathans

Internal gills are arranged in several different ways among fishes, and there are different patterns of ventilation. Agnathans have large, saccular branchial pouches that are lined with the primary gill lamellae (Fig. 18–2**A** and **B**). These are called **pouched gills.** Water normally is drawn into the pharynx through the mouth, enters the branchial pouches through pore-shaped **internal gill slits,** and leaves the pouches through pore-shaped **external gill slits.** One of the specializations of the lamprey for its sucking mode of feeding is a longitudinal division of the pharynx into a dorsal food passage, the "esophagus," and a ventral, blind **respiratory tube** from which the internal gill slits arise (see Fig. 17–4**A**). When the lamprey is feeding on another fish, water must be pumped in and out of the branchial pouches through the external gill slits.

Gills of Elasmobranchs

The branchial pouches of chondrichthyan fishes are narrow chambers, and the gill lamellae are borne on the interbranchial septa, which continue to the body surface (Figs. 18–1**B** and 18–2**C**). These are **septal gills.** A vertically elongated internal gill slit leads from the pharynx into each **branchial chamber. Gill rakers** at the bases of the interbranchial septa keep food in the pharynx. The structure of the gill rakers correlates with the type of food eaten. They are short processes in predacious sharks but form numerous long, thin filaments in the plankton-feeding basking and whale sharks. **Parabranchial chambers** lie between the branchial chambers and the small, slit-shaped external gill slits. The distal tips of the interbranchial septa act as valves that can close the external gill slits.

The embryonic aortic arches give rise to arteries that supply and drain the gills. Branches of **afferent branchial arteries** (L. *ad* = toward + *ferent-* = carrying) carry blood low in oxygen content into all of the primary gill lamellae. Tributaries of **pretrematic** and **posttrematic arteries** collect aerated blood from the gills. The pretrematic and posttrematic arteries lead to an **efferent branchial artery** (L. *ex* = out + *ferent-* = carrying) that carries blood to the dorsal aorta to be distributed to the body. Vascular beds where gas exchange occurs lie in the secondary lamellae between the afferent and the pretrematic and posttrematic arteries (Fig. 18–3**A, B**). Blood enters small vessels in the interbranchial septum from the afferent branchial artery and then flows through vascular spaces in the secondary lamellae. It is collected by small vessels and carried to the pretrematic and posttrematic arteries at the gill base. As it flows laterally from the internal to the external gill slits, most of the water passes between the secondary lamellae into **septal channels** beside the interbranchial septum from which it is discharged. Blood and water flow in opposite directions through and across the secondary lamellae (Fig. 18–3**B**). This **countercurrent flow** affords a considerably more efficient gas exchange than blood and water moving in the same direction, that is, concurrently (Fig. 18–4). In concurrent flow oxygen would diffuse from the water and enter the blood until an equilibrium was reached. Because of the presence of hemoglobin that binds with oxygen, the blood would finally contain more oxygen than the water, but considerable oxygen would remain in the water.

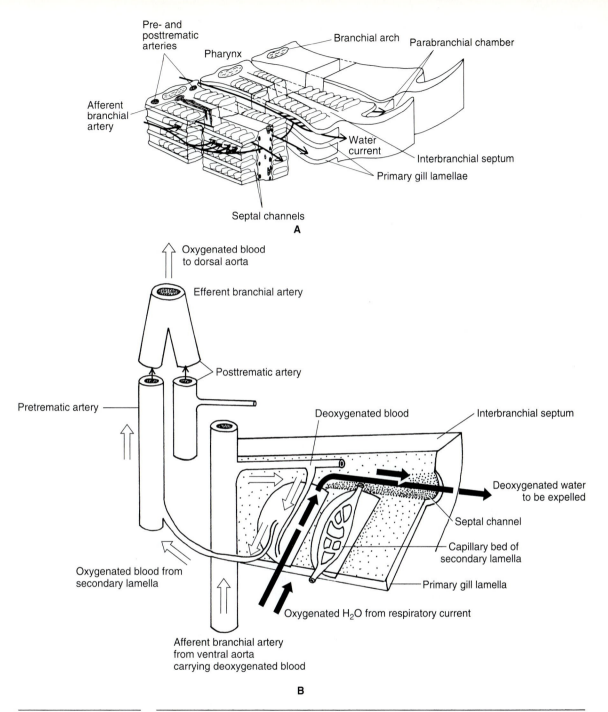

Pre- and posttrematic arteries

Pharynx

Branchial arch

Parabranchial chamber

Afferent branchial artery

Water current

Interbranchial septum

Primary gill lamellae

Septal channels

A

Oxygenated blood to dorsal aorta

Efferent branchial artery

Posttrematic artery

Pretrematic artery

Deoxygenated blood

Interbranchial septum

Deoxygenated water to be expelled

Septal channel

Capillary bed of secondary lamella

Primary gill lamella

Oxygenated blood from secondary lamella

Oxygenated H_2O from respiratory current

Afferent branchial artery from ventral aorta carrying deoxygenated blood

B

FIGURE 18–3   **A**, A stereodiagram of portions of the gills of a dogfish. Water flows between the second-ary lamellae to septal channels located beside the interbranchial septum, whence it is dis-charged (heavy arrows). Blood flow (not shown) is in the opposite direction through the secondary lamellae. **B**, Enlarged portion of primary and secondary gill lamellae with associ-ated blood vessels (open arrows) and water flow (solid arrows). (**A**, After Hughes 1984.)

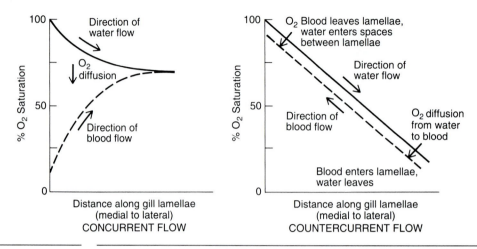

FIGURE 18–4 The results of concurrent flow (left) and countercurrent flow (right) of water beside the secondary lamellae and blood through them.

In countercurrent flow well-aerated blood leaving the secondary lamellae encounters water that has not yet crossed the secondary lamellae and so contains more oxygen than the blood. The opposite conditions prevail at the septal channel end of the secondary lamellae. There, water that has lost most of its oxygen is beside blood containing even less oxygen. Thus, a gradient for the diffusion of oxygen from the water to the blood exists all along the secondary lamellae, and most of the oxygen in the water enters the blood.

During inspiration in sharks, the mouth and valves that close the spiracles open, and the pharynx expands by contractions of the coracomandibularis and rectus cervicis (sternohyoideus) muscles (Fig. 18–5A and B). This reduces the pressure within the pharynx relative to that in the surrounding water. Pressure is reduced even more in the parabranchial chambers by the outward bowing of the valves closing the external gill slits. The pharynx and parabranchial chambers act together as a suction pump, creating a pressure gradient that draws water into the pharynx through the mouth and spiracles, across the gills, and into the parabranchial chambers. During expiration (Fig. 18–5C and D), the mouth and spiracles close, the pharynx and branchial chambers are compressed by the adductor mandibulae and preorbitalis muscles, the external gill slits open, and water is driven out. The pharynx and branchial chambers act together as a force pump. Branchial muscles are most active during expiration, compressing the pharynx and branchial chambers. The visceral skeleton also is compressed, and considerable energy is stored in the bent cartilages. Expansion of the system during inspiration occurs partly by the elastic recoil of the visceral skeleton, and partly by the contraction of somatic hypobranchial muscles that pull the pharynx floor ventrally. All this minimizes energy consumption. The shark conserves still more energy by moving water fairly steadily in only one direction, rather than in and out external gill slits. The mass of the water need not be alternately accelerated and decelerated. Most sharks rely on an equal balance of the force and suction pumps in their respiratory cycle. Some fast-swimming sharks, and also some teleosts, open their mouths after

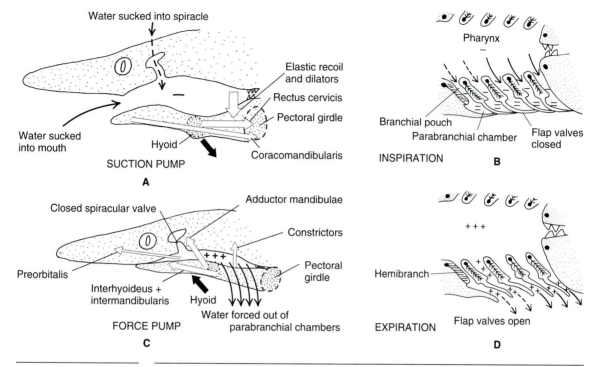

FIGURE 18–5 The mechanics of gill ventilation in the dogfish as seen in lateral views (left) and frontal sections (right) of the pharynx. Relative pressures are indicated by + and −. Solid arrows indicate the flow of water that entered through the mouth; dashed arrows, that of water that entered through the spiracle. Open arrows depict muscle activity, with the width expressing level of activity. Wide, solid arrows depict movement of the hyoid. **A** and **B** show inspiration; **C** and **D** expiration. (**B** and **D**, after Hughes 1963.)

reaching a certain speed and use the forward motion that is generated by their trunk and tail muscles to drive water across the gills. This is called **ram ventilation.** Sharks using ram ventilation have reduced or lost their spiracles. Skates and rays, which are bottom-dwelling fish with ventrally placed mouths, have exceptionally large spiracles through which most of the water enters the pharynx. Skates and rays ventilate their gills primarily by the suction pump, while the force pump plays only a minor role.

Gills of Bony Fishes

Bony fishes evolved a branchial apparatus independently of elasmobranchs, one that is somewhat different. A flap of body wall supported by bones, known as the **operculum,** extends from the hyoid arch region of the head laterally and caudally over the gills. There is a large, common **opercular cavity** for all the gills, and one valved, external gill slit, rather than a series of small parabranchial chambers, each with its own external gill slit (Figs. 18–2**D** and 18–6). The interbranchial septa are reduced, greatly so in teleosts, so the primary gill lamellae extend freely into the

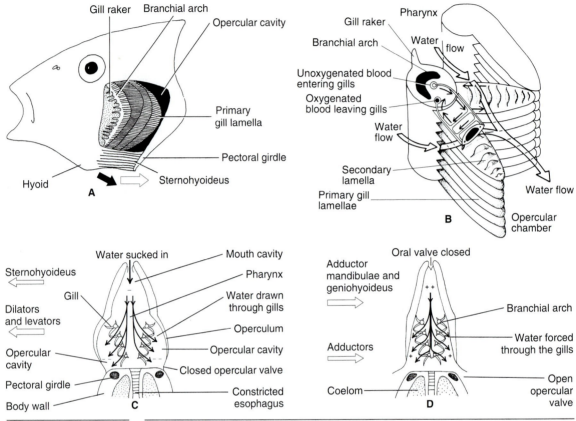

FIGURE 18-6 The gills in a teleost: **A,** A lateral view of the opened opercular chamber. **B,** The counter-current flow of blood and water. **C,** Mechanics of inspiration. **D,** Mechanics of expiration. Arrows indicate direction of water flow. Open arrows indicate muscle action, the width of which represents the level of activity. Wide, solid arrow depicts movement of hyoid. (**B,** after Dorit, Walker, and Barnes.)

opercular cavity. The gills are described as **aseptal.** With the emergence of the opercular system, teleosts have developed a respiratory cycle that ensures a continuous flow of water in one direction from the oropharynx to the opercular cavity over the gills. The cycle consists of two stages. The suction pump stage is activated by an expansion of the mouth cavity and pharynx by the sternohyoideus muscle (Fig. 18–6**A** and **C**). Because the volume of the oropharynx is increased, a pressure lower than the ambient surrounding pressure is created and water flows into the mouth. At the same time the opercular cavity is greatly enlarged by dilator muscles, which result in an even lower pressure causing the water from the oropharynx to be drawn over the gills into the opercular cavity (Fig. 18–6**B** and **C**). In the second stage, the force pump is activated by the adductors and geniohyoideus muscles (Fig. 18–6**D**). The oropharynx is compressed, creating a positive pressure that is even larger than the rising pressure in the opercular cavity. As a result, water continues to flow from the oropharynx through the gills into the opercular cavity and thence to the outside through the opened opercular cavity. The efficiency of the teleostean respiratory system is based on the maintenance of a differential pressure between the oropharynx and opercular cavity, which are separated by a curtain of gills (Fig. 18–6). It produces an uninterrupted unidirectional flow of respiratory water that runs in the opposite direction of the blood flow for an optimized countercurrent system.

The structure of the enormous number of secondary lamellae provides a large surface area (*A* in the diffusion equation) and an exceptionally short diffusion distance (*d* in the diffusion equation) between water and blood. The surface epithelium on each side of a secondary lamella often is only one cell thick (Fig. 18–7**A**).

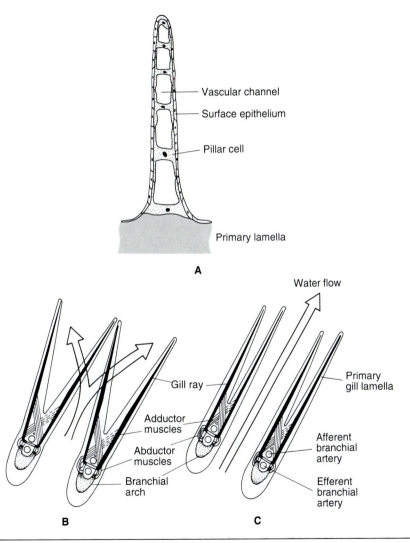

FIGURE 18–7   Gill structure in teleosts: **A**, A transverse section through a secondary lamella. **B**, Primary gill lamellae of adjacent gills meet when adductor muscles relax. **C**, Primary gill lamellae of adjacent gills separate when the adductor muscles contract. (**A**, after Hughes, 1984, **B** and **C**, after Bijtel.)

These epithelial layers are held apart by **pillar cells** bearing thin cytoplasmic flanges that spread out beneath them. Blood flows in the narrow spaces among the pillar cells rather than through typical capillary beds.

The total cross-sectional area of all the vascular channels within the secondary lamellae is considerably greater than the sum of the cross-sectional areas of the branchial arteries supplying them. Similarly, the cross-sectional area of all the spaces for water passage between the large number of secondary lamellae is greater than the cross-sectional area of the pharynx. Since the velocity of a liquid moving from a narrow area into a larger area decreases, as does water flowing from a stream into a pond, the rate of blood flow through the lamellae, and of water flow across them, is reduced. Adequate time for diffusion is available.

The combined surface area of the secondary lamellae varies greatly among phylogenetically unrelated species of teleosts and is probably adapted to their modes of life. Hughes (1984) has calculated that a sluggish, bottom-dwelling toad-fish has 151 mm^2 of gill surface area per gram of body weight, compared with 1241 mm^2/g for an active, pelagic menhaden. Diffusion distance across the gill lamellae is also less in active fishes, and the hemoglobin content of their blood is higher.

Gas exchange across the gills is accompanied by the diffusion of other small molecules. Considerable ammonia in most fishes and urea in some teleosts are excreted through the gills. Chondrichthyan fishes are an exception because their gills are impervious to the diffusion of urea, and a large quantity of this molecule accumulates in their body fluids and tissues. Water and salts also are lost or gained through the gills depending on their relative concentrations in the external environment and body fluids. Water, for example, is gained by osmosis through the gills of freshwater fishes but lost in most marine species whose environment is saltier than their body fluids (Chapter 20).

Because of the attendant osmotic problems, it is to a fish's advantage not to ventilate its gills, or perfuse blood through them, more rapidly than needed to meet its oxygen requirements at any particular time. Several mechanisms help match the rate of gas exchange with a fish's metabolic requirements.

1. Ventilation rate can be regulated.
2. The primary lamellae of adjacent gills can be brought together or moved farther apart (Fig. 18–7**B** and **C**). When adductor muscles within the gills are relaxed, the elasticity of the skeletal gill rays, assisted by the contraction of small abductor muscles, spreads the primary lamellae apart. The tips of adjacent gills meet, a damlike mechanism is formed, and most of the water passing across the gills must pass through the spaces between the secondary lamellae. When the adductor muscles contract, the tips of adjacent gills are pulled apart and much of the water leaves the branchial chambers without crossing the secondary lamellae.
3. The amount of blood perfusing the secondary lamellae can also be controlled. Vascular shunts that divert some of the blood directly from afferent to efferent branchial arteries without going through the secondary lamellae have been found in eels. In many other species the number of secondary lamellae being perfused is subject to control.

Accessory Respiratory Organs

The gills of bony fishes are such efficient respiratory organs that 80 to 95 percent of the limited oxygen available in the water is taken up by the blood. Yet there are situations where the gills alone cannot provide for a fish's needs. Oxygen levels can be very low in shallow, warm pools or in swamps where there is considerable decay, and in other bodies of stagnant water. A fish could not live in these habitats without an accessory respiratory organ that enables it to use the oxygen in the air. Gills are not effective for this purpose because the delicate gill lamellae cannot be supported in air; they collapse and clump together, greatly reducing the respiratory surface.

Many types of accessory respiratory organs enable bony fishes belonging to over 50 genera to obtain oxygen directly from the air. Freshwater eels often leave the water to migrate over land. They can use their scaleless and vascular skin as a respiratory membrane. Arborized and vascularized extensions from certain branchial arches allow walking catfish and climbing perch to breathe air (Fig. 18–8). The lining of the stomach of armored catfish and parts of the intestine of some loaches are vascularized and modified as lunglike structures. These fishes can obtain oxygen by swallowing air. Digestive functions are temporarily suspended when the gut is used in this way.

The buccal cavity and pharynx of actinopterygian fishes with accessory air-breathing organs initially expand and suck in air. Then these chambers act as a force or compressive pump forcing air into the cavity surrounding the accessory respiratory organ. The respiratory cycle is said to be driven by a **pulse pump,** reflecting the alternating suction and compression of the buccal cavity and pharynx.

Lungs

Lungs are but one of many accessory respiratory organs to have evolved in fishes. Many bony fishes have either lungs or swim bladders, except many bottom-dwelling species in which they have been secondarily lost, so these organs are a common character of the group. The organ is most lunglike in the primitive members of the group, so lungs appear to be the more primitive structure. One Devonian placoderm fossil (*Bothriolepis*) also has been discovered with impressions of what appears to be lungs.

Lungs evolved in certain placoderms and ancestral bony fishes that probably were living in unstable freshwater habitats during the late Silurian and Devonian periods (p. 74). Geological evidence suggests that the earth at this time was subject to pronounced fluctuations in rainfall. Dry and wet periods alternated as they do in some tropical countries today. During the hot, dry seasons, many bodies of fresh water would have become smaller, and water temperatures would have risen. Oxygen levels would have fallen. Some ponds probably became stagnant swamps or dried up completely. Chondrichthyan fishes and placoderms, most of which were marine, would not have been affected by these climatic fluctuations, but early freshwater bony fishes, including the rhipidistian ancestors of amphibians, would have been unable to survive unless they had accessory respiratory organs,

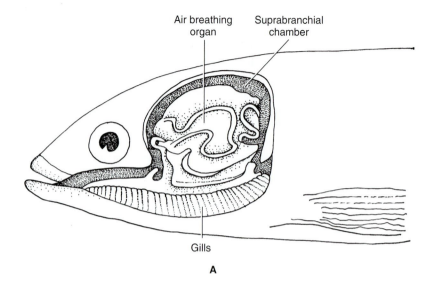

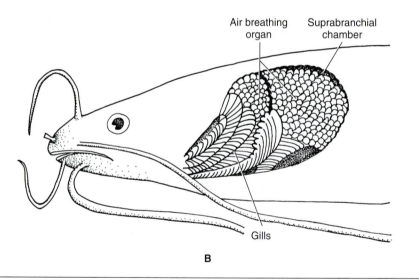

FIGURE 18–8   Air-breathing organs derived from labyrinthine expansions of gill arch elements in the climbing perch *Anabas* (**A**) and the "walking catfish" *Clarias* (**B**).

such as lungs. *Polypterus* and lungfishes have retained lungs because their ancestors continued to live in tropical freshwater environments (Fig. 18–9).

Lungs develop embryonically in bichirs (primitive chondrosteans), lungfishes, and amphibians as a ventral evagination from the floor of the digestive tract (pharynx or, in some cases, esophagus) just caudal to the last pair of pharyngeal pouches. The primordia of the lungs resemble a pair of displaced pharyngeal pouches in some amphibian larvae, and this has led some investigators to suggest a homology. In the bichir (*Polypterus*), the African lungfish (*Protopterus*), and amphibians, the

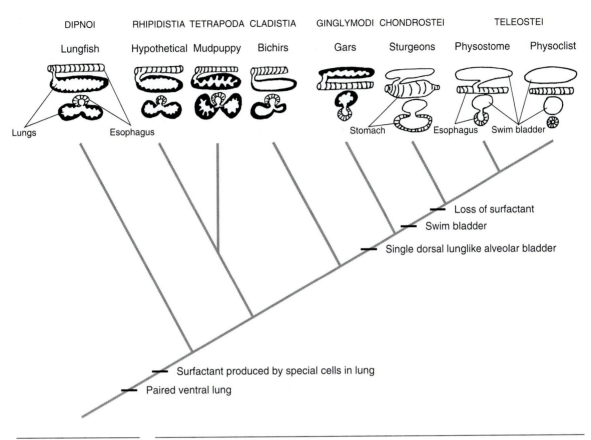

FIGURE 18–9   The evolution of lungs and the swim bladder.

bilobed lungs extend into the pleuroperitoneal cavity and grow dorsally, one on each side of the digestive tract (Fig. 18–9). In other lungfishes the single lung primordium extends caudally and dorsally on one side of the digestive tract, and may subsequently become bilobed. The similarity between the vascular connections of the lungs of lungfishes and those of *Polypterus* and amphibians indicates that they are homologous organs. Specialized cells in lungs secrete a surface film of lipoprotein, known as a **surfactant,** that reduces the surface tension of the lining. A surfactant has been demonstrated in the lungs of all pulmonate vertebrates that have been studied for this feature. Surfactants reduce the resistance to lung expansion and hence the energy needed to expand and fill them. A slit-shaped **glottis** on the floor of the digestive tract leads to the lungs.

When breathing air, a lungfish comes to the surface and first begins to inhale fresh air. It does so by expanding its buccal cavity and sucking in fresh air (Fig. 18–10). During this process the glottis opens and "spent" air is transferred from the lungs by elastic recoil of the pressurized lungs and contractions of smooth muscles that surround the lungs. As a result the spent air mixes with the just inhaled fresh air in the buccal cavity. Excess mixed air escapes from the open mouth. The fish closes its mouth and compresses its buccal cavity, which acts as

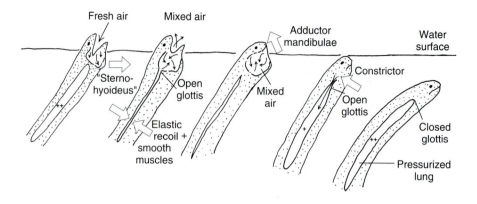

FIGURE 18–10 Ventilation of the lung by the buccal pulse pump in the South American (*Lepidosiren*) and African (*Protopterus*) lungfish. Small, black arrows indicate air flow. Wide, open arrows depict action of muscle, the width of which represents the level of activity.

a force pump driving the mixed air into the lungs. Thus the lungfish's ventilation pattern is typically a pulse pump. Because mixed air is delivered to the lung, this pattern is also called the **mixed-air buccal pump** system. This mode of ventilation is characteristic of sarcopterygian fishes. It has been verified by high-speed X-ray cinematography (Brainerd, Ditelberg, and Bramble, 1993) and direct measurements of the gas mixture. Lungfishes hold air in the lung for a considerable time as the oxygen is slowly used. Breathing is not continuous. Long periods of breath holding, or **apnea,** alternate with short periods of lung ventilation.

Some compartmentalization of the lung by internal septa provides a large surface area for gas exchange (Fig. 18–9). The Australian lungfish, *Neoceratodus*, uses its lung to supplement the gills in oxygen-poor water. It need not use them when oxygen supplies are normal. Internal lung septa are more numerous in *Protopterus*, which has lost its first two gills and is an obligate air breather. It will drown if prevented from reaching the surface to gulp air. The South American lungfish, *Lepidosiren*, is also an obligate air breather.

Carbon dioxide accumulates in the lungs during the periods of apnea. The hemoglobin of lungfishes, like that of terrestrial vertebrates, must be adapted to bind with oxygen in the presence of more carbon dioxide than normally is present in water. Accumulated carbon dioxide is released during expiration, but the frequency of lung ventilation is too low to unload all the carbon dioxide. Most continues to diffuse from the gills, since it is very soluble in water.

The Swim Bladder

The lungs of ancestral bony fishes have transformed into **swim bladders** in most neopterygian species. They live in oxygen-rich environments, and selection would favor the conversion of lungs into effective hydrostatic organs. In an early actinopterygian, the swim bladder arises as a dorsal outgrowth from the caudal end of the esophagus and remains connected to the esophagus by a **pneumatic duct** (Fig. 18–9). The pneumatic duct remains in **physostomous species** (Gr. *physa*

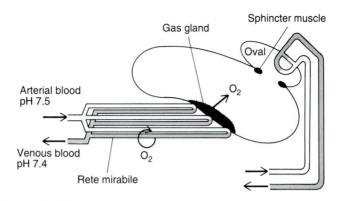

FIGURE 18–11   The operation of the swim bladder in a physoclistous teleost. (Modified after Alexander.)

= bladder + *stoma* = mouth) but is lost later in development by **physoclistous species** (Gr. *kleiein* = to close).

The swim bladder of an early actinopterygian continues to be an important site for the uptake of oxygen, but it also has acquired a hydrostatic function. In a teleost, it is primarily a hydrostatic organ, enabling the fish to attain neutral buoyancy and so maintain its position in the water with a minimum of muscular effort. Neutral buoyancy requires that the density of the fish equal that of the water. Density equals mass or weight divided by volume. A cubic centimeter of pure water weighs 1 g and so has a density of 1. But 1 cc of flesh weighs slightly more than 1 g. Since flesh alone has a density slightly greater than that of water, a fish without a swim bladder, such as many elasmobranchs, tends to sink. A sac of air in the fish reduces weight and can make its overall density equal to that of water. Problems arise, however, when a fish changes depth. When it goes deeper, water pressure increases, the swim bladder is compressed and becomes smaller, and the fish's density increases. The fish would sink faster except for the addition of enough gas to the swim bladder to maintain it at the appropriate size. When the fish rises, surrounding water pressure decreases, the swim bladder tends to expand, and gas must be removed from the bladder to maintain neutral buoyancy.

The gas in the swim bladder of most species is about 80 percent oxygen, and it is secreted into the bladder by a **gas gland** on one surface of the organ (Fig. 18–11). Often gas must be secreted against a considerable concentration gradient, for water pressure increases rapidly with depth. At a depth of 500 meters, for example, the fish and the gases in its swim bladder are subject to a pressure of 50 atmospheres, and the partial pressure of oxygen in the swim bladder is 40 atmospheres. The partial pressure, or tension, of oxygen in the water and in the fish's blood, however, is 0.2 atmosphere. (*Tension* is the term applied to the partial pressure of a gas when in solution.) Somehow, a fish secretes oxygen into the swim bladder against a pressure gradient and keeps it there. Oxygen is kept in the bladder by a **rete mirabile** (L. *rete* = net + *mirabile* = wonderful), a set of relatively long, parallel capillaries located just before the gas gland (Fig. 18–11). Together, the rete mirabile and gas gland are called the **red body.** The tension of the oxygen in the venous capillaries just leaving the swim bladder is high because it is in equilibrium with the partial pressure of oxygen in the swim bladder. Oxygen starts to be carried off, but the venous capillaries are surrounded by arterial capillaries carrying blood with a lower oxygen tension in the opposite direction. There is a countercurrent exchange: oxygen diffuses from the venous blood into the arterial blood and is carried back to the swim bladder. The rete mirabile acts as an oxygen gate, and the longer the capillaries in this organ, the less oxygen escapes.

Secretion of oxygen into the swim bladder requires a local increase in the acidity of the blood. As the blood becomes more acid, hemoglobin releases bound oxygen. There are two sources of acid. Carbon dioxide entering the blood combines with water to form carbonic acid. This reaction is accelerated greatly in the presence of the enzyme carbonic anhydrase in the red blood cells. In addition, considerable lactic acid is secreted by the gas gland. The release of oxygen by hemoglobin is a far faster reaction than its recombination, so considerable unbound oxygen accumulates in the blood, and the rete mirabile prevents most of it from escaping. Additional oxygen is continually being brought into the system by the arterial blood, and it, too, is released. Eventually, the tension of oxygen in the blood in the gas gland exceeds its partial pressure in the swim bladder, and oxygen diffuses into the swim bladder. The rete mirabile and gas gland together form a **countercurrent multiplier** system.

Removal of oxygen from the swim bladder as a fish rises in the water is a far simpler process. Oxygen is released through the pneumatic duct in physostomes, but it is reabsorbed from a special, vascularized compartment of the swim bladder known as the **oval** in physoclists. A sphincter that normally separates the oval from the main part of the swim bladder relaxes; oxygen enters the oval and, because of its high partial pressure, rapidly diffuses into the blood. Oxygen does not diffuse back into the blood through other parts of the swim bladder because its wall contains a diffusion barrier of guanine plates.

Although the swim bladder is primarily a hydrostatic organ, it has additional functions in certain species. It contains a store of oxygen that can be used in cellular respiration if needed. It is part of the hearing mechanism in ostariophysean fishes (p. 419). Drums and some other species can pluck on the swim bladder wall with specialized muscles and produce sounds.

Respiration in Early Tetrapods

Since dry, 20°C air at sea level contains 210 ml of oxygen per liter, adult terrestrial vertebrates have far more oxygen available to them than do fishes, but they need a moist respiratory membrane because the oxygen must be in solution to diffuse into the blood. Major problems for terrestrial vertebrates are how to expose and ventilate the respiratory membrane without an excessive loss of body water, and how to support it in air, which is less dense than water. Lungs, which terrestrial vertebrates inherited from fishes, are well adapted to meet these problems. They usually contain internal septa or divisions of the airways that increase the surface area of the respiratory membrane and provide the needed support. Air within the lungs contains a great deal of water vapor; in mammals it is normally saturated. The respiratory surface is kept moist without a great deal of water loss because

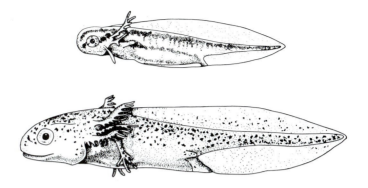

FIGURE 18–12   The development of the external gills in the larva of a salamander. (After Glaesner.)

the rate of ventilation is low, the air is conditioned by mucous glands in the airways prior to entering the lungs, and not all the air in the lungs is exchanged in each breathing cycle. The low rates of ventilation and exchange are made possible by the high oxygen content of the air in the lungs. Water vapor, which is a gas, reduces the partial pressure of oxygen in the lungs to about 100 mm Hg, compared with 160 mm Hg for dry air at sea level, but this is still ample.

Amphibian Respiratory Organs

The larvae of two of the three genera of contemporary lungfishes have external gills attached to the surface of their heads, and the larvae of rhipidistians probably had them, too. Amphibian larvae are aquatic and retain external gills. Fossil evidence indicates that larval labyrinthodonts had external gills. Salamanders retain external gills throughout their larval period (Fig. 18–12), and neotenic species, such as the mudpuppy, *Necturus*, have them throughout life. The external gills of young frog tadpoles become covered by an opercular fold that opens on the body surface through a single pore called the **spiracle.** (This term is unfortunate, since the pore is neither homologous nor analogous to the spiracle of a fish.) The **operculum** is extensive and even envelops the developing forelimbs. Older tadpoles lose the external gills as they develop deeper ones attached to the branchial arches. We are uncertain whether these deeper gills are homologous to the internal gills of fishes. Lungs develop and begin to function late in larval life. Gills are lost at metamorphosis, the forelimbs push through the operculum, and the remains of the operculum fuse with the body wall.

The frog respiratory system is representative of contemporary adult amphibians (Fig. 18–13). Valved, external nostrils lead to short nasal cavities that open through choanae at the front of the palate. Air passes through the buccopharyngeal cavity and enters the glottis, which is supported by small, **lateral laryngeal cartilages** derived from the sixth visceral arch. The glottis leads to a small, triangular **laryngotracheal chamber,** from which the lungs emerge. All these passages are lined by cilia and by mucous and serous secretory cells. These keep air moist and trap dirt particles and carry them away from the lungs. The lungs of frogs are sac-

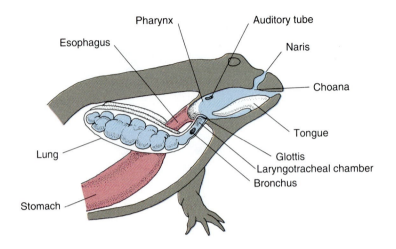

FIGURE 18-13   A sagittal section of the respiratory system of the frog. (After Dorit, Walker and Barnes.)

shaped organs; in salamanders they are elongated. The respiratory surface of frog lungs is increased by primary septa, and sometimes also by secondary ones, that form a series of pockets. The interior of the lungs is open.

Ventilation of the lungs of contemporary amphibians resembles that of lung-fishes, except that air enters and leaves through the nares and nasal cavities rather than through the mouth opening (Fig. 18–14). In a representative breathing cycle, a lowering of the pharynx floor draws fresh air through the nasal cavities into the buccopharyngeal cavity. The glottis then opens, and positive pressure in the lungs drives stale air out. Expiration is assisted by the elastic recoil of the lungs, and

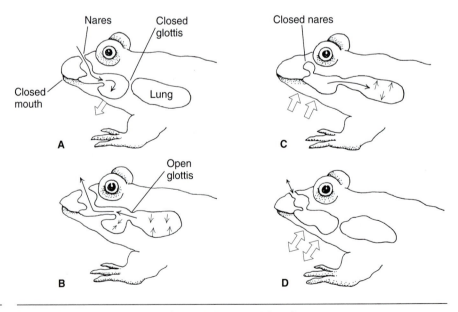

FIGURE 18-14   The ventilation of a frog's lungs. (After DeJongh and Gans.)

especially by the contraction of flank (hypaxial) muscles (DeJongh and Gans, 1969). The expired air is mainly expelled over the fresh air, which is sequestered to some degree in the floor of the buccopharyngeal cavity, but some mixing of air does occur. The nostrils are closed, the pharynx floor is raised, air is forced into the lungs, and the glottis closes. The pumping action of the pharynx may be repeated until the lungs are well inflated. Modifications of this sequence have been described (Vitalis and Shelton, 1990). Lung ventilation in amphibians is accomplished by a pulse pump and with the assistance of the hypaxial flank muscles. As in lungfish, amphibians can tolerate prolonged periods of apnea alternating with brief periods of ventilation. Because their metabolism and oxygen needs are low, a lung full of air can provide an amphibian with oxygen for many minutes. Because ventilation rates are low, eliminating carbon dioxide is more of a problem than obtaining oxygen.

In addition to lungs, the thin, moist, vascular skin of most amphibians acts as a respiratory membrane. Considerable **cutaneous gas exchange** occurs in most species, but there is much variation. Some arboreal, tropical frogs that live in dry habitats have very little cutaneous gas exchange. On the other hand, one large family of salamanders (the plethodontids) have lost their lungs and depend on cutaneous gas exchange. Reliance on cutaneous gas exchange restricts these animals to habitats where temperatures are cool and the oxygen supply dependable. Some live on the bottom of cool mountain streams; others live beneath rocks and logs in damp woods. Plethodontids are also small animals with a surface/mass ratio favorable for gas exchange through the skin, and they are not very active.

In species in which both the lungs and skin are used as respiratory membranes, the lungs are usually more important for oxygen uptake than carbon dioxide loss, and the skin is more important for carbon dioxide loss than oxygen uptake (Shoemaker et al., 1992). But the role of the lungs and skin varies with temperature and activity. As temperature rises and/or activity increases, the lungs assume a greater role in the exchange of both gases. Conversely, as temperature falls or activity decreases, the skin becomes more important. At very low temperatures, the lungs may not be used at all.

Cutaneous gas exchange also exposes amphibians to considerable water loss. Amphibians mitigate this by living in wet or moist habitats where they can make up for the lost water, and, in some species (toads), by being most active during the early morning and evening when humidity is high.

As part of their reproductive behavior, frogs have evolved a method of vocalization. **Vocal cords** are located in the laryngotracheal chamber and can be vibrated by passing air across them. Although both sexes have vocal cords, only the males have resonating **vocal sacs.** These are evaginations from the floor or lateral walls of the pharynx that can be filled with air. It is the distinctive call of the males in the spring that attracts conspecific females.

The Reptile Respiratory System

Gas exchange in the embryos of reptiles, birds, and egg-laying mammals is through the allantois, a vascularized extraembryonic membrane that extends from the hindgut to unite with the chorion just beneath the eggshell (p. 132). The chorioallantoic membrane contributes to the placenta in eutherians.

Neck length increases in reptiles, and the primitive laryngotracheal chamber becomes divided into a **larynx** (Gr. *larynx* = gullet) and **trachea** (Gr. *tracheia* = rough artery). The larynx wall is supported by a pair of **arytenoid cartilages** that flank the glottis and by a ring-shaped **cricoid cartilage.** These cartilages evolved from the lateral laryngeal cartilages of amphibians. Vocal cords are absent, but some reptiles can squeak, hiss, or bellow by rapidly expelling air. Cartilaginous rings, or partial rings, in the tracheal wall keep the trachea open for the free flow of air.

The horny skin of most reptiles precludes effective cutaneous gas exchange, so the lungs of adults are the primary or only site for gas exchange. Certain aquatic turtles exchange some gases through their skin and also the buccopharyngeal cavity and cloaca, in and out of which they can pump water. Some species of turtles have accessory, vascular bladders that evaginate from the cloaca for this purpose (Fig. 17–10). *Sphenodon* retains amphibian-like lungs with only a slight increase in internal surface area, but most reptiles have lungs that are relatively more compartmentalized and larger than those of amphibians (Fig. 18–15). A wide **central**

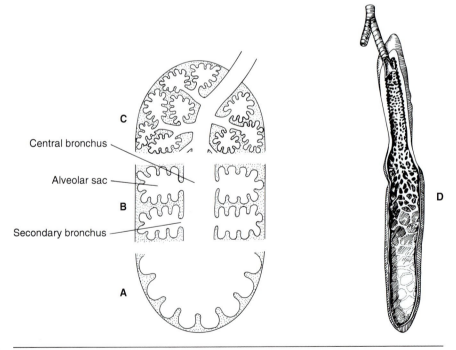

Central bronchus

Alveolar sac

Secondary bronchus

FIGURE 18–15 A diagram (left) of the compartmentalization of the lungs of tetrapods: **A,** An amphibian. **B,** A reptile. **C,** A mammal. **D,** The lung of a lizard, *Ophiosaurus*. (After Portmann.)

bronchus (Gr. *bronchos* = windpipe) leads from the trachea into each lung, **secondary bronchi** branch from it, and **alveolar sacs** of varying size bud off them. Lung structure varies considerably among species. The anterior part of the lung is more compartmentalized in lizards and snakes than the posterior part, which is frequently a simple sac with poorly vascularized walls. The posterior region appears to serve for air storage or acts as a bellows to help ventilate the rest of the lung. This region parallels the development of air sacs in birds. One lung usually is lost in amphisbaenians, snakes, and long-bodied lizards. As one would expect, large species have more compartmentalization of the lungs than smaller ones. This maintains an adequate ratio between respiratory surface and body mass.

Of particular significance is the reptilian method of ventilating the lungs. Reptiles do not ventilate their lungs by a pulse pump. Instead they use an **aspiration pump.** During inspiration, movements of the ribs enlarge the pleuroperitoneal cavity in which the lungs lie, pressure within the pleuroperitoneal cavity falls below atmospheric pressure, the lungs expand, and air is sucked into them, hence the term *aspiration pump.* The glottis is closed and air is held in the lungs until the next breathing cycle. Prolonged periods of apnea occur, as in lungfishes and amphibians. Expiration results from opposite rib movements and from the contraction of smooth muscle fibers in the lung wall. Aspiration by a suction pump mechanism is more efficient than a buccal force pump because air can be transferred into the lungs in one movement. The first documented aspiration pump in vertebrate evolution is found in *Polypterus* (Focus 18–1).

Crocodiles have evolved an unusual method of ventilation. Their lungs lie in separate pleural cavities anterior to the liver. The contraction of a unique **diaphragmatic muscle,** which extends from the liver to the pelvic girdle, pulls the liver caudad and enlarges the pleural cavities (Fig. 18–16). (This muscle is only analogous to the mammalian diaphragm.) Expiration occurs by the contraction of abdominal flank muscles, which increases intra-abdominal pressure and pushes

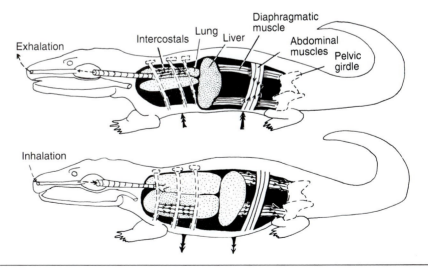

FIGURE 18–16 Muscle actions and movements of ribs, liver, and abdominal wall in the caiman. (Modified from Gans and Clark.)

| FOCUS 18-1 | The First Aspiration Pump in the Evolution of Vertebrates |

Virtually all air-breathing fishes and amphibians ventilate their lungs by using a pulse pump. In pulse pump systems air forces the lung to expand, whereas in aspiration systems air is sucked into the already expanding lung. Air-breathing fishes and amphibians begin to fill the lung only after the mouth is closed and the compressive muscles act on the buccal force pump. A remarkable exception is found in the polypterid fishes *Polypterus* (bichir) and *Calamoichthys* (reedfish). The first appearance of lung ventilation by aspiration in the evolutionary history of vertebrates is exhibited by the pattern of ventilation in these primitive actinopterygians. Polypterid fishes are encased in a stiff scale jacket analogous to the dermal armor of early tetrapods (Focus Figure Part **A**) Cineradiographic analysis, pressure recordings, and strain gauge recordings sensing deformations of the "dermal armor" by Brainerd, Liem, and Samper (1989) have elucidated an evolutionarily early aspiration pump. The fish exhales first by contraction of smooth muscles around the lung (Part **B**). Spent air is forced into the pharynx and exits from the opercular slit (Part **C**). The reduction in size of the lungs will cause a decrease in volume of the space occupied by organs within the peritoneal cavity, causing the ventral dermal armor to buckle inward (Part **C**). Such buckling is made possible by the nature of the peg and socket articulations between the thick scales.

These articulations have the capacity to undergo recoil after the initial deformation. As the dermal armor passively recoils to its original resting position, it creates a subatmospheric pressure in the pleuroperitoneal cavity, causing the lung to expand (Part **D**). The resulting subatmospheric pressure in the expanding lung will draw air in through the widely opened mouth very rapidly. The use of recoil aspiration by polypterid fishes demonstrates that a stiff body wall capable of storing energy can contribute to inhalation in vertebrates.

The ribs of polypterid fishes are confined to the dorsal part of the body and are not directly involved in the ventilation process even though they provide stiffness of the body, thus allowing deformation of the belly. The discovery of recoil aspiration in polypterid fishes has important implications for our attempts to understand the ventilation mechanics of early Paleozoic amphibians (Part **E**). Early Paleozoic amphibians retained ventral, bony scales from their air-breathing fish ancestors. These V-shaped scale rows are very similar to the rhomboid scale body armor of polypterid fishes. The configuration of their ribs suggests that they confer stiffness on the body wall. The presence of recoil aspiration in polypterid fishes encased in a thick dermal scale jacket suggests that the ventral dermal armor of early amphibians may well have played an important role in aspiration breathing.

---

the liver forward. Turtles ventilate their lungs by limb movements and by the contraction of muscles lining the limb pockets. Other parts of their body walls are firmly encased by the carapace and plastron.

Respiration in Birds

Because they are endothermic and flying animals, birds need an exceptionally efficient and compact respiratory system that will sustain a high level of metabolism and not add greatly to their weight. Bird lungs are relatively small organs that adhere to the dorsal wall of the pleural cavities and do not change appreciably in volume (Fig. 18–17**A**). The lungs themselves are only half the size of those of a

FOCUS 18–1 *continued*

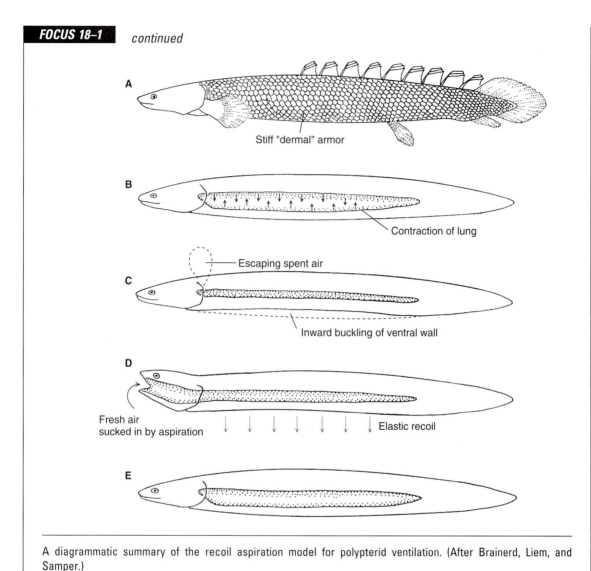

A diagrammatic summary of the recoil aspiration model for polypterid ventilation. (After Brainerd, Liem, and Samper.)

mammal of comparable size, but they connect to a system of **air sacs** that pass among the viscera and even extend into many of the bones. Together, lungs and air sacs have a volume two to three times that of the lungs of a comparable mammal. The air sac walls are not heavily vascularized, so they do not contribute to the gas exchange surface. Rather, the air sacs, in combination with a unique pattern of airways within the lungs, make possible a unidirectional flow of air through the lungs. In other pulmonate vertebrates, whose air moves in and out of the lungs through the same passages, a certain amount of stale air always mixes with inspired air. But in birds the one-way flow carries relatively fresh air, with more oxygen and less carbon dioxide, across the respiratory surfaces.

The morphology of the avian respiratory system is complex, but the essential features and the way they operate have been studied by Bretz and Schmidt-Nielsen (1971). By inserting minute probes into the airways, they have determined the composition of the air within them and the direction of its flow. The numerous air sacs can be grouped functionally into an anterior and a posterior set (Fig. 18–17**A**). The trachea bifurcates into a pair of **main bronchi,** each of which passes rather directly through the center of a lung to a **lateroventral bronchus,** which connects with the posterior air sacs (Fig. 18–18**A**). Air in the posterior sacs returns to a lung through a **mediodorsal bronchus,** which connects with thousands of small **parabronchi.** Innumerable short **air capillaries** bud off the parabronchi and branch and interconnect (Fig. 18–17**B**). They are surrounded by dense vascular capillary beds, and it is here that gas exchange occurs. Distances are short, and diffusion appears to keep the composition of the air in the air capillaries in equilibrium with that flowing unidirectionally through the parabronchi. The parabronchi lead to a **medioventral bronchus,** which in turn connects with the anterior air sacs and main bronchus (Fig. 18–18**B**). The parabronchi are parallel to each other throughout the lung in early birds and in most of the lung in other species. This part of the lung is called the **paleopulmo.** A small **neopulmo,** in which the parabronchi form a network, is also incorporated into the lung of most birds.

The lungs are ventilated by rocking movements of the sternum, which alternately expand and compress the bellows-like air sacs. Hinges between the dorsal and the ventral parts of the ribs allow the sternum to move (Fig. 8–14**A**). The uncinate processes on the dorsal parts of the ribs act as lever arms on which certain respiratory muscles attach. Because flight muscles also cause movements of the sternum, clavicle, and ribs, ventilation and the flight movements are coupled and occur in synchrony (Jenkins, Dial, and Goslow, 1988). The glottis is held open most of the time because bird lungs are ventilated continuously, and there are no prolonged periods of apnea. Two inspirations and expirations are needed to move a unit of air through the system (Fig. 18–18). During the first inspiration, fresh

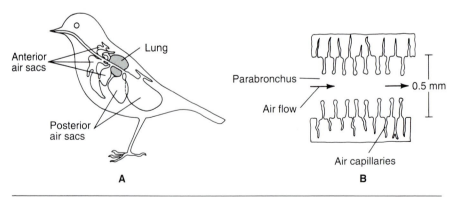

A

B

FIGURE 18–17   The anatomy of the respiratory system of a bird: **A,** A lateral view of the lungs and air sacs. **B,** A longitudinal section through a parabronchus and air capillaries. (**A,** after Salt. **B,** after Schmidt-Nielsen.)

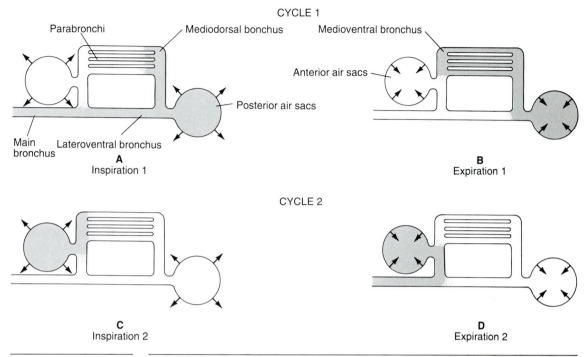

CYCLE 1

Parabronchi   Mediodorsal bonchus   Medioventral bronchus

Anterior air sacs

Posterior air sacs

Main
bronchus   Lateroventral bronchus
                    A   **B**
              Inspiration 1   Expiration 1

CYCLE 2

                    C   **D**
              Inspiration 2   Expiration 2

FIGURE 18–18   The movement of a volume of air (shaded) through the lungs and air sacs of a bird. Two
cycles of inspiration and expiration are needed to move the air through the system. (After
Bretz and Schmidt-Nielsen.)

air is drawn through the main bronchi and lateroventral bronchi into the posterior
air sacs (Part **A**). This air moves through the mediodorsal bronchi and parabronchi
during the first expiration (Part **B**). It is drawn into the anterior air sacs during
the second inspiration (Part **C**) and finally is expelled from the system during the
second expiration (Part **D**). Air, of course, enters and leaves the lungs and air sacs
during each cycle of inspiration and expiration, but it is a different volume of air.
During inspiration one volume of air passes through the lungs to the posterior air
sacs (Part **A**), and a volume of air inhaled previously is drawn out of the lungs into
the anterior air sacs (Part **C**). During expiration, air in the posterior sacs (Part **B**)
is driven into the lungs, and the stale air is expelled from the anterior air sacs
through the main bronchi and trachea to the outside (Part **D**).

Air flows through the system along the line of least resistance, and this is
determined primarily by the diameters and directions of the openings of the various bronchi and by other aerodynamic features. Smooth muscle fibers around the
openings of the medioventral bronchi also have a valvelike effect. The relatively
high carbon dioxide content of the air leaving the parabronchi relaxes these muscles, and this facilitates air entering the anterior sacs.

Blood in the vascular capillaries flows transversely to the air flow in the parabronchi and air capillaries in a pattern called **cross-current flow** (Fig. 18–19).
As air flows from right to left in the model, it loses oxygen to the blood; blood

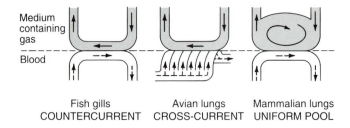

Fish gills
COUNTERCURRENT

Avian lungs
CROSS-CURRENT

Mammalian lungs
UNIFORM POOL

FIGURE 18–19   A comparison of the movement of blood and the external medium (water or air) through the respiratory systems of a fish, a bird, and a mammal. (After Piiper and Scheid.)

flowing across the air gains oxygen. Since air and blood flow across each other, a gradient for the transfer of oxygen exists at each crossing of parabronchi and vascular capillaries, but the gradient is different at each intersection. The net result of this system, as in the countercurrent flow of blood and water in a fish's gills, is that the blood leaving the system contains more oxygen than the exhaled air. In the lungs of other vertebrates, the oxygen content of the blood and air reach an equilibrium. The unidirectional flow of air through a bird's lungs, the cross-current flow of air and blood, and the very large gas exchange surface combine to produce an extraordinarily efficient respiratory system. Birds extract a greater proportion of oxygen from the air than mammals, and they can live and fly at high altitudes where mammals cannot survive.

Birds have also evolved a vocal apparatus that is associated with the airways. The calls and songs of birds are used in species recognition, for sounding alarms, in establishing territories, and in reproductive behavior. The vocal box is not the larynx, as in frogs and mammals, but a **syrinx** (Gr. *syrinx* = panpipe) that is located at the bifurcation of the trachea into the main bronchi (Fig. 18–20**A**). Syrinx structure varies among species, but usually one or more tympanum-like membranes lie between cartilaginous rings in its wall, and they are vibrated by air moving across

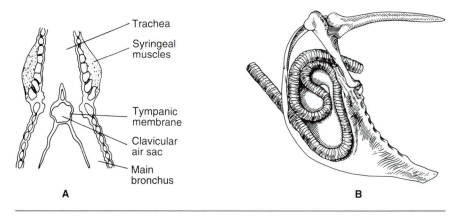

Trachea

Syringeal muscles

Tympanic membrane

Clavicular air sac

Main bronchus

A

B

FIGURE 18–20   Vocalization in birds: **A**, A frontal section through the syrinx of a male blackbird, *Turdus merula*. **B**, A lateral view of a dissection of the trachea and sternum of a whooping crane. (**A**, after Pettingill. **B**, after Portmann.)

them. Air vibrations are transformed into meaningful sounds by changes in tension on the tympani, by the configuration of the trachea and buccopharyngeal cavity, and by tongue movements. Trumpeter swans and whooping cranes have exceedingly long tracheae, half or more of which loops within the sternal keel (Fig. 18–20**B**). This gives their calls their deep, resonating, and somewhat trombone-like quality.

Respiration in Mammals

Mammals diverged from reptiles millions of years before the line to birds separated. Although they, too, evolved an efficient respiratory system, needed by endothermic animals, the mammalian system differs in many ways from that of birds because it evolved independently. The evolution of the secondary palate in synapsids and mammals made possible a separation of food and respiratory passages (p. 238). The paired **nasal cavities** of mammals, dorsal to the hard palate, are relatively larger than those of other vertebrates. Each contains three folds or scrolls of bones, known as the **turbinates** or **conchae,** which greatly increase the surface area (Fig. 18–21). The mucous membrane that covers the turbinates and lines the

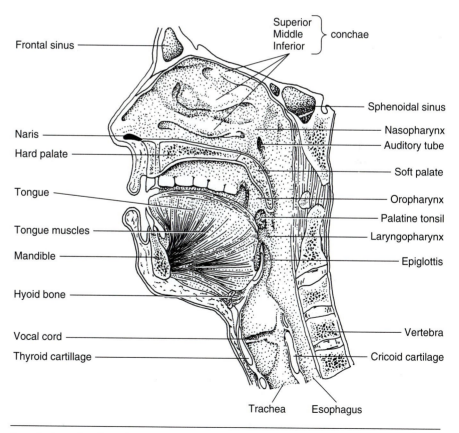

FIGURE 18–21   A sagittal section through part of the head and neck of a human being.

other respiratory passages as they continue into the lungs is very vascular, ciliated, and contains cells that produce mucous and serous secretions. Considerable conditioning of the air thus occurs as it passes across these surfaces to the lungs. The air is warmed and moistened, and dirt trapped in the mucus is carried by ciliary action to the throat to be swallowed or expectorated.

The nasal cavities connect through the **choanae** with the **nasopharynx,** which is separated from the oropharynx by the soft palate. Food and air passages cross in the laryngopharynx, but food normally does not enter the respiratory passages because the larynx and hyoid apparatus are pulled forward against the base of the tongue during swallowing. A troughlike fold, the **epiglottis,** flips back over the glottis or deflects food around it into the esophagus. The epiglottis is supported by a newly evolved cartilage that appears to have no homolog in other species. Except when food is swallowed, the **glottis,** which is bounded laterally by the **vocal cords,** is held open because mammalian lungs are ventilated continuously.

The mammalian **larynx** is more complex than that of other tetrapods (Fig. 7–23). It continues to be supported caudally by the ring-shaped **cricoid cartilage,** and its rostrodorsal wall is supported by the paired **arytenoid cartilages,** which also extend into the vocal cords. A new **thyroid cartilage** supports its lateroventral wall. The thyroid cartilage evolved from the fourth and fifth visceral arches as they became dissociated from the hyobranchial apparatus. The intrinsic muscles of the larynx that shape it and control the vocal cords are muscles of the fourth, fifth, and sixth visceral arches; as would be expected, they are innervated by a branch of the vagus nerve. Vibrations of air produced by the vocal cords are "shaped" by movements of the pharynx, soft palate, tongue, and lips.

The **trachea** continues down the neck and gives rise within the thorax to **primary bronchi,** which enter the lobes of the lungs. The internal passages within mammalian lungs are more finely compartmentalized than those of amphibians or reptiles, so there has been a great increase in internal surface area (Figs. 18–15**C** and 18–22). Whereas a frog has about 20 cm^2 of lung surface area per cc of lung tissue, a human being has 300 cm^2 per cc. The airways within the lungs branch and rebranch at least 20 times, forming a **respiratory tree** that terminates in **bronchioles, alveolar sacs,** and individual **alveoli.** The walls of the airways become progressively thinner along the course of the respiratory tree. Supporting cartilages, smooth muscle, and secretory and ciliated cells gradually disappear. The walls of the alveoli, through which gas exchange occurs, consist only of an exceedingly thin squamous epithelium that is nearly completely covered by vascular capillaries. Diffusion distance is about 0.2 micrometers (μm). Only in birds is the distance shorter, on the order of 0.1 μm. The alveoli are tightly packed, and the capillaries lie between them. This arrangement allows diffusion of gases to occur across two or more surfaces of the capillaries (Fig. 18–22**C**).

Mammalian lungs are ventilated by an aspiration pump. Air is moved in and out by changes in the size of the pleural cavities, alternately raising and lowering the pressure within the lungs relative to atmospheric pressure. Pleural cavity size is increased primarily by the contraction and caudal movement of the muscular **diaphragm** (Fig. 18–22**A**). During stronger inspiration, the external intercostal and other respiratory muscles pull the ribs rostrally and increase the dimensions

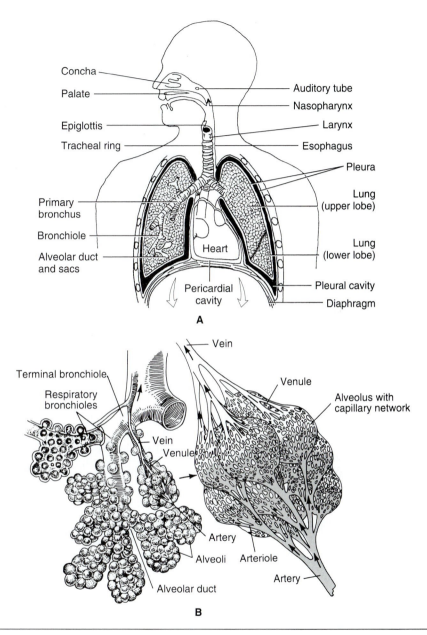

Concha

Palate

Epiglottis

Tracheal ring

Primary
bronchus

Bronchiole

Alveolar duct
and sacs

Auditory tube

Nasopharynx

Larynx

Esophagus

Pleura

Lung
(upper lobe)

Lung
(lower lobe)

Pleural cavity

Diaphragm

Heart

Pericardial
cavity

A

Vein

Terminal bronchiole

Respiratory
bronchioles

Venule

Alveolus with
capillary network

Vein
Venule

Artery

Alveoli

Arteriole

Artery

Alveolar duct

B

FIGURE 18–22   The mammalian respiratory system: **A,** An overview of the location and structure of the airways and lungs. **B,** An enlargement of the alveoli and their capillary bed. (*continued*)

of the thorax. Pressure within the lungs must be reduced enough during inspiration to overcome the frictional resistance of the numerous airways to the inflow of air, and also to overcome the surface tension of the walls of the many thousands of alveoli. This surface tension tends to collapse the alveoli. The resistance of the airways is reduced by a pattern of branching that maximizes the diameters of the

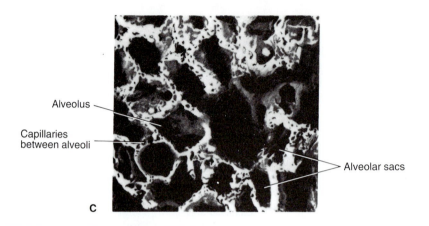

Alveolus

Capillaries
between alveoli

Alveolar sacs

C

FIGURE 18–22   *continued*
C, A drawing based on an electronmicrograph of a portion of a mammalian lung. Note that
most of the capillaries have alveoli on two of their surfaces. (**A** and **B,** after Dorit, Walker,
and Barnes. **C,** after Kessel and Kardon.)

major airways, by cartilaginous supports in their walls, and by the secretion of
surfactants. Surfactants are particularly important in mammals because of the
small size and large surface area of their alveoli. Expiration is largely a passive
process that depends on the elastic recoil of the lungs, ribs, and abdominal viscera,
which are pushed caudally by the contraction of the diaphragm. Internal intercostal, abdominal, and other respiratory muscles become active during forced expiration. Although certain intercostal muscles become active during strong inspiration and expiration, the primary function of these muscles is to maintain the
integrity of the chest wall so that the intercostal spaces do not bulge outward or
cave in as pressure within the thorax changes.

Evolution of Respiratory Patterns in Air-Breathing Vertebrates

The primitive pattern of the vertebrate breathing is the pulse pump found in
actinopterygian fishes (Fig. 18–23). When ventilating their lungs or lunglike air-
breathing organs, actinopterygian fishes first expire the spent air, after which they
fill their buccal cavity with fresh air, which is then compressed into the lung. In
the sarcopterygians and amphibians, fresh air and spent air are mixed in the buccal
cavity prior to being transferred by the buccal force pump into the lung. The
patterns of aerial ventilation in the actinopterygians, sarcopterygians, and amphibians are derived from the ancestral aquatic pulse pump of water-breathing fishes.
Actinopterygian and sarcopterygian air-breathing fishes rely solely on the buccal
force pump to drive respiratory air into the lung. Exhalation in actinopterygians
and sarcopterygians depends on suction by an expanding buccopharynx aided by
contractions of the muscles surrounding the lung.

In amphibians, the flank (hypaxial) muscles assume a respiratory role for the
first time in vertebrate history (Brainerd, Ditelberg, and Bramble, 1993). These

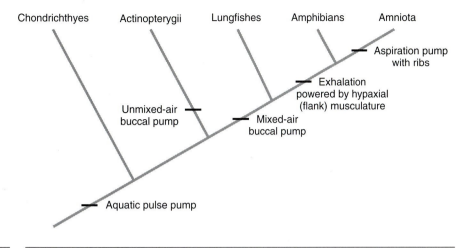

Chondrichthyes   Actinopterygii   Lungfishes   Amphibians   Amniota

Aspiration pump
with ribs

Exhalation
powered by hypaxial
(flank) musculature

Unmixed-air
buccal pump

Mixed-air
buccal pump

Aquatic pulse pump

FIGURE 18–23   Evolution of lung ventilation. (After Brainerd, Ditlberg, and Bramble.)

flank muscles play a role during exhalation and help empty the lung. Flank muscles continue their role in ventilating lungs in reptiles and mammals by moving the ribs.

In amniotes, the buccal pump is lost and an entirely new mode of respiration has emerged. Air is no longer compressed and forced into the lung but is sucked into the lung by aspiration. This is accomplished by an expanding cavity in which the lungs are housed. Expansion of the cavity is effected in various ways. In most reptiles, changes in the configuration of the rib cage by muscular actions create a sufficiently low pressure in the pleuroperitoneal cavity to make the lungs expand. Specialized reptiles such as the Crocodylia can move their livers back and forth through the action of a specialized diaphragmatic muscle. In mammals, a new structure, the diaphragm, is interposed between the pleural and peritoneal cavities. Contractions of the diaphragm assume the primary function in perfecting the aspiration pump.

SUMMARY

1. Gases diffuse between the external medium (water or air) and the blood in the body through a respiratory membrane. Vertebrates require a means of ventilating this membrane by the bulk flow of the medium across it.
2. The respiratory system of a fish must be adapted to the limited supply of oxygen available in the water and to the high density of this medium. Unloading carbon dioxide is not a problem for a fish because this gas is very soluble in water. Heat, salts, and water are also exchanged across the respiratory membrane.
3. Many larval fishes have external gills, but gas exchange in adults occurs through internal gills.
4. The gill pouches of fishes develop from endodermal pharyngeal pouches that meet ectodermal furrows extending inward from the body surface. The first pouch often is reduced to a spiracle or has been lost.

5. A complete gill, or holobranch, bears primary gill lamellae on each surface. These support numerous secondary gill lamellae through which gas exchange occurs. A skeletal branchial arch, gill rays, vascular derivatives of an embryonic aortic arch, muscles, and nerves lie within the holobranch, often in an interbranchial septum.

6. The pouched gills of agnathans line spherical gill pouches. The pouches are numerous.

7. Elasmobranchs have septal gills borne on interbranchial septa and facing narrow branchial chambers. Each chamber opens into a parabranchial chamber and to the surface through an external gill slit.

8. A countercurrent flow—whereby water flows across the secondary lamellae in the opposite direction that blood flows through them—increases the efficiency of gas exchange.

9. The valvular action of the mouth and external gill slits allows the pharynx and parabranchial chambers, by their concurrent expansion and contraction, to act together first as a suction pump and then as a force pump in ventilating the gills.

10. Teleosts have aseptal gills, which extend into a common opercular cavity that opens to the surface through a single gill slit. The absence of septa increases the efficiency of the countercurrent exchanges of gases between blood and water. The pharynx and opercular cavity act alternately and in sequence as suction and force pumps. The combination of suction and force is described as a pulse pump.

11. The structure of the secondary lamellae provides a large surface area and a short diffusion distance between water and blood.

12. Because of the attendant osmotic problems, a fish does not ventilate its gills more than necessary to meet its oxygen needs. Various mechanisms regulate the amount of ventilation.

13. Lungs probably evolved in ancestral bony fishes as accessory respiratory organs that permitted them to live in unstable freshwater environments.

14. Lungfishes also ventilate their lungs with a pulse pump. Some mixing of fresh and spent air occurs, and this mixed air is held in the lungs under positive pressure for a prolonged period of apnea. Some carbon dioxide accumulates in the lungs, but most is eliminated through the gills.

15. The lungs of ancestral bony fishes have transformed into hydrostatic swim bladders in most contemporary species.

16. Fishes attain neutral buoyancy by regulating the amount of air in the bladder. In most species oxygen is secreted into the bladder by a gas gland and held in the bladder by a countercurrent rete mirabile. Oxygen can be reabsorbed through the oval.

17. Terrestrial vertebrates must avoid an excess loss of body water through their lungs. The large amount of oxygen in air and a low metabolic rate allow amphibians to avoid water loss by ventilating their lungs at a low rate.

18. Contemporary adult amphibians ventilate their lungs by a buccopharyngeal pump and retain air in the lungs under positive pressure for a prolonged period of apnea. Oxygen also is gained and carbon dioxide is eliminated by cutaneous respiration.

19. Plethodontid salamanders have lost their lungs and rely on cutaneous respiration for the exchange of gases.

20. Lung surface area is relatively larger in reptiles than in amphibians. Reptiles use rib movements, limb movements, and (in crocodiles) an unusual muscle that pulls the liver caudad to ventilate the lungs. Reptiles continue to have prolonged periods of apnea. Reptiles have an aspiratory pump.

21. Endothermic birds have small lungs that are connected to an extensive system of air sacs. Their lungs are ventilated continuously by rib and sternal movements and a bellows-like action of the air sacs.

22. The airways of birds are arranged in such a way that a unidirectional flow of air passes through numerous parabronchi and across air capillaries. Little stale air is retained in the lungs. A cross-current flow of blood over the parabronchi and of air within them thoroughly aerates the blood.

23. The airways of mammals end in numerous, thin-walled alveoli within the lungs that are covered by dense capillary networks. A large surface area is available for gas exchange, but new air entering the lungs always mixes with some stale air. The lungs are ventilated continuously by a well-developed muscular diaphragm.

REFERENCES

Alexander, R. McN., 1966: Physical aspects of swim bladder function. *Biological Reviews,* 41:141–176.

Bijtel, H. J., 1949: The structure and the mechanism of the movement of the gill filaments in the Teleostei. *Archives Neerlandaises Zoologie,* 8:1–22.

Brainerd, E. L., Ditelberg, J. S., and Bramble, D. M., 1993: Lung ventilation in salamanders and the evolution of vertebrate air-breathing mechanisms. *Biological Journal of the Linnean Society,* 49:163–183.

Brainerd, E. L., Liem, K., and Samper, C. T., 1989: Air ventilation and recoil aspiration in Polypterid fishes. *Science,* 246:1593–1595.

Bramble, D. M., and Carrier, D. R., 1983: Running and breathing in mammals. *Science,* 291:251–256.

Bretz, W. L., and Schmidt-Nielsen, K., 1971: Bird respiration: Flow patterns in the duck lung. *Journal of Experimental Biology,* 54:103–118.

Burggren, W., and Roberts, J., 1991: Respiration and metabolism. *In* Prosser, C. L., editor: *Comparative Animal Physiology,* 4th edition, *Environmental and Metabolic Animal Physiology.* New York, Wiley-Liss.

Clements, J. A., Nellenbogen, J., and Trahan, H. J., 1970: Pulmonary surfactant and evolution of the lungs. *Science,* 169:603–604.

DeJongh, H. J., and Gans, C., 1969: On the mechanism of respiration in the bullfrog, *Rana catesbeiana*: A reassessment. *Journal of Morphology,* 127:259–290.

Dejours, P., 1981: *Principles of Comparative Respiratory Physiology,* 2nd edition. Amsterdam, North Holland Co.

Feder, M. E., and Burggren, W. W., 1985: Skin breathing in vertebrates. *Scientific American,* 235:(5):126–142.

Gans, C., 1970a: Strategy and sequence in the evolution of the external gas exchangers in ectothermal vertebrates. *Forma et Functio,* 3:61–104.

Gans, C., 1970b: Respiration in early tetrapods: The frog is a red herring. *Evolution,* 24:723–734.

Gatten, R. E., Jr., 1985: The use of anaerobiosis by amphibians and reptiles. *American Zoologist*, 25:945–954.

Gaunt, A. S., editor, 1973: Vertebrate sound production. *American Zoologist*, 13:1139–1255.

Gaunt, A. S., and Gans, C., 1969: Mechanics of respiration in the snapping turtle *Chelydra serpentina*. *Journal of Morphology*, 128:195–228.

Hughes, G. M., 1963: *Comparative Physiology of Vertebrate Respiration*. Cambridge, Harvard University Press.

Hughes, G. M., editor, 1974: *Respiration of Amphibious Vertebrates*, 2nd edition. London, Heinemann.

Hughes, G. M., 1984: General anatomy of the gills. *In* Hoar, W. S., and Randall, D. J., editors: *Fish Physiology.*, vol. 10. New York, Academic Press.

Hughes, G. M., and Morgan, M., 1973: The structure of fish gills in relation to their respiratory function. *Biological Reviews*, 48:419–475.

Jenkins, F. A., Jr, Dial, K. P., and Goslow, G. E., 1988: A cineradiographic analysis of bird flight: The wishbone in starlings is a spring. *Science*, 241:1495–1498.

Johansen, K., 1968: Air-breathing fishes. *Scientific American*, 219:102–111.

Johansen, K., 1971: Comparative physiology: gas exchange and circulation in fishes. *Annual Review of Physiology*, 33:569–599.

Jones, F. R. H., and Marshall, N. B., 1953: The structure and function of the teleostean swimbladder. *Biological Reviews*, 28:16–83.

Jones, J. D., 1972: *Comparative Physiology of Respiration*. London, Arnold Publishing Co.

Kessel, R. G., and Kardon, R. H., 1979: *Tissues and Organs, A Text–atlas of Scanning Electron Microscopy*. San Francisco, W. H. Freeman and Company.

Lagler, K. F., Bardach, J. E., Miller, R. E., 1962: *Ichthyology*. New York, John Wiley and Sons.

Liem, K. F., 1985: Ventilation. *In* Hildebrand, M., Bramble, D. M., Liem, K. F., and Wake, D. B., editors: *Functional Vertebrate Morphology*. Cambridge, Harvard University Press.

Liem, K. F., 1988: Form and function of lungs: The evolution of air breathing mechanisms. *American Zoologist*, 28:739–759.

McMahon, B. R., 1969: A functional analysis of the aquatic and aerial respiratory movements of the African lungfish, *Protopterus aethiopicus*, with reference to the evolution of lung ventilation mechanisms in vertebrates. *Journal of Experimental Biology*, 51:407–430.

Pettingill, O. S., Jr., 1985: *Ornithology in Laboratory and Field*, 5th edition. Orlando, Academic Press.

Piiper, J., editor, 1978: *Respiratory Function in Birds, Adult and Embryonic*. Berlin, Springer-Verlag.

Piiper, J., and Scheid, P., 1977: Comparative physiology of respiration: Functional analysis of gas exchange organs in vertebrates. *International Review of Physiology*, 14:219–253.

Randall, D. J., Burggren, W. W., Farrell, A. P., and Haswell, M. S., 1981: *The Evolution of Air Breathing in Vertebrates*. Cambridge, Cambridge University Press.

Scheid, P., and Piiper, J., 1972: Cross-current gas exchange in avian lungs: Effects of reversed parabronchial air flow in ducks. *Respiration Physiology*, 16:304–312.

Schmidt-Nielsen, K.: How birds breathe. *Scientific American*, 225:6:72–79, 1971.

Shoemaker, V. H., Hillman, S. S., Hillyard, S. D., Jackson, D. C., McClanahan, L. L., Withers, P. C., and Wygoda, M. L., 1992: Exchange of respiratory gases, ions, and water in amphibious and aquatic amphibians. *In* Feder, M. E., and Burggren, W. W., editors: *Environmental Physiology of the Amphibians*. Chicago, University of Chicago Press.

Vitalis, T. Z., and Shelton, G., 1990: Breathing in *Rana pipiens*: The mechanism of ventilation. *Journal of Experimental Biology*, 154:537–556.

Weibel, E. R., and Taylor, C. R., 1981: Design of the mammalian respiratory system. *Respiration Physiology*, 44:1–164.

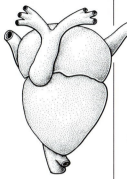

19 | The Circulatory System

PRECIS

The circulatory system is essential for the transport of materials throughout the body and for the maintenance of homeostasis. Blood performs the biochemical and physiological functions of the system, the heart propels the blood, and the vessels transport it. The heart and vessels change greatly as vertebrates move from water to land during their evolution and become endothermic animals.

Outline

The circulatory system is first of all the transport systems of the body: it picks up oxygen, nutrients, and other needed materials at sites of intake, transports them to the tissues, and then returns carbon dioxide, nitrogenous wastes, and other excess substances in the interstitial fluids to sites of removal from the body. Materials enter and leave the blood by diffusion, supplemented for some molecules by active transport. In essence, the circulatory system shortens diffusion distance by providing a means for the bulk flow of materials between various parts of the body. This is essential for animals to grow more than a few millimeters in diameter and to lead active lives. Beyond this, the circulatory system distributes heat between parts of the body, carries hormones, and generally plays a vital role in maintaining a constant internal environment, which is called **homeostasis** (Gr. *homoios* = alike + *stasis* = standing). Many components of the blood also are needed to prevent an excess loss of blood after an injury, heal wounds, and defend the body against disease organisms.

Components of the Circulatory System

The Spleen and other Hemopoietic Tissues

The biochemical and physiological functions of the circulatory system are performed by blood, which is composed of a liquid carrying cells of several types. In humans all blood cells are relatively short lived, with life spans ranging from one or two days for some leucocytes to 120 days for erythrocytes. **Hemopoietic tissues** (Gr. *haima* = blood + *poietikos* = producing) constantly produce new blood cells. Hemopoietic tissues are delicate, reticular connective tissues containing stem blood cells that multiply and differentiate. Hemopoietic tissues are widespread in embryos but become restricted in the adults, especially of the amniotes. The spleen, kidneys, and liver are important sites for hemopoiesis in adult anamniotes. Most of the blood cells of adult mammals develop in the red bone marrow.

The spleen is a conspicuous organ attached by a mesentery to the left side of the stomach in most vertebrates. Cyclostomes and lungfishes lack definitive spleens, but splenic tissue is embedded in the wall of part of their digestive tracts. Arteries enter the spleen through a network of dense, connective tissue **trabeculae** that extend into the organ from its connective tissue capsule (Fig. 19–1). Many of the small arterioles into which the arteries break up are surrounded by masses of lymphocytes that constitute the **white pulp.** Blood eventually is collected into **venous sinusoids** that lead to the veins draining the organ. A great deal of blood and many blood cells are present throughout the interstices of a delicate reticulum that constitutes the **red pulp,** but it is uncertain how blood enters and leaves these spaces. The "open theory" holds that the arterioles open directly into the interstices and drain into the venous sinusoids. Proponents of the "closed theory" believe that the arterioles lead to the venous sinusoids, and that blood passes back and forth through their walls to the interstices of the red pulp.

The spleen performs many functions. Lymphocytes multiply in the white pulp, and many enter the red pulp to participate in the body's immune responses. Numerous macrophages line the interstices of the red pulp and remove damaged or aging erythrocytes and other debris. Breakdown products of erythrocyte

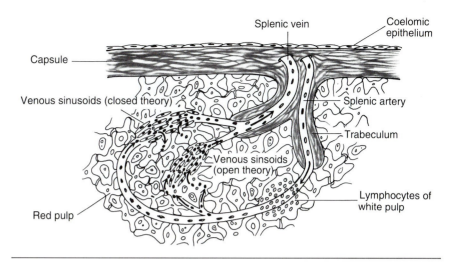

FIGURE 19–1   The structure of the spleen. Both the open and the closed theories of circulation through the organ have been shown. (Modified after Williams et al.)

destruction go to the liver, where iron is salvaged and other materials are excreted as bilirubin in the bile. Many hemopoietic cells occur in the red pulp of nonmammalian vertebrates and embryonic mammals. Finally, numerous erythrocytes are stored in the red pulp and venous sinusoids. They are released by the contraction of the spleen when there is a sudden need for an increase in the blood's oxygen-carrying capacity, as during vigorous exercise or a hemorrhage.

The Heart and Blood Vessels

The circulatory system of vertebrates is closed, for the blood is confined to blood vessels. By definition, **arteries** are vessels that carry blood from the heart to capillary networks in the tissues; **veins** are vessels that return blood to the heart. Usually, arteries carry blood high in oxygen content; veins carry blood low in oxygen. But this is not always the case, for the type of blood in vessels depends on the site of gas exchange. The pulmonary artery of a mammal, for example, carries blood low in oxygen content from the heart to the lungs.

The **heart** is a pump that receives blood low in hydrostatic pressure and increases the pressure sufficiently to drive the blood through the system. It is lined by a simple squamous epithelium that is called the **endothelium** and is covered by a coelomic epithelium, the **pericardium.** The **myocardium** between these epithelial layers is composed of a connective tissue "skeleton" and cardiac muscle. Cardiac muscle resembles skeletal muscle in having striated fibers that contract rapidly with considerable force (see Fig. 9–1); it resembles smooth muscle in its control. Cardiac muscle has an inherent rhythm of contraction that can be modified by autonomic nerve fibers, most of which terminate on a pacemaker. The pacemaker, called the **sinoatrial node** in mammals, is composed of modified cardiac muscle fibers and is located in the heart wall near where veins enter the

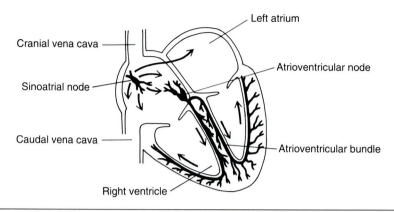

Cranial vena cava

Left atrium

Sinoatrial node

Atrioventricular node

Caudal vena cava

Atrioventricular bundle

Right ventricle

FIGURE 19–2   A ventral view of the conducting system of the mammalian heart. (After Dorit, Walker, and Barnes.)

heart (Fig. 19–2). Action potentials generated at the pacemaker spread from fiber to fiber, passing through specialized junctions. Birds and mammals also have a specialized conducting system in the ventricle composed of modified cardiac muscle known as **Purkinje fibers.** Once an action potential traveling through the atrial muscles reaches the **atrioventricular node,** located at the beginning of the conducting system, a wave of excitation is transmitted rapidly through Purkinje fibers to all parts of the ventricle.

The heart is not only a device to pump blood, but it also functions as an endocrine organ. The walls of one of the chambers, the atrium, produce **atrial natriuretic peptide,** a hormone that acts on the kidneys to increase excretion of salt and water.

Endothelium lines all the blood vessels. The rest of the wall of the vessels is composed of varying amounts of elastic fibers, smooth muscles, and collagen fibers (Fig. 19–3). The amount of each is determined by the forces acting on the vessels and their functions. Vessels are often described as having three layers of tissue. A **tunica intima** includes the endothelium and elastic fibers, if present; a **tunica media** is composed primarily of smooth muscles; and the **tunica externa** contains primarily collagen fibers. The larger arteries, and especially those near the heart, have a relatively large amount of elastic fibers. They absorb kinetic energy and stretch as blood is pumped into them from the heart. When the heart relaxes, the elastic recoil of these fibers keeps the blood flowing. The small **arterioles** preceding the capillaries have a relatively large amount of smooth muscle. Their degree of contraction regulates peripheral resistance and the amount of blood flowing through particular capillary beds.

The walls of **capillaries** consist only of endothelium. It is here that exchanges occur between the blood and the interstitial fluid bathing the cells. Diffusion is the primary factor in the exchanges, but **pinocytotic vesicles** are known to form on one surface of the endothelial cells, move across the cells, and discharge on the other side. In some organs, such as many endocrine glands and the kidneys of higher vertebrates, the capillaries are **fenestrated.** Small slits are present in the

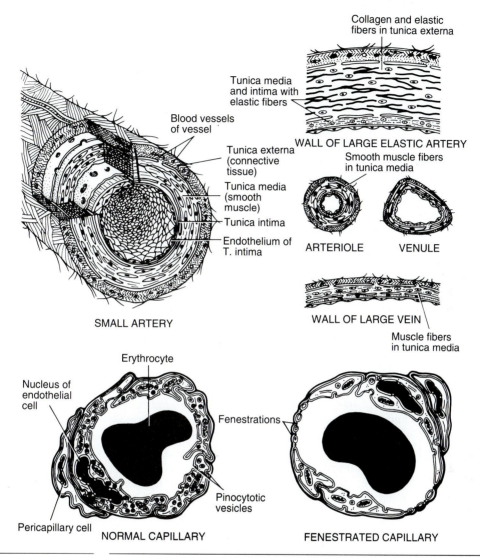

Collagen and elastic
fibers in tunica externa

Tunica media
and intima with
elastic fibers

WALL OF LARGE ELASTIC ARTERY

Blood vessels
of vessel

Tunica externa
(connective
tissue)

Tunica media
(smooth
muscle)

Tunica intima

Endothelium of
T. intima

Smooth muscle fibers
in tunica media

ARTERIOLE   **VENULE**

WALL OF LARGE VEIN

Muscle fibers
in tunica media

SMALL ARTERY

Erythrocyte

Nucleus of
endothelial
cell

Fenestrations

Pinocytotic
vesicles

Pericapillary cell   **NORMAL CAPILLARY**   **FENESTRATED CAPILLARY**

FIGURE 19–3   The structure of representative blood vessels. That of the capillaries is based on electronmicrographs, and they have been drawn on a much larger scale than the other vessels.
Capillaries have a diameter not much larger than the erythrocytes within them. (Modified
after Williams et al.)

endothelial cells, so blood and interstitial fluid are separated only by the delicate
basement membrane of the endothelial cells.

The **venules** that receive blood from the capillaries, and the larger veins to
which they lead, have much thinner walls than arteries of comparable size. Collagen and some smooth muscles are the walls' major components. Veins are also
much larger in diameter than the arteries they accompany. Calculations made on

TABLE 19–1 **Blood Vessels in the Mesenteric Vascular Bed of a Dog***

	Diameter in mm	Number of Vessels	Total Cross-Sectional Area in mm²	Circumference of Individual Vessels in mm	Total Circumference in mm
Aorta	10.0	1	78.5	31.4	31.4
Capillaries	0.008	1,200,000,000	60,319.0	0.025	30,159,289.0
Vena cava	12.5	1	122.7	39.3	39.3

*Dimensions and number of vessels from A. C. Burton.

a dog indicate that they contain nearly 70 percent of total blood volume. Veins are, therefore, an important reservoir of blood. A slight contraction of the veins and a decrease in their diameter can add greatly to the volume of blood flowing through the heart at any given time.

As arteries branch and lead to capillary beds in the tissues, the diameter of the vessels decreases. Because the number of vessels has increased, however, their combined cross-sectional area and circumference increase (Table 19–1). By the time the capillaries reach the intestine, for example, their combined cross-sectional area is about 770 times the cross-sectional area of the aorta that supplies them, and their combined circumference is nearly a million times the circumference of the aorta. As capillaries lead to veins and the veins coalesce into fewer but larger vessels, total cross-sectional area and circumference decrease. These changes have important functional consequences. A huge surface area (the circumference) is available in the capillary beds where exchanges occur. Also, since the total blood flow through the system at any one time is a constant, the velocity of flow changes greatly. It decreases as blood flows to the capillaries and total cross-sectional area increases. More time is available for the diffusion of materials between the blood and the interstitial fluid. Flow rate increases as the blood collects in the veins and total cross-sectional area decreases. An analogy is the decreasing velocity of a stream as it widens into a pond, and the increasing velocity as the pond flows into a narrowing outlet. Blood pressure, on the other hand, continues to decrease throughout the system because it is continuously reduced by friction. It drops particularly rapidly in capillary beds because of the very small diameter of the capillaries and their combined large surface area. It is very low in veins. Veins usually contain valves that prevent a backflow of blood. Such valves allow the surrounding skeletal muscles to squeeze the blood in the direction of the heart. Valves in veins assist the unidirectional flow of blood in both aquatic and terrestrial vertebrates.

All the vessels so far described constitute the **cardiovascular system.** Most vertebrates also have a **lymphatic system** of vessels. Blind lymphatic capillaries in most of the body tissues lead to progressively larger and larger lymphatic vessels that eventually discharge into the veins, usually near the heart, where venous pressure is very low. The lymphatic system is a supplementary drainage system that is needed in vertebrates with relatively high blood pressures to return excess liquid and plasma proteins, some of which escape from the blood. The return of the plasma proteins is essential to maintain the balance between the blood's osmotic and hydrostatic pressures that is needed for water exchanges.

The control and structure of the blood, the heart, and all the vessels enhance the flow of blood and maintain a relatively constant composition to the interstitial fluid, despite changes in the level of metabolism of particular tissues or changes in the external environment of the animals.

Embryonic Development of the Blood Vessels and Heart

Blood vessels develop within the mesenchyme of the embryo. Groups of mesenchymal cells first aggregate as small clusters called **blood islands** (Fig. 19–4A). The mesenchyme within the islands differentiates as blood cells; that on the periphery of the islands forms the vessel walls. The islands gradually coalesce to form small vessels. The first vessels develop in the splanchnic layer of the lateral plate mesoderm adjacent to the yolk-laden archenteron or yolk sac. Some of these vessels unite to form a pair of **vitelline** or **subintestinal veins** that extend forward beneath the developing pharynx. The vitelline veins unite and expand beneath the pharynx to form a tubular heart (Figs. 19–4**B** and 19–5**A**). A **ventral aorta** extends forward from the heart and gives rise to a series of paired aortic arches that extend dorsally through the branchial bars to the dorsal aortae. Six pairs of **aortic arches** eventually differentiate from anterior to posterior adjacent to the first six visceral arches, but the most anterior ones are lost or highly modified before the more posterior ones form. The **dorsal aortae** are paired above the pharynx but unite more posteriorly to form a single vessel that continues caudally, lying ventral to the developing vertebral column. The first branches to develop from the dorsal aorta are **vitelline arteries,** which return blood to the intestine or yolk sac. A visceral vitelline circulation is now established that enables the embryo to utilize energy reserves stored in the yolk (Fig. 19–4**B**).

As the liver develops from a ventral outgrowth of the archenteron just behind the heart, it expands around the vitelline veins, which break up within the liver into capillary-sized **hepatic sinusoids** (Fig. 19–5**A** and **B**). The sinusoids drain into a pair of **hepatic veins,** which differentiate from the anterior ends of the vitelline veins and lead to the heart. Caudal to the liver, the vitelline veins anastomose and parts atrophy as they form the **hepatic portal system** leading to the hepatic sinusoids (Fig. 19–5**C**). Portal systems are defined as groups of veins that drain one set of organs and lead to capillaries or sinusoids in another organ. The hepatic portal system brings blood from the intestinal region into close contact with the hepatic cells (p. 626).

As these changes are occurring, a somatic circulation begins to develop. Paired arteries leave the aorta to supply the body wall, appendages, kidneys, and developing gonads (Fig. 19–4**C**). A pair of **anterior cardinal veins** return blood from the head, and a pair of **posterior cardinal veins** return blood from most of the posterior parts of the body. Anterior and posterior cardinals on each side unite to form a **common cardinal,** which extends ventrally to the heart passing through the dorsal part of the transverse septum that it helped form (p. 558). The append-

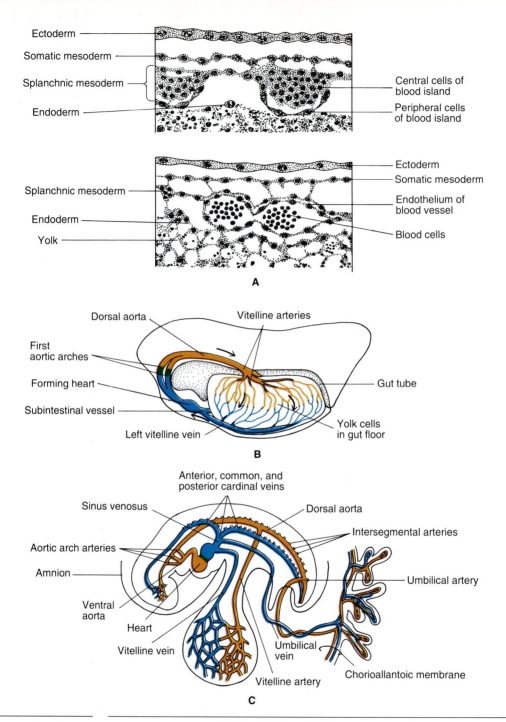

A

B

C

FIGURE 19–4   The development of the circulatory system: **A,** Two stages in the formation of blood cells and blood vessels from blood islands in the splanchnic mesoderm. **B,** A lateral view of the vitelline circulation in an early amphibian embryo. **C,** A lateral view of the major blood vessels in a 26-day-old human embryo. (**A,** after Corliss. **B,** after Romer and Parsons. **C,** after Moore.)

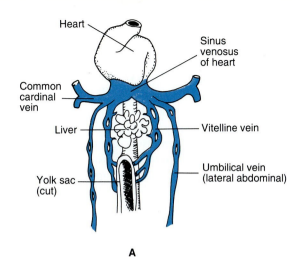

A

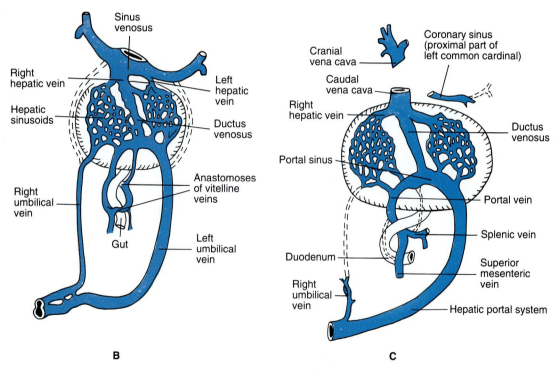

B   C

FIGURE 19–5   Ventral views of three stages in the development of the hepatic portal system and related veins in a mammalian embryo. (After Corliss.)

ages are drained primitively by a pair of **lateral abdominal veins** that enter the common cardinals (Fig. 19–5**A**).

The pattern of vessels now established resembles the pattern in adult fishes and amphibians. In the embryos of amniotes, some of these primitive channels are lost and replaced by others. Many of the differences are correlated with differences in the sites of nutrient, gas, and excretory exchanges. In the embryos of eutherian mammals, a pair of **allantoic** or **umbilical arteries** develop in the wall of the allantois and lead from the dorsal aorta to the placenta. **Allantoic** or **umbilical veins** return blood to the embryo (Fig. 19–4**C**). The umbilical veins develop from the primitive lateral abdominal veins and at first enter the common cardinals, but they soon establish a new passage through the liver, the **ductus venosus,** which leads into the right hepatic vein (Fig. 19–5**B**).

The Circulatory System of Early Fishes

The structure of the heart and circulatory patterns changed considerably in the course of vertebrate evolution as sites of exchanges between the body and the external environment changed, and as vertebrates became active, endothermic animals. A particularly important variable is the site for gas exchange, gills or lungs, so the evolution of the circulatory system is closely coupled to the evolution of the respiratory system.

The Heart

In early fishes without lungs—elasmobranchs are a good example—the heart and the pericardial cavity in which it lies are located beneath the posterior end of the pharynx floor. This position is close to the gills through which the heart must drive the blood before it is distributed to the body. Because the tubular embryonic heart grows in length more than the pericardial cavity, the adult heart has folded upon itself and forms an S-shaped loop that lies in the vertical plane (Fig. 19–6). The

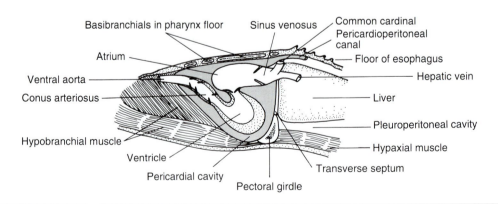

FIGURE 19–6   A sagittal section through the heart of a dogfish and the structures that surround it in the pharynx floor.

fish heart is not divided into left and right sides, as is a mammalian heart, because it receives and pumps only blood low in oxygen content. In most fishes the heart has differentiated into four chambers, lying in linear sequence. Valves between the chambers prevent backflow of blood. The most caudal and dorsal chamber is a thin-walled **sinus venosus,** which continuously receives blood very low in pressure from the common cardinal and hepatic veins. The sinus leads to the somewhat thicker walled **atrium,** a large chamber with valves at each end that accumulates blood. When it contracts, blood is pumped into the thicker walled **ventricle** with sufficient force to stretch slightly the ventricular muscles. This is advantageous because stretching the muscle maximizes the force with which it can contract. Contraction of the ventricle increases blood pressure sufficiently to drive the blood into the last heart chamber, the **conus arteriosus,** then through the capillary-like vascular beds in the gills, and on to the body. The conus arteriosus, which also has thick, muscular walls and several tiers of valves, acts as a buffer. It absorbs the abrupt increase in pressure during ventricular systole (Gr. *systole* = drawing together) and maintains pressure during ventricular diastole (Gr. *diastole* = drawing apart). It tends to even out and prolong blood flow to the gills. Teleost fishes have lost the muscular conus arteriosus. Instead they have developed an elastic chamber, the **bulbus arteriosus,** that lies in the pericardial cavity. The elastic properties of the bulbus arteriosus reduce the pulsations generated by the ventricle. Even so, blood flow through the gills is not perfectly even, but heartbeat and respiratory movements are so synchronized that the maximum flow of blood through the gills coincides with the maximum flow of water across them.

The pressure of blood in the ventral aorta of a dogfish ranges from 18 (diastole) to 29 (systole) mm Hg, according to Satchell (1971). Pressure drops considerably as blood passes through the gills to a range of 15 to 21 mm Hg in the dorsal aorta. This is adequate to drive the blood through the tissues, since a fish is supported by water and blood pressure need not also overcome high gravitational forces, as in terrestrial vertebrates. Additional pressure drops occur as blood passes through tissue capillaries and the renal and hepatic portal systems. Pressures in the large venous sinuses preceding the heart are very low, ranging from slightly less than ambient pressure to slightly more (− 0.4 mm Hg to + 0.1 mm Hg).

The pericardial cavity in which the heart lies is surrounded by stiff tissues (Fig. 19–6). A large basibranchial cartilage lies dorsal to the pericardial cavity, the pectoral girdle and liver lie posterior to it, and the rest of the cavity is surrounded by the thick hypobranchial musculature. Since the pericardial cavity is surrounded by stiff walls, it does not collapse when the ventricle contracts, the volume within the pericardial cavity increases slightly and pressure within the cavity drops. Concurrent contraction of the respiratory hypobranchial muscles, many of which extend between the visceral arches and the wall of the pericardial cavity, helps reduce pressure more. Reduced pressure in the pericardial cavity allows the thin-walled sinus venosus to expand and suck blood from the large venous sinuses. The heart acts first as a suction and then as a force pump. To work effectively, the system requires that secreted serous liquid not be permitted to accumulate in the pericardial cavity. Elasmobranchs and many early bony fishes have a **pericardioperitoneal canal** that passes through the transverse septum and continuously drains the pericardial cavity.

The Arterial System

The six embryonic aortic arches become interrupted in fishes by the capillary-like networks in the gills. The ventral part of the first aortic arch is lost in gnathostomes, but the ventral parts of the other five form the **afferent branchial arteries,** which lead into each of the gills (Fig. 19–7). The dorsal parts of most of the arches form **collector loops** around the internal gill slits. Utilization of the entire inner surface of the gill pouches is made possible by establishing complete loops around each gill pouch. The anterior and posterior limbs of the loop are called, respectively, the **pretrematic** and **posttrematic arteries** (Fig. 18–3**B**). The collector loops receive oxygenated blood from the gills and lead to **efferent branchial arteries** that continue to the dorsal aorta. The first embryonic gill pouch is reduced to a spiracle in many fishes and has been lost in others. When a spiracle is present, it frequently contains a rudimentary, gill-like structure, the pseudobranch (p. 637). The dorsal part of the first aortic arch contributes to a **spiracular artery** that carries aerated blood from the first collector loop through the pseudobranch to the internal carotid artery.

The paired **internal carotid arteries,** which supply blood to most of the head, are rostral extensions of the embryonically paired dorsal aortae. One branch of this vessel enters the skull to supply the brain. Another branch, the **stapedial**

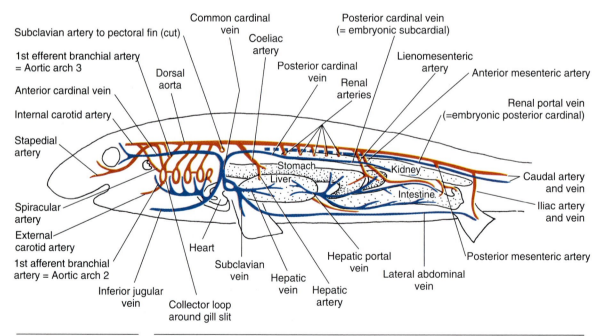

FIGURE 19–7   A lateral view of the major arteries and veins of an elasmobranch. Intersegmental vessels have not been shown except for the subclavian and iliac, which supply the paired fins. The dashed blue line above the kidney indicates the portion of the embryonic posterior cardinal that atrophies. This portion is replaced functionally by a subcardinal, located ventral to the kidney, that becomes a part of the adult posterior cardinal.

artery (so called because it runs through the stapes of embryonic and ancestral mammals), supplies the eye and most of the outside of the head. The **external carotid artery** of fishes extends forward from the ventral part of the first collector loop and supplies only tissues near the lower jaw.

The **dorsal aorta** continues caudally beneath the vertebral column to the tail, where it is known as the **caudal artery.** Several median visceral arteries, called **coeliac** and **mesenteric arteries,** extend through the mesenteries to most of the abdominal viscera. The gonads and kidneys are supplied by a series of small, paired **gonadal** and **renal arteries.** Larger paired **intersegmental arteries** extend between the myomeres of the trunk and tail. Two pairs of enlarged intersegmental arteries, the **subclavian** and **iliac arteries,** lead to the **brachial** and **femoral arteries** that enter the pectoral and pelvic fins.

The Venous System

Blood from the abdominal viscera, apart from the kidneys, drains into tributaries of the **hepatic portal vein.** After flowing through the capillary-like hepatic sinusoids in the liver, this blood is collected by a pair of **hepatic veins** that pass through the transverse septum to the sinus venosus. All these vessels develop from the embryonic vitelline veins (Figs. 19–5 and 19–7). Arterial blood reaches the hepatic sinusoids by a **hepatic artery,** a branch of the coeliac artery.

Most of the head is drained by the **anterior cardinal vein,** also known as the **lateral head vein.** After receiving superficial tributaries from the orbit and adjacent areas, and three deeper cerebral veins from within the skull, the anterior cardinal extends caudally to the **common cardinal.** The ventral surface of the head is drained by an **inferior jugular vein,** which enters the common cardinal independently.

The tail and trunk are drained by the **posterior cardinal system.** In jawless fishes and in the early embryos of other fishes, the **caudal vein** bifurcates at the posterior end of the trunk into a pair of posterior cardinal veins, which extend forward above the kidneys to the common cardinal. During the subsequent development of jawed fishes, new **subcardinal veins** form ventral to the kidneys and tap into the anterior part of the posterior cardinals. The midsections of the posterior cardinals, shown by the dashed line in Figure 19–7, then atrophy. The posterior parts of the early embryonic posterior cardinals have now been converted into **renal portal veins,** for they carry the blood they receive from the tail to capillary beds on the kidney tubules. The kidneys are drained by the subcardinals, which, with the anterior part of the embryonic posterior cardinals, now form the adult posterior cardinals.

The significance of the renal portal system is not entirely clear. The kidney tubules of vertebrates are supplied by two capillary beds. A knot of capillaries, known as the glomerulus, protrudes into the beginning of each tubule, and an artery from it leads to a second capillary bed lying over the rest of the tubule (see Fig. 20–2). The renal arteries lead first to the glomeruli, where a filtrate of the blood passes into the kidney tubules (p. 713); in contrast, the renal portal system leads directly to the tubular capillary beds, where materials may be added to or reabsorbed from the tubules. Since the renal arteries are very small in fishes, a

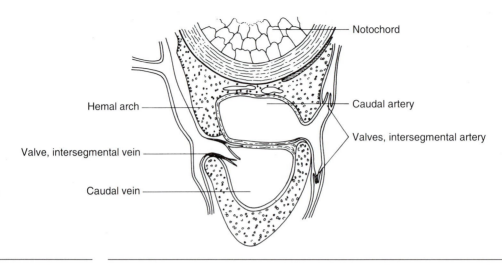

FIGURE 19–8   A transverse section through the hemal arch and the caudal artery and vein of a herring, *Clupea*. (After Satchell, 1971.)

renal portal system ensures the passage of a large volume of blood from the tail through the kidneys. Although pressure is often low in the caudal vein, Satchell (1971) has shown that the pressure can build up and equal that in the dorsal aorta by the operation of a "**tail pump**" (Fig. 19–8). The caudal artery and vein are completely encased by the hemal arches of the caudal vertebrae, which protect them from the waves of muscle contraction that sweep down the tail as the fish swims. Muscle contractions do, however, squeeze the tributaries of the caudal vein and drive blood into the caudal vein with considerable force. Valves at the junctions of the tributary veins and caudal vein prevent backflow. Unusual arterial valves at the origins of the intersegmental arteries to the tail muscles also prevent arterial blood from being forced back into the caudal artery. Arterial blood enters the tail muscles during periods of muscle relaxation.

In jawed fishes the paired fins are drained by veins that enter the **lateral abdominal system** (Fig. 19–7). The lateral abdominal veins are absent from teleosts. The femoral veins from the pelvic fins of teleosts enter the renal portal veins, and the subclavian veins from the pectoral fins enter the common cardinals.

Evolution of the Heart and Arteries

Few changes of evolutionary significance occur in the branching pattern of the dorsal aorta. All vertebrates have several **median visceral arteries** that supply the digestive tract and its derivatives, paired **lateral visceral arteries** to the gonads and kidneys, and paired **intersegmental arteries** to the trunk, tail, and paired appendages. Terrestrial vertebrates usually have fewer paired vessels than fishes—often only a single pair of gonadal arteries and another pair of renal arteries. Longitudinal anastomoses between the peripheral branches of the intersegmental arteries enable one artery to supply a more extensive area than in fishes. This is accompanied by a reduction in the number of points of origin of the intersegmental arteries from the dorsal aorta.

In contrast, major changes occur in the heart and aortic arches. These are correlated with the change from gills to lungs that accompanied the shift from water to land. Few changes in vertebrate evolution have been of greater significance.

Lungfishes

Branchial circulation is a very efficient means of gas exchange. The countercurrent flow of water and blood in the gills enables a fish to take up most of the limited amount of oxygen in the water and to unload carbon dioxide, which is very soluble in water. But fishes that live in water subject to a reduced oxygen tension or to periodic drought cannot rely solely on gills. They have evolved accessory respiratory organs, of which lungs are of particular interest because they also were present in the rhipidistian fishes ancestral to terrestrial vertebrates. Unfortunately, we do not know the nature of rhipidistian gas exchange, but we believe it was similar to that of contemporary lungfishes. The tropical habitat of lungfishes is subject to water stagnation and seasonal drought, not unlike the environment of the rhipidistian ancestors of tetrapods. Lungfishes take up oxygen from the air in the lungs by means of a newly evolved **pulmonary circulation,** but they have not dispensed with **branchial circulation.**

Gills continue to play an important role in oxygen uptake and the discharge of carbon dioxide. The gills are fully functional when oxygen tension in the water is greater than that in the blood. When lungfishes are in oxygen-poor stagnant water, blood is diverted to the lungs and away from the gills, from which oxygen could be lost. Experimental and cineradiographic studies have revealed that the lungfish vascular system permits considerable flexibility. By constricting certain segments of the aortic arches, it can direct the flow of blood. These arterial segments that can constrict and dilate function in an analogous way as valves in industrial plumbing. Lungfishes have evolved a heart and a pattern of aortic arches that allows for both branchial and pulmonary circulation and maintains a high degree of separation between oxygenated and nonoxygenated bloodstreams (Fig. 19–9). A new vessel, the **pulmonary vein,** has evolved and carries blood from the lung to a partially divided atrium (Fig. 19–9**A**). We will now review two extreme modes of the flexible circulation in lungfishes.

During the aquatic breathing mode, oxygen-poor blood from the body and head are collected in the sinus venosus (Fig. 19–9**A**). The deoxygenated blood is pumped sequentially into the atrium, ventricle, conus arteriosus, and the aortic arches. The oxygen-poor blood is directed to gills on the second, fifth, and sixth arches where the blood is oxygenated and collected by the efferent branchial arteries to the dorsal aorta. This pattern resembles the branchial circulation in water-breathing fishes (Fig. 19–7). In the aquatic breathing mode, the lungfish prevents blood already oxygenated in the gill of the sixth arch from entering the lung by constriction of the pulmonary artery (Fig. 19–9**A**). With this valve closed (constriction), oxygenated blood passes through the remaining segment of the sixth

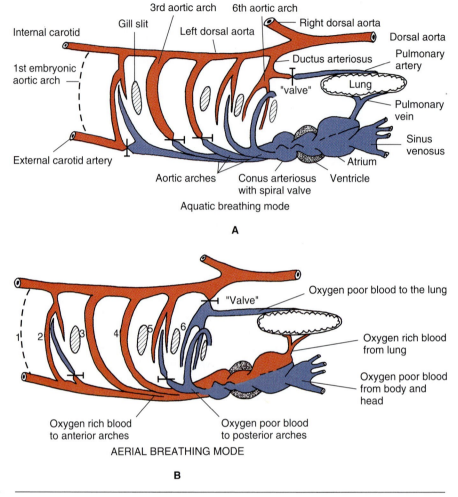

3rd aortic arch   6th aortic arch

Gill slit   Right dorsal aorta

Internal carotid   Left dorsal aorta   Dorsal aorta

1st embryonic   Ductus arteriosus   Pulmonary artery
aortic arch

"valve"   Lung

Pulmonary vein

External carotid artery   Sinus venosus

Atrium

Aortic arches   Conus arteriosus   Ventricle
                       with spiral valve

Aquatic breathing mode

A

"Valve"   Oxygen poor blood to the lung

1   2   3   4   5   6   Oxygen rich blood
                               from lung

Oxygen poor blood
from body and
head

Oxygen rich blood   Oxygen poor blood
to anterior arches   to posterior arches

AERIAL BREATHING MODE

B

FIGURE 19–9   Lateral views of heart and aortic arches of the African lungfish, *Protopterus*. The heart has been diagrammed in ventral view and with its chambers in linear sequence. **A,** Pattern when fish relies on gill irrigation in highly oxygenated water. **B,** Pattern when fish uses its lungs exclusively for air breathing.

efferent branchial artery, called the **ductus arteriosus,** into the dorsal aorta. Oxygen-poor blood is prevented from passing through the third and fourth aortic arches, which have no gills, by constrictions of the aortic bases. In this way all oxygen-poor blood in the anterior arches will pass through the gill on the second arch for oxygenation (Fig. 19–9**A**).

During the aerial breathing mode, the pattern changes drastically and foreshadows the circulation in tetrapods. Oxygen-poor blood from the sinus venosus enters the atrium and is propelled by the single ventricle through the conus arteriosus, which has a **spiral valve** (Fig. 19–9**B**). This valve directs the oxygen-poor blood into the sixth aortic arch. This oxygen-poor blood is shunted to bypass the gill and enters the lung via the pulmonary artery because the ductus arteriosus is closed off (Fig. 19–9**B**). Oxygenated blood from the lung is returned to the partially divided atrium via the pulmonary vein. Radiographic studies have shown that the oxygen-rich and oxygen-poor bloodstreams remain separated to a high degree as they pass through the partly divided atrium, ventricle, and conus arteriosus. Oxygen-rich blood is deflected to the anterior arches number 2 to 4. (The first embryonic arch has been lost.) Arches 3 and 4 form direct passages to the dorsal aorta. Oxygen-rich blood is distributed to the head and body by tributaries of the dorsal aorta. The aerial pattern of circulation of lungfishes foreshadows tetrapod circulation. For example, we will see that an open and subsequently closed ductus arteriosus form the basis of fetal and adult circulation in placental mammals.

The lungfish system does not only represent the ancestral condition from which tetrapod patterns evolve, but it also offers considerable functional flexibility. The two blood-streams are separated to a large extent, yet the possibility exists for different mixings and volumes of blood to be sent to different parts of the system according to environmental conditions and the extent to which gills or lungs are being used.

Lungs are accessory respiratory organs in the Australian lungfish, *Neoceratodus*, which retains gills on all the arches, but they are essential in *Protopterus* and *Lepidosiren*, whose gills are reduced.

Amphibians

The hearts of amphibians and most reptiles remain incompletely divided, and shunts permit some mixing of oxygen-rich and oxygen-poor blood. This appears to be an inefficiency, and for a long time biologists believed that the partial divisions were necessary transitional stages in the evolution of the double circulation of birds and mammals, with a completely divided heart and separation of the two blood-streams. Yet contemporary amphibians and reptiles have been evolving as long as birds and mammals, and there is no reason that the evolutionary processes would result in incomplete adaptations in some evolutionary lines but not in others. Recently, we have recognized that the shunts are adaptations to the unique modes of life of amphibians and reptiles.

Amphibians and reptiles resemble lungfishes in having long periods of apnea interspersed with brief periods of lung ventilation. They do not ventilate their lungs continuously, as do birds and mammals. As oxygen in the lungs is consumed, it would be energetically inefficient to continue to pump a large volume of blood through them. The shunts between the left and the right sides of the heart permit the volumes of blood flowing through the lungs and body to be unequal and to be altered as circumstances change. In birds and mammals, in contrast, the volume of blood passing through the two sides of the completely divided heart is equal at all times.

Larval amphibians retain gills and acquire a pulmonary circulation (Fig. 19–10**A**). At metamorphosis gills are lost, the heart migrates caudally closer to the lungs, and the animals rely on their lungs and pulmonary circulation for most of their oxygen uptake. Amphibians have a relatively low level of metabolism so their

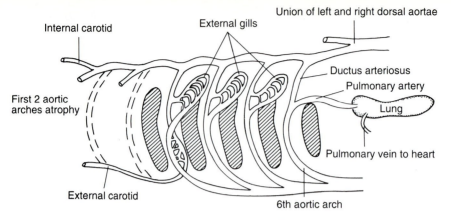

Internal carotid

External gills

Union of left and right dorsal aortae

Ductus arteriosus

Pulmonary artery

Lung

First 2 aortic arches atrophy

Pulmonary vein to heart

External carotid

6th aortic arch

URODELA, LARVA

A

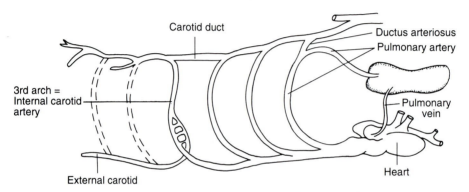

Carotid duct

Ductus arteriosus

Pulmonary artery

3rd arch = Internal carotid artery

Pulmonary vein

External carotid

Heart

URODELA, ADULT

B

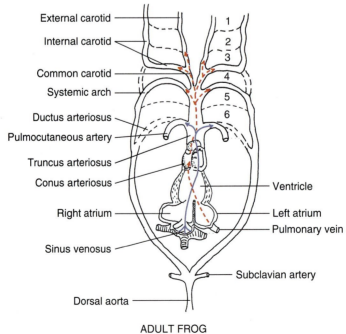

External carotid

1

Internal carotid

2

3

Common carotid

4

Systemic arch

5

Ductus arteriosus

6

Pulmocutaneous artery

Truncus arteriosus

Conus arteriosus

Ventricle

Right atrium

Left atrium

Pulmonary vein

Sinus venosus

Subclavian artery

Dorsal aorta

ADULT FROG

C

686

oxygen needs are not great. Some oxygen also is taken in through the skin, especially when the animals are beneath water.

The aortic arches of an adult amphibian are reduced to a greater extent than in a lungfish (Fig. 19–10**B** and **C**). The second embryonic arch is lost as well as the first, so the external carotid artery originates from the base of the third arch. The third arch itself is now a part of the internal carotid artery. The ventral aorta between the third and the fourth arches becomes the **common carotid artery.** The small enlargement at the branching of the external and internal carotid arteries is called the **carotid body.** Its function in amphibians is not certain. Some evidence suggests that it is a receptor that monitors the oxygen content of the blood; other evidence, that it detects pressure changes; and yet other evidence, that it secretes epinephrine. The segment of the dorsal aorta between the third and the fourth arches, which is called the **carotid duct,** has been lost in most adult amphibians, so the common carotid artery supplies both the external and the internal carotids. Adult salamanders usually retain both the fourth and the fifth arches as systemic arches leading to the dorsal aorta and trunk. The fifth arch frequently is smaller, and it has been lost in frogs. The sixth arch continues into the **pulmocutaneous artery** to the lungs and skin. Its embryonic connection to the dorsal aorta, known as the **ductus arteriosus,** usually disappears in adults.

The frog heart is representative (Fig. 19–11). Radiographic studies and analysis of the amount of oxygen and blood pressures in various vessels indicate that the two bloodstreams are separated to a considerable degree when the animals are on land and the lungs are functioning. The atrium is completely divided. Although the ventricle is not divided, the oxygen-rich and oxygen-poor bloodstreams appear to be separated to a large extent by slight differences in the time they enter and leave, and by the spongy wall of the ventricle. The streams are finally sorted out by a complex spiral valve in the conus arteriosus and by divisions within the bifurcated ventral aorta, now called the **truncus arteriosus.** Oxygen-rich blood returns from the lungs to the left atrium, and most of it is directed to the carotid and systemic arches. Blood entering the sinus venosus and right atrium has come from both the body and the skin, where some gas exchange occurs. Although this blood does not contain as much oxygen as blood from the lungs, it contains more than oxygen-poor blood that enters the sinus venosus of a fish. Most of this blood is directed to the pulmocutaneous and systemic arches. This pattern of circulation through the heart sends blood richest in oxygen to the head, a somewhat mixed blood to the rest of the body, and blood lowest in oxygen to the lungs.

The volume of blood flowing to the body, skin, and lungs can probably be altered to meet different circumstances. There is evidence that pulmonary resistance increases as oxygen in the lungs is consumed, especially when the animal is beneath water, and blood is diverted from the lungs to the skin and body. A

◀ *FIGURE 19–10*   The aortic arches of amphibians: **A** and **B,** Lateral views of the aortic arches in a larval and an adult urodele, respectively. **C,** Ventral view of the aortic arches and heart of an adult frog. The heart chambers are diagrammed in linear sequence. (Modified after Goodrich.)

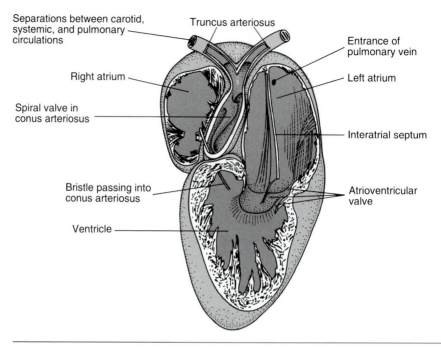

Separations between carotid, systemic, and pulmonary circulations

Truncus arteriosus

Entrance of pulmonary vein

Right atrium

Left atrium

Spiral valve in conus arteriosus

Interatrial septum

Bristle passing into conus arteriosus

Atrioventricular valve

Ventricle

FIGURE 19–11   A ventral dissection of a frog heart, *Rana*. (After Goodrich, from Kerr.)

muscular sphincter in the pulmonary artery just distal to the origin of the cutaneous artery probably regulates the amount of flow to the skin and lungs.

Reptiles

Embryonic reptiles never have a branchial circulation; gas exchange is through the chorioallantoic membrane. Aortic arches appear during development and are transformed into a pattern similar to that in adult frogs (Fig. 19–12). The third arches contribute to the internal carotid arteries, only the fourth arches form the systemic or aortic arches to the body, and the sixth arches are a part of the pulmonary arteries to the lungs. There are no branches from the pulmonary arteries to the skin because reptiles have a method of ventilating the lungs that is adequate for both oxygen uptake and carbon dioxide removal. The ductus arteriosus, present in embryos, disappears in adults, but the carotid duct often persists. The unique features of contemporary reptiles are the absorption during embryonic development of the conus arteriosus into the ventricle, and an unusual tripartite division of the ventral aorta that carries the origins of the pulmonary artery, left systemic arch, and right systemic arch to the ventricle. The right systemic arch gives rise to the entire carotid circulation to the head and then curves caudally to join the left systemic arch and form the dorsal aorta.

The atria are completely separated, as in most amphibians. Pulmonary veins return to the left atrium and systemic veins to the right one, which has largely

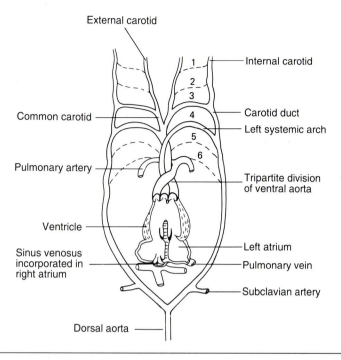

External carotid

Internal carotid

1

2

3

Common carotid

4

Carotid duct

Left systemic arch

5

6

Pulmonary artery

Tripartite division
of ventral aorta

Ventricle

Left atrium

Sinus venosus
incorporated in
right atrium

Pulmonary vein

Subclavian artery

Dorsal aorta

FIGURE 19–12   A ventral view of the aortic arches and heart of a reptile. The heart structure applies to most reptiles but not to crocodiles (see Fig. 19–13**D**). (After Goodrich.)

absorbed the sinus venosus in most reptiles. The ventricle is completely divided in crocodiles and partly divided in other reptiles. The condition in lizards is representative. The interventricular septum is complete caudally near the apex of the ventricle (Fig. 19–13**A**). More anteriorly, slightly caudal to the entrances from the atria, the interventricular septum is represented by a muscular ridge that extends into the ventricle from its ventral surface, curves to the right, and partly divides the ventricle into three functional compartments (Fig. 19–13**B**). Dorsally, a **cavum arteriosum** receives oxygen-rich blood from the left atrium, and a **cavum venosum** receives oxygen-poor blood from the right atrium. A wide **interventricular canal** connects these two cava, but it is closed off by the opening of the atrioventricular valves as the ventricle fills (Fig. 19–13**C**). A third compartment, the **cavum pulmonale,** lies ventral to the curvature of the muscular ridge and also receives oxygen-poor blood from its broad connection with the cavum venosum. The pulmonary artery to the lungs leaves from the cavum pulmonale, and both systemic arches leave from the cavum venosum. When the ventricle begins to contract, most of the oxygen-poor blood in the cavum venosum moves into the cavum pulmonale. This blood is sent out the pulmonary artery rather than the systemic arches because the lungs offer less peripheral resistance than the body. Concurrently, the atrioventricular valves close. This opens the interventricular canal, and oxygen-rich blood begins to move from the cavum arteriosum, where it has been sequestered, into the cavum venosum. Continued contraction of the ventricle

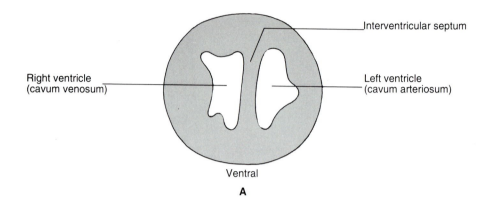

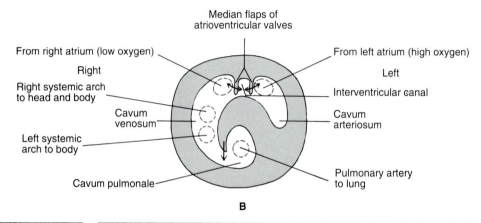

FIGURE 19–13 The reptilian heart: **A,** A transverse section through the apex of the ventricle of a lizard or chelonian (turtle) heart. **B,** A transverse section through the ventricle just caudal to the atria and the point of departure of the arteries. The location of the entrances from the atria and the exits to the arteries leaving the ventricle are shown by dotted circles.

brings its ventral wall up against the crest of the muscular ridge and prevents most of the oxygen-rich blood from entering the cavum pulmonale. Oxygen-rich blood, now in the cavum venosum, leaves through the systemic arches. In many cases both systemic arches receive oxygen-rich blood, but their exits from the cavum venosum are positioned in such a way that when there is some mixing of oxygen-rich and oxygen-poor blood, the distribution of the blood to the two systemic arches will not be the same. The right systemic arch, which leaves nearer the cavum arteriosum, will receive more oxygen-rich blood, whereas the left systemic arch, whose origin is closer to the cavum pulmonale, will receive more oxygen-poor blood.

The system is very labile in squamates and turtles, since the distribution of the two types of blood depends so much on the relative peripheral resistance in

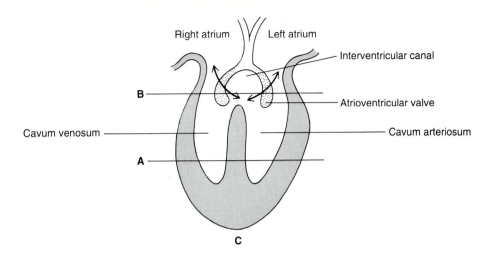

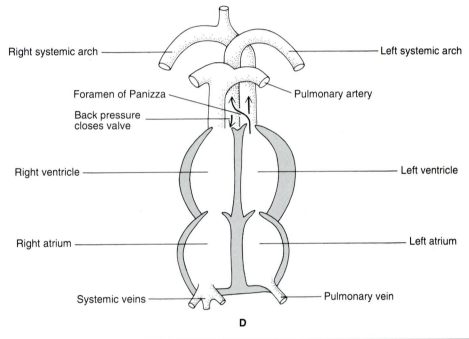

FIGURE 19–13 *continued*
C, A frontal section of a lizard or chelonian heart. Transverse lines indicate the level of the sections shown in **A** and **B. D,** The crocodile heart. The chambers have been diagrammed in linear sequence. (**A, B,** and **C** after White.)

the lungs and body. When the lungs are being ventilated and pulmonary resistance is low, as much as 60 percent of cardiac output goes to the lungs. Even some of the blood that has been through the lungs is recycled to the lungs. There is a **left-to-right shunt,** and the blood becomes thoroughly saturated with oxygen. When the lungs are not being ventilated, oxygen in the lungs is consumed, pulmonary arterioles constrict, and pulmonary resistance increases. The same thing occurs when the animals dive. Under these circumstances less blood is sent to the lungs, and some oxygen-poor blood is recycled to the body. There is a **right-to-left shunt.** The admixture of some oxygen-poor blood with oxygen-rich blood going to the body does not necessarily mean that less oxygen is delivered to the tissues. The increased carbon dioxide content of this mixed blood will cause hemoglobin to unload a greater amount of its bound oxygen than it normally would (the Bohr effect).

The crocodilian heart has a complete interventricular septum (Fig. 19–13**D**). The right systemic arch arises from the left ventricle and the left arch arises from the right ventricle along with the pulmonary artery. A **foramen of panizza** connects the two systemic arches at their base. Recent physiological studies (Jones and Shelton, 1993) show that in a resting alligator there is a right-to-left shunt through the foramen of panizza so that oxygen-poor blood, which is also acid blood, is directed into the left systemic arch. Branches of the left systemic arch supply much of the blood to the stomach and intestine. The ability to deliver shunted acidic blood to the gut is advantageous for secreting hydrochloric acid into the stomach of a resting animal after a meal. When the animal becomes agitated or active, systemic blood pressure rises. Thus there is no right-to-left shunt through the foramen of panizza because it is closed by a valve in the right systemic arch. There is a complete separation of arterial and venous blood when the animal is active and using its lungs. Likewise, during a dive, the animal maintains a high systemic blood pressure that prevents the development of a shunt.

The vascular arrangements in reptiles optimize the use of energy by matching the degree of blood flow through the lungs with the extent to which the lungs are being used. They permit a greater muscularization of the ventricles and the development of a blood pressure significantly higher than in fish and amphibians. Increased pressure and an adjustable separation of the bloodstreams enable reptiles to be considerably more active.

Birds

Birds are active endothermic vertebrates that evolved along with crocodiles from early archosaurian reptiles. Possibly, their ancestor had a heart and pattern of aortic arches similar to that of a crocodile. Since birds evolved endothermy and lungs that are continuously ventilated, there would have been no adaptive value to shunts bypassing the lungs. Equal volumes of blood are sent to the lungs and body at all times. This appears to have been accomplished simply by the loss of the left systemic arch (Fig. 19–14**A**). Birds have a single systemic or aortic arch, the right one; as in reptiles, it gives rise to the carotid circulation before turning caudad to the rest of the body. As might be expected, the subclavian arteries supplying the large flight and wing muscles are exceptionally large.

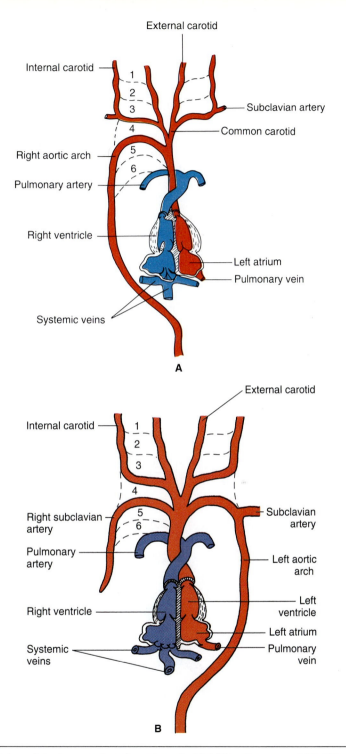

FIGURE 19–14 A ventral view of the heart and aortic arches of a bird (**A**) and a mammal (**B**). (After Goodrich.)

693

Adult Mammals

The synapsid line of evolution to mammals diverged from stem reptiles long before the lines of evolution to other reptiles and birds. The hearts of ancestral mammals probably were closer to the condition seen in lungfishes than to the specialized hearts and aortic arches of contemporary reptiles. As mammals became active, endothermic vertebrates with lungs that were ventilated continuously, they, too, evolved completely divided hearts. Oxygen-poor blood is received by the right atrium, which has absorbed the sinus venosus, and is pumped by the right ventricle to the lungs (Fig. 19–14**B**). Oxygen-rich blood returns to the left atrium and is pumped by the left ventricle to the body. Superficially, the mammalian heart resembles that of birds and some reptiles, but the interventricular septum evolved independently and develops embryonically in a slightly different way, so it is not homologous to the interventricular septum of crocodiles and birds.

Six embryonic aortic arches appear during the embryonic development of a mammal, but as in most other terrestrial vertebrates, only three persist (Figs. 19–14**B**, 19–15). The third arches contribute to the internal carotid arteries. The left fourth aortic arch alone forms the arch of the aorta. The right one is present but supplies only the subclavian artery on that side of the body. Its connection to the dorsal aorta is lost. The sixth arches contribute to the pulmonary arteries. The

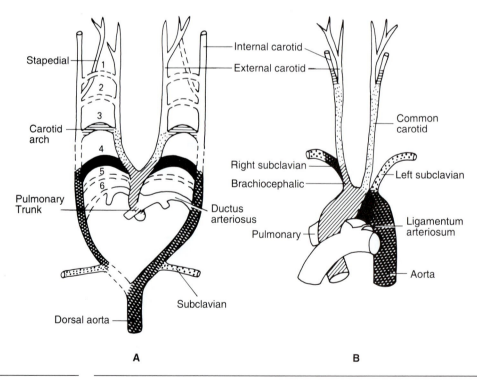

FIGURE 19–15 Ventral views of the aortic arches and carotid circulation of a mammal: **A,** An early developmental stage. **B,** The pattern in an adult human being. (After Barry.)

dorsal part of the left sixth arch, the ductus arteriosus, is present in the embryo. The ventral aorta and conus arteriosus divide during development, so the arch of the aorta arises directly from the left ventricle; the pulmonary arteries arise by a common pulmonary trunk from the right ventricle.

Embryonic and early mammals have a carotid circulation in which the stapedial branch of the internal carotid artery supplies most of the outside of the head, as in other vertebrates, but in most adult eutherian mammals the external carotid taps into the distal branches of the stapedial (Fig. 19–15). It takes over these branches when the proximal part of the stapedial artery (the segment passing through the stapes) atrophies. Thus, the adult external carotid artery supplies the entire outside of the head. The term *external* carotid, although used in all vertebrates, derives from the distribution of the artery in eutherian mammals. The internal carotid artery carries blood primarily to the cerebral hemispheres. The rest of the blood supply to the brain comes from derivatives of the intersegmental arteries that ascend with the spinal cord and from anastomoses with branches of the external carotid. These anastomoses usually are very small. Groups of specialized receptor cells are found at the bifurcation of the common carotid artery. These aggregations of receptors are called carotid bodies. Their function is to sense blood oxygen levels.

The complete division of the ventricle allows for different degrees of muscularization of its wall and the development of different systemic and pulmonary pressures. The left ventricular musculature is far more massive than that of the right side. The left ventricle develops a mean pressure of about 100 mm Hg—far higher than that developed by the comparable part of the reptilian ventricle, which is between 30 mm Hg and 40 mm Hg. A high systemic pressure, an efficient system for conducting action potentials through the ventricular musculature, and a rapid heartbeat combine to circulate blood quickly and efficiently through the body. Blood moves as rapidly through the lungs, but it would be undesirable for them to have such a high pressure. High pressure would cause an excessive filtration of water from the pulmonary capillaries into the alveoli. Pulmonary resistance is much lower, and pulmonary pressure is therefore only about one fifth the systemic pressure, on the order of 15 mm Hg to 20 mm Hg. This is comparable to the pulmonary pressures in reptiles.

The relatively massive and compact cardiac muscles of mammals necessitate a separate **coronary circulation** to supply their metabolic needs. A pair of coronary arteries leave the very base of the arch of the aorta. Coronary veins return to a coronary sinus that empties into the right atrium beside the entrances of the systemic veins. The coronary system is less well developed in amphibians and reptiles because their heart walls are thin and spongy enough to be supplied to a large extent by the blood flowing through the heart. Fishes also have coronary arteries that derive from the efferent branchial circulation and bring oxygenated blood to the heart. This is needed because only oxygen-poor blood flows through their hearts.

Fetal and Neonatal Mammals

Although an adult mammal has no shunts between the systemic and pulmonary circuits, a fetal mammal does, for the placenta, and not the lungs, is the site for gas exchange (Fig. 19–16**A**). Blood rich in oxygen returns to the fetus from the placenta in the umbilical vein. Most of this blood enters a direct passageway

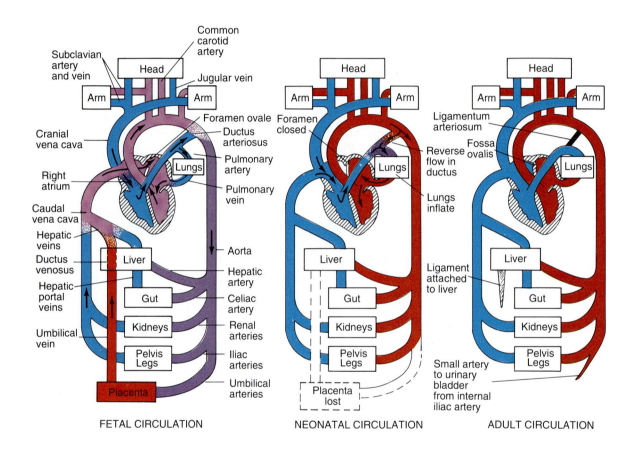

FETAL CIRCULATION   NEONATAL CIRCULATION   ADULT CIRCULATION

Degree of oxygen saturation of blood:

High

Low

FIGURE 19–16 Ventral views of the circulation in a fetal, neonatal, and adult mammal. Relative oxygen content in the vessels is shown by the color. (Modified after Dorit, Walker, and Barnes.)

through the liver, the **ductus venosus,** and joins venous blood in the caudal or inferior vena cava returning from the posterior parts of the fetus. The entrance of the caudal vena cava into the right atrium is situated in such a way that most of this blood passes through a valved opening, the **foramen ovale,** in the interatrial septum and enters the left atrium. This blood, which has the highest oxygen content of blood in the fetus, has bypassed the lungs. It is pumped by the left ventricle into the arch of the aorta, and much of it leaves through the first branches of the aorta to the head and shoulders. Oxygen-poor blood returns from the front of the body in the cranial or superior vena cava. Its entrance into the right atrium is positioned in such a way that this blood, together with a small admixture of blood from the caudal vena cava, enters the right ventricle to be pumped toward the lungs. Fetal lungs are collapsed and offer a higher resistance to blood passage than the systemic circuit, so most of this blood flows through the open **ductus arteriosus** to join the aorta caudal to its branches to the head and shoulders. The ductus arteriosus is a part of the left sixth aortic arch. A mixed blood is distributed to the rest of the body and, by the umbilical arteries, to the placenta.

In addition to diverting most of the blood from the lungs, these shunts provide both ventricles with a sufficient volume of blood to pump so that their musculature can develop normally. The ductus arteriosus is sometimes described as the "exercise channel" of the right ventricle because it permits the ventricle to pump considerable blood even though little of it can go through the lungs. Similarly, the foramen ovale is the "exercise channel" of the left ventricle, for it provides the left ventricle with a reasonable volume of blood and compensates for the small amount of blood that returns from the lungs.

Pressures within the circulatory system change abruptly at birth. The lungs inflate and pulmonary resistance drops below systemic resistance. Blood from the right ventricle now goes to the lungs rather than through the constricting ductus arteriosus to the body (Fig. 19–16**B**). The return of a large volume of aerated blood from the lungs raises pressure in the left atrium sufficiently to hold the valve in the foramen ovale closed. Accordingly, all of the blood in the right atrium now goes to the right ventricle and lungs. Soon the interatrial valve grows onto the interatrial septum, but a depression, the **fossa ovalis,** remains here throughout life. The ductus arteriosus remains open for a few hours or days after birth and continues to act as a shunt, but since pulmonary resistance is so low, it recirculates some aerated blood to the lungs rather than diverting blood from them. This is of considerable physiological importance during the **neonatal period,** when there is a changeover from fetal to adult types of hemoglobin. Fetal hemoglobin is adapted to take up oxygen at relatively low tensions because the fetus must be able to take oxygen away from the mother's blood, but fetal hemoglobin also does not unload oxygen in the tissues as easily as adult hemoglobin. As long as fetal hemoglobin still circulates in the newborn, the neonatal blood must become as thoroughly saturated with oxygen as possible to compensate for the greater tendency of fetal hemoglobin to hang onto its oxygen. A recirculation of some aerated blood through the lungs accomplishes this. After the hemoglobin changeover has occurred, the ductus arteriosus contracts fully and the adult circulatory pattern is established (Fig. 19–16**C**). Eventually, the lumen of the ductus arteriosus is filled with connective tissue, and the duct becomes the adult **ligamentum arteriosum.**

Shunts in fetal and neonatal mammals are performing the same functions as shunts in reptiles. They divert blood from the lungs when pulmonary resistance is high. They recirculate some blood through the lungs when pulmonary resistance is low and a more thorough oxygenation of the blood is needed.

Evolution of the Venous System

The **hepatic portal system** is conservative, and few changes occur from the condition seen in fishes. All vertebrates have hepatic portal systems, but the pattern of tributaries varies slightly with the configuration of the visceral organs that they drain. The liver sinusoids of amphibians and reptiles also receive some of the drainage from the hind legs because the veins draining their hind legs connect either with the hepatic portal system or directly with the liver (Fig. 19–17**C**). Vascular connections of the liver enable this important organ to process a large volume of blood.

Pulmonary veins appear in lungfishes and terrestrial vertebrates (p. 683) and carry oxygenated blood from the lungs to the heart.

The remaining changes in the venous system occur in the drainage of the appendages, in the renal portal and posterior cardinal systems, which are replaced by a **caudal vena cava,** and in the anterior cardinal, which is replaced by jugular veins and a **cranial vena cava.**

Drainage of the Appendages

Agnathans have no paired appendages to be drained (Fig. 19–17**A**), but a pair of **lateral abdominal veins** evolved and drain the appendages of early jawed fishes, such as elasmobranchs (Fig. 19–17**B**). In amphibians the lateral abdominal veins acquire a connection with the liver. The part of the lateral abdominal vein cranial to this connection then disappears in the adults, and the **brachial** and **subclavian veins** from the pectoral appendage enter the common cardinal directly. The part of the lateral abdominal system caudal to the liver remains paired in reptiles but fuses to form a single, median, **ventral abdominal vein** in amphibians (Fig. 19–17**C**). The **femoral veins** from the pelvic appendages of amphibians and reptiles acquire a new connection, the **external iliac veins**, with the renal portal system. Blood from the hind legs is now diverted either to the liver by the lateral abdominal system or to the kidneys by the renal portal system. These important organs receive more blood than they do in fishes. The external iliac connection to the renal portal system finally becomes dominant, and the lateral abdominal vein itself disappears in adult birds and mammals (Figs. 19–17**D** and 19–18). In embryonic birds and mammals, however, part of the primitive lateral abdominal system persists as the allantoic or umbilical veins, which return blood from the allantois in birds and from the placenta in mammals (Fig. 19–4**C**).

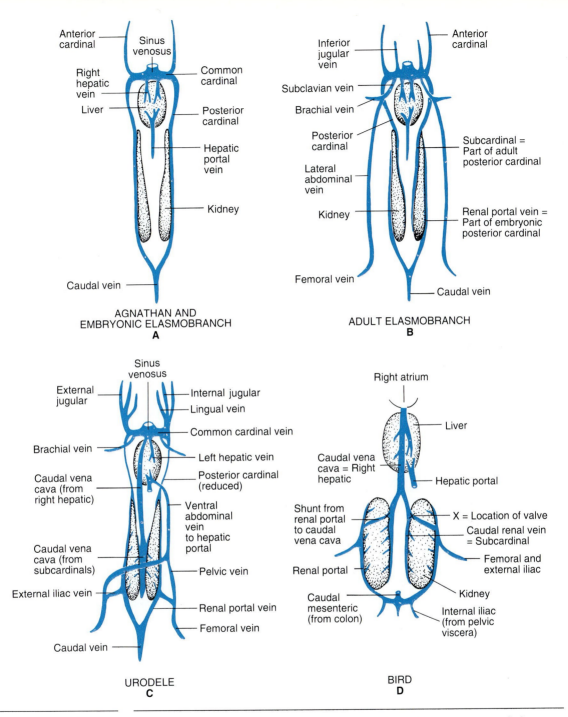

FIGURE 19-17 Ventral views showing important stages in the evolution of the venous system: **A,** An agnathan. **B,** An elasmobranch. **C,** A urodele. **D,** A bird. The reptilian system is similar to that of the bird.

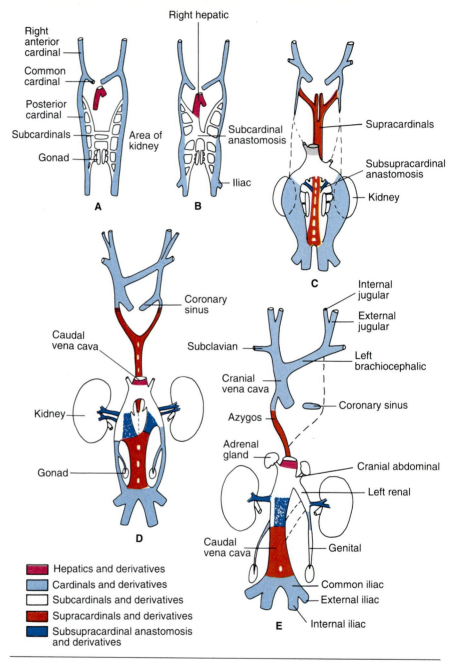

Right anterior cardinal

Common cardinal

Posterior cardinal

Subcardinals

Gonad

Area of kidney

A

Right hepatic

Subcardinal anastomosis

Iliac

B

Supracardinals

Subsupracardinal anastomosis

Kidney

C

Coronary sinus

Caudal vena cava

Kidney

Gonad

D

Internal jugular

External jugular

Subclavian

Left brachiocephalic

Cranial vena cava

Coronary sinus

Azygos

Adrenal gland

Cranial abdominal

Left renal

Caudal vena cava

Genital

Common iliac

External iliac

Internal iliac

E

Hepatics and derivatives

Cardinals and derivatives

Subcardinals and derivatives

Supracardinals and derivatives

Subsupracardinal anastomosis and derivatives

FIGURE 19–18 Ventral views of the development of the veins of a cat. A primitive cardinal system in an early embryo (**A**) is transformed through intermediate stages (**B–D**) into a cranial and caudal vena cava (**E**). Only the stump of the segment of the caudal vena cava that develops from the right hepatic vein is shown. (After Huntington and McClure.)

The Caudal Vena Cava

In most fishes the tail, kidneys, and trunk are drained by the renal portal and posterior cardinal systems (Fig. 19–17**A** and **B**). The beginning of an alternative course for the blood, the **caudal** or **inferior vena cava**, first appears in lungfishes and amphibians. It is a large vein that extends caudally from the sinus venosus and incorporates the now-fused embryonic subcardinal veins (Fig. 19–17**C**). (Recall that the subcardinals form much of the posterior cardinals in the adult jawed fishes.) The anterior part of the caudal vena cava develops in lungfishes by an enlargement of the anterior portion of the embryonic right posterior cardinal, but in all terrestrial vertebrates it forms as a caudal extension of the right hepatic vein from the liver (Fig. 19–17**C**).

The caudal vena cava of urodeles drains the kidneys and also receives most of the drainage from the tail, for most of this blood enters the kidneys through the renal portal system. Some blood from the tail continues forward in small posterior cardinal veins, which retain their connection with the renal portal system. (Recall that the renal portal system develops embryonically from the caudal portion of the posterior cardinals.) This connection has been lost in frogs, reptiles, and birds (Fig. 19–17**D**), so the caudal vena cava receives all the drainage of the kidney and tail, as well as most of the blood from the hind legs that enters the renal portal system. Not all this blood goes through the capillary beds over the kidney tubules, for reptiles and birds have one or more direct shunts from the renal portal system to the caudal vena cava. Valves within the shunts regulate the amount of blood going through the renal peritubular capillaries or directly to the caudal vena cava. When the valves are open, blood is returned more rapidly to the heart, but the factors that control the operation of these valves are not well known.

In the evolution of mammals, the caudal vena cava has extended caudally to unite with the most caudal section of the renal portal system, the part receiving the external iliac and caudal veins. The section of the renal portal system anterior to this connection is then lost, so blood from the tail and hind legs goes directly into the caudal vena cava rather than through the kidneys. How this occurs can best be seen in the embryonic development of the veins. The pattern of veins in an early mammalian embryo is similar to that of an elasmobranch (Fig. 19–18**A**). A bit later, a caudal vena cava develops from the right hepatic and taps into the subcardinals. The embryo's veins now resemble those of a urodele (Fig. 19–18**B**). The caudal extension of the caudal vena cava involves the development of a pair of **supracardinal veins** located dorsal to the kidneys. These tap into both the anterior and the posterior ends of the embryonic posterior cardinals (Fig. 19–18**C**). They also acquire a connection, known as the **subsupracardinal anastomosis**, with the partly fused subcardinals. Shortly after this the cranial and caudal sections of the supracardinals become disconnected from each other (Fig. 19–18**D**). Blood from the tail and hind legs now has two routes forward. The early tetrapod route includes the caudal segment of the posterior cardinal system (the adult renal portals), passages through the kidney, fused subcardinals, and the caudal extension of the right hepatic vein. The other route, which will become the adult mammal's caudal vena cava, includes the caudal segment of the posterior cardinal system, the caudal half of the supracardinals (the one on the right side usually becomes dominant), the right subsupracardinal anastomosis, the subcardinal (primarily the right one), and the caudal extension of the right hepatic (Fig. 19–18**D**). Other parts of the embryonic vessels form the renal, gonadal, and cranial abdominal veins.

Most of the thoracic wall of a mammal is drained by intercostal veins that lead into an **azygos vein** (Gr. *a* = without + *zygon* = yoke), which develops from the cranial half of the embryonic supracardinal plus a small segment of the embryonic posterior cardinal (Fig. 19–18**E**). Usually, the azygos vein receives the intercostal veins from both sides of the body, but in some species those of the left side enter a distinct **hemiazygos vein**, which enters the **coronary sinus** (the left common cardinal).

Why the renal portal system and much of the posterior cardinal system are replaced by the caudal vena cava is not entirely clear. Replacement may be related to an increase in arterial systemic blood pressure. The renal arteries provide the kidney tubules with a sufficient volume of blood to be cleared of nitrogenous and other wastes, so an additional venous contribution is not needed. The caudal vena cava is also a more direct passageway for blood from the caudal parts of the body and speeds up venous return to the heart.

The Cranial Vena Cava

The head of a jawed fish is drained by the anterior cardinals (also called the lateral head vein) and inferior jugulars, both of which enter the common cardinals (Figs. 19–17**B** and 19–19**A**). In a tetrapod the deeper tributaries of the anterior cardinal, which are within the skull, coalesce to form an **internal jugular vein** that emerges through the jugular foramen (Figs. 19–17**C** and 19–19**B**). In mammals this foramen is located near the back of the skull. The internal jugular vein is accompanied by the glossopharyngeal, vagus, and accessory nerves. Part of the internal jugular is homologous to the fish's posterior cerebral vein, which also emerges through a foramen accompanied by comparable nerves. The superficial tributaries of the anterior cardinal coalesce to form the **external jugular vein.** This vein is located in the head and neck more superficially than the anterior cardinal. In some species the external jugular also receives blood from within the skull by anastomoses in the orbital region with the internal jugular. The inferior jugular, now known as the **lingual vein,** acquires a new connection with the external jugular.

The evolution of the external and internal jugular veins is accompanied by the loss of the primitive anterior cardinal, except for its most caudal portion, which receives the jugulars and usually the subclavian vein as well. After the jugulars and subclavian veins unite, the base of the anterior cardinal and the common cardinal into which it leads are known as the **cranial** or **superior vena cava**. Many adult mammals retain the embryonic condition, in which both a left and a right cranial vena cava enter the right atrium (Fig. 19–18**C**). To reach the right atrium, the left cranial vena cava crosses the dorsal surface of the heart. As it does, it receives the coronary veins draining the heart, so this segment of the left cranial vena cava is also a **coronary sinus** (Fig. 19–18**D**). In many other mammals, including humans and cats, a cross connection called the **left brachiocephalic vein** extends from

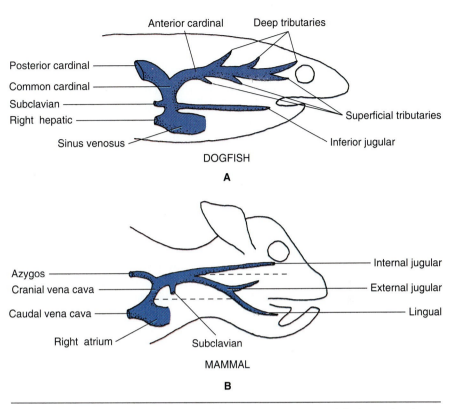

FIGURE 19–19   Lateral views of the veins of the head to illustrate the development of the internal and external jugular veins from the anterior cardinal and inferior jugular. **A,** A fish. **B,** A mammal. The broken lines in **B** represent atrophied vessels.

the left jugulars and subclavian to join the right cranial vena cava (Fig. 19–18**E**). After this connection has been established, the base of the left anterior cardinal and the common cardinal become very small or disappear, except for the segment of the left common cardinal serving as the coronary sinus.

Evolution of the Lymphatic System

Cyclostomes, chondrichthyans, and some other early fishes have networks of small, blind vessels that accompany the veins and help the cardiovascular capillaries drain the tissues. These vessels develop from the veins and empty into them at frequent intervals. Although similar to lymphatic vessels, they sometimes contain a few erythrocytes. They probably represent an early stage in the evolution of the lymphatic system.

Teleosts and tetrapods have independently evolved **lymphatic systems**, in which **lymphatic capillaries** (p. 674) help drain most of the tissues of the body. (Exceptions include the eye, inner ear, and central nervous system, which have their own drainage patterns: humors of the eye, endolymph, and cerebrospinal fluid, respectively. Bone marrow, cartilage, and the deep parts of the spleen and liver also lack lymphatic drainage.) The pattern of the large vessels into which the lymphatic capillaries enter varies considerably. Amphibians typically have **subcutaneous ducts** draining the skin and superficial muscles, **subvertebral ducts** draining deeper tissues, and **visceral plexuses** in the major internal organs (Fig. 19–20**A**). Large **lymph sinuses** are associated with many of these channels. Lymph is propelled by the contraction of surrounding body muscles and pulsating sections of the vessels, the **lymph hearts**, which keep it moving more or less in one direction. Valves are not present. The lymphatic vessels usually discharge into the anterior cardinal and renal portal veins.

Mammals have a single, large lymph sinus, the **cisterna chyli** (Gr. *chylos* = juice) located dorsal to the aorta just behind the diaphragm (Fig. 19–20**B**). It receives all the drainage caudal to the diaphragm. Valves are numerous in mammalian lymphatic vessels. Vessels coming from the intestinal region, the **lacteals** (L. *lac* = milk), often are prominent and have a whitish appearance, for they are the primary route for the passage of fat molecules absorbed in the intestine. A large subvertebral duct, which is called the **thoracic duct** in mammals, continues forward from the cisterna chyli, receives lymphatics from the thoracic wall, and joins the left subclavian vein near its union with the jugulars. Vessels draining the left side of the head, neck, and the left shoulder and arm join the thoracic duct near its entrance into the subclavian. Those draining comparable parts of the right side of the body form a short **right lymphatic duct**, which enters the right subclavian vein. Lymph hearts are absent, and lymph is moved by the contraction of surrounding body muscles. Numerous valves prevent a backflow.

Lymph nodes occur in a few aquatic birds and are abundant in mammals. Many are found in the neck, armpits, groin, and base of the mesentery, where many lymphatic vessels converge. Each node consists of many groups of lymphocytes, called **lymph follicles**, enmeshed in a network of reticular fibers (Fig. 19–21). The follicles are separated by dense connective tissue trabeculae that penetrate the node from its outer capsule. Several **afferent lymphatic vessels** enter the periphery of a node, and the lymph seeps through sinuses of reticular fibers around, between, and through the lymph follicles. Many macrophages line the sinuses and phagocytize foreign material in the lymph. The lymphocytes often initiate immune reactions to invading antigens. Lymph leaves through a single **efferent vessel**. Small arteries and veins also supply the nodes.

Additional aggregations of lymphatic tissue occur in other organs. Many lymph nodules are embedded in the mucosa and submucosa of the digestive tract, just beneath the lining epithelium. Among these are the **lingual, pharyngeal**, and **palatine tonsils** that form a ring of lymphatic tissue around the mammalian pharynx at the level of the embryonic second pharygeal pouch. Another large group, **Peyer's patches**, occur in the small intestine of young mammals; many lymph nodules occur in the wall of the **vermiform appendix**. Young birds have a distinct patch of lymph nodules in the **bursa of Fabricius**, a dorsal pouch of the cloaca. It plays a vital role in the development of the avian immune system.

The **thymus**, located in the base of the neck, is a lymphatic organ of particular importance. It develops in all vertebrates as epithelial buds from certain pharyngeal pouches (p. 607), but the epithelial tissue is invaded by and dominated by

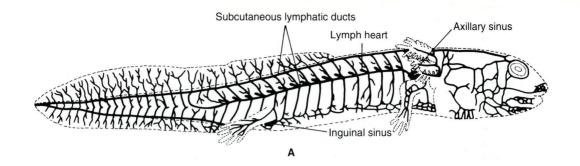

A

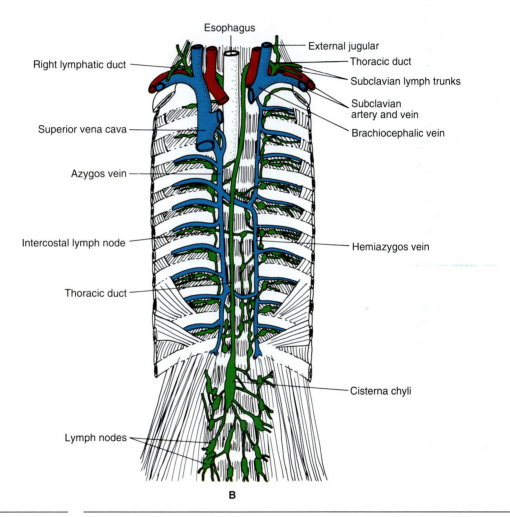

B

FIGURE 19–20 The lymphatic system: **A,** The superficial lymphatic vessels of a salamander larva. **B,** Deep lymphatic vessels and nodes of a human being. (**A,** after Hoyer and Udziela. **B,** after Williams et al.)

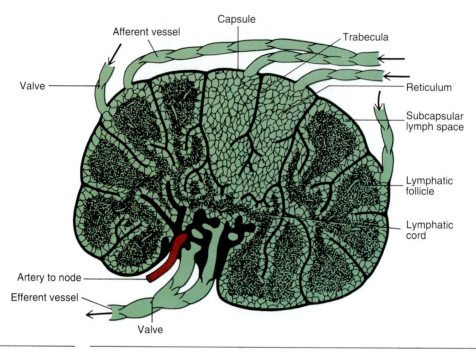

FIGURE 19–21   The structure of a lymph node. (After Fawcett.)

lymphocytes (p. 607). Lymphocytes enter the organ from other hemopoietic tissues, but they proliferate here. A population of lymphocytes known as the **T lymphocytes** matures in the thymus before spreading to the lymph nodes and other lymphatic tissues. The remaining lymphocytes of the body are known as **B lymphocytes**. In birds they mature in the bursa of Fabricius, and in mammals, possibly in some of the intestinal lymphoid nodules. On appropriate antigenic stimulation, B lymphocytes transform into cells that synthesize and release circulating antibodies in a **humoral immune response.**

SUMMARY

1. The circulatory system transports materials throughout the body and helps maintain homeostasis.
2. The biochemical and physiological functions of the system are performed by blood, which consists of a liquid plasma and several categories of cells.
3. The spleen functions as a hemopoietic organ, a site for the destruction of old erythrocytes, a site for phagocytosis, and a storage organ for erythrocytes.
4. The heart is the pump that creates the hydrostatic pressure that drives the blood through the vessels.
5. Cardiac muscle has an inherent rhythm that in advanced vertebrates is initiated at the sinoatrial node and modulated by autonomic nerves terminating on the node and heart. Birds and mammals have specialized systems for conducting impulses in the ventricles.

6. Arteries are defined as vessels that lead away from the heart, veins as vessels leading to the heart. Either may contain blood high or low in oxygen. Vessels are lined by endothelium and have a wall containing elastic fibers, smooth muscle, and connective tissue. Wall structure correlates with the forces within the vessels and their function.

7. Capillaries are very small vessels whose walls consist only of endothelium. Exchanges between the blood and the interstitial fluid occur here.

8. Because of changes in the combined cross-sectional area and circumference of vessels in different parts of the system, the rate of blood flow decreases as blood moves toward the capillaries and increases as blood flows from the capillaries into the veins. Pressure falls throughout the system. The fall is especially pronounced in the capillaries.

9. Blood vessels and cells begin their development as blood islands in the splanchnic mesoderm. An early vitelline circulation to and from the yolk is established before the somatic circulation. The pattern of the embryonic vessels in a mammal is similar to their pattern in an adult fish.

10. An early fish, such as an elasmobranch, has a four-chambered heart: sinus venosus, atrium, ventricle, and conus arteriosus. The heart receives low-pressure oxygen-poor blood and increases the pressure sufficiently to drive it through the gills, where aeration occurs, and on to the body.

11. The location of the heart in a pericardial cavity surrounded by stiff tissues that do not collapse enables the heart to act alternately as a suction and force pump.

12. The six embryonic aortic arches are transformed into a branchial circulation that consists of afferent branchial arteries that lead to the gills, collector loops that receive aerated blood, and efferent branchial arteries that lead to the dorsal aorta, from which blood is distributed to the body.

13. The branches of the aorta consist of median visceral (coeliac and mesenteric) arteries that supply the digestive tract and its derivatives, lateral visceral (gonadial and renal) arteries to the gonads and kidneys, and intersegmental arteries to the body wall and appendages.

14. A hepatic portal system carries blood from the abdominal viscera to the sinusoids in the liver. Two hepatic veins carry blood from the liver to the sinus venosus.

15. In jawed fishes a renal portal system has evolved that carries blood from the tail to capillaries over the renal tubules. A tail pump provides the necessary increase in pressure.

16. The kidneys and trunk are drained by a pair of posterior cardinals, the appendages by lateral abdominal veins, and the head by the paired anterior cardinals and inferior jugulars. All these vessels enter the paired common cardinals that lead to the sinus venosus.

17. Lungfishes have evolved a pulmonary circulation that carries blood to the lungs and back to the heart as well as retaining most of the branchial circulation.

18. Adult amphibians lose the branchial circulation; it is never present in reptiles. Aortic arches are reduced in number, but the third pair remains as part of the internal carotids to the head, the fourth and sometimes the fifth are sys-

temic arches to the dorsal aorta and body, and the sixth pair provides pulmonary arches to the lungs.

19. Blood returning to the heart from the body is separated from that returning from the lungs in a partially divided atrium (lungfishes) or a completely divided atrium (amphibians and reptiles). The division of the ventricle, conus arteriosus, and ventral aorta varies morphologically in these groups, but in general, oxygen-poor and oxygen-rich blood are physiologically separated to a high degree when the lungs are functioning. Blood richest in oxygen goes to the head and body; that lowest in oxygen, to the gills and lungs.

20. The system is very flexible, and when the lungs are not functioning, the incomplete division of the heart permits blood to be diverted from the lungs. The volume of the pulmonary and systemic flow need not be equal.

21. Birds and mammals have independently evolved completely divided hearts. The right side receives oxygen-poor blood from the body and sends it to the lungs, and the left side receives oxygen-rich blood from the lungs and sends it to the body. Blood flow through the pulmonary and systemic circuits is equal. The complete division of the heart permits the evolution of high pressures in the systemic circuit and the retention of modest pressures in the pulmonary circuit.

22. Both birds and mammals retain the paired third aortic arch as part of the internal carotids and the paired sixth arches as part of the pulmonary arteries. In mammals the left fourth arch is the systemic arch to the body; in birds the right fourth arch is the systemic arch.

23. Fetal mammals retain bypasses in the heart that divert blood from the lungs, and one of these is used temporarily in neonatal mammals to divert more blood to the lungs. Pulmonary and systemic volumes are not the same.

24. The hepatic portal system and pulmonary veins have changed little during vertebrate evolution.

25. The pectoral appendages of tetrapods drain into the common or anterior cardinals. A lateral (or ventral) abdominal vein continues to receive the drainage of the pelvic appendages in amphibians and reptiles, but it carries blood to the hepatic or renal portal system rather than to the common cardinals. The liver and kidney process a great deal of the blood returning from the hind legs.

26. In lungfishes, amphibians, reptiles, and birds a caudal vena cava evolves from part of the right hepatic vein and the subcardinals. It receives most or all of the drainage from the kidneys, posterior part of the trunk, and hind legs.

27. The caudal vena cava is extended caudal to the kidneys in mammals and receives directly the drainage of the pelvic region and hind legs. The renal portal system has been lost. The extension of the caudal vena cava results from the evolution of a supracardinal system of veins and a subsupracardinal anastomosis. Higher blood pressures in mammals delivers a large volume of blood to the kidneys by way of the renal arteries, and the loss of the renal portal system enables blood from the caudal parts of the body to return more rapidly to the heart.

28. By a fusion of its deep and superficial tributaries, the anterior cardinal system is converted to the external and internal jugulars and the cranial vena cava.

29. The lymphatic system is a subsidiary drainage system of the tissues that returns excess water and plasma proteins to the veins. It is absent from early fishes but evolved independently in teleosts and tetrapods.

30. Lymph nodes that are associated with the lymphatic system have evolved in mammals. With lymph nodules in the wall of the digestive system, they are part of the body's immune system.

REFERENCES

Adolph, E. F., 1967: The heart's pacemaker. *Scientific American*, 216:3:32–37.

Altman, P. L., and Dittmer, D. S., editors, 1971b: *Biological Handbooks: Respiration and Circulation*. Bethesda, Md., Federation of American Societies for Experimental Biology.

Baker, M. A., 1979: A brain-cooling system in mammals. *Scientific American*, 240:5:130–139.

Barnett, C. H., Harrison, R. J., and Tomlinson, J. D. W., 1958: Variations in the venous system of mammals. *Biological Reviews*, 33:442–487.

Barry, A., 1951: The aortic arch derivatives in the human adult. *Anatomical Record*, 111:221–238.

Bourne, G. H., editor, 1980: *Hearts and Heart-Like Organs*. New York, Academic Press.

Burggren, W. W., and Johansen, K., 1987: Circulation and respiration in lungfishes (Dipnoi). *In* Bemis, W. E., Burggren, W. W., and Kemp, N. E. editors: *The Biology and Evolution of Lungfishes*. New York, Alan R. Liss.

Burggren, W. W., and Johansen, K., 1982: Ventricular hemodynamics in the monitor lizard, *Varanus exanthematicus*: Pulmonary and systemic separation. *Journal of Experimental Biology*, 96:343–354.

Butler, P. J., and Jones, D. R., 1982: The comparative physiology of diving in vertebrates. *Advances in Comparative Physiology and Biochemistry*, 8:179–364.

Burton, A. C., 1972: *Physiology and Biophysics and the Circulation*, 2nd edition. Chicago, Yearbook Medical Publications.

Carey, F. G., 1973: Fishes with warm bodies. *Scientific American*, 228:3:36–44.

Caro, C. G., Pedley, T. J., Schroter, R. C., and Seed, W. A., 1978: *The Mechanics of Circulation*. Oxford, Oxford University Press.

DeLong, K. T., 1962: Quantitative analysis of blood circulation through the frog heart. *Science*, 138:693–694.

Farrell, A. P., 1991: Circulation of body fluids. *In* Prosser, C. L., editor: *Environmental and Metabolic Animal Physiology*. New York, Wiley-Liss.

Folkow, B., and Neil, E., 1971: *Circulation*. New York, Oxford University Press.

Foxon, G. E. H., 1955: Problems of the double circulation of vertebrates. *Biological Reviews*, 30:196–228.

Holmes, E. B., 1975: A reconsideration of the phylogeny of the tetrapod heart. *Journal of Morphology*, 147:209–228.

Huntington, G. S., and McClure, C. F. W., 1920: The development of the veins in the domestic cat. *Anatomical Record*, 20:1–31.

Johansen, K., and Burggren, W., 1980: Cardiovascular function in the lower vertebrates. *In* Bourne, G. H., editor: *Hearts and heartlike organs. Comparative Anatomy and Development*. Vol. 1. New York, Academic Press.

Johansen, K., and Burggren, W., editors, 1985: *Cardiovascular Shunts*. Copenhagen, Munksgaard.

Johansen, K., and Hanson, D., 1968: Functional anatomy of the hearts of lungfishes and amphibians. *American Zoologist*, 8:191–210.

Jones, D. R., and Shelton, G., 1993: The physiology of the alligator heart: left aortic flow patterns and right-to-left shunts. *Journal of Experimental Biology,* 176:247–269.

Kampmeier, O. F., 1969: *Evolution and Comparative Morphology of the Lymphatic System.* Springfield, Ill., Charles C. Thomas.

Lillywhite, H. B., 1988: Snakes, blood circulation and gravity. *Scientific American,* 259:92–98.

O'Donoghue, C. H., and Abbott, E., 1928: The blood-vascular system of the spiny dogfish, *Squalus acanthias Linne* and *Squalus sucklii Gill. Transactions of the Royal Society of Edinburgh,* 55:823–890.

Randall, D. J., 1968: Functional morphology of the heart of fishes. *American Zoologist,* 8:179–189.

Robinson, T. F., Factor, S. M., and Sonnenblick, E. H., 1986: The heart as a suction pump. *Scientific American,* 254:84–91.

Satchell, G. H., 1971: *Circulation in Fishes.* Cambridge, Cambridge University Press.

Satchell, G. H., 1991: *Physiology and Form of Fish Circulation.* Cambridge, Cambridge University Press.

Shelton, G., and Burggren, W. W., 1976: Cardiovascular dynamics of the chelonia during apnoea and lung ventilation. *Journal of Experimental Biology,* 64:323–343.

Szidon, J. P., Lahiti, S., Lev, M., and Fishman, A. P., 1969: Heart and circulation of the African lungfish. *Circulation Research,* 25:23–38.

White, F. N., 1968: Functional anatomy of the heart of reptiles. *American Zoologist,* 8:211–219.

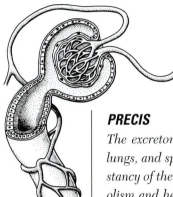

20 | The Excretory System and Osmoregulation

PRECIS

The excretory system includes the kidneys and their ducts. With the skin, gills, lungs, and special salt-excreting or salt-absorbing structures, it maintains the constancy of the internal environment by eliminating the nitrogenous wastes of metabolism and helping regulate the salt and water balances of the body. Its structure and evolution must be considered in the context of the environments in which vertebrates live and their level of metabolism.

The composition of the interstitial fluid that bathes the cells of vertebrates must be maintained within narrow limits. Cellular metabolism, however, produces carbon dioxide, water, and nitrogenous wastes that diffuse into the interstitial fluid. Other metabolic processes add organic acids, phosphates, sulfate ions, and other materials. The interstitial fluid may gain or lose water or salts depending on the vertebrate's environment: fresh water, salt water, or land. Despite these destabilizing factors, the composition of the interstitial fluid is held relatively constant by exchanges between it and the blood, and the composition of the blood, in turn, is maintained by carefully controlled exchanges between the animal and its external environment. These exchanges occur through many organs, including the skin, liver, gills, lungs, kidneys, and special structures that excrete or absorb salt. The liver eliminates bile pigments, and the respiratory system, assisted sometimes by the skin, removes carbon dioxide. Here we will examine the kidneys and other organs that remove the nitrogenous and other wastes of metabolism and concurrently maintain the water and salt balances of the body. The removal of waste products of metabolism is called **excretion.** Excretion should not be confused with defecation, which is the removal primarily of undigested residues and bacteria from the digestive tract. Only the bile pigments in the feces are byproducts of cellular metabolism.

Morphologically, the kidneys and their ducts are associated intimately with the reproductive system. Kidneys and gonads develop from adjacent tissues, and after the excretory or urinary ducts have developed, the reproductive system usually taps into them or their derivatives. These two systems are often treated together as the **urogenital system,** but we will first examine the excretory system and relate the reproductive system to it in the next chapter.

The Renal Tubules

Renal Tubule Structure and Function

The kidneys of vertebrates are of mesodermal origin. They develop from the pair of embryonic **nephric ridges** (Gr. *nephros* = kidney) or **mesomeres** that are located between the somites (epimeres) and the lateral plate (hypomere) (Figs. 4–7 and 20–1) and extend the length of the coelom. The segmentation of the somites extends into the nephric ridges of primitive vertebrates and divides them into a series of segmented **nephrotomes** (Gr. *tome* = cutting), each of which will differentiate into a **nephron** or **renal tubule** (L. *renes* = kidneys). The segmentation may be limited to the anterior part of the nephric ridges. The renal tubules are the structural and functional units of the kidneys. The proximal end of a representative tubule forms a two-layered, cup-shaped capsule of simple squamous epithelium that is known as **Bowman's capsule** or a **renal capsule** (Fig. 20–2A). The capsule envelops a tangled knot of capillaries known as a **glomerulus** (L. *glomus* = ball). The glomerulus receives an **afferent arteriole** from the renal artery and is drained by an **efferent arteriole.** Together, the renal capsule and glomerulus form a **renal corpuscle.** The rest of each tubule is composed of a simple epithelium, but the nature of the epithelial cells and the length and pattern of the nephron vary considerably among vertebrates according to the excretory

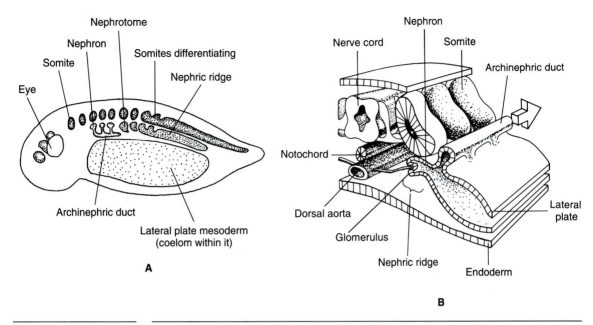

FIGURE 20–1   Development of nephrons: **A,** A diagram of a lateral view of an early embryo, showing the sequential differentiation (from anterior to posterior) of nephrons from the nephric ridge. The nephric ridge lies between the developing somites and the lateral plate mesoderm. **B,** A stereodiagram of a portion of the nephric ridge and adjacent structures. (**B,** after Williams et al.)

and osmoregulatory problems each group faces. Often, especially in anamniotes, a short **neck** of ciliated cells connects the renal corpuscle with a thicker **proximal tubule** of secretory or absorptive cells. A thin **intermediate tubule,** which varies in length and sometimes contains ciliated cells, may lie between the proximal and **distal tubules.** In amniotes, and some anamniotes, **collecting tubules** receive a number of renal tubules and, in turn, enter the excretory duct.

The renal corpuscles are **filtration** mechanisms from which a filtrate of the blood plasma enters the renal tubules. Most of the large plasma proteins are held back, but filtration is nonselective for smaller molecules, and they occur in the filtrate in the same proportion as in the blood. The filtrate contains nitrogenous wastes and other materials that must be eliminated, as well as glucose, amino acids, and other molecules that should be saved. Whether water or salts need to be conserved or eliminated depends on the animal's environment. The renal tubules and collecting tubules are surrounded by **peritubular capillaries** that receive their blood in different ways among the different groups of vertebrates (Fig. 20–2**B–D**). The peritubular capillaries of chondrichthyan fishes, reptiles, and birds receive blood from both the efferent renal arteriole and afferent renal vein leaving, respectively, the glomeruli and renal portal vein. Those of osteichthyan fishes and amphibians receive blood only from the renal portal vein because the efferent renal arteriole enters the renal vein directly. The renal portal system is lost in

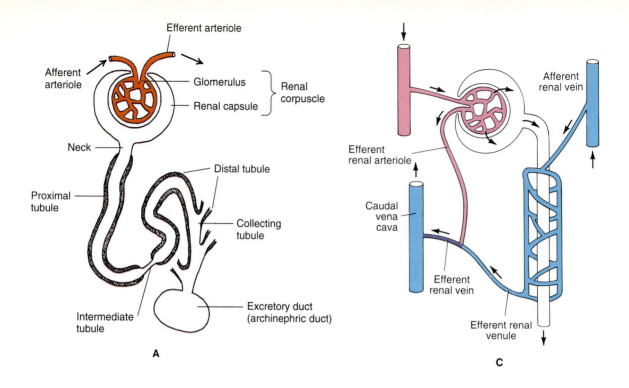

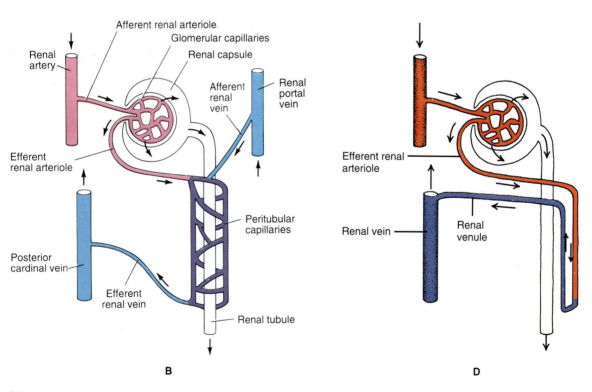

714

mammals, so their peritubular capillaries receive blood only from the efferent renal arterioles. In all cases blood leaving the peritubular capillaries drains into renal veins. As the filtrate passes down the tubules, substances that must be saved are selectively **reabsorbed** and enter the capillaries around the tubules. Reabsorption of some products is active and requires the tubular cells to do metabolic work; for other products reabsorption is by passive diffusion. In some species the amount of waste products in the filtrate is augmented by selective **secretion** by the tubular cells. Waste products are concentrated in this way as the filtrate passes down the tubules to be discharged as urine.

Renal Tubule Evolution

Most vertebrates have renal tubules of the type described, in which the renal tubules do not connect with the coelom and the glomeruli are surrounded by renal capsules. These are known as **internal glomeruli** (Fig. 20–3**D**). The first few tubules that develop at the anterior end of the nephric ridges of ammocoetes and some larval amphibians open into coelomic recesses through a ciliated **nephrostome.** These are **external glomeruli** (Fig. 20–3**B**). In larval vertebrates possessing external glomeruli, the filtrate is discharged into the coelomic recesses and is drawn by ciliary action into the nephrostomes. The glomeruli are internal in adult vertebrates, but many of the tubules of adult elasmobranchs, early actinopterygians, and many amphibians retain nephrostomes and the renal capsule connects to the coelom through a narrow coelomic funnel (Fig. 20–3**C**). Nephrostomes are absent from other adult vertebrates.

This variation in nephron types and their pattern of distribution suggests an evolutionary sequence. Ancestral vertebrate nephrons may have had external glomeruli and nephrostomes, as do the first few to develop in very primitive vertebrates. Many early vertebrates probably lived in fresh water. Excess water would enter their bodies by osmosis and would have to be eliminated. External glomeruli would add considerable water to the coelomic fluid along with nitrogenous waste products, and water and wastes could be drawn off through the nephrostomes and renal tubules. The mechanism would become more efficient if the coelomic recess into which each external glomerulus discharged became a part of the tubule, that is, grew around the glomerulus as a renal capsule. The nephrostomes were lost during subsequent evolution, leaving the type of tubule found in most vertebrates.

The vertebrate nephron is unique, but many groups of invertebrates drain wastes and other materials from the coelomic fluid through metanephridia. Amphioxus drains its dorsal coelomic canals with unusual cells called "solenocytes" (p. 37).

◀ *FIGURE 20–2* Vertebrate nephrons. **A,** Parts of a nephron as seen in a frog. **B–D,** Diagrams of the blood supply to the nephrons of different vertebrates: **B,** Chondrichthyans, reptiles, and birds; **C,** Osteichthyans and amphibians; **D,** Mammals. Peritubular capillaries cover all parts of the tubules including the collecting tubules, but only a small part of this capillary bed is shown. Arrows indicate the direction of blood and urine flow. Red = oxygen-rich blood, blue = oxygen-poor blood, purple = mixed blood. (From Walker and Homberger.)

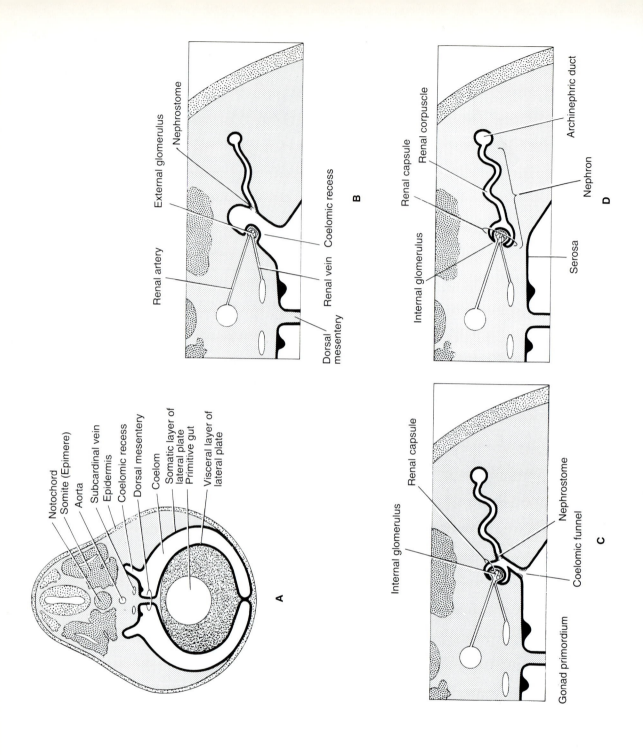

Kidney Development and Evolution

The Holonephros

Like somites and many other embryonic structures, nephrons differentiate sequentially from anterior to posterior during embryonic development along most of the nephric ridge, but not all of them become functional in most vertebrates (Fig. 20–1**A**). Analysis of the variation in the tubules that do become functional in vertebrate larvae and adults has led anatomists to postulate that in ancestral vertebrates, segmental tubules, all of which were functional, developed from the entire nephric ridge, and drained into an **archinephric duct.** This hypothetical ancestral kidney is called an **archinephros** or **holonephros** (Fig. 20–4**A**). Larval hagfishes and caecilians have kidneys that resemble the holonephros. Segmental tubules develop from the entire nephric ridge, but not all are functional concurrently.

The Pronephros

In all vertebrate embryos the kidney begins with the differentiation of a few renal tubules from the anterior end of the nephric ridge overlying the pericardial cavity. The distal ends of these tubules unite to form an archinephric duct that rapidly grows caudally to the cloaca. This early embryonic kidney is called the **pronephros** (Fig. 20–5**A**). A dozen or more pronephric tubules appear in some anamniotes, but most species have only four or five and even some of these soon regress. Only one to three pronephric tubules develop in amniotes. In anamniotes pronephric tubules are segmental and usually have coelomic funnels connecting to the coelom. Many of these tubules form a functional pronephric kidney in the embryos and larvae of cyclostomes, many bony fishes, and amphibians, but the pronephros is not functional in the embryos of chondrichthyan fishes or amniotes. Its only role in these vertebrates is to initiate the formation of the archinephric duct. Adult hagfishes and a few teleosts retain a pronephros, but the extent to which it continues to function is uncertain. It is called a **head kidney** because it is located far forward over the pericardial cavity and is separated from the more caudal and functional kidney by a gap in the nephric ridge (Fig. 20–4**B**). The pronephros has been lost in the adults of other vertebrates.

◀ *FIGURE 20–3*   **A,** A cross section of an early vertebrate embryo showing the location of the coelomic recess that contributes to the formation of the renal capsule. **B–D,** The probable sequence in the embryonic and evolutionary development of the renal capsule: **B,** External glomeruli seen in ammocoetes and some larval amphibians; **C,** The internal glomeruli of elasmobranchs and some actinopterygians retain a connection with the coelom; **D,** The external glomeruli of most vertebrates. (From Walker and Homberger.)

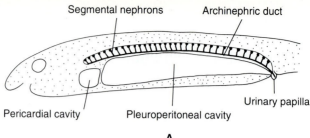

Segmental nephrons   Archinephric duct

Pericardial cavity

Pleuroperitoneal cavity

Urinary papilla

A

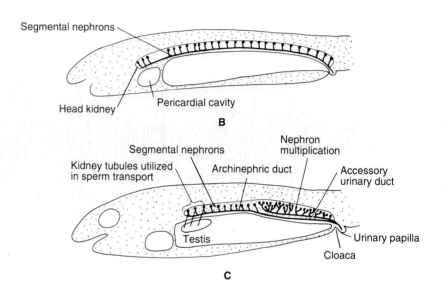

Segmental nephrons

Head kidney   Pericardial cavity

B

Nephron
multiplication

Segmental nephrons

Kidney tubules utilized
in sperm transport   Archinephric duct

Accessory
urinary duct

Testis

Urinary papilla

Cloaca

C

Archinephric duct = sperm duct

Kidney remnant utilized
in sperm transport   Metanephros

Ureter

Testis

Pericardial
cavity

Cloaca

D

FIGURE 20–4   Lateral views of the evolution of the kidney and its ducts in adult vertebrates: **A,** The theoretical holonephros. **B,** The primitive opisthonephros found in hagfishes. **C,** The advanced
opisthonephros characteristic of most fishes and amphibians. **D,** The metanephros of
amniotes.

The Mesonephros

Kidney tubules fail to differentiate in a few body segments caudal to the pronephros, so in all vertebrates there is a gap between the pronephros and more caudal renal tubules. As the pronephros regresses, the archinephric duct induces the sequential differentiation of tubules in the more caudal parts of the nephric ridge. These tubules tap into the archinephric duct and become functional (Fig. 20–5B). The new tubules at first are segmental, and those of anamniotes retain coelomic funnels. As development continues, additional tubules usually grow out from the base of each primary one, and the segmental nature of the kidney becomes obscure. Secondary and tertiary tubules lack coelomic funnels. Tubules having a common origin unite to form a **collecting duct** before entering the archinephric duct.

Tubules that differentiate in the middle part of the nephric ridge form a kidney called the **mesonephros.** This kidney functions in the embryos and larvae of all vertebrates, but the degree to which it functions in mammalian embryos is inversely related to the amount of excretion that occurs through the placenta. The mesonephric tubules are distinct from tubules that develop in adjacent parts of the nephric ridge in the embryos of amniotes. The cranial mesonephric tubules acquire a connection with the developing gonad in most vertebrates. Teleosts are a notable exception.

The Opisthonephros

In amniote embryos a gap is present between the mesonephros and tubules that differentiate in the caudal part of the nephric ridge, so the mesonephros is recognizable as a distinct entity (Fig. 20–5B). In contrast, in anamniotes the sequential differentiation of kidney tubules continues without interruption throughout the nephric ridge. New tubules function as soon as they tap into the archinephric duct. Adult hagfishes have functional, segmental renal tubules with coelomic funnels throughout most of the nephric ridge (Fig. 20–4B). This kidney in adult hagfishes approaches in its structure the holonephros postulated for ancestral vertebrates except for the specialization of the pronephric region and the gap in tubule sequence just caudal to the pronephros. Since this kidney includes the embryonic mesonephros and also tubules that develop in the caudal part of the nephric ridge, it is called an opisthonephros (Gr. *opisthen* = behind, at the back). More specifically, it is a **primitive opisthonephros** because of the primitive, segmental nature of the tubules.

Other fishes and amphibians have an **advanced opisthonephros** (Fig. 20–4C). Renal tubules multiply, especially in the caudal portion of the kidney, so they are no longer segmentally arranged. As a consequence, the caudal part of the kidney usually is enlarged, and most urine production occurs here. The cranial part of the organ, which is derived from the embryonic mesonephros, is slender in chondrichthyan fishes and urodeles, produces little or no urine, and in males receives sperm. The opisthonephric kidneys of teleosts are extremely variable.

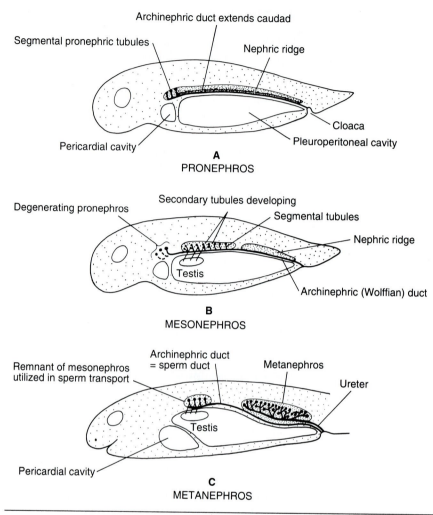

FIGURE 20–5 Lateral views of the sequence of kidneys that occur during the embryonic development of an amniote.

Some are long and slender and the two kidneys are partly united; some are divided into more or less separate anterior and posterior parts; yet others are short, compact organs confined to the caudal part of the trunk. Frogs have short trunks, and their kidneys also are short, compact organs. One or more **accessory urinary ducts** may bud off the archinephric duct in elasmobranchs and urodeles and enter the thickened, urinary portion of the opisthonephros (Fig. 20–4C). As accessory urinary ducts develop, they may separate completely from the archinephric duct and enter the cloaca independently. Accessory urinary ducts are more likely to be present in males than in females. When present, most of the

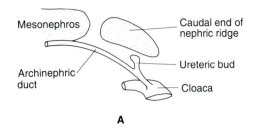

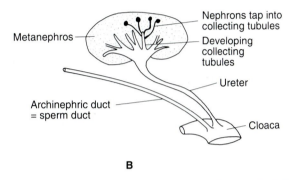

FIGURE 20–6   Lateral views of the induction by a ureteric bud of nephrons in the metanephric portion of the nephric ridge.

urine is transported in them, and the archinephric duct transports primarily sperm.

The Metanephros

Although the embryonic mesonephros contributes to the adult opisthonephros in most anamniotes, in amniotes it is a transitional kidney and functions only in their embryos (Fig. 20–5**B**). Later in the development of amniotes a **ureteric bud** extends from the caudal end of the archinephric duct, grows into the caudal end of the nephric ridge, and branches extensively (Fig. 20–6). The branching ureteric bud induces the differentiation of a great many renal tubules that tap into it. The ureteric bud itself forms the **collecting tubules** and the **ureter** that drain the adult kidney. This type of kidney, called a **metanephros,** occurs in all adult amniotes (Figs. 20–4**D** and 20–5**C**). It is homologous only to the posterior portion of the opisthonephros. A metanephros is always drained exclusively by the ureter, which in reptiles and birds separates from the archinephric duct and enters the cloaca independently. In mammals it enters the urinary bladder. The ureter develops in the same way as an accessory urinary duct and is its homolog. As the metanephros and ureter develop and become functional, the amniote mesonephros and archinephric duct regress, except for those parts in a male that are connected to the testis. The cranial mesonephric tubules and archinephric duct become part of the system of ducts that carry sperm (Chapter 21).

The kidneys develop dorsal to the coelom, but they frequently bulge into it as the tubules multiply during development. The dorsal surfaces of the kidneys lie against back muscles and are not covered by coelomic epithelium. This position is described as **retroperitoneal** (L. *retro* = backward). Although the metanephric kidneys develop from the caudal part of each nephric ridge, they migrate anteriorly during development and come to lie retroperitoneally just caudal to the liver. Much of this migration results from differential growth of parts of the body in this area.

In mammals the kidneys are usually compact, bean-shaped organs and tend to be subdivided into many lobes in large species, including many ungulates, bears, seals, and cetaceans. The subdivisions shorten the length of the collecting tubules and thus may facilitate the flow of urine.

During the evolution of the kidney, urinary functions tend to shift and become concentrated in the posterior part of the nephric ridge (Fig. 20–4). An ancestral holonephros has been replaced by an opisthonephros as the functional adult kidney of most anamniotes. Urinary functions frequently are concentrated in the caudal part of the opisthonephros, and this region may be drained by one or more accessory urinary ducts. A metanephros replaces the opisthonephros as the adult kidney in amniotes. There is also a posterior shift in urinary functions during the embryonic development of an amniote (Fig. 20–5). A very short and transitory pronephros forms the archinephric duct. The mesonephros utilizes this duct and becomes the functional embryonic kidney. It is replaced in the adult by a metanephros.

The Urinary Bladder and Cloaca

The caudal ends of either the archinephric ducts or the accessory urinary ducts are slightly enlarged in many fishes, and these areas are sometimes called **urinary bladders** or a **urogenital sinus** (Fig. 20–7A). In some species the caudal ends of the left and right ducts are conjoined, so the bladders are partly united. All these sacs are small, and their functional significance not known. A continuous urine discharge occurs in the fresh water environment. In elasmobranchs the ducts and sinus open into the cloaca dorsal and caudal to the opening of the digestive tract. In many fish species the cloaca is partially divided by folds on its lateral walls into a dorsal **urodaeum** receiving urine and genital products and a ventral **coprodaeum** (Gr. *kopros* = dung + *hodaion* = way) receiving feces.

Amphibians have large, often bilobed urinary bladders that evaginate from the midventral part of the cloaca into the body cavity (Fig. 20–7B). The epithelium of the urinary bladder is of endodermal origin, but the wall becomes muscularized and vascularized by mesoderm. The epithelium is an unusual type known as **transitional,** which allows for changes in the lining of structures that expand and contract. Transitional epithelium is composed of two layers of cells that are connected to the basement membrane and a third layer wedged in between the tops of the cells of the basal layer. The shapes of the cells and their distributional

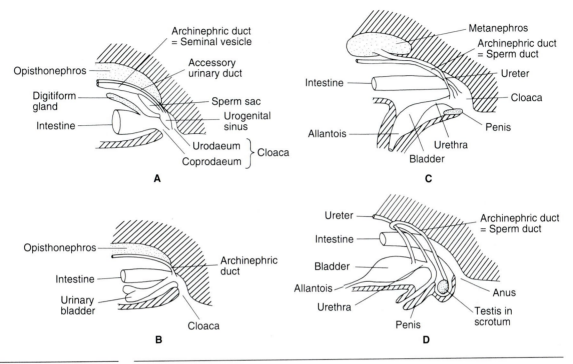

FIGURE 20–7 Lateral views of the cloacal region of representative male vertebrates, showing the termination of the intestine, urinary ducts, and associated structures. The way the cloaca becomes divided in eutherian mammals to continue the intestine and urogenital ducts independently to the body surface is considered in the next chapter. **A,** A dogfish. **B,** A salamander. **C,** A reptile. **D,** A eutherian mammal.

configuration give the epithelium the appearance of being stratified when the bladder is empty. When the bladder is fully distended the cells are spread apart resulting in two layers of flattened cells: one basal layer of triangular cells alternating with inverted triangles and a second layer of very flattened cells wedged in between the basal cells. The archinephric ducts open into the dorsal part of the cloaca, and urine flows into the bladder by gravity and the contraction of the cloacal wall. Urine discharge is not continuous in the terrestrial environment. Urine accumulates temporarily in the amphibian urinary bladder, and some water is reabsorbed from here.

A homologous cloacal evagination develops in the embryos of amniotes and expands beyond the confines of the body into the extraembryonic coelom (Fig. 20–7C). This outgrowth is the **allantois,** and with the embryonic chorion, it forms an important excretory and respiratory organ (p. 132). At hatching or birth the extraembryonic part of the allantois is lost along with the other extraembryonic membranes, but in many species the part of the allantois within the body enlarges to become the urinary bladder and its duct, the **urethra.** Some frogs, a few teleosts, most turtles, *Sphenodon,* and lizards have urinary bladders that, in addition to temporarily storing urine, are a site for water reabsorption. Certain aquatic turtles can pump water in and out of the bladder through the cloaca and utilize it in aquatic gas exchange. Indeed, some species have paired **accessory urinary bladders** that evaginate from the cloaca for this purpose (see Fig. 17–10). The urinary bladder has been lost in the adults of other reptiles and most birds, which excrete nitrogenous waste products as a semisolid paste. Mammals produce a liquid urine that is stored temporarily in a urinary bladder, although no water is reabsorbed.

The ureters of most reptiles, birds, and monotremes continue to open in the dorsolateral wall of the cloaca, and urine must cross the cloaca to enter the bladder in those species having one. Urinary and reproductive products and undigested residues become sorted out by the division of the cloaca in marsupials and eutherian mammals (Chapter 21). The ureters join the bladder, and the bladder is drained by a urethra that continues to the body surface or to a urogenital canal (Fig. 20–7**D**).

Excretion and Osmoregulation

Nitrogen Excretion

Although some nitrogenous wastes derive from the metabolism of nucleic acids, most come from the deamination of amino acids. This process occurs primarily in the liver. Each removed amino group (NH_2) picks up a hydrogen ion and transforms into an **ammonia** molecule (Fig. 20–8). Ammonia is toxic, so it must be either removed quickly from the tissues and body or converted to a less toxic substance. Since ammonia is very soluble in water, it can be flushed out of the tissues by a rapid water turnover, that is, by a rapid entry and removal of water from the body. This occurs in freshwater teleosts. If the requisite enzymes are available, ammonia can be converted to other materials. **Urea,** which contains two amino groups, is far less toxic than ammonia and need not be removed so rapidly. Alternatively, more amino groups can be combined to form the larger **uric acid** molecule. Uric acid has a very low solubility in water, so it precipitates from solution as minute crystals. It is chemically inert and not toxic. It can be stored in tissues, but it is discharged by adult vertebrates as a pastelike mass with little or no water being required to carry it off. Other forms of nitrogenous wastes occur in small amounts, but ammonia, urea, and uric acid are the primary ones. Since they require different amounts of water for their elimination, the molecular form

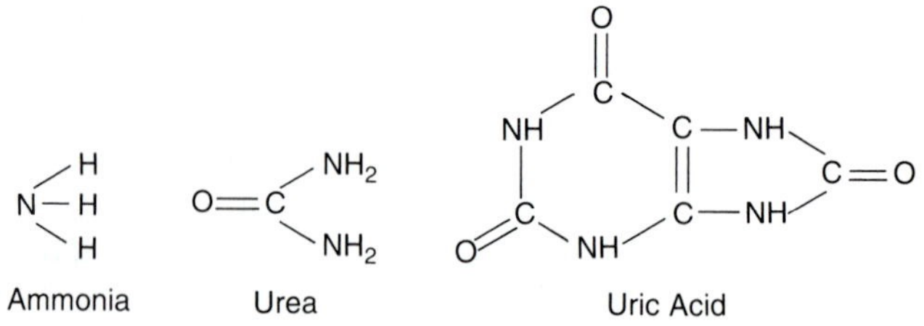

Ammonia   Urea   Uric Acid

FIGURE 20–8   The structural formulas of the three most common types of nitrogenous excretory products.

TABLE 20–1 **Osmotic Concentration of Inorganic Salts and Urea in Plasma***

	Habitat	Osmotic Concentration (mOsm/l)	Urea (mOsm/l)
Sea water		≈1000	
Hagfish *(Myxine)*	Marine	1152	
Lamprey *(Petromyzon)*	Marine	317	9
Dogfish *(Squalus)*	Marine	1000	354
Freshwater ray *(Potamotrygon)*	Freshwater	308	1+
Goldfish *(Carassius)*	Freshwater	259	
Toadfish *(Opsanus)*	Marine	392	
Eel *(Anguilla)*	Marine	371	
Eel *(Anguilla)*	Freshwater	323	
Coelacanth *(Latimeria)*	Marine	1181	355
Frog *(Rana)*	Freshwater	200	1+

*The amount of urea included in the osmotic concentration is shown separately if it is 1 milliosmol per liter or greater and therefore osmotically significant. (Data from K. Schmidt-Nielsen.)

in which nitrogen is eliminated by a vertebrate is closely linked to the availability of water and so to the problem of osmoregulation.

The Environment of Ancestral Vertebrates

Since excretion and osmoregulation are coupled, the structure of the renal tubules of primitive living vertebrates may help us understand the environment of ancestral vertebrates and the circumstances in which the renal tubules evolved. If vertebrates evolved from some protochordate group, as we believe, their remote ancestors must have resembled the protochordates in being marine. The protochordates, like other marine invertebrates, have an inorganic salt content in their body fluids that is similar to the salt water in which they live, so their body fluids have the same osmotic pressure as sea water. Their body fluids are **iso-osmotic** to their environment.

Possibly, the earliest vertebrates also were marine. Evidence is suggested by the contemporary but primitive hagfish, which is marine and whose body fluids are nearly iso-osmotic with sea water. The hagfish is the only vertebrate to have an inorganic salt content in its plasma equivalent to that of sea water (Table 20–1). It differs from the protochordates and resembles many vertebrates in having renal tubules with exceptionally large renal corpuscles that filter a great deal of water from the blood (Fig. 20–9). This water need not be reabsorbed because water easily reenters the iso-osmotic body fluids. Why an iso-osmotic animal needs to eliminate so much water through the kidneys when water diffuses so easily through the gills is not entirely clear. Kidneys do, however, supplement the gills in flushing out ammonia. Limited paleontological evidence also indicates that the earliest vertebrates may have been marine. Whether a fossil bed was formed in

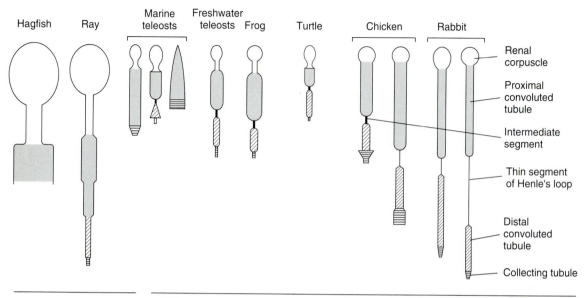

Hagfish Ray Marine teleosts Freshwater teleosts Frog Turtle Chicken Rabbit

Renal corpuscle

Proximal convoluted tubule

Intermediate segment

Thin segment of Henle's loop

Distal convoluted tubule

Collecting tubule

FIGURE 20–9   Nephron structures of representative vertebrates. (After Prosser.)

marine or freshwater deposits can usually be determined by the presence or absence of certain types of invertebrates. Echinoderms, for example, have been a marine group throughout their history, and their fossils are found only in marine deposits. The earliest known fragments of ostracoderm armor come from deposits that have been interpreted in this way as marine or brackish. But remember that the remains of freshwater organisms can be carried by rivers into marine deposits.

Whether the first vertebrates were marine or freshwater, it is clear that they adapted to a freshwater environment early in their evolutionary history. In freshwater teleosts, the body fluids have a higher osmotic pressure than the water they inhabit. Their body fluids are **hyperosmotic** to their environment so water enters the body. All other vertebrates have an *inorganic salt content* in their plasma that is substantially less than salt water (Table 20–1). Their body fluids are **hypoosmotic** to their environment, so in marine teleosts water tends to leave the body. This means that vertebrates have evolved physiological mechanisms to become ionically independent of their environment, whether this is salt water, fresh water, or land. Maintaining ionic independence (osmoregulation) requires the expenditure of considerable energy to eliminate or conserve water and salts. It is unlikely that ionic independence, and the adaptation of cellular enzymes to function at lower levels of inorganic salts than those present in sea water, would have evolved except as an adaptation to life in fresh water.

Renal corpuscles, which occur in nearly all vertebrates, are well suited to remove the excess water that floods the tissues in a freshwater fish. Although renal corpuscles may have evolved first in the marine environment as a means of ensuring a high water turnover for the elimination of ammonia, their presence would have facilitated the entry of vertebrates into fresh water. The fossil record also

TABLE 20–2   **Major Form of Nitrogen Excretion**

Animal	*Environment*	*Type of Nitrogen Excretion*
Teleost	Freshwater	Ammonia, some urea
Teleost	Marine	Ammonia, some urea
Elasmobranch	Marine	Urea
Larval amphibian	Freshwater	Ammonia, some urea
Adult amphibian	Terrestrial	Urea, some ammonia
Reptile	Terrestrial	Uric acid
Bird	Terrestrial	Uric acid, urea in some species
Mammal	Terrestrial	Urea, small amounts of ammonia, and sometimes uric acid

supports the hypothesis that vertebrates early adapted to fresh water: most ostra-coderms and other early fish fossils are found only in freshwater deposits.

Freshwater Fishes

About 90 percent of the nitrogenous wastes of freshwater fishes is excreted as ammonia; most of the rest, as urea (Table 20–2). The gills are the primary excretory organs, for about six times as much nitrogen is lost by diffusion through them as through the kidneys. The kidneys supplement the gills in nitrogen excretion and are essential in osmoregulation, especially in removing water. The body fluids of a freshwater fish have a salt content far greater than that of the environment in which it lives, so water enters the body by osmosis through any diffusible surface, such as the gills, gut lining, and sometimes the skin. A freshwater fish must produce a copious, dilute urine that is hypo-osmotic to its body fluids. The renal tubules with their large renal corpuscles are admirably suited for water removal because they produce a large filtrate volume. The tubule contains two ciliated segments, the narrow part immediately following the capsule and the intermediate segment (Fig. 20–9). The cilia work like water pumps, to move the filtrate at a high velocity.

Two mechanisms increase the amount of filtrate formed in the renal corpuscles compared with capillary beds in most other organs. First, the capillaries of the glomerulus lie between two arterioles (Fig. 20–2). The diameter of the efferent arteriole is smaller than that of the afferent arteriole, and its contraction under some conditions increases the hydrostatic pressure of the blood in the glomerular capillaries. This increases filtration pressure, which is the difference between the hydrostatic and the colloidal osmotic pressures of the blood. Second, diffusion distance is short because the walls of the capillaries and renal capsule are exceptionally thin (p. 672). Details have not been studied in all vertebrates, but electron microscope studies of mammalian renal corpuscles have shown that the endothelial cells that form the walls of the capillaries are perforated by small slits; they are **fenestrated** (Fig. 20–10). The epithelial cells that form the adjacent wall of the renal capsule are called **podocytes.** Podocytes do not lie tightly against the

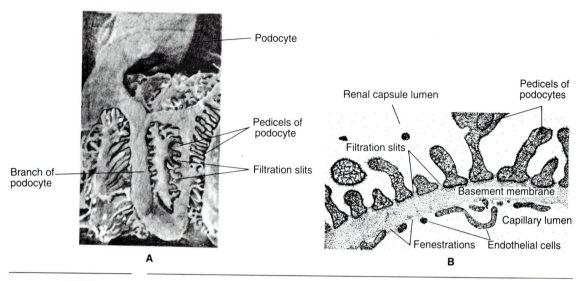

A   **B**

FIGURE 20–10   Drawings based on electronmicrographs of the relationship between the podocytes that line a renal capsule and the adjacent glomerular capillaries: **A,** A view from within the renal capsule lumen looking toward the capillaries. **B,** A vertical section extending from the lumen of a renal capsule to the lumen of a glomerular capillary. (After Kessel and Kardon.)

basement membrane that separates them from the capillary endothelial cells; rather, they are attached only by footlike pedicels with **filtration slits** between them. A similar condition was seen in the "solenocytes" of amphioxus (p. 37). The cavities of the glomerular capillaries and renal capsule are separated in many places only by the very thin basement membrane of the epithelial cells. This membrane is composed of delicate collagen fibers and glycoproteins that are believed to be secreted by both the endothelial cells and the podocytes.

Although a freshwater fish produces a copious and dilute urine that flushes out water and helps eliminate ammonia, water entry into the body must be limited so that more does not enter than can be removed (Fig. 20–11**A**). Mucus secreted by the skin reduces osmosis through the body surface. Although some water is unavoidably swallowed during feeding, the fish minimizes water uptake through the gut by not drinking extra water.

Body salts also are lost by a freshwater fish by diffusion through the gills. Additional salts are lost through the filtrate, but some salts are actively reabsorbed by distal parts of the renal tubules. Lost salts are regained in the food and by the active uptake of the limited amount of salt available in fresh water by special cells on the gills, called **ionocytes.** A freshwater fish, however, cannot obtain enough salt to maintain a salt level as high as that in sea water. Its cells must be able to function at lower salt levels.

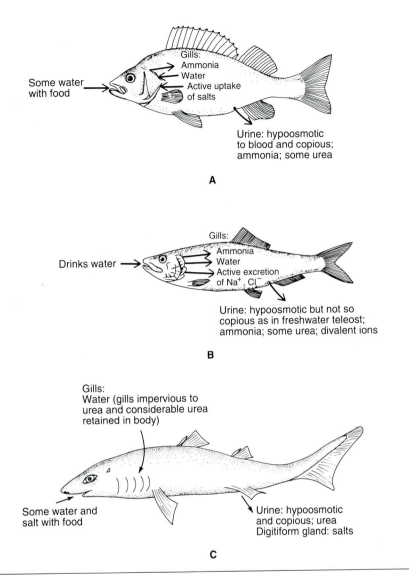

FIGURE 20–11 Osmoregulation and excretion in fishes. The major sites for exchanges of water, salts, and nitrogenous wastes are indicated. **A,** A freshwater teleost. **B,** A saltwater teleost. **C,** A saltwater shark.

Saltwater Teleosts

When early vertebrates entered the sea from the freshwater habitat to which they had become adapted, they were confronted with a different set of problems. The salt content of their body fluids was now less than salt water, so water would leave their bodies by osmosis. An additional problem would be taking in too much salt. Actinopterygians and cartilaginous fishes have evolved different solutions to these problems. Marine teleosts conserve some water by a reduction in the size of their renal corpuscles (Fig. 20–9), but because they continue to eliminate most of their nitrogen as ammonia, they need a high water turnover. Their hypo-osmotic urine is not as copious as that of freshwater fishes (Fig. 20–11**B**). The marine fish tubule has lost the ciliated segments (Fig. 20–9). Consequently the filtrate moves much slower since it is not acted on by cilia. In this way the volume of urine is reduced and water loss minimized. Saltwater teleosts make up for water lost in the urine and in osmosis by drinking sea water and then actively excreting the excess salts. Little sodium and chloride are lost in the urine, but ionocytes in their gills can actively excrete these ions. Note that ionocytes can move salts in either direction: they take up salts in the freshwater environment and excrete them in the marine environment. The divalent magnesium and sulfate ions taken in with sea water cannot be eliminated through the gills, but they are excreted actively by the renal tubules.

Some teleosts, such as eels and salmon, move between salt and fresh water during their life cycle. Their physiology and behavior resemble those of freshwater fishes in fresh water and marine fishes in salt water. In the presence of the pituitary hormone prolactin the kidney functions as that of a freshwater fish, producing copious amounts of dilute urine. When prolactin is absent, the renal tubule eliminates excess ions actively, and water is moved slowly because the cilia are inactivated. Eels do, however, have a slightly higher salt content in their body fluids when in the ocean than when in fresh water (Table 20–1). In the course of their evolution, many marine fishes have reentered fresh water and readapted to this habitat. Indeed, the fossil record and analyses of the distribution of lungs and swim bladders indicate that contemporary freshwater teleosts had a marine phase in their ancestry.

A few marine teleosts have lost their renal corpuscles (Fig. 20–9). This certainly lessens the problem of water loss through the kidneys, for it eliminates filtration. All of the blood reaching the peritubular capillaries now comes from the renal portal system (Fig. 20–2**C**). Nitrogen can be removed by renal tubule secretion and by diffusion through the gills. Most deep-sea fishes have lost the renal corpuscles resulting in considerable water conservation. The blood of some antarctic fish species contains many relatively small glycoprotein molecules that act as an antifreeze by lowering the blood's freezing point. It is hypothesized that these molecules are retained because of the loss of renal corpuscles.

Chondrichthyan Fishes

Marine chondrichthyans took a different evolutionary pathway. They convert ammonia to the less toxic urea and eliminate most of their nitrogenous wastes in this form (Table 20–2). They also retain sufficient urea in their body fluids to raise their internal osmotic pressure to a level comparable to that of sea water (Table 20–1). Their tissues have become adapted to function at urea levels most other vertebrates could not tolerate for long. Urea retention is possible because their gills have become impervious to its diffusion. Some urea enters the filtrate in their renal corpuscles, but much of this is reabsorbed more distally along the tubule. Urea retention gives marine cartilaginous fishes the same osmotic problem as freshwater fishes (Fig. 20–11**C**). Water diffuses into their bodies by osmosis and must be eliminated by the production of a copious and dilute urine. Their renal corpuscles are large, even larger than those of freshwater teleosts, most of whom evolved from marine ancestors (Fig. 20–9). Although cartilaginous fishes do not drink sea water, some salt water enters the gut during feeding, and additional salts enter the body through the gills. Excess salts are excreted by a special **digitiform** or **rectal gland** that discharges into the caudal end of the intestine. Very few cartilaginous fishes are able to re-enter fresh water for extended periods because the high urea level of their body fluids would draw in more water than could be eliminated. The few species that have adapted to fresh water retain less urea than marine species and have reduced digitiform glands (Table 20–1).

The retention of urea by adult cartilaginous fishes is an interesting solution to the problem of osmoregulation in the marine environment, but the synthesis of urea and the habituation of the tissues to high urea levels may have evolved first as embryonic adaptations. The young of cartilaginous fishes develop either in egg cases deposited in the sea or in the reproductive tracts of their mothers (p. 768). In either environment water turnover in the embryos is relatively low, and they could not survive unless ammonia was converted to the less toxic urea.

From Water to Land: Lungfishes and Amphibians

Unfortunately, we know nothing about excretion and osmoregulation in the freshwater rhipidistians that were the ancestors of terrestrial vertebrates. The related coelacanth, *Latimeria*, which has survived, tells us nothing; it is a marine species, gives birth to living young as do most cartilaginous fishes, and solves its osmotic problem by urea retention (Table 20–1)—an interesting example of convergence with chondrichthyans. Contemporary lungfishes, some of which live in bodies of fresh water that dry up, and amphibians inhabit environments like those of the ancestors of terrestrial vertebrates. They face similar problems and probably excrete and osmoregulate in similar ways.

Lungfishes and larval amphibians live in fresh water and have the same problems as freshwater fishes. Excess water enters their bodies by osmosis and must be eliminated by a copious urine that is hypo-osmotic to their body fluids. Although some salts are reabsorbed in the renal tubules, the net loss must be made up by obtaining salts from food or directly from the environment. Given a high water turnover, most nitrogen is eliminated as ammonia, but some urea is produced (Table 20–2).

When water is in short supply, lungfishes and amphibians conserve it by excreting most of their nitrogen as urea. The African lungfish, *Protopterus*, can estivate in a dried mud cocoon when its pond dries up. Since metabolism is very low during this period, the lungfish produces less waste nitrogen than usual, and all of it is converted to urea. The urea is not eliminated but accumulates in the tissues, often for several years, until water returns.

Most adult amphibians spend considerable time on land and must conserve water. They, too, convert most of their waste nitrogen to urea. A great deal of water and urea are filtered through the modest-sized renal corpuscles, and additional urea is added by tubular secretion (Fig. 20–9). Amphibian renal tubules

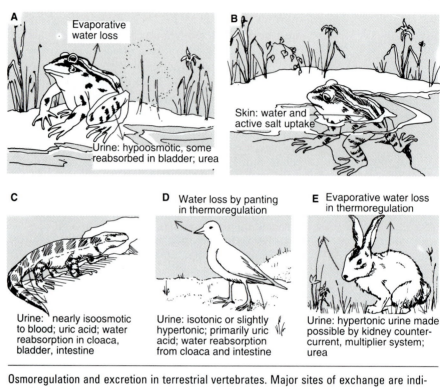

A Evaporative water loss

Urine: hypoosmotic, some reabsorbed in bladder; urea

B Skin: water and active salt uptake

C Urine: nearly isoosmotic to blood; uric acid; water reabsorption in cloaca, bladder, intestine

D Water loss by panting in thermoregulation
Urine: isotonic or slightly hypertonic; primarily uric acid; water reabsorption from cloaca and intestine

E Evaporative water loss in thermoregulation
Urine: hypertonic urine made possible by kidney counter-current, multiplier system; urea

FIGURE 20–12   Osmoregulation and excretion in terrestrial vertebrates. Major sites of exchange are indicated, but there is considerable variation within groups (see text). **A,** A frog on land. **B,** A frog in water. **C,** A lizard. **D,** A bird. **E,** A mammal.

resemble those of their freshwater fish ancestors. They have the same two ciliated segments (the narrow neck after the capsule and the intermediate segment) to move large amounts of filtrate rapidly. The urine remains hypo-osmotic to the body fluids, so a great deal of water is lost this way, although some is reabsorbed from the urinary bladder (Fig. 20–12**A**). Additional water is lost by evaporation through the thin and vascular skin, which functions as a respiratory membrane (p. 653). Evaporation through the skin of many amphibians occurs at the same rate as from a free water surface under similar conditions of temperature and humidity. Amphibians minimize this route of water loss by living in cool, moist habitats or being active during the cooler times of day, when the humidity is higher. Despite these water-saving mechanisms, amphibians lose a great deal, and most return to water periodically to regain water by osmosis through the skin (Fig. 20–12**B**). Their skin cells also can actively take up salts from the water.

Several anurans are able to live in environments where very little water is available by retaining a great deal of urea in their tissues. This solution is used by the crab-eating frog, *Rana cancrivora*, of Southeast Asia, which lives in saltwater mangrove swamps; by several toads; and by some salamanders that live in dry soil for several months during the year. A few arboreal frogs, including *Chiromantis* and *Phyllomedusa*, conserve water, as do most reptiles, by excreting their nitrogen primarily as uric acid. Another frog, *Cyclorana* (*Chirolepsis*), lives in the deserts of central Australia. It soaks up water during the rainy periods and produces a copious and very dilute urine that is stored in the urinary bladder. As much as a third of the animal's weight is represented by this urine. Additional water is stored in the bloated subcutaneous lymph sacs. During the dry periods, which may last two years or more, the frog estivates within a cocoon in deep burrows and gradually uses up the stored water.

Adaptations for Terrestrial Life: Reptiles

Vertebrates could not live and be active under truly terrestrial conditions unless they evolved mechanisms that enabled them to conserve water more successfully than most amphibians and to eliminate nitrogenous wastes at the same time. Water loss by respiration is reduced in all amniotes (Fig. 20–12**C**). They do not use cutaneous gas exchange or ventilate their lungs more than necessary (Chapter 18). Cornification of the skin of reptiles, birds, and mammals also reduces evaporative water loss. A desert tortoise, for example, loses as little as 3μg (micrograms) of water by cutaneous evaporation per square centimeter of body surface per hour, but reptiles that have readapted to an amphibious environment, such as crocodiles and some turtles, lose much more. The more terrestrial amniotes also have excretory adaptations that conserve considerable water.

In addition to being able to convert waste nitrogen into urea, many amniotes also can synthesize uric acid, which is chemically inert and requires little water for its removal. Although the enzymes needed for uric acid synthesis are present in several frogs, they may have first evolved in amniotes as an embryonic adaptation, as Joseph Needham suggested early in this century. Adaptation to land involved the evolution of a cleidoic egg that can be laid on land or retained in the body of the mother. The egg provides all the metabolic needs of the embryo, a free-living aquatic larval stage is suppressed, and the young hatch or are born as miniature adults. Waste nitrogen must be stored within the allantois as uric acid because there is not enough water turnover between the embryo and its environment to eliminate it as ammonia or urea.

Adult tortoises, lizards, snakes, and other terrestrial reptiles continue to eliminate most of their waste nitrogen as uric acid (Fig. 20–12**C**). Their renal corpuscles are exceptionally small (Fig. 20–9) and contain fewer glomerular capillaries than in most other vertebrates. Indeed, many tubules are aglomerular in some species of reptiles. Tubular filtration is reduced greatly, and with it, the loss of water. Some uric acid is filtered, but most enters the tubules by active tubular secretion. This is possible because the renal portal system provides the peritubular capillaries with a large volume of blood that is independent of the blood supply to the reduced glomeruli. The urine is iso-osmotic or slightly hypo-osmotic to the body fluids. Reptiles are unable to produce a urine that is more concentrated than their body fluids. Considerable water is reabsorbed in the cloaca and the urinary bladder, if one is present. Material in the cloaca can back up into the intestine, and water also is reabsorbed here. Uric acid leaves the body with the feces as a whitish, pastelike material.

Although all reptiles synthesize uric acid, the more aquatic species have a higher water turnover, produce a more dilute urine, and eliminate some of their nitrogen in other ways. Crocodiles and sea snakes remove a great deal of their waste nitrogen as ammonia; aquatic turtles eliminate it as urea. Reptiles that spend much or all of their lives in the sea also must have salt-excreting glands that eliminate excess salts. The salt-excreting glands of sea turtles discharge into the orbits; those of the marine iguana, into the nasal cavities; those of sea snakes, into the oral cavity.

High Metabolism and Endothermy: Birds and Mammals

The excretory and osmoregulatory problems of birds and mammals have been compounded by the evolution of a high level of metabolism and endothermy. Birds and mammals must cope with an increased volume of nitrogenous wastes and also conserve body water and salts. Water conservation is particularly important because birds and mammals unavoidably lose some water through the respiratory passages and skin in their thermoregulation. Birds and mammals have an increased number of renal tubules. Although some very primitive vertebrates have only one pair per body segment, and most ectothermic vertebrates have only a few hundred or thousand, estimates of the number of renal tubules in a human being range from 2 to 4 million. Filtration pressures also are higher in birds and mammals than in lower vertebrates because the complete division of the heart makes possible the development of a high systemic blood pressure. The increased number of renal tubules and high filtration pressures enable birds and mammals to clear nitrogenous wastes from a large volume of plasma. A chicken filters a volume of plasma four times that of its body fluids in 24 hours; a dog can filter eight times the volume of its body fluids. In reptiles and anamniotes filtration volume in a 24-hour period is less than body fluid volume.

Birds conserve water partly in the same way as reptiles. Renal corpuscles are small, and uric acid is the primary excretory product (Figs. 20–9 and 20–12**D**). Some uric acid is filtered, but much is added by tubular secretion from the blood that reaches the peritubular capillaries through the renal portal system. Birds, too, reabsorb water in the cloaca and from material that backs up into the intestine from the cloaca. In addition, some species of birds have evolved long, narrow loops of the renal tubules that dip into the medullary region of the kidney. This type of tubule enables the reabsorption of considerable water and the production of a very concentrated urine that is hyperosmotic to the body fluids; that is, it contains less water than the body fluids. Urea is the primary nitrogenous waste filtered in birds with many tubules of this type; precipitation of much uric acid would clog the narrow medullary loops.

The mechanism for the production of a hyperosmotic urine is understood most clearly in mammals, where a similar type of tubule evolved independently. The mechanism depends on the configuration of the renal tubules and their arrangement within the kidney. A gross section of the kidney shows distinct **cortical** and **medullary regions** (Fig. 20–13). The renal corpuscles and both the proximal and the distal convoluted tubules lie in the cortex (Fig. 20–14). A highly specialized new mammalian segment of the tubule forms a thin, hairpin loop, known as the **medullary loop of Henle,** that dips into the medulla. Most mam-

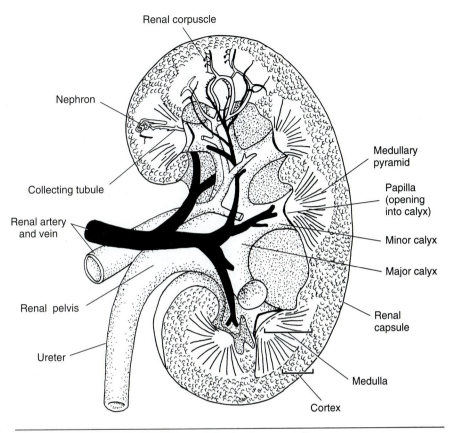

Renal corpuscle

Nephron

Collecting tubule

Renal artery
and vein

Renal pelvis

Ureter

Medullary
pyramid

Papilla
(opening
into calyx)

Minor calyx

Major calyx

Renal
capsule

Medulla

Cortex

FIGURE 20–13 A longitudinal section of a mammalian kidney. Highly schematized nephrons are drawn in the upper portion. (From Smith.)

mals have two types of tubules: some with short loops of Henle and others with long loops that dip far into the medulla. Long **collecting tubules** begin in the cortex, where they receive the distal convoluted tubules of many nephrons, and extend through the medulla to open into the **renal pelvis.** The renal pelvis is an expansion of the ureter with the kidney. The long loops of Henle and the collecting tubules are aggregated in the medulla to form one or more **renal pyramids,** whose apices form **renal papillae** that protrude into the renal pelvis or its subdivisions, called the **renal calices** (Fig. 20–13). Capillary venous and arterial loops, known as the **vasa recta,** follow the loops of Henle into the medulla.

The filtrate produced in the renal corpuscles contains a great deal of water and all the same types of smaller solute molecules present in the blood. Sugars, amino acids, vitamins, and other solutes needed by the body are actively reabsorbed from the filtrate by the cells of the proximal and distal convoluted tubules, as they are in other vertebrates. Excesses of any of these substances are allowed to stay in the urine. The ability to reabsorb water and concentrate the urine derives from the unique properties and topography of the loops of Henle, collecting

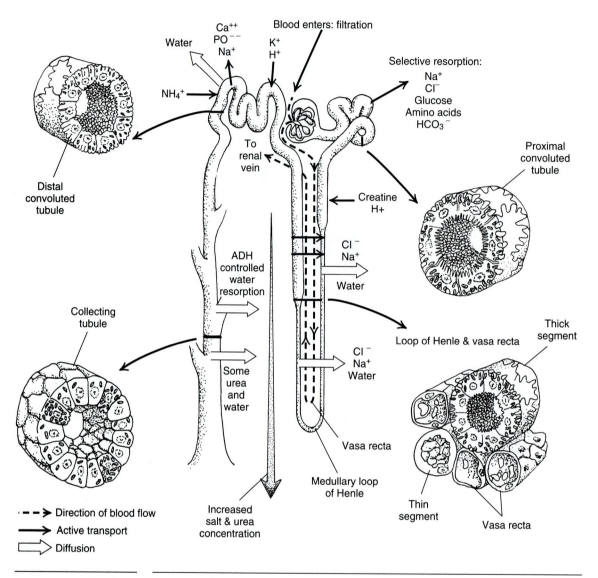

FIGURE 20–14 A mammalian nephron, showing the regions where materials are exchanged actively (solid arrows) or by diffusion (open arrows) and the operation of the countercurrent multiplying system, which permits the production of a hypertonic urine. (Modified slightly after Williams et al.)

tubules, and vasa recta (Fig. 20–14). Although the wall of most of the loop of Henle is composed of thin epithelial cells, cuboidal cells in the upper half of the ascending limb of the loop have the metabolic machinery to pump salt out of the tubular contents and into the interstitial fluid surrounding the loop. In particular, negatively charged chloride ions are actively pumped out, and positively charged sodium ions follow the chlorine. As more salts enter the filtrate from the blood, only to be pumped out again in the ascending limb of the loop of Henle, salts accumulate in the medullary interstitial fluid. Few are carried off in the medullary capillary loops because the blood flow is sluggish and the two limbs of the vasa recta form a countercurrent mechanism. The arterial and venous vasa recta also form loops in a countercurrent pattern: venous and arterial blood run in opposite directions. As a result the gradient in the medullary interstitial tissue is maintained. The venous component of the vasa recta returns blood to the renal vein with recaptured water and eliminated waste products.

The countercurrent flow of blood in the vasa recta and the pumping out of salts in the ascending limbs of the loops of Henle combine to form a very efficient **countercurrent multiplying system** that establishes a salt gradient, which becomes increasingly salty deeper and deeper in the medulla. The amount of salt that can be concentrated depends on the lengths of the loops of Henle and vasa recta.

This salt gradient enables water to diffuse passively out of the tubular contents. As the dilute tubular contents in the descending limbs of the loops of Henle enter an ever saltier environment, water moves out by osmosis. The water does not reenter the ascending limbs of the loops of Henle because their cells are impervious to its passage. As salts are pumped out of the ascending limbs of the loops of Henle, and as other solutes are actively reabsorbed in the distal convoluted tubules, the tubular contents again become so dilute that water passively diffuses from the distal convoluted tubules. As the tubular contents reenter the medulla in the collecting tubules, the contents again pass through a progressively saltier environment, and more water diffuses out. Water that enters the interstitial fluid diffuses into the vasa recta and is carried off. Water is not affected by the countercurrent flow in the vasa recta because the plasma proteins stay in the blood and exert sufficient osmotic pressure to retain the water. As the valuable solutes are actively reabsorbed from the tubular contents, and water diffuses out, the urea in the tubules becomes more and more concentrated. Most of the urea remains in the tubules because the cells of their walls are impervious to its diffusion, but the cells of the collecting tubules permit some urea to diffuse out. The presence of this urea in the interstitial fluid of the medulla intensifies the salt gradient that promotes the reabsorption of water.

The amount of water that must be conserved depends on the mammal's environment. Most or all of the tubules of desert rodents have very long loops of Henle, and some of these rodents can produce an extremely concentrated urine that has 25 times the salt content of their blood. These mammals need drink no water, for they lose very little, and they meet their needs with the water produced as a byproduct of metabolism. All the kidney tubules of a beaver, in contrast, have short medullary loops. A beaver produces a urine that is only twice as concentrated as its body fluids. About a third of our renal tubules have long loops of Henle, and our urine is about four times as concentrated as our blood.

The lengths of the loops of Henle are correlated with the capability to form concentrated urine. The longer the loops, the more concentrated urine can be produced. Mammals inhabiting dry environments, as well as those that live on diets rich in salts and other solutes, possess very long loops of Henle and collecting

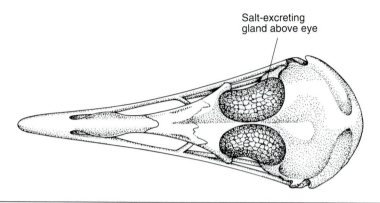

Salt-excreting
gland above eye

FIGURE 20–15 A dorsal view of the skull of a sea gull, showing the location of its salt-excreting gland. (After Schmidt-Nielsen.)

tubules. The extraordinary lengths of the collecting tubules are accommodated in very long papillae. The length and mass of the renal papillae are correlated with the power of the kidney to produce concentrated urine.

Urine concentration varies within limits in a particular species of mammal, depending on water intake. A major controlling factor is the level of **antidiuretic hormone** (Gr. *dia* = through + *ouron* = urine) in the blood. This hormone is synthesized by certain hypothalamic cells and released from the neural lobe of the pituitary gland (p. 545). A high level of antidiuretic hormone increases the permeability of the cells of the collecting tubules to water and so promotes a high concentration of urine. If water levels in the blood increase too much, antidiuretic hormone synthesis and release are decreased, water permeability of the collecting ducts decreases, and the urine becomes more dilute.

Birds and mammals that spend much of their lives in marine environments have the additional problem of salt balance. Salt intake can be a problem, especially if they feed on marine invertebrates and plants that are iso-osmotic with salt water. Fish-eating marine species have less of a problem, for they are ingesting food with a substantially lower salt content. (Recall that all vertebrates, except for the hagfishes and those that retain urea in their tissues, are hypo-osmotic to sea water.) However, some excess salt is taken in. Most birds have small nasal glands that excrete some salt. These glands are large in marine species and can excrete a very salty solution into the nasal cavities when the bird is under salt stress (Fig. 20–15). The kidneys of most mammals cannot excrete excess amounts of salt because they cannot produce a urine that is more concentrated than sea water, although they do produce one more concentrated than their blood. If a person drinks sea water, 1350 ml of urine are produced to remove the excess salts in every liter of salt water, and he or she would dehydrate rapidly. The kidneys of many marine species, however, can produce urine with a higher salt concentration than sea water. If a whale consumes a liter of sea water, only 650 ml of urine are produced to carry off the salts. There is a net gain of water. Marine mammals excrete excess salt in this way rather than by specialized salt-excreting glands.

SUMMARY

1. With the skin, lungs, and special salt-excreting or salt-absorbing cells, the excretory system maintains the stability of the internal environment by eliminating nitrogenous wastes and by helping regulate the salt and water balances of the body.
2. The structural and functional units of the excretory system are the kidney tubules, or nephrons, which develop from the mesodermal nephric ridge.
3. A filtration process occurs in the renal corpuscles; selective reabsorption and secretion occur in other parts of the tubules.
4. Ancestral kidney tubules may have had external glomeruli that filtered materials into the coelom; wastes then drained into the tubules through coelomic recesses. Glomeruli are internal in most tubules and filter directly into the tubules, but traces of the coelomic recesses remain in some primitive tubules as coelomic funnels.
5. An ancestral vertebrate may have had a holonephric kidney, in which segmental tubules developed along the entire nephric ridge. A larval hagfish has a kidney that approaches this.
6. The pronephros is an embryonic kidney that develops over the pericardial cavity and forms the archinephric duct. It is functional in the larvae or embryos of cyclostomes and many bony fishes and amphibians, but not in other groups.
7. The mesonephros is an embryonic kidney that differentiates from the portion of the nephric ridge that overlies the anterior half of the pleuroperitoneal cavity. Its tubules tap into the archinephric duct and are functional in the larvae or embryos of most vertebrates. The ducts from the testis acquire a connection with some of the cranial mesonephric tubules in most male vertebrates.
8. The adult kidney of a hagfish is a primitive opisthonephros. It includes the embryonic mesonephros and segmental tubules that develop from the caudal part of the nephric ridge.
9. Most adult fishes and amphibians have an advanced opisthonephros, which differs from the primitive opisthonephros by the multiplication of tubules, especially in the caudal part of the kidney. The caudal part may be drained by one or more accessory urinary ducts.
10. An adult amniote has a metanephros that develops only from the caudal part of the nephric ridge and is drained by a ureter. In a male, part of the embryonic mesonephros and the archinephric duct remain and transport sperm.
11. Urinary and genital ducts lead to a cloaca in most vertebrate groups. Most tetrapods, except for birds, have a urinary bladder where urine accumulates, and from which water is sometimes reabsorbed. The cloaca becomes divided in therian mammals, and the digestive and urogenital passages then lead directly to the body surface.
12. Nitrogen is eliminated primarily as ammonia if the animal has a high water turnover, or as less toxic urea and uric acid if water must be conserved.
13. Although the body fluids of the marine hagfish are iso-osmotic with sea water,

the animal has large renal corpuscles and a high water turnover. This may help flush out ammonia.

14. Ancestral vertebrates probably also were marine and iso-osmotic with their environment, as are protochordates. But vertebrates must have adapted early in their evolution to fresh water, because all contemporary species, except for the hagfishes, are ionically independent of their environment.

15. Freshwater fishes have an internal salt concentration greater than that in their environment. They must produce a copious urine that is hypo-osmotic to their body fluids to eliminate excess water that enters their bodies by osmosis. Salts must be conserved and replaced through salt-absorbing ionocytes on their gills.

16. Saltwater teleosts have a lower internal salt concentration than their environment, so they must conserve water and eliminate salts. Their urine, although hypo-osmotic to their body fluids, is less copious. Their renal tubules have very small glomeruli or are aglomerular. They drink sea water and excrete the excess salts through ionocytes on their gills and through their kidney tubules.

17. Because chondrichthyans retain considerable urea in their tissues, they have a salt concentration greater than the marine environment where most species live. They must eliminate excess water and salts.

18. Adult amphibians continue to produce a hypo-osmotic urine but conserve water by staying in moist microhabitats and reabsorbing some from their urinary bladders. They must return to water periodically to regain lost water and salts, partly by absorption through their skin.

19. The cornified skin of reptiles reduces water loss, their renal corpuscles are very small, and much nitrogen is excreted as uric acid. Most of the small amount of water used to remove uric acid from the kidneys is reabsorbed in the cloaca, urinary bladder, or caudal end of the intestine.

20. Since they are endothermic, birds and mammals produce a large volume of nitrogenous wastes, and they also lose some water in their thermoregulation. Most birds conserve water by excreting nitrogen as uric acid and reabsorbing water in the cloaca and caudal end of the intestine.

21. Mammals excrete nitrogen primarily as urea. Water is conserved by the production of a hyperosmotic urine. This is made possible by the evolution of a countercurrent multiplying system formed by the loops of Henle of the renal tubules and the accompanying vasa recta.

REFERENCES

Beeuwkes, R., 1982: The two solute model of the mammalian nephron. *In* Taylor, C. R., Johansen, K., and Bolis, L., editors: *A Companion to Animal Physiology*. Cambridge, Cambridge University Press.

Bentley, P. J., 1971: *Endocrines and Osmoregulation: A Comparative Account of the Regulation of Water and Salt in Vertebrates*. New York, Springer-Verlag.

Bentley, P. J., 1976: Osmoregulation. *In* Gans, C., and Dawson, W. R., editors: *Biology of the Reptilia*, vol. 5. New York, Academic Press.

Chase, S. W., 1923: The mesonephros and urogenital ducts of *Necturus maculosus* Rafinesque. *Journal of Morphology*, 37:457–532.

Dantzler, W. H., and Braun, E. J., 1980: Comparative nephron function in reptiles, birds, and mammals. *American Journal of Physiology*, 239:R197–R213.

Denison, R. H., 1956: A review of the habitat of the earliest vertebrates. *Fieldiana, Geology*, 11:359–457.

Dobbs, G. H., III, Lin, Y., and DeVries, A. L., 1974: Aglomerulism in antarctic fish. *Science*, 185:793–794.

Foskett, J. K., and Scheffey, C., 1982: The chloride cell: Definitive identification of the salt-secretory cell of teleosts. *Science*, 215:164–166.

Fox, H., 1963: The amphibian pronephros. *Quarterly Review of Biology*, 38:1–25.

Fox, H., 1977: The urogenital system of reptiles. *In* Gans, C., and Parsons, T. S., editors: *The Biology of the Reptilia*, vol. 6. New York, Academic Press.

Fraser, E. A., 1950: The development of the vertebrate excretory system. *Biological Reviews*, 25:159–187.

Gordon, M. S., Schmidt-Nielsen, K., and Kelly, H. M., 1961: Osmotic regulation in the crab-eating frog *(Rana cancrivora)*. *Journal of Experimental Biology*, 38:659–678.

Hickman, C. P., and Trump, B. F., 1969: The kidney. *In* Hoar, W. S., and Randall, D. J., editors: *Fish Physiology*, vol. 1. New York, Academic Press.

Kessel, R. G., and Kardon, R. H., 1979: *Tissues and Organs, A Text–atlas of Scanning Electron Microscopy*. San Francisco, W. H. Freeman and Company.

Kirschner, L. B., 1991: Water and ions. *In* Prosser, C. L., editor: *Environmental and Metabolic Animal Physiology*. New York, John Wiley and Sons.

Moffat, D. B., 1975: *The Mammalian Kidney*. Cambridge, Cambridge University Press.

Murrish, D. E., and Schmidt-Nielsen, K., 1970: Water transport in the cloaca of lizards: Active or passive? *Science*, 170:324–326.

Pang, P. K. T., Griffith, R. W., and Atz, J. W., 1977: Osmoregulation in elasmobranchs. *American Scientist*, 17:365–377.

Peaker, M., and Linzell, J. L., 1975: *Salt Glands in Reptiles and Birds*. Cambridge, Cambridge University Press.

Prosser, C. L., 1973: *Comparative Animal Physiology*, 3rd edition. Philadelphia, W. B. Saunders Company.

Rankin, J. C., and Davenport, J., 1981: *Animal Osmoregulation*. New York, John Wiley and Sons.

Riegel, J. A., 1972: *Comparative Physiology of Renal Excretion*. Edinburgh, Oliver and Boyd.

Schmidt-Nielsen, K., 1979: *Desert Animals: Physiological Problems of Heat and Water*. Reprinted 1964. New York, Dover Publications.

Schoemaker, V. H., and Nagy, K. A., 1977: Osmoregulation in amphibians and reptiles. *Annual Reviews of Physiology*, 39:449–471.

Skadhauge, E., 1981: *Osmoregulation in Birds*. Berlin, Springer-Verlag.

Smith, H. W., 1932: Water regulation and its evolution in fishes. *Quarterly Review of Biology*, 7:1–26.

21 The Reproductive System and Reproduction

PRECIS

Vertebrates reproduce and perpetuate their species by sexual reproduction. The method of sex determination, the pattern of reproduction, and the structure of the gonads and reproductive passages vary greatly among vertebrates and relate to their modes of life and the environment in which they live. Reproduction is integrated by gonadal hormones that interact with the hypothalamus and hypophyseal hormones.

Tunicates can reproduce asexually by budding, as well as sexually, but the reproduction of cephalochordates and vertebrates is exclusively sexual. Sexual reproduction ensures the recombination of genetic material necessary for the maintenance of genetic variation and evolutionary flexibility in a population. A few vertebrate species and populations within a species are entirely female. The eggs are activated in ways other than by normal fertilization, a chromosome doubling restores the diploid number, and the eggs develop **parthenogenetically** (Gr. *parthenos* = virgin + *genesis* = descent or birth). This would seem to reduce genetic recombination, but the known parthenogenetic populations have evolved by interspecific hybridization, so considerable genetic diversity is maintained by the way parental chromosomes become distributed during the processes of meiosis and subsequent chromosome doubling. Examples are seen in the whip-tailed lizards, *Cnemidophorus*, in some populations of the salamander *Ambystoma* (which are triploid), and in a tropical teleost species, the Amazon molly, *Poecilia*.

Sex Determination

Genetic recombination through sexual reproduction requires that some individuals in a population be female and others male. Determining sex is far more complicated than generally realized. Sex is determined in birds and mammals at the time of fertilization by the distribution of distinctive **sex chromosomes**. Male mammals are the heterogametic sex because their sperm-forming cells have X and Y chromosomes that segregate during meiosis to form two kinds of sperm, X and Y. Female mammals are homogametic (XX) and produce only one kind of egg, X. Although determined genetically at the time of fertilization, sex may be altered during development by hormonal influences. Sex is determined in birds, too, by the segregation of sex chromosomes, but the females are the heterogametic sex. To avoid confusion, different symbols are used in birds: females are WZ and males ZZ.

Sex is not so rigorously determined by sex chromosomes in most other vertebrates. A few have sex chromosomes; most do not. Sex chromosomes have been found in many lizards and snakes, but the male is the heterogametic sex in some species, and the female in others. Sex chromosomes are rare in turtles and unknown in crocodilians and *Sphenodon*. They are often absent in amphibians and fishes. When sex chromosomes are not present, sex appears to be determined primarily by environmental factors, although some autosomal genetic factors may play a role. Recent studies have shown that an important environmental determinant of sex, at least in reptiles without sex chromosomes, is the temperature at which the eggs are incubated. Turtle eggs incubated at temperatures above 30°C produce a preponderance of females; those incubated below 25°C hatch primarily as males. Although the first studies were done in the laboratory, field studies of the green sea turtle have confirmed these results. Mostly females emerge from nests on open beaches, whereas mostly males emerge from shaded nests. The position of an egg within a clutch can be a determinant when incubation temperatures are close to the thresholds. Eggs near the center of a clutch are exposed to more metabolic heat from their neighbors than peripheral eggs, and they hatch as females. Most of the peripheral eggs hatch as males.

Just how temperature determines sex is not clear, but it is likely that this was the primitive mechanism, at least in reptiles. **Temperature-dependent sex determination,** as the phenomenon is called, is found only among egg-laying reptiles. So far as we know, reptiles that retain the young in the body of the female until hatching have evolved sex chromosomes. This would appear to be necessary because the body temperature of reptiles, even though they are ectotherms, tends to be maintained at a high level during their diurnal period of activity. Reliance on temperature-dependent sex determination would produce so many individuals of the same sex that the continuation of the species would be endangered. The evolution of endothermy by birds and mammals may also have been a factor in the evolution of their sex chromosomes.

When sex is not rigorously determined by sex chromosomes, hermaphroditism may evolve. **Hermaphroditism** (Gr. mythology: the son of Hermes and Aphrodite became united in one body with a nymph) is the presence in one individual of both testes and ovaries. Many tunicates are **synchronous hermaphrodites** because both testes and ovaries function concurrently. A very few teleosts, such as certain sea bass, are also synchronous hermaphrodites and have a combined **ovotestis** (Fig. 21–1). Self-fertilization, possible in these cases, ensures reproduction in the absence of the other sex. This could be an advantage to sessile tunicates, but cross-fertilization normally occurs because of slight differences in the timing of gamete release or because spawning behavior is sometimes necessary to stimulate gamete release.

Individuals in five orders of teleost fishes are **sequential hermaphrodites,** for the gonad changes function during the life of an individual. **Protogynous** (Gr. *gyne* = woman) species are first female and then male. In protogynous species, the male usually is territorial and mates selectively with large females. **Protan-**

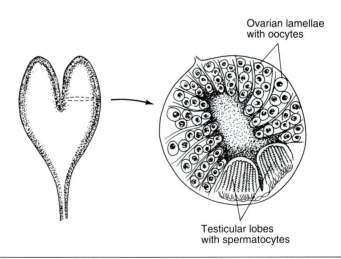

Ovarian lamellae
with oocytes

Testicular lobes
with spermatocytes

FIGURE 21–1   The ovotestis of a sea bass (*Serranidae*). (After D'Ancona.)

drous (Gr. *andr-* = man) species begin as functional males and change into females later in life. Protandrous species are typically non-territorial and mate randomly. Many sex reversals occur at some species-specific age when gonadal tissues and ducts are capable of responding to a shift in the balance of male and female hormones. But other factors, including social interactions, may cause a reversal. For example, in some species of coral reef fishes, in which a group of females breeds with a single male, the death of the male (or his experimental removal from the population) stimulates the dominant female to transform into a male. An advantage of sex reversals is that every individual is a potential female, and the population can increase rapidly if some misadventure occurs.

Reproductive Patterns

We considered the structure of mature gametes, fertilization, and early development in Chapter 4. Here we will examine other aspects of the reproductive system and the major patterns of reproduction. The reproductive system consists of the **gonads** (Gr. *gone* = seed), testes or ovaries, which produce the **gametes** (Gr. *gamet-* = spouse), sperm or eggs, and usually a set of ducts that carry them from the body. Sperm are small, motile cells surrounded only by their plasma membrane. Eggs are much larger cells that contain food reserves. They are surrounded by their plasma membrane, the **primary envelope,** and by an acellular **secondary envelope** that is deposited while they are in the ovary.

Most male vertebrates utilize certain excretory passages for the discharge of sperm. Parts of these ducts or associated glands produce secretions that help carry the sperm and sometimes assist in their nutrition and maturation. If fertilization is internal, the male usually has a penis or other type of copulatory organ by which sperm are transferred to the female.

Most female vertebrates have evolved a separate oviduct for the transport of eggs. Parts of it usually secrete **tertiary envelopes** around the eggs—jelly, albumenous layers, or shells that help nourish and protect the embryos. The structure and function of the oviduct are closely related to the pattern of reproduction of vertebrates. Most fishes and amphibians are **oviparous** (L. *ovum* = egg + *parere* = to bring forth). They lay macrolecithal or mesolecithal eggs containing sufficient yolk for the embryos to develop into free-swimming larvae that can provide for their own needs. Courtship behavior at the onset of the reproductive season ensures that the males and females are together, and sperm is discharged over the eggs as they are laid. Large numbers of eggs and sperm must be produced by oviparous species, for not all eggs are fertilized, and mortality of eggs and larvae is high. Some teleosts may lay many millions of eggs in a season. Some toads of the genus *Bufo* lay as many as 30,000 eggs in a season. Salamanders and frogs that brood their eggs, however, lay only a dozen or two. The probability of successful fertilization and embryonic development is greater in oviparous reptiles and birds because fertilization is internal, the eggs contain a great deal of yolk, and they are deposited in hidden or protected nests. Egg numbers are reduced. Turtles may lay nearly 100, but lizard and snake clutch size ranges from only 1 or 2 to about 40.

Other vertebrates, including most sharks, some teleosts, a few amphibians, some lizards and snakes, and all therian mammals, retain their embryos in an expanded part of the oviduct, called the uterus, for much or all of their embryonic development. These are **live-bearing** species, for the embryos are born as miniature adults. The source of energy and organic nutrients for the developing embryos ranges along a continuum from a complete dependence on material stored in the egg while it is in the ovary to a complete dependence on the transfer of materials in the uterus from the mother. The first extreme is called **ovoviviparity**; the last, **viviparity** (L. *vivus* = living + *parere* = to bring forth). There are no fully satisfactory terms for the intermediate conditions, and in many cases the degree of dependence is not known. For this reason, some investigators prefer to use the term **placental viviparity** for cases where a fully functional placenta is present, and **aplacental viviparity** for cases where there is less dependence on the mother, yet the young are born alive. Fertilization is internal in all live-bearing species and usually occurs in the upper part of the oviduct. Fewer sperm and eggs need be produced than in oviparous species because the chances of individual survival are much greater, but the eggs must have a great deal of yolk if few nutrients are transferred to them from the mother while they are in the uterus. In a sense, live-bearing species are mobile incubators.

In addition to the primary sex organs that produce and carry gametes, many vertebrates develop **secondary sex characters** that are important in sex recognition, as signals of sexual maturity and receptivity, in courtship behavior, and sometimes in combat among males for access to females (Fig. 21–2). Secondary sex characters include body colors, special ornamentation, antlers, beards, and so on. Another is body size. Male mammals frequently are larger than females, but the reverse is true in many other vertebrates. For instance, the females of many raptorial birds (hawks, kestrels) are larger than the males. Secondary sex characters develop at maturity under the influence of sex hormones. In cases where sex reversal occurs, the secondary sex characters may also change. Behavior also is strongly affected by the sex hormones. The biological purpose of life is reproduction, and much of an animal's structure and activity is devoted to this end. Biological fitness is commonly measured by the number of viable offspring produced.

Development of the Reproductive System

The early development of the gonads and reproductive ducts is the same in both sexes. The embryo passes through a **sexually indifferent stage** in which the primordia of both male and female structures are present. Potentially, the embryo may become either a male or a female. If subsequent differentiation is toward a male, male structures continue to develop and female ones atrophy. The opposite occurs during the differentiation of a female. Because of their similarity in early development, it is possible to homologize many male and female reproductive organs. Often, vestiges of female organs remain in males, and vice versa. The development of the reproductive system is best known in mammals but is similar in most other vertebrates. Teleosts are a notable exception and will be considered separately.

FIGURE 21–2   Secondary sex characters: **A,** The snout and body shape of a sexually mature male salmon (*Oncorhynchus*) differ from those of a female. **B,** The male creek chub (*Semotilus*) has nuptial tubercles on its head that are not present in a female. **C,** A male mule deer (*Odocoileus*). **D,** A man with a beard and mustache. (**A** and **B,** after Lagler, Bardach, and Miller. **C,** after Vaughan. **D,** after Austin and Short.)

Development of the Gonads

The gonads begin their development in all embryos as a thickening of the coelomic epithelium and mesenchyme adjacent to the medial border of the mesonephric kidneys. As mesenchyme accumulates, a **genital ridge** appears on each side of the body (Fig. 21–3**A**). Only the central part of the genital ridge forms the functional gonad in most vertebrates. The coelomic epithelium overlying the developing gonad thickens as a **germinal epithelium,** and **primordial germ cells,**

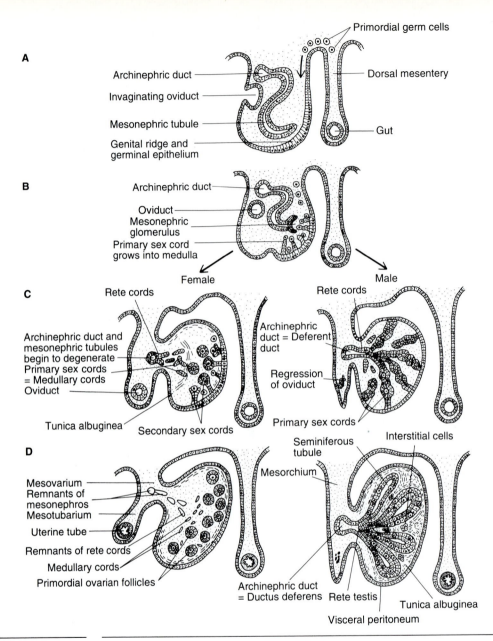

A

Primordial germ cells

Archinephric duct

Invaginating oviduct

Mesonephric tubule

Genital ridge and
germinal epithelium

Dorsal mesentery

Gut

B

Archinephric duct

Oviduct
Mesonephric
glomerulus

Primary sex cord
grows into medulla

Female

Male

C

Rete cords

Archinephric duct and
mesonephric tubules
begin to degenerate
Primary sex cords
= Medullary cords
Oviduct

Tunica albuginea

Secondary sex cords

Rete cords

Archinephric
duct = Deferent
duct

Regression
of oviduct

Primary sex cords

Interstitial cells

Seminiferous
tubule

Mesorchium

D

Mesovarium
Remnants of
mesonephros
Mesotubarium

Uterine tube

Remnants of rete cords

Medullary cords

Primordial ovarian follicles

Archinephric duct
= Ductus deferens   Rete testis

Tunica albuginea

Visceral peritoneum

FIGURE 21–3   Transverse sections through the mesonephros and developing gonad of a mammal: **A** and **B**, The development of the sexually indifferent gonad. **C**, An early stage in the differentiation of an ovary (left) and testis (right). **D**, A later stage in the differentiation of the ovary and testis. (After Williams, et al.)

which are the progenitors of the gametes, invade it. The primordial germ cells can be recognized first as large, distinctive cells in the wall of the yolk sac or gut next to the endoderm. The primordial germ cells migrate, primarily by ameboid motion, through the dorsal mesentery and into the germinal epithelium. Although few in number at first (20 to 30 in a human embryo), they multiply en route and in the germinal epithelium. By this stage, the indifferent gonad has a well-defined **medulla** of mesenchymal origin and a **cortex** of epithelial origin. Cords of germinal epithelium then grow into the medullary tissue to form the **primary sex cords** (Fig. 21–3**B**).

If the embryo becomes a male, the primary sex cords, probably under the influence of medullary tissue, multiply, enlarge, and differentiate into the sperm-producing **seminiferous tubules** of the testis (Fig. 21–3**C** and **D,** right column). All the primordial sex cells become incorporated into the primary sex cords in male amniotes, but some sperm-forming cells remain in other parts of the testis in anamniotes. Next, a peripheral extension of mesenchyme separates the cords from the cortex and, in mammals, thickens to form a dense, connective tissue **tunica albuginea** at the surface of the testis (Fig. 21–3**D,** right column). The testis is primarily an organ that develops in the medulla of the indifferent gonad. The cortex regresses, leaving only the visceral peritoneum that covers the surface of the testis. Some other parts of the medulla, and possibly some epithelial cells that do not contribute to the seminiferous tubules, form the **interstitial cells** of the testis that produce the male sex hormone. As the testis enlarges, it separates from the mesonephros or its derivative, to varying extents in different groups of vertebrates, but it remains connected by a mesentery known as the **mesorchium** (Gr. *mesos* = middle + *orchis* = testis).

A few primary sex cords also invade the medulla of the gonad during the differentiation of the female, and a rudimentary tunica albuginea separates them from the cortex (Fig. 21–3**C** and **D,** left column). Other primary cords remain in the cortical region, and they are joined in mammals and some other vertebrates by a proliferation of **secondary sex cords** that carry in the remaining primordial germ cells from the germinal epithelium. Sex cells remain in the germinal epithelium in most female vertebrates. As the cortex is thickening, blood vessels and connective tissue invade it from the medulla, and the medulla itself regresses, although traces of the primary sex cords may remain as **medullary cords.** The cords in the cortex break up into clusters of cells, each of which contains a developing egg surrounded by a layer of epithelial cells. These groups of cells are the **primordial follicles.** The ovary, unlike the testis, develops primarily in the cortex of the indifferent gonad. As the ovary enlarges, it, too, separates from the mesonephros, but it remains connected to the body wall by a mesentery known as the **mesovarium.**

Development of the Reproductive Ducts

Cyclostomes, as we shall see, lack reproductive ducts, but they are present in other vertebrates. As the testis or ovary is developing, a network of potential tubules begins to form deep in the medullary region of the indifferent gonad between the primary sex cords and the anterior mesonephric tubules. They are called **rete**

cords, or **cords of the urogenital union,** for they have the potential to connect the gonad and cranial mesonephric tubules (Fig. 21–3**C** and **D**). The rete cords canalize during the development of most male vertebrates, and the route through the mesonephros and its archinephric duct becomes the definitive sperm passage. The mesonephros and archinephric duct are also parts of the excretory system in male anamniotes, but not in amniotes that evolve a metanephros and ureter (Chapter 20). The rete cords regress in females or, at most, leave functionless vestiges. Female anamniotes also retain mesonephric tubules and the archinephric duct as parts of their excretory system, but in female amniotes these, too, regress or become vestigial.

Both sexes also begin to develop an **oviduct** (often called the **Müllerian duct** in embryos). The oviduct of chondrichthyans and some amphibians arises by a longitudinal splitting of the archinephric duct (Fig. 21–4). Its anterior opening into the coelom, known as the **ostium tubae** (L. *ostium* = door), appears to represent the coelomic funnels of one or more cranial renal tubules. This may be the way an oviduct evolved, but in other vertebrates it simply arises as a longitudinal, groovelike invagination of the coelomic epithelium on the lateral surface of the mesonephros (Fig. 21–3**A** to **D**). As the groove extends caudad, it deepens, folds on itself, and forms the oviduct, which continues to grow posteriorly to the cloaca. The oviduct regresses in males, sometimes leaving traces, but in females it continues to develop and becomes the passage for the removal of eggs. As the oviduct enlarges, it separates from the mesonephros but remains connected to the body wall by a mesentery known as the **mesotubarium.**

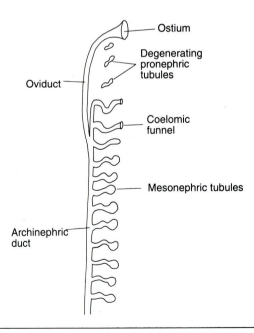

Ostium

Degenerating pronephric tubules

Oviduct

Coelomic funnel

Mesonephric tubules

Archinephric duct

FIGURE 21–4   A ventral view of the developing oviduct of a chondrichthyan, in which the oviduct forms by a longitudinal splitting of the archinephric duct. (Modified after van der Broek.)

The Mature Gonads

Although their structure varies to some extent among vertebrates, the gonads are relatively conservative organs, and those of mammals are representative. Important differences in other groups will be indicated.

Testes

Amniotes.   The testes of amniotes are compact organs composed of long, highly convoluted **seminiferous tubules** that lead to a system of ducts (cords of the urogenital union or rete testis) through which sperm leaves. Many tubules end blindly, but others branch and anastomose or form loops, both ends of which connect to the duct system of the testis. Humans may have as many as 600 seminiferous tubules in a testis, some of which are 80 cm long. The tubules are surrounded by a loose connective tissue that contains the **interstitial cells of Leydig,** which secrete the male sex hormone testosterone (Fig. 21–5A). Tubules often are grouped into lobules by dense connective tissue septa.

The tubule wall consists of sperm-forming cells and Sertoli cells (Fig. 21–5B). **Sertoli cells** rest on the basement membrane on the periphery of a seminiferous tubule and extend inward to the tubule lumen. They are large and irregularly shaped, with many recesses that surround groups of developing sperm cells, and probably contribute to their nutrition and maturation. Processes of Sertoli cells branch extensively in the wall of the tubule and contribute to its supporting framework.

The stem sperm-forming cells are diploid **spermatogonia** that lie in the periphery of the tubule (Fig. 21–5B). Spermatogonia divide mitotically near the beginning of a reproductive season or, in humans and other species without reproductive seasons, throughout reproductive life. Some progeny cells always remain as spermatogonia; others continue to differentiate. The division of the cytoplasm is incomplete in the last division of a spermatogonium. The two progeny cells enlarge to form a pair of conjoined **primary spermatocytes** (Fig. 21–6). The primary spermatocytes then divide once meiotically to form **secondary spermatocytes,** and these undergo a second meiotic division to form haploid **spermatids.** Cytoplasmic division also is incomplete in the meiotic divisions, so a cluster of eight spermatids arises from each pair of conjoined primary spermatocytes. During the process of their development, the sperm-forming cells move from basal compartments in the Sertoli cells to compartments next to the lumen. Spermatids do not divide again but undergo a complex transformation to become the motile **spermatozoa.** A flagellate tail and acrosome form, and much of the cytoplasm is lost (p. 117). The conjoined spermatids normally separate during this process, but it is not unusual to find abnormal double sperm. The whole process of sperm formation, or **spermatogenesis,** requires several weeks (64 days in humans). In species with extended reproductive periods, successive waves of spermatogenesis travel along the tubules, so different regions of the tubules are in different stages of spermatogenesis.

The testes lie within the body cavity close to their site of embryonic development in all vertebrates except for many mammals, in which the testes descend

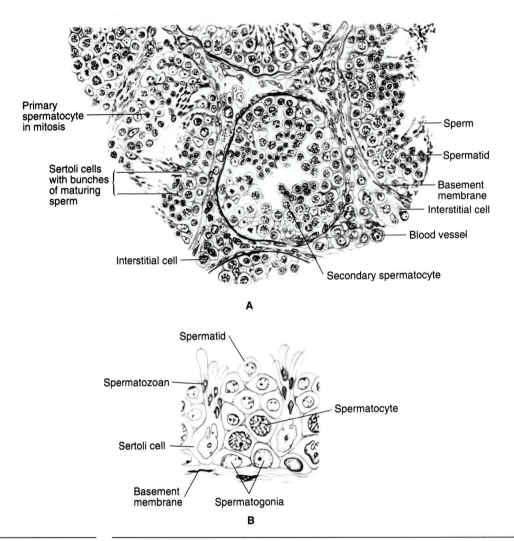

Primary spermatocyte in mitosis

Sertoli cells with bunches of maturing sperm

Interstitial cell

Sperm

Spermatid

Basement membrane

Interstitial cell

Blood vessel

Secondary spermatocyte

A

Spermatid

Spermatozoan

Spermatocyte

Sertoli cell

Basement membrane

Spermatogonia

B

FIGURE 21–5   Seminiferous tubules and spermatogenesis in a mammal: **A,** A transverse section of part of the testis, showing seminiferous tubules and interstitial cells. **B,** A detail of a portion of a tubule to show some of the stages in spermatogenesis. (After Fawcett.)

into a **scrotum** (L. *scrotum* = pouch). In a few mammals, including monotremes, edentates, elephants, sirenians, and cetaceans, the testes remain in the abdomen. The testes of bats, most rodents and lagomorphs, and some carnivores and ungulates descend into the scrotum during the breeding season and are withdrawn into the body cavity after reproduction (Focus 21–1).

The scrotum is a pair of sacs formed by the evagination of all the layers of the body wall. Early during development a cord of tissue known as the **gubernaculum** (L. *gubernaculum* = rudder) extends from the caudal end of each developing

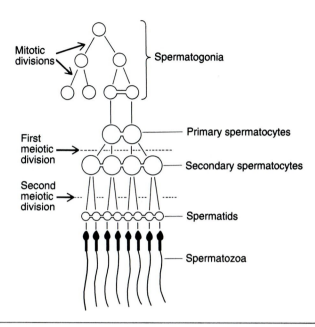

FIGURE 21–6   Mammalian spermatogenesis. (After Fawcett.)

testis, through the incompletely formed body wall in the region of the groin, and into the skin of the future scrotum (Fig. 21–7**A** and **B**). Each testis descends in a retroperitoneal position soon afterward. As a testis approaches the groin, its descent is accompanied by an evagination of the muscular and connective tissue layers of the body wall, and by a coelomic sac, the **vaginal sac** (or processus vaginalis) (L. *vagina* = sheath), that enfolds the testis with its sperm duct, nerve, and blood vessels. These layers and the skin constitute the wall of the scrotum, but the muscular and connective tissue layers form a well-defined sac that travels a short distance beneath the skin of the groin before entering the scrotal skin (Fig. 21–7**C**). It sometimes is convenient to refer to this sac as the **cremasteric pouch,** for the muscular tissue within its wall is the **cremasteric muscle** (Gr. *kremastos* = hung). The area of the body wall from which the cremasteric pouch arises is the **inguinal canal.** The cord of blood vessels, nerve, and sperm duct that travels to and from the testis within the vaginal sac is the **spermatic cord.** The vaginal sac remains open in most mammals, but its proximal part normally atrophies in human beings, leaving only a distal part surrounding the testis.

The mechanisms for the descent of the testis are not well known. The gubernaculum, being composed of soft tissue, may provide a route of low resistance. The failure of the gubernaculum to elongate after its formation and the continued growth of surrounding regions of the body may, in effect, pull the testis caudad. In species in which the testis migrates, intra-abdominal pressure contributes to its descent, and the contraction of the cremasteric muscles shortens the pouch and pulls the testis inward.

FOCUS 21–1 **The Descent of the Testes**

The evolutionary reasons for the descent of the testes into an external scrotum are not entirely clear. Temperatures are several degrees lower in the scrotum than intra-abdominal temperatures, 3°C in human beings. Moore (1926) proposed that the lower scrotal temperatures are critical for certain stages in spermatogenesis, especially for the transformation of the spermatids into spermatozoa. If the testes fail to descend in young boys, normal spermatogenesis does not occur at puberty. Conversely, if scrotal temperatures are elevated in experimental animals, sterility results. Several factors reduce scrotal temperatures. The scrotum is exposed, its skin is thin and often sparsely haired, sweat glands are abundant, and subcutaneous fat is absent. Moreover, each testicular artery and vein is closely entwined to form a **pampiniform plexus** (L. *pampinus* = tendrile + *forma* = shape) that acts as a heat exchanger. Much of the heat in the warm blood going to the testis in the artery is transferred to the cooler venous blood returning from the testis before the arterial blood reaches the testis. The cremasteric muscle and smooth muscle fibers in the dartos tunic, which insert into the scrotal skin, fine tune the mechanism. If temperatures become too cool, contraction of these muscles pulls the testes closer to the warmth of the body wall. Bedford (1977) has proposed that the critical thermosensitive stage is not a part of spermatogenesis in the testis, but the storage of spermatozoa in the adjacent epididymis. Experiments confirm that in some species, sperm cannot survive at the higher body core temperatures. The hypothesis that lower scrotal temperatures are needed for sperm development or survival may explain the presence of permanently descended testes in some species, and their migration in others. But this hypothesis does not fully account for the retention of the testes in the abdomen in many species. Many mammals with abdominal testes do have lower core temperatures than species with a scrotum, and they would not need a scrotum. However, this is not always the case. Elephants with intra-abdominal testes have a relatively high core temperature. (Birds, which do not have scrota, also have high core temperatures, although it is possible that air sacs lower the temperature in the immediate vicinity of their testes.)

Frey (1991) has proposed an alternate hypothesis that he believes explains why some mammals have descended testes and others do not. He points out that the last stage in spermatogenesis (when spermatids are transformed into spermatozoa) is highly sensitive to fluctuations in pressure. Oogenesis lacks this last pressure-sensitive stage. He further points out that intra-abdominal pressures fluctuate widely in mammals that gallop, jump, or whose vertebral column is strongly flexed and extended during locomotion. The testes of mammals with these patterns of locomotion must be shielded from the damaging fluctuations in intra-abdominal pressures. Frey reasons that a scrotum evolved as a protection against pressure fluctuations as many mammals adopted new locomotor strategies in relation to predation or predator avoidance. Mammals without these specialized patterns of locomotion had no need for a scrotum. Once the testes were descended, they may have become secondarily adapted to the cooler scrotal temperature. Frey also believes that the pam-

---

Anamniotes. Fertilization is external in most anamniotes, so enormous amounts of sperm must be produced to ensure fertilization. The testes of most fishes are large and often fill the body cavity prior to spawning. The paired testes unite to form a single organ in lampreys and many teleosts. The testes of frogs are composed of seminiferous tubules, but those of most anamniotes consist of **seminif-**

FOCUS 21–1 *continued*

piniform plexus is more a peripheral pump than a heat exchange device (See Focus Figure). Pulsations of the testicular artery help drive venous blood back into the caudal vena cava and thus prevent its peripheral accumulation, which could raise pressures in the testis. Not all agree

with Frey. Indian elephants, in which the testes are intra-abdominal, are used to move logs and do other heavy work, and this activity probably raises intra-abdominal pressures. Clearly more studies are needed before we unravel the mysteries of descended testes.

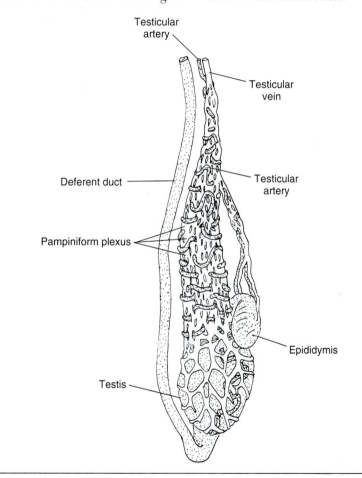

Blood vessels surrounding the testis in a human. (Modified from Frey.)

erous ampullae, or lobules, of developing sperm cells. Cells in the walls of the ampullae appear to function as Sertoli cells. The ampullae are completely evacuated when sperm are discharged, and the ampullae themselves may regress. Before the next reproductive season, stem sperm-producing cells move into old and newly formed ampullae from a permanent population of spermatogonia that reside in

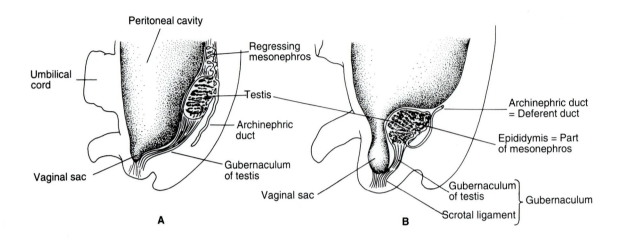

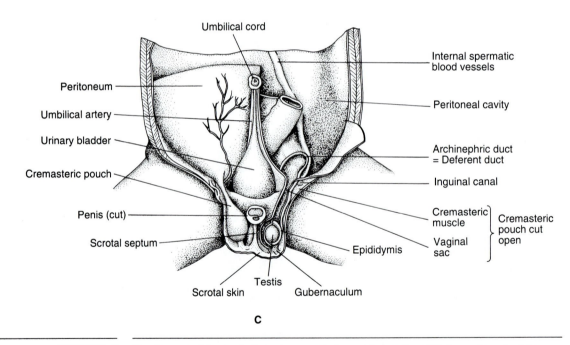

FIGURE 21–7 The descent of the testis, based on a male human being: **A** and **B,** An early and a later stage in lateral view. **C,** A ventral dissection of a nine-month fetus to show the descended testis and its relationship to surrounding structures. (After Corliss.)

other parts of the testis. Hormone-secreting interstitial cells are not present in the testes of all anamniotes, but they have been found in those of elasmobranchs and many amphibians. Branching, finger-shaped **fat bodies** are attached to the gonads of many amphibians and contain reserves of energy probably utilized in sperm and

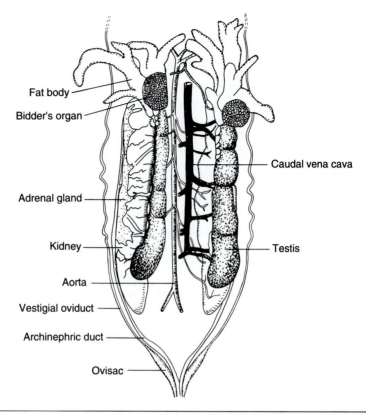

Fat body

Bidder's organ

Caudal vena cava

Adrenal gland

Kidney

Testis

Aorta

Vestigial oviduct

Archinephric duct

Ovisac

FIGURE 21–8   A ventral view of the urogenital system of a male toad. (After Turner and Bagnara.)

egg production (Fig. 21–8). In temperate species gamete production begins during the late winter period of dormancy, when the animals are not feeding.

Developmental stages are found in some sexually immature anamniotes, in which the anterior part of the gonad resembles an ovary while the posterior part more closely resembles a testis. This is not hermaphroditism but a reflection of the potential of the embryonic gonad to develop into either a testis or an ovary. Only one part of the gonad normally functions, but the potential for sex reversal exists. Male toads of the family Bufonidae have a **Bidder's organ** at the front of the testis that resembles an arrested ovary (Fig. 21–8). Seasonal changes in its size suggest an endocrine function, but when the testis ceases to function in old males, a Bidder's organ may become a functional ovary.

Ovaries

The eggs undergo their maturation, or **oogenesis,** within the ovaries. The ovaries of most vertebrates are paired organs, but in lampreys and many teleosts they unite to form a single organ. In some species either the left or the right ovary atrophies. Many viviparous sharks retain only the right one. The ovaries are paired in a few birds, including hawks, but most birds retain only the left one. Monotremes have both ovaries, but eggs mature only in the left one. The adaptive significance of the loss or reduction in function of an ovary is not clear.

Mammalian Ovaries. The ovaries of mammals are relatively smaller organs than in other vertebrates. Few eggs are produced in a reproductive period, and they contain less yolk than the eggs of reptiles or birds. Therian embryos undergo at least part of their development in a uterus, but monotremes are oviparous. The eggs of monotremes are about 4 mm in diameter when they leave the ovary; a human egg has a diameter of about 140 μm when it leaves the ovary. Mammalian ovaries, like the testes, become connected by a gubernaculum to the body wall during their development, but they undergo only a partial descent. The gubernaculum often persists in an adult female as a **round ligament** extending between the ovary and the groin.

Mature mammalian ovaries are dense organs consisting of a vascularized, connective tissue stroma in which the follicles are embedded (Fig. 21–9). Unlike spermatogonia, the oogonia cease multiplying during embryonic life. Eggs in the **primordial follicles,** which are located around the periphery of the ovary, have reached the **primary oocyte** stage of oogenesis. A human ovary contains approximately 2,000,000 of these follicles at birth, but most atrophy, leaving only about 40,000 at puberty. Most of these also atrophy and only 400 or so will ovulate. As the remaining eggs and follicles continue to develop after puberty, the follicular cells multiply and form many layers of **granulosa cells** around each egg. Surrounding layers of connective tissue contribute a sheath, or **theca,** to the wall of the follicle. The follicle protects and transfers nutrients to the egg. It is also an endocrine gland that secretes female sex hormones. As a follicle continues to develop, a **follicular liquor** accumulates in spaces among the granulosa cells. These spaces gradually enlarge, coalesce, and form a large central cavity, making the egg eccentric in position. A **mature** or **graafian follicle** may be 10 mm or more in diameter, and it bulges outward on the ovary surface. During these processes the primary oocyte enlarges and undergoes its first meiotic division to form a **secondary oocyte** and a minute, **first polar body.** Food reserves are kept within one functional cell. The second meiotic division, which will give rise to the mature **ootid** and a **second polar body,** begins shortly before ovulation but normally requires the stimulus of fertilization to be completed. As an egg cell matures, the follicular cells secrete around it the secondary envelope, known in mammals as the **zona pellucida.** The number of follicles that mature during a reproductive period depends on whether the species normally gives birth to single or multiple young.

Each mature follicle ruptures at the ovary surface at **ovulation,** and the eggs, still surrounded by some granulosa cells, are discharged into the coelom to be picked up by the reproductive tract. A ruptured follicle is converted to a hard, yellowish body, the **corpus luteum,** which also secretes female sex hormones, estrogen, and progesterone. Corpora lutea begin to atrophy before the next reproductive cycle.

Nonmammalian Ovaries. The ovaries of other vertebrates are relatively much larger than those of mammals. So many eggs are produced by oviparous anam-

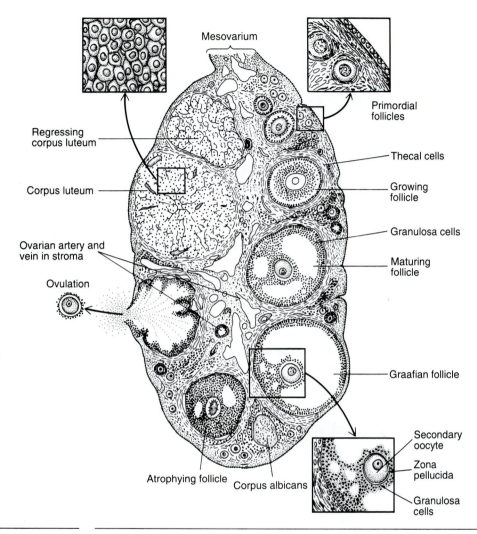

Mesovarium

Regressing corpus luteum

Corpus luteum

Ovarian artery and vein in stroma

Ovulation

Primordial follicles

Thecal cells

Growing follicle

Granulosa cells

Maturing follicle

Graafian follicle

Secondary oocyte

Zona pellucida

Granulosa cells

Atrophying follicle   Corpus albicans

FIGURE 21–9   A composite drawing of a longitudinal section of a mammalian ovary to show developing follicles, ovulation, and the development and regression of the corpus luteum. All these stages would not be present at the same time. (After Turner and Bagnara.)

niotes, and the eggs of oviparous amniotes contain so much yolk, that the ovaries fill all the available space in the body cavity prior to egg laying; they regress to a small size after the reproductive season. The ovaries do not have so extensive a stroma as in mammals (Fig. 21–10). Often they are hollow and contain a lymph space. Oogenesis occurs within follicles and passes through the same stages as in mammals. Oogonial divisions are completed during embryonic life in birds, some reptiles, and chondrichthyans, but the germinal epithelium remains active in most other vertebrates and they produce many eggs during their reproductive lives. The follicles of most vertebrates have very thin walls and consist of only one or a few layers of granulosa cells. As an egg accumulates yolk, it enlarges and fills the

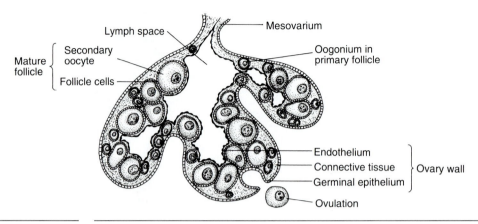

FIGURE 21–10   A section through an amphibian ovary.

follicle. A follicular cavity is not present. Usually, several generations of developing follicles are present at one time. The remains of the follicles usually are reabsorbed after ovulation. Functional corpora lutea may develop from the follicles in some species, but chiefly those that have some degree of viviparity: most sharks, several amphibians, and some lizards and snakes. Birds lack corpora lutea.

Evolution of the Reproductive Passages

Cyclostomes lack reproductive ducts. Both sperm and eggs are discharged into the coelom at the time of spawning. Gametes leave the caudal end of the coelom through a pair of **genital pores** that enter the caudalmost part of the archinephric ducts, where these ducts conjoin to form a urogenital papilla that protrudes into the cloaca (Fig. 21–11). The genital pores open under the influence of sex hormones only a few weeks before spawning. Genital pores may be the ancestral vertebrate method of gamete discharge. This certainly is the simplest way morphologically, and the discharge of gametes into the coelom also occurs in many invertebrate groups. Chondrichthyan fishes have a pair of **abdominal pores** of uncertain function that lead from the coelom to the cloaca. Some investigators regard them as vestiges of a primitive method of gamete discharge, but all gnathostomes have evolved reproductive ducts that remove the gametes.

Male Reproductive Passages

Amphibians.   The simplest and probably the ancestral pattern of male reproductive ducts is found in oviparous amphibians. The duct system of the testis, which develops embryonically from the rete testis, converges to a **testis canal** located in the center or along one margin of the testis (Fig. 21–12**A**). A variable number

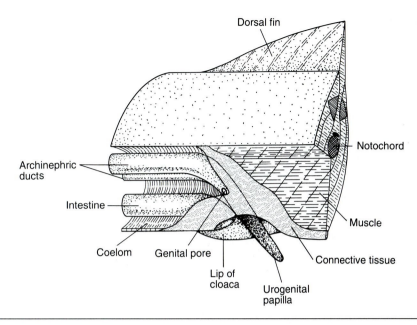

FIGURE 21–11   A stereodiagram of the cloacal region of an adult lamprey, showing a genital pore. (After Knowles.)

of **connecting tubules,** also called **efferent ductules** in amphibians (L. *ex* = out + *ferent-* = carrying) carry sperm from the testis canal, through the mesorchium, to the anterior end of the opisthonephros, where they connect to a **lateral kidney canal** or directly to kidney tubules. The lateral kidney canal, when present, develops from anterior mesonephric tubules. The connecting tubules appear to develop from the testicular duct system (rete testis), but contributions from mesonephric tubules cannot be ruled out. Sperm then pass through a few **anterior opisthonephric tubules** to the **archinephric duct,** through which they continue to the cloaca. In caecilians and some salamanders, sperm-carrying kidney tubules retain glomeruli, but in other groups the tubules lose them and their urinary function. In a few salamanders the adjacent part of the archinephric duct becomes highly coiled and stores sperm. This part of the duct is sometimes given the name of its mammalian homolog, the **duct of the epididymis.** The archinephric duct of most amphibians carries both sperm and urine, although not at the same time (Fig. 21–8). A few salamanders have evolved one or more accessory urinary ducts that transport most of the urine (Fig. 21–12**A**). The embryonic Müllerian ducts often remain as vestigial oviducts in males (Fig. 21–8).

Chondrichthyan Fishes.   The reproductive passages of male cartilaginous fishes represent a modification of the pattern, seen in amphibians, that evolved with internal fertilization. One or more accessory urinary ducts drain the caudal urinary portion of the kidney, and the rest of the kidney and its archinephric duct are taken over by the reproductive system (Fig. 21–12**B**). Most of the opisthonephric

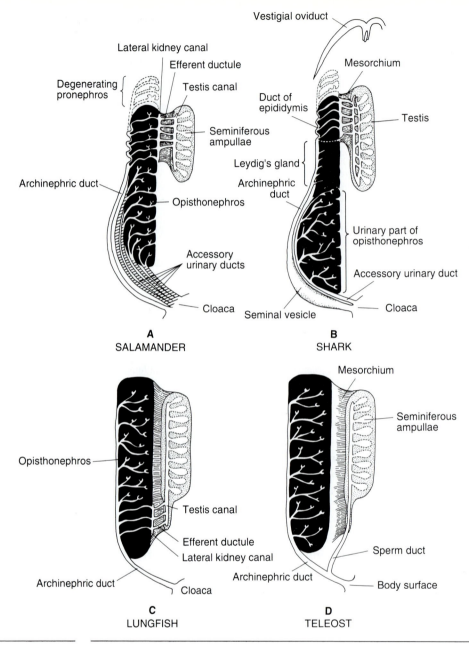

FIGURE 21–12 Ventral views of the male reproductive ducts and their relationship to excretory ducts in representative male anamniotes: **A,** A salamander. **B,** A shark. **C,** The South American lungfish, *Lepidosiren*. **D,** A teleost. (After Portmann.)

tubules between those receiving sperm and those carrying urine are modified as **Leydig's gland** and secrete a **seminal fluid.** Seminal fluid transports the sperm to the female and probably has some of the nutritive and activating functions that it has in mammals (p. 764). The archinephric duct is wide as it crosses the urinary part of the kidney and forms a **seminal vesicle** that contributes to the seminal fluid and helps store sperm. Part of each pelvic fin is modified as a copulatory **clasper** that transfers sperm to the female. A long **siphon** with muscular walls is associated with each clasper. Water taken into the siphon plus its own secretions help propel sperm into the female. Vestiges of the oviducts frequently occur in males.

Bony Fishes.   Lungfishes and early actinopterygians retain the primitive gnathostome pattern of male reproductive passages, except that the testis connects with caudal rather than cranial renal tubules in some lungfishes (Fig. 21–12**C**). Teleosts have evolved a unique **sperm duct** that develops as an extension of the testis canal, completely bypasses the kidneys, and continues caudad to open on the outside of the body adjacent to the excretory openings. It joins the caudal end of the archinephric duct in a few species (Fig. 21–12**D**).

Amniotes.   Except for the location of the testis within a scrotum, the pattern of reproductive ducts in all male amniotes is similar to that in mammals (Fig. 21–13). The kidney, now a metanephros, is drained by a ureter. The archinephric duct and the part of the embryonic mesonephros connecting to the testis function only as reproductive passages. The developing testis does not separate far from the mesonephros, so the rete testis leads directly into modified mesonephric tubules, now called the **efferent ductules.** It should be noted that the efferent ductules of amniotes develop from mesonephric tubules. They are not homologous to the connecting tubules (sometimes called efferent ductules) of anamniotes, which develop from the rete testis and pass through a rather wide mesorchium to the testis. The distal ends of the efferent ductules are highly coiled and become the head of a band of tissue, called the **epididymis** (Gr. *epi* = upon + *didymos* = testis), that lies on the surface of the testis. The body and tail of the epididymis are composed of an extensively convoluted part of the archinephric duct called the **duct of the epididymis.** The amniote epididymis is homologous to the cranial end of the anamniote opisthonephros and adjacent parts of the archinephric duct. The continuation of the archinephric duct, now called the **deferent duct** (L. *de* = from + *ferent-* = carrying), extends caudally to the cloaca, or to the part of the mammalian urethra that is derived from the cloaca (p. 775). In a mammal with a scrotum, each deferent duct travels in the spermatic cord and enters the body cavity through the inguinal canal. Vestiges of caudal mesonephric tubules (the **paradidymis**) and an embryonic Müllerian duct (**appendix testis, prostatic utricle**) frequently persist in adult males (Fig. 21–14 and Table 21–1).

Spermatozoa are stored in the epididymis and deferent duct, where secretions nourish them and appear to play a role in their maturation. Sperm removed from the epididymis before they have stayed there for a few hours are incapable of becoming motile and fertilizing eggs.

Several accessory sex glands that secrete the seminal fluid develop along the

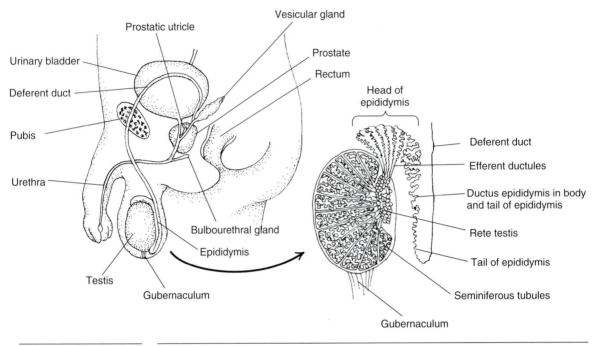

FIGURE 21–13   The reproductive system of a male amniote, based on a human being. The pattern of the accessory sex glands varies among mammals. In male reptiles and birds the testis would not descend, the ducts would open into a cloaca, and the penis would be embedded in the ventral cloacal wall.

sperm passages. They are particularly well developed in mammals and may include **prostate, vesicular gland,** and **bulbourethral gland** (Fig. 21–13). In addition to transporting sperm, the **seminal fluid** of mammals nourishes the sperm, helps make them motile, and neutralizes acids and other materials in the female reproductive tract that would interfere with their survival.

Most male amniotes have evolved a single or bifid intromittent organ with which sperm are transferred to the female. This organ, known in mammals as a **penis,** develops in the ventral wall of the cloacal region. It lies in the tail base in reptiles. When not engorged with blood and erect, it usually is withdrawn into a sheath. Most male birds transfer sperm to the females simply by a brief cloacal apposition, but an intromittent organ is present in ratites, ducks, geese, and a few other species.

Passages and Reproduction in Nonmammalian Females

Embryonic mesonephric tubules and the archinephric ducts are never used in gamete transport by females, rather the eggs take an independent route through the oviducts. The mesonephric tubules in female anamniotes are incorporated in the opisthonephros, which is drained by the archinephric duct. These structures

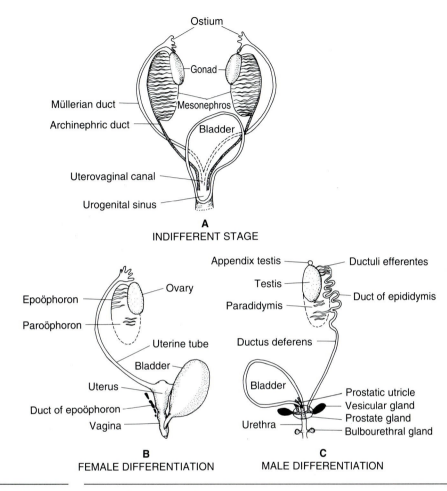

A
INDIFFERENT STAGE

B
FEMALE DIFFERENTIATION

C
MALE DIFFERENTIATION

FIGURE 21–14   Ventral views of the embryonic differentiation of female and male reproductive tracts in a eutherian mammal. The oviduct and its derivatives are shown in white; the archinephric duct and its derivatives are black. Table 21–1 summarizes the sexual homologies. (After Turner and Bagnara.)

are vestigial in female amniotes (**paroöphoron, epoöphoron**), for the metanephros and ureter are the functional excretory organs (Fig. 21–14 and Table 21–1).

Amphibians.   The reproductive passages of female oviparous amphibians, like those of the male, are closer to the ancestral vertebrate pattern than the more specialized passages of most fishes (Fig. 21–15A). A pair of long, convoluted **oviducts** extend along the dorsal body wall beneath the kidneys from the anterior end of the body cavity to the cloaca. The anterior end of each oviduct is expanded slightly to form a funnel-shaped **infundibulum** (L. *infundibulum* = funnel), which opens into the coelom through an **ostium tubae.** Its caudal end sometimes expands to form an **ovisac,** in which eggs accumulate for a brief period before

TABLE 21–1   **Embryology and Sexual Homology of Mammalian Reproductive Organs***

Embryo	Adult Male	Adult Female
Primary sex cords	Seminiferous tubules	(Medullary cords of ovary)
Secondary sex cords	—	Ovarian follicles
Gubernaculum	Gubernaculum	Round ligament of ovary
Rete testis	Rete testis	—
Mesonephros		
Cranial tubules	Efferent ductules in head of epididymis	(Epoöphoron)
Caudal tubules	(Paradidymis)	(Paroöphoron)
Archinephric duct	Duct of epididymis	(Duct of epoöphoron)
	Deferent duct	(Duct of epoöphoron)
Oviduct (Müllerian duct)		
Cranial end	(Appendix testis)	Uterine tube
		Uterus
Caudal end	(Prostatic utricle)	Vagina
Allantois	Urinary bladder	Urinary bladder
	Part of urethra	Urethra
Urogenital sinus	Part of urethra	Vaginal vestibule
Genital tubercle	Much of penis	Clitoris
Genital groove	Part of urethra	Vaginal vestibule
Genital folds	Ventral part of penis	Labia minora
Scrotal swellings	Cutaneous part of scrotum	Labia majora

*Parentheses identify vestigial structures; a dash indicates that the structure has been lost.

they are laid. Reproductive hormones stimulate the development of cilia on the peritoneum, so as the eggs are ovulated, they are swept toward the ostium. Ciliary action carries them into the ostium, and ciliary action within the oviduct and muscular contractions of its wall carry them caudad. As the eggs pass down the oviduct, they become coated with tertiary envelopes secreted by oviductal cells. The tertiary envelopes of amphibians are jelly-like layers that imbibe water and swell after the eggs have been laid and fertilized. These layers help protect and insulate the eggs.

Fertilization is external. Usually, sperm are sprayed over the eggs as they are laid. Before egg laying many male salamanders release their sperm in mucous clumps called **spermatophores** (Gr. *phoros* = bearing), which the female collects with her cloacal lips and stores for a short time in cloacal recesses. Only a few species build nests. Most eggs are simply deposited in the water, but they often clump to each other and to surrounding objects. A few salamanders that lay in damp locations on land wrap their bodies around their eggs and brood them. This helps maintain a high humidity.

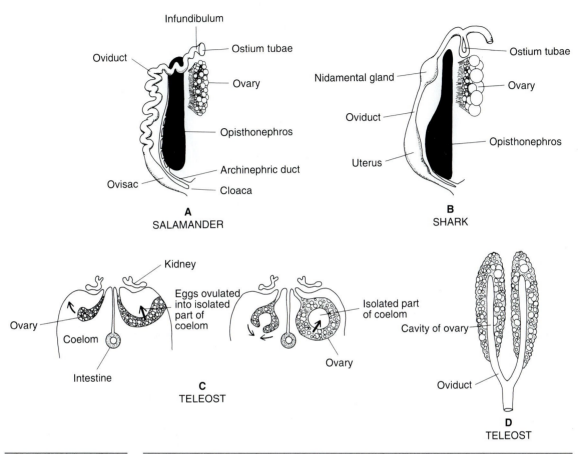

FIGURE 21–15 The reproductive tracts of female anamniotes: **A,** A salamander. **B,** A shark. **C,** Transverse sections through the developing ovaries of teleosts to show two ways the hollow ovary develops. **D,** A ventral view of the ovaries and oviduct of a teleost. (**A** and **B,** partly after Portmann.)

Some degree of viviparity has evolved in a few caecilians, two frogs, and some salamanders of the genus *Salamandra* that retain eggs in the oviducts. A number of tropical frogs have evolved other ways to protect and brood their tadpoles, including carrying them on their backs, in vocal sacs, or even in their stomachs.

Chondrichthyan Fishes. Primitive cartilaginous fishes are oviparous, but the male has a copulatory organ, and fertilization is internal. Although only a few eggs are produced, these contain enough yolk to provide for the young until they hatch. Eggs are large, 5 cm or more in diameter. Both oviducts curve ventrally anterior to the liver and unite in the falciform ligament (Fig. 21–15**B**). The ventral position and large size of the single ostium tubae are specializations for receiving the exceptionally large eggs. Tertiary envelopes, which take the form of a horny, protein-

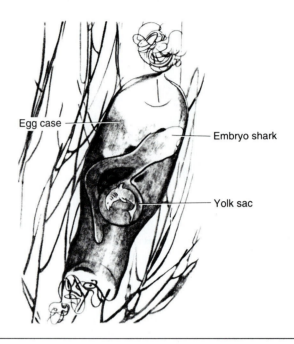

Egg case

Embryo shark

Yolk sac

FIGURE 21–16   The egg case of the oviparous swell shark, *Cephaloscyllium*. (After Springer.)

aceous egg case, are secreted by a **nidamental gland** (L. *nidamentum* = nesting material). The egg cases of oviparous species become entangled in seaweed or other material, and the young pass through a larval stage in the egg cases (Fig. 21–16). Embryos of live-bearing species are lodged in an expanded **uterus** that develops from the caudal part of the oviduct. Only thin-walled egg cases are secreted around the eggs in these species, and they are later reabsorbed. The degree of vascularity and the complexity of folding within the uterus correlate with the degree to which the embryos depend on the mother, from very little to full placental viviparity. The placenta, when present, is formed by the apposition of the embryo's vascular yolk sac to the uterine lining, so it is called a **yolk sac placenta.**

Bony Fishes.   Lungfishes and nearly all actinopterygians are oviparous. The oviducts of primitive species receive eggs that are ovulated into the general body cavity, but most teleosts have evolved a unique ovary and oviduct. The ovary contains a large central cavity representing a part of the coelom that has become separated from the main body cavity by a folding of the ovary against the body wall or on itself (Fig. 21–15**C**). Eggs are discharged into this cavity and are not free in the general coelom. The oviduct usually is a simple tubelike extension of the ovary, so it is not homologous to the oviducts of other vertebrates (Fig. 21–15**D**). Both oviducts fuse caudally and open through a genital papilla located near the anus and excretory openings. These modifications are correlated with the large number of eggs, sometimes numbering in the millions, that teleosts release in a season. Fertilization is external in oviparous species. Some teleosts build nests, and a few brood their young. One or both parents protect them from predators and help circulate water across them. The embryos of the few live-bearing species develop in the oviduct, in the coelomic space within the ovary, or in the follicles themselves.

Reptiles and Birds.   Reptiles have a pair of long, convoluted oviducts (Fig. 21–17**A**), but the right one is reduced or lost in most birds along with the right ovary (Fig. 21–17**B**). The eggs of reptiles enter the oviduct from the coelom, but in birds they directly enter an exceptionally large ostium tubae and infundibulum, which partially invest the ovary and mature follicles. The eggs are fertilized in the upper region of the oviduct, and the glandular oviduct wall secretes the tertiary membranes.

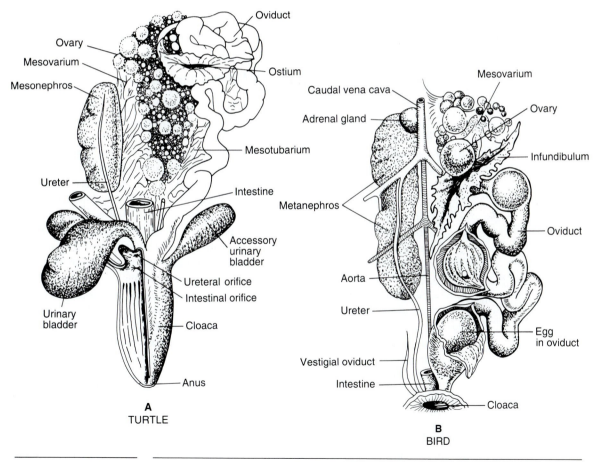

A
TURTLE

B
BIRD

FIGURE 21–17   Ventral views of the urogenital systems of female reptiles and birds: **A**, A turtle. **B**, A bird. Only the left ovary is shown for the turtle, but a right one is present; most birds have only the one ovary shown. (**A**, after Turner and Bagnara. **B**, after Portmann.)

Most reptiles and all birds are oviparous, but the evolution of a cleidoic egg permits them to bypass the aquatic larval stage and reproduce in a terrestrial environment (p. 131). Large amounts of yolk are stored in the eggs, and other materials are provided by the albumen that is secreted around the eggs by the oviduct. The egg is completely surrounded by a shell membrane and a shell, which are secreted by the lower end of the oviduct. The shell is composed of a meshwork of proteinaceous fibers impregnated with calcium salts. The reptilian eggshell is thin and parchment like; the bird eggshell is more heavily calcified. Gas exchange and, depending on the external environment, some loss or gain of water vapor occurs through it. The need for gas exchange precludes cleidoic eggs from being laid in the water. Most reptiles bury their eggs in nests on land where they will be incubated by environmental temperatures, but a few reptiles and all birds brood their eggs. As the embryo develops, its extraembryonic chorion and amnion help protect it, and its allantois allows gas exchange and stores excretory products. The eggs of reptiles and birds are large because they contain so much yolk. Some turtles and snakes may lay nearly 100 eggs, but most reptiles lay fewer, and birds lay only 1 to 20 in a season. Live bearing has evolved independently in many lizards and snakes, and all degrees of dependence on the mother are known.

Passages and Reproduction in Mammalian Females

Monotremes.   Although monotremes are oviparous, their oviducts have become differentiated into many of the distinctive regions that characterize other mammals (Fig. 21–18A). Each oviduct has an infundibulum that partly enfolds the ovary, and a narrow **uterine tube,** or **fallopian tube,** that continues to an expanded **uterus.** The two uteri independently enter a **urogenital sinus** that, after also receiving the excretory products, continues to the cloaca. As the fertilized eggs pass down the uterine tube and enter the uterus, the tertiary envelopes that characterize cleidoic eggs are secreted around them, but the shell is very thin, permeable, and expansile. The egg accumulates additional food from uterine secretions while in the uterus and grows from its ovulation diameter of 4 mm to 15 mm by the time it is laid. The shell thickens after an egg has reached its final size. Embryonic development begins while the egg is in the female reproductive tract. We see in monotremes, and also in some lizards and snakes, a shift in the dependence of the embryo from food stored in the egg while it was in the ovary to an increasing dependence on food provided by the mother by uterine secretions. This appears to be an intermediate stage in the evolution of placental viviparity. An advantage of this intermediate condition is less waste of maternal resources if the egg does not become fertilized or fails to develop. Eggs are brooded by the female after they are laid; echidnas develop a temporary skin pouch that folds over them. Hatched young lap up milk secreted by the mammary glands. Nipples are not present. The evolution of the female's ability to secrete milk after the hatching or birth of the young, and the maternal care associated with nursing, are particularly important features that have contributed greatly to the success of mammals.

Marsupials.   All therian mammals are placentally viviparous, but metatherians and eutherians have evolved different patterns. In marsupial females the caudal

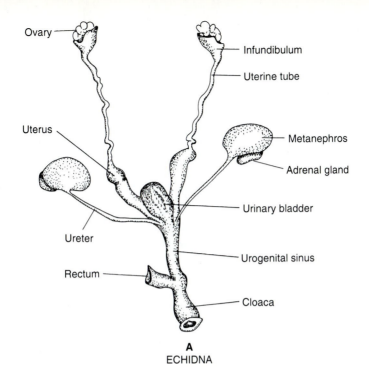

A
ECHIDNA

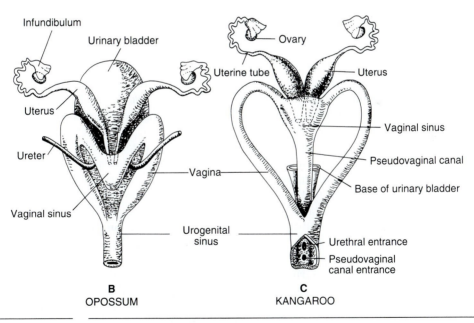

B
OPOSSUM

C
KANGAROO

FIGURE 21-18   The reproductive system of female monotremes and marsupials: **A,** A ventral view of an echidna. **B,** A dorsal view of an opossum. **C,** A dorsal view of a kangaroo. (**A,** after Griffiths. **B** and **C,** after Portmann.)

end of each oviduct becomes specialized as a **vagina,** and the vaginae enter the urogenital sinus (Fig. 21–18**B** and **C**). The urogenital sinus opens directly to the outside because the cloaca becomes divided in all therians (p. 775). The anterior ends of the vaginae unite to form a **vaginal sinus,** into which the uteri enter. Each uterus has its own neck, or **cervix,** so it is independent, a condition described as **duplex.** A uterine tube continues from each uterus to the infundibulum. The penis of many male marsupials is bifid, and sperm travel up both vaginae. Shortly before birth in primitive marsupials, such as the opossum, the vaginal sinus grows caudad and connects directly with the urogenital sinus to form a shorter and more direct birth canal. This pathway, known as the **pseudovaginal canal,** is a permanent structure in most marsupials.

Fertilized marsupial eggs receive a mucoid coat and a shell membrane as they travel down the uterine tubes to the uteri, but no shell is secreted. The embryos develop within the shell membrane during the first two thirds of their gestation period. Gestation periods are very brief, ranging from ten days in the opossum to about a month in some kangaroos. Nutrients secreted by the uterus are absorbed through the shell membrane, and gas exchange also occurs across it. After the reabsorption of the shell membrane, an expanded and vascular yolk sac reaches the chorionic ectoderm (Fig. 21–19**A**) and forms a **yolk sac placenta** that is applied loosely to the uterine lining. The early expansion of the yolk sac prevents the allantois from reaching the chorion in most species, but the allantois does reach it in a few marsupials, including the koala, and forms an additional **chorioallantoic placenta** (Fig. 21–19**B**). These placentas do not penetrate the uterine lining in most species, but in the bandicoot the chorioallantoic placenta is invasive.

The placental relationship of marsupials lasts only a short time before the young are born. The young complete their development attached to the nipples of the mammary glands, which often are located within a skin pouch, the **marsupium** (L. *marsupium* = pouch). A marsupium is particularly well developed in hopping and arboreal species. The reasons for an early birth are not entirely understood. Since half of the embryo's genes are paternal, the embryo is partly foreign tissue to the mother. It is hypothesized that the shell membrane, which is maternal tissue, prevents foreign embryonic antigens from reaching the mother and stimulating the synthesis of antibodies. Antibodies are synthesized, however, when the shell membrane is reabsorbed and a placenta is established, since the chorionic ectoderm, or trophoblast, does not function as an immunological barrier, as it appears to do in eutherian mammals (p. 133). Thus, the early birth is partly an immunological rejection.

The pattern of marsupial reproduction should not be regarded as a step toward the eutherian chorioallantoic placenta and relatively long intrauterine life. Marsupials and eutherians evolved independently from an ancestral therian group (p. 99). The reproductive success of marsupials is at least as high as in eutherians. Marsupial reproduction is particularly well adapted to the stressful and unpredictable environments where most species live. Food is plentiful on the Australian continent after a rainfall, but it becomes scarce and patchy during droughts, some of which last for years. Few maternal resources are invested in the young during their short intrauterine life, and young in a pouch can easily be abandoned when environmental resources fail. A marsupial mother lives to reproduce again, but a

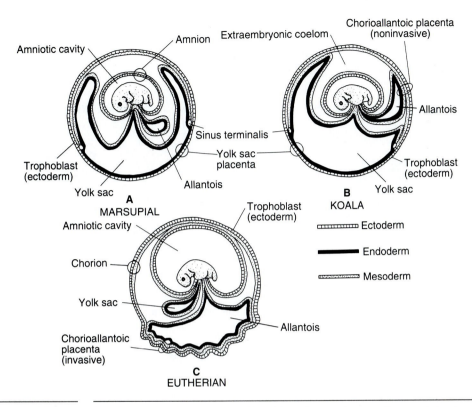

FIGURE 21–19   The placentas of therian mammals. **A,** The yolk sac placenta characteristic of most marsupials. **B,** The combined yolk sac and chorioallantoic placenta of the koala. **C,** The chorioallantoic placenta of a eutherian. (After Dawson.)

eutherian mother, drawing on her own stored reserves to support the embryos when food is in short supply, may weaken herself and jeopardize her life. Moreover, marsupials can reproduce very rapidly during favorable periods because pregnancy does not interrupt the cyclical production of eggs, as it does in eutherians (p. 782). A kangaroo may have one joey hopping about at heel, who occasionally crawls into the marsupium for milk, a younger offspring attached to a nipple and being supplied with a different type of milk, and a third in utero. The development of the intrauterine embryo is arrested in an early stage because the lactation for its older siblings prevents further development and additional pregnancies. When they are weaned, development of the intrauterine embryo is resumed and additional pregnancies may occur.

Eutherians.   An infundibulum, uterine tube, uterus, and vagina also differentiate along the oviducts of eutherian mammals, but the vaginal portions unite to form a single, median organ that leads to the urogenital sinus, or vaginal vestibule (Fig. 21–20**A**). The uteri are **duplex** in lagomorphs, many rodents, and elephants, for each enters the vagina independently and has its own cervix. The caudal ends of

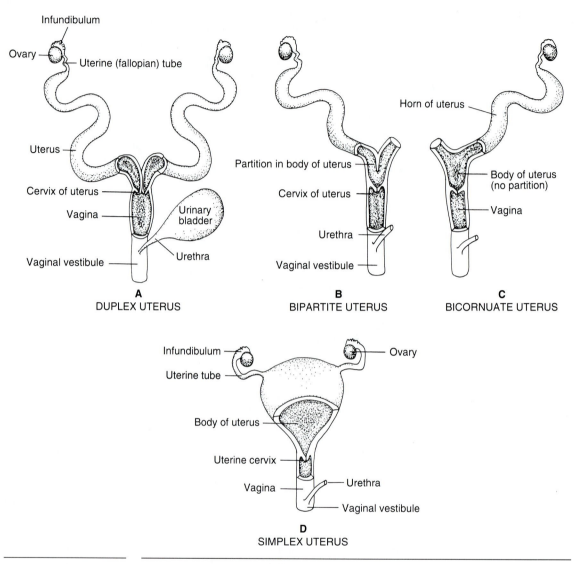

FIGURE 21–20 The reproductive tracts of female eutherian mammals, showing the types of uteri.

the uteri have fused in most eutherians to form a small **uterine body** from which a pair of long, convoluted **uterine horns** extend. **Uterine tubes** extend from the horns. Development occurs within the horns. A slight partition remains within the body of the **bipartite uterus** of most carnivores and many ungulates, but not in the **bicornuate uterus** of other ungulates, many carnivores, and whales (Fig. 21–20**B** and **C**). A complete fusion of the uteri into a single, large body occurs in armadillos and most primates, including humans (Fig. 21–20**D**). Although called a **simplex uterus,** this type represents the most advanced degree of uterine fusion.

Tertiary envelopes are not secreted around the fertilized eggs of eutherians during their passage down the uterine tubes to the uterus. Rather, each embryo quickly develops a surface trophoblast and establishes a placental relationship with the mother soon after reaching the uterus (p. 133). The placenta is **chorioallantoic,** for the allantois vascularizes the trophoblast, which is the mammalian homolog of the chorionic ectoderm (Fig. 21–19**C**). The degree of union between the embryo's chorioallantoic membrane and the mother's uterine lining, or **endometrium** (Gr. *metra* = womb), varies considerably among groups. The fetal component of the placenta does not penetrate the endometrium in a few ungulates, including pigs, so the uterine lining is not disrupted and is not shed at birth. This is described as a **nondeciduous placenta** (L. *decidere* = to fall down or off). In most eutherians, however, the early embryo becomes partly or completely embedded within the endometrium. Much of the uterine lining is cast off at birth with these **deciduous placentas.** The degree to which the endometrium is eroded by the fetal trophoblast also varies; the terms used describe the maternal and fetal tissues that are in contact. No disruption of the endometrium is seen in the **epitheliochorial placenta** of a pig (Fig. 21–21**A**). With a **hemochorial placenta,** found in insectivores, bats, many rodents, and anthropoid primates, maternal blood vessel walls break down so that villi on the surface of the fetal chorion are bathed by maternal blood (Fig. 21–21**B**). Lagomorphs and some rodents have a **hemoendothelial placenta,** for most of the trophoblast on the surface of the chorion also is lost, and maternal blood flows across the endothelial walls of fetal capillaries (Fig. 21–21**C**).

Division of the Cloaca; the External Genitalia

Most vertebrates have a common chamber, the **cloaca,** that receives the intestine and the excretory and reproductive ducts (see Fig. 20–7). Most actinopterygian fishes have lost the cloaca, and their excretory and reproductive ducts open on the body surface, often through a common papilla, just posterior to the anus. The urinary bladder of amphibians, and those amniotes that have one, is a ventral outgrowth from the cloaca. Monotremes retain the cloaca, but it has become partly divided into a dorsal portion, which receives the intestine, and a ventral **urogenital sinus,** which receives the urinary bladder, ureters, and female uteri (Fig. 21–22). The points of entrance of the ureters have shifted more ventrally than in reptiles, and they enter the urogenital sinus beside the neck of the bladder, or **urethra.** The deferent ducts of the male join the distal end of each ureter. The union of a ureter and deferent duct reflects the embryonic condition, in which the ureter is an outgrowth from the caudal end of the archinephric duct (now the deferent duct). The intestinal and urogenital parts of the cloaca enter a common chamber posteriorly.

The cloaca of an adult therian mammal is completely divided, so the intestine and urogenital passages have separate openings on the body surface. The way the division comes about can be seen clearly during embryonic development. An early, sexually indifferent embryo has a cloaca very similar to that of monotremes (Fig. 21–23**A**). A **urorectal fold** grows caudally and gradually divides the cloaca into a dorsal part, which becomes the **rectum,** and a ventral **urogenital sinus.** The allantois, whose base will form the urinary bladder, connects with the anteroventral

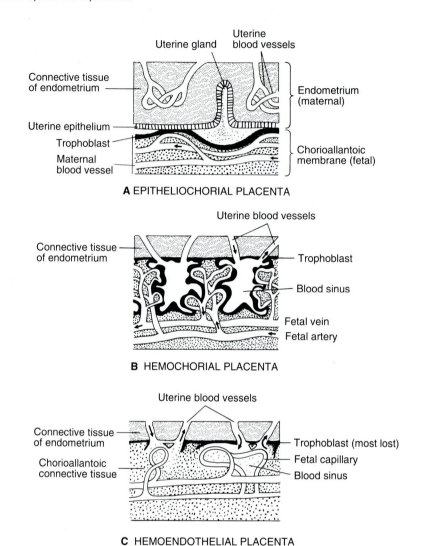

Uterine gland

Uterine
blood vessels

Connective tissue
of endometrium

Endometrium
(maternal)

Uterine epithelium
Trophoblast
Maternal
blood vessel

Chorioallantoic
membrane (fetal)

A EPITHELIOCHORIAL PLACENTA

Uterine blood vessels

Connective tissue
of endometrium

Trophoblast

Blood sinus

Fetal vein
Fetal artery

B HEMOCHORIAL PLACENTA

Uterine blood vessels

Connective tissue
of endometrium

Chorioallantoic
connective tissue

Trophoblast (most lost)
Fetal capillary
Blood sinus

C HEMOENDOTHELIAL PLACENTA

FIGURE 21–21   Representative chorioallantoic placentas, showing the degree of union between maternal and fetal tissues: **A**, Epitheliochorial placenta. **B**, Hemochorial placenta. **C**, Hemoendothelial placenta. (**A** and **B**, modified after Witschi.)

part of the urogenital sinus, and the still conjoined archinephric ducts and ureters attach close to the entrance of the allantois. The oviducts also enter the urogenital sinus close by. A **genital tubercle** begins to form on the anteroventral wall of the cloaca. In a later sexually indifferent stage, the ureters separate from the archinephric ducts and shift to the developing bladder (Fig. 21–23**B**). The neck of the bladder begins to narrow to form the urethra, which joins the urogenital sinus. A **genital groove** flanked by **genital folds** develops along the cloacal side of the enlarging genital tubercle.

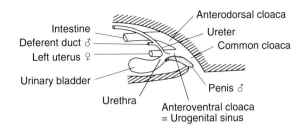

FIGURE 21–22   A composite diagram in lateral view of the cloaca of monotremes. Both male and female reproductive organs are shown, but an individual would have only one or the other. (After Grant.)

If the embryo differentiates into a male (Fig. 21–23**C**), the oviducts regress, but their distal ends often remain as a **prostatic utricle** within the prostate (see Fig. 21–13). The urogenital sinus has narrowed and continues the urethra through the prostate and pelvic canal. The genital tubercle enlarges further to form the **penis,** and the genital folds unite to enclose the genital groove within it as the penile part of the urethra. The male urethra is composed of three segments that develop from the neck of the urinary bladder, the urogenital sinus, and the genital groove. The scrotum first appears as a pair of **scrotal swellings.** As the testes descend into them, the scrotal swellings unite around the base of the penis. Several columns of spongy and vascular **erectile tissue** develop within the penis. The erectile tissue becomes engorged with blood during an erection. A **corpus spongiosum** forms around the penile urethra, and its distal end expands as the **glans penis.** A pair of **corpora cavernosa** form on the opposite side of the penis (Fig. 21–24). The glans penis is encased in a pocket of skin called the **prepuce.**

If the embryo differentiates as a female (Fig. 21–23**D**), the archinephric ducts are lost, and in eutherians the lower ends of the oviducts unite to form a median vagina. The vagina and urethra enter the urogenital sinus, which becomes the **vaginal vestibule** in an adult female. The female urethra is comparable to only the first segment of the male urethra, that is, to the part that develops from the neck of the bladder. The vaginal vestibule is a long, tubular passage in most female mammals, but it is a shallow depression in some, including humans, that is flanked by liplike skin folds. The **labia minora** develop from the genital folds, so they are the sexual homologs of the part of the male penis surrounding the urethra. The **labia majora** are homologous to the scrotal swellings. A **glans clitoridis** lies at the anterior junction of the labia minora and is comparable to the male glans penis. Small columns of erectile tissue form the **body of the clitoris** and are comparable to the male corpora cavernosa.

Since the male and female reproductive systems develop from common primordia present in a sexually indifferent embryo, they share many features. The pattern of development of the entire reproductive system of both male and female amniotes is shown in Figure 21–14, and a comparison of adult structures is given in Table 21–1.

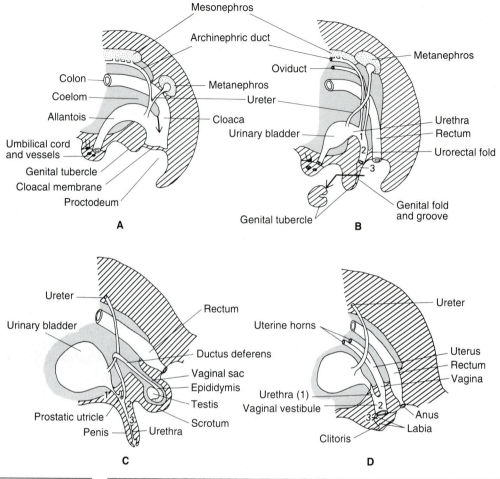

FIGURE 21–23 Lateral diagrams of the division of the cloaca in a eutherian mammal: **A** and **B**, Early and later sexually indifferent stages. **C**, Differentiation of the male. **D**, Differentiation of the female. Sexually homologous urethral segments are identified by the same number. (After Walker and Homberger.)

Reproductive Hormones

The Hormones and Their Effects

Besides producing eggs and sperm, the gonads also are endocrine glands that secrete sex hormones under the influence of the adenohypophyseal gonadotropins: **follicle-stimulating hormone** (FSH) and **luteinizing hormone** (LH). The major male sex hormones, known as androgens, are **cortical androgen,** secreted by the adrenal cortex (p. 550), and **testosterone,** secreted by the interstitial cells of the testis. The major female hormones are **estrogens** (chiefly **estradiol**) and **progesterone,** which are produced by the ovary. All these hormones have much

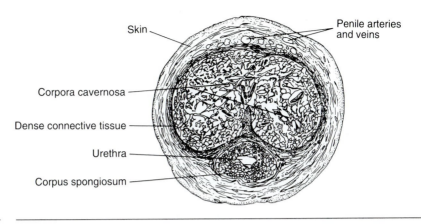

Skin

Penile arteries and veins

Corpora cavernosa

Dense connective tissue

Urethra

Corpus spongiosum

FIGURE 21–24   A cross section of a human penis. (After Fawcett.)

in common, for they are steroid derivatives of cholesterol, and the biochemical pathways by which they are synthesized overlap to a considerable extent. Enzymes needed for their synthesis occur in both the ovary and the testis, as well as in some other tissues, so it should not be surprising that some androgens are synthesized by females and some estrogens and progesterone by males. Males and females differ not in the types of hormones they produce but in their quantity. Mature human males have only about one tenth of the blood level of estrogens found in females at ovulation, and mature females have only one fifth of the amount of androgens as males.

The balance of these hormones is an important factor in controlling the direction of differentiation from the sexually indifferent stage of the embryo. Androgens promote protein synthesis and growth in both sexes. High levels of androgen are necessary for the maturation of the male gonads and reproductive passages and for the development of male secondary sex characters. In many species abnormally high androgen levels during the development can cause genetic females to differentiate in a male direction and become intersexes, that is, have morphological features that are intermediate between the two sexes. Unlike true hermaphrodites, intersexes usually either are sterile or function only as one sex. Normal differentiation in the female direction results from low androgen levels, supplemented in many species by an increase in the level of estrogens. Estrogens are needed for females to mature and develop their secondary sex characters.

Reproductive periods also are regulated to a large extent by gonadotropins and sex hormones. A few vertebrates, including human beings, rats, and domestic chickens, reproduce throughout the year. Mature males of these species produce sperm continuously, but egg production is cyclical. During a mammalian **ovarian cycle** the follicles and eggs enlarge, the eggs mature and ovulate, and the follicles transform into corpora lutea, which finally regress. Cyclical changes in the size of the female reproductive passages, in the amount of their glandular secretions, and often in behavior accompany the ovarian cycle. Many female mammals come into heat, or **estrus** (Gr. *oistros* = gadfly; frenzy), near the time of ovulation. Their behavior advertises their sexual condition, and they will permit copulation only at this time. This behavior increases the probability of fertilization. The length of an ovarian cycle varies greatly. It is 28 days in humans but only 4 to 5 days in rats. Species with repetitive cycles throughout the year are called **polyestrous.**

Reproduction is seasonal in most vertebrates. Birds and most ectotherms of the temperate region are spring breeders, but deer and many other mammals copulate in the autumn and give birth in the spring. Dogs and cats typically have two or three reproductive periods in a year. Seasonal breeders have one or more ovarian cycles during their reproductive season, and then they enter a long period of **anestrus.** Testes and ovaries become small and nonfunctional, and reproductive passages become smaller and lose their secretory state.

Integration of Reproduction

These seasonal and cyclical changes are controlled and integrated by the hypothalamus, which acts by the cyclical production of follicle-stimulating releasing hormone and luteinizing hormone releasing hormone (p. 545). These releasing hormones travel through the hypophyseal portal system to the adenohypophysis and stimulate the synthesis and release of follicle-stimulating and luteinizing hormones. These gonadotropins, in turn, stimulate the secretion of sex hormones by the gonads. The cyclical activity of the hypothalamus is controlled to a large extent by negative feedback from the fluctuating levels of the sex hormones, but it also is affected by nerve stimuli in many cases. It is possible that the hypothalamus of some species has an inherent rhythm, or biological clock, that is modulated by the sex hormones and nerve stimuli. Control mechanisms have been studied most thoroughly in birds and mammals, but considerable evidence indicates that these mechanisms operate in many other species as well. Much more research in this area is needed.

Seasonal breeders clearly respond to one or more seasonal environmental changes that affect the hypothalamus. Many environmental changes have been shown to act on the hypothalamus in some species: food quantity, day length, temperature, and even social interactions in colonial sea birds. Day length is particularly important. As the hours of light increase in late winter, spring breeders' gonadotropin and sex hormone production increases, and the gonads and reproductive passages enlarge. In fall breeders a decrease in day length triggers these events. Light, which can penetrate the skull and brain in small species, may stimulate the hypothalamus directly, but stimuli usually reach the hypothalamus from the image forming eyes or, in some species, via the median eye or pineal gland.

Females.   Under the influence of gonadotropin releasing hormones produced by the hypothalamus, the adenohypophysis of mature female mammals produces some FSH and LH throughout the reproductive period. But the amounts of these gonadotropins fluctuate. Increasing levels of both LH and FSH are needed for the growth and enlargement of the follicles, and the synthesis by the follicles of estrogen (Fig. 21–25). The thecal cells of the follicles only have receptors for LH. Under the influence of LH they produce androgen, the male sex hormone that is the biochemical precursor for estrogen. The granulosa cells of the follicles at first

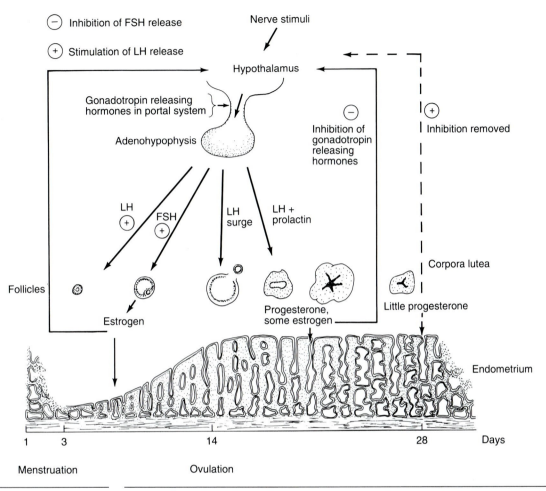

FIGURE 21–25 Cyclical changes during an ovarian cycle in a eutherian mammal. Day lengths and endometrial changes are based on the human cycle.

have receptors only for FSH, and FSH activates the enzymes that convert androgen to estrogen. Estrogen acting with LH and FSH promotes the rapid growth of one or more follicles. In some species of mammal, estrogen also causes the uterine lining to thicken in preparation for the reception of an embryo. Estrogen may also bring the female into heat. Estrogen causes the granulosa cells to develop receptors for LH. The positive feedback to the hypothalamus of increased and sustained levels of estrogen produced by one or more follicles nearly ready to ovulate promotes a surge in the secretion of LH. Because all of the follicle cells can now respond to LH, the high levels of LH also cause them to release a proteolytic enzyme that breaks down the follicle wall. Ovulation occurs. Most mammals are **spontaneous** or **cyclic ovulators** because ovulation will occur whether or not males are present. The shrew, mink, rabbit, cat, and a few other species are **induced ovulators.** The hypothalamus will not release the hormone needed for an LH surge unless nerve stimuli from copulation reach it. A single copulation will induce ovulation in a rabbit, but repetitive and frequent copulations are necessary in many induced ovulators.

Continued production of LH causes the ovulated follicles to transform into corpora lutea. **Prolactin,** which also is secreted by the adenohypophysis at this time, has luteinizing effects in rats and mice but not in all mammals. It does help maintain the corpora lutea. The corpora lutea secrete some estrogens, but their primary product is progesterone. Progesterone increases the vascularity and glandular secretion of the endometrium. It also feeds back to the hypothalamus and inhibits the release of gonadotropic releasing hormones in eutherian mammals. Follicle development stops when these releasing hormones fall to low levels. If pregnancy does not occur, the corpora lutea soon regress, progesterone secretion stops, and the inhibition of the hypothalamus on gonadotropic releasing hormone production no longer exists. Under the influence of their releasing hormones, FSH and LH production increase, follicles begin to develop, and the ovarian cycle starts anew, unless the animal goes into a period of anestrus. The uterine lining cannot be maintained in its receptive condition in the absence of progesterone. It sloughs off abruptly in most primates in a period known as **menstruation.**

If a eutherian becomes pregnant, the embryonic trophoblast of many species soon secretes **chorionic gonadotropin.** This hormone promotes the growth of the corpora lutea and the continued secretion of estrogens and progesterone, which are needed to maintain the uterine lining and form the placenta. Human pregnancy tests are based on the detection of this gonadotropin. A continued secretion of prolactin by the adenohypophysis, rather than chorionic gonadotropin, maintains the corpora lutea in other species. Corpora lutea persist until late in pregnancy in many mammals, but the placenta itself synthesizes estrogens and progesterone in humans and some other species. When this placental activity begins, chorionic gonadotropin secretion falls off and the corpora lutea regress. Since the continued high levels of progesterone block the ovarian cycles in eutherians, they cannot become pregnant again. This blockage does not occur in marsupials. As pregnancy continues, the ovaries and placentas of many mammals secrete **relaxin.** This hormone peaks just before birth and relaxes the pelvic symphysis and other pelvic ligaments. The increased flexibility of the pelvis facilitates the passage of the fetus through the pelvic canal. Many other hormonal changes occur at birth. Estrogen and progesterone secretion decrease, and prolactin production increases. The release of oxytocin from the neural lobe of the hypophysis helps bring about the uterine contractions that expel the fetus.

Estrogens, progesterone, and **placental lactogen** stimulate the proliferation of the ducts and alveoli of the mammary glands during pregnancy, but not the secretion of milk. Milk secretion is, in fact, blocked by progesterone. It does not begin until after birth, when progesterone secretion abruptly decreases and the adenohypophysis increases its output of prolactin. Prolactin is essential for milk production, and it also promotes maternal behavior in many species. The ejection of milk is caused by a combined neuronal and hormonal reflex (Fig. 21–26). When nerve impulses resulting from the tactile stimulus of the infant's sucking reach the hypothalamus, oxytocin is released from the neural lobe of the hypophysis. Oxytocin reaches the mammary glands through the circulatory system and causes

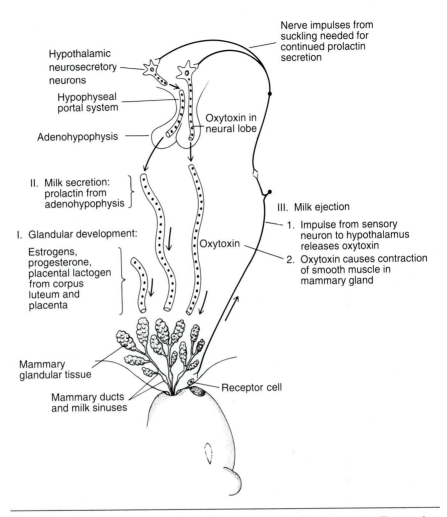

Nerve impulses from
suckling needed for
continued prolactin
secretion

Hypothalamic
neurosecretory
neurons

Hypophyseal
portal system

Adenohypophysis

Oxytoxin in
neural lobe

II. Milk secretion:
prolactin from
adenohypophysis

III. Milk ejection

1. Impulse from sensory
neuron to hypothalamus
releases oxytoxin

2. Oxytoxin causes contraction
of smooth muscle in
mammary gland

I. Glandular development:

Estrogens,
progesterone,
placental lactogen
from corpus
luteum and
placenta

Oxytoxin

Mammary
glandular tissue

Mammary ducts
and milk sinuses

Receptor cell

FIGURE 21–26   The hormones and nerve impulses needed for mammary gland development, milk secretion, and milk ejection.

smooth muscle contraction and milk ejection. Mammary gland growth, milk secretion, and milk ejection are separate processes mediated by different hormones. Prolactin secretion continues as long as the infant is suckling, and its presence inhibits the release of gonadotropic hormones and new ovarian cycles in both eutherians and marsupials. Most mammals cannot become pregnant during lactation. When nursing stops, prolactin secretion falls off and ovarian cycles begin again.

Males.   The control of male reproduction is diagrammed in Figure 21–27. Follicle-stimulating hormone and luteinizing hormone also are produced in males under the influence of hypothalamic releasing hormones. Their production is cyclical in seasonal breeders, though it is not so strongly cyclical as in polyestrous

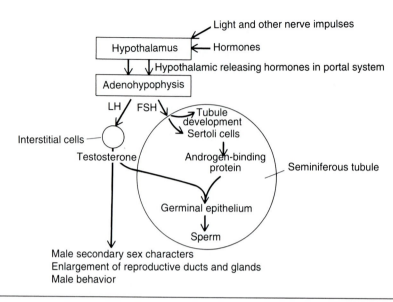

FIGURE 21–27 The factors that control male reproduction.

females. FSH is necessary for the development of the seminiferous tubules and for the enlargement of the testes at maturity and during the reproductive season in seasonal breeders. It also promotes the synthesis by the Sertoli cells of an **androgen-binding protein.** LH acts on the interstitial cells and causes them to secrete testosterone. Some testosterone enters the seminiferous tubules, unites with androgen-binding protein, enters the germinal epithelium, and promotes the development and maturation of the sperm. Testosterone traveling throughout the body causes the enlargement and secretion of the reproductive passages and accessory sex glands, the development of the secondary sex characters, and usually, by its effects on the nervous system, courtship and other behavioral changes.

SUMMARY

1. Vertebrates reproduce sexually. The eggs of a few hybrid populations of fishes, amphibians, and reptiles can develop parthenogenetically.
2. Sex is determined in mammals, birds, and a few other vertebrates by heteromorphic sex chromosomes, but the environmental conditions in which the eggs develop is an important determinant of sex in many vertebrates. Temperature-dependent sex determination is widespread among egg-laying reptiles.
3. A few fishes are hermaphroditic, but cross-fertilization normally occurs.
4. Primitive members of each vertebrate class are oviparous. Live bearing has evolved independently in some species of every vertebrate class except for agnathans and birds.
5. In live-bearing species the embryos depend on maternal food, which the mother transfers either to the eggs while the eggs are in the ovary (ovoviviparity or aplacental viviparity) or to the embryos while they are in the uterus

(placental viviparity). Many intermediate conditions between these extremes occur.

6. The primary sex cords that invade the medullary region of the gonad become the seminiferous tubules or ampullae of the testis. They regress in females, and secondary sex cords that invade the gonad cortex become the ovarian follicles.

7. Rete cords develop in the gonad between the primary sex cords and mesonephric tubules. They canalize in males and become part of the passageway for sperm; they regress in females.

8. An oviduct begins to develop in both sexes, usually by a folding of the coelomic epithelium adjacent to the mesonephros.

9. The testes of amniotes contain seminiferous tubules, in which spermatogenesis occurs, and interstitial cells, which secrete the male sex hormone. The testes of most anamniotes are composed of seminiferous ampullae. Spermatogonia multiply throughout sexual maturity.

10. The testes of most mammals lie permanently in a scrotum or descend to the scrotum during the reproductive season, but the testes of many mammals remain in the abdominal cavity. The reasons for the evolution of a scrotum are not entirely clear. Hypotheses have been proposed that the lower scrotal temperature is necessary for the production or storage of sperm, or that a scrotum removes the testes from the inhibiting effects on spermatogenesis of fluctuations in intra-abdominal pressure resulting from certain patterns of locomotion.

11. Oogonia cease dividing in mammalian ovaries prior to birth. The eggs mature within the ovarian follicles and are discharged into the coelom at ovulation. In mammals each follicle transforms into a corpus luteum after ovulation.

12. Mammalian ovaries are relatively small, but those of other vertebrates are large during the reproductive season because many eggs are produced, and in oviparous species they are laden with yolk. Oogonial divisions continue throughout sexual maturity in the ovaries of most vertebrates except for mammals, birds, a few reptiles, and chondrichthyan fishes. Corpora lutea develop from ovulated follicles primarily in species that have some degree of viviparity.

13. Cyclostomes lack reproductive ducts. Sperm as well as eggs are shed into the coelom. They leave through a genital pore that opens into the caudal end of the archinephric duct.

14. The sperm of amphibians, and probably ancestral jawed vertebrates, are carried from the testis by connecting tubules to cranial opisthonephric tubules, and thence down the archinephric duct to the cloaca. Urine also is carried by the archinephric duct in some species, but by accessory urinary ducts in others.

15. Fertilization is internal in chondrichthyans, and cranial parts of the kidney and much of the archinephric duct are specialized to produce a seminal fluid.

16. Teleosts have a unique sperm duct that leads directly from the testis to the outside, or to the caudal end of the archinephric duct. Kidney tubules have no role in sperm transport.

17. The passageway for sperm is the same in amniotes as in amphibians, but the terminology is different. Sperm leave the testis through the rete testis (con-

necting tubules), pass through efferent ductules (mesonephric tubules) to the duct of the epididymis (part of the archinephric duct), enter the deferent duct (the rest of the archinephric duct), and continue to the cloaca or, in therians, to the urethra. The efferent ductules and duct of the epididymis form the epididymis.

18. Specializations for internal fertilization in male amniotes include the evolution of accessory sex glands that produce the seminal fluid, and usually a penis. Most birds lack a penis. The accessory sex glands of mammals include the prostate, vesicular, and bulbourethral glands.

19. The eggs of amphibians and primitive jawed vertebrates are carried from the coelom to the cloaca by a pair of oviducts. Tertiary envelopes are usually secreted around the eggs as they traverse the oviducts.

20. The upper part of the oviduct of chondrichthyans is specialized to secrete a shell around the fertilized eggs, and the lower part forms a uterus in live-bearing species.

21. Most cartilaginous fishes are live bearing. The dependence of the embryos on maternal food transferred in the uterus ranges from very little to great. A yolk sac placenta is present in some species.

22. Teleosts discharge their eggs into a hollow ovary, part of which forms a unique oviduct that leads directly to the outside.

23. Reptiles and birds have oviducts (usually only the left one in birds) that secrete the tertiary envelopes that characterize the cleidoic egg: albumen, a shell membrane, and a shell.

24. Each oviduct of a monotreme has differentiated into a uterine tube and uterus. A shell membrane and shell are secreted around the eggs. The eggs continue to accumulate food from maternal secretions and begin to develop while in the uterus. Eggs are brooded after they are laid, and newly hatched young are fed milk.

25. The caudal end of each marsupial oviduct differentiates as a vagina, and these open independently into the urogenital sinus.

26. A shell membrane, but no shell, is secreted by marsupial mothers around their embryos. The embryos are nourished early in gestation by uterine secretions, which are absorbed through the shell membrane. When the membrane is reabsorbed, a brief placental relationship is established between the uterine lining and, usually, the embryonic yolk sac. Birth occurs soon after the placenta forms, apparently because the fetal trophoblast is not an immunological barrier. The embryos attach to nipples, which often are located in a marsupium, and complete their development outside the uterus.

27. A eutherian has only a single vagina, and the uterine portions of the oviducts usually unite to some degree.

28. Tertiary envelopes are not secreted around eutherian eggs, and a chorioallantoic placenta is established very early. The trophoblast appears to be an effective immunological barrier, so the embryos are not rejected.

29. Details of placental structure, including the intimacy of the union between maternal and fetal tissues, vary considerably among eutherian groups.

30. The cloaca becomes divided in therians, so the intestine and urogenital passages continue directly to the surface of the body.

31. The male sex hormones are cortical androgen, produced by the adrenal cortex, and testosterone, produced by the interstitial cells of the testis; the female hormones are estrogen and progesterone, secreted by the ovarian follicles and corpora lutea. All the sex hormones are chemically related steroids. Males have some female hormones; females have some male ones.

32. Androgens promote protein synthesis and growth in both sexes. High levels of cortical androgens and testosterone allow males to mature and develop their secondary sex characters. Low androgen levels and high estrogen levels permit females to mature and develop their secondary sex characters.

33. Reproductive periods, ovarian cycles, the production of sperm, and many aspects of reproductive behavior are integrated by complex interactions among hypothalamic releasing hormones, gonadotropins secreted by the adenohypophysis, and the gonadal hormones.

REFERENCES

Asdell, S. A., 1964: *Patterns of Mammalian Reproduction*, 2nd edition. Ithaca, N.Y., Cornell University Press.

Austin, C. R., and Short, R. V., editors, 1976: *Reproduction in Mammals*, book 6: *The Evolution of Reproduction*. Cambridge, Cambridge University Press.

Bedford, J. M., 1977: Evolution of the scrotum, The epididymis as prime mover. *In* Calaby, J. H., and Tyndale-Biscoe, C. H., editors: *Reproduction and Evolution*. Australian Academy of Science.

Bull, J. J., 1980: Sex determination in reptiles. *Quarterly Review of Biology*, 55:3–21.

Cowles, R. B., 1958: The evolutionary significance of the scrotum. *Evolution*, 12:417–418.

Duvall, D., Guillette, L. J., Jr., and Jones, R. E., 1982: Environmental control of reptilian reproductive cycles. *In* Gans, C., and Pough, F. H., editors: *Biology of the Reptilia*, vol. 13. New York, Academic Press.

Frey, Von R., 1991: Zur Ursache des Hodenabstiegs (Descensus testiculorum) bei Saügetieren. *Zeitschrift Zoologische Systematische Evolution forschung.*, 29:40–65.

Grant, T., 1984: *The Platypus*. Kensington, Australia, New South Wales University Press.

Griffiths, M., 1978: *The Biology of Monotremes*. New York, Academic Press.

Jameson, E. W., Jr., 1988: *Vertebrate Reproduction*. New York, John Wiley and Sons.

Jones, R. E., editor, 1978: *The Vertebrate Ovary: Comparative Biology and Evolution*. New York, Plenum Press.

Lagler, K. F., Bardach, J. E., Miller, R. E., 1962: *Ichthyology*. New York, John Wiley and Sons.

Lofts, B., 1974: Reproduction. *In* Lofts, B., editor: *Physiology of the Amphibia*. New York, Academic Press.

Moore, C. R., 1926: The biology of the mammalian testis and scrotum. *Quarterly Review of Biology*, 1:4–50.

Packard, G. C., 1977: The physiological ecology of reptilian eggs and embryos, and the evolution of viviparity within the class Reptilia. *Biological Reviews*, 52:71–105.

Parkes, A. S., editor, 1952–1966: *Marshall's Physiology of Reproduction*. London, Longmans Green.

Perry, J. S., 1972: *The Ovarian Cycle of Mammals*. New York, Nafner Publishing.

Potts, G. W., and Wootton, R. J., editors, 1984: *Fish Reproduction: Strategies and Tactics*. Orlando, Fla., Academic Press.

Sadleir, R. M. F. S., 1973: *The Reproduction of Vertebrates*. New York, Academic Press.

Standora, E. A., and Spotila, J. R., 1985: Temperature dependent sex determination in sea turtles. *Copeia*, 1985:711–722.

Taylor, D. H., and Guttman, S. I., editors, 1977: *The Reproductive Biology of Amphibians*. New York, Plenum Press.

Turner, C. D., and Bagnara, J. T., 1976: *General Endocrinology*, 6th edition. Philadelphia, W. B. Saunders Company.

Tyndale-Briscoe, C. H., and Renfree, M., 1987: *Reproductive Physiology of Marsupials*. Cambridge, Cambridge University Press.

Witschi, E., 1956: *Development of Vertebrates*. Philadelphia, W. B. Saunders Company.

Glossary

This glossary defines the most important technical terms that you will encounter. The pronunciation of a term is indicated by the phonetic spelling in brackets. The principal stressed syllables are indicated by a prime ('); other syllables are separated by hyphens (-). The macron (ˉ) is used for long vowels and the breve (˘) is used for short vowels.

For most terms, the classical derivation is given in parentheses because the derivation is descriptive of some aspect of the term and will help you learn and remember it. Because many word roots are repeated in different combinations in other terms, you will soon become familiar with the more common roots. The derivations are brief and typically include only three components: (1) an abbreviation of the language of the original word; (2) the original word in *italics;* and (3) the meaning of the original word. Usually only the nominative is given for Greek and Latin nouns, but the genitive (gen.) is included when this is necessary to recognize the root. For Greek and Latin verbs, the original word usually is shown in the first person, singular, present tense because this form of the word is closer to the root term than is the infinitive, but the past participle (pp.) or present participle (pres.p.) also is given when this is necessary to recognize the root. The English meaning is given in the infinitive. In some cases a noun or verb is given in the form it takes when used in combination with other words. This is indicated by a hyphen before or after the word (e.g., *odonto-* = tooth, as in **Odontoblast**).

When the English and classical terms are identical, only the classical meaning is given:

Abducens (L. = leading away)

When two or more successive terms use the same root, the derivation is given only for the first one:

Archicerebellum (Gr. *arche* = origin, beginning + L. *cerebellum* = small brain)

Archinephric duct (Gr. *nephros* = kidney)

The origin of many repetitive terms is given under the first entry of the term. For example, "ligamentum arteriosum" is defined the way this combination of words is used, but the derivation of "ligamentum" and "arteriosum" will be found under the terms "ligament" and "artery."

Pronunciations and derivations may not be given for common English words.

The pronunciations, derivations, and definitions of additional terms can be found in Dorland's or Stedman's Medical Dictionary or in unabridged dictionaries.

Abdomen [ab'dō-men] (L. *abdomen,* from *abdo* = to conceal). The part of the body cavity containing the visceral organs, in mammals limited to the part caudal to the diaphragm.

Abducens nerve [ab-dū'senz] (L. = leading away). The sixth cranial nerve, which innervates the lateral rectus muscle of the eyeball.

Abductor [ab-dŭk'ter] (L. *ab-* =

prefix meaning away from + *duco*, pp. *ductus*, = to lead). Describes a muscle that abducts, or moves a structure away from the midventral line of the body or some other point of reference, e.g., the abductor femoris.

Acanthodians [ak-an-thō′dē-anz] (Gr. *akanthodes* = spiny). A group of early bony fishes characterized by many spines, often considered to be a subclass of the Osteichthyes; the spiny sharks.

Acceleration [ak-sel′er-ā-shŭn]. The rate of increase in speed, often expressed as meters per second per second.

Accessory nerve. The eleventh cranial nerve of amniotes, which innervates the sternocleidomastoid and trapezius complex of muscles.

Acetabulum [as-ĕ-tab′yū-lŭm] (L. = vinegar cup). The socket in the pelvic girdle that receives the head of the femur.

Acinar [as′i-nar] (L. *acinus* = berry). A berry-shaped group of glandular cells.

Acoelous [ā-sē′lăs] (Gr. *a* = without +*koiloma* = hollow). A vertebral body that lacks cavities and is flat on each surface.

Acousticolateralis system [ă-kūs′ti-kō-lat-er-ā′lis] (Gr. *akoustikos*= related to hearing). The ear and lateral line system of fishes and larval amphibians; also called the octavolateralis system.

Acrodont tooth [ak′rō-dont] (Gr. *akron* = tip). A tooth that is loosely attached to the crest or inner edge of the jaw.

Acromion [ă-krō′mē-on] (Gr. *omos* = shoulder). The process on the scapula to which the clavicle articulates in species with a well-developed clavicle.

Acrosome [ak′rō-sōm] (Gr. *soma* = body). The cap at the apex of a sperm head that contains enzymes needed for the sperm to penetrate the egg.

Action potential. The electrical changes that occur across the plasma membrane of muscle and nerve cells when they become active.

Actinopterygians [ak′tin-op-te-rij′ē-anz] (Gr. *aktin*= ray + *pteryg-* = fin or wing). The largest subclass of the Osteichthyes; the ray-finned fishes.

Adaptation. A feature suited for a particular environment or mode of life; the evolutionary process by which organisms become fitted to their environment.

Adaptive radiation. The evolutionary process in which descendants from an ancestral species multiply and diverge to occupy many different habitats and modes of life.

Adductor [ă-dŭk′ter] (L. *ad-* = prefix meaning toward + *duco*, pp. *ductus* = to lead). Describes a muscle that adducts, or moves a structure toward the midventral line of the body or other point of reference, e.g., the adductor mandibulae.

Adenohypophysis [ad′ĕ-nō-hī-pof′i-sis] (Gr. *aden*= gland + *hypophysis* = under growth). The secretory portion of the pituitary gland, or hypophysis, that develops as an outgrowth from the stomodaeum; secretes hormones that regulate pigment production, growth, and the activity of the thyroid gland, adrenal gland, and gonads.

Adrenal gland [ă-drē′năl] (L. *ad-* = prefix meaning toward +*rene* = kidney). An endocrine gland next to the kidney consisting of distinct cortical and medullary parts. Major hormones of the cortex are cortisol, aldosterone, and cortical androgen; the major hormone of the medulla is adrenaline.

Adrenaline. The hormone produced by the adrenal medulla, resembles noradrenalin produced by the postganglionic sympathetic neurons and also helps the body adjust to stress.

Adrenergic fibers [ad-re-nūr′jik] (L. *ad* = toward + *rene* = kidney). Postganglionic sympathetic neurons that release noradrenaline (norepinephrine) at the neuroeffector junctions.

Adrenocorticotropic hormone [ă-drē-nō-kōr′ti-kō-trō′pik] (Gr. *trophe* = nurture). A hormone produced by the adenohypophysis that promotes the synthesis and release of adrenal cortical hormones.

Afferent [af′er-ent] (L. *ad-* = prefix meaning toward + *fero* = to carry). Describes structures that travel toward a point of reference, such as neurons toward the central nervous system or arteries toward the gills.

Agnatha [ag′nā-thă] (Gr. *a* = without + *gnathos*= jaw). The class of jawless vertebrates containing the ostracoderms and cyclostomes.

Aldosterone [al-dos′ter-ōn]. A hormone of the adrenal cortex that helps regulate mineral metabolism.

Allantois [ă-lan′tō-is] (Gr. *allas* = sausage + *eidos* = form). The extraembryonic membrane that develops as an outgrowth of the hind gut. It serves for respiration and excretion in reptile and bird embryos, contributes to the placenta in eutherians, and forms the urinary bladder and part of the urethra in adult amniotes.

Allometry [ă-lom′i-trē] (Gr. *allos* = other + *metron* = measure).

The study of relative growth in which the proportions of some part or activity changes as the size of the animal changes.

Alula [al′yu-la] (L. *ala* = wing + -*ule* = suffix denoting diminutive). The tuft of feathers borne by the first digit of a bird's wing.

Alveolus [al-vē′ō-lŭs] (L. = small pit). A small pit or cavity, such as a tooth socket or a small saclike structure in a lung where gas exchange occurs.

Ameloblasts [ă-mel′ō-blasts] (Middle English *amel* = enamel + Gr. *blastos* = bud). Cells that secrete enamel.

Ammocoete [am′ă-sēt] (Gr. *ammocoetes* = something bedded in sand). The larva of the lamprey.

Ammonia [ă-mō′nē-ă]. The first breakdown product of nitrogen metabolism; very toxic and soluble in water and requires a high water turnover for its elimination; often converted to the less toxic urea, and the nontoxic and stable uric acid.

Amnion [am′nē-on] (Gr. *amnion* = fetal membrane). The innermost of the extraembryonic membranes that surrounds the embryo and encases it in amniotic fluid.

Amniote [am′nē-ōt]. A vertebrate whose embryo has an amnion: a reptile, bird, or mammal.

Amphibians [am-fi′bē-ănz] (Gr. *amphi* = both, double + *bios* = life). Frogs, salamanders and other members of the class of ancestral terrestrial vertebrates; usually have aquatic larvae and terrestrial adults.

Amphicoelous [am-fi′sē-lăs] (Gr. *koilma* = hollow). A vertebral body that is concave on each surface.

Amphistylic suspension [am-fi′stī-lic] (Gr. *stylos* = pillar). A type of jaw suspension in fishes in which the upper jaw is supported by connections to both the chondrocranium and hyoid arch.

Ampulla [am-pul′lă] (L. = flask). A small membranous vesicle, such as that on the end of a semicircular duct.

Ampullary organ. Ampulla-shaped electroreceptors in the skin of some fishes and amphibians, e.g., the ampullae of Lorenzini of sharks.

Amygdala [ă-mig′dă-lă] (Gr. *amygdale* = almond). A subcortical nucleus of gray matter in the cerebral hemisphere.

Analogy [ă-nal′ō-je] (Gr. *analogia* = correspondence). A functional similarity among nonhomologous organs.

Anamniote [an-am′nē-ōt] (Gr. *an* = without + *amnion* = fetal membrane). A vertebrate without an amnion; a fish or amphibian.

Anapsid [ă-nap′-sid] (Gr. *a* = without + *aspid* = loop or bar). Without arches; a vertebrate skull with a complete roof of bone in the temporal region, or a reptilian subclass with such a skull.

Anastomosis [ă-nas′tō-mō′sis] (Gr. = opening, outlet). A peripheral union between blood vessels or other structures.

Androgen [an′drō-jen] (Gr. *aner*, gen. *andros* = male). A hormone that promotes the development of male characteristics.

Anestrus [an-es′trŭs] (Gr. *oistros* = gadfly, frenzy). The nonbreeding period of sexually mature animals.

Angle of attack. The angle at which the leading edge of a bird's wing is elevated above the horizontal; an increase in angle of attack increases lift up to the stalling point.

Antagonist [an-tag′ŏ-nist] (Gr. *anti* = against + *agona* = contest). A structure, usually a muscle, that opposes or resists the action of another.

Antebrachium [an-te-brā′kē-ŭm] (L. *ante* = prefix meaning before + *brachium* = arm). The forearm.

Anterior chamber. The space within the eyeball located between the iris and cornea.

Anterior commissure. An olfactory commissure within the cerebrum located just rostral to the columns of the fornix.

Antidiuretic hormone [an′tē-dī-yū-ret′ik] (Gr. *anti* = against + *dia* = through + *ouresis* = urination). A hormone produced in the hypothalamus and stored and released from the neurohypophysis; promotes water reabsorption from part of the kidney tubule and so concentrates the urine.

Antler [ant′ler] (L. *ante* = before + *oculus* = eye). One of the bony, branching, and deciduous horns of members of the deer family; usually restricted to males.

Anurans [an-yūr′anz] (Gr. *a* = without + *oura* = tail). The amphibian order to which frogs belong; also called Salientia.

Anus [ā′nŭs] (L. = anus). The caudal opening of the digestive tract.

Aorta [ā-ōr′tă] (Gr. *aorte* = great artery). A large artery; if unspecified, the dorsal aorta that carries blood from the heart to the body.

Aortic arches. Embryonic arteries that pass between the pharyngeal pouches as they carry blood from the ventral to the dorsal aorta.

Apnea [ap′nē-ă] (Gr. *a* = without

+ *pnoia* = breathing). The cessation of breathing, during which the breath is held and the lungs are not ventilated.

Apomorphic character.   See Derived character.

Aponeurosis   [ap-ō-nū-rō′sis] (Gr. *apo* = away from + *neuron* = nerve, sinew). A sheetlike tendon of a muscle.

Appendix   [ă-pen′diks] (L. *appendo* = to hang something on). A dangling extension of another organ, such as the vermiform appendix on the caecum.

Aqueduct of Sylvius   (*Franciscus Sylvius*, 1614–1672, Dutch anatomist). See Cerebral aqueduct.

Aqueous humor   [ā′kwē-ŭs hyū′mer] (L. *aqua* = water + *humor* = liquid). The lymphlike liquid filling the anterior and posterior chambers of the eye.

Arachnoid   [ă-rak′noyd] (Gr. *arachne* = spider + *eidos* = form). The weblike middle meninx around the central nervous system of mammals.

Arbor vitae   [ar′bōr vēt′ē] (L. = tree + *vita* = life). The treelike configuration of white fibers entering and leaving the mammalian cerebellum.

Archaeornithes   [ăr′kē-or′ne-thez] (Gr. *arche* = beginning + *ornis* = bird). The ancestral subclass of birds, includes *Archaeopteryx*.

Archenteron   [ark-en′ter-on] (Gr. *arche* = origin, beginning + *enteron* = intestine, gut). The embryonic gut cavity, lined with endoderm.

Archicerebellum   [ar′ki-ser-ĕ-bel′ŭm] (Gr. *cerebellum* = small brain). The part of the cerebellum that receives vestibular impulses from the ear and impulses from the lateral line

system; the flocculonodular lobes in mammals.

Archinephros.   See Holonephros.

Archinephric duct   [ar′ki-nef-rik] (Gr. *nephros* = kidney). The first-formed kidney duct, which drains the kidney of most anamniotes and becomes the ductus deferens of male amniotes.

Archipallium   [ar-ki-pal′ē-ŭm] (Gr. *pallium* = mantle). A primitive, medial olfactory area in the cerebrum; becomes the hippocampus of mammals.

Archipterygium   [ar′ki-tĕ-rij′ē-ŭm] (Gr. *ptery-* = fin or wing). The paired fins of lungfishes in which radials extend from each side of a central axis; once believed to be the ancestral type of paired fins.

Archosaurs   [ar-kō′sōrz] (Gr. *archon* = ruler + *sauros* = lizard). The reptilian subclass that includes the two extinct orders of dinosaurs, the extinct pterosaurs (flying reptiles), and the contemporary crocodiles.

Arcualia   [ar′kyū-ā-lia] (L. *arcus* = bow, arch). Small arches of cartilage or bone that often contribute to the formation of a vertebra.

Artery   [ar′ter-ē] (L. *arteria* = artery). A vessel that carries blood away from the heart. The blood may be high or low in oxygen content.

Articular   [ar-tik′yū-lăr] (L. *articulus* = joint). Pertaining to a joint.

Artiodactyls   [ar′ti-ō-dak′tilz] (Gr. *artios* = even + *daktylos* = finger or toe). The mammalian order that includes ungulates with an even number of toes: pigs, deer, cattle.

Arytenoid cartilage   [ar-i-tē′noyd] (Gr. *arytainoeides* = ladle-shaped). The ladle-shaped

cartilage of the mammalian larynx that attaches to and modifies the tension of the vocal cords.

Ascidians   [a-sid′ē-anz] (Gr. *askidion* = little wineskin). The most abundant class of tunicates, commonly called the sea squirts.

Aspiration pump.   A method of lung ventilation in which air is sucked into the lungs; occurs primarily in amniotes.

Astrocytes   [as′trō-sītz] (Gr. *astron* = star + *kytos* = hollow vessel or cell). Star-shaped nutritive and supportive glia cells of the central nervous system.

Atlas   [at′las] (Gr. mythology, a god supporting the earth upon his shoulders). The first cervical vertebra of terrestrial vertebrates, which articulates with the skull; nodding movements of the head occur between the atlas and skull.

Atrium   [ā′trē-um] (L. = entrance hall). A chamber, such as the atrium of the heart, which receives blood from the sinus venosus or veins.

Atrophy   [at′rō-fē] (Gr. *a* = without + *trophe* = nourishment). The decrease in size and sometimes loss of a structure.

Auditory   [aw′di-tōr-ē] (L. *audio* = to hear). Pertaining to the ear.

Auditory tube.   A tube that extends between the tympanic (middle ear) cavity and pharynx of most tetrapods and equalizes the air pressure on both sides of the tympanic membrane; homologous to the spiracle of fishes. Sometimes called the eustachian tube.

Auricle   [aw′ri-kl] (L. *auricula* = external ear). The external flap of the mammalian ear.

Autonomic nervous system   [aw-tō-nom′ik] (Gr. *autos* = self +

nomos = law). The part of the nervous system carrying visceral motor fibers to the viscera and glands.

Autostylic suspension [aw′tō-stī-lic] (Gr. *stylos* = pillar). A type of jaw suspension in which the upper jaw is attached to the rest of the skull by its own processes.

Aves [ā′vēz] (L. = birds). The vertebrate class that contains the birds.

Axillary [ak′sil-ār-ē] (L. *axilla* = armpit). Pertaining to the arm pit: axillary artery.

Axis [ak′sis] (L. = axle, axis). The second cervical vertebra of mammals; rotary movements of the head occur between the axis and atlas.

Axon [ak′son] (Gr. = axle, axis). The long, slender process of a neuron specialized for the transmission of nerve impulses.

Azygos vein [az′ī-gos] (Gr. *a* without + *zygon* = yoke). An unpaired vein that drains most of the intercostal spaces on both sides of the mammalian thorax.

Basal nuclei. A group of nuclei in the corpus striatum of the mammalian cerebrum.

Basapophysis [bā′să-pof′i-sis] (Gr. *basis* = base + *apo* = away from + *phsys* = growth). A transverse process low on a vertebral body to which a subperitoneal rib of a fish attaches; serially homologous to a hemal arch.

Basement lamina. A thin layer of delicate connective tissue that separates many epithelial tissues from deeper structures.

Biceps [bī′seps] (L. *bi* = two + *ceps* = head). A structure with two heads, such as the biceps muscle.

Bicornuate [bī-kōr′nū-āt] (L. *cornu* = horn). A structure with two horns, such as a bicornuate uterus.

Bile [bīl] (L. *bilis* = bile). The secretion of the liver, containing bile pigments and fat-emulsifying bile salts.

Biped [bī′ped] (L. *bi* = two + *pes*, gen. *pedis* = foot). A two-legged animal, such as a human being.

Bladder. A membranous sac filled with air or liquid.

Blastocoele [blas′tō-sēl] (Gr. *blastos* = bud + *koilos* = hollow). A cavity of the blastula that becomes obliterated during gastrulation and mesoderm formation.

Blastocyst [blas′tō-sist] (Gr. *kystis* = bladder). The modified blastula of a eutherian mammal.

Blastodisk [blas′tō-disk] (Gr. *diskos* = disk). The disk of cells formed during cleavage that lies on the top of the yolk of large-yolked eggs of fishes, reptiles, and birds, and on the top of the yolk sac of mammals.

Blastomere [blas′tō-mēr] (Gr. *meros* = part). One of the cells of the blastula.

Blastopore [blas′tō-pōr] (Gr. *poros* = pore). The opening into the archenteron that is formed during gastrulation.

Blastula [blas′tyū-lă] (L. diminutive of Gr. *blastos* = bud). The ball of cells formed during cleavage, usually containing a blastocoele.

Blood. The liquid circulating in the arteries, capillaries, and veins, consisting of a liquid plasma and cellular elements.

Blood-brain barrier. The structural and physiological barriers that regulate the exchange of materials between the blood, brain tissue, and cerebrospinal fluid.

Bone. The hard skeletal material of vertebrates, which consists of collagen fibers to which calcium phosphate crystals are bound, usually arranged in alternating layers of matrix and bone-forming cells.

Boundary layer. The layer of water or air surrounding a moving aquatic or flying animal in which shear forces occur; causes frictional drag.

Bony fish. See Osteichthyes.

Bowman's capsule. (*Sir William Bowman*, British anatomist, 1816–1892). The dilated end of a kidney tubule that surrounds a knot of capillaries.

Brachial [brā′kē-ăl] (L. *brachium* = upper arm). Pertaining to the upper arm.

Brachium conjunctivum [kon-jŭnk-tī′-vum] (L. *conjungo*, pp. *conjunctus* = to join together). The most cranial cerebellar peduncle—an armlike neuronal tract of mammals through which impulses enter and leave the cerebellum.

Brachium pontis [pon′tis] (L. *pons*, gen. *pontis*, bridge). The middle cerebellar peduncle carrying impulses into the cerebellum from the pons.

Brain. The cranial portion of the central nervous system enclosed by the cranium; the major integrative center of the nervous system.

Braincase. The cartilages and bones that encase the brain.

Brainstem. The brain exclusive of the cerebellum and forebrain (diencephalon and cerebrum).

Branchial [brang′kē-ăl] (Gr. *branchia* = gills). Pertaining to the gills.

Branchial arches. Those visceral arches (numbers 3 through 7) that support the gills in fishes. Compare with Visceral arches.

Branchiomeric [brang′kē-o-

mēr′ik] (Gr. *meros* = part).
Pertaining to muscles and other
structures associated with the
visceral arches.

Bronchus [brong′kŭs] (Gr.
bronchos = windpipe). A branch
of the trachea that enters the
lungs.

Buccal [bŭk′ăl] (L. *bucca* =
cheek). Pertaining to the mouth,
as in buccal cavity.

Bulbourethral glands [bŭl′bō-
yū-rē-thrăl] (L. *bulbus* = a
bulbous root + Gr. *ourethra* =
urethra). Accessory sex glands of
male mammals that are located
near the base of the penis and
discharge into the urethra.

Bulla [bul′ă] (L. = bubble). A
bubble-like expansion of some
structure, such as the tympanic
bulla on the temporal bone.

Bunodont [bū′nō-dont] (Gr.
bounos = mound +*odont-* =
tooth). Molar teeth with low,
rounded cusps.

Bursa [ber′să] (L. = purse). A
saclike cavity.

Bursa of Fabricius (*Giralamo
Fabricius,* Italian anatomist and
embryologist, 1533–1619). A
dorsal cloacal diverticulum of
birds, site of the maturation of B
lymphocytes.

Caecilians [sē-sil′ē-ănz] (L.
caecilia = blindworm). Tropical
wormlike, burrowing amphibians
of the order Gymnophiona.

Caecum [sē′kŭm] (L. *caecus* =
blind). A blind-ending pouch
attached to part of the intestine,
such as the one at the beginning
of the mammalian large intestine.

Calcaneus [kal-kā′nē-ŭs] (L. =
heel). The large proximal tarsal
bone that forms the "heel bone"
of mammals.

Calcitonin [kal-si-tō′nin] (L. *calx*
= lime + Gr. *tonos* = tension).
A hormone produced by the C

cells of the ultimobranchial
bodies or thyroid gland; its
actions oppose those of
parathormone, for it promotes
the deposition of calcium in bone
and reduces its level in the
blood.

Calyx, pl. **calyces** [kā′liks, kal′i-
sēz] (Gr. *kalyx* = cup). A cuplike
compartment, such as the renal
calyces or subdivisions of the
renal pelvis.

Canaliculi [kan-ă-lik′yū-lī] (L.
canaliculi = small channels).
Small canals in bone matrix that
contain the processes of the
osteocytes.

Canine [kā′nīn] (L. *canis* = dog).
The mammalian tooth next
behind the incisors, usually
longer than other teeth.

Cantilever. A projecting beam or
other structure that is supported
only at one end.

Capillary [kap′i-lār-ē] (L. *capillus*
= hair). One of the minute blood
vessels between arteries and
veins through which exchanges
between the blood and tissue
fluids occur.

Capitulum [kă-pit′yū-lŭm] (L. =
small head). A small, articulating
knob on the end of a bone such
as a rib.

Captorhinida [kap′tō-rī′nid-ă] (L.
capus = capture + Gr. *rhis* =
nose). The ancestral order of
reptiles.

Carapace [kar′ă-pās] (Spanish
carapacho = covering). The
dorsal shell of a turtle.

Cardiac [kar′dē-ak] (Gr. *kardia* =
heart). Pertaining to the heart.

Cardinal veins [kar′di-năl] (L.
cardinalis = principal). One of
the principal veins of embryonic
vertebrates and adult
anamniotes.

Carnassials [kar-nas′ē-ălz] (L.
caro, gen. *carnis*= flesh). The
specialized shearing teeth of

carnivores; the fourth upper
premolar and first lower molar.

Carnivores [kar′ni-vōrz] (L. -
vorous = devouring). Animals
specialized to feed on other
animals.

Carotid [ka-rot′id] (Gr. *karotides*
= large neck artery, from *karoo*
= to put to sleep, because
compressing the artery causes
unconsciousness). Pertaining to a
large artery in the neck or to
nearby structures.

Carpal [kar′păl] (Gr. *karpos* =
wrist). One of the small bones of
the wrist.

Cartilage [kar′ti-lij] (Gr. *cartilago*
= cartilage). A firm but elastic
skeletal tissue whose matrix
contains proteoglycan molecules
that bind with water. Occurs in
all embryos, in adult cartilaginous
fishes, and in parts of the
skeleton of other vertebrates
where firmness as well as
flexibility are needed.

**Cartilage replacement
bone.** Bone that develops
within and around the embryonic
endoskeleton.

Cartilaginous fish. See
Chondrichthyes.

Caudal [kaw′dăl] (L. *cauda* =
tail). Pertaining to the tail.

Caudata [kaw′dă-tă] (L. *cauda* =
tail). The amphibian order that
includes the salamanders.

Cecum. See Caecum.

Center of buoyancy. The point
in the body of an aquatic
vertebrate through which the
resultant force of buoyancy acts.

Center of gravity. The point in
the body of an animal through
which the resultant force of
gravity acts.

Central nervous system. That
part of the nervous system
located in the longitudinal axis of
the body, consists of the brain
and spinal cord.

Central pattern generator.
Groups of neurons in the spinal
cord and brain whose activity is
responsible for innate, cyclical
movements of body parts,
as occur in swimming and
walking.

Centrum [sen'trŭm] (Gr. *kentron*
= center). The vertebral body
that lies ventral to the vertebral
arch.

Cephalic [se-fal'ik] (Gr. *kephale*
= head). Pertaining to the head.

Cephalization [sef'ăl-ĭ-zā-shŭn].
The development during
evolution of a well-defined head.

Cephalochordata [sef'ă-lō-kōr-
dătă] (L. *chordata* = string). The
subphylum of chordates that
includes amphioxus.

Ceratotrichia [ser'ă-tō-trik'i-ă]
(Gr. *kerat-* = horn + *trich-* =
hair). The horny fin rays of
Chondrichthyes.

Cerebellum [ser-ĕ-bel'um] (L. =
small brain). The dorsal part of
the metencephalon, which is a
center for motor coordination.

Cerebral aqueduct [se-rē'-brăl]
(L. *cerebrum* = brain). The
narrow passage within the brain
that extends between the third
and fourth ventricle; also called
the aqueduct of sylvius.

Cerebral hemispheres (Gr.
hemi- prefix meaning one-half +
sphaira = globe, ball). The pair
of hemispheres that form most of
the telencephalon. They are the
major integrating centers of the
brain in mammals.

Cerebrospinal fluid. A lymphlike
fluid that circulates within and
around the central nervous
system, which it helps to protect
and nourish.

Cervical [ser'vĭ-kal] (L. *cervix* =
neck). Pertaining to the neck.

Cervix [ser'viks]. The necklike
portion of an organ, such as the
neck of the uterus.

Cheek teeth. A collective term
for the premolar and molar teeth
of mammals.

Cheiropterygium [kī-rō-tĕ-rij'ē-
ŭm] [Gr. *chiro-* = hand +
pteryg- = fin or wing). The
paired appendage of a terrestrial
vertebrate.

Chelonia [kē-lō'-nē-ă] (Gr.
chelone = tortoise). The reptilian
order to which turtles belong.

Chimaera [kī-mēr'-ă] (Gr.
chimaira = monster). A
cartilaginous fish belonging to the
subclass Holocephali.

Choana [kō'an-ă] (Gr. *choane* =
funnel). One of the paired
openings from the nasal cavities
into the pharynx; an internal
nostril.

Chondrichthyes [kon-drik'thi-ēz]
(Gr. *chondros* = cartilage +
ichthyos = fish). The class of
cartilaginous fishes, including
sharks, skates, and chimaeras.

Chondrocranium [kon-drō-
krā'nē-um] (Gr. *chondros*=
cartilage + *kranion* = skull).
Cartilages that encase the brain
and major sense organs in
embryos and the adults of some
vertebrates, also called the
neurocranium.

Chondrocyte [kon'drō-sīt] (Gr.
kytos = hollow vessel or cell). A
mature cartilage cell, develops
from a chondroblast.

Chondrosteans [kon'drōs-te-anz]
(Gr. *osteon* = bone). The
infraclass of ancestral
actinopterygians; includes the
contemporary sturgeon and
related species. Considerable
cartilage is retained in the adult
skeleton, although some bone is
present.

Chordamesoderm [kōr-dă-mes'ō-
derm]. The longitudinal,
middorsal group of mesodermal
cells that moves into the roof of
the archenteron during

gastrulation and gives rise to the
notochord.

Chordates [kōr'-dātz]. The
phylum to which protochordates
and vertebrates belong;
characterized by having a
notochord, at least at some stage
of their life cycle.

Chorion [ko'rē-on] (Gr. *chorion*
= skinlike membrane enclosing
the fetus). The outermost
extraembryonic membrane of
amniotes.

Choroid [kō'royd] (Gr.
chorioeides = like a membrane).
The highly vascularized middle
tunic of the eyeball, which lies
between the fibrous tunic and
the retina.

Choroid plexus [plek'sŭs] (L.
plexus = network). The vascular
network of the tela choroidea
that protrudes into certain brain
ventricles and secretes the
cerebrospinal fluid.

Chromaffin cells [krō'maf-in]
(Gr. *chromo-* = color + L.
affinis = affinity). Cells in the
medulla of the adrenal gland of
neural crest origin that secrete
noradrenaline and adrenaline and
have an affinity for chromic
stains.

Chromatophore [krō-mat'ō-fōr]
(Gr. *phoros* = bearing from
pherein = to bear). A vertebrate
cell of neural crest origin that
carries pigment or reflective
granules.

Cilia [sil'ē-ă] (L. *cilia* = hairs).
Minute, movable processes of
some epithelial cells that contain
a characteristic pattern of nine
peripheral and two central
microtubules.

Ciliary body. A part of the
vascular tunic of the eyeball that
secretes the aqueous humor and
contains muscle fibers used in
focusing the eye.

Circadian rhythm [ser-kā'dē-ăn].

(L. *circa* = about + *dies* = day). A metabolic or behavioral pattern with a cycle of approximately 24 hours.

Cisterna chyli [sis-ter′nă kĭl′ē] (L. *cisterna* = an underground reservoir, cistern + Gr. *chylos* = juice). The sac that receives lymph from the abdominal viscera and caudal parts of the body.

Cladistics [klă-dis′tikz] (Gr. *clados* = branch). Pertaining to an analysis of evolution that seeks to determine the points at which various lineages separated; all lineages are monophyletic.

Clasper. The modified part of the pelvic fin of male chondrichthyan fishes used to transfer sperm to the female.

Clavicle [klav′i-kl] (L. *clavicula* = small key, nail). A dermal bone of the pectoral girdle extending medially from the scapula to the interclavicle or sternum.

Cleavage [klĕv′ij]. The mitotic divisions by which the single-celled zygote is converted to a multicellular blastula of the same size.

Cleidoic egg [klī-dō′ik] (Gr. *kleid-* = clavicle, key). The self-contained eggs of amniotes in which a free larval stage is bypassed; modified in viviparous species.

Cleithrum [kli′thr-ŭm] (Gr. *kleithron* = bar). A bar-shaped dermal element of the pectoral girdle of some fishes and early tetrapods; located dorsal to the clavicle.

Clitoris [klit′ō-ris] (Gr. *kleitoris* = hill). The small erectile organ of a female mammal that corresponds to the male glans penis and corpora cavernosa penis.

Cloaca [klo-ā′kă] (L., *cloaca* = sewer). The posterior chamber of most fishes, nonmammalian tetrapods, and monotreme mammals into which the digestive tract and urogenital passages discharge.

Coccyx [kok′siks] (Gr. *kokkyx* = cuckoo). Several fused caudal vertebrae of humans; does not reach the body surface, but serves for the attachment of certain muscles.

Cochlea [kok′lē-ă] (L. = snail shell). The snail-shaped part of the mammalian inner ear, consisting of the cochlear duct and the scala vestibuli and scala tympani.

Cochlear duct. The duct within the cochlea that is a part of the membranous labyrinth and contains the receptive cells for sound.

Coelacanths [se′lă-kanthz] (Gr. *koilos* = hollow + *akantha* = spine). An order of sarcopterygian fishes that specialized early for a marine life.

Coelom [sē′lom] (Gr. *koiloma* = a hollow). A body cavity that is completely lined by an epithelium of mesodermal derivation.

Collagen [kol′lă-jen] (Gr. *kolla* = glue + *genos* = descent). A protein produced by fibroblasts; forms most of the extracellular fibers of connective and skeletal tissues.

Collecting ducts. The small tubules that receive material from the kidney tubules and lead to the renal pelvis or urinary duct.

Colliculus [ko-lik′yū-lŭs] (L. = little hill). One of the small elevations on the dorsal surface of the mesencephalon of mammals that is a center for certain optic (superior colliculus) or auditory (inferior colliculus) reflexes.

Colon [kō′lon] (Gr. *kolon* = colon). The large intestine of tetrapods exclusive of the caecum and rectum.

Columella auris [kol-ū-mel′ă aw′ris] (L. *columella* = small column). The single rod-shaped auditory ossicle of nonmammalian tetrapods that transmits vibrations from the tympanic membrane to the inner ear; often called the stapes.

Commissure [kom′i-syūr] (L. *commissura* = seam). A band of nervous tissue, usually fibers, that crosses the midline of the body.

Common bile duct. The principal duct carrying bile to the intestine, formed by the confluence of hepatic ducts from the liver and, when present, the cystic duct from the gallbladder.

Compression. A stress that results when two parallel forces move toward each other.

Concha [kon′kă] (Gr. *konkhe* = sea shell). One of several folds within the mammalian nasal cavities that increase their surface area; also called a turbinate bone.

Condyle [kon′dĭl] (Gr. *kondylos* = knuckle). Any rounded articular surface, such as the occipital condyles or mandibular condyles.

Conjunctiva [kon-jūnk-tī′va] (L. *conjunctus* = joined together). The epithelial layer that lines the eyelids and reflects over the cornea.

Connective tissue. A widespread body tissue characterized by an extensive extracellular matrix of fibers. It connects other tissues and supports the body; includes fibrous tissue, fat, cartilage, and bone.

Contralateral. Descriptive of a structure that is located on the opposite side of the body from the point of reference.

Conus arteriosus [kō′năs ar-tēr-ē-ō′săs]. The fourth chamber of the heart of most fishes, which extends between the ventricle and the ventral aorta.

Convergent evolution. The evolution of homoplastic structures by distantly related animals as they adapt to a common environment; e.g., the evolution of wings by birds and bats.

Coprodaeum [kŏ-prō-dē′ŭm] (Gr. *kopros* = dung + *hodaion* = way). The portion of the cloaca that receives the feces.

Coprophagy [kŏ-prof′ă-je] (Gr. *phagein* = to eat). The reingestion of feces; characteristic behavior of many rodents and lagomorphs.

Coracoid [kōr′ă-koyd] (Gr. *korax* = crow + *eidos* = form). A cartilage replacement bone that forms the posteroventral part of the pectoral girdle, reduced to a small process shaped like a crow's beak in therians.

Cornea [kōr′nē-ă] (L. *corneus* = horny). The transparent part of the fibrous tunic at the front of the eyeball.

Cornua [kōr′nū-ă] (L. *cornua* = horns). Hornlike processes of a structure, such as the cornua of the hyoid bone.

Corpora quadrigemina [kōr′pōr-ă kwah-dri-jem′i-nă] (L. *corpus*, plural, *corpora*, = body +*quadrigeminus* = fourfold). A collective term for the paired superior and inferior colliculi on the roof of the mesencephalon of mammals.

Corpus cavernosum penis [kōr′pŭs kav-er-nō-sum pē′nis] (L. *caverna* = hollow place). One of a pair of columns of erectile

tissue that forms much of the penis.

Corpus callosum [ka-lō′sum] (L. *callosus* = hard). The large commissure interconnecting the two cerebral hemispheres.

Corpus luteum [lu-te′um] (L. *luteus* = yellow). The hard, yellowish body that develops from an ovulated follicle and acts as an endocrine gland.

Corpus spongiosum penis. A column of erectile tissue that surrounds the penile portion of the urethra.

Corpus striatum [strī-ā′tum] (L. *striatus* = striped). A group of basal nuclei in the base of the cerebrum through which white fibers pass.

Cortex [kōr′teks] (L. = bark). A layer of distinctive tissue on the surface of an organ, such as the adrenal cortex or the cerebral cortex.

Cortisol [kōr′ti-sol]. A hormone produced by the adrenal cortex that helps regulate carbohydrate metabolism.

Cosmine [koz′mēn] (Gr. *kosmios* = well ordered). A type of dentine found in certain bony scales in which the dentine tubules are grouped into radiating tufts.

Cosmoid scale. A thick, bony scale with a conspicuous layer of cosmine, characteristic of early sarcopterygians.

Costal [kos′tăl] (L. *costa* = rib). Pertaining to the ribs.

Cowper's gland (*William Cowper*, British anatomist, 1666–1709). See Bulbourethral gland.

Cranial kinesis [ki-nē′sis] (Gr. *kinesis* = movement). Movement of parts of the skull, exclusive of the lower jaw, relative to each other, occurs during feeding in many nonmammalian vertebrates.

Cranium [krā′nē-ŭm] (Gr.

kranion = skull). The skull, especially the part encasing the brain.

Cremasteric pouch [krē-mas-ter′ik] (Gr. *kremaster* = suspender). Layers of the body wall that suspend the testis; the scrotal wall apart from the skin.

Cribriform plate [krib′ri-fōrm] (L. *cribrum* = sieve + *forma* = shape). The perforated portion of the sphenoid bone through which groups of olfactory neurons pass.

Cricoid cartilage [krī′koyd] (Gr. *krikos* = ring +*eidos* = form). Ring-shaped cartilage of the mammalian larynx.

Crop. The distal part of the esophagus of certain birds, especially grain-eating species, that stores food.

Crossopterygians [kro-sop-te-rij′ē-ănz] (Gr. *krossoi* = tassels + *pteryg-* = fin or wing). A collective name sometimes used for two groups of sarcopterygian fishes, the coelacanths and rhipidistians. The rhipidistians were ancestral to terrestrial vertebrates.

Crus, pl. **crura** [krūs, krū′ră] (L. = leg). The lower leg, shank, or shin of a tetrapod.

Crypt of Lieberkühn (*Johann N. Lieberkühn*, German anatomist, 1711–1756). Glandlike invaginations from the small intestine of mammals; epithelial cells multiply here, spread over the intestinal lining to replace worn-out cells, and some release digestive enzymes.

Ctenoid scale [tē′noid] (Gr. *ktenoeides* = like a cock's comb). A thin, bony scale having comblike processes on its outer part and a serrate margin.

Cupula [kū′pū-la] (L. a small tub). A cup-shaped, jelly-like secretion that caps the group of hair cells in a neuromast.

Cursorial [kūr-sor'ē-ăl] (L. *cursor* = runner). Pertaining to a vertebrate specialized for running.

Cutaneous [kyū-tā'nē-ŭs] (L. *cutis* = skin). Pertaining to the skin.

Cycloid scale [sī'kloyd] (Gr. *kyklos* = circle). A thin, bony scale having a smooth surface and rounded margins.

Cyclostome [sī'klǎstōm] (Gr. *stoma* = mouth). A collective term for the two orders of contemporary agnathans: Petromyzontiformes (lampreys) and Myxiniformes (hagfishes).

Cystic duct [sis'tik] (Gr. *kystis* = bladder). The duct of the gallbladder.

Deciduous [dē-sid'yū-ŭs] (L. *deciduus* = falling off). Teeth that are shed; the first set of teeth of a mammal, which are replaced by the permanent teeth.

Decussation [dē-kŭ-sā'shŭn] (L. *decusso*, pp. *decussatus* = to divide crosswise in an X). The crossing of neuronal tracts in the midline of the central nervous system.

Defecation [def'ě-kā'shŭn] (L. *defaeco*, pp. *defaecatus* = to remove the dregs). The elimination of undigested residue and bacteria from the digestive tract.

Deferent duct [def'er-ent] (L. *defero*, pres.p. *deferens* = to carry away). The sperm duct of amniotes, homologous to the archinephric duct of anamniotes.

Delamination [dē-lam-i-nā'shŭn] (L. *de* = from + *lamina* = small plate). The splitting off of cells to form a new layer.

Dendrite [den'drīte] (Gr. *dendrites* = relating to a tree). Branching neuronal processes that receive nerve impulses.

Density. The mass or weight of a body divided by its volume.

Dentine [den'tēn] (L. *dens*, gen. *dentis* = tooth). Bonelike material that forms the substance of a tooth deep to the superficial enamel.

Derived character. A unique, evolved character that distinguishes one group from another; also called an apomorphic character.

Dermal bones [der'măl] (Gr. *derma* = skin). Superficial bones that lie in or just beneath the skin and develop from the direct deposition of bone in connective tissue; also called membrane bones.

Dermal denticle [den'ti-kl] (L. *denticulus* = small tooth). A small, toothlike scale often found in the skin of cartilaginous fishes; also called a placoid scale.

Dermatocranium [der-mǎ-tō-krā'nē-ŭm] (Gr. *kranion* = skull). The portion of the skull composed of dermal bones.

Dermatome [der'mǎ-tōm] (Gr. *tome* = a cutting). The lateral portion of an epimere, which will form the dermis of the skin.

Dermis [der'mis]. The dense connective tissue layer of the skin deep to the epidermis.

Deuterostome [dū'ter-ō-stōm] (Gr. *deuteros* = second + *stoma* = mouth). The group of coelomate animals in which the stomodaeum rather than the blastopore forms the adult mouth; includes echinoderms and chordates.

Diaphragm [dī'ǎ-fram] (Gr. *dia* = through, across + *phragma* = a partition wall). The membranous and muscular partition between the thoracal and the abdominal cavities in mammals.

Diaphysis [dī-af'i-sis] (Gr. *physis* = growth). The shaft of a limb bone.

Diapophysis [dī-ă-pof'ĭ-sis] (Gr. *apo* = away from + *physis* = growth). A transverse process that extends from the vertebral arch and receives the tuberculum of a rib.

Diapsid [dī-ap'sid] (Gr. *di-* = two + *apsis* = arch). Pertaining to a reptilian skull in which there are two temporal fenestrae and two arches of bone, or to a reptile with such a skull.

Diarthrosis [dī-ar-thrō'sis] (Gr. *arthron* = joint). A joint allowing considerable movement between the elements.

Diastole [di-as'tō-lē] (Gr. *diastole* = dilation). The period during which the ventricle of the heart relaxes and fills with blood.

Diencephalon [dī-en-sef'ă-lon] (Gr. *dia* = through, across + *enkephalos* = brain). The region of the brain between the telencephalon and mesencephalon, consisting of the epithalamus, thalamus, and hypothalamus.

Digit [dij'it] (L. *digitus* = digit). A finger or toe.

Digitigrade [dij'i-ti-grad] (L. *gradus* = step). Walking with the heel and ankle raised off the ground so only the digits bear the body weight.

Dinosaur [dī'nō-sōr] (Gr. *deinos* = terrible + *sauros* = lizard). A member of the archosauran orders Saurischia or Ornithischia.

Diphycercal tail [di-fi'ser'kl] (Gr. *diphyes* = twofold + *kerkos* = tail). A caudal fin in which the vertebral axis runs straight to the tip and divides the fin into symmetrical upper and lower lobes.

Diphyodont [dif'ē-ō-dont] (Gr. *di-* = two + *phyo* = to produce

+ *odont-* = tooth). Pertaining to mammals with two sets of teeth, deciduous and permanent.

Diplospondyly [dip′lō-spon-di-le] (Gr. *diploos* = double + *spondylos* = vertebra). A condition in which there are two vertebral bodies per body segment; found in some early tetrapods and the tail region of some fishes.

Dipnoans [dip′nō-ănz] (Gr. *di-* + *pnoe* = breath). Lungfishes that breathe through both gills and lungs.

Divergent evolution. The division of an ancestral group into two or more lineages that share homologous structures.

Drag. The resistance to the movement of an animal through the water or air in which it lives.

Duct [dŭkt] (L. *duco*, pp. *ductus* = to lead). A small, tubular passage.

Duct of Cuvier (*Baron Georges Cuvier*, 18th-century French scientist). The common cardinal vein.

Ductus arteriosus [dŭk′tŭs ar-tēr′-ē-ō-sus]. The dorsal part of the sixth aortic arch, may serve as a bypass of the lungs in larval or fetal stages.

Ductus venosus [ve′nō-sŭs]. An embryonic connection between the umbilical vein and the caudal vena cava; bypasses the hepatic sinusoids.

Duodenum [dū-ō-dē′nŭm] (L. *duodeni* = 12 each). The first portion of the tetrapod small intestine, which is 12 fingerbreadths long in human beings.

Dura mater [dū-ră mā′ter] (L. = hard mother). The dense outer meninx surrounding the mammalian central nervous system.

Ear [ēr] (Anglo-Saxon *eare* = ear). The organ of hearing.

Ectoderm [ek′tō-derm] (Gr. *ektos* = outside + *derma* = skin). The outermost of the three embryonic germ layers; forms the epidermis, nervous system, and neural crest.

Ectothermy [ek′to-thŭrm-ē] (Gr. *thermos* = heat). A condition in which an animal derives its body heat primarily from the external environment, so its body temperature is about the same as the ambient temperature.

Effector [ē-fek′tōr] (L. = producer). Any organ or cell that responds in some way to a stimulus.

Efferent [ef′er-ent] (L. *ex* = out + *fero*, pres.p. *ferens* = to carry). Pertaining to structures that carry something away from a point of reference, such as efferent neurons leading from the central nervous system.

Efferent ductules. Minute, sperm-transporting ducts; the cords of the urogenital union in anamniotes and mesonephric tubules in the head of the epididymis in amniotes.

Egest [ē-jest′] (L. *egestus* = taken out). The elimination of material from the caudal end of the digestive tract; also called defecation.

Elasmobranchs [ē-las′mō-brankz] (Gr. *elasmos* = thin plate + *branchia* = gills). The subclass of cartilaginous fishes that includes the sharks and skates.

Electric organ. An organ composed of modified muscle or glandular tissue that produces electric currents. Electric organs are used for electrolocation or defense, found primarily in certain fishes.

Electroplaque [ē-lek′trō-plak] (Gr. *electron* = amber, from

which electricity can be produced by friction + French *plaque* = plate). The plates of modified muscular tissue that form the electric organs of some fishes.

Embryo [em′brē-ō] (Gr. *embryon* = ingrowing). An early stage in the development of an organism that is dependent for energy and nutrients on materials stored within itself or obtained from a mother; embryos are not free living.

Enamel [ē-nam′ĕl] (Middle English *amel* = enamel). The very hard material on the surface of teeth and some bony scales; consists almost entirely of crystals of hydroxyapatite.

Endocrine glands [en′dō-krin] (Gr. *endon* = within + *krino* = to separate). Ductless glands that discharge their secretions (hormones) into the blood.

Endoderm [en′dō-derm] (Gr. *derma* = skin). The innermost of the three germ layers; forms the lining of most of the digestive and respiratory tracts and glandular cells derived from these structures.

Endolymph [en′dō-limf] (L. *lympha* = liquid). The liquid within the membranous labyrinth.

Endometrium [en′dō-mē-trē-um] (Gr. *metra* = womb). The mucous membrane lining the uterus.

Endoskeleton [en-dō-skel′ĕ-tŏn]. The part of the skeleton that lies deep within the body wall, appendages, and pharynx; composed of cartilage or cartilage replacement bone.

Endostyle [en-dō′stīl] (Gr. *stylos* = pillar). An elongated, ciliated groove in the pharynx floor of protochordates.

Endothelium [en-dō-thē′lē-ŭm]

(Gr. *thele* = delicate skin). Delicate epithelium lining blood vessels and the heart.

Endothermy [en′dō-ther′mē] (Gr. *therme* = heat). A condition in which an animal derives its body heat from internal metabolic processes, so it maintains a high and relatively constant body temperature despite variations in ambient temperature; also known as homorothermic.

Enterocoele [en′ter-o-sēl] (Gr. *enteron* = gut +*koilos* = hollow). A coelom that develops primitively as buds from the gut cavity.

Epaxial [ep-ak′sē-ăl] (Gr. *epi* = upon + *axon*= axle, axis). Pertaining to structures that lie above or beside the vertebral axis.

Ependymal epithelium [ep-en′di-măl] (Gr. *ependyma* = garment). The epithelial layer that lines the central nervous system.

Epiboly [ē-pib′o-lē] (Gr. *epibole* = act of throwing on). The spreading of animal hemisphere cells over vegetal hemisphere cells during the gastrulation of some vertebrates.

Epidermis [ep-i-derm′is] (Gr. *epi* = upon + *derma* = skin). The epithelial layer that forms the surface of the skin.

Epididymis [ep-i-did′i-mis] (Gr. *didymoi* = testes). A band of tissue on the amniote testis that is homologous to the cranial part of the opisthonephros and part of the archinephric duct of anamniotes.

Epiglottis [ep-i-glot′is] (Gr. *glottis* = entrance to the windpipe). The flap of fibrocartilage that deflects food around the entrance of the mammalian larynx.

Epimere.   See Somite.

Epinephrine.   See Adrenaline.

Epiphysis [e-pif′i-sis] (Gr. *physis* = growth). The end of a mammalian long bone; a threadlike outgrowth from the roof of the diencephalon of cartilaginous fishes.

Epithalamus [ep′i-thal′ă-mŭs] (Gr. *thalamos* = chamber, bedroom). The roof of the diencephalon lying above the thalamus; part of it is an olfactory center.

Epithelium [ep-i-thē-′lē-um] (Gr. *thele* = delicate skin). The delicate cellular tissue that covers surfaces and lines cavities.

Epoophoron [ep′ō-of′ŏ-ron] (Gr. *oon* = egg + *phero* = to bear). A vestigial organ near the ovary of amniotes that is homologous to the male epididymis.

Erectile tissue.   A tissue containing cavernous vascular spaces that swell when they become filled with blood.

Esophagus [ē-sof′ă-gŭs] (Gr. *oisophagos* = gullet). The part of the digestive tract between the pharynx and stomach, or between the pharynx and intestine if a stomach is absent.

Estivation [es-ti-vā′-shŭn] (L. *aestivus* = summer). A period of torpor during hot, dry weather.

Estradiol [es-tră-di′ol]. The primary hormone produced by the ovarian follicle; promotes the development of female secondary sex characters, the development of the uterine lining during an ovarian cycle, and its feedback to the hypothalamus promotes the LH surge needed for ovulation in many mammals.

Estrus [es′trŭs] (Gr. *oistros* = gadfly, frenzy). A period in some female mammals of increased sexual excitement about the time of ovulation during which copulation may occur.

Euryapsid [yū-rē-ap′sid] (Gr. *eurys* = wide + *apsis* = arch). Pertaining to a reptilian skull in which there is a single temporal fenestra high on the skull and a wide arch of bone beneath it; a subclass of reptiles with such a skull, includes the ichthyosaurs and plesiosaurs.

Eustachian tube   (*Bartolomeo Eustachio,* a 16th-century Italian anatomist). See Auditory tube.

Eutherians [yū-thē′-rē-ănz] (Gr. *eu* = true, good + *therion* = wild beast). The group of therian mammals with a relatively long gestation period; the placental mammals.

Evagination [ē-vag-i-nā′-shŭn] (L. *e* = out + *vagina* = sheath). An outgrowth from another structure, or the process that gives rise to the outgrowth.

Evolutionary homology. Fundamentally similar parts in different organisms that have evolved from a common precursor in an ancestral species; they may or may not resemble each other superficially or functionally.

Excretion [eks-krē′shŭn] (L. *ex* = out + *cretus*= separated). The elimination of nitrogenous wastes.

Exocrine glands [ek′sō-krin] (Gr. *ex* = out + *krino*= to separate). Glands whose secretions are discharged through a duct onto some surface or into a cavity.

Extension [eks-ten′shŭn] (L. *tendere* = to stretch). A movement that carries a distal limb segment away from the next proximal segment, retracts a limb at the shoulder or hip, or moves the head or a part of the trunk toward the middorsal line.

External acoustic meatus [ă-kūs′tik mē-ā′tŭs]. The external ear canal of amniotes extending from the body surface to the tympanic membrane.

External nostrils.   See Nares.

Extrinsic   [eks-trin'sik] (L. *extrinsicus* = from without). Acting from outside the organ in question; applied to muscles that are not within or a part of the organ to which they attach.

Extrinsic ocular muscles   [ok'yū-lăr]. The group of small muscles that extend from the wall of the orbit to the eyeball and control the movements of the eyeball.

Facial   [fā'shăl] (L. *facies* = face). Pertaining to the face; applied to muscles, the seventh cranial nerve, and other structures.

Facial nerve.   The seventh cranial nerve; innervates facial and other muscles associated with the second visceral arch, some salivary glands, and taste receptors on the front of the tongue.

Fallopian tube.   (*Gabriele Fallopio*, 16th-century Italian anatomist). See Uterine tube.

Falx cerebri   [falks ser'ĕ-brē] (L. *falx* = sickle + *cerebrum* = brain). The sickle-shaped fold of dura mater that projects between the cerebral hemispheres.

Fascia   [fash'ē-ă] (L. = band, bandage). Sheets of connective tissue that lie beneath the skin (superficial fascia) or ensheathe groups of muscles (deep fascia or perimysium).

Fasciculus   [fă-sik'yū-lŭs] (L. = small bundle). A small bundle of muscle or nerve fibers.

Feathers   [feth'arz] (Old English *fether* = feather). Skin derivatives, characteristic of birds, that consist primarily of keratinized epidermal cells, provide insulation, and form the flying surfaces of the wing and tail.

Femur   [fē'mŭr] (L. = thigh). The thigh or the bone within the thigh.

Fenestra   [fe-nes'tră] (L. = window). A relatively large opening, such as a temporal fenestra in the skull.

Fenestra cochleae   [kok'lē-ē] (L. *cochlea* = snail shell). The opening in the wall of the otic capsule through which pressure waves are released from the cochlea to the tympanic cavity; also called the round window.

Fenestra vestibuli   [ves-ti'būl-ē] (L. *vestibulum* = antechamber). The opening in the wall of the otic capsule through which vibrations of the auditory ossicles establish pressure waves in the cochlea; also called the oval window.

Fibroblast   [fi'brō-blast] (L. *fibra* = fiber + Gr. *blastos* = bud). An irregularly shaped connective tissue cell that produces the extracellular matrix, including collagen fibers.

Fibrous tunic   [tū'nik] (L. *tunica* = coat). The dense connective tissue forming the outer layer of the eyeball; divided into the transparent cornea and opaque sclera.

Fibula   [fib'yū-la] (L. = buckle). The slender bone on the lateral side of the shin of tetrapods.

Filtration   [fil-trā'shŭn]. The nonselective passage of molecules in the blood, other than plasma proteins, from the glomerulus into the renal tubule.

Fissure   [fish'ŭr] (L. *fissura* = cleft). A deep groove or cleft in certain organs, such as the brain and skull.

Flexion   [flek'shŭn] (L. *flexus* = bending). A movement that brings a distal limb segment toward the next proximal segment, advances a limb at the

shoulder or hip, or bends the head or a part of the trunk toward the midventral line.

Follicle-stimulating hormone.   A hormone of the adenohypophysis that promotes the development of the ovarian follicles.

Foramen,   pl. **foramina** [fō-rā'men, -ram'i-nă] (L. = opening). A perforation of an organ, usually a small opening.

Foramen magnum   [mag'nŭm] (L. *magnus* = large). The large opening through which the spinal cord enters the skull.

Foramen of Monro   (*Alexander Monro, Secundus*, 1759–1808, Scottish anatomist). See Interventricular foramen.

Foramen of Panizza.   An opening between the bases of the left and right systemic arches in crocodilians; shunts blood.

Foramen ovale   [ō-vāl'ē]. A valved opening in the interatrial septum of fetal mammals that allows some blood to pass from the right to the left atrium, thereby bypassing the lungs; becomes the adult fossa ovalis.

Force.   The product of mass and acceleration.

Forebrain.   See Prosencephalon.

Fornix   [fōr'niks] (L. = vault, arch). An arch-shaped neuronal tract deep in the cerebrum that carries impulses from the hippocampus to the hypothalamus.

Fossa   [fos'ă] (L. = ditch). A groove or depression in an organ.

Fossa ovalis   [ō-vah'lis]. A depression in the interatrial septum that represents the fetal foramen ovale.

Fossorial   [fo-sōr'ē-ăl] (L. *fossorius* = adapted for digging). Descriptive of an animal adapted for digging, such as a mole.

Fovea   [fō've̅-ă] (L. = a pit). A small depression, such as the fovea in the retina that contains a concentration of cones.

Friction. The resistance to motion of an object resulting from its contact with the surface on which it is moving or the medium through which it is moving.

Frontal [frŭn'tăl] (L. *frons*, gen. *frontis* = forehead). Pertaining to the forehead, such as the frontal bone.

Fulcrum [ful'krŭm] (L. = bedpost). The point of rotation or pivot in a lever system.

Funiculus [fyū-nik'yū-lŭs] (L. = slender cord). A bundle or column of white matter in the spinal cord.

Furcula [fer'kyū-lă] (L. = small fork). The united clavicles or wishbone of a bird.

Fusiform [fyū'zi-fōrm] (L. *fusus* = spindle + *forma* = shape). A spindle-shaped or streamlined object.

Gait. The repetitive sequence for moving and placing the feet on the ground during locomotion of tetrapods.

Gallbladder [gawl'blad-er] (Old English *galla* = bile). A small sac attached to the liver in which bile accumulates before its discharge into the intestine.

Gametes [gam'ētz] (Gr. *gamet-* = spouse). The haploid germ cells: the sperm and eggs.

Ganglion [gang'glē-on] (Gr. = little tumor, swelling). A group of neuron cell bodies that lie peripheral to the central nervous system in vertebrates.

Ganoid scale [gan'oid] (Gr. *ganos* = sheen). A bony scale with a thick layer of surface ganoine, characteristic of the scales of early actinopterygians.

Ganoine [gan'ō-in]. Enamel or enamel-like material deposited in layers on the surface of some bony scales.

Gastralia [gas-tral'jē-ă] (Gr. *gaster* = stomach). Riblike structures in the ventral abdominal wall of some reptiles.

Gastric [gas'trik] (Gr. *gaster* = stomach). Pertaining to or resembling the stomach.

Gastrulation [gas-trū-lā'shŭn] (Gr. *gastrula* = little stomach). The process by which a single-layered blastula is converted into a two-layered gastrula with an archenteron; mesoderm formation often accompanies gastrulation.

Gear ratio. An expression of the relationship between force and velocity; determined by dividing the length of the out-lever by the length of the in-lever.

Genus [jē'nŭs] (L. *genus* = race). The taxon that comprises very closely related species, and the first term in the binomial name for a species.

Germ layers (L. *germen* = bud). The three tissue layers (ectoderm, mesoderm, endoderm) in an early embryo from which all organs will arise.

Gills. The respiratory organs of aquatic vertebrates, consisting of platelike or filamentous outgrowths from a surface across which water flows.

Girdles. The skeletal elements in the body wall that support the pectoral and pelvic appendages.

Gizzard [giz'ărd] (Old French *gezier* = gizzard). A muscular compartment of the stomach that usually contains swallowed stones with which food is ground up.

Gland [gland] (L. *glans* = acorn). A group of secretory cells.

Glans clitoridis [glanz kli-tō'ri-dis] (L. = acorn + Gr. *kleitoris* = hill). The small mass of erectile tissue at the distal end of the clitoris of a female mammal.

Glans penis [pē'nis] (L. *penis* = tail, penis). The bulbous distal end of the penis of a mammal.

Glenoid fossa [glen'oyd] (Gr. *glene* = socket + *eidos* = form). The socket in the pectoral girdle of tetrapods that receives the head of the humerus.

Glia. See Neuroglia.

Glide. A controlled descent at a low angle to the horizontal.

Glomerulus [glō-mār'yū-lŭs] (L. *glomus* = ball). A ball-like network of capillaries that is surrounded by Bowman's capsule at the proximal end of a renal tubule.

Glossal [glos'ăl] (Gr. *glossa* = tongue). Pertaining to the tongue; also used to describe certain muscles, such as the genioglossus.

Glossopharyngeal nerve [glos'ō-fă-rin-jē-ăl] (Gr. *pharynx* = throat). The ninth cranial nerve, which innervates muscles of the third visceral arch and returns sensory fibers from the part of the pharynx near the base of the tongue.

Glottis [glot'is] (Gr. *glottis* = opening of the windpipe). The opening near the base of the tongue that leads from the pharynx to the larynx.

Glucagon [glū'kă-gōn] (Gr. *glykys* = sweet). A hormone produced by the islet cells of the pancreas that promotes the breakdown of glycogen and the release of sugar from the liver; increases blood sugar level.

Gnathostomes [nă-thos'tōmz] (Gr. *gnathos* = jaw + *stoma* = mouth). A collective term for all the jawed vertebrates as opposed to the jawless agnathans.

Gonads [gō'nadz] (Gr. *gone* = seed). The gamete-producing reproductive organs, the ovaries and testes.

Graafian follicle [grä′fē-ăn] (*Rijnier de Graaf*, Dutch anatomist, 1641–1673). A mature ovarian follicle.

Gray matter. Tissue in the central nervous system consisting of neuron cell bodies and unmyelinated nerve fibers.

Gubernaculum [gū′ber-nak′yū-lŭm] (L. = small rudder). A cord of tissue that extends between the embryonic testis of therian mammals and the developing scrotum and guides the descent of the testis.

Gustatory [gŭs′tă-tōr-ē] (L. *gusto,* pp. *gustatus* = to taste). Pertaining to the sense of taste.

Gymnophiona [jim′nō-fi′on-ă] (Gr. *gymnos* = naked + *ophidion* = snake). The order of tropical amphibians that includes the wormlike, burrowing caecilians.

Gyrus [jī′rŭs] (Gr. *gyros* = circle). One of the folds on the surface of the cerebrum.

Habenula [ha-ben′yū-lă] (L. = small strip). A small, epithalamic nucleus with olfactory connection.

Hair. A filamentous skin derivative of mammals that consists primarily of keratinized epidermal cells, helps provide insulation.

Hair cells. The receptive cells of the ear and lateral line system, so called because they bear superficial cytoplasmic processes, most of which are modified microvilli.

Halecomorphi [hal′e-kō-mor-fī] (L. *halec* = herring + Gr. *morphe* = form). The division of the neopterygians that includes *Amia.*

Hallux [hal′ŭks] (Gr. = big toe). The first or most medial digit of the foot.

Hard palate. A shelf of bone in mammals that separates the oral cavity from the nasal cavities; together with the soft palate it forms the secondary palate.

Harderian gland (*Johann Harder,* a 17th-century Swiss anatomist). A tear gland present in certain mammals and located rostral to and beneath the eyeball; also called the gland of the nictitating membrane.

Haversian system. See Osteon.

Head kidney. A group of pronephric renal tubules that persists in the adults of hagfishes and some teleosts.

Heart. A hollow muscular organ that pumps blood through the body.

Heliothermy [hē′lē-ō-thūrm-ē] (Gr. *helios* = sun + *thermos* = heat). The maintenance of a high body temperature by regulation of the body's exposure to the sun; characteristic of many reptiles.

Hemal arch [hē′măl] (Gr. *haima* = blood). A skeletal arch on the ventral surface of a caudal vertebra that forms a canal around the caudal artery and vein.

Hemibranch [hem′ē-brangk] (Gr. *hemi* = half + *branchia* = gills). A gill of fishes with gill filaments or lamellae present on only one surface of the interbranchial septum, often the first gill.

Hemichordate [hem′ē-kōr-dāt] (Gr. *chorde* = string). A member of the phylum that contains the acorn worms, shows some affinity to the chordates.

Hemopoietic tissue [hē′mō-poy-et′ik] (Gr. *haima* = blood + *poietikos* = producing). A tissue in which blood cells are formed.

Hepatic [he-pat′ik] (Gr. *hepar,* gen. *hepatikos* = liver).

Pertaining to blood vessels, ducts, or other structures associated with the liver.

Hepatic portal system. A system of veins that drain the abdominal digestive organs and lead to sinusoids within the liver.

Hepatic vein. One of the veins that receives blood from the hepatic sinusoids and leads to the heart or caudal vena cava.

Herbivores [hēr′ba-vōrz] (L. *herba* = herb +-*vorous* = devouring). Animals specialized to feed on plant material.

Hermaphrodite [her-maf′rō-dīt] (Greek mythology: the son of Hermes and Aphrodite who became united in one body with a nymph). An animal with both male and female reproductive organs.

Heterocercal tail [het-ăr-ō-ser′kl] (Gr. *heteros* = other + *kerkos* = tail). A caudal fin of fishes in which the vertebral axis turns upward into the enlarged dorsal lobe.

Heterochrony [het′ă-rok-ră-nē] (Gr. *chronos* = time). A genetic shift in the timing of the development of a body part or process relative to the ancestral condition.

Heterodont [het′er-ō-dont] (Gr. *odous, odont-* = tooth). Pertaining to dentition in which the teeth are differentiated and perform different functions, as in mammals.

Hibernation. (L. *hibernus* = wintery). The period of torpor in which some vertebrates pass the winter.

Hindbrain. See Rhombencephalon.

Hippocampus [hip-ō-kam′pŭs] (Gr. = seahorse). The archipallium of mammals, which has shifted medially and protrudes into the lateral ventricle; part of the limbic system.

Holoblastic cleavage [hol-ō-blas'tik] (Gr. *holos* = whole + *blastos* = bud). A pattern of cleavage in which the cleavage furrows pass through the entire egg.

Holocephalians [hol'o-sef'ăl-i-anz] (Gr. *holos* = whole + *kephale* = head). The subclass of cartilaginous fishes that includes the chimaeras.

Holonephros [hol-ō-nef'ros] (Gr. *nephros* = kidney). The hypothetical ancestral vertebrate kidney consisting of segmented renal tubules that develop along the full length of the nephric ridge; also called an archinephros.

Homeostasis [hō'mē-ō-stā'sis] (Gr. *homoios* = alike + *stasis* = standing). The condition in which a constant internal environment is maintained despite factors that tend to destabilize it.

Homoiothermic [hō'mē-ō-ther'mik] (Gr. *homios* = like + *therme* = heat). Pertains to vertebrates in which the body temperature remains relatively constant despite variations in ambient temperature; endothermic.

Homocercal tail [hō'mō-ser'kl] (Gr. *homos* = same + *kerkos* = tail). A caudal fin that is superficially symmetrical but retains a slight uptilt in the skeleton of the vertebral axis; characteristic of teleosts.

Homodont [hō'mō-dont] (Gr. *odous, odont-* = tooth). Pertaining to dentition in which all the teeth are essentially alike, differing only in size.

Homology [hŏ-mol'ō-jē] (Gr. *homologia* = agreement). The fundamental similarity among organs in different organisms based on their evolution from a precursor organ in a common ancestor; homologous organs may or may not resemble each other superficially and functionally.

Homoplasty [hō'mō-plas'tē] (Gr. *plastos* = molded). Morphological resemblance among organs that are not homologous.

Hormones [hōr-mōnz] (Gr. *hormono*, pres.p. *hormon* = to rouse or set in motion). The secretions of the endocrine glands.

Horn (Anglo-Saxon = horn). A bony projection from the skull of many ruminants that is covered by layers of keratinized epidermis and is not shed, usually occurs in both sexes.

Hox genes. Short sections of DNA that occur in clusters called homeoboxes; nearly identical sequences have been found in many invertebrate and vertebrate groups and regulate the expression of genes that determine the features characteristic of each body segment.

Humerus [hyū'mer-ŭs] (L. = upper arm). The bone of the upper arm.

Hyaline cartilage [hī'ă-lin] (Gr. *hyalos* = glass). Cartilage with a clear, translucent matrix.

Hyobranchial apparatus. The group of visceral arches that support the tongue and larynx of tetrapods; includes the hyoid arch and one or more caudal arches.

Hyoid [hī'oyd] (Gr. *hyoeides* = shaped like the letter *ypsilon* = Y). Pertaining to structures associated with the second visceral arch, known as the hyoid arch.

Hyoid apparatus. See Hyobranchial apparatus.

Hyomandibular cartilage [hī-ō-man-dib'yū-lăr] (L. *mandibula* = jaw). The dorsal element of the hyoid arch of fishes that extends from the otic capsule to the caudal end of the upper jaw.

Hyostylic suspension [hī'ō-stī-lic] (Gr. *stylos* = pillar). A type of jaw suspension in fishes in which the upper jaw is attached to the skull by the hyomandibular.

Hypaxial [hī-pak'sē-ăl] (Gr. *hypo* = under + *axon* = axle, axis). Pertaining to structures that lie ventral to the vertebral axis.

Hyperosmotic [hī'per-oz-mot'ik] (Gr. *hyper* = above + *osmos* = action of pushing). A condition in which the concentration of osmotically active solutes in the liquid in question is greater than in the comparison liquid.

Hypobranchial [hī-pō-brang'kē-ăl] (Gr. *hypo* = under + *branchia* = gills). Pertaining to muscles or other structures located ventral to the gills.

Hypocercal tail [hī'pō-ser'kl] (Gr. *kerkos* = tail). A caudal fin in which the vertebral axis turns into an enlarged ventral lobe.

Hypoglossal nerve [hī-pō-glos'ăl] (Gr. *glossa* = tongue). The 12th cranial nerve of amniotes, which innervates muscles in the tongue; homologous to the hypobranchial nerve of anamniotes.

Hypomere. See Lateral plate.

Hypo-osmotic [hī'pō-oz-mot'ik] (Gr. *osmos* = action of pushing). A condition in which the concentration of osmotically active solute in the liquid in question is less than that in the comparison liquid.

Hypophysis [hī-pof'i-sis] (Gr. *physis* = growth). The pituitary gland.

Hypothalamus [hī-pō-thal'ă-mŭs] (Gr. *thalamos* = chamber, bedroom). The ventral part of the diencephalon that lies beneath the thalamus; an important center for visceral integration.

Hypsodont [hip′sō-dont] (Gr. *hypsos* = height +*odont*- = tooth). A high crowned tooth.

Ileum [il′ē-ŭm] (L. = small intestine; from Gr. *eileo* = to roll up, twist). The caudal portion of the small intestine of tetrapods.

Iliac [il′ē-ak]. Pertains to structures near or supplying the ilium, such as the iliac artery.

Ilium [il′ē-ŭm] (L. = groin, flank). The dorsal bone of the tetrapod pelvic girdle that attaches onto the sacrum.

Incisor [in-sī′zŏr] (L. = the cutter; from *incido*= to cut into). One of the front teeth of mammals lying rostral to the canine; used for cutting or cropping food.

Incus [ing′kŭs] (L. = anvil). The anvil-shaped middle auditory ossicle of mammals, homologous to the quadrate bone.

Induction [in-dŭk′shŭn] (L. *inductus* = led in). An embryonic process whereby a tissue causes an adjacent tissue to differentiate in a characteristic way.

Inertia [in-er′shăh] (L. *iners* = sluggish). The tendency of a body at rest to remain at rest, or of one in motion to remain in motion.

Infundibulum [in-fŭn-dib′yū-lŭm] (L. = little funnel). A funnel-shaped structure, such as the coelomic entrance of the oviduct; also a ventral evagination of the hypothalamus that forms the neurohypophysis.

Ingest [in-jest′] (L. *ingestus* = taken in). To take material into the mouth.

Inguinal [ing′gwi-năl] (L. *inguen*, gen. *inguinis* = groin.) A term used to describe structures in or near the groin.

Inguinal canal. A passage through the body wall that leads from the abdominal cavity into the vaginal cavity of the scrotum; the ductus deferens as well as the blood vessels and nerves supplying the testis pass through it.

In-lever. The lever arm through which a force is delivered into a lever system; it is the perpendicular distance from line of action of the in-force to the axis of rotation of the lever system.

Innate behavior [i′nāt] (L. *innatus* = inborn). Those aspects of behavior that are inherited or instinctive and not learned.

Inner ear. That portion of the ear that lies within the otic capsule of the skull and contains the receptive cells for equilibrium and hearing.

Insectivore [in-sek′tă-vōr] (L. *insectum* = insect + -*vorous* = devouring). An insect-eating animal, specifically the order of eutherian mammals that includes the shrews and moles; they retain many ancestral eutherian features.

Insertion [in-ser′shŭn] (L. *insertio* = a planting). That point of attachment of a muscle that moves the most when the muscle shortens; it is the most distal end of limb muscles.

Insulin [in′sŭ-lin] (L. *insula* = island). The hormone produced by the pancreatic islets that decreases blood sugar by promoting the uptake of glucose by cells and its conversion into glycogen in liver and muscle cells.

Integument [in-teg′yū-ment] (L. *integumentum* = covering). The skin.

Integumentary skeleton. Hard structures such as plates of dermal bone, bony scales, and teeth that develop in or just beneath the skin. (See also Dermal bone.)

Intercentrum [in-ter-sen′trŭm] (L. *inter* = between + Gr. *kentron* = center). The vertebral body of rhipidistians, labyrinthodonts, and ancestral reptiles that lies between the pleurocentra; lost in most amniotes.

Interclavicle [in-ter-klav′i-kl] (L. *clavicula* = small key). The ventromedian element of the pectoral girdle of early tetrapods that lies between the clavicles.

Internal capsule. A sheet of white fibers passing through the corpus striatum, which carries most impulses to and from the cerebral cortex.

Internal nostrils. See Choanae.

Interneurons [in′ter-nū′ronz] (L. *inter* = between + Gr. *neuron* = nerve, sinew). Neurons within the central nervous system that lie between the motor and sensory neurons. Their connections are responsible for most of the integrative activity of the central nervous system.

Interstitial cells [in-ter-stish′ăl] (L. *interstitium*= space between). Cells of the testis between the seminiferous tubules that produce testosterone.

Interstitial fluid. A lymphlike fluid that lies in the minute spaces between the cells of the body.

Interventricular foramen [in-ter-ven-trik′yū-lar fō-rā′men] (L. *ventriculus* = belly + *foramen* = hole). The opening between the lateral ventricles and third ventricle of the brain; also called the foramen of Monro.

Intervertebral disk [in-ter-ver'te-brăl] (L. *vertebratus*= jointed). Disks of fibrocartilage that lie between the vertebral bodies of mammals and some other vertebrates.

Intervertebral foramen [in-ter-ver'te-brăl]. An opening between successive vertebral arches through which a spinal nerve passes.

Intestine [in-test'tin] (L. *intestinus* = the intestine). The portion of the digestive tract between the stomach and cloaca or anus; site of most digestion and absorption.

Intrinsic [in-trin'sik] (L. *intrinsicus* = on the inside). A structure that is an inherent part of an organ, such as the ciliary muscles of the eyeball.

Invagination [in-vaj'i-nā-shŭn] (L. in = into + *vagina* = sheath). An ingrowth or the process that gives rise to an ingrowth.

Involution [in-vō-lū-shŭn] (L. *involutus* = rolled up). A process that occurs during gastrulation of some vertebrates by which surface cells roll over the lip of the blastopore and move into the archenteron.

Ipsilateral [ip-si-lat'er-ăl] (L. *ipse* = the same + *latus* = side). Pertaining to structures on the same side of the body.

Iris [ī'ris] (Gr. *iris* = rainbow). The part of the vascular tunic of the eyeball that lies in front of the lens, with the pupil in its center.

Ischium [is'kē-ŭm] (Gr. *ischion* = hip). The ventral and posterior element of the pectoral girdle.

Islets of Langerhans (*Paul Langerhans*, 19th-century German physician). See Pancreatic islets.

Isometric contraction [ī-sō-met'rik] (Gr. *isos* = equal + *metron* = measure). A muscle contraction in which force is developed but the muscle does not shorten.

Iso-osmotic [ī-sō-os-mo'tik] (Gr. *osmos* = action of pushing). A condition in which the concentration of osmotically active solutes in the liquid in question is the same as in the comparison liquid.

Isotonic contraction [ī-sō-ton'ik] (Gr. *tonos* = tension). A muscle contraction in which the tension developed remains the same and the muscle shortens.

Jacobson's organ (*Ludwig L. Jacobson*, 19th-century Danish surgeon and anatomist). See Vomeronasal organ.

Jejunum [jĕ-jū'nŭm] (L. *jejunus* = empty). Approximately the first half of the mammalian postduodenal small intestine; usually found to be empty at autopsies.

Jugular veins [jŭg'yū-lar] (L. *jugulum* = throat). Major veins in the neck of mammals that drain the head.

Keratin [ker'ă-tin] (Gr. *keras* = horn). A horny protein synthesized by the epidermal cells of many vertebrates.

Kidney [kid'nē]. The organ that removes waste products, especially nitrogenous wastes, from the blood and produces urine.

Kinetic skull [ki-net'ik] (Gr. *kinein* = to move). A skull in which the upper jaw and palate can move relative to other parts, found in many fishes, squamates, and birds.

Labia [lā'bē-ă] (L. = lips). Liplike structures.

Labyrinth. (Gr. *labyrinthos* = labyrinth). An intricate system of connecting pathways, such as the membranous labyrinth of the inner ear.

Labyrinthodonts [lab-i-rin'thō-dont] (Gr. *odous, odont-* = tooth). The ancestral subclass of amphibians, now extinct; named for a labyrinthine infolding of the enamel layer on their teeth.

Lacrimal apparatus [lak'ri-māl] (L. *lacrima* = tear). Pertaining to glands and associated structures that produce and transport the tears.

Lactation [lak-tā'shŭn] (L. *lac* = milk). The production of milk.

Lacuna [lă-kū'nă] (L. = pit). A small cavity, such as one in bone that contains an osteocyte.

Lagena [lă-jē'nă] (Gr. *lagenos* = flask). A posteroventral evagination of the sacculus; homologous to the cochlear duct.

Lamella [lă-mel'ă] (L. = small plate). A thin plate or layer of tissue, such as the lamellae in fish gills where gas exchange occurs.

Laminar flow. The smooth, nonturbulent flow of water or air across the surface of the body.

Larva [lar'vă] (L. = mask). A free-living developmental stage that is markedly different from the adult.

Larynx [lar'ingks] (Gr. = larynx). A chamber at the entrance to the trachea; contains the vocal cords in many tetrapods, but not in birds.

Lateral line system. A sensory system of fishes and larval amphibians that detects low-frequency water disturbances; parts are sometimes modified as electroreceptors.

Lateral plate. The most lateral or ventral portion of the mesoderm that contains the coelomic cavity; also called the hypomere.

Lens [lenz] (L. = lentil). The part of the eyeball that focuses light on the retina.

Lepidosaurs [lep′i-dō-sōrz] (Gr. *lepis* = scale + *sauros* = lizard). The subclass of primitive diapsid reptiles, including *Sphenodon*, lizards, snakes, and amphisbaenians.

Lepidotrichia [lep′i-dō-trĭ-kī-ă] (Gr. *trich-* = hair). Bony fin rays of fishes composed of rows of small, modified scales.

Lepospondyls [lep′ō-spon-dĭlz] (Gr. *lepos* = husk + *spondylos* = vertebra). A subclass of early amphibians characterized by a vertebral body that ossified as a spool-shaped structure around the notochord.

Levers. Rodlike mechanical devices that exert a force by turning about a pivot or fulcrum.

Leydig cells. (*Franz von Leydig*, German anatomist, 1821–1908). See Interstitial cells.

Lift. An upward force generated by a stream of water or air flowing across a fin or wing. The line of action of the lift force is perpendicular to the stream.

Ligament [lig′ă-ment] (L. *ligamentum* = band, bandage). Strong connective tissue band that extends between structures, usually skeletal elements; also describes certain mesenteries.

Ligamentum arteriosum [lig′ă-men-tŭm artēr-ē-ō′sŭm]. The connective tissue band extending between the pulmonary artery and aorta; a remnant of the embryonic ductus arteriosus.

Limbic system [lim′bik] (L. *limbus* = border). A brain region that encircles the diencephalon and leads to the hypothalamus; includes the hippocampus, fornix, and cingulate cortex. Important

in behaviors related to survival of the species, such as feeding and sexual activity.

Lingual [ling′gwăl] (L. *lingua* = tongue). Pertaining to the tongue, such as lingual muscles.

Lissamphibians (Gr. *lissos* = smooth + *amphibianz*). The subclass of contemporary amphibians.

Liver [liv′er] (Anglo-Saxon *lifer* = liver). A large gland that develops from the floor of the archenteron just behind the stomach; secretes bile and processes blood brought to it in the hepatic portal system.

Loop of Henle (*Friedrich G. J. Henle*, German anatomist, 1809–1885). That portion of the renal tubule of mammals and some birds and reptiles that loops into the medulla of the kidney, essential in establishing the interstitial salt gradient needed for the production of a concentrated urine.

Lophodont [lof′ō-dont] (Gr. *lophos* = crest + *odont-* = tooth). A cheek tooth whose cusps have united to form ridges.

Lumbar [lŭm′bar] (L. *lumbus* = loin). Descriptive of structures in the back between the thorax and pelvis, such as lumbar vertebrae.

Lung [lŭng]. One of a pair of respiratory organs of terrestrial vertebrates that develops as an outgrowth from the floor of the pharynx.

Luteinizing hormone [lū′tē-in-īz-ing] (L. *luteus* = yellow). A hormone produced by the adenohypophysis that promotes maturation of ovarian follicles, ovulation, and the growth of the corpus luteum.

Lymph [limf] (L. *lympha* = clear water). A clear liquid derived from interstitial fluid that flows through the lymphatic vessels.

Lymph heart. Muscular sections

of lymphatic vessels of some amphibians and reptiles whose contractions help propel the lymph.

Lymph node. Nodules of lymphatic tissue along the course of the lymphatic vessels; the contained lymphocytes respond to invading antigens and initiate immune responses.

Macrolecithal [mak′rō-les-i-thăl] (Gr. *makros* = large + *lekithos* = yolk). An egg with a large amount of yolk, found in many fishes, reptiles, and birds.

Macrophage [mak′rō-fāj] (Gr. *phagein* = to eat). Large cells that phagacytose, or ingest, foreign material.

Macula [mak′yū-lă] (L. = spot). Spot or patch, specifically clusters of hair cells in the sacculus and utriculus of the inner ear.

Malleus [mal′ē-ŭs] (L. = hammer). The outermost of the three mammalian auditory ossicles; homologous to the articular bone.

Mammals [măm′ălz] (L. *mamma* = breast). The vertebrate class characterized by mammary glands and hair.

Mammary glands. Cutaneous glands that secrete milk.

Mandibular arch [man-dib′yū-lăr] (L. *mandibula* = lower jaw). The first visceral arch of jawed vertebrates.

Mandibular cartilage. The ventral part of the mandibular arch; forms the lower jaw of cartilaginous fishes.

Mandibular gland. A mammalian salivary gland that is located near the caudal end of the mandible.

Manus [mā′nŭs] (L. = hand). The hand.

Marsupials [mar-sū-′pē-ălz] (L. *marsupium* = pouch). The pouched mammals; see also Metatheria.

Marsupium [mar-sū′pē-ŭm]. The pouch of a marsupial in which the young are carried.

Mass.   The quantity of material an object contains, usually measured by weight.

Matrix [mā′triks] (L. = a female set aside for bearing). The medium in which a substance is embedded, specifically the extracellular material in connective tissues.

Meatus [mē-ā′tŭs] (L. = passage). A passage such as the external auditory meatus, which leads to the tympanic membrane.

Meckel's cartilage.   (*Johann F. Meckel,* 18th-century German anatomist). See Mandibular cartilage.

Mediastinum [me′dē-as-tī′nŭm] (L. *mediastinus* = medial, from *medius* = middle). The area between the two pleural cavities of mammals that contains the pericardial cavity, thymus, and other structures.

Medulla [me-dūl′ă] (L. = core, marrow). The central part of an organ, often as opposed to its periphery or cortex.

Medulla oblongata.   The caudal region of the brain that is continuous with the spinal cord.

Melanophore [mel′ă-nō-fōr] (Gr. *melas* = black + *-phore* = bearing). A cell of neural crest origin in the skin that produces and carries the black pigment melanin.

Melanophore-stimulating hormone.   Hormone produced by the intermediate part of the adenohypophysis, causes the dispersal of melanin granules in some animals.

Melatonin [mel-ă-tōn′in] (Gr. *tonos* = stain). A hormone produced by the pineal gland in

inverse proportion to the amount of light received; may be important in regulating sexual development and biorhythms.

Membrane bone.   See Dermal bone.

Membranous labyrinth [mem′bră-nŭs lab′i-rinth]. The sacs and ducts of the inner ear that are filled with endolymph and contain the receptive cells for balance and hearing.

Meninges [mě-nin′jēz] (Gr. *meninx,* pl., *meninges* = membrane). Connective tissue membranes that surround the central nervous system, namely, the dura mater, arachnoid, and pia mater.

Meniscus [mě-nis′kūs] (Gr. *meniskos* = crescent). A crescent-shaped disk of fibrocartilage found in some joints, including the knee joint.

Meroblastic cleavage [mer′ō-blas-tik] (Gr. *meros* = part + *blastos* = bud). The partial cleavage of macrolecithal eggs.

Mesectoderm [mez-ek′tō-derm] (Gr. *mesos* = middle + *ektos* = outside). Mesoderm-like tissue in the head of vertebrates that arises from neural crest cells.

Mesencephalon [mez-en-sef′ă-lon] (Gr. *mesos* = middle + *enkephalos* = brain). The midbrain, which dorsally forms the optic lobes or corpora quadrigemina.

Mesenchyme [mez′en-kīm] (Gr. *enchein* = to pour in). An embryonic tissue that consists of star-shaped, wandering cells and gives rise to most adult tissues, except for epithelium.

Mesentery [mez′en-ter-ē] (Gr. *enteron* = intestine). Any fold of coelomic epithelium that suspends visceral organs or extends between them, carrying blood vessels and nerves; in a

limited sense, the membrane that suspends the small intestine.

Meso- [mez′ō] (Gr. *mesos* = middle). A term that when combined with the name of a visceral organ denotes a mesentery suspending that organ, such as the mesocolon suspending the colon.

Mesoderm [mez′ō-derm] (Gr. *derma* = skin). The central germ layer of an early embryo; gives rise to most of the connective tissue, muscles, and blood.

Mesolecithal [mez′ō-les-i-thăl] (Gr. *lekithos* = yolk). An egg, such as that of an amphibian, with a moderate amount of yolk.

Mesomere.   See Nephric ridge.

Mesonephric duct.   See Archinephric duct.

Mesonephros [mez′ō-nef′ros] (Gr. *nephros* = kidney). An embryonic kidney that develops in the central part of the nephric ridge; contributes to the adult kidney of anamniotes and the epididymis of male amniotes.

Metacarpal [met′ă-kar′păl] (Gr. *meta* = after + *karpos* = wrist). One of the skeletal elements in the palm of the hand.

Metamerism [me-tam′er-ism] (Gr. *meros* = part). The condition in which the body is divided into similar segments.

Metamorphosis [met-ă-mōr′fō-sis] (Gr. = transformation). The rapid change in form from a larva to an adult.

Metanephros [met-ă-nef′ros] (Gr. *nephros* = kidney). The adult kidney of amniotes, which develops from the caudal part of the mesomere.

Metatarsal [met′ă-tar′săl] (Gr. *tarsos* = sole of the foot). One of the skeletal elements of the sole of the foot.

Metatheria [met′ă-thē′rē-a] (Gr. *therion* = wild beast). The infraclass of therian mammals that includes the marsupials.

Metencephalon [met′en-sef′ă-lon] (Gr. *enkephalos* = brain). The brain region that includes the cerebellum and, in birds and mammals, the pons.

Microglia [mī-krog′lē-ă] (Gr. *micros* = small + *glia* = glue). Small neuroglial cells of mesodermal origin, some of which are phagocytic.

Microlecithal [mī-krō′les-i-thăl]. An egg with a small amount of yolk.

Microvilli [mī-krō-vil′ī] (L. *villus* = shaggy hair). Minute, nonmotile cytoplasmic processes on the surface of many epithelial cells; they greatly increase surface area.

Midbrain. See Mesencephalon.

Middle ear. That portion of the ear of tetrapods that usually contains the tympanic cavity and one or more auditory ossicles that transmits vibrations from the body surface (usually from a tympanic membrane) to the inner ear.

Middle ear cavity. See Tympanic cavity.

Modulus of elasticity. A measure of the elastic properties of a material; equals stress divided by strain.

Molar [mō′lăr] (L. *mola* = millstone). One of the teeth in the most posterior group of mammalian teeth, usually adapted for crushing or grinding.

Moment. The product of a force times the perpendicular distance from the line of action of the force to an axis of rotation; also called a torque.

Monophyletic [mon′ō-fī-let′ik] (Gr. *monos* = single + *phyle* = tribe). An ancestral group and all its descendants.

Monotremes [mon′ō-trēmz] (Gr.

monos = single + *trema* = hole). The contemporary order of prototherians; includes the platypus and spiny anteaters.

Morphogenesis [mōr-fō-jen′ē-sis] (Gr. *morphe* = form + *genesis* = production). The development of form.

Morphology [mōr-fol′ō-jē] (Gr. *morphe* = form + *logos* = discourse). The study of structure.

Motor unit. A motor neurone and the muscle fibers it supplies.

Mucosa [myū-kō′să] (L. *mucosus* = mucous, slimy). The lining of the gut or other visceral organs, consisting of epithelium and associated connective tissue.

Mucus [myū′kŭs]. (L. = slime). A slimy material produced by some epithelial cells that is rich in the glycoprotein mucin.

Muscle [mŭs-ĕl] (L. *musculus* = muscle). A contractile tissue primarily responsible for the movement of an animal or its parts; discrete groups of muscle cells with a common origin and insertion.

Myelencephalon [mī′el-en-sef′ă-lon] (Gr. *myelos* = core, marrow + *enkephalos* = brain). The most caudal region of the brain; consists of the medulla oblongata and leads to the spinal cord.

Myelin sheath [mī′ĕ-lin]. A sheath around most axons, composed of lipid materials.

Myocardium [mī-ō-kar′dē-ŭm] (Gr. *my-* = muscle + *kardia* = heart). The muscular layer of the heart.

Myoepithelial cells [mī′ō-ep-i-thē-lē-al]. Elongated epithelial cells with contractile properties, such as those associated with sweat glands.

Myofilaments [mī′ō-fil′-ă-mentz]. Ultramicroscopic filaments of actin and myosin that form the

contractile mechanism of muscle cells.

Myoglobin [mī-ō-glō′bin] (L. *globus* = globe). A hemoglobin-like molecule in red muscle.

Myomere [mī′ō-mēr] (Gr. *meros* = part). A muscle segment, usually applied to adult segments.

Myometrium [mī′-ō-mē′trē-ŭm] (Gr. *metra*= uterus). The muscular layer of the uterus.

Myoseptum [mī-ō-sep-tŭm]. A connective tissue septum between myomeres.

Myotome [mī′ō-tōm] (Gr. *tome* = cutting). A muscle segment, usually applied to embryonic segments.

Myxiniformes [mix′in-i-fōr-mēz] (Gr. *myxa* = slime + L. *forma* = form). The order of contemporary agnathans that includes the hagfishes.

Nares [nā′res] (L. *naris*, pl., *nares* = nostrils). The paired openings from the outside into the nasal cavities; external nostrils.

Nasal [nā′zăl] (L. *nasus* = nose). Pertaining to the nose, as in nasal bone.

Neocerebellum [nē′ō-ser-ĕ-bel′ŭm] (Gr. *neos*= new + L. *cerebellum* = small brain). The portion of the mammalian cerebellum that has connections with the cerebrum; includes the cerebellar hemispheres and part of the vermis.

Neocortex. See Neopallium.

Neognathous birds [nē′o-nath-ŭs] (Gr. *gnathos* = jaw). The superorder of birds with a nonreptilian type of palate; includes most orders of birds.

Neonatal [nē-ō-nā′tăl] (L. *natus* = born). Newborn.

Neopallium [nē-ō-pal′ē-ŭm] (Gr. *pallium* = mantle). The newly expanded portion of the mammalian cerebral cortex, encompassing all the cortex except for the olfactory paleopallium and archipallium.

Neopterygians [nē-ōp-tē-rij'e-anz] (Gr. *neos* = new + *pteryg-* = fin or wing). The infraclass of actinopterygian fishes that includes *Amia* and the teleosts.

Neornithes [nē-or'ne-thez] (Gr. *neos* = new + *ornis* = bird). The subclass of birds that has lost many of the primitive features of the Archaeornithes (including the long tail); essentially modern birds.

Neoteny [nē-ot'ĕ-nē] (Gr. *teinein* = to extend). Paedomorphosis that results from the slowing down of somatic development relative to reproductive development; it occurs in many salamanders.

Nephric ridge [nef'rik] (Gr. *nephros* = kidney). The region of the mesoderm between the somite and lateral plate that gives rise to the kidneys and gonads; also called nephrogenic ridge and mesomere.

Nephron [nef'ron]. A renal tubule, the structural and functional unit of the kidneys.

Nerve [nerv] (L. *nervus* = nerve). A cordlike group of axons and associated connective tissue that lies outside the brain and spinal cord; nerves connect the central nervous system with peripheral organs.

Neural arch. See Vertebral arch.

Neural crest [nūr'ăl] (Gr. *neuron* = nerve, sinew). A pair or ridges of ectodermal cells that develop along the top of the neural tube as the neural folds close; this derived character of vertebrates gives rise to many of their distinctive features, including the visceral skeleton, pigment cells, sensory and postganglionic neurons, the dentine-producing cells of teeth, and certain bony scales.

Neural tube. The tube formed in the embryo by the joining of the pair of neural folds; the precursor of the brain and spinal cord.

Neurilemma [nūr-i-lem'ă] (Gr. *lemma* = husk). The thin sheath formed by cells of neural crest origin that surrounds an unmyelinated axon, or, after having myelinated an axon, lies on the surface of the myelin sheath.

Neuroectoderm [nūr-o-ek'tō-derm] (Gr. *ektos* = outside + *derma* = skin). That portion of the ectoderm that give rise to the neural tube and neural crest.

Neuroglia [nū-rog'lē-ă] (Gr. *glia* = glue). Cells in the central nervous system that help support, protect, and maintain the neurons; they include astrocytes, oligodendrocytes, and microglia.

Neurohemal organ [nūr-ō-hē'măl] (Gr. *haima* = blood). An organ, such as the neurohypophysis, formed by the termination of a group of neurosecretory neurons and the blood vessels into which they discharge their products.

Neurohypophysis [nūr'ō-hī-pof'i-sis] (Gr. *hypo* = under + *physis* = growth). The posterior part of the pituitary gland that develops from the infundibulum of the brain; its hormones promote the reabsorption of water and smooth muscle contraction.

Neuromast [nūr'ō-mast] (Gr. *mastos* = knoll, breast). An aggregation of sensory hair cells and supporting cells in the lateral line system that is overlain by a gelatinous cupula.

Neuron [nūr'on]. A nerve cell, the structural and functional unit of the nervous system.

Neurosecretory cells [nūr'ō-sē'krē-tōr-ē]. Neurons that secrete hormones.

Neurotransmitters [nūr'ō-trans-mit'erz]. Substances released by neurons at synapses and neuro-effector junctions that activate or inhibit the target cells.

Nictitating membrane [nik-ti-tā'ting] (L. *nicto*, pp. *nictatus* = to wink). A third eyelid of many amniotes that helps to protect and cleanse the surface of the eyeball.

Nidamental gland [nīd-ă-men'tal] (L. *nidamentum* = nesting material). An aggregation of glands in the oviduct that secrete coverings for the eggs.

Nipple [nip'l] (Old English *neb* = small nose). A papilla that bears the openings of the ducts from the mammary glands.

Noradrenaline [nor-ă-dren'ă-lin] (L. *nor* = short for normal + *ad* = toward + *ren* = kidney). The hormone produced by postganglionic sympathetic fibers and by chromaffin cells of the adrenal medulla.

Norepinephrine. See Noradrenaline.

Notochord [nō'tō-kōrd] (L. *notos* = back + *chorda* = string, cord). A rod of vacuolated cells encased by a firm sheath that lies ventral to the neural tube in vertebrate embryos and some adults.

Nucleus [nū'klē-ŭs] (L. = kernel). An organelle within a cell that contains the genetic material; a group of neuron cell bodies within the brain.

Obturator foramen [ob'tū-rā-tŏr] (L. *obturo*, pp. *obturatus* = to stop up). A foramen in the pubis of reptiles, or an opening between the pubis and ischium in mammals; the obturator muscles arise from the periphery of the obturator foramen and close it.

Occipital nerves [ok-sip′i-tăl] (L. *occiput* = back of the head). Nerves that emerge from the occipital region of the skull, or just behind it, in fishes and some amphibians; they become the hypoglossal nerve of amniotes.

Occlusion [ŏ-klū-shŭn] (L. *occludo,* pp. *occlusus* = to shut up). The closing of a passage; the coming together of the surfaces of the teeth of upper and lower jaws.

Octavolateralis system [ok-tā′vō-lat-er-ā′lis] (L. *octavus* = the eighth + *latus* = side, flank). The combined vestibuloauditory and lateral line systems of fishes and amphibians; fibers from the ear return in the eighth nerve and those from the lateral line system return in the adjacent seventh, ninth, and tenth nerves.

Oculomotor nerve [ok′yū-lō-mō′tŏr] (L. *oculus* = eye + *motorius* = moving). The third cranial nerve, which innervates most of the extrinsic muscles of the eyeball and carries autonomic fibers into the eyeball.

Odontoblasts [ō-dont′tō-blastz] (Gr. *odont-* = tooth + *blastos* = bud). Cells of neural crest origin that produce the dentine of teeth or certain bony scales.

Olecranon [ō-lek′ră-non] (Gr. *olene* = elbow + *kranion* = head). A process on the proximal end of the ulna to which the triceps muscle attaches.

Olfactory [ol-fak′tŏ-rē] (L. *olfacio,* pp. *olfactus* = to smell). Pertaining to the nose.

Olfactory bulb. A rostral enlargement of the brain in which the olfactory nerve terminates.

Olfactory nerve. The first cranial nerve, consisting of neurons

returning from the nose to the olfactory bulb.

Oligodendrocytes [ol′i-gō-den′drō-sītz] (Gr. *oligos* = few + *dendron* = tree + *kytos* = hollow vessel or cell). Neuroglial cells of ectodermal origin that myelinate axons in the central nervous system.

Omentum [ō-men′tŭm] (L. = fatty membrane). The peritoneal fold, often containing a great deal of fat, which extends between the body wall and stomach (greater omentum), or between the stomach and liver and duodenum (lesser omentum).

Omnivores [om-niv′vōrz] (L. *omnis* = all + *-vorous* = devouring). An animal that eats a wide variety of food, both plant and animal.

Ontogeny [on-toj′ĕ-nē] (Gr. *on* = being + *genesis* = birth or descent). The development of an individual.

Oogenesis [ō-ō-jen′ĕ-sis] (Gr. *oon* = egg). The development and maturation of an egg.

Operculum [ō-per′kyū-lŭm] (L. = covering). The gill covering of fishes and some amphibian larvae; also an auditory ossicle in contemporary amphibians.

Ophthalmic nerve [of-thal′mik] (Gr. *ophthalmos* = eye). One of the main branches of the trigeminal nerve.

Opisthocoelous [ō-pis′thō-sē′las] (Gr. *opisthen* = behind + *kolima* = hollow). A vertebral body that is concave on the caudal surface.

Opisthonephros [ō-pis′thō-nef-ros] (Gr. *nephros* = kidney). The adult kidney of most anamniotes; kidney tubules are concentrated caudally.

Optic [op′tik] (Gr. *optikos* = pertaining to the eyes). Pertaining to the eyes.

Optic chiasma [kī-az′ma] (Gr.

chiasma = cross, from the Greek letter *chi* = X). The complete or partial decussation of the optic nerves on the floor of the diencephalon.

Optic lobes. A pair of enlargements of the roof of the mesencephalon that are important integration centers for sight and other senses in anamniotes.

Optic nerve. The second cranial nerve, which carries impulses from the retina.

Oral cavity [or′ăl] (L. *os,* gen. *oris* = mouth). The mouth cavity, also called the buccal cavity.

Orbit [or′bit] (L. *orbis* = circle, eye). The cavity in the skull for the eyeball.

Organ of Corti [(*Marquis Alfonso Corti,* Italian anatomist, 1822–1888). The sound receptive organ in the mammalian cochlea.

Origin [or′i-jin] (L. *origio* = beginning). The starting point of a structure; that end of a muscle that attaches to the more fixed part of the skeleton, which is the proximal end in limb muscles.

Osmosis [os-mō′sis] (Gr. *osmos* = action of pushing). The movement of water through a semipermeable membrane, through which solute molecules do not pass, from an area of high water concentration to one with a lower concentration.

Osmotic pressure. The pressure that results from the movement of water by osmosis into a solution surrounded by a semipermeable membrane.

Ossicle [os′i-kl] (L. *ossiculum* = small bone). Any small bone, such as one of the auditory ossicles.

Osteichthyes [os-tē-ik′thē-ēz] (Gr. *osteon* = bone + *ichthyes* = fishes). The class of fishes in which all or part of the endoskeleton ossifies; includes most fishes.

Osteoblast   [os′tē-ō-blast] (Gr. *blastos* = bud). A cell that produces the bone matrix.

Osteoclast   [os′tē-ō-klast] (Gr. *klastos* = broken). A cell that removes bone and calcified cartilage during the process of bone remodeling and growth.

Osteocyte   [os′tē-ō-sīt] (Gr. *kytos* = hollow vessel or cell). A mature osteoblast that is surrounded by the matrix it has produced.

Osteoderm   [os′tē-o-derm] (Gr. *derma* = skin). Small bones embedded in the skin of some vertebrates.

Osteon   [os′tē-on]. A cylindrical unit of bone consisting of concentric layers that have developed around a central cavity containing blood vessels; also called a haversian system.

Ostium   [os′tē-ŭm] (L. = entrance, mouth). The entrance to an organ, such as the oviduct.

Ostracoderms   [os-trā′kō-dermz] (Gr. *ostrakon* = shell + *derma* = skin). A term applied to several orders of Paleozoic agnathans that are characterized by the extensive development of bone in the skin.

Otic capsule   [ō′tik] (Gr. *otikos* = pertaining to the ear). The portion of the chondrocranium that houses the inner ear.

Otolith   [ō′tō-lith] (Gr. *oto-* = ear + *lithos* = stone). A calcareous structure found in the sacculus and utriculus of vertebrates; its movement with respect to gravity stimulates underlying hair cells and allows an animal to detect its position and movement.

Out-lever.   The lever arm through which a force is delivered out of a lever system to its point of application; it is the perpendicular distance from the line of action of the out-force to the axis of rotation of the lever system.

Oval window.   See Fenestra vestibuli.

Ovarian follicles   [ō-var′ē-an] (L. *ovarium* = ovary). Groups of epithelial and connective tissue cells in the ovary that invest and nourish maturing eggs.

Ovary   [ō′vă-rē] (L. *ovarium* = ovary). One of a pair of female reproductive organs containing the ovarian follicles and eggs.

Oviduct   [ō′vi-dŭkt] (L. *ovum* = egg + *ducere*, pp. *ductus* = to lead). The tube that carries eggs from the coelomic cavity to the outside.

Oviparous   [ō-vip′ă-rŭs] (L. *pario* = to bear). A pattern of reproduction in which eggs are laid and then develop outside the body of the mother.

Ovoviviparous   [ō′vō-vī-vip′ă-rŭs] (L. *viviparus* = bringing forth alive). A pattern of reproduction in which the eggs are retained within the uterus and the embryos are born as miniature adults. The term is often limited to aplacental viviparity for all or most of the needed nutrients and energy are contained within the egg.

Ovulation   [ov′yū-lā′shŭn]. The rupture of the ovarian follicle and the discharge of the eggs from the ovary into the coelomic cavity.

Ovum   [ō′vum] (L. = egg). The mature egg cell.

Oxytocin   [ok-sē-tō′sin] (Gr. *okytokos* = swift birth). A hormone produced by the neurohypophysis that promotes the contraction of uterine muscles at birth and the release of milk during lactation.

Paedomorphosis   [pē-dō-mōr-fō′sis] (Gr. *paid-* from *pais* = child + *morphe* = shape). The retention of juvenile characters into the adult stage.

Palate   [pal′ăt] (L. *palatum* = palate). The roof of the mouth.

Palatoquadrate cartilage   [pal-ă-tō-kwah′drāt] (L. *quadratus* = square). The dorsal part of the mandibular arch.

Paleocerebellum   [pā′lē-ō-ser′ĕ-bel′ŭm] (Gr. *palaios* = ancient + *cerebellum* = small brain). The part of the cerebellum that receives proprioceptive impulses; the flocculonodular lobes in mammals.

Paleognathous birds   [pā′lē-ō-nă-thos] (Gr. *gnathos* = jaw). Birds that retain a primitive reptile-like palate; the kiwi, emu, ostrich, and similar birds, most of which are flightless.

Paleopallium   [pā′lē-ō-pal′ē-ŭm] (Gr. *pallium* = mantle). The primary olfactory region of the cerebrum; forms the piriform lobes of mammals.

Pampiniform plexus   [pam-pin′i-fōrm] (L. *pampinus* = tendril + *forma* = shape). A convoluted network of veins in mammals that surrounds the spermatic artery.

Pancreas   [pan′krē-as] (Gr. *pan* = all + *kreas* = flesh). A large glandular outgrowth of the duodenum that secretes many digestive enzymes; also contains the pancreatic islets.

Pancreatic islets.   Small clusters of endocrine cells in the pancreas that produce hormones that regulate sugar metabolism; also called the islets of Langerhans.

Papilla   [pă-pil′ă] (L. = nipple). A small conical protuberance.

Paradidymis   [par′ă-did′i-mis] (Gr. *para* = beside + *didymoi* = testes). A small group of vestigial mesonephric tubules in mammals located beside the epididymis and testis.

Paraganglia [par-ă-gang′glē-ă] (Gr. *ganglion* = little tumor). Small groups of chromaffin cells that lie beside the sympathetic ganglia.

Parallel evolution. The independent evolution of homoplastic structures by two or more closely related groups.

Paraphyletic [par-ă-fī-let′ik] (Gr. *phyle* = tribe). Pertains to a group that contains an ancestral species and some but not all of its descendants.

Parapophysis [par-ă-pof′i-sis] (Gr. *apo* = away from + *physis* = growth). A transverse process on a vertebral body to which the head of a rib attaches, or the facet on a vertebral body for such an attachment.

Parasympathetic nervous system [par-ă-sim-pa-thet′ik] (Gr. *syn* = with + *pathos* = feeling). The portion of the autonomic nervous system that, in mammals, leaves the central nervous system through certain cranial and sacral nerves; promotes metabolic processes that produce and store energy.

Parathormone [par-ă-thor′mōn] (Gr. *horma*, pres. p. *hormon* = to rouse or set in motion). The hormone of the parathyroid gland; helps regulate mineral metabolism.

Parathyroid glands [par-ă-thī′royd] (Gr. *thyreos* = oblong-shaped shield + *eidos* = form). Endocrine glands of tetrapods located dorsal to or near the thyroid gland; their hormone regulates calcium and phosphate metabolism.

Paraxial mesoderm. That portion of the mesoderm that lies just lateral to the neural tube,

differentiates into somites in the trunk and caudal part of the head and into somitomeres more rostrally.

Parietal [pă-rī′ĕ-tăl] (L. *paries* = wall). Pertaining to the wall of some structure, such as the parietal bone, e.g., parietal peritoneum.

Parietal eye. A median, photoreceptive eye of some fishes and reptiles; lies between the parietal bones.

Parotid gland [pă-rot′id] (Gr. *para* = beside + *otikos* = pertaining to the ear). A mammalian salivary gland located caudal to the ear.

Parthenogenesis [par′the-nō-jen′ĕ-sis] (Gr. *parthenos* = virgin + *genesis* = descent or birth). Activation and development of an egg without fertilization.

Patella [pa-tel′ă] (L. = small plate). The kneecap.

Pectoral [pek′tŏ-răl] (L. *pectoralis*, pertaining to the breast; from *pectus* = breastbone). Pertaining to the chest, as in pectoral appendage, pectoral muscles.

Pelvic [pel′vik] (L. *pelvis* = basin). Pertaining to basin-shaped structures, such as the human pelvic girdle, or to structures near the pelvic girdle.

Pelycosaurs [pel′i-ko-sōrz] (Gr. *pelyx*, gen. *pelykos*= bowl, axe + *sauros* = lizard). The ancestral order of synapsids, most of which have narrow, deep, axe-shaped skulls.

Penis [pē′nis] (L. = tail, penis). The male copulatory organ.

Pericardial cavity [per-i-kar′dē-ăl] (Gr. *peri* = around + *kardia* = heart). The portion of the coelom that surrounds the heart.

Perichondrium [per-i-kon′drē-ŭm] (Gr. *chondros* = cartilage).

The connective tissue covering of a cartilage.

Perilymph [per′i-limf] (L. *lympha* = a clear liquid). The lymphlike fluid that surrounds the membranous labyrinth of the inner ear.

Periosteum [per-ē-os′tē-ŭm] (Gr. *osteon* = bone). The connective tissue covering of a bone.

Peripheral nervous system. The portion of the nervous system lying peripheral to the brain and spinal cord; the cranial and spinal nerves.

Perissodactyls [pĕ-ris′ō-dak-tilz] (Gr. *perissos* = odd + *daktylos* = finger or toe). The mammalian order that includes the ungulates with an odd number of digits (three or one): the rhinoceros, tapir, horse.

Peritoneal cavity [per′i-tō-nē′al] (Gr. *peritonaion* = to stretch over). The part of the mammalian coelom that surrounds the abdominal viscera.

Peritoneum [per′-i-tō-nē-ŭm]. The connective and epithelial layer that lines the peritoneal cavity, forms mesenteries, and covers the abdominal viscera.

Permanent teeth. The teeth of mammals that replace the milk, or deciduous, teeth.

Pes [pez] (L. = foot). Foot.

Petromyzontiformes [pe′trō-mī-zon-ti-fōr′mēz] (Gr. *petros* = stone + *myzo* = to suck in + L. *forma* = form). The order of agnathans that includes the contemporary lampreys.

Phagocytosis [fag′ō-sī-tō-sis] (Gr. *phagein* = to eat + *kytos* = hollow vessel or cell). The ingestion and breaking down of foreign particles by a cell.

Phalanges [fă-lan′jēz] (Gr. *phalanx*, pl. *phalanges* = battle line of soldiers). Bones of the digits that extend beyond the palm or sole.

Pharynx [far′ingks] (Gr. = throat). The portion of the digestive tract from which the pharyngeal pouches develop in an embryo; lies between the oral cavity and esophagus; the crossing place of digestive and respiratory tracts.

Pheromones [fer′ō-mōnz] (Gr. *pherein* = to bear + *horma*, pres.p. *hormon* = to rouse or set in motion). Chemical secretions that act as signals for another individual of the same species.

Phylogeny [fī-loj′ĕ-nē] (Gr. *phylon* = race + *genesis* = birth or descent). The study of the evolutionary development of a group.

Physoclistous [fī-sō-kli′stŭs] (Gr. *physa* = bladder + *kleien* = to close). Pertaining to fishes in which the swim bladder is not connected to the digestive tract.

Physostomous [fī-sō-stō′mŭs] (Gr. *stoma* = mouth). Pertaining to the fishes in which the swim bladder remains connected to the digestive tract by a pneumatic duct.

Pia mater [pī′ă mā′ter] (L. = tender mother). The delicate vascular membrane that invests the brain and spinal cord; the innermost of the three mammalian meninges.

Pineal eye [pin′ē-ăl] (L. *pineus* = relating to pine; from *pinus* = pine tree). A dorsal outgrowth of the diencephalon that forms a light-sensitive eye in some fishes and amphibians and becomes the pineal gland in mammals.

Pineal gland. An endocrine gland that produces melatonin, especially in the dark. The functions of melatonin are not clearly understood, but melatonin has been implicated in adjusting

physiological processes to diurnal and seasonal cycles.

Pisces [pis′ēz] (Gr. = fishes). A collective term for all fishes.

Pitch. The vertical rotation of a swimming or flying vertebrate about its transverse axis.

Pituitary gland [pi-tū′i-tār-ē]. See Hypophysis.

Placenta [plă-sen′tă] (L. = flat cake). The apposition or union of parts of the uterine lining and fetal extraembryonic membranes through which exchanges between mother and embryo occur.

Placental mammals. See Eutherians.

Placode [plak′ōd] (Gr. *placodes* from *plax* = plate + *eidos* = like). A thickened disk of ectoderm that gives rise to certain sense organs and nerves.

Placoderms [plak′ō-dermz] (Gr. *derma* = skin). A class of Paleozoic jawed fishes characterized by the extensive development of bone in the head and thorax.

Plantigrade [plan′ti-grād] (L. *planta* = sole of the foot + *gradus* = step). Walking with the sole of the foot on the ground.

Plastron [plas′tron] (French = breastplate). The ventral shell of a turtle.

Plesiomorphic character [plē′sē-ō-mōr-fik] (Gr. *plesios* = near + *morphe* = shape). A primitive or ancestral character.

Pleura [plūr′ă] (Gr. = side, rib). The coelomic epithelium in the pleural cavities.

Pleural cavities. The coelomic spaces that enclose the lungs of mammals.

Pleurapophysis [plūr′ă-pof′i-sis] (Gr. *apo* = away + *physis* = growth). A vertebral transverse process that incorporates a rib.

Pleurocentrum [plūr′ō-sen′trŭm]

(L. *centrum* = center). A dorsolateral element of the vertebral body of rhipidistians and labyrinthodonts that becomes the definitive vertebral body of amniotes.

Pleurodont tooth [plūr′ō-dont] (Gr. *odont-* = tooth). A tooth that is loosely attached to the outside edge of the jaw.

Pleuroperitoneal cavity [plūr′ō-per-i-tō-nē′al]. The peritoneal cavity and potential pleural cavities of anamniotes and some reptiles; contains the abdominal viscera and lungs (if present).

Plexus [plek′sŭs] (L. = a braid). A network of nerves or blood vessels.

Pneumatic duct [nū-mat′ik] (Gr. *pneuma* = air). The duct that connects the swim bladder with the pharynx in physostomous fishes.

Poikilothermic [poy′ki-lō-ther′mik] (Gr. *poikilos* = varied + *thermos* = heat). Pertains to vertebrates in which the body temperature varies with the ambient temperature; ectothermic.

Pollex [pol′eks] (Gr. = thumb). The thumb.

Polyphyletic [pol-ē-fī-let′ik] (Gr. *polys* = many + *phyle* = race). Pertaining to a group that has evolved from more than one ancestral group.

Polyphyodont [pol-ē-fī′ō-dont] (Gr. *polyphyes* = manifold + *odont-* = tooth). Pertaining to many successive sets of teeth.

Pons [ponz] (L. = bridge). The ventral part of the metencephalon of birds and mammals; has a conspicuous, superficial band of transverse fibers.

Portal veins [pōr′tăl] (L. *porta* = gate). Veins that drain one capillary bed and lead to another one in a different organ, such as the hepatic portal and hypophyseal portal systems.

Posterior chamber. The cavity within the eyeball located between the iris and the ciliary body.

Postganglionic fiber [pōst′gang-glē-on′ik] (Gr. *ganglion* = small tumor). A neuron of the autonomic nervous system with its cell body in a peripheral ganglion and its axon extending to the effector organ.

Posttrematic [pōst-trē-mat′ik] (Gr. *trema* = hole). Pertaining to blood vessels or nerves that lie caudal to a branchial pouch.

Power. The rate of doing work.

Preadaptation. The evolution of a feature, usually as an adaptation to the existing environment, that enables an animal to exploit a new environment, e.g., the evolution of lungs in certain fishes.

Preganglionic fibers [pre′gang-glē-on′-ik] (Gr. *ganglion* = small tumor). A neuron of the autonomic nervous system with its cell body in the brain or spinal cord and its axon extending to a peripheral ganglion.

Premolars [prē-mō′lǎrz] (L. *molaris* = millstone). Cheek teeth that lie rostral to the molars and may be specialized for cutting or grinding.

Pressure. Force per unit area, e.g., grams per square centimeter.

Pretrematic [prē-trē-mat′ik] (Gr. *trema* = hole). Pertaining to blood vessels or nerves that lie rostral to a branchial pouch.

Primates [prī-ma′tez] (L. *primus* = one of the first). The eutherian order that includes lemurs, monkeys, apes, and human beings.

Primitive character. A character that is ancestral for a group; also called a plesiomorphic character.

Primitive streak. A longitudinal thickening of cells on the blastoderm of large-yolked eggs, through which prospective chordamesoderm and mesoderm cells move inward; homologous to the blastopore.

Primordium [prī-mōr′dē-ŭm] (L. = beginning). The first indication of the formation of a structure in an embryo.

Processus vaginalis [prō-ses′ŭs va-ji′nal′is] (L. = process + *vagina* = sheath). A sac that contains the mammalian testis and its sperm duct and blood vessels, as well as the coelomic vaginal cavity; located in the scrotum, also called vaginal sac.

Procoelous [prō-sē′lŭs] (Gr. *koilios* = hollow). A vertebral body with a concavity on its cranial surface.

Proctodaeum [prok-tō-dē′ŭm] (Gr. *proktos* = anus + *hodaion* = way). An ectodermal invagination near the caudal end of the embryo that contributes to the cloaca.

Progenesis [prō-jen′i-sis] (Gr. *pro* = before + *genesis* = origin). Paedomorphosis that results from the acceleration of reproductive maturity relative to somatic development.

Progesterone [prō-jes′ter-ōn] (L. *gesto*, pp. *gestatus* = to bear). A hormone produced by the corpus luteum and later by the placenta, prepares the uterus for the reception of a fertilized egg and maintains the uterine lining during pregnancy.

Prolactin [prō-lak′tin] (L. *lac*, *lact-* = relating to milk). A hormone produced by the adenohypophysis that promotes maternal behavior and milk production.

Pronephros [prō-nef′ros] (Gr. *pro* = before + *nephros* = kidney). The first formed kidney of a vertebrate embryo, which lies dorsal to the pericardial cavity and forms the archinephric duct before it atrophies.

Proprioceptor [prō′prē-ō-sep′ter] (L. *proprius* one's own + *capio*, pp. *ceptus* = to take). A receptor in muscles, tendons, and joints that monitors the activity of muscles.

Prosencephalon [pros-en-sef′ǎ-lon] (Gr. *pro* = before + *enkephalos* = brain). The embryonic forebrain, which gives rise to the telencephalon and diencephalon.

Prostate [pros′tāt] (Gr. *prostates* = one who stands before). An accessory sex gland of male mammals that surrounds the urethra just before the urinary bladder.

Protandry [prō-tan′drē] (Gr. *protos* = first + *andr-* = man). Sequential hermaphroditism in which the gonad functions first as a testis before it acts as an ovary.

Protochordates [prō-tō-kor′dātz] (L. *chorda* = string, cord). A collective term for the subphyla of nonvertebrate chordates: the tunicates and cephalochordates.

Protogyny [prō-tō-gī′nē] (Gr. *gyne* = woman). Sequential hermaphroditism in which the gonad functions first as an ovary before it acts as a testis.

Protostomes [prō-tō′stōmz] (Gr. *stoma* = mouth). The group of coelomate animals in which the blastopore forms or contributes to the mouth; includes molluscs, annelids, and arthropods.

Prototherians [prō-tō-the′rē-anz] (Gr. *therion* = wild beast). The ancestral subclass of mammals;

includes the contemporary, egg-laying monotremes.

Protraction   [prō-trak'shŭn] (L. *pro* = before + *traho,* pp. *tractus* = to pull). Muscle action that moves the entire appendage of a quadruped forward.

Proventriculus   [prō-ven-trik'yū-lus] (L. *ventriculus* = small belly). The anterior, glandular portion of the stomach of birds.

Pseudobranch   [sū-dō-brank] (Gr. *pseudes* = false + *branchia* = gills). A small first gill of some fishes without a respiratory function.

Pterosaur   [ter'ō-sōr] (Gr. *pteryg-* = fin or wing + *sauros* = lizard). An extinct order of flying reptiles.

Pterygiophores   [tĕ-rij'iō-fōrz] (Gr. *phoros* = bearing). The proximal supporting cartilages or bones of a fin.

Pubis   [pyū'bis] (L. *pubes* = genital hair). The cranioventral bone of the pelvis of tetrapods.

Pulmonary   [pŭl'mō-nār-ē] (L. *pulmo* = lung). Pertaining to the lungs, as the pulmonary artery.

Pupil   [pyū'pĭl] (L. *pupilla* = pupil). The central opening through the iris of the eye.

Pygostyle   [pī-gō'stĭl] (Gr. *pyge* = rump +*stylos* = pillar). The fused, caudal vertebrae of a bird, which support the tail feathers.

Pylorus   [pī-lōr'ŭs] (Gr. *pyloros* = gatekeeper). The caudal end of the stomach, which contains a sphincter muscle.

Pyramidal system   [pi-ram'i-dal] (Gr. *pyramis* = pyramid). The direct motor pathway in mammals from the cerebrum to the motor nuclei and columns.

Radius   [rā'dē-ŭs] (L. = ray). A bone of the antebrachium of tetrapods that rotates around the ulna; located on the thumb side when the hand is supine.

Ramus   [rā'mŭs] (L. = branch). A branch such as those of a spinal nerve.

Rathke's pouch   [(*Martin H. Rathke,* German anatomist, 1793–1860). A dorsal evagination of the stomodaeum that forms the adenohypophysis.

Ray-finned fishes.   See Actinopterygians.

Receptor   [rē-sep'tōr] (L. = receiver). A specialized cell or tissue that responds to a specific stimulus and initiates a nerve impulse.

Rectum   [rek'tŭm] (L. *rectus* = straight). The terminal segment of the intestine that leads to the anus.

Reflex   [rē'flekz] (L. *refecto,* pp. *reflexus* = to bend backward). An innate reaction in response to a peripheral stimulus.

Releasing hormones.   Hormones produced by the hypothalamus that travel in the hypophyseal portal system and promote the release of specific adenohypophyseal hormones. In several cases inhibiting hormones are also known.

Renal   [rēnăl] (L. *ren* = kidney). Pertaining to the kidneys.

Renal tubule.   A kidney tubule or nephron.

Renal portal system.   A system of veins that drains the tail and hind legs of most nonmammalian vertebrates and leads to the peritubular capillaries of the kidneys.

Reptiles   [rep'tĭlz] (L. *reptilis* = creeping). The class of amniotes that includes turtles, lizards, snakes, and crocodiles.

Resultant of force.   A vector that expresses the interaction between two or more vectors.

Rete cords   [rē'tē] (L. *rete* = net). Minute cords in the embryo that interconnect the primary sex

cords and the cranial mesonephric tubules; they contribute to the sperm passages in males and regress in females.

Rete mirabile   [rē'tē mē-ră'bi-le'] (L. = wonderful net). A network of small arteries or capillaries such as that associated with the gas gland of the swim bladder.

Reticular formation   [re-tik'yū-lăr] (L. *reticulum* = small net). A network of short interneurons in the brainstem that forms a primitive integrating system. In mammals it also projects to the cerebrum and helps maintain the level of arousal.

Retina   [ret'i-nă]. The innermost layer of the eyeball; contains pigment cells, photoreceptive cells, and neurons.

Retraction   [rē-trak'shŭn] (L. *retractio* = a drawing back). Muscle action that moves the entire appendage of a quadruped backward.

Retroperitoneal   [re'trō-per'i-tō-nē'ăl] (L. *retro* = backward + Gr. *peritonaion* = to stretch over). Pertaining to structures, such as the kidneys, that lie dorsal to the peritoneal cavity.

Rhinal   [rī'năl] (Gr. *rhin-* = nose). Pertaining to the nose.

Rhipidistians   [ri-pi-dis'tē-anz] (Gr. *rhipis* = fan). The order of freshwater sarcopterygian fishes that gave rise to the amphibians.

Rhombencephalon   [rom-ben-sef'ă-lon] (Gr. *rhombos* = lozenge-shaped + *enkephalos* = brain). The hindbrain, the most posterior of the three primary divisions of the developing brain; subdivides into the metencephalon and myelencephalon.

Roll.   Rotation of a swimming or flying vertebrate around its longitudinal axis.

Round window.   See Fenestra cochleae.

Rudiment   [rū′di-ment] (L. *rudimentum* = first attempt). An early stage in the development of an organ; a primordium.

Rumen   [rū′men] (L. *rumen* = gullet). The first and largest chamber of the ruminant stomach.

Ruminants   [rū-mi-năntz] (L. *rumino* = to chew the cud). Those artiodactyls with chambered stomachs, including deer, sheep, and cattle.

Sacculus   [sak′yū-lŭs] (L. = small sac). The most ventral chamber of the membranous labyrinth.

Sacral vertebrae,
   Sacrum   [sā′krăl, sa′krŭm] (L. *sacrum* = sacred). The vertebrae, or the union of two or more vertebrae and their ribs, by which the pelvis articulates with the vertebral column.

Salientia.   See Anurans.

Salivary gland   [sal′i-vār-ē] (L. *saliva* = saliva). A gland that produces the saliva; the major ones in mammals are the parotid, mandibular, and sublingual glands.

Salt gland.   A gland or secretory cells that secrete excess salt, found near the nose and eye in certain marine reptiles and birds and on the gill of certain marine fishes.

Saltatorial   [sal′tă-tōr′ē-ăl] (L. *saltatio* = to dance). Adapted for leaping.

Sarcopterygians   [sar′kōp-te-rij′ē-ănz] (Gr. *sarkodes* = fleshy + *pteryg-* = fin or wing). The subclass of Osteichthyes with fleshy paired fins: the coelacanths, rhipidistians, and lungfishes.

Scala tympani   [skālă′ tim′pă-nē] (L. *scala* = ladder + *tympanum* = drum). The perilymphatic duct through which pressure waves pass from the cochlea to the tympanic cavity.

Scala vestibuli   [ves-tib′yū-lē] (L. *vestibulum* = antechamber). The perilymphatic duct through which pressure waves enter the cochlea from the auditory ossicle.

Scaling.   Analyzing the relationship between the size of a structure, or level of activity of a process, and body size.

Scapula   [skap′yū-lā] (L. = shoulder blade). The dorsal element of the pectoral girdle that ossifies from cartilage.

Schizocoele   [skiz′ō-sēl] (Gr. *schizo* = to cleave + *koilos* = hollow). A coelom formed by cavitation of the mesoderm rather than by enterocoelic pouches, characteristic of protostomes.

Schwann cells   (*Theodor Schwann*, German histologist, 1810–1882). See Neurilemma.

Sclera   [sklēr′ă] (Gr. *skleros* = hard). The opaque, "white" portion of the fibrous tunic of the eyeball.

Sclerotic bones   [sklĕ-rot′ik] (Gr. *oto-* = ear). A ring of bones that develops in the sclera of some vertebrates and strengthens the eyeball wall.

Sclerotome   [sklēr′ō-tōm] (Gr. *tomos* = a cutting). The medial portion of a somite that forms the vertebrae and much of the chondrocranium.

Scrotum   [skrō′tŭm] (L. = pouch). The sac that encases the mammalian testes; it includes all the layers of the body wall.

Sebaceous glands   [sē-bā′shŭs] (L. *sebum* = tallow). Mammalian cutaneous glands that secrete oily and waxy materials.

Secondary palate.   A palate that separates the food and air passages; in mammals it consists of a bony hard palate that separates the oral and nasal cavities and a fleshy soft palate that separates the oral pharynx from the nasal pharynx.

Secretin   [se-krē′tin] (L. *secerno*, pp. *secretus* = to secrete). A hormone produced by the duodenal mucosa that promotes the secretion of the aqueous portion of the pancreatic juice.

Segmentation   [seg′men-tā-shŭn]. Refers to the division of the body into a longitudinal series of segments.

Selachian   [si-lā′kē-an] (Gr. *selachios* = resembling a shark). A member of one of the orders of sharks.

Selenodont   [sĕ-lē′nō-dont] (Gr. *selene* = crescent + *odont-* = tooth). Mammalian cheek teeth with crescent-shaped cusps.

Semicircular duct   [sem′ē-sir′kyū-lăr]. One of the ducts, shaped like a half circle, of the membranous labyrinth; semicircular ducts are located within a set of semicircular canals in the otic capsule of the skull.

Seminal fluid   [sem-i-năl] (L. *semen* = seed). The fluid secreted by male reproductive ducts and accessory sex glands that carries the sperm.

Seminal vesicle.   See Vesicular gland.

Seminiferous tubules   [sem-i-nif′er-ŭs]. The tubules within the testis that produce the sperm.

Sense organ.   An aggregation of receptive cells and associated cells that support them and may amplify a stimulus.

Serial homologs.   Structures in different parts of the body of the same individual that develop from a precursor that is repeated in different body segments.

Sertoli cells   (*Enrico Sertoli,*

Italian histologist, 1842–1910). Epithelial cells of the seminiferous tubules that play a role in the maturation of the sperm.

Sesamoid bone   [ses'ă-moyd] (Gr. *sesamon* = sesame seed + *eidos* = form). A bone that develops in the tendon of a muscle near its insertion and facilitates the movement of a muscle across a joint or alters its direction of pull; the patella and pisiform are examples.

Sessile   [ses'il] (L. *sessilus* = fit for sitting). Describes an animal that lives attached to its substratum.

Sex cords.   Embryonic cords of epithelium and primordial germ cells that give rise to the seminiferous tubules or ovarian follicles.

Sexual homology.   Parts in different sexes of the same species that develop from the same type of primordium.

Shear.   A stress that results from two parallel but not directly opposite forces that are moving toward each other.

Sinus   [sī'nŭs] (L. = a cavity). A cavity or space within an organ.

Sinus venosus   [vē-nō'sus]. The most caudal chamber of the heart of anamniotes and some reptiles; receives the systemic veins.

Sinusoids   [sī'nŭ-soydz] (Gr. *eidos* = form). Capillary-sized blood spaces in the liver or other organs that are not completely lined by endothelial cells.

Skin.   See Integument.

Skull   [skŭl] (Old English *skulle* = bowl). The group of bones and cartilages that encase the brain and major sense organs and form the jaws; the lower jaw sometimes is not considered to be a part of the skull.

Soft palate.   A fleshy palate in mammals that separates the nasal and oral pharynx; part of the secondary palate.

Somatic   [sō-mat'ik] (Gr. *somatikos* = bodily). Refers to structures that develop in the body wall or appendages as opposed to those in the gut tube, such as the somatic muscles, somatic skeleton.

Somite   [sō'mīt]. One of the series of dorsal segments, or divisions of the paraxial mesoderm, in the trunk and caudal part of the head in a developing embryo; also called an epimere.

Somitomere   [sō'mi-tō-mēr] (Gr. *meros* = part). One of the partial divisions of the paraxial mesoderm in the rostral part of the head of a developing embryo.

Soaring.   A type of flight in which the wings are held stationary and the animal remains aloft by utilizing upward air currents (static soaring) or differential air speeds at different elevations (dynamic soaring).

Specialization.   Adaptations to a particular habitat and mode of life.

Species   [spē'shēz] (L. = particular kind). A group of inbreeding, or potentially inbreeding, individuals that in nature are normally reproductively isolated from other such groups.

Sperm   (Gr. *sperma* = seed). The mature male gametes, also called spermatozoa.

Spermatogenesis   [sper'mă-tō-jen'ě-sis] (Gr. *genesis* = birth, descent). The formation and maturation of the sperm.

Spermatophore   [sper'mă-tō-fōr] (Gr. *phoros* = bearing). A clump of sperm encapsulated in mucoid material; deposited by some male salamanders.

Sphincter   [sfingk'ter] (Gr. *sphinkter* = band, lace). A circular muscle that closes the opening of an organ or surrounds another structure, e.g., the pyloric sphincter, sphincter colli muscle.

Spinal column   [spī'năl] (L. *spina* = thorn, backbone). The vertebral column.

Spinal cord.   The central nervous system caudal to the brain.

Spiracle   [spī'ră-kl] (L. *spiraculum* = air hole). The reduced first gill pouch of some fishes through which water may enter the pharynx; also, the opening from the gill chamber of frog tadpoles.

Spiral valve   (L. *spira* = coil). A helical coil in the intestine of early fishes; also a fold within the conus arteriosus and ventral aorta of lungfishes and some amphibians and reptiles that helps separate pulmonary and systemic bloodstreams.

Splanchnic   [splangk'nik] (Gr. *splanchnon* = gut, viscus). Descriptive of structures that supply the gut, such as the splanchnic nerves.

Splanchnocranium   [splangk-nō-krā'nē-ŭm] (Gr. *kranion* = skull). The portion of the head skeleton composed of the visceral arches.

Spleen   [splēn] (Gr. *splen* = spleen). A vascular organ near the stomach in which blood cells may be produced, stored, and eliminated.

Squamates   [skwā'māts] (L. *squama* = scale). The reptilian order that includes the lizards, amphisbaenians, and snakes.

Stall.   Sudden loss of lift by the wings.

Stapes   [stā'pēz] (L. = stirrup). The single auditory ossicle of nonmammalian tetrapods and the most medial of the three auditory ossicles of mammals; homologous to the hyomandibular of fishes.

Step.   The distance a tetrapod

moves forward by the action of one leg and foot.

Sternum [ster′nŭm] (Gr. *sternon* = chest). The breastbone of tetrapods.

Stomach [stŭm′ŭk] (Gr. *stomachos* = stomach). The part of the digestive tract where food is stored temporarily and where digestion is usually initiated.

Stomodaeum [stō-mō-dē′ŭm] (Gr. *stoma* = mouth + *hodaion* = on the way). An ectodermal invagination at the front of the embryo that forms the oral cavity.

Strain. The deformation in a material that results from stress.

Stratum [strat′ŭm] (L. = layer). A layer of tissue, such as the stratum corneum on the skin surface.

Stress. The force per unit area that is applied to a material.

Stride. The distance a tetrapod moves forward from the placement of one foot on the ground to the next placement of the same foot; equivalent to four steps in a quadruped.

Sulcus [sūl′kŭs] (L. = groove). A groove on the surface of an organ, such as the sulci on the cerebrum of a mammal.

Sulcus limitans [lim′i-tăns]. A groove in the central canal of the nervous system that delineates the dorsal sensory areas of gray matter from the ventral motor ones.

Summation [sūm-ā′shŭn]. The addition of successive events that come in rapid sequence to produce a response, or a response of greater magnitude.

Suprarenal gland. See Adrenal gland.

Suprasegmental control. A level of integration by parts of the brain that is superimposed on the basic pattern of activity of lower centers.

Surfactant [ser-fak′tănt] (L. *superficius* = superficial + *actio*, pp. *actus* = to do). A surface tension depressant found on the lining of the lungs.

Suture [sū′chūr] (L. = seam). An immovable joint in which the bones are separated by a septum of connective tissue, such as those between dermal bones of the skull.

Sweat glands. Mammalian cutaneous glands that secrete a watery solution that sometimes contains odoriferous materials.

Swim bladder. A sac of gas, located dorsally in the body cavity of most actinopterygians, that has a hydrostatic function.

Sympathetic nervous system [sim-pă-thet′ik] (Gr. *syn* = with + *pathos* = feeling). The part of the autonomic nervous system that, in mammals, leaves the central nervous system from parts of the spinal cord; its activity helps an animal adjust to stress by promoting physiological processes that increase the energy available to tissues.

Symphysis [sim′fi-sis] (Gr. *physis* = growth). A joint between bones that permits limited movement by the deformation of the fibrocartilage between them, as the pelvic symphysis; usually occurs in the midline of the body.

Synapse [sin′aps] (Gr. *synapsis* = union). The junction at which an impulse passes from one neuron to another.

Synapsid [si-nap′sid] (Gr. *apsid* = loop or bar). A skull with a single laterally placed temporal fenestra, or a group of vertebrates with such a skull, such as the synapsid reptiles and mammals.

Synapsid reptiles. The subclass of reptiles that led to mammals.

Synarthrosis [sin′ar-thrō-sis] (Gr. *arthron* = joint). A joint with fibrous or cartilaginous material between the adjacent elements; growth can occur here but only limited movement.

Synchondrosis [sin′chon-drō-sis] (Gr. *chondros* = cartilage). A synarthrosis in which cartilage separates two bony elements, found between bones that ossify in the chondrocranium.

Synergy [sin′er-jē] (Gr. *ergon* = work). Pertaining to different muscles or other organs that interact to produce a common effect.

Synovial fluid [si-nō′ve-ăl] (L. *synovia* = joint oil). A clear fluid that serves as a lubricant in movable joints.

Synovial joint. See Diarthrosis.

Synsacrum [sin-sā′krŭm] (L. *sacrum* = sacred). The group of fused vertebrae and their ribs in birds that articulates with the pelvis.

Syrinx [sir′ingks] (Gr. = panpipe). The voice box of birds, located at the distal end of the trachea.

Systemic circulation [sis-tem′ik] (Gr. *systema* = a whole composed of several parts). The circulation through the body as a whole, exclusive of the circulation through the respiratory organs (branchial or pulmonary circulation), or heart (coronary circulation).

Systole [sis′tō-lē] (Gr. = a drawing together). The period during which the ventricle of the heart contracts and expels blood.

Tapetum lucidum [tā-pē′tŭm lū′sid-ŭm] (L. *tapete* = carpet + *lucidus* = clear, shining). A layer within or behind the retina of some vertebrates that reflects light back onto the photoreceptive cells.

Tarsal [tar′săl] (Gr. *tarsos* = sole of the foot). One of the bones in the ankle.

Taxon [tak′son] (Gr. *taxis* = arrangement). Any unit in the classification of organisms.

Tectum [tek′tŭm] (L. = roof). A roof, specifically the roof of the mesencephalon.

Tegmentum [teg-men′tŭm] (L. *tegmen* = covering). The floor of the mesencephalon or metencephalon.

Tela choroidea [tē′la kō-roy′dē-a] (L. *tela* = web + Gr. *chorion* = membrane enclosing the fetus + *eidos* = form). A thin membrane composed of the ependymal epithelium and the vascular meninx that forms the roof or wall of some ventricles.

Telencephalon [tel-en-sef′ă-lon] (Gr. *telos* = end + *enkephalos* = brain). The rostral part of the forebrain from which the olfactory bulbs and cerebral hemispheres develop.

Teleosts [tel′ē-ostz] (Gr. *osteon* = bone). The most advanced group of neopterygian fishes.

Telodendria [tel-ō-den′driă] (Gr. *dendria* = trees). The terminal branches of an axon.

Tendon [ten′dŏn] (L. *tendo* = to stretch). A cord of dense connective tissue that extends between a muscle and its attachment.

Tendon organ. A proprioceptor in tendons that is stimulated by tension developed by muscle contraction.

Tension. The stress that results from two parallel forces pulling directly away from each other.

Tentorium [ten-tō′rē-um] (L. = tent). The septum of dura mater, ossified in some species, that

extends between the cerebrum and cerebellum.

Terminal nerve. A small nerve present beside the olfactory nerve in most vertebrates; its function is uncertain, but it may have a role in detecting pheromones and regulating reproductive functions.

Testis [tes′tis] (L. = witness, originally an adult male, testis). The male reproductive organ, which produces sperm and male sex hormones.

Testosterone [tes-tos′tĕ-rōn]. The male sex hormone produced by the testis; promotes the development of male secondary sex characteristics and the development of sperm.

Tetrapods [tet′ră-podz] (Gr. *tetra* = four + *pous, podos* = foot). A collective term for the terrestrial vertebrates; they have four feet unless some have been secondarily lost or converted to other uses.

Thalamus [thal′ă-mŭs] (Gr. *thalamos* = chamber, bedroom). The lateral walls of the diencephalon; an important center between the cerebrum and other parts of the brain.

Thecodont teeth [thē′kō-dont] (Gr. *theke* = case + *odous, odont-* = tooth). Teeth that are set in sockets.

Therapsid [the-rap′sid] (Gr. *therion* = wild beast + *apsis* = arch). The order of advanced synapsids that had many mammalian features and includes the ancestors of mammals.

Therians [thē′rē-anz]. The subclass of mammals that includes the marsupials and eutherians.

Thoracic duct [thō-ras′ik] (Gr. *thorax* = chest). The large lymphatic duct of mammals that passes through the thorax and

enters the large veins near the heart.

Thorax [thō′raks]. The region of the mammalian body encased by the ribs.

Thymus [thī′mŭs] (Gr. *thymos* = thyme, thymus; so called because it resembles a bunch of thyme). The lymphoid organ that develops from certain pharyngeal pouches, necessary as the site where certain T lymphocytes mature.

Thyroid gland. [thī′royd] (Gr. *thyroides* = resembling an oblong shield). An endocrine gland that develops from the floor of the pharynx, and in humans is located adjacent to the thyroid cartilage of the larynx; its hormones increase the rate of metabolism.

Thyroid-stimulating hormone. A hormone produced by the adenohypophysis that promotes the secretion of the thyroid gland.

Thyroxine [thī-rok′sēn] (Gr. *oxo-* = oxygen). One of the hormones released by the thyroid gland.

Tibia [tib′ē-ă] (L. = the large shinbone). The bone on the medial side of the lower leg, in line with the first digit.

Tissue [tish′ū] (Old French *tissu* = woven). An aggregation of cells that together perform a similar function.

Tongue [tŭng] (Old English *tunge*). A muscular mobile organ on the floor of the oral cavity of tetrapods that often helps gather food and manipulates it within the mouth.

Tonsil [ton′sil] (L. *tonsilla* = tonsil). One of the lymphoid organs that develops in the wall of the pharynx near the level of the second pharyngeal pouch.

Torque. A turning force equal to the product of the force and the perpendicular distance between the line of action of the force and the fulcrum about which it acts; also called a moment.

Trabeculae [tră-bek′yū-lē] (L. = little beams). Small rodlike skeletal structures, such as the trabeculae within bones.

Trachea [trā′kē-ă] (L. *tracheia arteria* = rough artery, windpipe). The respiratory tube between the larynx and the bronchi.

Tract [trakt] (L. *tractus* = a drawing out). A group of axons of similar function traveling together in the central nervous system.

Transverse septum. The partition of coelomic epithelium that separates the pericardial from the pleuroperitoneal cavity.

Trigeminal nerve [trī-jem′i-năl] (L. *trigeminus* = threefold). The fifth cranial nerve, which has three branches in mammals; it innervates the muscles of the mandibular arch and returns sensory fibers from cutaneous receptors over most of the head.

Tri-iodothyronine [trī-ī′ō-dō-thi′rō-nēn] (Gr. *tri* = three + *iodo* = violet-like or iodine + *thyroides* = resembling an oblong shield). A hormone produced by the thyroid gland.

Trochanter [trō-kan′ter] (Gr. = a runner). One of the processes on the proximal end of the femur to which thigh muscles attach.

Trochlear nerve [trok′lē-ăr] (L. *trochlea* = pulley). The fourth cranial nerve, which innervates the superior oblique muscle; the mammalian muscle passes through a connective tissue pulley before inserting on the eyeball.

Trophoblast [trof′ō-blast] (Gr. *trophe* = nourishment + *blastos* = bud). The outer layer of the mammalian blastocyst; initiates placenta formation; homologous to the chorionic ectoderm.

Tunic [tū′nik] (L. *tunica* = coating). Describes a layer of an organ, such as the layers of the eyeball.

Tunicates [tū′ni-kātz]. The subphylum of chordates that includes the sea squirts and their allies; also called urochordates.

Turbinate bones [ter′bi-nāt] (L. *turbinatus* = top-shaped, whirlwind). Scroll-shaped bones in the nasal cavities of mammals that increase the surface area of the cavities; also called conchae.

Turbulent flow. A disrupted flow of fluid along the surface of a swimming or flying vertebrate.

Tympanic cavity (L. *tympanum* = drum). The middle ear cavity, which lies between the tympanic membrane and the otic capsule containing the inner ear.

Tympanic membrane. The eardrum.

Ulna [ŭl′nă] (L. = elbow). The bone of the antebrachium of tetrapods that extends behind the elbow, lying on the side adjacent to the little finger when the hand is supine.

Ultimobranchial bodies [ŭl′ti-mō-brang′kē-ăl] (L. *ultimus* = farthest + Gr. *branchia* = gills). Derivatives of the caudal surface of the last branchial pouch; in fishes they contain the C cells whose hormone, calcitonin, helps regulate mineral metabolism.

Ungulates [ung′gyū-lātz] (L. *ungula* = hoof). A collective term for the hoofed mammals: artiodactyls and perissodactyls.

Unguligrade [ung′gyū-li-grād] (L. *gradus* = step). Walking on the toe tips.

Urea [yū-rē′ă] (Gr. *ouron* = urine). A breakdown product of nitrogen metabolism; occurs in elasmobranchs, some amphibians, and mammals.

Ureter [yū-rē′ter] (Gr. *oureter* = ureter, from *ouron* = urine). The duct of amniotes that carries urine from a metanephric kidney to the urinary bladder.

Urethra [yū-rē′thră] (Gr. *ourethra* = urethra). The duct in amniotes that carries urine from the urinary bladder to the cloaca or outside; part of it also carries sperm in males.

Uric acid [yūr′ik] (Gr. *ouron* = urine). A breakdown product of nitrogen metabolism; occurs chiefly in reptiles and birds, requires that little water be removed from the body.

Urinary bladder [yūr′i′nār-ĕ]. A saccular organ in which urine accumulates before discharge from the body.

Urodels [yū′rō-dēlz] (Gr. *oura* = tail + *delos* = visible). The amphibian order that includes the salamanders; also called Caudata.

Urogenital [yū-rō′jen-i-tăl] (Gr. *ouron* = urine + L. *genitalis* = genital). Pertains to structures that are common to the urinary and genital systems, such as certain urogenital ducts.

Urophysis [yū-rō-fi-sis] (Gr. *oura* = tail + *physis* = growth). A neurosecretory organ on the caudal end of the spinal cord in elasmobranchs and teleosts.

Uropygeal gland [yū-rō-pij′ē-al] (Gr. *pyge* = rump). An oil-secreting gland of birds located dorsal to the tail base.

Urostyle [yū′rō-stīl] (Gr. *stylos* = pillar). An elongated bone of anurans that represents fused caudal vertebrae.

Uterine tube [yū′ter-in] (L. *uterus* = womb). The portion of the mammalian oviduct that carries eggs from the coelom to the uterus; also called the fallopian tube; site of fertilization.

Uterus   [yū′ter-ŭs]. The portion of an oviduct in which embryos develop in live-bearing species.

Utriculus   [yū′trik′yū-lŭs] (L. = small sac). The upper chamber of the membranous labyrinth to which the semicircular ducts attach.

Vagina   [vă-jī′na] (L. = sheath). The passage in female therians that leads from the uterus to the vaginal vestibule.

Vaginal vestibule   [ves′ti-būl] (L. *vestibulum* = antechamber). The passage or space in female therians that receives the vagina and urethra; also called the urogenital sinus.

Vagus nerve   [vā′gŭs] (L. = wandering). The tenth cranial nerve; carries motor fibers to the muscles of the last four visceral arches, autonomic fibers to the heart and abdominal viscera, returns sensory fibers from these areas, and supplies the lateral line canal in fishes and larval amphibians.

Vas deferens   [vas]. See Deferent duct.

Vasa efferentia   [vā′să]. See Efferent ductules.

Vascular tunic   [vas′kyū-lăr] (L. *vasculum* = small vessel). The middle layer of the eyeball; it forms the choroid, ciliary body, and iris.

Vasopressin.   See Antidiuretic hormone.

Vector   [vek′tōr] (L. *vector* = bearer). A quantity, such as a force, that has both a magnitude and a direction.

Vein   [vān] (L. *vena* = vein). A vessel that conveys blood toward the heart; most veins contain blood low in oxygen content, but

pulmonary veins from the lungs are rich in oxygen.

Velocity   [vĕ-los′i-tē]. Distance traveled divided by the time units.

Vena cava   [vē′nă cā′vă] (L. *vena cava* = hollow vein). One of the major veins of lungfishes, amphibians, and amniotes; leads directly to the heart.

Ventral aorta   [ā-ōr′tă]. An artery that leads from the heart to the aortic arches and their derivatives; contributes to the arch of the aorta and base of the pulmonary artery in mammals.

Ventricle   [ven′tri-kl] (L. *ventriculus* = small belly). The chamber of the heart that greatly increases blood pressure and sends blood to the arteries; also a chamber within the brain.

Vermis   [ver′mis] (L. = worm). The "segmented" medial portion of the amniote cerebellum.

Vertebra   [ver′te-bră] (L. = joint, vertebra). One of the skeletal units that make up the spinal column.

Vertebral arch.   The arch of a vertebra that surrounds the spinal cord; also called a neural arch.

Vertebral body.   The main supporting component of a vertebra, lying ventral to the vertebral arch; sometimes called the centrum.

Vertebrates   [ver′tĕ-brāts]. The subphylum of chordates that contains the backboned animals.

Vesicular gland   [vĕ-sik′yū-lăr] (L. *vesicus* = small bladder). One of the accessory sex glands of males that contributes to the seminal fluid.

Vestibular apparatus   [ves-tib′yū-lăr] (L. *vestibulum* = entrance). The portion of the inner ear that detects changes in position and acceleration.

Vestibulocochlear nerve   [ves-tib′yū-lō-kok′ lē-ăr] (L. *cochlea* = snail shell). The eighth cranial nerve, which returns fibers from the parts of the inner ear related to equilibrium and sound detection; often called the statoacoustic nerve in anamniotes.

Vestige   [ves′tij] (L. *vestigium* = trace). A remnant in one organism of a structure that is well developed in another organism and has no function or a different function from that of its well-developed homologue.

Villi   [vil′i] (L. = shaggy hairs). Multicellular but minute, often finger-shaped projections of an organ that increase its surface area, e.g., the intestinal villi.

Viscera   [vis′er-ă] (L. = internal organs). A collective term for the internal organs.

Visceral arches.   The skeletal arches that develop in the wall of the pharynx; includes the mandibular, hyoid, and branchial arches.

Vitelline   [vī-tel′in] (L. *vitellus* = yolk). Pertains to structures associated with the embryonic yolk sac, such as the vitelline arteries and veins.

Vitreous body   [vit′rē-ŭs] (L. *vitreus* = glassy). The clear, viscous material in the eyeball between the lens and retina.

Viviparity   [viv′i-pār′i-tē] (L. *vivus* = living + *pario* = to bring forth). A pattern of reproduction in which the embryos are born as miniature adults. The term is often limited to placental viviparity, in which the embryos are completely dependent on materials transferred from the mother.

Vocal cords   [vō′kăl] (L. *vocalis* = pertaining to the voice). Folds of mucous membrane within the larynx of many tetrapods whose vibrations produce sound.

Vomeronasal organ   [vō′mer-ō-nā′zăl]. An accessory olfactory organ located between the palate and nasal cavities of most tetrapods, important in feeding and sexual behavior. See also Jacobson's organ.

Vulva   [vŭl′vă] (L. = covering). The external genitalia of a female.

Weberian ossicles   [os′i-klz] (*E. H. Weber*, German anatomist, 1795–1878 + L. *ossiculum* = small bone). A set of small bones that transmit sound waves from the swim bladder to the inner ear in some teleosts.

White matter.   Tissue in the central nervous system that consists primarily of myelinated axons.

Wing loading.   The weight of a bird divided by the area of its wings.

Wolffian duct.   (*K. F. Wolff*, 18th-century German embryologist). A term often applied to the embryonic archinephric duct.

Work.   The product of a force and the distance through which it acts.

Yaw.   The tendency for the head of swimming or flying vertebrates to move from left to right about its longitudinal axis.

Yolk sac   [yok] (Anglo-Saxon *geula* = yellow). The yolk-containing sac attached to the ventral surface of embryos that develops from macrolecithal eggs.

Zygapophysis   [zī′gă-pof′i-sis] (Gr. *zygon* = yoke + *apo* = away from + *physis* = growth). A process of a vertebral arch that articulates with a comparable process on an adjacent arch; also called articular process.

Zygomatic arch   [zī′gō-mat′ik] (Gr. *zygoma* = bar, yoke). The arch of bone in a mammalian skull that lies beneath the orbit and connects the facial and cranial regions of the skull.

Zygote   [zi′got] (Gr. *zygotes* = yoked together). The cell formed by the union of a sperm and egg.

Index

Page references to figures are followed by (f); to tables by (t).